Linear Algebra
with Applications

Linear Algebra with Applications

George Nakos
U.S. Naval Academy

David Joyner
U.S. Naval Academy

Brooks/Cole Publishing Company

I(T)P® An International Thomson Publishing Company

Pacific Grove • Albany • Belmont • Bonn • Boston • Cincinnati • Detroit
Johannesburg • London • Madrid • Melbourne • Mexico City • New York
Paris • Singapore • Tokyo • Toronto • Washington

 A GARY W. OSTEDT BOOK

Publisher: *Gary W. Ostedt*
Marketing Team: *Caroline Croley, Christine Davis*
Editorial Associate: *Carol Benedict*
Production Coordinator: *Kirk Bomont*
Manuscript Editor: *Linda Thompson*
Interior Design: *Merry Obrecht Sawdey*
Interior Illustration: *Scientific Illustrators*

Cover Design: *Craign Hanson, George Nakos*
Project Management and Typesetting: *Integre Technical Publishing Company, Inc.*
Cover Printing: *Phoenix Color Corporation, Inc.*
Printing and Binding: *R.R. Donnelley, Crawfordsville*

COPYRIGHT © 1998 by Brooks/Cole Publishing Company
A division of International Thomson Publishing Inc.
I(T)P The ITP logo is a registered trademark under license.

For more information, contact:

BROOKS/COLE PUBLISHING COMPANY
511 Forest Lodge Road
Pacific Grove, CA 93950
USA

International Thomson Publishing Europe
Berkshire House 168–173
High Holborn
London WC1V 7AA
England

Thomas Nelson Australia
102 Dodds Street
South Melbourne, 3205
Victoria, Australia

Nelson Canada
1120 Birchmount Road
Scarborough, Ontario
Canada M1K 5G4

International Thomson Editores
Seneca 53
Col. Polanco
11560 México, D.F., México

International Thomson Publishing GmbH
Königswinterer Strasse 418
53227 Bonn
Germany

International Thomson Publishing Asia
221 Henderson Road
#05–10 Henderson Building
Singapore 0315

International Thomson Publishing Japan
Hirakawacho Kyowa Building, 3F
2-2-1 Hirakawacho
Chiyoda-ku, Tokyo 102
Japan

Printed in the United States of America

10 9 8 7 6 5 4 3

Library of Congress Cataloging-in-Publication Data
Nakos, George
    Linear algebra with applications / George Nakos, David Joyner.
      p. cm.
    Includes index.
    ISBN 0-534-95526-6 (alk. paper)
    1. Algebras, Linear. I. Joyner, David. II. Title.
QA184.N34   1998
512′.5—dc21
                                                        97-47341
                                              CIP

This book is dedicated to
Constantine, Helen, Debra,
Constantine, and David Nakos,
and to
Elva Joyner

Contents

Preface

Linear algebra is one of the most useful courses a student of science or mathematics will ever take. It is the first course where *concepts* are at least as important as *calculations*, and *applications* are motivating and mind opening. This multiple role of linear algebra is emphasized throughout the book.

Applications of linear algebra to science and real life are numerous. The solutions to many problems in physics, engineering, biology, chemistry, medicine, computer graphics, image processing, economics, and sociology require tools from linear algebra. So do all main branches of modern mathematics.

A good introduction to this subject can help students think clearly and precisely, and give them practice with interesting and useful calculations.

A General Goal By designing the book for maximum flexibility, we try to address the needs of the student and instructor faced with rapid changes in technology. The instructor may use the theoretical, numerical, applied, and computer material in any number of ways. However, we do believe that the mathematical theory should be combined with computer exploration.

The Level The material largely agrees with the recommendations in the *Summary Report of the Linear Algebra Study Group* of the NSF-sponsored workshop held in August 1990 at the College of William and Mary.

The book was written primarily for a one-semester course at the sophomore level of mathematics or science majors. However, there is enough material for a complete two-semester course, if desired. Also, with appropriate choices, the book can be used at the freshman or even at the junior level.

Simple, Direct Style This book is easy to read. We proceed from the particular to the general with examples that build understanding, but we avoid lengthy discussions.

Many Examples   The book's unusually high number of examples (over 420) is perhaps its most distinctive feature. The examples are carefully chosen and presented in sufficient detail to teach the material thoroughly. They can be used in class and in self-study or group-study. The instructor can choose which examples to discuss depending on what needs to be emphasized.

Wide Variety of Applications   The applications interlaced throughout the text represent a broad spectrum. Also, there is at least one separate applications section per chapter. The applications never intrude into the basic material. They can be used or, if a more theoretical course is desired, they can be ignored. There are applications not only to physics, engineering, mechanics, chemistry, economics, business, sociology, and psychology, but also to mathematics in the areas of graph theory, analytic geometry, fractal geometry, coding theory, wavelet theory, dynamical systems, and solutions of polynomial systems.

Emphasis on Geometry   Often linear algebra books are lacking in geometric insight. In this book, there has been a substantial effort to emphasize the geometric understanding of the material. With over 180 figures, we try to illuminate the basic concepts both geometrically and algebraically. As an example of this approach, we introduce dot products, cross products, orthogonality, lines, and planes very early on (in Chapter 2).

Numerical Methods   Sections on numerical methods show what can go wrong when we deal with real-life problems. Many of these problems involve large-scale calculations that require reliability and precision. The numerical sections, which are independent of the basic material, address these issues. Sample methods include iterative solutions, LU factorization, numerical computations of eigenvalues, QR factorization, the singular value decomposition, and least squares solutions.

New Emphasis on Determinants   In recent years texts on linear algebra have tended to minimize the role of determinants, because determinants have mostly theoretical use and are overshadowed by Gaussian elimination in numerical calculations. This trend may change. Determinants are again becoming increasingly important. One of their most interesting uses is in solving polynomial systems with resultants. The renewed attention to determinants is reflected in the book.

Miniprojects   A miniprojects section is a distinctive feature in each chapter. The projects are fairly short and not very involved. Some are lengthy exercises, some extend the basic theory, some emphasize a particular application, and some entertain with a small lesson from history.

Optional Computer Exercises   Each chapter ends with computer exercises. These exercises are all generic: They can be used with home-grown software or with commercially available packages. The computer exercises are not just regular exercises with messy numbers. They are designed to help the student understand how to use the available computer program. This material is independent of the basic theory. It can be used or ignored, depending on whether technology is desired.

Maple, Mathematica, and MATLAB Another unique feature is the complete solutions for over 90% of the computer exercises using Maple, Mathematica, and MATLAB. Most basic linear algebra commands of these programs are illustrated and explored. We also offer a generous supply of comments and explanations.

CAS Subsections Short computer algebra systems (CAS) subsections display the input and output of basic linear algebra commands for Maple, Mathematica, and MATLAB.

Statements of Reader's Goals All sections start with brief statements of the reader's goals. These are especially useful at the time of a second reading, or in reviewing for a quiz or test.

Historical and Modern Examples Often texts discuss only modern examples and applications. It is our opinion that the student needs *some* historical examples. We make an effort to show that the roots of linear algebra are deep and interesting. So, we include some simple but motivating problems from the past (such as Fibonacci's money pile, Archimedes' oxen of the sun, and Newton's cows and fields). At the same time we discuss many modern applications in fractals, wavelets, codes, and so on.

Supplements The following supplements are currently available:

- Instructor's Manual: Includes solutions to all problems and projects.
- Student's Manual: Includes solutions to all odd-numbered problems.

An Overview

Chapter 1: Linear Systems Gauss elimination is emphasized as an important computational tool; nearly all subsequent material depends on it. A thorough discussion of Sections 1.1 and 1.2 is desirable. Section 1.3, on numerical solutions, can be outlined or omitted if time is limited. Section 1.4 offers a variety of independent applications. A selection of examples here helps the student appreciate the applicability of linear systems.

Chapter 2: Vectors The most important sections are 2.1–2.5. Section 2.1 discusses vector operations. Section 2.2 is about the dot product and orthogonality in $\mathbf{R}^n$, with a brief subsection on orthogonal projections. The next two sections are very important. Section 2.3 introduces the span of vectors and ties it to linear systems. Section 2.4 discusses the fundamental notion of linear independence. Sections 2.3 and 2.4 are intentionally brief, but they should not be covered in the same class meeting. Section 2.5 introduces the matrix-vector product $A\mathbf{x}$ and relates it to linear combinations and linear systems.

Section 2.6 is on the cross product. Section 2.7 is on lines and planes. Some of the material in Sections 2.6 and 2.7 may be familiar from multivariate calculus. In this case, these sections can be discussed very quickly. There are several applications in Section 2.8, with a notable introduction to dynamical systems. Also, there are applications to statics geared toward an engineering audience.

Chapter 3: Matrices

The basic material is covered in Sections 3.1–3.4. Section 3.1 is on matrix operations, including matrix multiplication. Section 3.2 introduces the inverse of a matrix and how to compute it. It ends with brief applications to heat conduction and related markets. Section 3.3 is on elementary matrices. Here we justify the matrix inversion algorithm and we characterize invertible matrices in many different ways. Section 3.4 is about the important LU factorization of any size matrix. The last subsection, on interchanges, is optional. Section 3.5 has detailed applications to graph theory with adjacency and incidence matrices and line and dominance graphs. It also covers stochastic matrices, Markov chains, and the Leontief input-output models.

Chapter 4: Vector Spaces

The student is well-prepared by now for the important abstraction. Section 4.1, on subspaces of $\mathbf{R}^n$, is preparatory. Section 4.2 has the definitions of vector space and subspace and many examples that should be discussed thoroughly. The student should make an effort to supply the details of the verification of the axioms in these examples. Section 4.3 is on the important concepts of spanning sets, linear independence, and basis. The first two concepts are already familiar. What needs to be stressed here is how to rely more on the definition of linear independence than on the direct use of matrix row reduction. Section 4.4 is on dimension. Several theorems connect dimension, linear independence, spanning sets, and bases together. Finally, there are two optional proofs of the exchange theorem. Section 4.5 is on coordinates and change of bases. Section 4.6 is basic; it covers the null space, the column space, and the row space. It also introduces the rank and nullity and the very important rank theorem. It then discusses the connection between rank and linear systems and ends with the optional proof of the uniqueness of reduced row echelon form. Section 4.7 is on applications to coding theory, especially on the linear Hamming (7, 4)-code.

Chapter 5: Linear Transformations

Sections 5.1–5.3 are important. Section 5.1 is introductory and geometric. It discusses the basic matrix transformations of the plane. Section 5.2 defines the general linear transformation. This section has a variety of examples that should be studied thoroughly. Section 5.3 is on the basic notions of kernel, range, and isomorphism. The highlight here is the important dimension theorem. The relation of this material with that of Section 4.6 should be emphasized. Section 5.4 is on the matrix of a linear transformation and how it changes when the bases are changed. Section 5.5 is on the operations of linear transformations and how they relate to matrix operations. The applications in Section 5.6 are on affine transformations and fractals.

Chapter 6: Determinants

Determinants are introduced in Section 6.1 by cofactor expansion. This is not the mathematical method of choice but it is direct and works with the students. Section 6.2 is on the basic properties of determinants. Computing determinants by correct row reduction is a point that should be emphasized here. Section 6.3 is on the adjoint and Cramer's rule. Section 6.4 discusses how to define and compute determinants by permutations. There is an optional subsection on crossing diagrams for the sign of a permutation. In Section 6.5 we have applications of determinants on equations of geometric objects, on elimination theory, Sylvester resultant and solutions of polynomial systems, and on electrical circuits and the counting of their spanning trees.

Chapter 7: Eigenvalues and Eigenvectors

Sections 7.1 and 7.2 are the most important sections here. In Section 7.1 we define eigenvectors and eigenvalues and study several examples. We conclude with eigenvalues of linear transformations. Section 7.2 is on diagonalization of matrices and linear transformations. We also discuss the computation of powers of diagonalizable matrices. Section 7.3 is on the various numerical computations of eigenvalues. This section is quite explicit on the different numerical methods. Section 7.4 consists of a very detailed discussion of dynamical systems, including long-term behavior. Section 7.5 is devoted to Markov chains, probability vectors, and limits of stochastic matrices.

Chapter 8: Dot and Inner Products

Section 8.1 is on orthogonality and orthogonal matrices. This material is important and should be learned well. Section 8.2 includes a detailed discussion on orthogonal projections and the Gram-Schmidt process. Section 8.3 is on the very useful QR factorization. Section 8.4, on least squares, is also important. In practice, most systems seem to be inconsistent and we use least squares to best fit solutions. Section 8.5 is on how to diagonalize a symmetric matrix using orthogonal matrices. Here we prove the spectral theorem using the Schur decomposition theorem. Section 8.6 is on quadratic forms and conic sections. Section 8.7 is on the numerically important singular value decomposition and on pseudoinverses. Section 8.8 is devoted to general inner product spaces and to how the different notions and processes generalize. In Section 8.9 we have an example from the NFL ratings, we discuss least squares polynomials, and, for readers who know basic integration, we have Fourier polynomials and an introduction to wavelets.

Appendices

Appendix A consists of a review of complex numbers and how to do linear algebra with them. It also discusses Hermitian and unitary matrices. *Appendix B* consists of the basic linear algebra commands for Maple, Mathematica, and MATLAB.

Answers to Selected Exercises

There are answers to selected odd-numbered exercises. In general, the proofs in the theoretical exercises are omitted.

Acknowledgments

Our first thanks go to our families. The second author wishes to thank his wife, Elva, for all her support and help. The first author wishes to thank his parents, Constantine and

Helen, his wife, Debra, and his sons, Constantine and David, for all the support and help over his six long years of writing.

We thank the U.S. Naval Academy faculty, especially Professor William P. Wardlaw, for the interest and the helpful conversations. We have adopted Bill's long-standing position on the distinction between sets and sequences of linearly dependent vectors. To Professor T. S. Michael we owe the information on some of our references.

The first author also wishes to thank Dr. Robert M. Williams for introducing him to resultants (discussed in Chapter 6) and for the long collaboration on linear algebra related research. He also wishes to thank Professor Peter R. Turner of the Naval Academy for all the good advice and for making his notes on wavelets available for use in the book. Also, some projects were modeled after Peter's material in numerical analysis.

We wish to thank all the reviewers for their help, criticism, and advice. These include: Michael Ward, Bucknell University; Thomas Moore, Bridgewater State College; David Johnson, Lehigh University; Lala Krishna, University of Akron; and Larry Grove, University of Arizona. Our contact with Brooks/Cole has been wonderful. Our thanks go to Gary Ostedt, Carol Benedict, Kirk Bomont, Kelly Shoemaker, Tim Spurlock, and Janna Miser. We also thank Integre Technical Publishing Co. for the excellent typesetting, and Scientific Illustrators for the impressive artwork. Last, but not least, we thank Steve Quigley, Art Minsberg, and Barbara Lovenvirth.

George Nakos
David Joyner

Linear Systems

Thou hast ordered all things in measure and number and weight.
—Wisdom of Solomon, Chapter 11, Verse 20

Introduction

M any questions in engineering, physics, mathematics, economics, and other sciences can be reduced to the problem of solving a linear system. The interest in solving such systems is very old, as Archimedes's *Cattle Problem* (studied in Section 1.5) demonstrates. Let us briefly look at a problem involving a linear system that occupied mathematicians about eight centuries ago. Its solution is discussed in Section 1.4.

Fibonacci

Our story concerns the medieval Italian mathematician Leonardo of Pisa (c. 1175–1250), better known as Fibonacci. During his travels Fibonacci learned the Arabic "new arithmetic" he later introduced to the West in his famous book *Liber abaci*. The story goes that Emperor Frederick II of Sicily invited Fibonacci and other scholars to participate in a sort of mathematical tournament, where several mathematical problems were posed. One of the problems was as follows:

> *Three men possess a single pile of money, their shares being $\frac{1}{2}$, $\frac{1}{3}$, and $\frac{1}{6}$. Each man takes some money from the pile until nothing is left. The first man then returns $\frac{1}{2}$ of what he took, the second man $\frac{1}{3}$, and the third $\frac{1}{6}$. When the total so returned is divided equally among the men it is discovered that each man then possesses what he is entitled to. How much money was there in the original pile and how much did each man take from the pile?*

Fibonacci came up with the solution: 47 for the total amount and 33, 13, and 1 for the amounts each of these men took from the pile. Was he correct?

1.1   Introduction to Linear Systems

> **Reader's Goal for This Section**
>
> To recognize and use elimination to solve a simple linear system.

In this section we introduce the notions of a linear equation and a system of linear equations. We discuss how to solve "small" systems by elimination. Typical real-life systems are solved by computer. They usually involve hundreds—or even thousands—of equations and unknowns.

The set of all real numbers is denoted by **R**. Unless stated otherwise, by a *scalar* we mean a real number. We usually abbreviate the phrase x is an element of a set A by

$$x \in A$$

Thus, $x \in \mathbf{R}$ means x is a scalar or x is a real number.

Linear Equations

DEFINITION

(Linear Equation)

An equation in n **variables** $x_1, \ldots, x_n$, is **linear** if it can be written in the form

$$a_1x_1 + a_2x_2 + \cdots + a_nx_n = b \qquad (1.1)$$

The a_is are the **coefficients**, and b is the **constant term** of the equation. The variables are also called **unknowns**, or **indeterminants**. If $b = 0$, the equation is called **homogeneous**. The equation obtained from (1.1) by replacing b with 0 is the homogeneous equation **associated** with (1.1). If we order the variables, the first variable with nonzero coefficient is called the **leading variable**. The remaining variables are called **free**.

▪ EXAMPLE 1   The equation

$$x_1 + x_2 + 4x_3 - 6x_4 - 1 = x_1 - x_2 + 2$$

is linear because it can be written in the *standard form* (1.1):

$$0x_1 + 2x_2 + 4x_3 - 6x_4 = 3$$

If the variables are ordered from x_1 to x_4, then x_2 is the leading variable and x_1, x_3, x_4 are the free ones. The coefficients are $0, 2, 4, -6$ and the constant term is 3.

▪ EXAMPLE 2   (Celsius to Fahrenheit)   The standard conversion of Celsius degrees, C, into Fahrenheit, F, is a linear equation in C and F.

$$F = \frac{9}{5}C + 32 \qquad (1.2)$$

■ EXAMPLE 3   The following linear equations are homogeneous.

$$x_1 + 2x_2 - \sqrt{5}x_3 - x_4 = 0 \qquad x - y + z = (\sin 4)w$$

■ EXAMPLE 4   These equations are not linear, or **nonlinear**:

$$xy - 3 = 2x \qquad x^2 - y = 1 \qquad \sin x + y = 0$$

A (**particular**) **solution** of a linear equation is a sequence of numbers that, when substituted for the variables, yields an equation that is an identity.

For example, $C = 5°$ and $F = 41°$ is a solution of (1.2), because $\frac{9}{5} \cdot 5 + 32 = 41$. On the other hand, $C = 5°$ and $F = 40°$ is not a solution, because $\frac{9}{5} \cdot 5 + 32 \neq 40$.

The set of all particular solutions is called the **solution set**. The solution set is obtained by solving for the leading variable in terms of the free variables and letting each free variable take on any scalar value. This results in a generic element of the solution set called a **general solution**.

■ EXAMPLE 5   Find the general solution of the equation

$$2x_1 + 0x_2 - 4x_3 = -2$$

SOLUTION   We solve for the leading variable, x_1, to get $x_1 = 0x_2 + 2x_3 - 1$. The free variables x_2 and x_3 can take on any value, say $x_2 = s$ and $x_3 = r$. Hence, the general solution is given by

$$x_1 = 2r - 1, \qquad x_2 = s, \qquad x_3 = r \qquad \text{for all } r, s \in \mathbf{R}$$

The letters r and s used to denote the free variables are called **parameters**. The solution set just found is a *two-parameter set*. All particular solutions can be obtained from the general solution by assigning values to the parameters. For instance, $r = -1$ and $s = 2$ yields the particular solution $x_1 = -3, x_2 = 2,$ and $x_3 = -1$.

Linear Systems

A linear system is a set of linear equations, such as

$$\begin{aligned} 3x + 2y + z &= 39 \\ 2x + 3y + z &= 34 \\ x + 2y + 3z &= 26 \end{aligned} \qquad (1.3)$$

This system and its solution are in the third century B.C. Chinese mathematical book *Nine Chapters of Mathematical Art.*[1]

---

[1] See Carl Boyer's *A History of Mathematics*, p. 219 (New York: Wiley).

DEFINITION

> ## (Linear System)
>
> A **linear system** of m equations in n variables (or unknowns) $x_1, \ldots, x_n$, is a set of m linear equations of the form:
>
> $$\begin{aligned} a_{11}x_1 + a_{12}x_2 + \cdots + a_{1n}x_n &= b_1 \\ a_{21}x_1 + a_{22}x_2 + \cdots + a_{2n}x_n &= b_2 \\ &\vdots \\ a_{m1}x_1 + a_{m2}x_2 + \cdots + a_{mn}x_n &= b_m \end{aligned} \qquad (1.4)$$
>
> The numbers $a_{11}, a_{12}, \ldots, a_{1n}, a_{21}, \ldots, a_{2n}, \ldots, a_{m1}, \ldots, a_{mn}$ are the **coefficients** of the system, and $b_1, b_2, \ldots, b_n$ are the **constant terms**. If all constant terms are zero, the system is called **homogeneous**. The homogeneous system that has the same coefficients as system (1.4) is said to be **associated** with (1.4).

Consider the system

$$\begin{aligned} x_1 + 2x_2 \phantom{{}+ 2x_3} &= -3 \\ 2x_1 + 3x_2 - 2x_3 &= -10 \\ -x_1 \phantom{{}+ 2x_2} + 6x_3 &= 9 \end{aligned} \qquad (1.5)$$

Its coefficients are, in order, $1, 2, 0, 2, 3, -2, -1, 0, 6$. The constant terms are $-3, -10, 9$. The associated homogeneous system is

$$\begin{aligned} x_1 + 2x_2 \phantom{{}+ 2x_3} &= 0 \\ 2x_1 + 3x_2 - 2x_3 &= 0 \\ -x_1 \phantom{{}+ 2x_2} + 6x_3 &= 0 \end{aligned}$$

A linear system can be abbreviated by recording only its coefficients and constant terms, provided that the names and an order of the variables have been specified. The rectangular arrangement of the coefficients and constant terms of a system is called its **augmented matrix**. For example, the augmented matrix of (1.5) is

$$\begin{bmatrix} 1 & 2 & 0 & -3 \\ 2 & 3 & -2 & -10 \\ -1 & 0 & 6 & 9 \end{bmatrix} \quad \text{or} \quad \begin{bmatrix} 1 & 2 & 0 & : & -3 \\ 2 & 3 & -2 & : & -10 \\ -1 & 0 & 6 & : & 9 \end{bmatrix}$$

The second form involves a separator to indicate where the column of the constant terms is. In general, a **matrix** is a rectangular arrangement of numbers. The matrix that consists of the coefficients of a system is its **coefficient matrix**. The one-column matrix that displays the constant terms is the **vector of constants**. The coefficient matrix and the vector of constants of system (1.5) are, respectively:

$$\begin{bmatrix} 1 & 2 & 0 \\ 2 & 3 & -2 \\ -1 & 0 & 6 \end{bmatrix} \quad \text{and} \quad \begin{bmatrix} -3 \\ -10 \\ 9 \end{bmatrix}$$

▪ EXAMPLE 6   Write a system with augmented matrix.

$$\begin{bmatrix} 1 & 2 & 0 & -4 \\ 0 & 3 & -2 & -1 \end{bmatrix}$$

SOLUTION   Since the augmented matrix has four columns, the system has three variables. If we choose names x_1, x_2, and x_3 for the variables, then

$$\begin{aligned} x_1 + 2x_2 \qquad &= -4 \\ 3x_2 - 2x_3 &= -1 \end{aligned}$$

is a system with the given augmented matrix.

DEFINITION

(Solution of a Linear System)

A sequence $r_1, r_2, \ldots, r_n$ of scalars is a **(particular) solution** of system (1.4) if all the equations are satisfied when we substitute $x_1 = r_1, \ldots, x_n = r_n$. The set of all possible solutions is the **solution set**. Any generic element of the solution set is called the **general solution**.

▪ EXAMPLE 7   Show that $x_1 = -15, x_2 = 6$, and $x_3 = -1$ is a particular solution of system (1.5).

SOLUTION   The substitution $x_1 = -15, x_2 = 6$, and $x_3 = -1$ yields the following true statements.

$$\begin{aligned} -15 + 2 \cdot 6 \qquad\qquad &= -3 \\ 2 \cdot (-15) + 3 \cdot 6 - 2 \cdot (-1) &= -10 \\ -(-15) \qquad + 6 \cdot (-1) &= 9 \end{aligned}$$

If a system has solutions it is called **consistent**; otherwise, it is called **inconsistent**. System (1.5) is consistent. The system $x + y = 1, x + y = -1$ is inconsistent.

A linear system can have one solution, infinitely many solutions, or no solutions. This is illustrated geometrically for

$$\begin{array}{ccc} y + x = 2 & y + x = 2 & y + x = 2 \\ y - x = 0 & 2y + 2x = 4 & y + x = 1 \end{array}$$

for which the equation lines intersect, coincide, or are parallel (Fig. 1.1).

Graphs of Solution Sets

We know that the graph of the equation $ax + by = c$ is a straight line (except in the extreme cases $0x + 0y = 0$ and $0x + 0y = c \neq 0$). So, in general, the graph of the solution set of a two-variable system is the intersection of several straight lines.

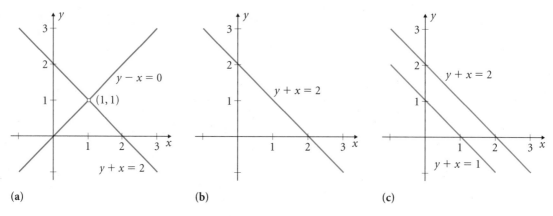

Figure 1.1 Solutions: (a) exactly one, (b) infinitely many, (c) none.

Although planes are discussed in Chapter 2, we should mention that the graph of the equation $ax + by + cz = d$ is a plane (except in the extreme cases $0x + 0y + 0z = 0$ and $0x + 0y + 0z = d \neq 0$). Hence, the graph of the solution set of a three-variable system is, in general, the intersection of several planes.

Note that $x_1 = 0, \ldots, x_n = 0$ is always a solution of a homogeneous system. This is called the **trivial**, or **zero, solution**. Any other solution is called **nontrivial**. For example, $x = 1, y = 1$ is a nontrivial solution of the homogeneous system

$$
\begin{aligned}
x - \ y &= 0 \\
-2x + 2y &= 0
\end{aligned}
$$

Two linear systems with the same solution sets are called **equivalent**. The systems

$$
\begin{aligned}
-2x \quad\ \ = -2 \qquad &2x - 2y = -2 \\
x - y = -1 \qquad &\qquad\ \ y = 2
\end{aligned}
$$

are equivalent. Their common (and only) solution is $x = 1, y = 2$.

Solving a Linear System

The easiest systems to solve are those in triangular, or **echelon**, form. In these systems, the leading variable in each equation occurs to the right of the leading variable of the equation above it. We solve such a system by starting at the bottom and moving upward. First we solve the last equation, then substitute the values found into the equation above it and solve it, and then repeat. This method is called **back-substitution**.

■ EXAMPLE 8   Use back-substitution to solve the system

$$
\begin{aligned}
x_1 - x_2 + \ x_3 - x_4 + 2x_5 - x_6 &= 1 \\
-2x_3 \qquad\quad + 2x_5 \qquad\ \ &= 2 \\
-x_5 + x_6 &= 3
\end{aligned}
\tag{1.6}
$$

SOLUTION The last equation implies that if x_6 is any number, such as r, then $x_5 = r - 3$. By the second equation, $x_3 = x_5 - 1 = r - 3 - 1 = r - 4$. Solved for x_1, the first equation yields $x_1 = 1 + x_2 - x_3 + x_4 - 2x_5 + x_6$. x_2, x_4 can be any numbers, say, $x_4 = s, x_2 = t$. Hence, $x_1 = 1 + t - (r - 4) + s - 2(r - 3) + r = -2r + s + t + 11$. So the general solution is

$$
\begin{aligned}
x_1 &= -2r + s + t + 11 \\
x_2 &= t \\
x_3 &= r - 4 \\
x_4 &= s \\
x_5 &= r - 3 \\
x_6 &= r
\end{aligned}
\qquad r, s, t \in \mathbf{R}
$$

The solution set is a *three-parameter* infinite set.

To solve a system in general, we eliminate unknowns to get an equivalent system in echelon form. Then we use back-substitution to solve the latter. This idea is refined and discussed more thoroughly in Section 1.2.

We eliminate unknowns by using a sequence of the following equation operations in such a way that the resulting system is in echelon form. *Application of any of these operations results in equivalent systems.*

DEFINITION

(Elementary Equation Operations)

The **elementary equation operations** of a linear system consist of the following.

(Elimination) *Adding a constant multiple of one equation to another.*
$$\boxed{E_i + cE_j \longrightarrow E_i}$$
(Scaling) *Multiplying an equation by a nonzero constant.* $\boxed{cE_i \longrightarrow E_i}$
(Interchange) *Interchanging two equations.* $\boxed{E_i \leftrightarrow E_j}$

Because the augmented matrix of a system serves as an abbreviation of the system, we can save time and avoid notational mistakes by working with the augmented matrix. The matrix operations that correspond to the elementary equation operations are called elementary row operations. Actually, these operations can be applied to *any* matrix.

DEFINITION

(Elementary Row Operations)

The **elementary row operations** of a matrix consist of the following.

(Elimination) *Adding a constant multiple of one row to another.* $\boxed{R_i + cR_j \longrightarrow R_i}$
(Scaling) *Multiplying a row by a nonzero constant.* $\boxed{cR_i \longrightarrow R_i}$
(Interchange) *Interchanging two rows.* $\boxed{R_i \leftrightarrow R_j}$

▪ EXAMPLE 9   Solve the system by elimination.

$$
\begin{aligned}
x_1 + 2x_2 &= -3 \\
2x_1 + 3x_2 - 2x_3 &= -10 \\
-x_1 \quad\quad + 6x_3 &= 9
\end{aligned}
$$

SOLUTION   We have

$$
\begin{aligned}
x_1 + 2x_2 &= -3 \\
2x_1 + 3x_2 - 2x_3 &= -10 \quad\text{or} \\
-x_1 \quad\quad + 6x_3 &= 9
\end{aligned}
\qquad
\begin{bmatrix}
1 & 2 & 0 & : & -3 \\
2 & 3 & -2 & : & -10 \\
-1 & 0 & 6 & : & 9
\end{bmatrix}
$$

Multiplying the first equation by -2 and adding to the second equation will eliminate x_1 from the second equation. Adding the first equation to the third one will also eliminate x_1 from the third equation. This can be abbreviated by $E_2 - 2E_1 \to E_2$ and $E_3 + E_1 \to E_3$ or by $R_2 - 2R_1 \to R_2$ and $R_3 + R_1 \to R_3$ on the augmented matrix.

$$
\begin{aligned}
x_1 + 2x_2 &= -3 \\
-x_2 - 2x_3 &= -4 \quad\text{or} \\
2x_2 + 6x_3 &= 6
\end{aligned}
\qquad
\begin{bmatrix}
1 & 2 & 0 & : & -3 \\
0 & -1 & -2 & : & -4 \\
0 & 2 & 6 & : & 6
\end{bmatrix}
$$

To eliminate x_2 from the third equation, we perform $E_3 + 2E_2 \to E_3$ (or $R_3 + 2R_2 \to R_3$ on the augmented matrix).

$$
\begin{aligned}
x_1 + 2x_2 &= -3 \\
-x_2 - 2x_3 &= -4 \quad\text{or} \\
2x_3 &= -2
\end{aligned}
\qquad
\begin{bmatrix}
1 & 2 & 0 & : & -3 \\
0 & -1 & -2 & : & -4 \\
0 & 0 & 2 & : & -2
\end{bmatrix}
$$

The system is now in echelon form. Starting at the bottom, we work upward to eliminate unknowns *above* the leading variables of each equation (back-substitution). To eliminate x_3 from the second equation, we perform $E_2 + E_3 \to E_2$ (or $R_2 + R_3 \to R_2$).

$$
\begin{aligned}
x_1 + 2x_2 &= -3 \\
-x_2 &= -6 \quad\text{or} \\
2x_3 &= -2
\end{aligned}
\qquad
\begin{bmatrix}
1 & 2 & 0 & : & -3 \\
0 & -1 & 0 & : & -6 \\
0 & 0 & 2 & : & -2
\end{bmatrix}
$$

To eliminate x_2 from the first equation, we perform $E_1 + 2E_2 \to E_1$ (or $R_1 + 2R_2 \to R_1$).

$$
\begin{aligned}
x_1 \quad\quad &= -15 \\
-x_2 &= -6 \quad\text{or} \\
2x_3 &= -2
\end{aligned}
\qquad
\begin{bmatrix}
1 & 0 & 0 & : & -15 \\
0 & -1 & 0 & : & -6 \\
0 & 0 & 2 & : & -2
\end{bmatrix}
$$

Finally, by $\frac{1}{2}E_3 \to E_3$ (or $\frac{1}{2}R_3 \to R_3$) and $(-1)E_2 \to E_2$ (or $(-1)R_2 \to R_2$), we get

$$
\begin{aligned}
x_1 \quad\quad &= -15 \\
x_2 &= 6 \quad\text{or} \\
x_3 &= -1
\end{aligned}
\qquad
\begin{bmatrix}
1 & 0 & 0 & : & -15 \\
0 & 1 & 0 & : & 6 \\
0 & 0 & 1 & : & -1
\end{bmatrix}
$$

Hence, $x_1 = -15$, $x_2 = 6$, and $x_3 = -1$ is the only solution of the system.

■ EXAMPLE 10   Solve the system

$$x + 3y - z = 4$$
$$-2x + y + 3z = 9$$
$$4x + 2y + z = 11$$

SOLUTION   Working with the augmented matrix we have

$$
\begin{bmatrix}
1 & 3 & -1 & : & 4 \\
-2 & 1 & 3 & : & 9 \\
4 & 2 & 1 & : & 11
\end{bmatrix}
\quad
\boxed{\begin{array}{l} R_2 + 2R_1 \rightarrow R_2 \\ R_3 - 4R_1 \rightarrow R_3 \end{array}}
\quad
\begin{bmatrix}
1 & 3 & -1 & : & 4 \\
0 & 7 & 1 & : & 17 \\
0 & -10 & 5 & : & -5
\end{bmatrix}
$$

$$
\boxed{R_3 + \tfrac{10}{7}R_2 \rightarrow R_3}
\quad
\begin{bmatrix}
1 & 3 & -1 & : & 4 \\
0 & 7 & 1 & : & 17 \\
0 & 0 & \frac{45}{7} & : & \frac{135}{7}
\end{bmatrix}
\quad
\boxed{\tfrac{7}{45}R_3 \rightarrow R_3}
$$

$$
\begin{bmatrix}
1 & 3 & -1 & : & 4 \\
0 & 7 & 1 & : & 17 \\
0 & 0 & 1 & : & 3
\end{bmatrix}
\quad
\boxed{\begin{array}{l} R_2 - R_3 \rightarrow R_2 \\ R_1 + R_3 \rightarrow R_1 \end{array}}
\quad
\begin{bmatrix}
1 & 3 & 0 & : & 7 \\
0 & 7 & 0 & : & 14 \\
0 & 0 & 1 & : & 3
\end{bmatrix}
$$

$$
\boxed{\tfrac{1}{7}R_2 \rightarrow R_2}
\quad
\begin{bmatrix}
1 & 3 & 0 & : & 7 \\
0 & 1 & 0 & : & 2 \\
0 & 0 & 1 & : & 3
\end{bmatrix}
\quad
\boxed{R_1 - 3R_2 \rightarrow R_1}
\quad
\begin{bmatrix}
1 & 0 & 0 & : & 1 \\
0 & 1 & 0 & : & 2 \\
0 & 0 & 1 & : & 3
\end{bmatrix}
$$

Hence $x = 1$, $y = 2$, and $z = 3$. The *geometric* solution is the point $P(1, 2, 3)$, which is the intersection of the three planes defined by the equations of the system (Fig. 1.2).

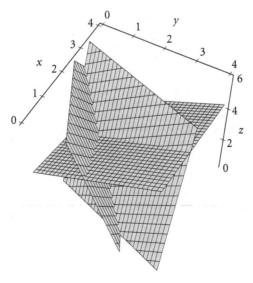

Figure 1.2   Exactly one solution: three planes intersecting at one point.

■ EXAMPLE 11   (Infinitely Many Solutions)   Solve the system

$$x + 2y - z = 4$$
$$2x + 5y + 2z = 9$$
$$x + 4y + 7z = 6$$

SOLUTION   By using elementary row operations on the augmented matrix of the

system, we get $\begin{bmatrix} 1 & 0 & -9 & : & 2 \\ 0 & 1 & 4 & : & 1 \\ 0 & 0 & 0 & : & 0 \end{bmatrix}$. This is the augmented matrix of the system

$$x_1 \quad\quad - 9x_3 = 2$$
$$x_2 + 4x_3 = 1$$

Hence, if $x_3 = r$ for any scalar r, then $x_2 = -4r + 1$ and $x_1 = 9r + 2$ by the first two equations. Therefore, the general solution is

$$x_1 = 9r + 2$$
$$x_2 = -4r + 1 \quad\quad r \in \mathbf{R}$$
$$x_3 = r$$

In this case the intersection of the planes defined by each of the equations is a straight line (Fig. 1.3). (Straight lines are studied in Chapter 2.)

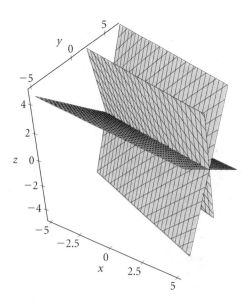

Figure 1.3   The three planes have a common line.

■ EXAMPLE 12 (No Solutions) Solve the system

$$y - 2z = -5$$
$$2x - y + z = -2$$
$$4x - y \quad\quad = -4$$

SOLUTION The augmented matrix of the system reduces to

$$\begin{bmatrix} 2 & -1 & 1 & : & -2 \\ 0 & 1 & -2 & : & -5 \\ 0 & 0 & 0 & : & 5 \end{bmatrix}$$

The last row corresponds to the false expression $0x_3 = 5$. Hence, the system is inconsistent.

The three graphical cases of an inconsistent three-variable system are (Fig. 1.4):

1. Three planes parallel to each other
2. Two planes parallel and the third one intersecting them
3. Three intersecting planes but with no common intersection

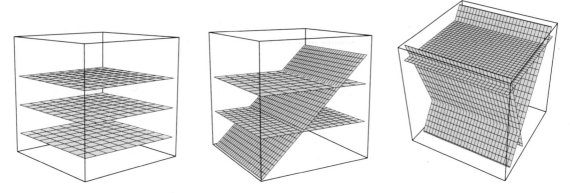

Figure 1.4 **No solutions.**

■ EXAMPLE 13 (Book Packing) A student is assigned a new dormitory room. While packing her books in boxes, she notices that if she puts seven books in each box she ends up with one book unpacked. If, on the other hand, she puts eight books in each box then the last box contains only one book. How many books and how many boxes does she have?

SOLUTION Let x be the number of books and y be the number of boxes. Then $7y = x - 1$ in the first case, and $8y = x + 7$ in the second. Solving this system of two equations yields $x = 57$ books and $y = 8$ boxes.

Linear Systems with CAS

Maple

```
>solve({3*x+2*y=1,2*x-y=-3},{x,y});
```
$$\{x = \frac{-5}{7}, y = \frac{11}{y}\}$$

Mathematica

```
In[1]:=  Solve[{3x+2y==1,2x-y==-3},{x,y}]
```

```
Out[1]=          5        11
          {{x -> -(-), y -> --}}
                  7        7
```

MATLAB (Symbolic Toolbox)

```
>> solve('3*x+2*y=1','2*x-y=-3')
ans =
x = -5/7, y = 11/7
```

Exercises 1.1

Linear Equations

For Exercises 1–4 consider the following equations:

(a) $3x - 5 - x = 2x + 2y + 5$
(b) $2x + 3y - x = x + 3y - 1$
(c) $1 + x + y + z = 1$
(d) $x + y + z = 1 + y$
(e) $x + y + z = 1 + y - w + t$
(f) $xy + z = x - y$

1. Identify each equation as linear or nonlinear. If an equation is linear, classify it as homogeneous or non-homogeneous. (b)

2. For each linear equation, write its coefficients, its constant term, and its associated homogeneous equation.

3. For each linear equation, order its variables and find the leading variable and the free variables.

4. For each linear equation, find, if possible, the general solution and two particular solutions.

5. Which of the points $P(2, -3, 0)$, $Q(2, -3, -1)$, $R(\frac{1}{2}, -\frac{1}{2}, -2)$, $S(\frac{1}{2}, \frac{1}{2}, -2)$ are in the plane with equation $x - y + z = -2$?

6. Which of the points $P(1, -1, 1)$, $Q(-2, 5, 3)$, $R(\frac{1}{2}, \frac{1}{7}, 10)$, $S(0, 0, 0)$ are in the plane with equation $2x - 7y + z = 10$?

7. Find all the values of a such that each of the following equations has

   (i) Exactly one solution;
   (ii) Infinitely many solutions;
   (iii) No solutions.

   a. $a^2x - 2 = 4x + a$

   b. $(a^2 - 4)x = 3$

   c. $(a^2 - 4)x = 0$

   d. $ax - a^2y = 3a$

8. Find all the *real* values of a such that each of the following equations has

(i) Infinitely many solutions;

(ii) No solutions.

a. $a^2(x + y) - x - y - a + 1 = 0$

b. $a(x + y) - x - y - a + 1 = 0$

Linear Systems

9. Rewrite the linear system in standard form.

$$2x + 4z + 1 = 0$$
$$2z + 2w - 2 = x$$
$$-2x - z + 3w = -3$$
$$y + z + t = w + 4$$

Find:

a. The coefficient matrix.

b. The vector of constants.

c. The augmented matrix.

d. The associated homogeneous system.

10. Use back-substitution to solve the system

$$x_1 + 2x_2 + x_3 \quad\quad + x_5 \quad\quad = -1$$
$$-2x_3 \quad\quad + 4x_6 = \quad 2$$
$$4x_4 - 2x_5 \quad\quad = \quad 0$$

11. Use back-substitution to solve the homogeneous system associated with the system of Exercise 10.

12. Let

$$M = \begin{bmatrix} 1 & -1 & 1 & -5 & 6 & -1 & 1 \\ 0 & 0 & 0 & 0 & -1 & 1 & 0 \\ 0 & 0 & -2 & 0 & 2 & 0 & 0 \end{bmatrix}$$

a. Write a system whose augmented matrix is M.

b. Write the associated homogeneous system with the linear system of part (a).

c. Apply one elementary row operation to M so that the resulting matrix corresponds to a system in echelon form.

13. Find the general solution of the system whose augmented matrix is matrix M, defined in Exercise 12.

14. Find the general solution of the associated homogeneous system with the system of Exercise 13.

15. Find the intersection of the straight lines $y + x = 1$ and $y - 2x = \frac{1}{2}$.

16. Find the intersection of the straight lines $2y - 3x = 5$ and $y + 2x = 20$.

17. Without actually solving the systems, show that they are equivalent.

$$x - y + z = 1 \quad\quad 4x - 4y + 4z = \quad 4$$
$$2x + 2y - 3z = -2 \quad\quad 2x + 2y - 3z = -2$$
$$-3x + 4y + 4z = -1 \quad\quad 5y + 2z = -2$$

In Exercises 18–28 find the consistent systems and compute their general solutions.

18. $\quad -x + y - z = 1$
$$-2x + y + 3z = 10$$
$$3x + y + 2z = 3$$

19. $\quad\quad\quad y + 2z = \quad 6$
$$3x - 3y - 3z = -15$$
$$x + 3y + 3z = \quad 11$$

20. $\quad\quad 3y + z = -9$
$$3x + y \quad\quad = -8$$
$$3x + 7y + 2z = -26$$

21. $\quad 3x + y + 3z = \quad 15$
$$-x + 3y - z = -5$$
$$2x + 4y + 2z = \quad 9$$

22. $\quad -x + 3y - 2z = -17$
$$-2x - 3y \quad\quad = \quad 14$$
$$-3x - y - 2z = \quad 1$$

23. $\quad x + y - 3z = 2$
$$-3x + y + z = 6$$

24. $\quad \frac{1}{3}x + \frac{1}{3}y + z = \quad 5$
$$-\frac{1}{3}x - \frac{1}{2}y - \frac{1}{3}z = -\frac{5}{3}$$

25. $\quad\quad 3y + z - w = 3$
$$x + y - 2z \quad\quad = 6$$
$$-2x + y + 2z - w = 9$$

26. $\quad x + 3y + z - w = \quad 0$
$$3x + y + 3z \quad\quad = -2$$
$$2x + 6y + 2z - 2w = \quad 2$$

27. $x + y = 1$
$$y + z = 1$$
$$z + w = 1$$
$$x + w = 1$$

28. $x + y = 1$
$$y + z = 1$$
$$z + w = 1$$
$$y + w = 1$$

29. Solve the ancient Chinese system mentioned in Section 1.1.

In Exercises 30–34, solve the systems with the given augmented matrices.

30. $\begin{bmatrix} -1 & 2 & 0 & -6 \\ 3 & -2 & -1 & 10 \\ 3 & 2 & 2 & -14 \end{bmatrix}$

31. $\begin{bmatrix} -1 & -1 & 0 & -1 \\ 0 & 2 & 1 & 3 \\ -2 & 1 & 3 & 7 \end{bmatrix}$

32. $\begin{bmatrix} -1 & -2 & -1 & 1 \\ -1 & -1 & 1 & 4 \\ 1 & 1 & -1 & 4 \end{bmatrix}$

33. $\begin{bmatrix} 2 & -1 & 3 & -4 \\ -1 & -3 & -1 & 7 \\ 2 & -1 & 0 & 8 \end{bmatrix}$

34. $\begin{bmatrix} -3 & -1 & 1 & 2 \\ 6 & 2 & -2 & -4 \\ 3 & 1 & -1 & -2 \end{bmatrix}$

35. Solve the system for θ:

$$\sin \theta - 4 \cos \theta = 4$$
$$4 \sin \theta - 4 \cos \theta = 4$$

36. Solve the system for θ:

$$2 \sin \theta + \sqrt{2} \tan \theta = 2\sqrt{2}$$
$$4 \sin \theta - 3\sqrt{2} \tan \theta = -\sqrt{2}$$

37. Consider the homogeneous system

$$a_1 x + b_2 y = 0$$
$$a_2 x + b_2 y = 0$$

   a. Show that if $x = x_0, y = y_0$ is a solution of this system, so is $x = kx_0, y = ky_0$, for any constant k.

   b. Show that if $x = x_1, y = y_1$ and $x = x_2, y = y_2$ are two solutions, so is $x = x_1 + x_2, y = y_1 + y_2$.

For Exercises 38 and 39 consider the system

$$a_1 x + b_1 y = c_1$$
$$a_2 x + b_2 y = c_2$$

38. Let $a_1 b_2 - b_1 a_2 \neq 0$. Show that

   a. The system has exactly one solution. Compute this solution.

   b. The associated homogeneous system has only the trivial solution.

39. Let $a_1 b_2 - b_1 a_2 = 0$. Show that

   a. The system has either infinitely many solutions or no solutions.

   b. The associated homogeneous system has nontrivial solutions.

Applications

40. Peter and Pam start a game of baseball cards, each having the same number of cards. During the first round Peter won 20 cards, but during the second round he lost $\frac{2}{3}$ of the cards he had started out with. At the end of the second round, Pam had three times as many cards as Peter did. What was the total number of cards?

41. A math teacher assigns three exercises and asks $\frac{1}{4}$ of her students to solve the first exercise, $\frac{3}{8}$, the second one, and $\frac{5}{16}$, the third one. Given that 2 students were absent, what is the total number of students in the class?

42. A bookshelf contains books whose number is $\frac{3}{5}$ of the number of books that are in the bookshelf next to it. If we move 10 books from the first bookshelf to the second one, the latter will have twice as many books as the first one. How many books were there in each bookshelf?

43. Forty-five U.S. Naval Academy midshipmen aboard a boat with 15 crew members are supplied with food to last 30 days. Twelve days later they rescue 10 people in a small inflatable boat with food supplies for 4 days. How long will the total available food supply last, provided that all people aboard the boat are given the same amount of daily food?

44. One-third of a mathematics textbook consists of exercise sections, with a total of 90 such sections. Given that each exercise section averages 25 exercises and that 15 exercises will fit on one page, how many pages are there in the book?

45. Determine the angles of a parallelogram with the property that two consecutive angles differ by 20°.

46. Compute the length of each side of a parallelogram with perimeter 10 ft and length difference of two consecutive sides of 1 ft.

47. Compute the length of each side of an isosceles triangle with perimeter 16 ft and length difference of two of the unequal sides of 2 ft.

48. When we add an extra hard disk to a personal computer, the new system costs $1400. It is known that

$\frac{1}{3}$ of the computer value plus $\frac{1}{5}$ of the hard disk total $400. What is the cost of the hard disk?

49. A 20-column crossword puzzle's black squares number $\frac{1}{8}$ of the total number of squares, and they are 5 less than the $\frac{1}{7}$ of them. What is the number of rows of the crossword puzzle?

In Exercises 50 and 51, although the resulting systems are nonlinear, they can be both reduced to linear.[2]

50. Half an hour after one of the two authors left home to drive to work, one lane of the freeway was closed due to an accident. This resulted in the author's driving, on the average, at $\frac{2}{5}$ of his former speed for the rest of the way and made him 1 h 3 min late for work. Had the accident occurred 15 mi farther down the road, he would have arrived 27 min sooner. How far does the author live from work? What is his regular traveling speed?

51. Two sailboats leave Annapolis at different times, which sets them 4 mi apart, with the same destination and the same speed. Half an hour after the departure of the second boat, the wind drops significantly. This slows the boats, which now travel at $\frac{2}{3}$ of their previous speed. The second boat reaches its destination 45 min later than the predicted time. The first boat is only 35 min late. What is the distance traveled by the boats?

1.2   Gauss Elimination

Reader's Goals for This Section

1. To recognize a matrix in row echelon and in reduced row echelon form.
2. To master the Gauss Elimination process.

We now take a closer look at the method we used to solve linear systems. It is called *Gauss elimination*,[3] although it far precedes the times of Gauss. In fact, it was used to solve the system in the ancient Chinese book mentioned in Section 1.1. We also discuss a variant of this elimination, *Gauss-Jordan elimination*.

Row Echelon Form; Gauss Elimination

Matrices are discussed in detail in Chapter 3. In this paragraph we introduce only the necessary notation to solve linear systems efficiently.

A matrix is a rectangular arrangement of numbers, called **entries**. The rows of a matrix are numbered top to bottom and the columns, left to right. So by the last column, we mean the rightmost column. The entries are numbered according to their (row, column) position. If a matrix has m rows and n columns it is called **size** $m \times n$.

---

[2] Both exercises are modeled after a problem due to Sam Loyd, one of the best puzzle and chess problem writers of all time.

[3] **Karl Friedrich Gauss** (1777–1855) is considered to be one of the greatest mathematicians of all time. He was born in Germany, and he was a child prodigy. At the age of 18 he succeeded in constructing a 17-sided regular polygon by ruler and compass, solving a 2000-year-old problem. He wrote *Disquisitiones Arithmeticae*, a masterpiece in number theory, and proved the *fundamental theorem of algebra*. In his time Gauss was known as the prince of mathematicians.

M, N, and P are matrices. M and N are size 3×4, but P is size 2×3. The $(2, 3)$ entry of P is 1. The $(1, 2)$ entry of N is -6.

$$M = \begin{bmatrix} -1 & 0 & 1 & 2 \\ 0 & 0 & -3 & 4 \\ 0 & 0 & 0 & 0 \end{bmatrix} \qquad N = \begin{bmatrix} 1 & -6 & 0 & 2 \\ 0 & 0 & 1 & 4 \\ 0 & 0 & 0 & 0 \end{bmatrix} \qquad P = \begin{bmatrix} 0 & 2 & -6 \\ 0 & -1 & 1 \end{bmatrix}$$

A **zero row** of a matrix is a row that consists entirely of zeros. A **nonzero row** is a row that has at least one nonzero entry. In the same manner, we can talk about zero and nonzero *columns*. The first nonzero entry of a nonzero row is called a **leading entry**. If a leading entry happens to be 1, we call it a **leading 1**.

For the matrix M we have the following: Its first two rows are nonzero. Its third row is a zero row. Its leading entries are -1 and -3. Its second column is a zero column. The remaining columns are nonzero. Note that none of the leading entries of M are leading 1s. In contrast, all leading entries of N are leading 1s.

DEFINITION

(Row Echelon Form)

Consider the following conditions on a matrix:

1. *All zero rows are at the bottom of the matrix.*
2. *The leading entry of each nonzero row after the first occurs to the right of the leading entry of the previous row.*
3. *The leading entry in any nonzero row is 1.*
4. *All entries in the column above and below a leading 1 are zero.*

If a matrix satisfies the first two conditions, we say that it is (in) **row echelon form** (or simply, **echelon form**). If a matrix satisfies all four conditions, we say that it is (in) **reduced row echelon form** (or just **reduced echelon form**). A reduced echelon form matrix is always echelon form.

M and N are echelon form matrices. N is also reduced echelon form. M is not, because condition 3 fails. Matrix P is not echelon form, because condition 2 fails. To study a few more examples, let

$$A = \begin{bmatrix} 1 & 0 & 0 & 0 \\ 0 & 0 & 1 & 0 \\ 0 & 0 & 0 & 0 \end{bmatrix} \qquad B = \begin{bmatrix} 1 & 0 & 0 & -6 \\ 0 & 1 & 0 & 0 \\ 0 & 0 & 1 & -1 \end{bmatrix} \qquad C = \begin{bmatrix} 1 & 0 & 1 \\ 0 & 0 & 1 \\ 0 & 0 & 1 \end{bmatrix}$$

$$D = \begin{bmatrix} 1 & 1 & 0 & 0 & 2 \\ 0 & 0 & 1 & 0 & 3 \\ 0 & 0 & 0 & 1 & 4 \end{bmatrix} \qquad E = \begin{bmatrix} 0 & 0 \\ 1 & 0 \end{bmatrix} \qquad F = \begin{bmatrix} 1 & 7 & 0 & 9 & 0 \\ 0 & 0 & 1 & -8 & 0 \\ 0 & 0 & 0 & 0 & 1 \end{bmatrix}$$

$$G = \begin{bmatrix} 1 & 0 & -1 & 0 \\ 0 & 1 & 0 & 0 \\ 0 & 0 & 1 & 0 \end{bmatrix} \qquad \cdot H = \begin{bmatrix} 1 & 0 & 0 & 0 \\ 0 & 0 & 1 & 0 \\ 0 & 0 & 0 & -2 \end{bmatrix}$$

A, B, D, F, G, H are in echelon form, because the first two conditions hold. Of these, A, B, D, F are in reduced echelon form, because all four conditions hold. G and H are not in reduced echelon form. For G condition 4 fails. For H condition 3 fails. C and E are not in echelon form. For C condition 2 fails. For E condition 1 fails.

DEFINITION   **(Equivalent Matrices)**

Two matrices are **(row) equivalent** if one can be obtained from the other by a finite sequence of elementary row operations. Sometimes we use the abbreviation

$$A \sim B$$

for the statement "matrix A is equivalent to B."

■ EXAMPLE 14

$$A = \begin{bmatrix} 0 & 3 \\ 1 & 2 \\ -1 & 1 \end{bmatrix} \quad \text{and} \quad B = \begin{bmatrix} 1 & 2 \\ 0 & 3 \\ 0 & 0 \end{bmatrix}$$

are equivalent, because

$$A \boxed{R_1 \leftrightarrow R_2} \begin{bmatrix} 1 & 2 \\ 0 & 3 \\ -1 & 1 \end{bmatrix} \boxed{R_3 + R_1 \to R_3} \begin{bmatrix} 1 & 2 \\ 0 & 3 \\ 0 & 3 \end{bmatrix} \boxed{R_3 - R_2 \to R_3} B$$

We say that a matrix *reduces to (reduced) echelon form* if it is equivalent to a matrix in (reduced) echelon form.

■ EXAMPLE 15   Show that the following matrix reduces to reduced echelon form.

$$\begin{bmatrix} 1 & 1 & 0 & 0 & 0 \\ 3 & 3 & 1 & 0 & -1 \\ -2 & -2 & 1 & 1 & 0 \end{bmatrix}$$

SOLUTION   This can be seen by using the following sequence of elementary row operations.

$$\begin{bmatrix} 1 & 1 & 0 & 0 & 0 \\ 3 & 3 & 1 & 0 & -1 \\ -2 & -2 & 1 & 1 & 0 \end{bmatrix} \boxed{\begin{array}{c} R_2 - 3R_1 \to R_2 \\ R_3 + 2R_1 \to R_3 \end{array}} \begin{bmatrix} 1 & 1 & 0 & 0 & 0 \\ 0 & 0 & 1 & 0 & -1 \\ 0 & 0 & 1 & 1 & 0 \end{bmatrix}$$

$$\boxed{R_3 - R_2 \to R_3} \begin{bmatrix} 1 & 1 & 0 & 0 & 0 \\ 0 & 0 & 1 & 0 & -1 \\ 0 & 0 & 0 & 1 & 1 \end{bmatrix}$$

The last matrix is in reduced echelon form.

The solution method in Example 15 is a special case of the following important algorithm (process) that enables us to reduce any matrix to echelon or reduced echelon form.

Algorithm 1

(Gauss Elimination)

To reduce any matrix to reduced row echelon form, apply the following steps:

Step 1. Find the leftmost nonzero column.

Step 2. If the first row has a zero in the column of Step 1, interchange it with one that has a nonzero entry in the same column.

Step 3. Obtain zeros below the leading entry by adding suitable multiples of the top row to the rows below that.

Step 4. Cover the top row and repeat the same process starting with Step 1 applied to the leftover submatrix. Repeat this process with the rest of the rows. (At this stage the matrix is already in echelon form.)

Step 5. Starting with the last nonzero row, work upward: For each row obtain a leading 1 and introduce zeros above it by adding suitable multiples to the corresponding rows.

The following example illustrates this process.

▪ **EXAMPLE 16**   Apply Gauss elimination to find a reduced echelon form of the matrix.

$$\begin{bmatrix} 0 & 3 & -6 & -4 & -3 & -5 \\ -1 & 3 & -10 & -4 & -4 & -2 \\ 4 & -9 & 34 & 0 & 1 & -21 \\ 2 & -6 & 20 & 2 & 8 & -8 \end{bmatrix}$$

SOLUTION

Step 1. Find the leftmost nonzero column: This is the first column here.

Step 2. If the first row has a zero in the column of Step 1, interchange it with one that has a nonzero entry in the same column.

$$R_1 \leftrightarrow R_2 \qquad \begin{bmatrix} -1 & 3 & -10 & -4 & -4 & -2 \\ 0 & 3 & -6 & -4 & -3 & -5 \\ 4 & -9 & 34 & 0 & 1 & -21 \\ 2 & -6 & 20 & 2 & 8 & -8 \end{bmatrix}$$

Step 3. Obtain 0s below the leading entry by adding suitable multiples of the top row to the rows below that:

$$\begin{matrix} R_3 + 4R_1 \rightarrow R_3 \\ R_4 + 2R_1 \rightarrow R_4 \end{matrix} \qquad \begin{bmatrix} -1 & 3 & -10 & -4 & -4 & -2 \\ 0 & 3 & -6 & -4 & -3 & -5 \\ 0 & 3 & -6 & -16 & -15 & -29 \\ 0 & 0 & 0 & -6 & 0 & -12 \end{bmatrix}$$

Step 4. Cover the top row and repeat the same process starting with Step 1 applied to the leftover submatrix. Repeat this process with the rest of the rows.

$$\begin{bmatrix} -1 & 3 & -10 & -4 & -4 & -2 \\ 0 & 3 & -6 & -4 & -3 & -5 \\ 0 & 3 & -6 & -16 & -15 & -29 \\ 0 & 0 & 0 & -6 & 0 & -12 \end{bmatrix}$$

$$R_3 - R_2 \rightarrow R_3 \quad \begin{bmatrix} -1 & 3 & -10 & -4 & -4 & -2 \\ 0 & 3 & -6 & -4 & -3 & -5 \\ 0 & 0 & 0 & -12 & -12 & -24 \\ 0 & 0 & 0 & -6 & 0 & -12 \end{bmatrix}$$

$$R_4 - \tfrac{1}{2}R_3 \rightarrow R_4 \quad \begin{bmatrix} -1 & 3 & -10 & -4 & -4 & -2 \\ 0 & 3 & -6 & -4 & -3 & -5 \\ 0 & 0 & 0 & -12 & -12 & -24 \\ 0 & 0 & 0 & 0 & 6 & 0 \end{bmatrix}$$

Step 5. *Starting with the last nonzero row, work upward: For each row, obtain a leading 1 and introduce 0s above it by adding suitable multiples to the corresponding rows.*

$$\tfrac{1}{6}R_4 \rightarrow R_4 \quad \begin{bmatrix} -1 & 3 & -10 & -4 & -4 & -2 \\ 0 & 3 & -6 & -4 & -3 & -5 \\ 0 & 0 & 0 & -12 & -12 & -24 \\ 0 & 0 & 0 & 0 & 1 & 0 \end{bmatrix} \begin{array}{l} R_3 + 12R_4 \rightarrow R_3 \\ R_2 + 3R_4 \rightarrow R_2 \\ R_1 + 4R_4 \rightarrow R_1 \end{array}$$

$$\begin{bmatrix} -1 & 3 & -10 & -4 & 0 & -2 \\ 0 & 3 & -6 & -4 & 0 & -5 \\ 0 & 0 & 0 & -12 & 0 & -24 \\ 0 & 0 & 0 & 0 & 1 & 0 \end{bmatrix} \quad -\tfrac{1}{12}R_3 \rightarrow R_3 \quad \begin{bmatrix} -1 & 3 & -10 & -4 & 0 & -2 \\ 0 & 3 & -6 & -4 & 0 & -5 \\ 0 & 0 & 0 & 1 & 0 & 2 \\ 0 & 0 & 0 & 0 & 1 & 0 \end{bmatrix}$$

$$\begin{array}{l} R_2 + 4R_3 \rightarrow R_2 \\ R_1 + 4R_3 \rightarrow R_1 \end{array} \quad \begin{bmatrix} -1 & 3 & -10 & 0 & 0 & 6 \\ 0 & 3 & -6 & 0 & 0 & 3 \\ 0 & 0 & 0 & 1 & 0 & 2 \\ 0 & 0 & 0 & 0 & 1 & 0 \end{bmatrix} \quad \tfrac{1}{3}R_2 \rightarrow R_2$$

$$\begin{bmatrix} -1 & 3 & -10 & 0 & 0 & 6 \\ 0 & 1 & -2 & 0 & 0 & 1 \\ 0 & 0 & 0 & 1 & 0 & 2 \\ 0 & 0 & 0 & 0 & 1 & 0 \end{bmatrix} \quad R_1 - 3R_2 \rightarrow R_1 \quad \begin{bmatrix} -1 & 0 & -4 & 0 & 0 & 3 \\ 0 & 1 & -2 & 0 & 0 & 1 \\ 0 & 0 & 0 & 1 & 0 & 2 \\ 0 & 0 & 0 & 0 & 1 & 0 \end{bmatrix}$$

$$(-1)R_1 \rightarrow R_1 \quad \begin{bmatrix} 1 & 0 & 4 & 0 & 0 & -3 \\ 0 & 1 & -2 & 0 & 0 & 1 \\ 0 & 0 & 0 & 1 & 0 & 2 \\ 0 & 0 & 0 & 0 & 1 & 0 \end{bmatrix}$$

Remarks

1. The first four steps of the algorithm are known as the **forward pass** of the Gauss elimination. They bring the matrix to echelon form. Step 5 is the **backward pass** (or back-substitution). This step reduces the matrix to reduced echelon form.
2. Often the algorithm is described with the introduction of leading 1s in Step 3, which tends to introduce fractions at an early stage of the calculation if the entries were integers. This variant is preferred by programmers in floating-point arithmetic. In addition, in Step 2 the row with the nonzero entry of largest absolute value is moved

to the top. This is called *partial pivoting*, and it is discussed in Section 1.3. Partial pivoting helps control the round-off errors.[4]

3. Sometimes it is convenient to obtain leading 1s at the end of the entire process.

Two Things to Avoid

1. Avoid changing the order of steps in the algorithm (you may end up with the matrix you started out with). Note that the next step in the reduction of $\begin{bmatrix} 1 & -1 & 0 \\ 0 & 1 & -1 \\ 0 & 1 & 0 \end{bmatrix}$

   is $\begin{bmatrix} 1 & -1 & 0 \\ 0 & 1 & -1 \\ 0 & 0 & 1 \end{bmatrix}$ and *not* $\begin{bmatrix} 1 & 0 & -1 \\ 0 & 1 & -1 \\ 0 & 1 & 0 \end{bmatrix}$. In other words, reduce to echelon form first, then use Step 5. The only liberty you are allowed to take is on when to obtain the leading 1s.

2. It is bad practice to combine several elementary operations into one and it may lead to errors. For example, operations of the form $cR_i + dR_j \rightarrow R_i$ and $cR_i + R_j \rightarrow R_i$ with $c \neq 1$ are non-elementary and they should be avoided. The correct operation is $R_i + cR_j \rightarrow R_i$. In other words, *the row that gets replaced should not be multiplied by a scalar $\neq 1$.*

Uniqueness of Reduced Row Echelon Form: Pivots

Gauss elimination implies that each matrix reduces to echelon and reduced echelon form. A matrix can be equivalent to several echelon form matrices but to only one reduced echelon form. This is the claim of the next theorem, whose proof is discussed in Section 4.6.

THEOREM 1

(Uniqueness of Reduced Row Echelon Form)

Every matrix is equivalent to one and only one matrix in reduced echelon form.

Let A, E, and R be three equivalent matrices such that E is in echelon form and R is in reduced echelon form. We call E **an echelon form** of A. By Theorem 1 we can call R **the reduced echelon form** of A.

Note that in any echelon form of a matrix A, the leading entries occur at the same columns. This follows from the uniqueness of the reduced echelon form and the fact that after Step 4 the positions of the leading entries do not change. These fixed positions are called the **pivot positions** of A. The fixed columns that contain pivot positions are called the **pivot columns** of A. A **pivot** is any nonzero entry of a pivot position.

---

[4]Also, unlike people, the computer needs no physical interchange of rows to keep track of the performed operations. Interchanging rows is costly and slow. Programmers use a kind of "renaming rows" called the *permutation vector method*.

For example, consider matrix A. Matrices E and R are two stages of the reduction of A, with E being an echelon form and R being the reduced echelon form of A.

$$A = \begin{bmatrix} 1 & -2 & 0 & 0 & 1 \\ 3 & -6 & -1 & 1 & 1 \\ 4 & -8 & 5 & -1 & 14 \end{bmatrix} \rightarrow E = \begin{bmatrix} 1 & -2 & 0 & 0 & 1 \\ 0 & 0 & -1 & 1 & -2 \\ 0 & 0 & 0 & 4 & 0 \end{bmatrix}$$

$$\rightarrow R = \begin{bmatrix} 1 & -2 & 0 & 0 & 1 \\ 0 & 0 & 1 & 0 & 2 \\ 0 & 0 & 0 & 1 & 0 \end{bmatrix}$$

Columns 1, 3, and 4 are the pivot columns of A, E, and R. The pivot positions for A, E, and R are the positions $(1, 1)$, $(2, 3)$, and $(3, 4)$. The pivots of E are $1, -1, 4$, and the pivots of R are $1, 1, 1$.

In general, *equivalent matrices have the same pivot column positions and pivot positions*.

■ EXAMPLE 17   Columns 1, 2, 4, and 5 are the pivot columns of the matrix in Example 16.

Solution of Linear Systems

Let us see now how to use Gauss elimination to solve any linear system. The process is applied to the augmented matrix of the system. It yields a reduced echelon form matrix whose corresponding system is equivalent to the given one. This last system is easy to solve: First we separate the variables into **leading** and **free**. The leading variables are the ones that correspond to pivot positions. The remaining variables, if any, are the free variables. Then we write the leading variables in terms of the free and/or constants. It is customary to assign new names to the free variables and call them **parameters**. The parameters can take on any scalar value.

■ EXAMPLE 18   (Infinitely Many Solutions)   Solve the system.

$$3x_2 - 6x_3 - 4x_4 - 3x_5 = -5$$
$$-x_1 + 3x_2 - 10x_3 - 4x_4 - 4x_5 = -2$$
$$2x_1 - 6x_2 + 20x_3 + 2x_4 + 8x_5 = -8$$

SOLUTION   By Gauss elimination to the augmented matrix of the system (the actual reduction is left as an exercise), we get

$$\begin{bmatrix} 1 & 0 & 4 & 0 & 1 & : & -3 \\ 0 & 1 & -2 & 0 & -1 & : & 1 \\ 0 & 0 & 0 & 1 & 0 & : & 2 \end{bmatrix}$$

Therefore, the original system reduces to the equivalent system

$$
\begin{aligned}
x_1 \quad\quad + 4x_3 \quad\quad + x_5 &= -3 \\
x_2 - 2x_3 \quad\quad - x_5 &= 1 \\
x_4 \quad\quad &= 2
\end{aligned}
$$

Because the pivot columns are columns 1, 2, and 4, x_1, x_2, and x_4 are the leading variables, and x_3, x_5 are the free variables. The free variables can take on any value. Let $x_5 = r, x_3 = s$. Next, we write the leading variables x_1, x_2 in terms of r and s to get

$$
\begin{aligned}
x_1 &= -4s - r - 3 \\
x_2 &= 2s + r + 1 \\
x_3 &= s \quad\quad\quad\quad\quad \text{for any } r, s \in \mathbf{R} \\
x_4 &= 2 \\
x_5 &= r
\end{aligned}
$$

This is the general solution of the system. The solution set is a two-parameter infinite set.

■ **EXAMPLE 19** (One Solution) Determine whether the five planes defined by the following equations pass through the same point.

$$
\begin{aligned}
- x \quad\quad + z &= -2 \\
2x - y + z &= 1 \\
- 3x + 2y - 2z &= -1 \\
x - 2y + 3z &= -2 \\
5x + 2y + 6z &= -1
\end{aligned}
$$

SOLUTION Row reduction of the augmented matrix yields

$$
\begin{bmatrix}
1 & 0 & 0 & : & 1 \\
0 & 1 & 0 & : & 0 \\
0 & 0 & 1 & : & -1 \\
0 & 0 & 0 & : & 0 \\
0 & 0 & 0 & : & 0
\end{bmatrix}
$$

Therefore, $x = 1$, $y = 0$, and $z = -1$. Hence, all five planes pass through the point with coordinates $(1, 0, -1)$.

Note that as soon as the pivot columns are known (end of Step 4 of Algorithm 1) we can tell whether our system is consistent or not: If the last column is a pivot column, then the system is inconsistent. This is because one of the rows of the echelon form augmented matrix will then be of the form

$$
\begin{bmatrix} 0 & 0 & 0 & \cdots & 0 & : & c \end{bmatrix}
$$

with pivot c (hence $c \neq 0$). But this corresponds to the equation

$$0x_1 + \cdots + 0x_n = c$$

which is false for any values of the variables, because $c \neq 0$. If the system is inconsistent, there is no need to continue with Step 5.

■ EXAMPLE 20   (No Solutions)   Determine whether the system is consistent or not.

$$
\begin{aligned}
x_1 \quad\quad + 2x_3 - 2x_4 &= 1 \\
-x_1 + x_2 \quad\quad + x_4 &= -2 \\
x_2 + 2x_3 - x_4 &= 1
\end{aligned}
$$

SOLUTION   Reduction yields

$$
\begin{bmatrix}
1 & 0 & 2 & -2 & : & 1 \\
0 & 1 & 2 & -1 & : & -1 \\
0 & 0 & 0 & 0 & : & 2
\end{bmatrix}
$$

Because the last column of the augmented matrix is a pivot column the system is inconsistent.

Of course, during *any* stage of the reduction the presence of a row like this one implies that the system is inconsistent. Often, we do not need to complete Gauss elimination, as the reduction $\begin{bmatrix} 1 & 1 & : & 0 \\ 1 & 1 & : & 2 \\ 3 & 4 & : & 5 \end{bmatrix} \rightarrow \begin{bmatrix} 1 & 1 & : & 0 \\ 0 & 0 & : & 2 \\ 3 & 4 & : & 5 \end{bmatrix}$ shows.

Algorithm 2

(Solution of Linear System)

To solve any linear system:

Step 1. *Apply Gauss elimination to the augmented matrix of the system (forward pass). If during any stage of this process it is found that the last column is a pivot column, stop. In this case the system is inconsistent. Otherwise, continue with Step 2.*

Step 2. *Complete the Gauss elimination. Write the system that corresponds to the reduced echelon form augmented matrix, ignoring any zero equations.*

Step 3. *Separate the variables of the reduced system into leading and free (if any). Write the leading variables in terms of the free variables or constants.*

Let us now draw a few important conclusions from studying Algorithm 2.

THEOREM 2

(Existence of Solutions)

A linear system is consistent if and only if the last column of its augmented matrix is **not** a pivot column, or, equivalently, if any echelon form of the augmented matrix does **not** have a row of the form

$$
\begin{bmatrix} 0 & 0 & 0 & \cdots & 0 & : & c \end{bmatrix} \qquad \text{with } c \neq 0
$$

Step 1 is sufficient not only for telling whether the system is consistent or not, but for finding the number of solutions as well. If a consistent system has free variables, then Step 3 shows that there are infinitely many solutions (by letting the parameters take on any value). If there are no free variables, then the leading variables are constants, so we get exactly one solution. Because leading variables correspond to pivot columns, we see that a consistent system has exactly one solution if all columns but the last are pivot columns. Let us summarize.

THEOREM 3

(Uniqueness of Solutions)

1. A consistent linear system has exactly one solution if and only if it has no free variables.
2. A consistent linear system has exactly one solution if and only if each column of the augmented matrix other than the last one is a pivot column and the last column is not a pivot column.

We should keep in mind that the presence of free variables does not guarantee infinitely many solutions, because the system may be inconsistent, as Example 20 shows.

■ EXAMPLE 21   What can you say about the systems whose augmented matrices reduce to the given echelon form matrices? (We need not specify the letter constants or draw the separator lines.)

$$\begin{bmatrix} 1 & a & b & d & g \\ 0 & 2 & c & e & h \\ 0 & 0 & 3 & f & i \\ 0 & 0 & 0 & 4 & j \end{bmatrix} \quad \begin{bmatrix} 1 & a & b & c & e \\ 0 & 0 & 2 & d & f \\ 0 & 0 & 0 & 3 & g \end{bmatrix} \quad \begin{bmatrix} 1 & a & b & d & f \\ 0 & 2 & c & e & g \\ 0 & 0 & 0 & 0 & 3 \end{bmatrix} \quad \begin{bmatrix} 1 & a & b & c \\ 0 & 0 & 0 & 2 \\ 0 & 0 & 0 & 0 \end{bmatrix}$$

SOLUTION   The first two systems are consistent, because their last columns are nonpivot columns (Theorem 2). The last two systems are inconsistent, because their last columns are pivot columns (Theorem 2). The first system has exactly one solution, because each column but the last is a pivot column (Theorem 3). The second system has infinitely many solutions, because there is a nonpivot column (the second one) other than the last column (Theorem 3).

A COMMON MISTAKE   The second system has a solution even if $g = 0$. Do not confuse $\begin{bmatrix} 0 & 0 & 0 & 3 & 0 \end{bmatrix}$ with $\begin{bmatrix} 0 & 0 & 0 & 0 & 3 \end{bmatrix}$.

Theorems 2 and 3 imply the following.

THEOREM 4

(Number of Solutions)

For any linear system, only one of the following is true:

1. The system has exactly one solution.
2. The system has infinitely many solutions.
3. The system has no solutions.

Finally, let us specialize to the case of a homogeneous linear system.

THEOREM 5   **(Solutions of Homogeneous Linear Systems)**

1. A homogeneous linear system has either only the trivial solution or infinitely many solutions.
2. A homogeneous linear system has infinitely many solutions if and only if it has free variables.
3. If a homogeneous system has more unknowns than equations, it has infinitely many solutions.   ▾

PROOF   Any homogeneous linear system is consistent, because it has the trivial solution as a solution. So Parts 1 and 2 follow from Theorem 3. To prove Part 3 we observe that because the system of the reduced echelon form augmented matrix has more unknowns than equations, there should be free variables. Hence, the system has infinitely many solutions by Part 2.   ■

Note that for homogeneous systems, the presence of free variables *does* guarantee infinitely many solutions.

■ EXAMPLE 22   Show that the system has nontrivial solutions.

$$x_1 + x_2 + x_3 = 0$$
$$x_1 - x_2 - x_3 = 0$$

SOLUTION   Because the system is homogeneous with more unknowns than equations, it has infinitely many solutions; hence, it has infinitely many *nontrivial* solutions.   ■

Gauss-Jordan Elimination

An interesting variant of Gauss elimination occurs if, during the forward pass, we first produce leading 1s and then 0s below and **above** them. So, by the time the forward pass is done, the matrix is already in reduced row echelon form. This is known as **Gauss-Jordan elimination.**[5]

■ EXAMPLE 23   Find the reduced echelon form of A by using Gauss-Jordan elimination.

$$A = \begin{bmatrix} 1 & 1 & 0 \\ 0 & 2 & -2 \\ 0 & 2 & 1 \\ 0 & -1 & 0 \end{bmatrix}$$

---

[5] **Wilhelm Jordan** (1842–1899), German engineer, wrote the popular *Pocket Book of Practical Geometry*. According to Gewirtz, Sitomer, and Tucker "he devised the pivot reduction algorithm, known as Gauss-Jordan elimination, for geodetic reasons."

SOLUTION   First we scale the second row to get a leading 1. Then we obtain 0s below and *above* this leading 1. Then we repeat the process.

$$A \sim \begin{bmatrix} 1 & 1 & 0 \\ 0 & \boxed{1} & -1 \\ 0 & 2 & 1 \\ 0 & -1 & 0 \end{bmatrix} \sim \begin{bmatrix} 1 & 0 & 1 \\ 0 & \boxed{1} & -1 \\ 0 & 0 & 3 \\ 0 & 0 & -1 \end{bmatrix} \sim \begin{bmatrix} 1 & 0 & 1 \\ 0 & 1 & -1 \\ 0 & 0 & \boxed{1} \\ 0 & 0 & -1 \end{bmatrix} \sim \begin{bmatrix} 1 & 0 & 0 \\ 0 & 1 & 0 \\ 0 & 0 & 1 \\ 0 & 0 & 0 \end{bmatrix}$$

Row Reduction by CAS

Maple

```
> with(linalg):
> rref(matrix([[1,2,3],[2,2,3],[3,3,3]]));
```

$$\begin{bmatrix} 1 & 0 & 0 \\ 0 & 1 & 0 \\ 0 & 0 & 1 \end{bmatrix}$$

Mathematica

```
In[1]:=
      RowReduce[{{1,2,3},{2,2,3},{3,3,3}}]
Out[1]=
      {{1, 0, 0}, {0, 1, 0}, {0, 0, 1}}
```

MATLAB

```
>> rref([1 2 3; 2 2 3; 3 3 3])
ans =
      1     0     0
      0     1     0
      0     0     1
```

Exercises 1.2

Echelon Form

In Exercises 1–5 place each matrix into one of the following categories:

1. Echelon but *not* reduced row echelon form.
2. Reduced row echelon form.
3. Not echelon form.

1. a. $\begin{bmatrix} 0 & 0 \\ 0 & 1 \end{bmatrix}$ b. $\begin{bmatrix} 1 & 0 \\ 0 & 0 \end{bmatrix}$

c. $\begin{bmatrix} 0 & 1 \\ 1 & 0 \end{bmatrix}$ d. $\begin{bmatrix} 2 & -1 \\ 0 & 1 \end{bmatrix}$

2. a. $\begin{bmatrix} 1 & 0 & 0 \\ 0 & 0 & 1 \\ 0 & 1 & 0 \end{bmatrix}$ **b.** $\begin{bmatrix} 1 & 1 \\ 0 & 0 \\ 0 & 0 \end{bmatrix}$

c. $\begin{bmatrix} -1 & 0 & -1 \\ 0 & 0 & 1 \\ 0 & 0 & 1 \end{bmatrix}$ **d.** $\begin{bmatrix} 1 & 0 & 0 & -6 \\ 0 & 0 & 1 & 5 \\ 0 & 0 & 0 & 0 \end{bmatrix}$

3. a. $\begin{bmatrix} 1 & 0 & 0 \\ 0 & 1 & 0 \\ 0 & 0 & -1 \end{bmatrix}$ **b.** $\begin{bmatrix} 1 & 4 & 2 & 1 & 0 \\ 0 & 0 & 1 & 0 & 0 \\ 0 & 0 & 0 & 0 & 1 \end{bmatrix}$

c. $\begin{bmatrix} 1 & 0 & 0 & 4 \\ 0 & 0 & 1 & 8 \\ 0 & 0 & 0 & 0 \end{bmatrix}$ **d.** $\begin{bmatrix} 0 & 0 & 0 \\ 1 & 0 & 0 \\ 0 & 0 & 1 \end{bmatrix}$

4. a. $\begin{bmatrix} 1 & 9 & 0 & -7 & 0 \\ 0 & 0 & 1 & 8 & 0 \\ 0 & 0 & 0 & 0 & 1 \end{bmatrix}$

b. $\begin{bmatrix} 0 & 1 & 8 & 6 & 0 & -1 \\ 0 & 0 & 0 & 0 & 1 & 0 \\ 0 & 0 & 0 & 0 & 0 & 1 \\ 0 & 0 & 0 & 0 & 0 & 0 \end{bmatrix}$

c. $\begin{bmatrix} 1 & 0 & 0 & -1 & 0 \\ 0 & 1 & 0 & 8 & 0 \\ 0 & 0 & 1 & 6 & 0 \\ 0 & 0 & 0 & 0 & 1 \end{bmatrix}$

d. $\begin{bmatrix} 1 & 6 & 0 & -1 & 0 & 2 \\ 0 & 0 & 1 & -1 & 0 & 3 \\ 0 & 0 & 0 & 0 & 1 & 4 \end{bmatrix}$

5. a. $\begin{bmatrix} 0 & 1 & 0 & 0 & 0 \\ 0 & 0 & 0 & 0 & 1 \\ 0 & 0 & 1 & 0 & 0 \\ 0 & 0 & 0 & 1 & 0 \end{bmatrix}$ **b.** $\begin{bmatrix} 1 & 2 & 0 & 7 & 0 \\ 0 & 1 & 0 & 0 & 1 \\ 0 & 0 & 1 & 0 & 0 \\ 0 & 0 & 0 & 1 & 0 \end{bmatrix}$

c. $\begin{bmatrix} 1 & 5 & 0 & -1 & 0 \\ 0 & 0 & 1 & 9 & 0 \\ 0 & 0 & 0 & 0 & 1 \\ 0 & 0 & 0 & 0 & 0 \end{bmatrix}$

d. $\begin{bmatrix} 1 & 4 & 0 & 5 & 0 & 6 \\ 0 & 0 & 1 & 4 & 0 & 4 \\ 0 & 0 & 0 & 0 & 1 & 2 \end{bmatrix}$

Equivalent Matrices; Row Reduction

6. Show that each of the elementary row operations is *reversible*. In other words, if an operation is used to produce matrix B from matrix A, then there is an elementary row operation that will reverse the effect of the first one and transform B back to A.

7. Use Exercise 6 to show that

$$\text{If } A \sim B, \text{ then } B \sim A.$$

8. Show that

$$\text{If } A \sim B \text{ and } B \sim C, \text{ then } A \sim C.$$

9. Use Exercises 7 and 8 to show that $A = \begin{bmatrix} 1 & 1 & 1 \\ 0 & 1 & -1 \\ -1 & 0 & 0 \end{bmatrix}$ and $B = \begin{bmatrix} 1 & 2 & 3 \\ 2 & 2 & 3 \\ 3 & 3 & 3 \end{bmatrix}$ are equivalent by first showing that each is equivalent to $I = \begin{bmatrix} 1 & 0 & 0 \\ 0 & 1 & 0 \\ 0 & 0 & 1 \end{bmatrix}$.

10. Show that

$$\begin{bmatrix} 0 & -1 & 0 & -1 \\ 1 & 0 & 1 & 1 \\ 1 & 1 & 1 & 1 \end{bmatrix} \sim \begin{bmatrix} 1 & 1 & 1 & 0 \\ 0 & 1 & 0 & 0 \\ 0 & 0 & 0 & 1 \end{bmatrix}$$

11. Are the following matrices equivalent?

$$\begin{bmatrix} 1 & 2 & 3 \\ 4 & -1 & 2 \end{bmatrix} \quad \begin{bmatrix} 1 & 0 & \frac{7}{9} \\ 0 & 1 & \frac{10}{9} \end{bmatrix}$$

12. Show that the matrices are *not* equivalent.

$$\begin{bmatrix} 1 & 2 & 3 \\ 4 & -1 & 2 \end{bmatrix} \quad \begin{bmatrix} 1 & 0 & 7 \\ 0 & 1 & 10 \end{bmatrix}$$

13. Show that the matrices are *not* equivalent.

$$\begin{bmatrix} 1 & 2 & 3 \\ 0 & 1 & 1 \end{bmatrix} \quad \cdot \quad \begin{bmatrix} 1 & 1 & -1 \\ 0 & 1 & 1 \end{bmatrix}$$

Let $I = \begin{bmatrix} 1 & 0 \\ 0 & 1 \end{bmatrix}$

14. Show that if $ad - bc \neq 0$, then the reduced row echelon form of $A = \begin{bmatrix} a & b \\ c & d \end{bmatrix}$ is I.

15. Use Exercise 14 to show that for any θ, the matrix reduces to I.

$$\begin{bmatrix} \cos\theta & -\sin\theta \\ \sin\theta & \cos\theta \end{bmatrix}$$

16. Reduce the following matrices to echelon form.

a. $\begin{bmatrix} 0 & -1 & 0 & -1 \\ 2 & 0 & 1 & 1 \\ 0 & 1 & 1 & 1 \end{bmatrix}$ **b.** $\begin{bmatrix} -1 & 0 & -1 \\ 0 & 1 & 1 \\ 1 & 1 & 1 \\ 0 & 0 & -1 \end{bmatrix}$

17. Find two different echelon forms for each of the matrices.

a. $\begin{bmatrix} 1 & 1 & 1 \\ 0 & 1 & -1 \\ -1 & 0 & 0 \end{bmatrix}$ **b.** $\begin{bmatrix} 0 & -1 & 0 & -1 \\ 1 & 0 & 1 & 1 \\ 1 & 1 & 1 & 1 \end{bmatrix}$

In Exercises 18–22 find the reduced row echelon forms of the matrices.

18. $\begin{bmatrix} 0 & 0 & 0 & 0 & 0 \\ 1 & -1 & 1 & -1 & 0 \\ 0 & 0 & 1 & 0 & 1 \\ 0 & 0 & 0 & -1 & 0 \end{bmatrix}$

19. $\begin{bmatrix} 0 & 0 & 0 & 0 & 0 \\ 1 & -1 & 0 & -1 & 0 \\ 0 & 0 & 1 & 0 & 1 \\ 1 & 0 & 0 & -1 & 0 \end{bmatrix}$

20. $\begin{bmatrix} -1 & -6 & 0 & 1 & 0 & -2 \\ 5 & 30 & 1 & -6 & 0 & 13 \\ 4 & 24 & 1 & -5 & 1 & 15 \end{bmatrix}$

21. $\begin{bmatrix} -1 & -4 & 0 & -5 & 0 & -6 \\ 5 & 20 & 1 & 29 & 0 & 34 \\ 4 & 16 & 1 & 24 & 1 & 30 \end{bmatrix}$

22. $\begin{bmatrix} 0 & -1 & -1 & 1 & 0 \\ 0 & -1 & -1 & 1 & 0 \\ -1 & 0 & 1 & 0 & 1 \\ 0 & 0 & -1 & 0 & 1 \\ 0 & 0 & 1 & 0 & 0 \end{bmatrix}$

Linear Systems

In Exercises 23–33 solve the systems.

23.
$$\begin{aligned} x \quad + z + w &= -5 \\ x \quad - z + w &= -1 \\ x + y + z + w &= -3 \\ 2x \quad + 2z \quad &= -2 \end{aligned}$$

24.
$$\begin{aligned} x_1 - 8x_2 \quad + 7x_4 \quad &= 9 \\ -2x_1 + 16x_2 - x_3 - 20x_4 \quad &= -24 \\ 2x_1 - 16x_2 + 6x_3 + 50x_4 + x_5 &= 51 \end{aligned}$$

25.
$$\begin{aligned} x_1 \quad + 4x_3 + 5x_4 \quad &= 0 \\ -2x_1 - x_2 - 10x_3 - 16x_4 \quad &= -6 \\ 2x_1 + 6x_2 + 20x_3 + 46x_4 + x_5 &= 33 \end{aligned}$$

26.
$$\begin{aligned} 6x_1 \quad - x_3 + 4x_4 &= 0 \\ 2x_1 - x_2 - x_3 - 6x_4 &= -6 \\ 16x_1 - 2x_2 - 4x_3 - 4x_4 &= 12 \end{aligned}$$

27.
$$\begin{aligned} x - y + z + 2w &= 0 \\ x \quad + \quad w &= -1 \\ y - z - w &= 1 \\ x + 2y \quad &= -3 \end{aligned}$$

28.
$$\begin{aligned} x \quad + 2z - w &= -1 \\ y + z + 2w &= 0 \\ x - y - 2z - w &= 4 \\ y + z \quad &= -2 \end{aligned}$$

29.
$$\begin{aligned} 3x \quad + z + w &= -3 \\ 2y + 3z + 3w &= -14 \\ z - 2w &= 6 \\ w &= -4 \end{aligned}$$

30.
$$\begin{aligned} x + y + z + w - t &= 1 \\ y \quad &= -1 \\ -2z - w + t &= -3 \\ w - 3t &= -1 \\ t &= 1 \end{aligned}$$

31.
$$\begin{aligned} x \quad - t &= -2 \\ y \quad - z + t &= 5 \\ -y \quad + z - t &= -5 \\ y \quad - z + t &= 5 \\ -y - w \quad &= -1 \end{aligned}$$

32.
$$\begin{aligned} x + 2y + 3z + 4w &= 0 \\ 2x + 2y + 3z + 4w &= 0 \\ 3x + 3y + 3z + 4w &= 0 \end{aligned}$$

33.
$$\begin{aligned} x + 2y + 3z + 4w &= 0 \\ 2x + 2y + 3z + 4w &= 0 \\ 3x + 3y + 3z + 4w &= 0 \\ 4x + 4y + 4z + 4w &= 0 \end{aligned}$$

34. Solve the system for x, y, z.

$$\begin{aligned} x + y \quad &= a \\ x - y \quad &= b \\ x + y + z &= c \end{aligned}$$

35. Solve the system for x, y, z.

$$\begin{aligned} x + y \quad &= a \\ x - y \quad &= b \\ x - y + z &= 0 \end{aligned}$$

36. Show that the values of λ for which the system

$$(a - \lambda)x + by = 0$$

$$cx + (d - \lambda)y = 0$$

has nontrivial solutions should satisfy the quadratic equation $\lambda^2 - (a + d)\lambda + ad - bc = 0$.

In Exercises 37–41 solve the systems with the given augmented matrices.

37.
$$\begin{bmatrix} 1 & -1 & -2 & 2 & -7 \\ -2 & 2 & 0 & 2 & -2 \\ -1 & 1 & 1 & -2 & 6 \\ -2 & 2 & -1 & 2 & -3 \end{bmatrix}$$

38.
$$\begin{bmatrix} 0 & 2 & -2 & -2 & 10 \\ -2 & -2 & 2 & 2 & -2 \\ -1 & -1 & 0 & 0 & 3 \\ -1 & -1 & 1 & -2 & -1 \end{bmatrix}$$

39.
$$\begin{bmatrix} -1 & 0 & 1 & 1 & -1 & -1 \\ 0 & 1 & 0 & -1 & -1 & 0 \\ 0 & 1 & -1 & -1 & 1 & -6 \\ 0 & 1 & 1 & -1 & 0 & 3 \\ 1 & -1 & -1 & 0 & 1 & 2 \end{bmatrix}$$

40.
$$\begin{bmatrix} 0 & -1 & -1 & -1 & -1 & 8 \\ 0 & 0 & -1 & -1 & 1 & 1 \\ -1 & 1 & -1 & 0 & -1 & -2 \\ 0 & 1 & 0 & 0 & 1 & -5 \\ -1 & 1 & -1 & 1 & 0 & -3 \end{bmatrix}$$

41.
$$\begin{bmatrix} 1 & -1 & -1 & 1 & 1 & 0 \\ 0 & 0 & -1 & -1 & 1 & 0 \\ 1 & -1 & 0 & 1 & 1 & 0 \\ -1 & -1 & 1 & 1 & -1 & 0 \\ -1 & 0 & 0 & 0 & 0 & 0 \end{bmatrix}$$

42. Solve the homogeneous system with each coefficient matrix.

a.
$$\begin{bmatrix} 1 & 2 & 3 \\ 2 & 2 & 3 \\ 3 & 3 & 3 \end{bmatrix}$$

b.
$$\begin{bmatrix} 1 & \frac{1}{2} & \frac{1}{3} \\ \frac{1}{2} & \frac{1}{3} & \frac{1}{4} \\ \frac{1}{3} & \frac{1}{4} & \frac{1}{5} \end{bmatrix}$$

43. Show that the systems with the following augmented matrices are equivalent.

$$A = \begin{bmatrix} -1 & -4 & 0 & -5 & 0 & -6 \\ 5 & 20 & 1 & 29 & 0 & 34 \\ 4 & 16 & 1 & 24 & 1 & 30 \end{bmatrix}$$

$$B = \begin{bmatrix} -1 & -4 & 0 & -5 & 0 & -6 \\ 2 & 8 & -1 & 6 & 0 & 8 \\ -2 & -8 & 3 & 2 & 1 & 2 \end{bmatrix}$$

44. The solution of the system with augmented matrix A codifies a message that can be obtained as follows: Each letter of the alphabet is numbered by its alphabetical position. What is the message?

$$A = \begin{bmatrix} 1 & -1 & -1 & -1 & 0 & 0 & 0 & -27 \\ 0 & 0 & -1 & 0 & -1 & 0 & -1 & -27 \\ 0 & -1 & 0 & 1 & -1 & 0 & 0 & -21 \\ 1 & 1 & 1 & 0 & -1 & 0 & 0 & 27 \\ 1 & 1 & 0 & 1 & -1 & 0 & 0 & 16 \\ 0 & 0 & 1 & 1 & -1 & -1 & -1 & -8 \\ 0 & 1 & 0 & 0 & 0 & 0 & -1 & 13 \end{bmatrix}$$

In Exercises 45–46 consider linear systems whose augmented matrices reduce to the following echelon forms. What can you say about the systems?

45. a.
$$\begin{bmatrix} 2 & a & b & d & f \\ 0 & 2 & c & e & g \\ 0 & 0 & 0 & 2 & h \end{bmatrix}$$
b.
$$\begin{bmatrix} 2 & a & b & d \\ 0 & 2 & c & e \\ 0 & 0 & 2 & f \\ 0 & 0 & 0 & 2 \end{bmatrix}$$

46. a.
$$\begin{bmatrix} 1 & a & b & c & d \\ 0 & 0 & 0 & 2 & 0 \\ 0 & 0 & 0 & 0 & 0 \end{bmatrix}$$
b.
$$\begin{bmatrix} 2 & a & b & c \\ 0 & 0 & 0 & 2 \\ 0 & 0 & 0 & 0 \end{bmatrix}$$

In Exercises 47–48 each row of the table gives the size and the number of pivot columns of the augmented matrix of some system. What can you say about the system?

47.

Size	Number of Pivot Columns
3×5	3
4×4	4
4×4	3
5×3	3

48.

Size	Number of pivot columns
6×4	4
5×5	5
5×5	4
4×6	4

49. Show that if a matrix has size $m \times n$, then the number of pivot columns is less than or equal to m and less than or equal to n.

50. Show that if the linear system with augmented matrix $[A : \mathbf{c}]$ is inconsistent, then the linear system with augmented matrix $[A : \mathbf{b}]$ is either inconsistent or it has infinitely many solutions.

51. Show that if the linear system with augmented matrix $[A : \mathbf{c}]$ has exactly one solution, then the linear system with augmented matrix $[A : \mathbf{b}]$ has also exactly one solution, if A has no more rows than columns.

In Exercises 52–53 find the values of a such that the system with augmented matrix shown has (a) exactly one solution, (b) infinitely many solutions, and (c) no solutions.

52. a. $\begin{bmatrix} 2 & 3 & 4 \\ 4 & a & 8 \end{bmatrix}$ **b.** $\begin{bmatrix} 2 & 3 & 4 \\ 4 & 6 & a \end{bmatrix}$

53. a. $\begin{bmatrix} 2 & 3 & 4 \\ a & 6 & 8 \end{bmatrix}$

b. $\begin{bmatrix} 1 & 2 & 1 & 3 \\ 1 & 3 & -1 & 4 \\ 1 & 2 & a^2 - 8 & a \end{bmatrix}$

1.3   Numerical Solutions

Reader's Goals for This Section

1. To understand some of the numerical considerations in solving systems.
2. To see how systems are often solved in practice.

Most linear systems appearing in applications cannot be solved by hand, because a typical such system consists of hundreds, or thousands, of equations and unknowns. The use of computers with efficient programs is absolutely necessary. Although Gauss and Gauss-Jordan eliminations are very important methods of solving a system, they are not always the most efficient, even when we use a computer. In this section we discuss two new methods that are widely used in practice. We also discuss some of the numerical considerations that we may encounter in solving particular systems.

First let us compare our two elimination methods. Note these methods are **direct**, which means that the solution is always obtained in finitely many steps. In fact, the number of steps can be well estimated.

Choosing Between Gauss and Gauss-Jordan Elimination

In Section 1.2 we spent some time studying Gauss elimination, whereas Gauss-Jordan elimination was barely illustrated. The reason is that Gauss-Jordan, though seemingly more efficient (since there is no backward pass), requires more arithmetic operations. In fact, for a system of n equations in n unknowns (for large n), it can be shown that Gauss elimination requires approximately $2n^3/3$ arithmetic operations. Gauss-Jordan, on the other hand, requires approximately n^3 operations. This is 50% more operations than Gauss elimination. So for a medium-size system, say, 500 equations with 500 unknowns, it takes approximately 125 million operations using Gauss-Jordan and only about 83 million operations using Gauss elimination. This is mainly why we prefer Gauss elimination.

Having said all this, we should not rush to discard Gauss-Jordan elimination. In fact, in today's parallel computing a Gauss-Jordan parallel algorithm is slightly more efficient than a corresponding Gauss one.

Iterative Methods

In addition to the direct methods, we also have **iterative methods**, where we try to approximate the solution of a system by using iterations, starting with an initial guess. If the successive iterations approach the solution, we say that the iteration **converges**. Otherwise, we say that it **diverges**. The procedure ends when two consecutive iterations yield the same answer within a desired accuracy. Unlike the direct methods, the number of steps needed is not known beforehand. We discuss two iterative methods, Jacobi iteration[6] and the Gauss-Seidel iteration.[7]

Jacobi Iteration

Jacobi iteration applies to **square systems**, i.e., systems with as many equations as unknowns. Suppose we have a system with n equations in n unknowns $x_1, \ldots, x_n$, such as

$$
\begin{aligned}
5x + y - z &= 14 \\
x - 5y + 2z &= -9 \\
x - 2y + 10z &= -30
\end{aligned}
\tag{1.7}
$$

Step 1. Solve the ith equation of the system for x_i.

$$
\begin{aligned}
x &= -0.2y + 0.2z + 2.8 \\
y &= 0.2x + 0.4z + 1.8 \\
z &= -0.1x + 0.2y - 3.0
\end{aligned}
\tag{1.8}
$$

Step 2. Start with an initial guess $x_1^{(0)}, x_2^{(0)}, \ldots, x_n^{(0)}$ for the solution. In the absence of any information, initialize all variables at zero, $x_1^{(0)} = 0, x_2^{(0)} = 0, \ldots, x_n^{(0)} = 0$.
In our example let $x^{(0)} = 0$, $y^{(0)} = 0$, and $z^{(0)} = 0$.

Step 3. Substitute the values $x_1^{(k-1)}, x_2^{(k-1)}, \ldots, x_n^{(k-1)}$ obtained after the $(k-1)$st iteration into the right side of (1.8) to obtain the new values $x_1^{(k)}, x_2^{(k)}, \ldots, x_n^{(k)}$.
In our example, the substitution $x = 0, y = 0, z = 0$ on the right side of (1.8) yields $x = 2.8, y = 1.8, z = -3$. Then, substitution of these new values back into the right side of (1.8) again yields $x = 1.84, y = 1.16, z = -2.92$. We continue in the same manner.

Step 4. Stop the process when a desired accuracy has been achieved. Usually we stop when two consecutive iterations yield the same values up to this accuracy.

In our example, we iterated using accuracy to four decimal places and stopped when two consecutive answers were the same.

The iterations suggest that $x = 2, y = 1$, and $z = -3$ is the solution of the system, correct at least to four decimal places. In fact, this is the exact solution in this case.

---

[6] **Karl Gustav Jacobi** (1804–1851) was an eminent German mathematician. Professor at the University of Könisberg, he made fundamental contributions in the theory of elliptic functions and the theory of differential equations.
[7] **Philipp Ludwig Seidel** (1821–1896) was a German mathematician. He taught at Munich and did research in analysis and astronomy.

Iteration	x	y	z
Initial guess	0.0000	0.0000	0.0000
1	2.8000	1.8000	-3.0000
2	1.8400	1.1600	-2.9200
3	1.9840	1.0000	-2.9520
4	2.0096	1.0160	-2.9984
5	1.9971	1.0026	-2.9978
6	1.9999	1.0003	-2.9992
7	2.0001	1.0003	-2.9999
8	2.0000	1.0001	-3.0000
9	2.0000	1.0000	-3.0000
10	2.0000	1.0000	-3.0000

Gauss-Seidel Iteration

Gauss-Seidel iteration also applies to square systems. We have the following steps.

Step 1. Same as the Jacobi iteration.

Step 2. Same as the Jacobi iteration.

Step 3. Substitute the *most recently calculated* unknown into the right side of the equations obtained in Step 1 to get the new approximation, $x_i^{(k)}$.

In our example, substitution of $y = 0$, $z = 0$ into the first equation yields $x = 2.8$. In the second equation we substitute $z = 0$ and $x = 2.8$ (the most recent value of x) to get $y = 2.36$. In the third equation we substitute $x = 2.8$ and $y = 2.36$ (the latest x and y) to get $z = -2.808$. We continue in the same manner.

Step 4. Same as the Jacobi iteration.

In our example we obtained the following.

Iteration	x	y	z
Initial guess	0.0000	0.0000	0.0000
1	2.8000	2.3600	-2.8080
2	1.7664	1.0301	-2.9706
3	1.9999	1.0117	-2.9976
4	1.9981	1.0006	-2.9997
5	1.9999	1.0001	-3.0000
6	2.0000	1.0000	-3.0000
7	2.0000	1.0000	-3.0000

Note that Gauss-Seidel iteration required fewer iterations than Jacobi iteration. This *appears* to be true in most cases, but it is not always true. Unfortunately, we do not know which method is more efficient in advance.

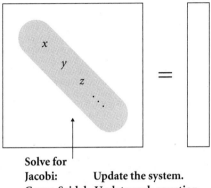

Solve for
Jacobi: **Update the system.**
Gauss-Seidel: Update each equation.

Convergence

A sufficient condition for Jacobi and the Gauss-Seidel iterations to converge is when the coefficient matrix of the system is **diagonally dominant**. This means that (1) the matrix is **square**, i.e., it has the same number of rows and columns and (2) each (i, i) entry (called a diagonal entry) has absolute value larger than the sum of the absolute values of the other entries in the same row.

For example, System (1.7) has coefficient matrix

$$\begin{bmatrix} 5 & 1 & -1 \\ 1 & -5 & 2 \\ 1 & -2 & 10 \end{bmatrix}$$

which is diagonally dominant, because $|5| > |1| + |-1|$, $|-5| > |1| + |2|$ and $|10| > |1| + |-2|$. So we are guaranteed that both iterations will converge in this case.

The matrix

$$\begin{bmatrix} 4 & 2 & -1 \\ 3 & -5 & 2 \\ 1 & -2 & 10 \end{bmatrix}$$

is not diagonally dominant, because in the second row $|-5| = 5$ is *not* greater than $|3| + |2| = 5$.

Note the Jacobi and Gauss-Seidel iterations may converge even if the coefficient matrix of the system is not diagonally dominant (see Exercise 15).

Sometimes a rearrangement of equations may result to a diagonally dominant coefficient matrix. For example, the system

$$2x + 4y - z = 1$$
$$x - 5y + 2z = 2$$
$$x - 2y + 10z = 3$$

has a coefficient matrix that is not diagonally dominant. However, if we interchange the first and second equations, the new coefficient matrix

$$\begin{bmatrix} 1 & -5 & 2 \\ 2 & 4 & -1 \\ 1 & -2 & 10 \end{bmatrix}$$

is diagonally dominant.

Choosing Between Gauss Elimination and Gauss-Seidel Iteration

Let us see how our seemingly most efficient direct and iterative methods compare with each other. It can be shown that for large n, the Gauss-Seidel method requires approximately $2n^2$ arithmetic operations per iteration. If we use fewer than $n/3$ iterations, the total amount of operations will be less than $2n^3/3$, and Gauss-Seidel will be more efficient than Gauss. For example, for a square system of 500 equations no more than 166 iterations make Gauss-Seidel a better choice.

In practice we often prefer Gauss-Seidel over Gauss even if we have to perform more operations, because during Gauss elimination the computer round-off errors with each elementary row operation accumulate and affect the final answer. In Gauss-Seidel there is only one round-off error, due to the last iteration. Indeed, we may consider the iteration before the last as an excellent initial guess.

Another virtue of Gauss-Seidel is that it is a **self-correcting** method. If at any stage there was a miscalculation, the answer is still usable; it is simply considered as a new initial guess.

Finally, both Jacobi and Gauss-Seidel are excellent choices if the coefficient matrix is **sparse**, i.e., if it has many zero entries. This is because the same coefficients are used in each stage, so the zeros remain throughout the process.

Numerical Considerations: Ill-Conditioning and Pivoting

Some systems, (even small-size ones) exhibit behavior that requires a careful numerical analysis. Consider the almost identical systems,

$$\begin{array}{ccc} x + y = 1 & & x + y = 1 \\ & \text{and} & \\ 1.01x + y = 2 & & 1.005x + y = 2 \end{array}$$

The exact solution of the first one is $x = 100$, $y = -99$, whereas the solution of the second is $x = 200$, $y = -199$. So, a small change in the coefficients resulted into a dramatic change in the solution. Such a system is called **ill-conditioned**.

If, for example, we were using floating-point arithmetic with accuracy of two decimal places, then our approximation for the solution of the second system would be off by 50%. In this case the reason for such a behavior is that the two lines defined by the first system are almost parallel. So, a small change in the slope of one may move the intersection point quite some distance (Fig. 1.5).

Another type of problem occurs when we use Gauss or Gauss-Jordan elimination with floating-point arithmetic and the entries of the augmented matrix of a system have vastly different sizes.

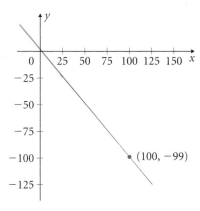

Figure 1.5 Almost parallel lines of an ill-conditioned system.

Consider the system

$$10^{-3}x + y = 2$$

$$2x - y = 0$$

It is easy to see that the exact solution is $x = 2000/2001$ and $y = 4000/2001$. Suppose now we want to solve the system numerically but we can perform floating point arithmetic to only three significant digits.

SOLUTION 1

$$\begin{bmatrix} 10^{-3} & 1 & 2 \\ 2 & -1 & 0 \end{bmatrix} \boxed{R_2 - 2 \cdot 10^3 R_1 \rightarrow R_2} \begin{bmatrix} 10^{-3} & 1 & 2 \\ 0 & \boxed{-2 \cdot 10^3} & -4 \cdot 10^3 \end{bmatrix}$$

The actual $(2, 2)$ entry of the last matrix was -2001, which was rounded to -2000 because we are working with three significant digits. The remaining of the reduction is as usual:

$$\sim \begin{bmatrix} 10^{-3} & 1 & 2 \\ 0 & 1 & 2 \end{bmatrix} \sim \begin{bmatrix} 10^{-3} & 0 & 0 \\ 0 & 1 & 2 \end{bmatrix}$$

So we get $x = 0$ and $y = 2$. We see that the approximation for x is quite poor.

SOLUTION 2   Suppose now we interchange equations 1 and 2 and scale the first row to get a leading 1.

$$\begin{bmatrix} 2 & -1 & 0 \\ 10^{-3} & 1 & 2 \end{bmatrix} \boxed{\tfrac{1}{2}R_1 \rightarrow R_1} \begin{bmatrix} 1 & -\tfrac{1}{2} & 0 \\ 10^{-3} & 1 & 2 \end{bmatrix}$$

Then

$$\boxed{R_2 - 10^{-3}R_1 \rightarrow R_2} \begin{bmatrix} 1 & -\tfrac{1}{2} & 0 \\ 0 & \boxed{1} & 2 \end{bmatrix} \sim \begin{bmatrix} 1 & 0 & 1 \\ 0 & 1 & 2 \end{bmatrix}$$

The actual $(2, 2)$ entry of the first matrix was $1 + (1/2) \cdot 10^{-3}$, which simplifies to 1 in our arithmetic. Hence, $x = 1$ and $y = 2$, a much better approximation this time.

What went wrong during Solution 1? The small coefficient, 10^{-3}, at the first pivot position forced large-size coefficients in the second row, which resulted in a slight error for y, due to rounding. This small error, however, caused a substantial error in estimating x in $10^{-3}x + y = 2$, because the coefficient of x was overpowered by that of y.

The second solution did not suffer from this problem because the row with the larger-size leading entry was moved to the pivot position. So, elimination did not yield large coefficients. And although y suffered from the same small rounding error, the value of x was only slightly affected.

In practice, during Gauss or Gauss-Jordan elimination, we always move the row with the *largest absolute value* leading entry to the pivot position before we eliminate. This is called **partial pivoting**, and it helps us keep the round-off errors under control. There is also a variant in which we pick the largest-size entry in the entire matrix as pivot. This forces interchanging of *columns* in addition to rows, which means we have to change the variables as well. This method is called **full pivoting**, and it yields better numerical results but can be quite slow. Partial pivoting is the most popular modification of Gauss or Gauss-Jordan elimination.

Numerical Solutions by CAS

Maple

```
> fsolve({3*x+2*y=1,2*x-y=-3},{x,y});
```

$$\{y = 1.571428572, x = -.7142857147\}$$

Mathematica

```
In[1]:=
        NSolve[{3x+2y==1,2x-y==-3},{x,y}]
Out[1]=
        {{x -> -0.714286, y -> 1.57143}}
```

MATLAB

```
>> rref([3 2 1; 2 -1 -3])
ans =
       1.0000         0     -0.7143
            0    1.0000      1.5714
```

Exercises 1.3

In Exercises 1–4 determine whether the matrix is diagonally dominant.

1. $\begin{bmatrix} -4 & 1 \\ 0 & 5 \end{bmatrix}$

2. $\begin{bmatrix} 4 & 2 \\ -5 & 5 \end{bmatrix}$

3. $\begin{bmatrix} 2 & 1 & 1 \\ 0 & 5 & 1 \\ 0 & 1 & 4 \end{bmatrix}$

4. $\begin{bmatrix} 5 & 1 & -1 \\ -1 & 5 & 1 \\ 1 & 1 & 5 \end{bmatrix}$

In Exercises 5–6 rewrite the system so that its (new) coefficient matrix *is* diagonally dominant.

5. $x - 2y = -6$
   $5x + y = 14$

6. $x + y + 5z = 15$
   $-x + 5y + z = -9$
   $5x + y - z = 5$

7. Use Jacobi's method with four iterations and initial values $x = 1, y = 1$ to approximate the solution of the system. Compare your answer with the exact solution.

$$5x + y = 14$$
$$x - 2y = -6$$

8. Repeat Exercise 7 with Gauss-Seidel iteration.

In Exercises 9–11 find approximate solutions of the system using Jacobi's method with four iterations. Initialize all variables at 0.

9. $7x \quad - z = 9$
   $-x + 4y \quad = 19$
   $y - 9z = 23$

10. $4x + y + z = 17$
    $x + 4y + z = 2$
    $x + y + 4z = 11$

11. $5x + y - z = 5$
    $-x + 5y + z = -9$
    $x + y + 5z = 15$

In Exercises 12–14 find approximate solutions of the system using the Gauss-Seidel method with four iterations. Initialize all variables at 0.

12. The system of Exercise 9.

13. The system of Exercise 10.

14. The system of Exercise 11.

15. The coefficient matrices of the following systems are not diagonally dominant. Apply Gauss-Seidel iteration initializing $x = 0, y = 0$, and use five iterations. Show that (a) The iteration for the first system *diverges*. (b) The iteration for the second system *converges* to two decimal places. (The difference between the last two iterates of each variable is $< .005$.)

a. $x - y = 2$ b. $4x - y = -3$
   $x + y = 0$ $x + y = 0$

In Exercises 16–19 use partial pivoting in Gauss elimination to solve the system. Use four-significant-digit arithmetic.

16. $x - 3y = -11$
    $10x + 5y = 30$

17. $1.2x - 4.5y = -1.23$
    $-5.5x + y = -15.95$

18. $x + 2y + 2z = 6$
    $2x + 4y + z = 9$
    $8x + 2y + z = 19$

19. $1.5x + 2.2y + 2.4z = 3.2$
    $2.5x + 4.2y + 1.5z = 2.3$
    $-8.4x + 2.2y + 1.5z = -17.5$

20. (**Scaling**) In the following system, all the coefficients of x are of different order of magnitude than the rest. In such cases the calculations are simplified if we scale the variable. In this case let $x' = 0.001x$. Write the system in the variables x', y, z and solve it using Gauss elimination. Then compute x.

$$0.004x + y - z = 15.8$$
$$0.001x + 5y + z = 14.2$$
$$0.001x + y + 5z = -29.8$$

1.4   Applications

Reader's Goal for This Section

To get a flavor of the many applications of linear systems.

In this section we discuss applications of linear systems to old and new problems. You will be pleased to know that even with the few tools you have acquired so far, you can solve—or tackle—a variety of real-life problems from several areas of science.

Now that you are rather experienced in solving linear systems, we shall *skip*, for the most part, the solution of a linear system.

Manufacturing, Social, and Financial Issues

■ EXAMPLE 24   (Manufacturing)   R.S.C.L.S. and Associates manufactures three types of personal computers: The Cyclone, the Cyclops, and the Cycloid. It takes 10 h to assemble the Cyclone, 2 h to test its hardware, and 2 h to install its software. The hours required for the Cyclops are 12 h to assemble, 2.5 h to test, and 2 h to install. The Cycloid, being the lower end of the line, requires 6 h to assemble, 1.5 h to test, and 1.5 h to install. If the company's factory can afford 1560 labor-hours per month for assembling, 340 h for testing, and 320 h for installation, how many PCs of each kind can be produced in a month?

SOLUTION   Let x, y, z be the number of Cyclones, Cyclops, and Cycloids produced each month. Then it takes $10x + 12y + 6z$ hours to assemble the computers. Hence $10x + 12y + 6z = 1560$. Similarly, we get equations for testing and installing. The resulting system is

$$10x + 12y + 6z = 1560$$
$$2x + 2.5y + 1.5z = 340$$
$$2x + 2y + 1.5z = 320$$

with solution $x = 60$, $y = 40$, $z = 80$. Hence 60 Cyclones, 40 Cyclops and 80 Cycloids can be manufactured monthly.

■ EXAMPLE 25   (Foreign Currency Exchange)   An international businessperson needs, on the average, fixed amounts of Japanese yen, English pounds, and German marks during each of her business trips. She traveled three times this year. The first time she exchanged a total of $2550 at the following rates: the dollar was 100 yen, 0.6 pounds, and 1.6 marks. The second time she exchanged a total of $2840 at these rates: the dollar was 125 yen, 0.5 pounds, and 1.2 marks. The third time she exchanged a total of $2800 at these rates: the dollar was 100 yen, 0.6 pounds, and 1.2 marks. How many yen, pounds, and marks did she buy each time?

SOLUTION   Let x, y, z be the fixed amounts of yen, pounds, and marks she purchases each time. Then the first time she spent $(1/100)x$ dollars to buy yen, $(1/0.6)y$ dollars to buy pounds, and $(1/1.6)z$ dollars to buy marks. Hence $(1/100)x+(1/0.6)y+(1/1.6)z = 2550$. The same reasoning applies to the other two purchases, and we get the system

$$\tfrac{1}{100}x + \tfrac{1}{0.6}y + \tfrac{1}{1.6}z = 2550$$

$$\tfrac{1}{125}x + \tfrac{1}{0.5}y + \tfrac{1}{1.2}z = 2840$$

$$\tfrac{1}{100}x + \tfrac{1}{0.6}y + \tfrac{1}{1.2}z = 2800$$

Gauss elimination yields $x = 80000$, $y = 600$, and $z = 1200$. Therefore, each time she bought 80,000 yen, 600 pounds, and 1,200 marks for her trips.

■ EXAMPLE 26   (Inheritance)   A father plans to distribute his estate, worth $234,000, among his four daughters as follows: $\tfrac{2}{3}$ of the estate is to be split equally among the daughters. For the rest, each daughter is to receive $3000 for each year that remains until her 21st birthday. Given that the daughters are all 3 years apart, how much would each receive from her father's estate? How old are the daughters now?

SOLUTION   Let x, y, z, w be the amount of money that each daughter will receive from the splitting of $\tfrac{1}{3}$ of the estate, according to age, starting with the oldest one. Then $x + y + z + w = \tfrac{1}{3} \cdot 234{,}000 = 78{,}000$. On the other hand, $w - z = 3 \cdot 3{,}000$, $z - y = 3 \cdot 3{,}000$, and $y - x = 3 \cdot 3{,}000$. Hence we have the system

$$x + y + z + w = 78{,}000$$

$$w - z = 9{,}000$$

$$z - y = 9{,}000$$

$$y - x = 9{,}000$$

with solution $x = 6{,}000$, $y = 15{,}000$, $z = 24{,}000$, $w = 33{,}000$. One-quarter of two-thirds of the estate is worth $\tfrac{1}{4} \cdot (\tfrac{2}{3} \cdot 234{,}000) = \$39{,}000$. So, the youngest daughter will receive $33{,}000 + 39{,}000 = \$72{,}000$, the next one $24{,}000 + 39{,}000 = \$63{,}000$, the next one $15{,}000 + 39{,}000 = \$54{,}000$, and the first one $6{,}000 + 39{,}000 = \$45{,}000$. The oldest daughter will receive $6{,}000 = 2 \cdot 3{,}000$ so she is currently $21 - 2 = 19$. The second one is 16, the third one is 13, and the last one is 10 years old.

■ EXAMPLE 27   (Weather)   The average of the temperatures for the cities of New York, Washington, D.C., and Boston was $88°$ during a given summer day. The temperature in Washington was $9°$ higher than the average of the temperatures of the other two cities. The temperature in Boston was $9°$ lower than the average temperature of the other two cities. What was the temperature in each one of the cities?

SOLUTION   Let x, y, z be the temperatures in New York, Washington, and Boston, respectively. The average temperature between all three cities is $(x + y + z)/3$ which is 88. On the other hand the temperature in Washington exceeds the average temperature between New York and Boston, $(x + z)/2$, by 9°. So $y = (x + z)/2 + 9$. Likewise, we have $z = (x + y)/2 - 9$. So, the system is

$$\frac{x + y + z}{3} = 88$$

$$y = \frac{x + z}{2} + 9$$

$$z = \frac{x + y}{2} - 9$$

After rewriting the system in standard form, we use Gauss elimination to get $x = 88°$, $y = 94°$, and $z = 82°$.

Economics

One of the most important functions in manufacturing, which concerns manufacturers, economists, marketing specialists, etc., is the **demand function**. It expresses the number of items, D, of a certain commodity that will be sold according to the demand for that commodity. The demand function D (or Q_d to economists) depends on several variables, such as the price, P, of the commodity, the income, I, of the consumers, the price, C, of a competing commodity, etc. Sometimes (actually rather often) the demand function, D, and its variables form a linear equation. For example, $D = -15P + 0.05I + 2.5C$. Note that as the price of the commodity increases by 1 unit, the demand drops by 15 units. Likewise, if the income of the consumer increases or if a competing commodity's price goes up, the demand increases.

■ EXAMPLE 28   (Computation of a Demand Function)   Bikey Inc. plans to manufacture a new kind of low-priced athletic shoe and researches the market for demand. It is found that if a pair of the new shoes costs $20 in a $20,000 average-family-income area and if their competitor, Triceps Inc., prices their competing shoes at $20 a pair, then 660 pairs will be sold. If, on the other hand, the price remains the same and Triceps drops their price to $10 a pair, then in a $30,000 income area, 1130 pairs will be sold. Finally, if the shoes are priced at $15 a pair and the competition remains at $20 a pair, then in an $25,000 income area, 1010 pairs will be sold. Compute the demand function by assuming that it depends linearly on its variables.

SOLUTION   Let $D = aP + bI + cC$. We need a, b, c. According to the first research case, $20a + 20,000b + 20c = 660$. Likewise, considering the other two cases we get the linear system

$$20a + 20,000b + 20c = 660$$

$$20a + 30,000b + 10c = 1130$$

$$15a + 25,000b + 20c = 1010$$

By Gauss elimination we get $a = -20, b = 0.05$, and $c = 3$. Hence, the demand function is given by $D = -20P + 0.05I + 3C$.

Chemistry

■ EXAMPLE 29   (Chemical Solutions)   It takes three different ingredients, A, B, and C, to produce a certain chemical substance. A, B, and G have to be dissolved in water separately before they interact to form the chemical. The solution containing A at 1.5 g per cubic centimeter (g/cm^3) combined with the solution containing B at 3.6 g/cm^3 combined with the solution containing C at 5.3 g/cm^3 makes 25.07 g of the chemical. If the proportions for A, B, C in these solutions are changed to 2.5, 4.3, and 2.4 g/cm^3, respectively (while the volumes remain the same), then 22.36 g of the chemical is produced. Finally, if the proportions are changed to 2.7, 5.5, and 3.2 g/cm^3, respectively, then 28.14 g of the chemical is produced. What are the volumes in cubic centimeters of the solutions containing A, B, and C?

SOLUTION   Let x, y, z cubic centimeters be the corresponding volumes of the solutions containing A, B, and C. Then $1.5x$ is the mass of A in the first case, $3.6y$ is the mass of B, and $5.3z$ is the mass of C. Added together, the three masses should give 25.07. So $1.5x + 3.6y + 5.3z = 25.07$. The same reasoning applies to the other two cases, and we get the system

$$1.5x + 3.6y + 5.3z = 25.07$$

$$2.5x + 4.3y + 2.4z = 22.36$$

$$2.7x + 5.5y + 3.2z = 28.14$$

with solution $x = 1.5$, $y = 3.1$, and $z = 2.2$. Hence the volumes of the solutions containing A, B, and C are, respectively, 1.5 cm^3, 3.1 cm^3, and 2.2 cm^3. ■

Another typical application of systems to chemistry is **balancing a chemical reaction**. We need to insert *integer* coefficients in front of each one of the reactants so that the number of atoms of each element is the same on both sides of the equation. For example, consider the reaction of the burning of methane:

$$a\,CH_4 + b\,O_2 \rightarrow c\,CO_2 + d\,H_2O \qquad (1.9)$$

Let us compute the coefficients a, b, c, d that will balance the reaction. Note that guessing is easy in this case but not in general.

■ EXAMPLE 30   (Balancing Chemical Reactions)   Balance Reaction (1.9).

SOLUTION $a = c$, because the number of carbon atoms should be the same on both sides. Likewise, we have

$$a = c$$

$$4a = 2d$$

$$2b = 2c + d$$

The solution of this homogeneous system is $a = \frac{1}{2}d$, $b = d$, $c = \frac{1}{2}d$. If $d = 2$, then $a = 1, b = 2, c = 1$.

Physics and Engineering

Suppose we have an electrical network such as the one pictured in Fig. 1.6(a). The currents and the voltage drops around the network satisfy Kirchhoff's first law.

THEOREM 6   **(Kirchhoff's Current Law)**

The algebraic sum of all currents at any branch point is zero.

THEOREM 7   **(Kirchhoff's Voltage Law)**

The algebraic sum of all voltage changes around a loop is zero.

A typical application of these laws is when we are given the voltage of the electromotive force (usually a battery or a generator) and the resistances of the resistors and we are asked to compute the currents.

Note that for each element of a circuit, we have to choose a positive direction for measuring the current through that element. The choice is indicated by arrows. For the voltage source we choose as positive direction the direction from the negative sign to the positive sign. The voltage source adds voltage; hence, the voltage change is positive, whereas the voltage change through the resistors is negative due to voltage dropping.

▪ **EXAMPLE 31**   (Electrical Circuits)   Find the currents i_1, i_2, i_3 in the electrical circuit of Fig. 1.6(a) if the voltage of the battery is $E = 6$ V and the resistances are $R_1 = 2$, $R_2 = 2, R_3 = 1\Omega$ each.

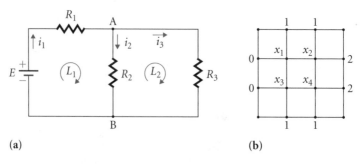

(a)   (b)

Figure 1.6   (a) Electrical circuit. (b) Heat transfer.

SOLUTION   By the first law we have $i_1 - i_2 - i_3 = 0$ from branch point A. Applying the second law to loop L_1 yields, $6 - i_1R_1 - i_2R_2 = 0$ hence $2i_1 + 2i_2 = 6$. Likewise, loop L_2 yields $i_3R_3 - i_2R_2 = 0$, or $-2i_2 + i_3 = 0$. Hence,

$$
\begin{aligned}
i_1 - i_2 - i_3 &= 0 \\
2i_1 + 2i_2 \quad\ &= 6 \\
-2i_2 + i_3 &= 0
\end{aligned}
$$

and by Gauss elimination we easily get $i_1 = 2.25$, $i_2 = 0.75$, $i_3 = 1.5$ A.

Another typical application of linear systems is in the heat-transfer problems in physics and engineering.

Suppose we have a thin rectangular metal plate whose edges are kept at fixed temperatures. As an example, let the left edge be at $0°$C, the right edge at $2°$C, and the top and bottom edges at $1°$C (Fig. 1.6(b)). We want to know the temperature inside the plate.

There are several ways of approaching this kind of problem, some of which require more advanced mathematics. The approach that will interest us will be the following type of approximation: We shall overlay our plate with finer and finer grids, or meshes (Fig. 1.6(b)). The intersections of the mesh lines are called *mesh points*. Mesh points are divided into *boundary* and *interior* points, depending on whether they lie on the boundary or the interior of the plate. We may consider these points as *heat elements*, such that each influences its neighboring points. We need the temperature of the interior points, given the temperature of the boundary points. It is obvious that the finer the grid, the better the approximation of the temperature distribution of the plate. To compute the temperature of the interior points, we use the following principle.

THEOREM 8

(Mean Value Property for Heat Conduction)

The temperature at any interior point is the average of the temperatures of its neighboring points.

Suppose, for simplicity, we have only four interior points with unknown temperatures x_1, x_2, x_3, x_4 and 12 boundary points (not named) with temperatures indicated in Fig. 1.6(b).

■ EXAMPLE 32   (Heat Conduction)   Compute x_1, x_2, x_3, x_4.

SOLUTION   According to the mean value property, we have

$$
\begin{aligned}
x_1 &= \tfrac{1}{4}(x_2 + x_3 + 1) \\
x_2 &= \tfrac{1}{4}(x_1 + x_4 + 3) \\
x_3 &= \tfrac{1}{4}(x_1 + x_4 + 1) \\
x_4 &= \tfrac{1}{4}(x_2 + x_3 + 3)
\end{aligned}
$$

We can certainly use Gauss elimination, but it is rather tedious for this particular system. Instead, we shall exploit the symmetry of the graph and the equations for a quick solution. By symmetry, we have $x_1 = x_3$ and $x_2 = x_4$. Let $r = x_1 + x_2 + x_3 + x_4 = 2x_1 + 2x_4$. Then by adding all the equations together, we get $r = \frac{1}{4}(2r + 4)$ which implies $r = 4$. Hence, $x_1 + x_4 = 2$. Therefore, the second equation implies $x_2 = \frac{5}{4}$. So we immediately conclude that $x_1 = \frac{3}{4}, x_2 = \frac{5}{4}, x_3 = \frac{3}{4}, x_4 = \frac{5}{4}$.

We should note at this point that systems arising from heat-conduction problems, such as the one above, are usually quite tedious to solve. Using Maple, Mathematica, or MATLAB could be of great assistance.

Statics and Weight Balancing

Let us now study a typical weight-balancing lever problem in statics. We need the following.

THEOREM 9

(Archimedes' Law of the Lever[8])

Two masses on a lever balance when their weights are inversely proportional to their distances from the fulcrum.

▪ EXAMPLE 33   Find weights w_1, w_2, w_3, w_4 to balance the levers in Fig. 1.7(a).

SOLUTION   To balance the two small levers, according to Archimedes' law, we have $2w_1 = 6w_2$ for the lever on the left and $2w_3 = 8w_4$ for the lever on the right. To balance the main lever, we need $5(w_1 + w_2) = 10(w_3 + w_4)$. Hence we have the following

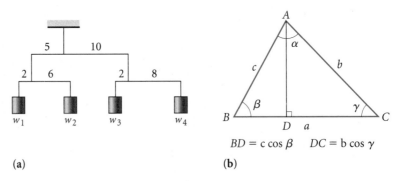

$BD = c \cos \beta$ $DC = b \cos \gamma$

(a)   (b)

Figure 1.7   (a) Balancing weights. (b) The law of cosines.

---

[8] Although this law is also found earlier, in works of Aristotle, it seems that Archimedes was the first to base the law on statics as opposed to kinematics. It is a special case of Archimedes' axiom of symmetry of a system in equilibrium.

homogeneous system of three equations and four unknowns:

$$5w_1 + 5w_2 - 10w_3 - 10w_4 = 0$$
$$2w_1 - 6w_2 \qquad\qquad = 0$$
$$2w_3 - 8w_4 = 0$$

The solution set is a one-parameter infinite solution set described by $w_1 = 7.5r$, $w_2 = 2.5r$, $w_3 = 4r$, and $w_4 = r, r \in \mathbf{R}$. So, infinitely many weights can balance this system, as we know from experience, as long as the weights, in the considered order, are multiples of the numbers 7.5, 2.5, 4, and 1.

Applications to Geometry

■ EXAMPLE 34   (Law of Cosines)   Prove the law of cosines in geometry; i.e., for any triangle ABC (Fig. 1.7(b)) we have

$$\cos\alpha = \frac{b^2 + c^2 - a^2}{2bc}, \qquad \cos\beta = \frac{a^2 + c^2 - b^2}{2ac}, \qquad \cos\gamma = \frac{a^2 + b^2 - c^2}{2ab}$$

SOLUTION   Side a is the sum $BD + DC$. But $BD = c\cos\beta$ and $DC = b\cos\gamma$. Hence, $c\cos\beta + b\cos\gamma = a$. Likewise, we obtain the other two equations of the system:

$$c\cos\beta + b\cos\gamma = a$$
$$c\cos\alpha + a\cos\gamma = b$$
$$a\cos\beta + b\cos\alpha = c$$

This is a linear system in $\cos\alpha$, $\cos\beta$, $\cos\gamma$. The augmented matrix of this system reduces as

$$\begin{bmatrix} 0 & c & b & : & a \\ c & 0 & a & : & b \\ b & a & 0 & : & c \end{bmatrix} \sim \cdots \sim \begin{bmatrix} 1 & 0 & 0 & : & (b^2 + c^2 - a^2)/2bc \\ 0 & 1 & 0 & : & (a^2 + c^2 - b^2)/2ac \\ 0 & 0 & 1 & : & (a^2 + b^2 - c^2)/2ab \end{bmatrix}$$

which proves the law.

■ EXAMPLE 35   (Quadratic Through Three Points)   Find the equation of the parabola with vertical axis, in the xy-plane, passing through the points $P(1, 4)$, $Q(-1, 6)$, and $R(2, 9)$.

SOLUTION   Let $y(x) = ax^2 + bx + c$ be the equation of the parabola. We need to determine the coefficients a, b, c. Because the point P belongs to the parabola, we should have $4 = y(1) = a \cdot 1 + b \cdot 1 + c$. Likewise, by using the other two points we get the system

$$
\begin{aligned}
a + b + c &= 4 \\
a - b + c &= 6 \\
4a + 2b + c &= 9
\end{aligned}
$$

Gauss elimination yields $a = 2$, $b = -1$, and $c = 3$. Hence the equation of the parabola is $y = 2x^2 - x + 3$.

▪ **EXAMPLE 36**   (Plane Through Three Points)   Find the equation of the plane in xyz-space passing through the points $P(1, 1, 2)$, $Q(1, 2, 0)$, $R(2, 1, 5)$.

SOLUTION   Let $ax + by + cz + d = 0$ be the equation of the plane. We need to determine the coefficients a, b, c and the constant d. Because the point P belongs to the plane, we should have $a \cdot 1 + b \cdot 1 + c \cdot 2 + d = 0$. Likewise, by using the other two points we get the homogeneous system

$$
\begin{aligned}
a + b + 2c + d &= 0 \\
a + 2b + d &= 0 \\
2a + b + 5c + d &= 0
\end{aligned}
$$

Solving yields the one-parameter infinite set $a = 3r$, $b = -2r$, $c = -r$, and $d = r$. If we let $d = r = 1$, then we get the plane equation $3x - 2y - z + 1 = 0$. (Any other value of r gives a constant multiple of this equation, which represents the same plane.)

Algebra

Systems are used in nearly all areas of algebra, from the study of polynomials and partial fractions to the proof of identities. Let us highlight some of these applications.

Recall that a *polynomial*, $f(x)$, in x, is an expression of the form

$$
f(x) = a_n x^n + a_{n-1} x^{n-1} + \cdots + a_1 x + a_0
$$

where $a_n, a_{n-1}, \ldots, a_0$ are fixed numbers, called the *coefficients* of $f(x)$, and x is a variable or indeterminate. If $a_n \neq 0$, we say that $f(x)$ has *degree* n. Two polynomials are *equal* if their corresponding coefficients are equal. The polynomial whose coefficients are all zero is the *zero polynomial*. A typical application of systems is involved when we compare polynomials.

▪ **EXAMPLE 37**   (Polynomial Equality)   Compute a, b, c such that the polynomials $ax^2 + 3x^2 + 2ax - 2cx + 10x + 6c$ and $-2bx^2 - 3bx + 9 + a - 4b$ are equal.

SOLUTION   The coefficients of the corresponding powers of x should be equal. So

$$(a + 3)x^2 + (2a - 2c + 10)x + 6c = -2bx^2 - 3bx + 9 + a - 4b$$

implies the system

$$a + 3 = -2b$$
$$2a - 2c + 10 = -3b$$
$$6c = 9 + a - 4b$$

You can easily verify that $a = 1, b = -2, c = 3$ by Gauss elimination.

■ EXAMPLE 38   (Partial Fractions)   Compute constants A and B such that

$$\frac{1}{(x - 1)(x - 2)} = \frac{A}{x - 1} + \frac{B}{x - 2}$$

SOLUTION   We have

$$\frac{1}{(x - 1)(x - 2)} = \frac{A}{x - 1} + \frac{B}{x - 2} = \frac{A(x - 2) + B(x - 1)}{(x - 1)(x - 2)}$$

Therefore, $1 = A(x - 2) + B(x - 1)$ or $(A + B)x - 2A - B = 1$, because the equal first and last fractions have the same denominators. So we need to solve

$$A + B = 0$$
$$-2A - B = 1$$

to get $A = -1$ and $B = 1$.

■ EXAMPLE 39   (Sum of Squares)   Derive a formula for the sum of squares, shown below, by assuming that the answer is a polynomial of degree 3 in n.

$$1^2 + 2^2 + \cdots + n^2$$

SOLUTION   Let $f(x) = ax^3 + bx^2 + cx + d$ be a polynomial with the property: $f(n) = 1^2 + 2^2 + \cdots + n^2$. Since $1 = f(1) = a1^3 + b1^2 + c1 + d$, we must have $a + b + c + d = 1$. On the other hand,

$$n^2 = f(n) - f(n - 1) = an^3 + bn^2 + cn + d - (a(n - 1)^3 + b(n - 1)^2 + c(n - 1) + d)$$

Hence, by expanding the left-hand side we get

$$n^2 = a(3n^2 - 3n + 1) + b(2n - 1) + c \quad \text{or} \quad n^2 = 3an^2 + (-3a + 2b)n + a - b + c$$

Comparing the coefficients of the powers of n on both sides and taking into account $a + b + c + d = 1$, we have

$$3a = 1$$
$$-3a + 2b = 0$$
$$a - b + c = 0$$
$$a + b + c + d = 1$$

This system can easily be solved: $a = \frac{1}{3}$, $b = \frac{1}{2}$, $c = \frac{1}{6}$, $d = 0$. Hence, $f(n) = \frac{1}{3}n^3 + \frac{1}{2}n^2 + \frac{1}{6}n$. By factoring the last expression we conclude

$$1^2 + 2^2 + \cdots + n^2 = \sum_{i=1}^{n} i^2 = \frac{n(n+1)(2n+1)}{6}$$

The Fibonacci Money Pile Problem

Now let us return to that famous money pile problem solved by Fibonacci a few centuries ago.

▪ EXAMPLE 40   (The Fibonacci Problem)   Three men possess a single pile of money, their shares being $\frac{1}{2}$, $\frac{1}{3}$, and $\frac{1}{6}$. Each man takes some money from the pile until nothing is left. The first man then returns $\frac{1}{2}$ of what he took, the second man, $\frac{1}{3}$, and the third, $\frac{1}{6}$. When the total so returned is divided equally among the men, it is discovered that each man then possesses what he is entitled to. How much money was there in the original pile and how much did each man take from the pile?

SOLUTION   Let x, y, z denote the amount taken from the pile of money by the three men, respectively, and let w denote the amount of money in the pile originally. Since no money is left after the three take their money, we have

$$x + y + z = w$$

The three men return a total of $x/2 + y/3 + z/6$, because they return, respectively, $\frac{1}{2}$, $\frac{1}{3}$, and $\frac{1}{6}$ of what each took. This amount is equally divided among them, so each receives $\frac{1}{3}(x/2 + y/3 + z/6)$ back.

The first man has $x - x/2 = x/2$ left after he returned $x/2$, plus $\frac{1}{3}(x/2 + y/3 + z/6)$. The total should be what he is entitled to, i.e., $w/2$. So,

$$\frac{x}{2} + \frac{1}{3}\left(\frac{x}{2} + \frac{y}{3} + \frac{z}{6}\right) = \frac{w}{2}$$

Likewise, the second man has $y - y/3 = 2y/3$ left after he returned $y/3$, plus $\frac{1}{3}(x/2 + y/3 + z/6)$. The total should be what he is entitled to, i.e., $w/3$. So,

$$\frac{2y}{3} + \frac{1}{3}\left(\frac{x}{2} + \frac{y}{3} + \frac{z}{6}\right) = \frac{w}{3}$$

Finally, the third man has $z - z/6 = 5z/6$ left after he returned $z/6$, plus $\frac{1}{3}(x/2 + y/3 + z/6)$. The total should be what he is entitled to, i.e., $w/6$. So,

$$\frac{5z}{6} + \frac{1}{3}\left(\frac{x}{2} + \frac{y}{3} + \frac{z}{6}\right) = \frac{w}{6}$$

and we have the homogeneous system:

$$x + y + z - w = 0$$

$$\frac{2x}{3} + \frac{y}{9} + \frac{z}{18} - \frac{w}{2} = 0$$

$$\frac{x}{6} + \frac{7y}{9} + \frac{z}{18} - \frac{w}{3} = 0$$

$$\frac{x}{6} + \frac{y}{9} + \frac{16z}{18} - \frac{w}{6} = 0$$

It is worth noticing that the sum of the last three equations equals the first one. This means that if we can find a simultaneous solution to the last three equations, we will automatically have a solution to the first one. So, in essence, we have three equations with four unknowns. Hence, we expect infinitely many solutions.

Reduction of the augmented matrix yields

$$\begin{bmatrix} 1 & 0 & 0 & -\frac{33}{47} & : & 0 \\ 0 & 1 & 0 & -\frac{13}{47} & : & 0 \\ 0 & 0 & 1 & -\frac{1}{47} & : & 0 \\ 0 & 0 & 0 & 0 & : & 0 \end{bmatrix}$$

Hence, $x - \frac{33}{47}w = 0$, $y - \frac{13}{47}w = 0$, and $z - \frac{1}{47}w = 0$. We have infinitely many solutions that can expressed as $x = \frac{33}{47}r$, $y = \frac{13}{47}r$, $z = \frac{1}{47}r$, $w = r$, $r \in \mathbf{R}$.

It is not known whether Fibonacci computed the entire solution set. We do know that he found the particular solution $w = 47$, $x = 33$, $y = 13$, and $z = 1$, which is obtained by letting $r = 47$.

Magic Squares

A **magic square** of size n is a $n \times n$ matrix whose entries consist of all integers between 1 and n^2 such that the sum of the entries of each column, row, or diagonal is the same. The sum of the entries of any row, column, or diagonal, of a magic square of size n is $n(n^2 + 1)/2$. (To see this use the identity $1 + 2 + \cdots + k = k(k + 1)/2$ with $k = n^2$.)

■ EXAMPLE 41   (Size 2 Magic Squares)   Show that magic squares of size 2 do not exist.

SOLUTION   Let $\begin{bmatrix} a & b \\ c & d \end{bmatrix}$ be a hypothetical magic square. Then,

$$a + b = 5$$

$$c + d = 5$$

$$a + c = 5$$

$$b + d = 5$$

$$a + d = 5$$

$$b + c = 5$$

We can use Gauss elimination or simply observe that these relations imply that $b = c$, by the first and the third; hence $2b = 5$, by the last equation. This is a contradiction, because b is supposed to be an integer. So this system has no *integer* solutions. Hence there are no magic squares of size 2.

■ EXAMPLE 42   (Size 3 Magic Squares)   Find the magic square of size 3 whose first row is the vector $(8, 1, 6)$.

SOLUTION   Let the magic square be of the form

$$\begin{bmatrix} 8 & 1 & 6 \\ a & b & c \\ d & e & f \end{bmatrix}$$

for unknowns a, b, c, d, e, f. According to the definition, we have a system of seven equations and six unknowns:

$$a + b + c = 15$$

$$d + e + f = 15$$

$$8 + a + d = 15$$

$$1 + b + e = 15$$

$$6 + c + f = 15$$

$$8 + b + f = 15$$

$$6 + b + d = 15$$

The Gauss elimination is quite tedious, but it can be avoided: If we eliminate a from the first and third equations, we get $b + c = 8 + d$. This combined with the last equation yields $2b + c = 17$. Hence, according to first equation $b - a = 2$. Let us see now what a can be. It cannot be 1, since 1 has already been used. If $a = 2$, then $b = 4$ and $c = 9$, but then the third column's sum would exceed 15. If $a = 3$, then $b = 5$ and $c = 7$; hence $d = 4, e = 9, f = 2$ by using the three columns. It is easy to show that these values form a solution (the only solution) of the system.

Therefore, the magic square is

$$\begin{bmatrix} 8 & 1 & 6 \\ 3 & 5 & 7 \\ 4 & 9 & 2 \end{bmatrix}$$

This square was mentioned in the ancient Chinese book *Nine Chapters of the Mathematical Art*.[9] Can you find another square based on this one?

---

[9]See Carl Boyer's *A History of Mathematics*, p. 219 (New York: Wiley).

Exercises 1.4

1. Suppose the numbers of bacteria of types A and B interdepend on each other according to the following experimental table. Is there a linear relation (equation) between A and B?

A	B
500	500
1,000	2,000
5,000	14,000
10,000	29,000

2. QuickInk Publishers publishes three different qualities of books: paperbacks, hardcovers, and leathercovers. For paperbacks the company spends, on the average, $5 for paper, $2 for illustrations, and $3 for the cover. For hardcovers the spending is $10 for paper, $4 for illustrations, and $8 for the cover; for the deluxe leathercovers, they spend $20 for paper, $12 for illustrations, and $24 for the cover. If the budget allows $235,000 for paper, $110,000 for illustrations, and $205,000 for covers, how many books of each kind can be produced?

3. An international businessperson needs, on average, fixed amounts of Japanese yen, French francs, and German marks during each of his business trips. He traveled three times this year. The first time he exchanged a total of $2400 at the following rates: the dollar was 100 yen, 1.5 francs, and 1.2 marks. The second time he exchanged a total of $2350 at these rates: the dollar was 100 yen, 1.2 francs, and 1.5 marks. The third time he exchanged a total of $2390 at these rates: the dollar was 125 yen, 1.2 francs, and 1.2 marks. What were the amounts of yen, francs and marks that he bought each time?

4. A mother plans to distribute her estate, worth $400,000 between her four sons as follows: $\frac{3}{4}$ of the estate is to be split equally among the sons. For the rest, each son is to receive $3,000 for each year that remains until his 25th birthday. Given that the sons are all 4 years apart, how much would each receive from their mother's estate?

5. The average of the temperatures for the cities of Boston, New York and Montreal was 30° during a given winter day. The temperature in New York was 9° higher than the average of the temperatures of the other two cities. The temperature in Montreal was 9° lower than the average temperature of the other two cities. What was the temperature in each one of the cities?

6. ToysOnDemand Inc. plans to manufacture a new kind of toy train and researches the toy market for demand. It is found that if the train costs $40 in a $30,000 average family income area and if their competitor, ToysSupplies Inc., prices their competing toy train at $30, 1160 trains will be sold. If on the other hand, the price remains the same and ToysSupplies raises their price to $50 a train, then in a $40,000 income area 1700 trains will be sold. Finally, if the train is priced at $30, but the competition remains at $40, then in a $35,000 income area, 1530 trains will be sold. Compute the demand function by assuming that it depends linearly on its variables.

7. It takes three different ingredients, A, B, and C, to produce a certain chemical substance. A, B, and C have to be dissolved in water separately before they interact to form the chemical. The solution containing A at 1.5 g/cm^3 combined with the solution containing B at 1.8 g/cm^3 , combined with the solution containing C at 3.2 g/cm^3 makes 15.06 g of the chemical. If the proportions for A, B, C in these solutions are changed to 2.0, 2.5, and 2.8 g/cm^3, respectively, (while the volumes remain the same), then 17.79 g of the chemical is produced. Finally, if the proportions are changed to 1.2, 1.5, and 3.0 g/cm^3, respectively, then 13.05 g of the chemical is produced. What are the volumes in cubic centimeters of the solutions containing A, B, and C?

In Exercises 8–10 find the currents i_1, i_2, i_3 in the electrical circuits given that in all cases the voltage of the battery is $E = 6$ V.

8. Referring to Fig. 1.8 it is known that $R_1 = 2$, $R_2 = 2$, $R_3 = 1$, and $R_4 = 2\ \Omega$.

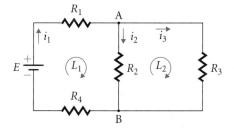

Figure 1.8

9. Referring to Fig. 1.9, it is given that $R_1 = 2$, $R_2 = 3$, $R_3 = 4$, and $R_4 = 2 \, \Omega$.

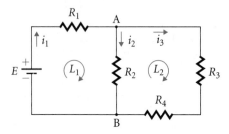

Figure 1.9

10. Referring to Fig. 1.10, it is known that $R_1 = 3$, $R_2 = 2$, $R_3 = 2$, $R_4 = 1$, and $R_5 = 2 \, \Omega$.

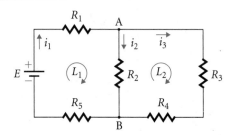

Figure 1.10

11. Find the temperatures at x_1, x_2, and x_3 of the triangular metal plate shown in Fig. 1.11 given that the temperature of each interior point is the average of its four neighboring points.

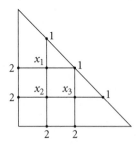

Figure 1.11

12. Find the temperatures at x_1, x_2, x_3, and x_4 of the triangular metal plate shown in Fig. 1.12 given that the temperature of each interior point is the average of its four neighboring points.

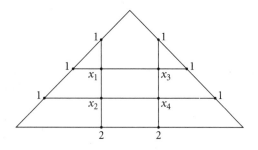

Figure 1.12

13. Balance the lever-weight system shown in Fig. 1.13.

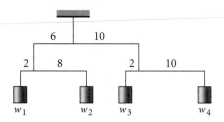

Figure 1.13

14. Find the equation of the parabola in the xy-plane passing through the points $P(1,0)$, $Q(-1,6)$, and $R(2,0)$.

15. Find the equation of the cubic curve in the xy-plane passing through the points $P(1,1)$, $Q(-1,5)$, $R(0,1)$, and $S(-2,7)$.

16. Find the equation of the plane in xyz-space passing through the points $P(1,1,-1)$, $Q(2,1,2)$, and $R(1,3,-5)$.

17. Compute a, b, c such that the quadratics $(a-b)x^2 + (a-c)x + b + c$ and $(3-c)x^2 - ax - b - 2c$ are equal.

18. Compute a, b, c, d such that the cubics $(a+b+c)x^3 + (a+b)x^2 + 2bx$ and $(-d)x^3 + (2-d)x^2 + (1+a)x + b + c$ are equal.

19. Compute constants A and B such that

$$\frac{1}{(x+1)(x-2)} = \frac{A}{x+1} + \frac{B}{x-2}$$

20. Find the magic squares of size 3 of the form

$$\begin{bmatrix} 4 & a & b \\ c & 5 & d \\ e & f & 6 \end{bmatrix}$$

1.5   Miniprojects

1 ■ Sets of Systems

Suppose you have to solve several systems with the same coefficient matrix. Find a way of solving the systems, simultaneously, so that the reduction of the coefficient matrix is done only once. Apply your notation to solve the following four linear systems.

$$x + 2y + 3z + 4w = 1, \ = 0, \ = 0, \ = 0$$

$$2x + 2y + 3z + 4w = 0, \ = 1, \ = 0, \ = 0$$

$$3x + 3y + 3z + 4w = 0, \ = 0, \ = 1, \ = 0$$

$$4x + 4y + 4z + 4w = 0, \ = 0, \ = 0, \ = 1$$

2 ■ Animal Intelligence

A set of experiments in psychology deals with the study of teaching tasks to various animals such as pigs, rabbits, rats, etc. One such experiment involves the search for food. An animal is placed somewhere in a square mesh of corridors that may lead to food (points labeled 1) or to a dead-end (points labeled 0) (Fig. 1.14). It is assumed that the probability of an animal occupying position x_i is the *average* of the probabilities of occupying the neighboring positions directly above, below, to the left, and to the right of it. If a neighboring position is one with food, this being success, its probability is $100\% = 1$. If a neighboring position is dead-end, this being failure, its probability is $0\% = 0$. For example, for (a) and (b) of Fig. 1.14 we have, respectively:

$$x_1 = \tfrac{1}{4}(1 + 0 + x_3 + x_2) \qquad x_1 = \tfrac{1}{4}(0 + 0 + x_4 + x_2)$$

$$x_2 = \tfrac{1}{4}(0 + x_1 + x_4 + 1) \qquad x_2 = \tfrac{1}{4}(1 + x_1 + x_5 + x_3)$$

$$x_3 = \tfrac{1}{4}(x_1 + 1 + 0 + x_4) \qquad x_3 = \tfrac{1}{4}(0 + x_2 + x_6 + 0)$$

$$\vdots \qquad\qquad\qquad \vdots$$

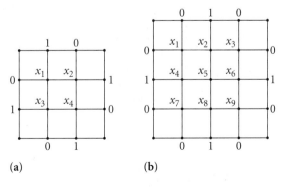

(a)   (b)

Figure 1.14   *Animal intelligence experiments.*

Compute:

(a) The probabilities x_1, x_2, x_3, x_4 for Fig. 1.14(a).
(b) The probabilities $x_1, x_2, x_3, x_4, x_5, x_6, x_7, x_8, x_9$ for Fig. 1.14(b).

Exploiting any *symmetries* of the data might help avoid lengthy calculations.

3 ■ Approximations to Curves; Linear Splines

One of the most elementary and basic ways of approximating the graph of a curve is
by connecting consecutive known points of the curve by straight-line segments. The
resulting polygonal line is a *linear spline* of the curve. Often, in practice, we are given
a set of values of an unknown function (usually, some experimental data), and we are
confronted with the task of identifying the function. A first graphical approximation
would be to draw the linear spline connecting these points. Suppose, for instance, that
we are given the following two sets of points, which belong to two different curves.

x	-1	-0.5	0.	0.5	1.	1.5	2.	2.5	3
$f(x)$	-5	-0.875	1.	1.375	1.	0.625	1.	2.875	7

x	-4.5	-3.5	-2.5	-1.5	-0.5
$f(x)$	3519.14	301.641	-32.4844	14.7656	-24.6094

x	0.5	1.5	2.5	3.5	4.5
$f(x)$	-24.6094	14.7656	-32.4844	301.641	3519.14

(a) Sketch the graphs of the two linear splines connecting the consecutive points.
(b) If the first curve is a cubic—i.e., of the form $ax^3 + bx^2 + cx + d$—and the second is a
sixth-degree polynomial—i.e., of the form $ax^6 + bx^5 + cx^4 + dx^3 + ex^2 + fx + g$—use
the tables to compute the coefficients of each polynomial curve. Round the answers
to the nearest integer.
(c) Sketch the graphs of the two curves.

4 ■ Archimedes' Cattle Problem

This is a famous problem sent by the ancient Greek mathematician Archimedes of
Syracuse[10] to Eratosthenes in Alexandria. Its original form was a collection of epigrams
in ancient Greek. It is generally believed that Archimedes worked on this problem, but it
is not known whether he is the author of it. We shall outline the part of the problem that
is relevant to the current project. For a nice translation see Sir Thomas L. Heath's *The
Works of Archimedes* (Dover Edition, 1953), pp. 319–326. The epigrams start as follows:

---

[10] **Archimedes** (c. 287–212 B.C.) is generally considered the greatest mathematician and physicist
of antiquity and one of the greatest mathematicians who ever lived. He grew up in Syracuse, a
Greek settlement in Sicily. His father, Pheidias, was an astronomer. After he studied mathematics
in Alexandria, Egypt, he returned to Syracuse, where he spent the rest of his life. He was killed by
a Roman soldier during the city's capture by the Romans.

Compute, O stranger, the number of the oxen of the Sun which once grazed upon the fields of the Sicilian isle of Thrinacia and which were divided, according to color, into four herds, one white, one black, one yellow, and one dappled...

Then the manuscript goes on to describe the relations between the cows and the bulls of the four herds. Let W, B, D, Y be the numbers of the bulls in the white, black, dappled, and yellow herds, respectively. Likewise, let w, b, d, y be the numbers of the cows in the same order. Then $W + w, B + b, D + d, Y + y$ are the numbers of the oxen in the white, black, dappled, and yellow herds, respectively. The manuscript gives us the following relationships (1) for the bulls:

$$W = \left(\frac{1}{2} + \frac{1}{3}\right) B + Y$$

$$B = \left(\frac{1}{4} + \frac{1}{5}\right) D + Y$$

$$D = \left(\frac{1}{6} + \frac{1}{7}\right) W + Y$$

and (2) for the cows:

$$w = \left(\frac{1}{3} + \frac{1}{4}\right) (B + b)$$

$$b = \left(\frac{1}{4} + \frac{1}{5}\right) (D + d)$$

$$d = \left(\frac{1}{5} + \frac{1}{6}\right) (Y + y)$$

$$y = \left(\frac{1}{6} + \frac{1}{7}\right) (W + w)$$

Solve this system of seven equations with eight unknowns. Since the system is homogeneous, with more unknowns than equations, you expect to find infinitely many solutions. Show that the smallest integer solution gives $B = 7,460,514$. Find the remaining of the unknowns and the total number of oxen in this case.

1.6   Computer Exercises

The first computer exercises are designed to help you learn basic commands of your software, related to Chapter 1, and also highlight some of the material. An exercise designated as [S] requires symbolic manipulation.[11] In all questions you are supposed to use some mathematics program.

---

[11] Skip if only numerical calculation is available.

1. Solve the system numerically. Display both in default and in higher accuracy. If your program supports rational arithmetic, find the exact answer. Finally, verify your answer.

$$\frac{1}{5}x + \frac{1}{6}y + \frac{1}{7}z = \frac{241}{1260}$$

$$\frac{1}{6}x + \frac{1}{7}y + \frac{1}{8}z = \frac{109}{672} \qquad (1.10)$$

$$\frac{1}{7}x + \frac{1}{8}y + \frac{1}{9}z = \frac{71}{504}$$

2. [S] Solve the following system for x and y.

$$a_1 x + b_1 y = c_1$$

$$a_2 x + b_2 y = c_2$$

3. Enter the augmented matrix of system (1.10) and find (a) a row echelon (if available), (b) the reduced echelon form. What is the solution of the system?

4. Let A be the coefficient matrix of system (1.10). Is A row-equivalent to B?

$$B = \begin{bmatrix} 1 & 2 & 3 \\ 2 & 2 & 3 \\ 3 & 3 & 3 \end{bmatrix}$$

5. Consider the following system. Use your program to show that if $c = -\frac{250}{3}$, then the system has infinitely many solutions. If $c \neq -\frac{250}{3}$, then the system has no solutions.

$$\frac{1}{5}x - \frac{1}{6}y = 100$$

$$-\frac{1}{6}x - \frac{5}{36}y = c$$

6. Consider matrix B in Exercise 4. Use your software to display the first column of B, the second row, the first 2 columns, the last two rows, and the portion $\begin{bmatrix} 1 & 2 \\ 2 & 2 \end{bmatrix}$.

7. If your program supports random numbers, generate and solve a random system of three equations and three unknowns. If you repeat this several times do you mostly get consistent or inconsistent systems?

8. Use your program to sketch the lines defined by a system of two equations and two unknowns on the same graph.

9. Use your program to sketch the planes defined by a system of three equations and three unknowns on the same graph.

10. Find the temperatures of the nine interior points of a square plate that has been subdivided by three equally spaced parallel vertical lines and three equally spaced parallel horizontal lines. Assume that the two vertical sides of the square are kept at 85° and the two horizontal sides are at 110°. Use the mean value property for heat conduction.

Selected Solutions with Maple

function	Maple	function	Maple
End of comand	; (semicolon)	continue to next line	\ (Backslash)
surpressing output	: (colon)	Solving	solve, linsolve
help	?topic or help(topic);	Solving numerically	fsolve
comments	# comment to end of line	Lin alg package	with(linalg);
last result	" (double quotes)	Row echelon form	gausselim, ffgausselim
assignment	:= (colon equal)	Reduced echelon	rref, gaussjordan
argument lists	() (parentheses)	Back-substitution	backsub

```
# EXERCISE 1 - Partial
sys:={1/5*x+1/6*y+1/7*z =241/1260, 1/6*x+1/7*y+1/8*z = 109/672,
     1/7*x+1/8*y+1/9*z = 71/504};      # Assigining a name to the system.
solve(sys, {x,y,z});                   # Rational number arithmetic.
evalf(");                              # Evaluation of the last output
                                       # to default accuracy.
evalf("",15);                          # Evaluation to higher accuracy.
fsolve(sys, {x,y,z});                  # A one-step alternative.
solve({1./5*x+1/6*y+1/7*z=241/1260,    # Also, forcing floating point
1/6*x+1/7*y+1/8*z = 109/672,           # arithmetic with 1./5 .
1/7*x+1/8*y+1/9*z = 71/504},{x,y,z});
# Another important way of solving a linear system is using linsolve. First
with(linalg);                  # load the linear algebra package, linalg. Then use
A:=matrix([[1/5,1/6,1/7],[1/6,1/7,1/8],[1/7,1/8,1/9]]);  # the coefficent matrix
b:=vector([241/1260,109/672,71/504]);           # and the constant vector
linsolve(A,b);                                  # in linsolve.
# EXERCISE 2
solve({a1*x + b1*y = c1, a2*x + b2*y = c2},{x,y});
# EXERCISE 3
m:=matrix([[1/5,1/6,1/7,241/1260],     # Entering the augmented matrix.
[1/6,1/7,1/8,109/672],[1/7,1/8,1/9,71/504]]);
gausselim(m);                          # Gauss Elimination; A row echelon form.
rref(m);                               # The reduced row echelon form.
gaussjord(m);                          # Same. The last column is the solution.
# EXERCISE 4
A:=matrix(3,3,[1/5,1/6,1/7,            # Entering A in a slightly
1/6,1/7,1/8,1/7,1/8,1/9]);             # different format.
B:=matrix(3,3,[[1,2,3],[2,2,3],[3,3,3]]);       # Yet another format.
rref(A);                               # The 2 reduced echelon forms are
rref(B);                               # the same, so A and B are equivalent.
```

```
# EXERCISE 6
col(B,1);row(B,2);col(B,1..2);row(B,2..3); submatrix(B,1..2,1..2);
# EXERCISE 7 - Hint
rin := rand(1..1000):    # generates a random integer between 1 and 1000.
a1 := evalf(rin()/1000); # Division by 1000 generates a random real in [0,1].
# Also related: randmatrix and randvector.
# EXERCISE 8 - Hint
plot(2*x-1, x=0..3);          # Plots 2x-1 as x varies from 0 to 3.
plot({2*x-1,x+2}, x=0..3);    # Plots 2x-1 and x+2 on the same graph.
# EXERCISE 9 - Hint
plot3d(x-y, x=0..3,y=0..2);        # 3D plot of x-y on [0,3]x[0,2]
plot3d({x-y,x+y}, x=0..3,y=0..2);  # 3D plots in one graph.
```

Selected Solutions with Mathematica

function	Mathematica	function	Mathematica
End of comand	(Return)	argument lists	[] (square brackets)
surpressing output	; (semicolon)	continue to next line	\ (Backslash)
help	?topic or ??topic	multiplication	* or (space)
comments	(* comment *)	Solving	Solve, LinearSolve
last result	% (percent sign)	Solving numerically	NSolve
assignment	= (equal)	Reduced echelon form	RowReduce
A linear algebra package:		<<LinearAlgebra`MatrixManipulation`	

```
(* EXERCISE 1 - Partial *)
sys={1/5 x+1/6 y+1/7 z == 241/1260, 1/6 x+1/7 y+1/8 z == 109/672,
     1/7 x+1/8 y+1/9 z == 71/504}    (* Assigining a name to the system. *)
Solve[sys, {x,y,z}]             (* Rational number arithmetic.      *)
N[%]              (* Evaluation of the last output to default accuracy. *)
N[%%,15]                        (* Evaluation to higher accuracy.   *)
NSolve[sys, {x,y,z}]            (* A one-step alternative.          *)
Solve[{1./5 x+1/6 y+1/7 z==241/1260,  (* Also, forcing floating point  *)
1/6 x+1/7 y+1/8 z == 109/672,   (* arithmetic with 1./5 .           *)
1/7 x+1/8 y+1/9 z == 71/504},{x,y,z}]
LinearSolve[{{1/5,1/6,1/7},{1/6,1/7,1/8},   (* Also! Use LinearSolve with the *)
{1/7,1/8,1/9}},{241/1260,109/672,71/504}]   (* coeff. matrix and constant vector. *)
(* EXERCISE 2 *)
Solve[{a1 x + b1 y == c1, a2 x + b2 y == c2},{x,y}]
Simplify[%]                     (* Answer needs simplification.     *)
```

```
(* EXERCISE 3 *)
m={{1/5,1/6,1/7,241/1260},{1/6,1/7,1/8,109/672},{1/7,1/8,1/9,71/504}}
RowReduce[m]   (* The reduced row echelon form. The soln is the last coln.*)
(* EXERCISE 4 *)
A={{1/5,1/6,1/7},{1/6,1/7,1/8},{1/7,1/8,1/9}}
B={{1,2,3},{2,2,3},{3,3,3}}
RowReduce[A]                    (* The 2 reduced echelon forms are      *)
RowReduce[B]                    (* the same, so A and B are equivalent.*)
(* EXERCISE 6 *)
<<LinearAlgebra'MatrixManipulation'      (* Load a linear algebra package. *)
TakeColumns[B,{1}]
TakeRows[B,{2}]
TakeColumns[B,{1,2}]
TakeRows[B,{2,3}]
TakeMatrix[B,{1,1},{2,2}]
(* EXERCISE 7 - Hint *)
Random[]                        (* A random real in [0,1].               *)
(* EXERCISE 8 - Hint *)
Plot[2*x-1, {x,0,3}]            (* Plots 2x-1 as x varies from 0 to 3.  *)
Plot[{2*x-1,x+2}, {x,0,3}]      (* Plots 2x-1 and x+2 on the same graph.*)
(* EXERCISE 9 - Hint *)
p1=Plot3D[x-y, {x,0,3},{y,0,2}]         (* 3D plot of x-y on [0,3]x[0,2]. *)
p2=Plot3D[x+y, {x,0,3},{y,0,2}]         (* A second plot.                 *)
Show[{p1,p2}]                           (* Displayed together.            *)
```

Selected Solutions with MATLAB

function	MATLAB	function	MATLAB
End of comand	(Return)	argument lists	() (parentheses)
surpressing output	; (semicolon)	continue to next line	. . . (ellipses)
help	help topic	Solving equation	roots
comments	% comment to end of line	Solving system $Ax = b$	A\b
last result	ans	Reduced echelon form	rref
assignment	= (equal)	Reduction in stages	rrefmovie

Note   The comment notation (ST) means that the Symbolic Toolbox should be available.

```
% EXERCISE 1 - Partial
A = [1/5 1/6 1/7; 1/6 1/7 1/8; 1/7 1/8 1/9]  % To solve a square system
b = [241/1260; 109/672; 71/504]   % form the coefficient matrix A, then
A\b                               % the constant vector b and type A\b.
format long                       % For hihger diplayed accuracy switch to
ans                               % long format and call the last output.
```

```
format short                    % Back to short format.
linsolve(A,b)                   % We may also use linsolve (ST).
% EXERCISE 2
solve('a1*x + b1*y = c1', 'a2*x + b2*y = c2', 'x,y')        % (ST)
% EXERCISE 3
m=[1/5 1/6 1/7 241/1260;1/6 1/7 1/8 109/672; 1/7 1/8 1/9 71/504]
rref(m)  % The reduced row echelon form. The last column is the solution.
% EXERCISE 4
B=[1 2 3; 2 2 3; 3 3 3]      % Matrix A was entered in Exer. 1.
rref(A)                      % The 2 reduced echelon forms are
rref(B)                      % the same so A and B are equivalent.
% EXERCISE 6
B(:,1)                       % Column 1.
B(2,:)                       % Row 2.
B(:,1:2)                     % Columns 1 and 2.
B(2:3,:)                     % Rows 2 and 3.
B(1:2,1:2)                   % Upper left 2-block.
% EXERCISE 7 - Hint
rand                         % A random real in [0 1].
% Also related: randn .
% EXERCISE 8 - Hint
fplot('[2*x-1,x+2]',[0 3]) % Plots 2x-1 and x+2 on [0,3] in one graph. Also:
x = 0:.1:3;                  % A more important way is to define an x-vector.
y1 = 2*x-1; y2 = x+2;        % then apply the functions to get the y-vectors
plot(x,y1,x,y2)              % and plot.
                             % Also see ezplot, from (ST).
% EXERCISE 9 - Hint
x = 0:1/4:3;    % To plot x-y and x+y on [0,3]x[0,2] on the same graph:
y = 0:1/6:2;    % Create vectors for the x- and y-coordinates of the points.
[X,Y]=meshgrid(x,y); % Builds an array for x and y suitable for 3-d plotting.
Z=[X-Y,X+Y];            % Define Z in terms of the two functions in
mesh(Z);                % X and Y and use mesh to plot.
% WARNING: x and y should have the same dimensions. Note x has (3-1)*4+1=13
% components and y has (2-1)*6+1=13 components.
% Related: Explore the command linspace!
```

2

Vectors

If survival is a measure of quality, the "Elements" of Euclid and the "Conics" of Appolonius were the best works in their fields.

—Carl C. Boyer

Introduction

Apollonius and Descartes

Quantities such as length, area, volume, temperature, mass, and potential can be determined by their magnitude only. Consider, however, displacement, velocity, and force. With these quantities we need both magnitude and direction to completely determine them. Displacement, velocity, and force are examples of *free vectors*. We are primarily interested in free vectors that start at the *origin*. These we simply call *vectors*.

In this chapter we study vectors along with their arithmetic and geometry, because they play an important role in mathematics, physics, engineering, image processing, computer graphics, and many other areas of science and everyday life.

In the plane and in space, vectors have a dual existence: they are both algebraic and geometric objects. This kind of duality enables us to study geometry by algebraic means. Two men more than others are credited with the realization of this, Apollonius and Descartes.

Apollonius[1] ranks along with Euclid and Archimedes as one of the three greatest geometers of antiquity. He proved many geometric properties of conic sections by using lines of reference to measure distances between points on curves. This is the first known systematic use of coordinates. In 1637 Descartes used one of Apollonius' theorems to test his new analytic geometry.

---

[1] **Apollonius of Perga** was born in the Greek city of Perga in Asia Minor. He spent some time in Alexandria and Pergamum. It is believed that he lived between 262 and 200 B.C. Most of his works, such as *Quick Delivery*, *Cutting-off of a Ratio*, *Tangencies*, and *Inclinations*, were lost. Seven books of his great treatise *Conics* survived. His theorems from the *Conics* have been used in astronomy for two millenia.

Descartes[2] is the true father of analytic geometry and the first person to systematically study geometry by purely algebraic methods. Unlike Apollonius, Descartes introduced coordinate systems independently of curves. He emphasized the idea that a curve is *defined* by an equation and not that the curve determines the equation. His most celebrated work is a philosophical treatise known as *Discours de la Méthode*. One of the appendices in this work is titled *La Geométrié* and introduces what today is known as analytic geometry.

2.1   Vector Operations

Reader's Goals for This Section

1. To do basic vector arithmetic and understand its geometry.
2. To know what a linear combination of vectors is.
3. To understand the relation between linear systems and vector equations.

We devote this section to vectors and their arithmetic. We also introduce the important concept of a *linear combination* of a sequence of vectors.

Addition and Scalar Multiplication

A **vector** is a one-column matrix. An *n*-**vector** is an $n \times 1$ matrix. For example,

$$\mathbf{u} = \begin{bmatrix} 1 \\ -1 \end{bmatrix}, \qquad \mathbf{v} = \begin{bmatrix} 1 \\ 2 \\ 3 \end{bmatrix}, \qquad \mathbf{w} = \begin{bmatrix} 0.5 \\ 1 \\ 0 \\ -0.2 \end{bmatrix}$$

are 2-, 3-, and 4-vectors, respectively. The value n is sometimes called the **size** of the vector. The entries of a vector are also called **components**. The components of $\mathbf{w}$ are 0.5, 1, 0, and -0.2. The set of all *n*-vectors is denoted by $\mathbf{R}^n$.

$$\mathbf{R}^n = \{\mathbf{x}, \mathbf{x} \text{ is an } n\text{-vector}\}$$

$\mathbf{u}$, $\mathbf{v}$, and $\mathbf{w}$ are elements of $\mathbf{R}^2$, $\mathbf{R}^3$, and $\mathbf{R}^4$, respectively.

Two- and 3-vectors can be interpreted geometrically as points in the plane or in space. Any 2-vector, say, $\mathbf{x} = \begin{bmatrix} x_1 \\ x_2 \end{bmatrix}$, can be graphically represented by the *point* with coordinates (x_1, x_2) in a Cartesian coordinate plane. Often $\mathbf{x}$ is viewed as the *arrow* starting at the origin $(0, 0)$ and with tip the point with coordinates (x_1, x_2). Figure 2.1(a) and 2.1(b) shows vectors $\mathbf{u} = \begin{bmatrix} 1 \\ -1 \end{bmatrix}$, $\mathbf{v} = \begin{bmatrix} 2 \\ 2 \end{bmatrix}$, and $\mathbf{w} = \begin{bmatrix} -2 \\ 1 \end{bmatrix}$ both as points and as arrows starting at the origin. Because all 2-vectors can be represented this way, $\mathbf{R}^2$ is the

---

[2] **René Descartes** was a philosopher, soldier, and mathematician. Born in 1596 in Tourraine, France, of a well-to-do family, he was educated at a Jesuit college and studied law. He traveled about and participated in several military campaigns. His works include *Le Monde* and *Discours de la Méthode*. He died in 1650 in Stockholm, where he was spending time tutoring Queen Christina of Sweden.

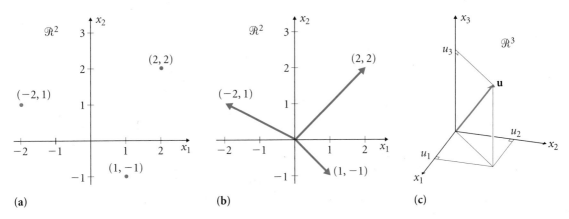

Figure 2.1   Geometric representation of 2- and 3-vectors.

entire plane. Three-vectors can be graphed in a similar fashion (Fig. 2.1(c)), and $\mathbf{R}^3$ is the entire 3-space.

We call two vectors $\mathbf{u}$ and $\mathbf{v}$ of the same size **equal**, and we write $\mathbf{u} = \mathbf{v}$, if their corresponding components are equal. Vectors of different sizes are never equal.

■ EXAMPLE 1   We have

$$\begin{bmatrix} 1 \\ a + b \end{bmatrix} = \begin{bmatrix} a \\ -1 \end{bmatrix}$$

only if $a = 1$ and $b = -2$.

Vectors of the same size can be added componentwise:

$$\begin{bmatrix} 1 \\ -1 \end{bmatrix} + \begin{bmatrix} -4 \\ 2 \end{bmatrix} = \begin{bmatrix} -3 \\ 1 \end{bmatrix}, \qquad \begin{bmatrix} 1 \\ 2 \\ 3 \end{bmatrix} + \begin{bmatrix} 4 \\ -2 \\ -7 \end{bmatrix} = \begin{bmatrix} 5 \\ 0 \\ -4 \end{bmatrix}$$

This operation is called **vector addition**.

The sum $\mathbf{u} + \mathbf{v}$ of two 2-vectors or two 3-vectors $\mathbf{u}$ and $\mathbf{v}$ can be represented geometrically as the diagonal arrow of the parallelogram with sides $\mathbf{u}$ and $\mathbf{v}$ (Fig. 2.2(a)). This rule is known as the **parallelogram law of addition**.

An n-vector can be multiplied by a scalar componentwise:

$$7 \begin{bmatrix} 2 \\ -1 \end{bmatrix} = \begin{bmatrix} 14 \\ -7 \end{bmatrix}, \qquad -2 \begin{bmatrix} 1 \\ 2 \\ -3 \end{bmatrix} = \begin{bmatrix} -2 \\ -4 \\ 6 \end{bmatrix}$$

This operation is called **scalar multiplication**. The vector $(-1)\mathbf{v}$ is called the **opposite** of $\mathbf{v}$, and it is denoted by $-\mathbf{v}$.

$$(-1)\mathbf{v} = -\mathbf{v}$$

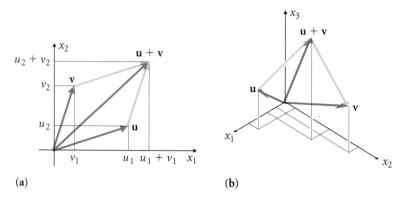

(a)   (b)

Figure 2.2   The parallelogram law of addition.

We usually write $\mathbf{u} - \mathbf{v}$ for $\mathbf{u} + (-1)\mathbf{v}$ and call this result the **difference** between $\mathbf{u}$ and $\mathbf{v}$.

$$\mathbf{u} - \mathbf{v} = \mathbf{u} + (-1)\mathbf{v}$$

A vector whose entries are all zero is called a **zero vector**, and it is denoted by $\mathbf{0}$.

$$\mathbf{0} = [0] \qquad \mathbf{0} = \begin{bmatrix} 0 \\ 0 \end{bmatrix} \qquad \mathbf{0} = \begin{bmatrix} 0 \\ 0 \\ 0 \end{bmatrix}$$

Geometrically, the scalar product $c\mathbf{u}$ is the arrow $\mathbf{u}$ scaled by a factor of c. If $c > 0$, then $c\mathbf{u}$ is in the same direction as $\mathbf{u}$. If $c < 0$, $c\mathbf{u}$ is in the opposite direction. If $|c| > 1$, then $c\mathbf{u}$ is $\mathbf{u}$ stretched by a factor of c. If $|c| < 1$, then $c\mathbf{u}$ is a contraction of $\mathbf{u}$ (Fig. 2.3(a) and 2.3(b)).

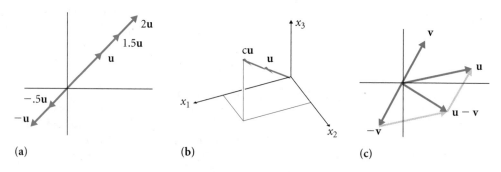

(a)   (b)   (c)

Figure 2.3   (a) and (b): Scalar products, (c) difference.

Note that the difference $\mathbf{u} - \mathbf{v}$ can be sketched as the sum $\mathbf{u}+(-1)\mathbf{v}$ (Fig. 2.3(c)).

The n-vector with 1 for its ith component and 0s everywhere else is denoted by $\mathbf{e}_i$. The vectors $\mathbf{e}_1, \mathbf{e}_2, \ldots, \mathbf{e}_n$ are called the **standard basis vectors** of $\mathbf{R}^n$. For example, the standard basis vectors of $\mathbf{R}^2$ are $\mathbf{e}_1 = \begin{bmatrix} 1 \\ 0 \end{bmatrix}$ and $\mathbf{e}_2 = \begin{bmatrix} 0 \\ 1 \end{bmatrix}$, whereas those of $\mathbf{R}^3$ are

$$\mathbf{e}_1 = \begin{bmatrix} 1 \\ 0 \\ 0 \end{bmatrix}, \qquad \mathbf{e}_2 = \begin{bmatrix} 0 \\ 1 \\ 0 \end{bmatrix}, \qquad \mathbf{e}_3 = \begin{bmatrix} 0 \\ 0 \\ 1 \end{bmatrix}$$

(See Fig. 2.4.)

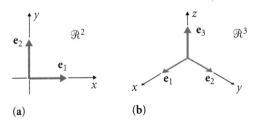

Figure 2.4   The standard basis vectors.

Vector addition and scalar multiplication satisfy a few basic properties, which are summarized in the next theorem.

THEOREM 1

(Rules for Vector Addition and Scalar Multiplication)

Let $\mathbf{u}$, $\mathbf{v}$, $\mathbf{w}$ be any n-vectors and let a and b be any scalars. Then we have the following equalities between n-vectors.

1. $(\mathbf{u} + \mathbf{v}) + \mathbf{w} = \mathbf{u} + (\mathbf{v} + \mathbf{w})$ **Associative law**
2. $\mathbf{u} + \mathbf{v} = \mathbf{v} + \mathbf{u}$ **Commutative law**
3. $\mathbf{u} + \mathbf{0} = \mathbf{0} + \mathbf{u} = \mathbf{u}$
4. $\mathbf{u} + (-\mathbf{u}) = (-\mathbf{u}) + \mathbf{u} = \mathbf{0}$
5. $a(\mathbf{u} + \mathbf{v}) = a\mathbf{u} + a\mathbf{v}$ **Distributive law**
6. $(a + b)\mathbf{u} = a\mathbf{u} + b\mathbf{u}$ **Distributive law**
7. $(ab)\mathbf{u} = a(b\mathbf{u}) = b(a\mathbf{u})$
8. $1\mathbf{u} = \mathbf{u}$
9. $0\mathbf{u} = \mathbf{0}$

PROOF   We prove only Part 1 and leave the remaining proofs as exercises. Because $\mathbf{u} + \mathbf{v}$ and $\mathbf{w}$ are n-vectors, so is their sum $(\mathbf{u} + \mathbf{v}) + \mathbf{w}$. Likewise, $\mathbf{u} + (\mathbf{v} + \mathbf{w})$ is an n-vector. Hence, $(\mathbf{u} + \mathbf{v}) + \mathbf{w}$ and $\mathbf{u} + (\mathbf{v} + \mathbf{w})$ have the same size. Let u_i, v_i, and w_i be the ith components of $\mathbf{u}$, $\mathbf{v}$, and $\mathbf{w}$, respectively. Then $u_i + v_i$ is the ith component of $\mathbf{u} + \mathbf{v}$, so $(u_i + v_i) + w_i$ is the ith component of $(\mathbf{u} + \mathbf{v}) + \mathbf{w}$. Likewise, $u_i + (v_i + w_i)$ is the ith component of $\mathbf{u} + (\mathbf{v} + \mathbf{w})$. Since $(u_i + v_i) + w_i = u_i + (v_i + w_i)$, the corresponding components of $(\mathbf{u} + \mathbf{v}) + \mathbf{w}$ and $\mathbf{u} + (\mathbf{v} + \mathbf{w})$ are equal. This statement is true for all $i = 1, \dots, n$. We conclude that $(\mathbf{u} + \mathbf{v}) + \mathbf{w}$ and $\mathbf{u} + (\mathbf{v} + \mathbf{w})$ are equal.

Theorem 1 can be used in solving simple vector equations.

■ EXAMPLE 2   Find the vector **x** such that $2\mathbf{x} - 4\mathbf{v} = 3\mathbf{u}$.

SOLUTION   Add 4**v** to both sides of the equation to get

$$
\begin{aligned}
(2\mathbf{x} - 4\mathbf{v}) + 4\mathbf{v} &= 3\mathbf{u} + 4\mathbf{v} \\
\Leftrightarrow 2\mathbf{x} + (-4\mathbf{v} + 4\mathbf{v}) &= 3\mathbf{u} + 4\mathbf{v} && \text{by 1 of Theorem 1} \\
\Leftrightarrow 2\mathbf{x} + \mathbf{0} &= 3\mathbf{u} + 4\mathbf{v} && \text{by 4} \\
\Leftrightarrow 2\mathbf{x} &= 3\mathbf{u} + 4\mathbf{v} && \text{by 3}
\end{aligned}
$$

Multiply both sides of the last equation by $\frac{1}{2}$:

$$
\begin{aligned}
\Leftrightarrow \tfrac{1}{2}(2\mathbf{x}) &= \tfrac{1}{2}(3\mathbf{u} + 4\mathbf{v}) \\
\Leftrightarrow \left(\tfrac{1}{2} \cdot 2\right)\mathbf{x} &= \tfrac{1}{2}(3\mathbf{u}) + \tfrac{1}{2}(4\mathbf{v}) && \text{by 7, 5} \\
\Leftrightarrow 1\mathbf{x} &= \left(\tfrac{1}{2} \cdot 3\right)\mathbf{u} + \left(\tfrac{1}{2} \cdot 4\right)\mathbf{v} && \text{by 7} \\
\Leftrightarrow \mathbf{x} &= \tfrac{3}{2}\mathbf{u} + 2\mathbf{v} && \text{by 8}
\end{aligned}
$$

Matrices as Sequences of Vectors

We often view matrices as sequences of vectors. For example, the matrix

$$
\begin{bmatrix} 1 & 3 & 1 \\ 2 & 4 & 2 \end{bmatrix}
$$

can be viewed as $\begin{bmatrix} \mathbf{v}_1 & \mathbf{v}_2 & \mathbf{v}_3 \end{bmatrix}$, where $\mathbf{v}_1 = \begin{bmatrix} 1 \\ 2 \end{bmatrix}$, $\mathbf{v}_2 = \begin{bmatrix} 3 \\ 4 \end{bmatrix}$, and $\mathbf{v}_3 = \begin{bmatrix} 1 \\ 2 \end{bmatrix}$. We said *sequence* instead of *set* for two reasons: In contrast with sets, (1) the elements of a sequence have a distinct order and (2) the same element is allowed to be repeated in different positions. Clearly, a matrix can have repeated columns, and the order of columns is important. Once we are aware of this, saying *set* instead of *sequence* does not usually cause trouble.

Linear Combinations

The associative and commutative laws allow us to drop parentheses from multiple sums and simplify the notation. For example, each of the equal expressions $(\mathbf{u} + \mathbf{v}) + (\mathbf{w} + \mathbf{r})$, $\mathbf{u} + ((\mathbf{v} + \mathbf{w}) + \mathbf{r})$, $\mathbf{u} + (\mathbf{v} + (\mathbf{w} + \mathbf{r}))$ and $\mathbf{v} + (\mathbf{u} + (\mathbf{w} + \mathbf{r}))$ is simply written as $\mathbf{u} + \mathbf{v} + \mathbf{w} + \mathbf{r}$. Also, we may write without ambiguity expressions such as

$$
\mathbf{v}_1 - 3\mathbf{v}_2 + 5\mathbf{v}_3 - 2\mathbf{v}_4
$$

where $\mathbf{v}_1, \mathbf{v}_2, \mathbf{v}_3$, and $\mathbf{v}_4$ are vectors of the same size. Such sums of scalar multiples of vectors are called *linear combinations*.

DEFINITION

(Linear Combination)

Let $v_1, v_2, \ldots, v_k$ be n-vectors and let $c_1, c_2, \ldots, c_k$ be scalars. The n-vector of the form

$$c_1 v_1 + c_2 v_2 + \cdots + c_k v_k$$

is called a **linear combination** of $v_1, \ldots, v_k$. The scalars $c_1, \ldots, c_k$ are called the **coefficients** of the linear combination.

■ EXAMPLE 3   Compute and sketch the linear combination $\frac{1}{2}v_1 - 3v_2$, where

$$v_1 = \begin{bmatrix} 2 \\ 4 \end{bmatrix}, \qquad v_2 = \begin{bmatrix} -1 \\ 1 \end{bmatrix}$$

SOLUTION $\frac{1}{2}\begin{bmatrix} 2 \\ 4 \end{bmatrix} - 3\begin{bmatrix} -1 \\ 1 \end{bmatrix} = \begin{bmatrix} 1 \\ 2 \end{bmatrix} + \begin{bmatrix} 3 \\ -3 \end{bmatrix} = \begin{bmatrix} 4 \\ -1 \end{bmatrix}$. This process is illustrated geometrically in Fig. 2.5.

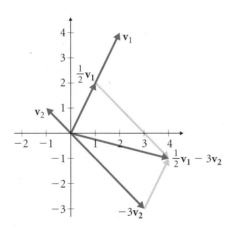

Figure 2.5   The linear combination $\frac{1}{2}v_1 - 3v_2$.

■ EXAMPLE 4   Determine whether each of the vectors

$$u = \begin{bmatrix} 0 \\ 2 \\ 1 \end{bmatrix} \quad \text{and} \quad v = \begin{bmatrix} 0 \\ 1 \\ 2 \end{bmatrix}$$

is a linear combination of $v_1, v_2,$ and v_3, where

$$v_1 = \begin{bmatrix} -1 \\ 1 \\ 0 \end{bmatrix}, \qquad v_2 = \begin{bmatrix} 2 \\ 0 \\ 1 \end{bmatrix}, \qquad v_3 = \begin{bmatrix} 1 \\ 1 \\ 1 \end{bmatrix}$$

SOLUTION   We are seeking scalars c_1, c_2, and c_3 such that $\mathbf{u} = c_1\mathbf{v}_1 + c_2\mathbf{v}_2 + c_3\mathbf{v}_3$.

$$\begin{bmatrix} 0 \\ 2 \\ 1 \end{bmatrix} = c_1 \begin{bmatrix} -1 \\ 1 \\ 0 \end{bmatrix} + c_2 \begin{bmatrix} 2 \\ 0 \\ 1 \end{bmatrix} + c_3 \begin{bmatrix} 1 \\ 1 \\ 1 \end{bmatrix}$$

Therefore,

$$\begin{bmatrix} -c_1 + 2c_2 + c_3 \\ c_1 + c_3 \\ c_2 + c_3 \end{bmatrix} = \begin{bmatrix} 0 \\ 2 \\ 1 \end{bmatrix}, \quad \text{or} \quad \begin{aligned} -c_1 + 2c_2 + c_3 &= 0 \\ c_1 + c_3 &= 2 \\ c_2 + c_3 &= 1 \end{aligned}$$

This system is linear with unknowns c_1, c_2, and c_3. Its augmented matrix has columns $\mathbf{v}_1$, $\mathbf{v}_2$, $\mathbf{v}_3$, and $\mathbf{u}$ and reduces as

$$\begin{bmatrix} -1 & 2 & 1 & : & 0 \\ 1 & 0 & 1 & : & 2 \\ 0 & 1 & 1 & : & 1 \end{bmatrix} \longrightarrow \cdots \longrightarrow \begin{bmatrix} 1 & 0 & 1 & : & 2 \\ 0 & 1 & 1 & : & 1 \\ 0 & 0 & 0 & : & 0 \end{bmatrix}$$

Hence, the solution of the system is $c_1 = -r + 2$, $c_2 = -r + 1$, $c_3 = r$ for any scalar r. In this case there are infinitely many scalars such that $\mathbf{u} = c_1\mathbf{v}_1 + c_2\mathbf{v}_2 + c_3\mathbf{v}_3$. For example, if $r = 0$, then $c_1 = 2$, $c_2 = 1$, $c_3 = 0$ and

$$\begin{bmatrix} 0 \\ 2 \\ 1 \end{bmatrix} = 2 \begin{bmatrix} -1 \\ 1 \\ 0 \end{bmatrix} + 1 \begin{bmatrix} 2 \\ 0 \\ 1 \end{bmatrix} + 0 \begin{bmatrix} 1 \\ 1 \\ 1 \end{bmatrix}$$

We are also seeking scalars d_1, d_2, and d_3 such that $\mathbf{v} = d_1\mathbf{v}_1 + d_2\mathbf{v}_2 + d_3\mathbf{v}_3$, or

$$\begin{bmatrix} 0 \\ 1 \\ 2 \end{bmatrix} = d_1 \begin{bmatrix} -1 \\ 1 \\ 0 \end{bmatrix} + d_2 \begin{bmatrix} 2 \\ 0 \\ 1 \end{bmatrix} + d_3 \begin{bmatrix} 1 \\ 1 \\ 1 \end{bmatrix}$$

The augmented matrix of the corresponding system in d_1, d_2, and d_3 has columns $\mathbf{v}_1$, $\mathbf{v}_2$, $\mathbf{v}_3$, and $\mathbf{v}$ and reduces as

$$\begin{bmatrix} -1 & 2 & 1 & : & 0 \\ 1 & 0 & 1 & : & 1 \\ 0 & 1 & 1 & : & 2 \end{bmatrix} \longrightarrow \cdots \longrightarrow \begin{bmatrix} 1 & 0 & 1 & : & 0 \\ 0 & 1 & 1 & : & 0 \\ 0 & 0 & 0 & : & 1 \end{bmatrix}$$

Therefore, the system is inconsistent. So, $\mathbf{v}$ is *not* a linear combination of $\mathbf{v}_1$, $\mathbf{v}_2$, and $\mathbf{v}_3$.

Linear Systems as Vector Equations

In the solution of Example 4 we saw that the vector equation $c_1\mathbf{v}_1 + c_2\mathbf{v}_2 + c_3\mathbf{v}_3 = \mathbf{u}$ yielded a system whose augmented matrix had columns $\mathbf{v}_1$, $\mathbf{v}_2$, $\mathbf{v}_3$, and $\mathbf{u}$. Conversely, the system with this augmented matrix implies the same vector equation. This important notational equivalence applies to any linear system.

Relation Between Linear Systems and Linear Combinations

Consider the system with unknowns $x_1, x_2, \ldots, x_n$, coefficient matrix A, and constant terms the components of a vector $\mathbf{b}$. If $\mathbf{a}_1, \mathbf{a}_2, \ldots, \mathbf{a}_n$ are the columns of A, then we have the equivalent notations

$$[A : \mathbf{b}] \Leftrightarrow x_1\mathbf{a}_1 + x_2\mathbf{a}_2 + \cdots + x_n\mathbf{a}_n = \mathbf{b} \qquad (2.1)$$

■ EXAMPLE 5   Write the system as a vector equation.

$$x - 5y = 1$$
$$-x + 6y = 3$$

SOLUTION

$$x \begin{bmatrix} 1 \\ -1 \end{bmatrix} + y \begin{bmatrix} -5 \\ 6 \end{bmatrix} = \begin{bmatrix} 1 \\ 3 \end{bmatrix}$$

■ EXAMPLE 6   Write the vector equation

$$x_1 \begin{bmatrix} a_{11} \\ a_{21} \end{bmatrix} + x_2 \begin{bmatrix} a_{12} \\ a_{22} \end{bmatrix} = \begin{bmatrix} b_1 \\ b_2 \end{bmatrix}$$

as a linear system.

SOLUTION

$$a_{11}x_1 + a_{12}x_2 = b_1$$
$$a_{21}x_1 + a_{22}x_2 = b_2$$

QUESTION   What expression is equivalent to 'the vector $\mathbf{b}$ is a linear combination of $\mathbf{v}_1, \mathbf{v}_2, \ldots, \mathbf{v}_n$'?

ANSWER   The system with augmented matrix

$$\begin{bmatrix} \mathbf{v}_1 & \mathbf{v}_2 & \cdots & \mathbf{v}_n & : & \mathbf{b} \end{bmatrix}$$

is consistent. Or, equivalently, 'the last column of this augmented matrix is *not* a pivot column.'

The equivalence between linear systems and linear combinations allows us to write the solutions of linear systems in vector form. For example, the solution $x = 21, y = 4$ of the system in Example 5 can be written as $\begin{bmatrix} x \\ y \end{bmatrix} = \begin{bmatrix} 21 \\ 4 \end{bmatrix}$ or just as $\begin{bmatrix} 21 \\ 4 \end{bmatrix}$ if the order of the variables is unambiguous.

AN IMPORTANT QUESTION   Often, both in theory and in practice (see Example 11), we need to answer the following question: can the system with augmented matrix $[A : \mathbf{b}]$ be consistent for *all* $\mathbf{b}$? The key to answering this is, once more, row reduction.

0 0 0 : c   row without pivot.
if c = 0, then consistent, But not for all vectors (need to be 0)

THEOREM 2

Let A be an $m \times n$ matrix. The following are equivalent:

1. The linear system with augmented matrix $[A : b]$ is consistent for all vectors $b \in R^m$.
2. Every vector $b \in R^m$ is a linear combination of the columns of A.
3. A has m pivot positions. (Or, each row has a pivot position.)

PROOF   By (2.1) Statements 1 and 2 are equivalent, so it suffices to prove the equivalence between 1 and 3. Suppose that the system is consistent for all $b \in R^m$. If one row of A is not a pivot row, then any echelon of $[A : b]$ will have a row of the form $\begin{bmatrix} 0 & 0 & 0 & \cdots & 0 & : & b \end{bmatrix}$. Because b is any n-vector, we may choose components for b so that $b \neq 0$. But then $[A : b]$ will be inconsistent for the particular b. This contradicts our assumption that $[A : b]$ is consistent for all m-vectors b. Therefore, all rows have a pivot position. Conversely, suppose that each row of A has a pivot position. Then the last entry of the last row of A is nonzero. Hence, the last column of the augmented matrix is never a pivot column, no matter what b is. So the system $[A : b]$ is consistent for all m-vectors b.

NOTE

1. To say that the $m \times n$ matrix A has m pivot positions automatically implies that $m \leq n$. (Why?)
2. Parts 2 and 3 of Theorem 2 refer to the *coefficient* matrix of a system and *not* the augmented matrix. For example, the augmented matrix $\begin{bmatrix} 1 & : & 0 \\ 0 & : & 1 \end{bmatrix}$ has two pivot columns, but the corresponding system has no solutions.

■ **EXAMPLE 7**   Consider the systems with augmented matrices $[A : b]$, $[B : b]$, and $[C : b]$, where

$$A = \begin{bmatrix} -1 & 3 & 2 & 0 \\ 0 & 2 & -2 & 4 \\ 0 & -1 & 1 & 2 \end{bmatrix}, \quad B = \begin{bmatrix} -1 & 3 & 2 & 0 \\ 0 & 2 & -2 & 4 \\ 0 & -1 & 1 & -2 \end{bmatrix}, \quad C = \begin{bmatrix} -1 & 3 \\ 0 & 2 \\ 0 & -1 \end{bmatrix}$$

Which systems are consistent for all vectors $b \in R^3$? What can you say about the systems that are not?

SOLUTION   Because

$$A \sim \begin{bmatrix} -1 & 3 & 2 & 0 \\ 0 & 2 & -2 & 4 \\ 0 & 0 & 0 & 4 \end{bmatrix} \quad B \sim \begin{bmatrix} -1 & 3 & 2 & 0 \\ 0 & 2 & -2 & 4 \\ 0 & 0 & 0 & 0 \end{bmatrix} \quad C \sim \begin{bmatrix} -1 & 3 \\ 0 & 2 \\ 0 & 0 \end{bmatrix}$$

A has one pivot position in each row; hence, $[A : b]$ is solvable for all $b \in R^3$, by Theorem 2. The third rows of B and C have no pivots; hence, the systems $[B : b]$ and $[C : b]$ are not solvable for all $b \in R^3$. According to Exercise 50 of Section 1.2, each of

$[B : \mathbf{b}]$ and $[C : \mathbf{b}]$ has either no solutions or infinitely many solutions, depending on what $\mathbf{b}$ is.

The $(x_1, x_2, \ldots, x_n)$ Notation

Occasionally, in order to save space, we use the notation $(x_1, x_2, \ldots, x_n)$ for the vector with components $x_1, x_2, \ldots, x_n$. This should not be confused with the matrix $\begin{bmatrix} x_1 & x_2 & \cdots & x_n \end{bmatrix}$, which has size $1 \times n$ and not $n \times 1$. Thus $(1, 2)$ is the same as $\begin{bmatrix} 1 \\ 2 \end{bmatrix}$ but not the same as $\begin{bmatrix} 1 & 2 \end{bmatrix}$.

Free Vectors

A **free vector** is a quantity that can be determined by its magnitude and its direction. Just as with vectors, geometrically, we think of free vectors as directed line segments or as arrows, and we denote them by $\mathbf{a}, \mathbf{b}, \ldots$, or by $\overrightarrow{PQ}, \overrightarrow{RS}, \ldots$. The length and the tip of the arrow describe the magnitude and the direction of the vector. In contrast with 2- and 3-vectors, free vectors can start at any point in the plane or space. In fact, every free vector with initial point the origin of the coordinate system is a vector.

Two free vectors $\mathbf{a}$ and $\mathbf{b}$ are **equal**, written $\mathbf{a} = \mathbf{b}$, if they have the same magnitude and the same direction (Fig. 2.6(a)). *A vector can be used to represent all the free vectors equal to it* (Fig. 2.6(b)). We call the vector that corresponds to a set of equal free vectors the **vector** (or **position vector**) of the set.

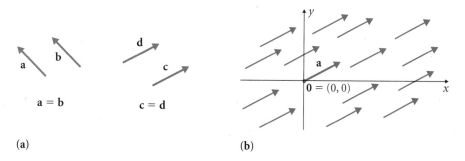

(a)   (b)

Figure 2.6   (a) Equal free vectors, (b) the vector of equal free vectors.

We add two free vectors by adding their corresponding vectors. Likewise, scalar multiplication is performed on the corresponding vector. The components of a free vector are the components of its vector. In general, *we study free vectors via their corresponding vectors.*

A free plane vector $\overrightarrow{PQ}$ with origin $P(p_1, p_2)$ and terminal point $Q(q_1, q_2)$ has components

$$\overrightarrow{PQ} = (q_1 - p_1, q_2 - p_2)$$

because

$$\overrightarrow{PQ} = \overrightarrow{OQ} - \overrightarrow{OP} = (q_1, q_2) - (p_1, p_2) = (q_1 - p_1, q_2 - p_2)$$

(Fig. 2.7(a)). The analogous formula holds for free vectors in space. If $P(p_1, p_2, p_3)$ and $Q(q_1, q_2, q_3)$, then the components of $\overrightarrow{PQ}$ are given by

$$\overrightarrow{PQ} = (q_1 - p_1, q_2 - p_2, q_3 - p_3)$$

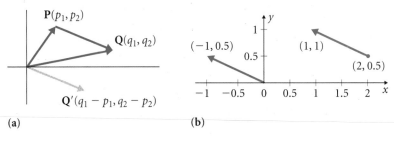

(a)   (b)

Figure 2.7   Components of free vectors.

▪ **EXAMPLE 8** (Components of a Free Vector)   Find the components of each vector.

(a) $\overrightarrow{PQ}$ if $P(2, 0.5)$ and $Q(1, 1)$ (Fig. 2.7(b)).
(b) $\overrightarrow{RS}$ if $R(-2, -1, 4)$ and $S(-3, 1, 0)$

SOLUTION

(a) $\overrightarrow{PQ} = \begin{bmatrix} 1 - 2 \\ 1 - 0.5 \end{bmatrix} = \begin{bmatrix} -1 \\ 0.5 \end{bmatrix}$ (b) $\overrightarrow{RS} = \begin{bmatrix} -3 - (-2) \\ 1 - (-1) \\ 0 - 4 \end{bmatrix} = \begin{bmatrix} -1 \\ 2 \\ -4 \end{bmatrix}$

Applications

▪ **EXAMPLE 9** (Olympic Diving Team)   The 10-member U.S. Olympic diving team participated in three international meets this year. Find the averages of the divers if the scores in the meets are given by the vectors $\mathbf{u}$, $\mathbf{v}$, $\mathbf{w}$. What is the average score of the tenth member of the team?

$$\mathbf{u} = (8.5, 9.5, 8, 9.2, 9.9, 10, 8.8, 6.5, 9.4, 9.8)$$
$$\mathbf{v} = (9.5, 7.5, 8.2, 8.2, 8.9, 7.9, 7.8, 8.5, 9.4, 9.6)$$
$$\mathbf{w} = (8.5, 8.5, 8.9, 9.2, 8.6, 9.9, 9.8, 9.5, 9.1, 8.9)$$

SOLUTION   The average vector is

$$\frac{1}{3}(\mathbf{u} + \mathbf{v} + \mathbf{w}) = (8.83, 8.50, 8.36, 8.86, 9.13, 9.26, 8.80, 8.16, 9.3, 9.43)$$

All entries were truncated to two decimal places. The tenth member of the team averaged 9.43 points.

Manufacturing

▪ EXAMPLE 10   A sports company owns two factories, each making aluminum and titanium mountain bikes. The first factory makes 150 aluminum and 15 titanium bikes a day. For the second factory the numbers are 220 and 20, respectively. If $\mathbf{v}_1 = \begin{bmatrix} 150 \\ 15 \end{bmatrix}$ and $\mathbf{v}_2 = \begin{bmatrix} 220 \\ 20 \end{bmatrix}$, compute and discuss the meaning of a–d:

(a) $\mathbf{v}_1 + \mathbf{v}_2$
(b) $\mathbf{v}_2 - \mathbf{v}_1$
(c) $10\mathbf{v}_1$
(d) $a\mathbf{v}_1 + b\mathbf{v}_2$, for $a, b > 0$.
(e) How many days should each factory operate if the company is to deliver 2600 aluminum and 250 titanium bikes?

SOLUTION

(a) $\mathbf{v}_1 + \mathbf{v}_2 = \begin{bmatrix} 370 \\ 35 \end{bmatrix}$ represents the total number of aluminum (370) and titanium (35) bikes produced by the two factories in one day.

(b) $\mathbf{v}_2 - \mathbf{v}_1 = \begin{bmatrix} 70 \\ 5 \end{bmatrix}$ shows how many more bikes the second factory makes a day over the first one.

(c) $10\mathbf{v}_1 = \begin{bmatrix} 1500 \\ 150 \end{bmatrix}$ represents how many bikes the first factory makes in 10 days.

(d) $a\mathbf{v}_1 + b\mathbf{v}_2 = \begin{bmatrix} 150a + 220b \\ 15a + 20b \end{bmatrix}$ represents the total number of bikes produced if the first factory operates for a days and the second for b days.

(e) Let x_1 and x_2 be the respective days of operation. Then $x_1\mathbf{v}_1 + x_2\mathbf{v}_2 = \begin{bmatrix} 2600 \\ 250 \end{bmatrix}$. Hence,

$$x_1 \begin{bmatrix} 150 \\ 15 \end{bmatrix} + x_2 \begin{bmatrix} 220 \\ 20 \end{bmatrix} = \begin{bmatrix} 2600 \\ 250 \end{bmatrix}$$

The corresponding system has augmented matrix $\begin{bmatrix} 150 & 220 & : & 2600 \\ 15 & 20 & : & 250 \end{bmatrix}$ with reduced row echelon form $\begin{bmatrix} 1 & 0 & : & 10 \\ 0 & 1 & : & 5 \end{bmatrix}$. Therefore, the first factory needs to operate 10 days and the second, 5 days.

Heat Transfer

As an example of application of Theorem 2, let us consider the heat-transfer problem studied in Section 1.4. A square plate has been given boundary temperature distribution $b_1, b_2, \ldots, b_8$ (Fig. 2.8). We need the temperature in the interior. For simplicity, we assign heat elements at a choice of mesh points. We then use the *mean value property for heat conduction*, namely, that the temperature of an interior mesh point is the average of the temperatures of its neighboring mesh points, to get a linear system. In this case,

$$
\begin{aligned}
x_1 &= \tfrac{1}{4}(x_2 + x_3 + b_1 + b_3) \\
x_2 &= \tfrac{1}{4}(x_1 + x_4 + b_2 + b_4) \\
x_3 &= \tfrac{1}{4}(x_1 + x_4 + b_5 + b_7) \\
x_4 &= \tfrac{1}{4}(x_2 + x_3 + b_6 + b_8)
\end{aligned}
\tag{2.2}
$$

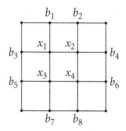

Figure 2.8   Heat transfer.

■ **EXAMPLE 11**   (An Engineer's Intuition)   Any engineer knows (intuitively) that any temperature values $b_1, \ldots, b_8$ on the boundary of the plate yield unique temperatures at $x_1, \ldots, x_4$. Prove this fact mathematically.

SOLUTION   We need to prove that system (2.2) can be solved for all $b_1, \ldots, b_8$. First, we rewrite (2.2) in standard form and read off the augmented matrix:

$$
[A : \mathbf{b}] =
\left[\begin{array}{cccc:c}
4 & -1 & -1 & 0 & b_1 + b_3 \\
-1 & 4 & 0 & -1 & b_2 + b_4 \\
-1 & 0 & 4 & -1 & b_5 + b_7 \\
0 & -1 & -1 & 4 & b_6 + b_8
\end{array}\right]
$$

Because

$$
A \sim
\begin{bmatrix}
4 & -1 & -1 & 0 \\
0 & -1 & -1 & 4 \\
0 & 0 & 4 & -2 \\
0 & 0 & 0 & 12
\end{bmatrix}
$$

A has four pivots. Hence, $[A : \mathbf{c}]$ has a solution for all 4-vectors $\mathbf{c}$ by Theorem 2. In particular, if $\mathbf{c}$ is given components $b_1 + b_3$, $b_2 + b_4$, $b_5 + b_7$, and $b_6 + b_8$, we see that (2.2) can be solved for any choice of temperatures $b_1, \ldots, b_8$. ■

Linear Combinations with CAS

Maple

```
> vector(2*[1,2,3]-5*[-3,2,-1]);
```

$$[17, -6, 11]$$

Mathematica

```
In[1]:=
    2*{1,2,3}-5*{-3,2,-1}
Out[1]=
    {17, -6, 11}
```

MATLAB

```
>> 2*[1 2 3]-5*[-3 2 -1]
ans =
       17      -6      11
```

Exercises 2.1

Vectors and Vector Arithmetic

In Exercises 1–4 perform, if possible, the indicated vector operations.

1. a. $\begin{bmatrix} -1 \\ 3 \end{bmatrix} + \begin{bmatrix} 2 \\ -4 \end{bmatrix}$ **b.** $\begin{bmatrix} 5 \\ -3 \end{bmatrix} - \begin{bmatrix} 10 \\ 6 \end{bmatrix}$

2. a. $\frac{1}{3}\begin{bmatrix} 9 \\ -3 \end{bmatrix}$ **b.** $3\begin{bmatrix} 4 \\ 1 \end{bmatrix} + 2\begin{bmatrix} -6 \\ -2 \end{bmatrix}$

3. a. $\begin{bmatrix} 5 \\ -3 \\ 7 \end{bmatrix} - \begin{bmatrix} 1 \\ 6 \\ 3 \end{bmatrix}$ **b.** $2\begin{bmatrix} 5 \\ -3 \\ 7 \end{bmatrix} - \begin{bmatrix} 1 \\ 4 \\ 7 \end{bmatrix}$

4. a. $3\begin{bmatrix} a \\ 1 \end{bmatrix} + 2\begin{bmatrix} -1 \\ b \end{bmatrix}$ **b.** $2\begin{bmatrix} 1 \\ -2 \\ 3 \end{bmatrix} - \begin{bmatrix} 1 \\ -1 \end{bmatrix}$

5. Let $\mathbf{a} = \begin{bmatrix} -5 \\ 8 \end{bmatrix}$, $\mathbf{b} = \begin{bmatrix} 2 \\ -4 \end{bmatrix}$, $\mathbf{c} = \begin{bmatrix} -7 \\ 0 \end{bmatrix}$. Compute:

a. $2\mathbf{a} - 4\mathbf{b} + 3\mathbf{c}$

b. $\mathbf{a} - 0\mathbf{b} - 6\mathbf{c} + 3\begin{bmatrix} 1 \\ -1 \end{bmatrix}$

6. Find a vector $\mathbf{x}$ such that

a. $3\mathbf{x} + \begin{bmatrix} 2 \\ -1 \end{bmatrix} = \frac{1}{2}\begin{bmatrix} -2 \\ 4 \end{bmatrix}$

b. $\mathbf{x} + 2\begin{bmatrix} 0 \\ -1 \\ 4 \end{bmatrix} = -\begin{bmatrix} -7 \\ 4 \\ 4 \end{bmatrix}$

Let

$$\mathbf{v}_1 = \begin{bmatrix} -1 \\ 1 \end{bmatrix}, \quad \mathbf{v}_2 = \begin{bmatrix} 2 \\ 0 \end{bmatrix}, \quad \mathbf{v}_3 = \begin{bmatrix} -2 \\ 3 \end{bmatrix}$$

$$\mathbf{w}_1 = \begin{bmatrix} -1 \\ 1 \\ 1 \end{bmatrix}, \quad \mathbf{w}_2 = \begin{bmatrix} 2 \\ 0 \\ -1 \end{bmatrix}$$

In Exercises 7–9 sketch the vectors.

7. $\mathbf{v}_1, \mathbf{v}_2, \mathbf{v}_3, \mathbf{v}_1 + \mathbf{v}_2, \mathbf{v}_1 - \mathbf{v}_3, \mathbf{v}_3 - \mathbf{v}_1$

8. $2\mathbf{v}_1 - \mathbf{v}_2, 3\mathbf{v}_2 + 2\mathbf{v}_3, 3\mathbf{v}_1 - 2\mathbf{v}_3, -\mathbf{v}_1 - \mathbf{v}_2$

9. $\mathbf{w}_1, \mathbf{w}_2, \mathbf{w}_1 + \mathbf{w}_2, 3\mathbf{w}_1, \mathbf{w}_2 - \mathbf{w}_1$

In Exercises 10–12 find the value(s) (if any) of a and b that make the equalities true.

10. a. $\begin{bmatrix} a - 1 \\ a + b \end{bmatrix} = \begin{bmatrix} 1 \\ 1 \end{bmatrix}$ b. $\begin{bmatrix} a - b \\ a + 2b \end{bmatrix} = \begin{bmatrix} b \\ 0 \end{bmatrix}$

11. a. $\begin{bmatrix} a - b \\ a + b \end{bmatrix} - \frac{1}{2}\begin{bmatrix} b \\ a \end{bmatrix} = \begin{bmatrix} 0 \\ 0 \end{bmatrix}$

    b. $\begin{bmatrix} b \\ a \\ 4 \end{bmatrix} = \begin{bmatrix} 1 \\ 1 \\ 3 \end{bmatrix}$

12. a. $\begin{bmatrix} a - b \\ a + b \end{bmatrix} - \frac{1}{2}\begin{bmatrix} b \\ a \end{bmatrix} = \begin{bmatrix} 0 \\ 1 \end{bmatrix}$

    b. $\begin{bmatrix} b \\ a \\ 0 \end{bmatrix} = \begin{bmatrix} 1 \\ 1 \end{bmatrix}$

Let $\mathbf{a}$ and $\mathbf{b}$ be 3-vectors.

13. Find the 3-vector $\mathbf{x}$ such that $2\mathbf{x} - 4\mathbf{b} = 3\mathbf{a}$.

14. Find the 3-vector $\mathbf{x}$ such that $4\mathbf{x} + 3\mathbf{b} = 2\mathbf{a}$.

15. Find the 3-vectors $\mathbf{x}$ and $\mathbf{y}$ such that

$$4\mathbf{x} - 3\mathbf{y} = 2\mathbf{a}$$
$$\mathbf{x} - \mathbf{y} = \mathbf{a} + \mathbf{b}$$

16. Find the 2-vectors $\mathbf{x}$ and $\mathbf{y}$ such that:

$$\mathbf{x} + \mathbf{y} = \begin{bmatrix} 5 \\ 0 \end{bmatrix}$$

$$2\mathbf{x} - 3\mathbf{y} = \begin{bmatrix} -10 \\ 15 \end{bmatrix}$$

17. Draw the free vector PQ and find its components.

    a. $P = (1, 1), Q = (1, -1)$

    b. $P = (2, -1, 1), Q = (-1, 1, 3)$

Linear Combinations

In Exercises 18–25 determine whether the first vector is a linear combination of the remaining vectors.

18. $\begin{bmatrix} -a - 2b \\ 4a + 3b \end{bmatrix}, \begin{bmatrix} -1 \\ 4 \end{bmatrix}, \begin{bmatrix} 2 \\ -3 \end{bmatrix}$

19. $\begin{bmatrix} a + b + c \\ a + b - 2c \end{bmatrix}, \begin{bmatrix} 1 \\ 1 \end{bmatrix}, \begin{bmatrix} -1 \\ 2 \end{bmatrix}$

20. $\begin{bmatrix} 1 \\ -3 \end{bmatrix}, \begin{bmatrix} 1 \\ 1 \end{bmatrix}, \begin{bmatrix} -1 \\ 2 \end{bmatrix}$

21. $\begin{bmatrix} 1 \\ -3 \end{bmatrix}, \begin{bmatrix} 1 \\ 1 \end{bmatrix}, \begin{bmatrix} -1 \\ -1 \end{bmatrix}$

22. $\begin{bmatrix} 0 \\ -2 \\ 0 \end{bmatrix}, \begin{bmatrix} 1 \\ 1 \\ 1 \end{bmatrix}, \begin{bmatrix} 1 \\ -1 \\ 1 \end{bmatrix}$

23. $\begin{bmatrix} 1 \\ -2 \\ 0 \end{bmatrix}, \begin{bmatrix} 1 \\ 1 \\ 1 \end{bmatrix}, \begin{bmatrix} 0 \\ -1 \\ 1 \end{bmatrix}$

24. $\begin{bmatrix} 1 \\ -2 \\ 0 \end{bmatrix}, \begin{bmatrix} 1 \\ 1 \\ 1 \end{bmatrix}, \begin{bmatrix} 1 \\ 0 \\ -1 \end{bmatrix}, \begin{bmatrix} 0 \\ -1 \\ 1 \end{bmatrix}$

25. $\begin{bmatrix} -3 \\ 2 \\ 4 \end{bmatrix}, \begin{bmatrix} 1 \\ 1 \\ 1 \end{bmatrix}, \begin{bmatrix} 1 \\ 0 \\ -1 \end{bmatrix}, \begin{bmatrix} 0 \\ -1 \\ 1 \end{bmatrix}$

In Exercises 26–32 determine whether $\mathbf{b}$ is a linear combination of the columns of A.

26. $A = \begin{bmatrix} 1 & -1 \\ 0 & -2 \end{bmatrix}, \mathbf{b} = \begin{bmatrix} 3 \\ -4 \end{bmatrix}$

27. $A = \begin{bmatrix} 1 & -1 \\ 3 & -2 \end{bmatrix}, \mathbf{b} = \begin{bmatrix} x \\ y \end{bmatrix}$

28. $A = \begin{bmatrix} 1 & -1 & 0 \\ 0 & 1 & 1 \end{bmatrix}, \mathbf{b} = \begin{bmatrix} 2 \\ 1 \end{bmatrix}$

29. $A = \begin{bmatrix} 1 & -1 \\ 2 & 1 \\ 0 & -2 \end{bmatrix}, \mathbf{b} = \begin{bmatrix} -1 \\ 2 \\ 4 \end{bmatrix}$

30. $A = \begin{bmatrix} 1 & -1 & 6 \\ 2 & 1 & 3 \\ 0 & -2 & 3 \end{bmatrix}, \mathbf{b} = \begin{bmatrix} -1 \\ 1 \\ 0 \end{bmatrix}$

31. $A = \begin{bmatrix} 1 & -1 & 0 \\ 0 & 1 & 0 \\ 0 & 0 & 0 \end{bmatrix}, \mathbf{b} = \begin{bmatrix} 2 \\ 1 \\ 0 \end{bmatrix}$

32. $A = \begin{bmatrix} 1 & 0 & 0 \\ 0 & 1 & 0 \\ -1 & 1 & 1 \end{bmatrix}, \mathbf{b} = \begin{bmatrix} x \\ y \\ z \end{bmatrix}$

33. Show that any 3-vector $\begin{bmatrix} a \\ b \\ c \end{bmatrix}$ is a linear combination

of $\begin{bmatrix} 1 \\ 1 \\ 1 \end{bmatrix}, \begin{bmatrix} 1 \\ 0 \\ -1 \end{bmatrix}$, and $\begin{bmatrix} 0 \\ -1 \\ 1 \end{bmatrix}$.

34. Show that any 3-vector is a linear combination of the columns of $\begin{bmatrix} 1 & 0 & 1 \\ 1 & 1 & 0 \\ -1 & 0 & 1 \end{bmatrix}$.

35. Find a 3-vector that is *not* a linear combination of $\begin{bmatrix} 1 \\ 1 \\ 1 \end{bmatrix}$, $\begin{bmatrix} 1 \\ 0 \\ -1 \end{bmatrix}$, and $\begin{bmatrix} 2 \\ 1 \\ 0 \end{bmatrix}$.

36. Show that any 4-vector is a linear combination of the standard basis vectors of $\mathbf{R}^4$.

In Exercises 37–41 describe the set of all linear combinations of the given vectors:

37. $\begin{bmatrix} 1 \\ 0 \end{bmatrix}$, $\begin{bmatrix} 0 \\ 1 \end{bmatrix}$ **38.** $\begin{bmatrix} 1 \\ 0 \end{bmatrix}$, $\begin{bmatrix} -1 \\ 0 \end{bmatrix}$

39. $\begin{bmatrix} 1 \\ 0 \\ 0 \end{bmatrix}$, $\begin{bmatrix} 0 \\ 1 \\ 0 \end{bmatrix}$ **40.** $\begin{bmatrix} 1 \\ 0 \\ 1 \end{bmatrix}$, $\begin{bmatrix} -1 \\ 0 \\ 2 \end{bmatrix}$

41. $\mathbf{e}_1, \mathbf{e}_2, \mathbf{e}_3$ in $\mathbf{R}^3$.

42. Find the value(s) of k such that $\begin{bmatrix} k \\ 2 \\ -2k \end{bmatrix}$ is a linear combination of $\begin{bmatrix} 0 \\ 2 \\ 1 \end{bmatrix}$ and $\begin{bmatrix} 1 \\ 0 \\ k \end{bmatrix}$.

43. Find the value(s) of k such that $\begin{bmatrix} 0 \\ -1 \end{bmatrix}$ is a linear combination of the columns of $\begin{bmatrix} 1 & -k & 1 \\ 0 & k & 0 \end{bmatrix}$.

Vector Equations and Linear Systems

44. Write the systems as vector equations.

a. $\begin{aligned} x - 4y &= 1 \\ -2x + y &= 0 \end{aligned}$

b. $\begin{aligned} x - 4y + z &= 2x + 1 \\ -2x + y - z &= -y \end{aligned}$

45. Write the systems as vector equations.

a. $\begin{aligned} x - 2y - z &= 1 \\ -x + y &= 0 \\ y - z &= 2 \end{aligned}$

b. $\begin{aligned} x - 2y - z &= 1 \\ -x + 2z &= -1 \\ y - 2 &= 0 \end{aligned}$

46. Write the vector equations as linear systems.

a. $x \begin{bmatrix} 3 \\ -2 \end{bmatrix} - y \begin{bmatrix} -1 \\ 5 \end{bmatrix} = \begin{bmatrix} 1 \\ 2 \end{bmatrix}$

b. $x \begin{bmatrix} -3 \\ 2 \end{bmatrix} = y \begin{bmatrix} 4 \\ 5 \end{bmatrix}$

In Exercises 47–48 say whether the system $[A : \mathbf{b}]$ is consistent for all vectors $\mathbf{b} \in \mathbf{R}^3$ if A is

47. a. $\begin{bmatrix} 0 & 2 \\ 1 & -1 \\ -1 & 1 \end{bmatrix}$ **b.** $\begin{bmatrix} -1 & 0 & 2 \\ 1 & 1 & -1 \\ 0 & -1 & 1 \end{bmatrix}$

48. a. $\begin{bmatrix} -1 & 0 & 2 \\ 1 & 1 & -1 \end{bmatrix}$ **b.** $\begin{bmatrix} 1 & 1 & 2 \\ 2 & 1 & 3 \\ 4 & -1 & 3 \end{bmatrix}$

$[A|b]$ consistent for all vectors $b \in \mathbb{R}^2$ —

Theoretical Exercises

49. Prove Parts 2–5 of Theorem 1.

50. Prove Parts 6–9 of Theorem 1.

51. Let A be an $m \times n$ matrix. Suppose the system $[A : \mathbf{b}]$ is consistent for all n-vectors $\mathbf{b}$. [m-vectors] True or false? (Explain.)

a. $\mathbf{b}$ is a linear combination of the columns of A.

b. A has m pivots.

c. A has n pivots.

d. Each row of A has a pivot.

e. Each column of A has a pivot.

f. A can have more than m pivots.

g. A can have more than n pivots.

h. $m \leq n$.

52. Suppose that the system $[A : \mathbf{b}]$ is inconsistent.

a. What can you say about the number of pivots of A?

b. Can $\mathbf{b}$ be a scalar multiple of the first column of A?

Applications

53. A sailboat traveling at 10 mi/h in an easterly direction is subjected to a south-north cross wind of 20 mi/h. What is the true velocity vector of the boat?

54. During takeoff a plane is rising at a speed of 20 mi/h. Its eastbound speed is 150 mi/h, and its northbound speed is 200 mi/h. Find the total velocity vector.

55. An airline buys food supplies for three of its planes. The average dollar cost per trip is given by the following matrix A with columns $\mathbf{a}_1, \mathbf{a}_2,$ and $\mathbf{a}_3$.

Class	Plane 1	Plane 2	Plane 3
First	350	300	450
Business	500	600	700
Economy	800	700	900

Compute and explain the meaning of each.

a. $\mathbf{a}_1 + \mathbf{a}_2 + \mathbf{a}_3$ **b.** $\mathbf{a}_3 - \mathbf{a}_2$

c. $10\mathbf{a}_3$ **d.** $7\mathbf{a}_1 + 8\mathbf{a}_2 + 9\mathbf{a}_3$

56. Referring to Exercise 55, how many trips did each plane make if the airline spent $23,000 for first class, $38,000 for business, and $49,000 for economy? (*Hint:* If x_1, x_2, and x_3 are the number of trips each plane made, then $x_1\mathbf{a}_1 + x_2\mathbf{a}_2 + x_3\mathbf{a}_3 = \begin{bmatrix} 23,000 \\ 38,000 \\ 49,000 \end{bmatrix}$).

2.2   Dot Product

Reader's Goals for This Section

1. To compute the length of a vector and the dot product and angle between vectors.
2. To compute the orthogonal projection of a vector along another vector.

In this section we define the dot product and length in $\mathbf{R}^n$. These concepts are basic in the theory and applications of vectors. In two and three dimensions they have familiar geometric interpretations.

DEFINITION

(Dot Product)

Let $\mathbf{u} = (u_1, \ldots, u_n)$ and $\mathbf{v} = (v_1, \ldots, v_n)$ be any two n-vectors. The **dot product** of $\mathbf{u}$ and $\mathbf{v}$ is the following *number*:

$$\mathbf{u} \cdot \mathbf{v} = u_1 v_1 + \cdots + u_n v_n$$

■ EXAMPLE 12   Let

$$\mathbf{u} = \begin{bmatrix} -1 \\ 2 \\ 3 \\ 0 \end{bmatrix}, \quad \mathbf{v} = \begin{bmatrix} -2 \\ 0 \\ 2 \\ 0 \end{bmatrix}, \quad \mathbf{w} = \begin{bmatrix} -2 \\ 0 \\ -2 \\ 1 \end{bmatrix}$$

Find $\mathbf{u} \cdot \mathbf{v}, \mathbf{u} \cdot \mathbf{w}, \mathbf{v} \cdot \mathbf{w}$.

SOLUTION

$$\mathbf{u} \cdot \mathbf{v} = (-1) \cdot (-2) + 2 \cdot 0 + 3 \cdot 2 + 0 \cdot 0 = 8$$
$$\mathbf{u} \cdot \mathbf{w} = (-1) \cdot (-2) + 2 \cdot 0 + 3 \cdot (-2) + 0 \cdot 1 = -4$$
$$\mathbf{v} \cdot \mathbf{w} = (-2) \cdot (-2) + 0 \cdot 0 + 2 \cdot (-2) + 0 \cdot 1 = 0$$

Example 12 shows that the dot product can be less than 0, greater than 0, or equal 0. In fact, if the dot product of two vectors is 0, we call the vectors **orthogonal**. So, $\mathbf{v}$ and $\mathbf{w}$ are orthogonal, but $\mathbf{u}$ and $\mathbf{v}$ are not.

The **norm**, **length**, or **magnitude** of an n-vector $\mathbf{u}$ is the positive square root:

$$\|\mathbf{u}\| = \sqrt{\mathbf{u} \cdot \mathbf{u}} = \left(u_1^2 + \cdots + u_n^2\right)^{1/2}$$

The norm is always defined, because

$$\|\mathbf{u}\|^2 = \mathbf{u} \cdot \mathbf{u} \tag{2.3}$$

is ≥ 0.

We also define the (**Euclidean**) **distance** between $\mathbf{u}$ and $\mathbf{v}$ by

$$d = \|\mathbf{u} - \mathbf{v}\|$$

■ EXAMPLE 13   Find the norm of $\mathbf{w}$, $\|\mathbf{w}\|$, and find the distance d between $\mathbf{u}$ and $\mathbf{v}$.

SOLUTION   The norm of $\mathbf{w}$ is

$$\|\mathbf{w}\| = \left((-2)^2 + 0^2 + (-2)^2 + 1^2\right)^{1/2} = \sqrt{9} = 3$$

The distance between $\mathbf{u}$ and $\mathbf{v}$ is

$$d = \|\mathbf{u} - \mathbf{v}\| = \|(1, 2, 1, 0)\| = \sqrt{6}$$

Note that the norm of a scalar product $c\mathbf{u}$ is given by

$$\|c\mathbf{u}\| = |c|\,\|\mathbf{u}\| \tag{2.4}$$

because $\sqrt{c^2} = |c|$. For example,

$$\|-5(1, 2)\| = \|(-5, -10)\| = 5\sqrt{5} = |-5|\,\|(1, 2)\|$$

The norm of a 2- or 3-vector is exactly what, geometrically, we call the length. If $\mathbf{u} = (u_1, u_2)$, then $\|\mathbf{u}\| = \sqrt{u_1^2 + u_2^2} = \sqrt{(OP')^2 + (PP')^2} = OP$ by the Pythagorean theorem (Fig. 2.9(a)). Likewise, if $\mathbf{u} = (u_1, u_2, u_3)$, then $\|\mathbf{u}\|$ is seen to be the geometric length by using the Pythagorean theorem on OPP' and $OP'P''$ of Fig. 2.9(b).

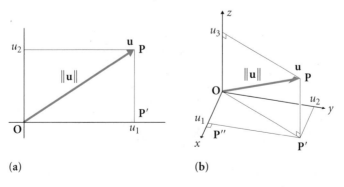

(a)   (b)

Figure 2.9   The norm for plane and space vectors.

If $\overrightarrow{PQ}$ is a free vector with $P(p_1, p_2, p_3)$ and $Q(q_1, q_2, q_3)$, then the length $\|\overrightarrow{PQ}\|$ of $\overrightarrow{PQ}$ is the length of the corresponding vector, i.e.,

$$\left\|\overrightarrow{PQ}\right\| = \sqrt{(q_1 - p_1)^2 + (q_2 - p_2)^2 + (q_3 - p_3)^2}$$

This formula also gives us the **distance** between the points P and Q in space. So, the distance between two vectors defined above is the geometric distance in space between the tips of the vectors. If $p_3 = q_3 = 0$, we get the distance between two points in the plane.

■ EXAMPLE 14

(a) Compute the length of $\mathbf{v} = (1, -2, 2)$.
(b) Find the distance between $P(2, 3, -1)$ and $Q(-1, 0, -2)$.

SOLUTION

$$\|\mathbf{v}\| = \sqrt{1^2 + (-2)^2 + 2^2} = 3$$
$$\left\|\overrightarrow{PQ}\right\| = \sqrt{(-3)^2 + (-3)^2 + (-1)^2} = \sqrt{19}$$

A vector with length 1 is called a **unit** vector.

■ EXAMPLE 15 $\mathbf{u} = (\frac{1}{2}, -\frac{1}{2}, \frac{1}{2}, -\frac{1}{2})$ is a unit vector, because

$$\|\mathbf{u}\|^2 = \left(\frac{1}{2}\right)^2 + \left(-\frac{1}{2}\right)^2 + \left(\frac{1}{2}\right)^2 + \left(-\frac{1}{2}\right)^2 = 1$$

So, $\|\mathbf{u}\| = 1$.

We are often interested in obtaining the unit vector in the direction of a given vector.

THEOREM 3   **(Unit Vector in a Given Direction)**

Let $\mathbf{v} = (v_1, \ldots, v_n)$ be a nonzero vector and let $\mathbf{u}$ be the unit vector in the direction of $\mathbf{v}$. Then,

$$\mathbf{u} = \frac{1}{\|\mathbf{v}\|}\mathbf{v} = \left(\frac{v_1}{\|\mathbf{v}\|}, \ldots, \frac{v_n}{\|\mathbf{v}\|}\right)$$

PROOF   Because $\mathbf{u}$ is a positive scalar multiple of $\mathbf{v}$, it has the same direction as $\mathbf{v}$. Also, $\mathbf{u}$ has unit length, because $\left\|\frac{1}{\|\mathbf{v}\|}\mathbf{v}\right\| = \frac{1}{\|\mathbf{v}\|}\|\mathbf{v}\| = 1$, by (2.4).

■ EXAMPLE 16   Find the unit vector in the direction of $\mathbf{v} = (1, -2, 1)$.

SOLUTION   By Theorem 3,

$$\mathbf{u} = \frac{1}{\|\mathbf{v}\|}\mathbf{v} = \frac{1}{\sqrt{6}}(1, -2, 1) = \left(\frac{1}{\sqrt{6}}, -\frac{2}{\sqrt{6}}, \frac{1}{\sqrt{6}}\right)$$

The unit 3-vectors along the coordinate axes are $\mathbf{e}_1$, $\mathbf{e}_2$, and $\mathbf{e}_3$. They are also denoted by $\mathbf{i}$, $\mathbf{j}$, and $\mathbf{k}$, respectively (Fig. 2.10(b)). This notation is common in physics and engineering.

$$\mathbf{i} = \mathbf{e}_1 = \begin{bmatrix} 1 \\ 0 \\ 0 \end{bmatrix}, \qquad \mathbf{j} = \mathbf{e}_2 = \begin{bmatrix} 0 \\ 1 \\ 0 \end{bmatrix}, \qquad \mathbf{k} = \mathbf{e}_3 = \begin{bmatrix} 0 \\ 0 \\ 1 \end{bmatrix}$$

In the plane we have (Fig. 2.10(a)),

$$\mathbf{i} = \mathbf{e}_1 = \begin{bmatrix} 1 \\ 0 \end{bmatrix}, \qquad \mathbf{j} = \mathbf{e}_2 = \begin{bmatrix} 0 \\ 1 \end{bmatrix}$$

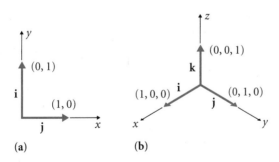

Figure 2.10   The unit vectors along the coordinate axes.

The dot product satisfies the properties summarized in Theorem 4.

THEOREM 4

(Properties of Dot Product)

Let $\mathbf{u}$, $\mathbf{v}$ and $\mathbf{w}$ be n-vectors and let c be any scalar.

1. $\mathbf{u} \cdot \mathbf{v} = \mathbf{v} \cdot \mathbf{u}$	**Symmetry**
2. $\mathbf{u} \cdot (\mathbf{v} + \mathbf{w}) = \mathbf{u} \cdot \mathbf{v} + \mathbf{u} \cdot \mathbf{w}$	**Additivity**
3. $c(\mathbf{u} \cdot \mathbf{v}) = (c\mathbf{u}) \cdot \mathbf{v} = \mathbf{u} \cdot (c\mathbf{v})$	**Homogeneity**
4. $\mathbf{u} \cdot \mathbf{u} \geq 0$ Furthermore, $\mathbf{u} \cdot \mathbf{u} = 0$ if and only if $\mathbf{u} = \mathbf{0}$.	**Positive definiteness**

PROOF   We prove Parts 2 and 4 and leave the remaining cases as exercises. Let $\mathbf{u} = (u_1, \ldots, u_n)$, $\mathbf{v} = (v_1, \ldots, v_n)$, and $\mathbf{w} = (w_1, \ldots, w_n)$. Part 2 follows from

$$\mathbf{u} \cdot (\mathbf{v} + \mathbf{w}) = \begin{bmatrix} u_1 \\ \vdots \\ u_n \end{bmatrix} \cdot \begin{bmatrix} v_1 + w_1 \\ \vdots \\ v_n + w_n \end{bmatrix}$$

$$= u_1(v_1 + w_1) + \cdots + u_n(v_n + w_n)$$

$$= (u_1 v_1 + \cdots + u_n v_n) + (u_1 w_1 + \cdots + u_n w_n)$$

$$= \mathbf{u} \cdot \mathbf{v} + \mathbf{u} \cdot \mathbf{w}$$

For Part 4 we have

$$\mathbf{u} \cdot \mathbf{u} = \|\mathbf{u}\|^2 \geq 0$$

and

$$\mathbf{u} \cdot \mathbf{u} = 0 \Leftrightarrow u_1^2 + \cdots + u_n^2 = 0$$

$$\Leftrightarrow u_1 = \cdots = u_n = 0$$

$$\Leftrightarrow \mathbf{u} = \mathbf{0}$$

Equation (2.3) and Theorem 4 can be combined to form many new identities. As an example, consider Theorem 5.

THEOREM 5

For any n–vectors $\mathbf{u}$ and $\mathbf{v}$ we have

$$\|\mathbf{u} + \mathbf{v}\|^2 = \|\mathbf{u}\|^2 + \|\mathbf{v}\|^2 + 2\mathbf{u} \cdot \mathbf{v}$$

$$\|\mathbf{u} - \mathbf{v}\|^2 = \|\mathbf{u}\|^2 + \|\mathbf{v}\|^2 - 2\mathbf{u} \cdot \mathbf{v}$$

PROOF   Because

$$\|\mathbf{u} + \mathbf{v}\|^2 = (\mathbf{u} + \mathbf{v}) \cdot (\mathbf{u} + \mathbf{v}) \qquad \text{by Equation (2.3)}$$

$$= (\mathbf{u} + \mathbf{v}) \cdot \mathbf{u} + (\mathbf{u} + \mathbf{v}) \cdot \mathbf{v} \qquad \text{by Theorem 4, Part2}$$

$$= \mathbf{u} \cdot \mathbf{u} + \mathbf{u} \cdot \mathbf{v} + \mathbf{v} \cdot \mathbf{u} + \mathbf{v} \cdot \mathbf{v} \qquad \text{by Parts 1 and 2}$$

$$= \mathbf{u} \cdot \mathbf{u} + 2\mathbf{u} \cdot \mathbf{v} + \mathbf{v} \cdot \mathbf{v} \qquad \text{by Part 1}$$

$$= \|\mathbf{u}\|^2 + \|\mathbf{v}\|^2 + 2\mathbf{u} \cdot \mathbf{v} \qquad \text{by Equation (2.3)}$$

By replacing $\mathbf{v}$ with $-\mathbf{v}$, we get the second identity.

One of the most useful consequences of Theorem 4 is the Cauchy-Schwarz inequality.[3]

THEOREM 6

(Cauchy-Schwarz Inequality)

For any n-vectors $\mathbf{u}$ and $\mathbf{v}$,

$$|\mathbf{u} \cdot \mathbf{v}| \leq \|\mathbf{u}\| \|\mathbf{v}\|$$

Furthermore, equality holds if and only if $\mathbf{u}$ and $\mathbf{v}$ are scalar multiples of each other.

---

[3] Proper credits and biographies are discussed in Chapter 8.

PROOF By Theorem 5,

$$0 \le (x\mathbf{u} + \mathbf{v}) \cdot (x\mathbf{u} + \mathbf{v}) = x^2(\mathbf{u} \cdot \mathbf{u}) + x(2\mathbf{u} \cdot \mathbf{v}) + \mathbf{v} \cdot \mathbf{v} \qquad (2.5)$$

for all scalars x. This is a quadratic polynomial $p(x) = ax^2 + bx + c$ with $a = \mathbf{u} \cdot \mathbf{u}$, $b = 2\mathbf{u} \cdot \mathbf{v}$ and $c = \mathbf{v} \cdot \mathbf{v}$. Because $a \ge 0$ and $p(x) \ge 0$ for all x, the graph of $p(x)$ is a parabola in the upper half-plane that opens upwards. Hence, the parabola is either above the x-axis, in which case $p(x)$ has two complex roots, or is tangent to the x-axis, in which case $p(x)$ has a repeated real root. Therefore, $b^2 - 2ac \le 0$. So,

$$(2\mathbf{u} \cdot \mathbf{v})^2 - 4(\mathbf{u} \cdot \mathbf{u})(\mathbf{v} \cdot \mathbf{v}) \le 0, \quad \text{or} \quad 4(\mathbf{u} \cdot \mathbf{v})^2 - 4\|\mathbf{u}\|^2\|\mathbf{v}\|^2 \le 0$$

which implies the Cauchy-Schwarz inequality.

Equality holds if and only if $b^2 - 2ac = 0$ or if and only if $p(x)$ has a double real root, say r. Hence, by Equation (2.5), with $x = r$ we have

$$(r\mathbf{u} + \mathbf{v}) \cdot (r\mathbf{u} + \mathbf{v}) = 0 \Leftrightarrow \|r\mathbf{u} + \mathbf{v}\| = 0$$

$$\Leftrightarrow r\mathbf{u} + \mathbf{v} = \mathbf{0}$$

$$\Leftrightarrow \mathbf{v} = -r\mathbf{u}$$

So $\mathbf{v}$ is a scalar product of $\mathbf{u}$, which proves the last claim of the theorem.

■ EXAMPLE 17 Verify the Cauchy-Schwarz inequality for $\mathbf{u} = (-1, 2, 0, -1)$ and $\mathbf{v} = (4, -2, -1, 1)$.

SOLUTION

$$|\mathbf{u} \cdot \mathbf{v}| = |(-1, 2, 0, -1) \cdot (4, -2, -1, 1)| = |-9| = 9$$

$$\|\mathbf{u}\|\,\|\mathbf{v}\| = \|(-1, 2, 0, -1)\|\,\|(4, -2, -1, 1)\|$$

$$= \sqrt{6}\,\sqrt{22} \cong 11.489 \ge 9$$

As an application of Theorem 4 and the Cauchy-Schwarz inequality we have the following useful theorem.

THEOREM 7

(The Triangle Inequality)

For any n-vectors $\mathbf{u}$ and $\mathbf{v}$, we have

$$\|\mathbf{u} + \mathbf{v}\| \le \|\mathbf{u}\| + \|\mathbf{v}\|$$

PROOF

$$\begin{aligned}
\|\mathbf{u} + \mathbf{v}\|^2 &= \|\mathbf{u}\|^2 + \|\mathbf{v}\|^2 + 2\mathbf{u} \cdot \mathbf{v} && \text{by Theorem 5} \\
&\le \|\mathbf{u}\|^2 + \|\mathbf{v}\|^2 + 2\|\mathbf{u}\|\,\|\mathbf{v}\| && \text{by Cauchy-Schwarz} \\
&= \left(\|\mathbf{u}\| + \|\mathbf{v}\|\right)^2
\end{aligned}$$

Therefore, $\|\mathbf{u} + \mathbf{v}\| \le \|\mathbf{u}\| + \|\mathbf{v}\|$.

In two and three dimensions the triangle inequality can also be verified geometrically: each side of a triangle has length less than the sum of the lengths of the other two sides (Fig. 2.11).

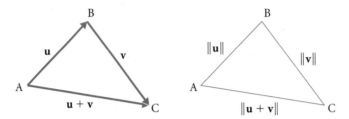

Figure 2.11   The triangle inequality in the plane.

Angle Between Two n-Vectors

The Cauchy-Schwarz inequality implies

$$\frac{|\mathbf{u} \cdot \mathbf{v}|}{\|\mathbf{u}\|\,\|\mathbf{v}\|} \leq 1, \quad \text{or} \quad -1 \leq \frac{\mathbf{u} \cdot \mathbf{v}}{\|\mathbf{u}\|\,\|\mathbf{v}\|} \leq 1$$

Because any number between -1 and 1 can be written as $\cos\theta$ for a unique $0 \leq \theta \leq \pi$, the last inequality allows us to define the angle between two n-vectors.

The **angle** between two nonzero n-vectors $\mathbf{u}$ and $\mathbf{v}$ is the unique number θ such that

$$\cos\theta = \frac{\mathbf{u} \cdot \mathbf{v}}{\|\mathbf{u}\|\,\|\mathbf{v}\|}, \qquad 0 \leq \theta \leq \pi \tag{2.6}$$

We can also write the dot product in terms of the angle:

$$\mathbf{u} \cdot \mathbf{v} = \|\mathbf{u}\|\,\|\mathbf{v}\| \cos\theta \tag{2.7}$$

▪ EXAMPLE 18   Find the angle θ between the vectors.

(a) $(1,1)$ and $(3,0)$
(b) $(1,1,1,1)$ and $(1,0,0,0)$

SOLUTION

(a) $\theta = \arccos \dfrac{(1,1) \cdot (3,0)}{\|(1,1)\|\,\|(3,0)\|} = \arccos \dfrac{1}{\sqrt{2}} = \dfrac{\pi}{4} = 45°$

(b) $\theta = \arccos \dfrac{(1,1,1,1) \cdot (1,0,0,0)}{\|(1,1,1,1)\|\,\|(1,0,0,0)\|} = \arccos \dfrac{1}{2} = \dfrac{\pi}{3} = 60°$

Our geometric notion of the angle agrees with the preceding definition in two and three dimensions (Fig. 2.12). In fact, (2.7) can be verified using the law of cosines. Here is how.

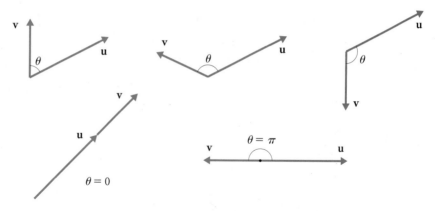

Figure 2.12   The angle θ in radians between two vectors is $0 \le \theta \le \pi$.

By the law of cosines on the triangle OPQ (Fig. 2.13), we have

$$\|\mathbf{u}\| \, \|\mathbf{v}\| \cos \theta = \frac{1}{2} \left(\|\mathbf{u}\|^2 + \|\mathbf{v}\|^2 - \|PQ\|^2 \right)$$

But $\|\mathbf{u}\|^2 + \|\mathbf{v}\|^2 = \|PQ\|^2 + 2\mathbf{u} \cdot \mathbf{v}$, by Theorem 5. Hence,

$$\|\mathbf{u}\| \, \|\mathbf{v}\| \cos \theta = \frac{1}{2} \left(\|PQ\|^2 + 2\mathbf{u} \cdot \mathbf{v} - \|PQ\|^2 \right) = \mathbf{u} \cdot \mathbf{v}$$

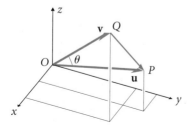

Figure 2.13

Equation (2.7) implies that $\mathbf{u} \cdot \mathbf{v}$ and $\cos \theta$ have the same sign and one is zero only if the other is. Because $\cos \theta$ is positive for $0 \le \theta < \frac{\pi}{2}$, negative for $\frac{\pi}{2} < \theta \le \pi$, and zero for $\theta = \frac{\pi}{2}$, we have

1. θ is **acute** if and only if $\mathbf{u} \cdot \mathbf{v} > 0$.
2. θ is **obtuse** if and only if $\mathbf{u} \cdot \mathbf{v} < 0$.
3. θ is **right** if and only if $\mathbf{u} \cdot \mathbf{v} = 0$.

If $\theta = \frac{\pi}{2}$, then $\mathbf{u}$ and $\mathbf{v}$ are called **perpendicular** and we write $\mathbf{u} \perp \mathbf{v}$. Because $\theta = \frac{\pi}{2}$ is equivalent to $\mathbf{u} \cdot \mathbf{v} = 0$ for nonzero vectors, we see that $\mathbf{u}$ and $\mathbf{v}$ are *perpendicular if and only if they are orthogonal*, i.e.,

$$\mathbf{u} \perp v \Leftrightarrow \mathbf{u} \cdot \mathbf{v} = 0 \qquad \text{for } \mathbf{u} \ne \mathbf{0} \text{ and } \mathbf{v} \ne \mathbf{0}$$

We also have a generalization of the Pythagorean theorem (Fig. 2.14).

$$\|\mathbf{u} + \mathbf{v}\|^2 = \|\mathbf{u}\|^2 + \|\mathbf{v}\|^2$$

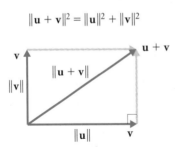

Figure 2.14   The Pythagorean theorem.

THEOREM 8   **(Pythagorean Theorem)**

The *n*-vectors **u** and **v** are orthogonal if and only if

$$\|\mathbf{u} + \mathbf{v}\|^2 = \|\mathbf{u}\|^2 + \|\mathbf{v}\|^2$$

PROOF   Exercise. (*Hint:* Use Theorem 5.)

Orthogonal Projections

Dot products can be used to write any *n*-vector as a sum of orthogonal vectors. Let **u** and **v** be given nonzero vectors. We want to write **u** as

$$\mathbf{u} = \mathbf{u}_{\text{pr}} + \mathbf{u}_{\text{c}}$$

where $\mathbf{u}_{\text{pr}}$ is a scalar multiple of **v** and $\mathbf{u}_{\text{c}}$ is orthogonal to $\mathbf{u}_{\text{pr}}$ (Fig. 2.15). As we see, such a decomposition of **u** is always possible, and it is *unique*. We call $\mathbf{u}_{\text{pr}}$ the **orthogonal projection of u on v** and $\mathbf{u}_{\text{c}}$ the **vector component of u orthogonal to v**.

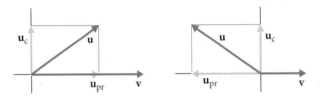

Figure 2.15   The orthogonal projection of **u** on **v**.

Let us now compute $\mathbf{u}_{\text{pr}}$ and $\mathbf{u}_{\text{c}}$ in terms of **u** and **v**. Because $\mathbf{u}_{\text{pr}}$ and **v** have the same direction, $\mathbf{u}_{\text{pr}} = c\mathbf{v}$ for some scalar c. In addition, since $\mathbf{u}_{\text{c}}$ and **v** are orthogonal

$\mathbf{u}_c \cdot \mathbf{v} = 0$. Hence,

$$\mathbf{u} \cdot \mathbf{v} = (\mathbf{u}_{pr} + \mathbf{u}_c) \cdot \mathbf{v}$$
$$= \mathbf{u}_{pr} \cdot \mathbf{v} + \mathbf{u}_c \cdot \mathbf{v}$$
$$= (c\mathbf{v}) \cdot \mathbf{v} + 0$$
$$= c(\mathbf{v} \cdot \mathbf{v})$$
$$\Rightarrow c = \frac{\mathbf{u} \cdot \mathbf{v}}{\mathbf{v} \cdot \mathbf{v}}$$

Therefore,

$$\mathbf{u}_{pr} = \frac{\mathbf{u} \cdot \mathbf{v}}{\mathbf{v} \cdot \mathbf{v}}\mathbf{v} \qquad \text{orthogonal projection of } \mathbf{u} \text{ on } \mathbf{v} \qquad (2.8)$$

and

$$\mathbf{u}_c = \mathbf{u} - \frac{\mathbf{u} \cdot \mathbf{v}}{\mathbf{v} \cdot \mathbf{v}}\mathbf{v} \qquad \text{vector component of } \mathbf{u} \text{ orthogonal to } \mathbf{v} \qquad (2.9)$$

■ **EXAMPLE 19** (Orthogonal Projection)   Let $\mathbf{u} = (1, 1, 1)$ and $\mathbf{v} = (2, 2, 0)$. Find the orthogonal projection of $\mathbf{u}$ on $\mathbf{v}$ and the vector component of $\mathbf{u}$ orthogonal to $\mathbf{v}$.

SOLUTION

$$\mathbf{u}_{pr} = \frac{(1, 1, 1) \cdot (2, 2, 0)}{(2, 2, 0) \cdot (2, 2, 0)}(2, 2, 0) = \frac{4}{8}(2, 2, 0) = (1, 1, 0)$$

and

$$\mathbf{u}_c = \mathbf{u} - \mathbf{u}_{pr} = (1, 1, 1) - (1, 1, 0) = (0, 0, 1)$$

(Verify geometrically.)

Dot Product with CAS

Maple

```
> with(linalg):
> dotprod([1,2,3],[-3,2,-1]);
```

$$-2$$

Mathematica

```
In[1]:=
    Dot[{1,2,3},{-3,2,-1}]
Out[1]=
    -2
```

MATLAB

```
>> dot([1 2 3], [-3 2 -1])
ans =
      -2
```

Exercises 2.2

Let $\mathbf{u} = (-1, 2, -2)$, $\mathbf{v} = (4, -3, 5)$, $\mathbf{w} = (-4, -2, 0)$, and $\mathbf{d} = (-1, -2, 1, \sqrt{3})$.

1. Find the lengths of the following vectors.

$\mathbf{u}$	$\mathbf{v}$	$\mathbf{w}$
$\mathbf{u} + \mathbf{v}$	$\mathbf{u} - \mathbf{v}$	$\mathbf{u} - \mathbf{v} + \mathbf{w}$
$\mathbf{d}$	$10\mathbf{d}$	$\|\mathbf{d}\|\mathbf{d}$

2. Compute the following expressions.

$\|\mathbf{u} + \mathbf{v}\|$ $\|\mathbf{u}\| + \|\mathbf{v}\|$

$\|\mathbf{u}\| - \|\mathbf{v}\|$ $\|\mathbf{u} - \mathbf{v}\|$

$\|\mathbf{v}\|\mathbf{v} + \|\mathbf{w}\|\mathbf{w}$ $(1/\|\mathbf{d}\|)\mathbf{d}$

3. Compute the following expressions.

$\mathbf{u} \cdot \mathbf{v}$	$\mathbf{w} \cdot \mathbf{u}$
$\mathbf{u} \cdot (\mathbf{v} + \mathbf{w})$	$\mathbf{v} \cdot \mathbf{u} + \mathbf{w} \cdot \mathbf{u}$
$\mathbf{d} \cdot \mathbf{d}$	$(\mathbf{d} \cdot \mathbf{d})\mathbf{d}$

4. Which of the following expressions are *undefined*, and why?

$\mathbf{u} \cdot \mathbf{v} \cdot \mathbf{w}$	$\mathbf{u} \cdot (\mathbf{v} \cdot \mathbf{w})$
$(\mathbf{u} \cdot \mathbf{v})\mathbf{w}$	$(\mathbf{u} \cdot \mathbf{v})(\mathbf{v} \cdot \mathbf{w})$
$\mathbf{u} \cdot (3\mathbf{v})$	$\mathbf{u} \cdot (3 + \mathbf{v})$
$(\mathbf{d} \cdot \mathbf{d})^3$	$\mathbf{d} \cdot \mathbf{d} + 2$

5. Find the unit vector in each direction.

   a. $\mathbf{u}$ b. $\mathbf{v}$ c. $\mathbf{w}$ d. $\mathbf{d}$

6. Find a vector of length 2 in the direction of $\mathbf{u}$.

7. Find a vector of length 2 in the opposite direction of that of $\mathbf{u}$.

8. Find a vector in the direction of $\mathbf{d}$ with length 9 times the length of $\mathbf{d}$.

9. Find a vector of length 9 in the direction of $\mathbf{d}$.

10. Find the distance between the points P and Q.

   a. $P(1, 1, -1)$, $Q(-2, 3, 4)$
   b. $P(4, -3, 2, 0)$, $Q(0, 2, -6, 4)$

11. Which pairs of vectors are orthogonal?

   a. $(1, 1)$, $(1, -1)$
   b. $(1, -1, -1)$, $(0, 1, -1)$
   c. $(4, -2, -1)$, $(-2, 3, 4)$
   d. $(-7, -3, 1, 0)$, $(0, 2, 6, 4)$

12. Find two vectors each orthogonal to $(1, -2, 4)$.

13. Find a unit vector orthogonal to $(-2, 3, 1)$.

In Exercises 14–15 compute the angle between $\mathbf{u}$ and $\mathbf{v}$. (You may need a calculator.)

14. a. $\mathbf{u} = (1, 1)$, $\mathbf{v} = (1, -1)$
    b. $\mathbf{u} = (-1, 1, 1)$, $\mathbf{v} = (2, -2, 1)$
    c. $\mathbf{u} = (1, -1, 1, -1)$, $\mathbf{v} = (0, 1, 1, 0)$

15. a. $\mathbf{u} = (3, 4)$, $\mathbf{v} = (5, 12)$
    b. $\mathbf{u} = (\sqrt{3}, 1, -2)$, $\mathbf{v} = (1, \sqrt{3}, -2)$
    c. $\mathbf{u} = (1, -1, -1, -1)$, $\mathbf{v} = (1, 1, 1, 1)$

16. Compute the orthogonal projection of $\mathbf{u}$ onto $\mathbf{v}$.

   a. $\mathbf{u} = (2, 3)$, $\mathbf{v} = (-2, 1)$
   b. $\mathbf{u} = (0, -1, 6)$, $\mathbf{v} = (-1, -3, 5)$
   c. $\mathbf{u} = (-2, -1, 0, 1)$, $\mathbf{v} = (0, 0, -1, 3)$

17. Referring to Exercise 16 find the vector component of $\mathbf{u}$ orthogonal to $\mathbf{v}$.

18. Verify the Cauchy-Schwarz inequality for the pairs $\mathbf{u}$ and $\mathbf{v}$ of Exercise 16.

19. Prove Theorem 8.

20. Is it true that if $\mathbf{u} \cdot \mathbf{v} = 0$, then either $\mathbf{u} = \mathbf{0}$ or $\mathbf{v} = \mathbf{0}$?

21. Is it true that if $\mathbf{u} \cdot \mathbf{v} = \mathbf{u} \cdot \mathbf{w}$ and $\mathbf{u} \neq \mathbf{0}$, then $\mathbf{v} = \mathbf{w}$?

22. (**Parallelogram Law**) Prove the following identity (Fig. 2.16).

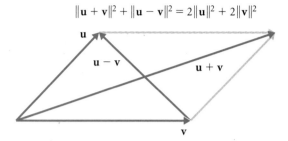

$$\|\mathbf{u} + \mathbf{v}\|^2 + \|\mathbf{u} - \mathbf{v}\|^2 = 2\|\mathbf{u}\|^2 + 2\|\mathbf{v}\|^2$$

Figure 2.16   Parallelogram law in the plane.

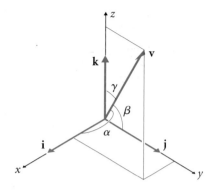

Figure 2.17   Direction angles.

23. (**Polarization Identity**) Prove the identity that gives us the dot product in terms of the norm.

$$\mathbf{u} \cdot \mathbf{v} = \frac{1}{4}\|\mathbf{u} + \mathbf{v}\|^2 - \frac{1}{4}\|\mathbf{u} - \mathbf{v}\|^2$$

24. Show that if $\mathbf{u}$ is orthogonal to $\mathbf{v}$ and to $\mathbf{w}$, then it is orthogonal to any linear combination $c\mathbf{v} + d\mathbf{w}$.

25. Describe geometrically the set of all 2-vectors $\mathbf{u}$ such that $\|\mathbf{u}\| = 1$.

26. Describe geometrically the set of all 3–vectors $\mathbf{v}$ such that $\|\mathbf{v}\| = 1$.

Polar Decomposition of a Vector

The angles α, β, and γ between a nonzero 3-vector $\mathbf{v}$ and the unit basis vectors $\mathbf{i}$, $\mathbf{j}$, and $\mathbf{k}$ are called the **direction angles** of $\mathbf{v}$ (Fig. 2.17). The cosines $\cos\alpha$, $\cos\beta$, and $\cos\gamma$ of the direction angles are called the **direction cosines** of $\mathbf{v}$.

27. Prove that $(\cos\alpha, \cos\beta, \cos\gamma)$ is a unit vector.

28. Prove that

$$\mathbf{v} = \|\mathbf{v}\|(\cos\alpha, \cos\beta, \cos\gamma) \qquad (2.10)$$

Equation (2.10) is called the **polar decomposition** of $\mathbf{u}$.

29. Let $\mathbf{v} = c\mathbf{u}$ for some scalar $c > 0$ and some unit vector $\mathbf{u}$. Show that

$$c = \|\mathbf{v}\| \quad \text{and} \quad \mathbf{u} = (\cos\alpha, \cos\beta, \cos\gamma)$$

For the given vectors in Exercises 30–31 find each.

(a) The polar decomposition
(b) The direction cosines
(c) The direction angles (You may need a calculator.)

30. $(1, 0, 1)$ $(-1, 1, 0)$
    $(1, -1, 1)$ $(2, -1, 0)$

31. $(1, 1, 1)$ $(-1, 1, -1)$
    $(\sqrt{3}, 1, 0)$ $(2, -2, 0)$

2.3   Span

Reader's Goal for This Section

To understand the span of vectors algebraically and geometrically.

The material in this section is important. We learn about the *span* of a sequence of n-vectors and its geometric interpretation for 2- and 3-vectors. This is a vast generalization of the notions of a line or a plane passing through the origin.

DEFINITION   (Span)

The set of all linear combinations of the n-vectors $\mathbf{v}_1, \ldots, \mathbf{v}_k$ is called the **span** of $\mathbf{v}_1, \ldots, \mathbf{v}_k$ and is denoted by $\mathrm{Span}\{\mathbf{v}_1, \ldots, \mathbf{v}_k\}$. If $V = \mathrm{Span}\{\mathbf{v}_1, \ldots, \mathbf{v}_k\}$ we say that $\mathbf{v}_1, \ldots, \mathbf{v}_k$ **span** V and that $\{\mathbf{v}_1, \ldots, \mathbf{v}_k\}$ is a **spanning set** of V.

▪ EXAMPLE 20   Show that the following vectors are in $\mathrm{Span}\{\mathbf{v}_1, \mathbf{v}_2\}$.

$$\mathbf{0}, \qquad \mathbf{v}_1, \qquad \mathbf{v}_2, \qquad \mathbf{v}_1 + \mathbf{v}_2, \qquad 3\mathbf{v}_1, \qquad 3\mathbf{v}_1 - 2.5\mathbf{v}_2$$

SOLUTION   Each of these vectors is a linear combination of $\mathbf{v}_1$ and $\mathbf{v}_2$, because

$$\mathbf{0} = 0\mathbf{v}_1 + 0\mathbf{v}_2 \qquad \mathbf{v}_1 = 1\mathbf{v}_1 + 0\mathbf{v}_2 \qquad \mathbf{v}_2 = 0\mathbf{v}_1 + 1\mathbf{v}_2$$
$$\mathbf{v}_1 + \mathbf{v}_2 = 1\mathbf{v}_1 + 1\mathbf{v}_2 \qquad 3\mathbf{v}_1 = 3\mathbf{v}_1 + 0\mathbf{v}_2 \qquad 3\mathbf{v}_1 - 2.5\mathbf{v}_2 = 3\mathbf{v}_1 + (-2.5)\mathbf{v}_2$$

▪ EXAMPLE 21   Show that $\mathrm{Span}\{\mathbf{v}\}$ is the set of all scalar multiples of $\mathbf{v}$.

SOLUTION   Any linear combination of $\mathbf{v}$ is of the form $c\mathbf{v}$ for some scalar c. Conversely, any scalar multiple $c\mathbf{v}$ of $\mathbf{v}$ is a linear combination of $\mathbf{v}$. Hence, $\mathrm{Span}\{\mathbf{v}\}$ is the set of all scalar multiples of $\mathbf{v}$.

▪ EXAMPLE 22   Is

$$\begin{bmatrix} 2 \\ 3 \end{bmatrix}$$

in $\mathrm{Span}\left\{ \begin{bmatrix} 1 \\ 2 \end{bmatrix}, \begin{bmatrix} 3 \\ 5 \end{bmatrix} \right\}$?

SOLUTION   The vector is in the span if and only if there are scalars c_1 and c_2 such that $\begin{bmatrix} 2 \\ 3 \end{bmatrix} = c_1 \begin{bmatrix} 1 \\ 2 \end{bmatrix} + c_2 \begin{bmatrix} 3 \\ 5 \end{bmatrix}$. This is equivalent to saying that the system with augmented matrix $\begin{bmatrix} 1 & 3 & : & 2 \\ 2 & 5 & : & 3 \end{bmatrix}$ is consistent. From the echelon form $\begin{bmatrix} 1 & 3 & : & 2 \\ 0 & -1 & : & -1 \end{bmatrix}$, we see that the last column is *not* a pivot column, so the system is consistent and the vector *is* in the span.

Although not necessary, we can easily solve the system to get $c_1 = -1$, $c_2 = 1$. Thus,

$$\begin{bmatrix} 2 \\ 3 \end{bmatrix} = -1 \begin{bmatrix} 1 \\ 2 \end{bmatrix} + 1 \begin{bmatrix} 3 \\ 5 \end{bmatrix}$$

▪ EXAMPLE 23   Show that $\mathrm{Span}\{\mathbf{e}_1, \mathbf{e}_2\} = \mathbf{R}^2$.

SOLUTION   Any linear combination in $\mathbf{e}_1$, $\mathbf{e}_2$ is a vector of the form $c_1\mathbf{e}_1 + c_2\mathbf{e}_2$ for scalars c_1 and c_2. Because

$$c_1\mathbf{e}_1 + c_2\mathbf{e}_2 = c_1 \begin{bmatrix} 1 \\ 0 \end{bmatrix} + c_2 \begin{bmatrix} 0 \\ 1 \end{bmatrix} = \begin{bmatrix} c_1 \\ c_2 \end{bmatrix}$$

we see that $c_1\mathbf{e}_1 + c_2\mathbf{e}_2$ can be any 2-vector and, conversely, any 2-vector can be written as a linear combination $c_1\mathbf{e}_1 + c_2\mathbf{e}_2$. Therefore, Span$\{\mathbf{e}_1, \mathbf{e}_2\} = \mathbf{R}^2$.

▪ EXAMPLE 24   Show that Span$\{\mathbf{e}_1, \mathbf{e}_2, \ldots, \mathbf{e}_n\} = \mathbf{R}^n$.

SOLUTION   This verification is left as an exercise.

▪ EXAMPLE 25   Compute Span$\{\mathbf{e}_1, \mathbf{e}_3\}$ in $\mathbf{R}^3$.

SOLUTION 1   Because

$$c_1\mathbf{e}_1 + c_2\mathbf{e}_3 = c_1 \begin{bmatrix} 1 \\ 0 \\ 0 \end{bmatrix} + c_2 \begin{bmatrix} 0 \\ 0 \\ 1 \end{bmatrix} = \begin{bmatrix} c_1 \\ 0 \\ c_2 \end{bmatrix}$$

Span$\{\mathbf{e}_1, \mathbf{e}_3\}$ is the set of all 3-vectors whose second component is zero.

SOLUTION 2   We need to find all $\mathbf{b}$ in $\mathbf{R}^3$ such that $c_1\mathbf{e}_1 + c_2\mathbf{e}_3 = \mathbf{b}$ can be solved for c_1 and c_2. Or, we can find all $\mathbf{b}$ such that the system $\begin{bmatrix} 1 & 0 & : & b_1 \\ 0 & 0 & : & b_2 \\ 0 & 1 & : & b_3 \end{bmatrix}$ is consistent.

Because this matrix is equivalent to $\begin{bmatrix} 1 & 0 & : & b_1 \\ 0 & 1 & : & b_3 \\ 0 & 0 & : & b_2 \end{bmatrix}$, its last column is nonpivot if and only if $b_2 = 0$. Hence, Span$\{\mathbf{e}_1, \mathbf{e}_3\}$ consists of all 3-vectors whose second component is zero.

Let

$$\mathbf{v}_1 = \begin{bmatrix} 1 \\ 2 \end{bmatrix}, \qquad \mathbf{v}_2 = \begin{bmatrix} 3 \\ 5 \end{bmatrix}, \qquad \mathbf{v}_3 = \begin{bmatrix} 2 \\ 4 \end{bmatrix}$$

▪ EXAMPLE 26   Find Span$\{\mathbf{v}_1, \mathbf{v}_2\}$.

SOLUTION   We need to find all $\mathbf{b} = \begin{bmatrix} b_1 \\ b_2 \end{bmatrix}$ such that the system $\begin{bmatrix} 1 & 3 & : & b_1 \\ 2 & 5 & : & b_2 \end{bmatrix}$ is consistent. Because $\begin{bmatrix} 1 & 3 \\ 2 & 5 \end{bmatrix} \sim \begin{bmatrix} 1 & 3 \\ 0 & -1 \end{bmatrix}$ the last column of the *augmented matrix* is never a pivot column. So, the system is consistent for all $\mathbf{b}$ in $\mathbf{R}^2$. Hence,

$$\text{Span}\{\mathbf{v}_1, \mathbf{v}_2\} = \mathbf{R}^2$$

▪ EXAMPLE 27   Find Span$\{\mathbf{v}_1, \mathbf{v}_3\}$.

SOLUTION   Because

$$\begin{bmatrix} 1 & 2 & b_1 \\ 2 & 4 & b_2 \end{bmatrix} \sim \begin{bmatrix} 1 & 2 & b_1 \\ 0 & 0 & b_2 - 2b_1 \end{bmatrix}$$

the system $\begin{bmatrix} 1 & 2 & : & b_1 \\ 2 & 4 & : & b_2 \end{bmatrix}$ is consistent if and only if $b_2 - 2b_1 = 0$ or $b_2 = 2b_1$. Therefore, $\mathbf{b}$ is in Span$\{\mathbf{v}_1, \mathbf{v}_3\}$ if and only if $b_2 = 2b_1$ or if and only if $\mathbf{b}$ is of the form $r \begin{bmatrix} 1 \\ 2 \end{bmatrix}$, for any scalar r. Hence, the span consists of all scalar multiples of $\mathbf{v}_1$.

$$\mathrm{Span}\{\mathbf{v}_1, \mathbf{v}_2\} = \{r\mathbf{v}_1, r \in \mathbf{R}\} = \mathrm{Span}\{\mathbf{v}_1\}$$

■ EXAMPLE 28   Find a spanning set for

$$V = \left\{ \begin{bmatrix} 3a - b \\ a + 5b \\ a \end{bmatrix}, a, b \in \mathbf{R} \right\} \subseteq \mathbf{R}^3$$

SOLUTION   Because $\begin{bmatrix} 3a - b \\ a + 5b \\ a \end{bmatrix} = a \begin{bmatrix} 3 \\ 1 \\ 1 \end{bmatrix} + b \begin{bmatrix} -1 \\ 5 \\ 0 \end{bmatrix}$ for all scalars a and b, V is

spanned by $\left\{ \begin{bmatrix} 3 \\ 1 \\ 1 \end{bmatrix}, \begin{bmatrix} -1 \\ 5 \\ 0 \end{bmatrix} \right\}.$

$[\underline{v_1 \, v_2 \, v_3}] \rightarrow \begin{bmatrix} ① & 0 & a \\ 0 & ① & b \end{bmatrix}$ RRAF   then, $\underline{v_3} = a\underline{v_1} + b\underline{v_2}$
v_3 linear combo of the rest.

THEOREM 9   (Reduction of Spanning Set)

If one of the m-vectors $\mathbf{v}_1, \mathbf{v}_2, \ldots, \mathbf{v}_k$ is a linear combination of the rest, then the span remains the same if we remove this vector.

PROOF   For notational convenience we may assume that $\mathbf{v}_k$ is a linear combination of $\mathbf{v}_1, \ldots, \mathbf{v}_{k-1}$ (by renaming the vectors, if necessary). Therefore,

$$\mathbf{v}_k = c_1 \mathbf{v}_1 + \cdots + c_{k-1} \mathbf{v}_{k-1} \qquad (2.11)$$

for some scalars $c_1, \ldots, c_{k-1}$. Let V and V' be the spans of $\mathbf{v}_1, \ldots, \mathbf{v}_k$ and $\mathbf{v}_1, \ldots, \mathbf{v}_{k-1}$, respectively. We need to show that $V = V'$. Because any linear combination of $\mathbf{v}_1, \ldots, \mathbf{v}_{k-1}$ is a linear combination of $\mathbf{v}_1, \ldots, \mathbf{v}_k$ (by adding $0\mathbf{v}_k$), we have $V' \subseteq V$. It suffices to show that $V \subseteq V'$. Let $\mathbf{u} \in V$. Then $\mathbf{u} = d_1 \mathbf{v}_1 + \cdots + d_k \mathbf{v}_k$ for some scalars $d_1, \ldots, d_k$. We have

$$\mathbf{u} = d_1 \mathbf{v}_1 + \cdots + d_{k-1} \mathbf{v}_{k-1} + d_k(c_1 \mathbf{v}_1 + \cdots + c_{k-1} \mathbf{v}_{k-1})$$

$$= (d_1 + d_k c_1)\mathbf{v}_1 + \cdots + (d_{k-1} + d_k c_{k-1})\mathbf{v}_{k-1}$$

which is a linear combination in $\mathbf{v}_1, \ldots, \mathbf{v}_{k-1}$. Therefore, $\mathbf{u} \in V'$. Hence, $V \subseteq V'$, as asserted.

■ EXAMPLE 29   Show that

$$\text{Span}\left\{\begin{bmatrix} 1 \\ -3 \end{bmatrix}, \begin{bmatrix} 0 \\ 2 \end{bmatrix}\right\} = \text{Span}\left\{\begin{bmatrix} 1 \\ -3 \end{bmatrix}, \begin{bmatrix} 0 \\ 2 \end{bmatrix}, \begin{bmatrix} 10 \\ -28 \end{bmatrix}\right\}$$

SOLUTION   Because $\begin{bmatrix} 10 \\ -28 \end{bmatrix} = 10\begin{bmatrix} 1 \\ -3 \end{bmatrix} + \begin{bmatrix} 0 \\ 2 \end{bmatrix}$, the two spans are the same according to Theorem 9.

Perhaps the two most important properties of the span of a set of vectors are described in the following theorem.

THEOREM 10

Let $V = \text{Span}\{\mathbf{v}_1, \ldots, \mathbf{v}_n\}$. Then for any $\mathbf{u}$ and $\mathbf{w}$ in V and any scalar c: $\underline{u}, w \in V$

1. $\mathbf{u} + \mathbf{w}$ is in V; $u + w \in V$
2. $c\mathbf{u}$ is in V. $c\underline{u} \in V$.

PROOF   Exercise.

Theorem 10 implies that any linear combination of elements in V is in V. So, not only is the span the set of linear combinations of vectors but also any linear combination of its own vectors is in it.

Relation Between Span and Linear Systems

To say that a vector $\mathbf{b} \in \mathbf{R}^m$ is a linear combination of vectors $\mathbf{v}_1, \ldots, \mathbf{v}_n$ is equivalent to saying that $\mathbf{b}$ is in $\text{Span}\{\mathbf{v}_1, \ldots, \mathbf{v}_n\}$. Therefore,

1. The system with augmented matrix $[A : \mathbf{b}]$ is consistent if and only if $\mathbf{b}$ is in $\text{Span}\{\mathbf{a}_1, \ldots, \mathbf{a}_n\}$.
2. $[A : \mathbf{b}]$ is consistent for all vectors $\mathbf{b} \in \mathbf{R}^m$ if and only if $\text{Span}\{\mathbf{a}_1, \ldots, \mathbf{a}_n\} = \mathbf{R}^m$. This is true only if each row of A has a pivot position (Theorem 2, Section 2.1).

■ EXAMPLE 30   Is

$$\text{Span}\left\{\begin{bmatrix} 1 \\ -3 \end{bmatrix}, \begin{bmatrix} 0 \\ 2 \end{bmatrix}\right\} = \mathbf{R}^2?$$

What can you say about the system $\begin{bmatrix} 1 & 0 & : & b_1 \\ -3 & 2 & : & b_2 \end{bmatrix}$?

SOLUTION   Because each row of $\begin{bmatrix} 1 & 0 \\ -3 & 2 \end{bmatrix}$ has a pivot (why?), $\begin{bmatrix} 1 & 0 & : & b_1 \\ -3 & 2 & : & b_2 \end{bmatrix}$
is consistent for all scalars b_1 and b_2, and $\text{Span}\left\{\begin{bmatrix} 1 \\ -3 \end{bmatrix}, \begin{bmatrix} 0 \\ 2 \end{bmatrix}\right\} = \mathbf{R}^2$.

Geometric Interpretation of Span{**u**} and Span{**u**, **v**}

Let **u** and **v** be either both 2-vectors or both 3-vectors.

Span{**u**}, **u** ≠ **0**

As c takes on any scalar value, $c\mathbf{u}$ runs through all points of the line through the origin with direction **u**. Because, according to Example 21, Span{**u**} is the set of all scalar multiples of **u**, we conclude that, geometrically, Span{**u**} is the line l through the origin with direction **u** (Fig. 2.18(a)). On the other hand, any straight line through the origin can be written as Span{**u**} by letting **u** be any nonzero vector on the line.

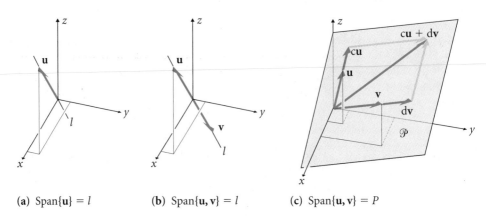

(a) Span{**u**} = l (b) Span{**u**, **v**} = l (c) Span{**u**, **v**} = P

Figure 2.18   The span of one and two vectors.

Span{**u**, **v**}, **u** ≠ **0**

Case 1. If **v** is a scalar multiple of **u**, then Span{**u**, **v**} = Span{**u**} = l (Fig. 2.18(b)).
Case 2. If **v** is not a scalar multiple of **u**, then **v** ≠ **0** and Span{**u**, **v**} is the plane through the origin that contains both **u** and **v** (Fig. 2.18(c)).

In fact, any plane through the origin can be represented as Span{**u**, **v**} by letting **u** and **v** be any two noncolinear vectors in that plane.

Exercises 2.3

1. Let $\mathbf{v} = \begin{bmatrix} 9 \\ -3 \end{bmatrix}$, $\mathbf{w} = \begin{bmatrix} -3 \\ 1 \end{bmatrix}$, and $S = \{\mathbf{w}\}$. True or false?

   a. **v** is in S.   **b.** **w** is in S.

   c. **v** is in Span(S).   **d.** **w** is in Span(S).

2. Let S be a finite set of n-vectors with at least one nonzero vector. Explain why Span(S) has *infinitely* many vectors.

Let

$$\mathbf{a} = \begin{bmatrix} -1 \\ 3 \end{bmatrix} \qquad \mathbf{b} = \begin{bmatrix} 0 \\ -2 \end{bmatrix}$$

$$\mathbf{c} = \begin{bmatrix} 6 \\ 3 \end{bmatrix} \qquad \mathbf{d} = \begin{bmatrix} -3 \\ 9 \end{bmatrix}$$

3. **a.** Is **c** in Span{**a**, **b**}?   **b.** Is **b** in Span{**a**, **c**}?

   c. Is **a** in Span{**b**, **c**}?   **d.** Is **d** in Span{**a**}?

e. Is **a** in Span {**d**}?   **f.** Is **d** in Span {**c**}?

g. Is **d** in Span {**b, c**}?

4. a. Is it true that Span{**a, b**} = $\mathbf{R}^2$?

b. Is it true that Span{**a, d**} = $\mathbf{R}^2$?

c. Is it true that Span{**a, c**} = Span{**b, c**}?

d. Compare Span{**a**} and Span{**d**}.

e. What is Span{**a, b, c**}?

f. What is Span{**a, c, d**}?

g. What is Span{**a, b, c, d**}?

h. What is the span of the columns of A = $\begin{bmatrix} \mathbf{a} & \mathbf{a} & \mathbf{b} & \mathbf{d} \end{bmatrix}$?

i. What is the span of the columns of B = $\begin{bmatrix} \mathbf{a} & \mathbf{a} & \mathbf{d} \end{bmatrix}$?

Let

$$\mathbf{v}_1 = \begin{bmatrix} -1 \\ 3 \\ 1 \end{bmatrix} \quad \mathbf{v}_2 = \begin{bmatrix} 2 \\ -3 \\ 0 \end{bmatrix}$$

$$\mathbf{v}_3 = \begin{bmatrix} 1 \\ 0 \\ 0 \end{bmatrix} \quad \mathbf{v}_4 = \begin{bmatrix} -1 \\ 0 \\ 4 \end{bmatrix}$$

5. a. Is $\mathbf{v}_4$ in Span{$\mathbf{v}_1, \mathbf{v}_2, \mathbf{v}_3$}?

b. Is $\mathbf{v}_4$ in Span{$\mathbf{v}_1, \mathbf{v}_2$}?

c. Is $\mathbf{v}_4$ in Span{$\mathbf{v}_2, \mathbf{v}_3$}?

6. a. Show that Span{$\mathbf{v}_1, \mathbf{v}_2, \mathbf{v}_3$} = $\mathbf{R}^3$.

b. Show that Span{$\mathbf{v}_1, \mathbf{v}_2, \mathbf{v}_4$} = $\mathbf{R}^3$.

c. Show that Span{$\mathbf{v}_1, \mathbf{v}_3, \mathbf{v}_4$} = Span{$\mathbf{v}_2, \mathbf{v}_3, \mathbf{v}_4$}.

In Exercises 7–13 determine whether or not the columns of the given $m \times n$ matrix span $\mathbf{R}^m$.

7. a. $\begin{bmatrix} -1 & 2 \\ 3 & -4 \end{bmatrix}$ **b.** $\begin{bmatrix} -1 & 3 \\ 3 & -9 \end{bmatrix}$

8. a. $\begin{bmatrix} a & 2a \\ b & 2b \end{bmatrix}$ **b.** $\begin{bmatrix} a & b \\ 2a & 2b \end{bmatrix}$

9. a. $\begin{bmatrix} 1 & -1 & 0 \\ 7 & 3 & -1 \end{bmatrix}$ **b.** $\begin{bmatrix} 0 & 5 \\ 1 & -1 \\ 1 & 3 \end{bmatrix}$

10. a. $\begin{bmatrix} a & b & 0 \\ a & b & 0 \end{bmatrix}$ **b.** $\begin{bmatrix} a & d \\ b & e \\ c & f \end{bmatrix}$

11. a. $\begin{bmatrix} 1 & 2 & -3 & 2 & -5 & 5 \\ 0 & 0 & 0 & 2 & -5 & 1 \\ 0 & 0 & 0 & -2 & 5 & -1 \end{bmatrix}$

b. $\begin{bmatrix} a & e & i \\ b & f & j \\ c & g & k \\ d & h & l \end{bmatrix}$

12. a. $\begin{bmatrix} 2 & 0 & 5 \\ 0 & -2 & 0 \\ -1 & 1 & 2 \end{bmatrix}$ **b.** $\begin{bmatrix} a & a & 0 \\ a & a & 0 \\ 1 & 0 & 1 \end{bmatrix}$

13. $\begin{bmatrix} 1 & 2 & -3 & 2 & -5 & 5 \\ 0 & 0 & 0 & 2 & -5 & 1 \\ 0 & 0 & 0 & -1 & 0 & -1 \\ 0 & 0 & 0 & 0 & 1 & 2 \end{bmatrix}$

14. Under what restriction(s) on a and b will the columns

of $\begin{bmatrix} a & b & 0 \\ a & 0 & 0 \\ a & 0 & 1 \end{bmatrix}$ span $\mathbf{R}^3$?

15. True or false?

a. $\mathbf{R}^{10}$ can be spanned by exactly nine 10-vectors.

b. $\mathbf{R}^{10}$ can be spanned by at least nine 10-vectors.

c. $\mathbf{R}^{10}$ can be spanned by ten 9-vectors.

d. $\mathbf{R}^{10}$ can be spanned by ten 10-vectors.

e. $\mathbf{R}^{10}$ can be spanned by eleven 10-vectors.

f. Any twenty 10-vectors span $\mathbf{R}^{10}$.

g. Twenty 10-vectors can span $\mathbf{R}^{10}$.

16. Show that

$$\text{Span}\{\mathbf{u}, \mathbf{v}\} = \text{Span}\{\mathbf{u} + \mathbf{v}, \mathbf{u} - \mathbf{v}\}$$

17. Show that

$$\text{Span}\{\mathbf{u}, \mathbf{v}, \mathbf{w}\} = \text{Span}\{\mathbf{u}, \mathbf{u} + \mathbf{v}, \mathbf{u} + \mathbf{v} + \mathbf{w}\}$$

18. Show that $S_1 = S_2$, where

$$S_1 = \text{Span}\left\{ \begin{bmatrix} 1 \\ 1 \end{bmatrix}, \begin{bmatrix} 1 \\ -1 \end{bmatrix} \right\}$$

$$S_2 = \text{Span}\left\{ \begin{bmatrix} 1 \\ -2 \end{bmatrix}, \begin{bmatrix} 0 \\ 5 \end{bmatrix} \right\}$$

19. Prove Theorem 10.

20. Let A be a matrix whose columns span $\mathbf{R}^{10}$. What can you say about the following?

a. The size of A

b. The system $[A : \mathbf{b}]$

21. Let A be a 10×9 matrix.

    a. Can the columns of A span $\mathbf{R}^{10}$?

    b. Is it true that there is a 10-vector **b** such that the system $[A : \mathbf{b}]$ is inconsistent?

22. Find a finite spanning set for

$$V = \left\{ \begin{bmatrix} a - b \\ 2a + 4b \end{bmatrix}, a, b \in \mathbf{R} \right\}$$

23. Find a finite spanning set for

$$V = \left\{ \begin{bmatrix} 3a - b \\ 4b \\ -a \end{bmatrix}, a, b \in \mathbf{R} \right\}$$

24. Suppose each row of the $m \times n$ matrix A has a pivot position. What can you say about the system $[A : \mathbf{b}]$?

25. If **b** is in the span of the columns of A, what can you say about the system $[A : \mathbf{b}]$?

26. Find all values of x such that

$$\text{Span} \left\{ \begin{bmatrix} 5 \\ 1 \end{bmatrix}, \begin{bmatrix} -10 \\ x \end{bmatrix} \right\} = \mathbf{R}^2$$

27. Find all values of x such that

$$\text{Span} \left\{ \begin{bmatrix} 1 \\ 1 \\ 0 \end{bmatrix}, \begin{bmatrix} -1 \\ 0 \\ -1 \end{bmatrix}, \begin{bmatrix} 0 \\ 1 \\ x \end{bmatrix} \right\} = \mathbf{R}^3$$

28. Draw the following sets.

    a. $\text{Span} \left\{ \begin{bmatrix} 1 \\ 1 \end{bmatrix}, \begin{bmatrix} -2 \\ -2 \end{bmatrix} \right\}$ **b.** $\text{Span} \left\{ \begin{bmatrix} 1 \\ 1 \end{bmatrix}, \begin{bmatrix} -2 \\ 2 \end{bmatrix} \right\}$

29. Draw the following sets.

    a. $\text{Span} \left\{ \begin{bmatrix} 1 \\ 0 \\ 0 \end{bmatrix}, \begin{bmatrix} 0 \\ 1 \\ 0 \end{bmatrix} \right\}$ **b.** $\text{Span} \left\{ \begin{bmatrix} 1 \\ 1 \\ 0 \end{bmatrix}, \begin{bmatrix} 0 \\ 1 \\ 1 \end{bmatrix} \right\}$

2.4   Linear Independence

Reader's Goals for This Section

1. To know what linearly dependent and linearly independent vectors are.
2. To understand the geometric interpretation of linear independence.

The material in this section is very important. We learn about *linearly dependent* and *independent n*-vectors. In addition, we discuss the geometric interpretation of these notions for 2- and 3-vectors.

A linear combination of n-vectors $\mathbf{v}_1, \ldots, \mathbf{v}_k$

$$c_1 \mathbf{v}_1 + c_2 \mathbf{v}_2 + \cdots + c_k \mathbf{v}_k$$

is called **nontrivial** if not all $c_1, \ldots, c_k$ are zero. A linear combination with all coefficients zero is called **trivial**. *(scalars could be zeros)*

scalars does not have to be all zero to obtain 0

DEFINITIONS

A *sequence* of n-vectors $\mathbf{v}_1, \ldots, \mathbf{v}_k$ is **linearly dependent** if **0** is a nontrivial linear combination of these vectors. In other words, there are scalars $c_1, \ldots, c_k$ not all zero such that

$$c_1 \mathbf{v}_1 + c_2 \mathbf{v}_2 + \cdots + c_k \mathbf{v}_k = \mathbf{0} \tag{2.12}$$

A *set* of n-vectors $\mathbf{v}_1, \ldots, \mathbf{v}_k$ is linearly dependent if it is linearly dependent as a sequence.[4] A relation of the form (2.12) with not all $c_1, \ldots, c_k$ zero is called a **linear dependence relation**.

---

[4]Once more, we use the term *sequence* when we want to allow repetitions of vectors.

scalars must be all zero to obtain $\underline{0}$

A *set* of n-vectors $\mathbf{v}_1, \ldots, \mathbf{v}_k$ is **linearly independent** if it is not linearly dependent. This is the same as saying that <u>the only linear combination of $\mathbf{0}$ in terms of $\mathbf{v}_1, \ldots, \mathbf{v}_k$ is the trivial one</u> or that (2.12) implies $c_1 = 0, \ldots, c_k = 0$.

■ EXAMPLE 31   The set

$$\left\{ \begin{bmatrix} 1 \\ -1 \end{bmatrix}, \begin{bmatrix} 1 \\ 2 \end{bmatrix}, \begin{bmatrix} 2 \\ 1 \end{bmatrix} \right\}$$

is linearly dependent, because

$$1 \begin{bmatrix} 1 \\ -1 \end{bmatrix} + 1 \begin{bmatrix} 1 \\ 2 \end{bmatrix} - 1 \begin{bmatrix} 2 \\ 1 \end{bmatrix} = \begin{bmatrix} 0 \\ 0 \end{bmatrix}$$

■ EXAMPLE 32   The sequence

$$\begin{bmatrix} 1 \\ 2 \end{bmatrix}, \begin{bmatrix} 2 \\ 1 \end{bmatrix}, \begin{bmatrix} 1 \\ 2 \end{bmatrix}$$

is linearly dependent, because

$$1 \begin{bmatrix} 1 \\ 2 \end{bmatrix} + 0 \begin{bmatrix} 2 \\ 1 \end{bmatrix} + (-1) \begin{bmatrix} 1 \\ 2 \end{bmatrix} = \begin{bmatrix} 0 \\ 0 \end{bmatrix}$$

■ EXAMPLE 33   Is the set

$$\left\{ \begin{bmatrix} 0 \\ -2 \end{bmatrix}, \begin{bmatrix} 1 \\ 2 \end{bmatrix}, \begin{bmatrix} 2 \\ 1 \end{bmatrix} \right\}$$

linearly dependent? If yes, find a linear dependence relation.

SOLUTION   Let c_1, c_2, c_3 such that

$$c_1 \begin{bmatrix} 0 \\ -2 \end{bmatrix} + c_2 \begin{bmatrix} 1 \\ 2 \end{bmatrix} + c_3 \begin{bmatrix} 2 \\ 1 \end{bmatrix} = \begin{bmatrix} 0 \\ 0 \end{bmatrix}$$

Hence,

$$c_2 + 2c_3 = 0$$

$$-2c_1 + 2c_2 + c_3 = 0$$

We need to solve this homogeneous linear system for c_1, c_2, c_3. Because the augmented matrix $\begin{bmatrix} 0 & 1 & 2 & : & 0 \\ -2 & 2 & 1 & : & 0 \end{bmatrix}$ reduces to $\begin{bmatrix} 1 & 0 & \frac{3}{2} & : & 0 \\ 0 & 1 & 2 & : & 0 \end{bmatrix}$, we get $c_1 = -\frac{3}{2}r, c_2 = -2r$, $c_3 = r$. Thus, there are nontrivial solutions and the set is linearly dependent. To get a particular linear dependence relation, we assign a value to the parameter r. For example, if $r = 2$, then

$$-3 \begin{bmatrix} 0 \\ -2 \end{bmatrix} - 4 \begin{bmatrix} 1 \\ 2 \end{bmatrix} + 2 \begin{bmatrix} 2 \\ 1 \end{bmatrix} = \begin{bmatrix} 0 \\ 0 \end{bmatrix}$$

This is one of the (infinitely many) linear dependence relations.

Note that if we had to check only linear dependence, an echelon form of the augmented matrix would be sufficient. For example, the echelon form $\begin{bmatrix} -2 & 2 & 1 & : & 0 \\ 0 & 1 & 2 & : & 0 \end{bmatrix}$ tells us that the *homogeneous* system has free variables. Hence, there are nontrivial solutions and the set is linearly dependent.

THEOREM 11   Consider two nonzero n-vectors. The following statements are equivalent:

1. The vectors are linearly dependent.
2. One vector is a scalar multiple of the other.
3. The angle between the two vectors is either 0 or π. for $\mathbb{R}^2$

PROOF   Exercise.

▪ **EXAMPLE 34**   Show that $\{e_1, e_2\}$ is linearly independent in $\mathbf{R}^2$.

SOLUTION   Let c_1 and c_2 be scalars such that $c_1 e_1 + c_2 e_2 = \mathbf{0}$. In other words,

$$c_1 \begin{bmatrix} 1 \\ 0 \end{bmatrix} + c_2 \begin{bmatrix} 0 \\ 1 \end{bmatrix} = \begin{bmatrix} 0 \\ 0 \end{bmatrix}$$

Then $\begin{bmatrix} c_1 \\ c_2 \end{bmatrix} = \begin{bmatrix} 0 \\ 0 \end{bmatrix}$. Hence, $c_1 = 0$ and $c_2 = 0$. Therefore, the set is linearly independent.

▪ **EXAMPLE 35**   Show that $\{e_1, e_2, \ldots, e_n\}$ in $\mathbf{R}^n$ is linearly independent.

SOLUTION   Exercise.

▪ **EXAMPLE 36**   Show that

$$\left\{ \begin{bmatrix} 2 \\ 3 \\ 2 \end{bmatrix}, \begin{bmatrix} 8 \\ -6 \\ 5 \end{bmatrix}, \begin{bmatrix} -4 \\ 3 \\ 1 \end{bmatrix} \right\}$$

is linearly independent.

SOLUTION   If $c_1 \begin{bmatrix} 2 \\ 3 \\ 2 \end{bmatrix} + c_2 \begin{bmatrix} 8 \\ -6 \\ 5 \end{bmatrix} + c_3 \begin{bmatrix} -4 \\ 3 \\ 1 \end{bmatrix} = \begin{bmatrix} 0 \\ 0 \\ 0 \end{bmatrix}$, then the corresponding system has coefficient matrix $\begin{bmatrix} 2 & 8 & -4 \\ 3 & -6 & 3 \\ 2 & 5 & 1 \end{bmatrix}$, which is equivalent to the echelon form

$\begin{bmatrix} 2 & 8 & -4 \\ 0 & -3 & 5 \\ 0 & 0 & -21 \end{bmatrix}$. There are no free variables, so the homogeneous system has only the

trivial solution and the set is linearly independent.

As the previous examples indicate, row reduction is an effective way of determining linear dependence or independence.

THEOREM 12   **(Criterion for Linear Independence)**

The following are equivalent:

1. The set of m-vectors $\{\mathbf{v}_1, \mathbf{v}_2, \ldots, \mathbf{v}_n\}$ is linearly independent.
2. The system has only the trivial solution

$$[\, \mathbf{v}_1 \quad \mathbf{v}_2 \quad \cdots \quad \mathbf{v}_n \ : \ \mathbf{0} \,]$$

3. The matrix has n pivot positions (or, each column is a pivot column):

$$[\, \mathbf{v}_1 \quad \mathbf{v}_2 \quad \cdots \quad \mathbf{v}_n \,]$$

PROOF $1 \Leftrightarrow 2$, because $c_1 \mathbf{v}_1 + \cdots + c_n \mathbf{v}_n = \mathbf{0}$ and $[\mathbf{v}_1 \cdots \mathbf{v}_n : \mathbf{0}]$ are equivalent expressions. On the other hand, $2 \Leftrightarrow 3$ by Part 2 of Theorem 3, Section 1.2.

■ EXAMPLE 37   Are the columns of A linearly independent?

$$A = \begin{bmatrix} -3 & 3 & 3 \\ 2 & 2 & 2 \\ 0 & 1 & 0 \end{bmatrix}$$

SOLUTION A has three pivot columns, because $A \sim \begin{bmatrix} -3 & 3 & 3 \\ 0 & 4 & 4 \\ 0 & 0 & -1 \end{bmatrix}$. Hence, its

columns are linearly independent by Theorem 12.

■ EXAMPLE 38   Are the columns of A linearly independent?

$$A = \begin{bmatrix} -5 & 5 & 5 \\ 9 & 0 & 9 \\ 4 & 6 & 16 \end{bmatrix}$$

SOLUTION   Because $A \sim \begin{bmatrix} -5 & 5 & 5 \\ 0 & 9 & 18 \\ 0 & 0 & 0 \end{bmatrix}$, the third column of A is not a pivot column. So the columns of A are not linearly independent, by Theorem 12. (They are linearly dependent.)

One useful consequence of Theorem 12 is Theorem 13.

THEOREM 13   If the set of m-vectors $\{\mathbf{v}_1, \mathbf{v}_2, \ldots, \mathbf{v}_n\}$ is linearly independent, then $n \leq m$.

PROOF   Because the set is independent, the matrix $[\, \mathbf{v}_1 \quad \mathbf{v}_2 \quad \cdots \quad \mathbf{v}_n \,]$ has n pivot columns by Theorem 12. But the number of pivots cannot exceed either the number of columns or the number of rows of a matrix, by Exercise 49, Section 1.2. Therefore, $n \leq m$ as asserted.

Theorem 13 says that if we have *more vectors than components, the vectors are linearly dependent.*

■ EXAMPLE 39   Are the columns of the matrix linearly dependent?

$$\begin{bmatrix} 45 & -80 & -93 & 92 \\ 43 & -62 & 77 & 66 \\ 54 & 55 & -99 & -61 \end{bmatrix}$$

SOLUTION   Yes; by Theorem 13, four 3-vectors cannot be linearly independent.

The following theorem is quite useful in showing that a sequence or set of vectors is linearly dependent.

THEOREM 14   **(Test for Linear Dependence)**

Let S be a finite set or sequence of m-vectors. We have

1. If S consists of one vector $\mathbf{v}$, then S is linearly dependent if and only if $\mathbf{v} = \mathbf{0}$.
2. If S consists of two or more vectors $\mathbf{v}_1, \ldots, \mathbf{v}_k$, then S is linearly dependent if and only if at least one vector is a linear combination of the remaining vectors.
3. If S consists of two or more vectors $\mathbf{v}_1, \ldots, \mathbf{v}_k$, with $\mathbf{v}_1 \neq \mathbf{0}$, then S is linearly dependent if and only if at least one vector, say, $\mathbf{v}_i$ ($i \geq 2$), is a linear combination of the vectors that precede it, i.e., $\mathbf{v}_1, \ldots, \mathbf{v}_{i-1}$.

PROOF   We prove only Part 3 and leave the remaining proofs as exercises. If $\mathbf{v}_i$ is a linear combination of $\mathbf{v}_1, \ldots, \mathbf{v}_{i-1}$, then there are scalars $c_1, \ldots, c_{i-1}$ such that

$$\mathbf{v}_i = c_1 \mathbf{v}_1 + \cdots + c_{i-1} \mathbf{v}_{i-1}$$

Therefore,

$$c_1\mathbf{v}_1 + \cdots + c_{i-1}\mathbf{v}_{i-1} + (-1)\mathbf{v}_i + 0\mathbf{v}_{i+1} + \cdots + 0\mathbf{v}_k = \mathbf{0}$$

This is a nontrivial linear combination of $\mathbf{0}$ in terms of $\mathbf{v}_1, \ldots, \mathbf{v}_k$, because the coefficient of $\mathbf{v}_i$ is $-1 \neq 0$. Hence, $\mathbf{v}_1, \ldots, \mathbf{v}_k$ is linearly dependent.

Conversely, let $\mathbf{v}_1, \ldots, \mathbf{v}_k$ be linearly dependent. Then there are scalars $c_1, \ldots, c_k$ not all zero such that

$$c_1\mathbf{v}_1 + \cdots + c_k\mathbf{v}_k = \mathbf{0}$$

Let c_i be the last nonzero scalar in the preceding equation. Then

$$c_1\mathbf{v}_1 + \cdots + c_i\mathbf{v}_i + 0\mathbf{v}_i + \cdots + 0\mathbf{v}_k = \mathbf{0}, \quad \text{or} \quad c_1\mathbf{v}_1 + \cdots + c_i\mathbf{v}_i = \mathbf{0}$$

with $c_i \neq 0$. We see $i \geq 2$, because if $i = 1$, the nontrivial combination $c_1\mathbf{v}_1 = \mathbf{0}$ would imply $\mathbf{v}_1 = \mathbf{0}$. Therefore,

$$c_i\mathbf{v}_i = (-c_1)\mathbf{v}_1 + \cdots + (-c_{i-1})\mathbf{v}_{i-1} \quad \text{or} \quad \mathbf{v}_i = \left(-\frac{c_1}{c_i}\right)\mathbf{v}_1 + \cdots + \left(-\frac{c_{i-1}}{c_i}\right)\mathbf{v}_{i-1}$$

Hence, $\mathbf{v}_i$ is a linear combination of $\mathbf{v}_1, \ldots, \mathbf{v}_{i-1}$.

■ EXAMPLE 40   The vectors

$$\begin{bmatrix} 1 \\ 2 \\ 0 \\ 0 \end{bmatrix}, \begin{bmatrix} 0 \\ 1 \\ 0 \\ 0 \end{bmatrix}, \begin{bmatrix} 1 \\ 5 \\ 0 \\ 0 \end{bmatrix}, \begin{bmatrix} 35 \\ 67 \\ 33 \\ 88 \end{bmatrix}$$

are linearly dependent by Theorem 14, since $\begin{bmatrix} 1 \\ 5 \\ 0 \\ 0 \end{bmatrix} = 1\begin{bmatrix} 1 \\ 2 \\ 0 \\ 0 \end{bmatrix} + 3\begin{bmatrix} 0 \\ 1 \\ 0 \\ 0 \end{bmatrix}.$

REMARK   Theorem 14 does *not* say that *every* vector is a linear combination of the remaining (or the preceding) vectors. For example, $\left\{ \begin{bmatrix} 1 \\ 0 \end{bmatrix}, \begin{bmatrix} 2 \\ 0 \end{bmatrix}, \begin{bmatrix} 0 \\ 1 \end{bmatrix} \right\}$ is linearly dependent, because $\begin{bmatrix} 2 \\ 0 \end{bmatrix} = 2\begin{bmatrix} 1 \\ 0 \end{bmatrix} + 0 \cdot \begin{bmatrix} 0 \\ 1 \end{bmatrix}$. However, $\begin{bmatrix} 0 \\ 1 \end{bmatrix}$ is not a linear combination of $\begin{bmatrix} 1 \\ 0 \end{bmatrix}$ and $\begin{bmatrix} 2 \\ 0 \end{bmatrix}$.

Let us collect now a few basic facts, whose proofs are left as exercises.

THEOREM 15
1. Any finite set or sequence of vectors that contains $\mathbf{0}$ is linearly dependent.
2. A finite sequence with repeated vectors is linearly dependent.
3. Any finite set (sequence) of vectors that contains a linearly dependent set (sequence) is itself linearly dependent.
4. Any subset of a finite linearly independent set is itself linearly independent.

We also have the following theorem.

THEOREM 16

Let $S = \{v_1, \ldots, v_n\}$ be a linearly independent set of m-vectors.

1. Any vector v in the span of S is uniquely expressible as a linear combination of vectors in S. That is, the relations

$$v = c_1 v_1 + \cdots + c_n v_n \quad \text{and} \quad v = d_1 v_1 + \cdots + d_n v_n$$

imply

$$c_1 = d_1, \ldots, c_n = d_n$$

2. If v is not in the span of S, then the set $\{v_1, \ldots, v_n, v\}$ is linearly independent.

PROOF

1. Because $v = c_1 v_1 + \cdots + c_n v_n = d_1 v_1 + \cdots + d_n v_n$ we have

$$(c_1 - d_1) v_1 + \cdots + (c_n - d_n) v_n = 0$$

Hence $c_1 - d_1 = \cdots = c_n - d_n = 0$, because S is linearly independent. Therefore, $c_1 = d_1, \ldots, c_n = d_n$, as asserted.

2. Suppose, on the contrary, that $S' = \{v_1, \ldots, v_n, v\}$ is linearly dependent. Then there is a nontrivial linear combination of 0,

$$c_1 v_1 + \cdots + c_n v_n + c v = 0$$

If $c \neq 0$, we may solve the equation for v to get $v = (c^{-1}c_1) v_1 + \cdots + (c^{-1}c_n) v_n$. But then v would be a linear combination in S. So it would be in the span of S, which contradicts our assumption. If, on the other hand, $c = 0$, then the equation reduces to

$$c_1 v_1 + \cdots + c_n v_n = 0$$

with at least one $c_i \neq 0$. (Why?) This, too, contradicts the assumption that S is linearly independent. We conclude that S' has to be linearly independent.

Geometric Interpretation of Linear Independence in $\mathbf{R}^2$ and $\mathbf{R}^3$

Using Theorem 15 and the geometric definitions of addition and scalar multiplication discussed in Section 2.1, we have the following observations (Fig. 2.19):

1. Two nonzero vectors in $\mathbf{R}^2$ or in $\mathbf{R}^3$ are linearly dependent if and only if they are on the same line passing through the origin.
2. Three vectors in $\mathbf{R}^3$ are linearly dependent if and only if they are on the same plane passing through the origin.
3. The span of two vectors in $\mathbf{R}^2$ or in $\mathbf{R}^3$ is a line through the origin if the vectors are linearly dependent or the plane defined by the vectors if they are linearly independent.
4. The span of two linearly independent 2-vectors is $\mathbf{R}^2$.

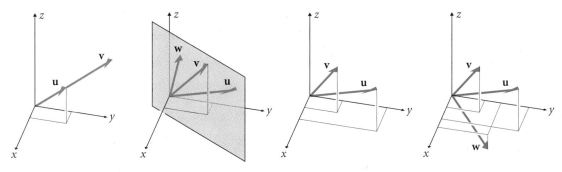

Linearly dependent

Linearly independent

Figure 2.19

5. The span of three linearly independent 3-vectors is $\mathbf{R}^3$.
6. Any linearly independent set in $\mathbf{R}^3$ has at most three vectors.
7. Any linearly independent set in $\mathbf{R}^2$ has at most two vectors.
8. Any set that spans $\mathbf{R}^2$ has at least two vectors.
9. Any set that spans $\mathbf{R}^3$ has at least three vectors.

Exercises 2.4

In Exercises 1–5 determine whether the vectors are linearly independent.

1. $\begin{bmatrix} 1 \\ -2 \\ 0 \end{bmatrix}, \begin{bmatrix} 1 \\ 1 \\ 1 \end{bmatrix}, \begin{bmatrix} 0 \\ -1 \\ 1 \end{bmatrix}$

2. $\begin{bmatrix} 1 \\ -2 \\ 0 \end{bmatrix}, \begin{bmatrix} 1 \\ 1 \\ 1 \end{bmatrix}, \begin{bmatrix} 1 \\ 0 \\ -1 \end{bmatrix}, \begin{bmatrix} 0 \\ -1 \\ 1 \end{bmatrix}$

3. $\begin{bmatrix} a \\ 1 \end{bmatrix}, \begin{bmatrix} 10a \\ 100 \end{bmatrix}$, for $a \neq 0$

4. $\begin{bmatrix} a \\ a \\ 1 \end{bmatrix}, \begin{bmatrix} b \\ b \\ 1 \end{bmatrix}, \begin{bmatrix} 0 \\ 0 \\ 1 \end{bmatrix}$

5. $\begin{bmatrix} a \\ a \\ 1 \end{bmatrix}, \begin{bmatrix} b \\ b \\ 1 \end{bmatrix}, \begin{bmatrix} 1 \\ 0 \\ 0 \end{bmatrix}$, for $a \neq b$

In Exercises 6–9 determine whether the columns of the matrix are linearly independent.

6. $\begin{bmatrix} -1 & 2 & 1 \\ 3 & -3 & 0 \\ 1 & 0 & 0 \end{bmatrix}$

7. $\begin{bmatrix} 15 & 0 \\ 20 & 0 \\ 25 & 0 \end{bmatrix}$

8. $\begin{bmatrix} -3 & 1 & 1 & 0 \\ 2 & 1 & 0 & -1 \\ 4 & 1 & -1 & 1 \end{bmatrix}$

9. $\begin{bmatrix} 1 & 1 & 2 & 3 \\ 0 & 1 & 1 & 0 \\ 0 & 1 & 0 & -1 \\ 0 & 0 & 0 & 1 \end{bmatrix}$

In Exercises 10–18 tell by inspection whether or not the vectors are linearly independent.

10. $\begin{bmatrix} 555 \\ 123 \end{bmatrix}, \begin{bmatrix} 55{,}500 \\ 12{,}300 \end{bmatrix}$

11. $\begin{bmatrix} 555 \\ 123 \end{bmatrix}, \begin{bmatrix} 334 \\ 654 \end{bmatrix}, \begin{bmatrix} 446 \\ 667 \end{bmatrix}$

12. $\begin{bmatrix} 5 \\ 3 \\ 1 \end{bmatrix}, \begin{bmatrix} -1 \\ 0 \\ -2 \end{bmatrix}, \begin{bmatrix} 6 \\ 3 \\ 3 \end{bmatrix}$

13. $\begin{bmatrix} 0 \\ 1 \\ 1 \\ -1 \end{bmatrix}, \begin{bmatrix} 1 \\ 0 \\ 1 \\ 0 \end{bmatrix}, \begin{bmatrix} 20 \\ 10 \\ 30 \\ -10 \end{bmatrix}$

14. $\begin{bmatrix} a \\ a \end{bmatrix}, \begin{bmatrix} b \\ b \end{bmatrix}$

15. $\begin{bmatrix} a \\ a \end{bmatrix}, \begin{bmatrix} a \\ b \end{bmatrix}, \begin{bmatrix} 0 \\ a - b \end{bmatrix}$

16. $\begin{bmatrix} a \\ 1 \\ 0 \end{bmatrix}, \begin{bmatrix} a \\ 0 \\ 1 \end{bmatrix}$ **17.** $\begin{bmatrix} 1 \\ 0 \\ 0 \end{bmatrix}, \begin{bmatrix} a \\ 1 \\ 0 \end{bmatrix}, \begin{bmatrix} a \\ 0 \\ 1 \end{bmatrix}$

18. $\begin{bmatrix} 1 \\ 0 \\ 0 \end{bmatrix}, \begin{bmatrix} a \\ 1 \\ 0 \end{bmatrix}, \begin{bmatrix} a \\ a \\ 0 \end{bmatrix}$

19. For which values of a is the set $\left\{ \begin{bmatrix} a \\ 1 \end{bmatrix}, \begin{bmatrix} a+2 \\ a \end{bmatrix} \right\}$ linearly dependent?

20. For which values of a is the set $\left\{ \begin{bmatrix} a \\ 2 \end{bmatrix}, \begin{bmatrix} a-2 \\ a \end{bmatrix} \right\}$ linearly dependent?

21. Is it true that the set with elements $\begin{bmatrix} a \\ 1 \end{bmatrix}, \begin{bmatrix} 2 \\ 1 \end{bmatrix},$ $\begin{bmatrix} a+2 \\ 2 \end{bmatrix}$ must be linearly dependent?

22. True or false?

  a. Any two distinct n-vectors are linearly independent.

  b. Any three distinct 2-vectors span $\mathbf{R}^2$.

  c. Any two linearly independent 2-vectors span $\mathbf{R}^2$.

  d. Any n linearly independent n-vectors span $\mathbf{R}^n$.

23. Let $\{v_1, v_2, v_3\}$ be a linearly independent set of n-vectors. Find c_1, c_2, and c_3 if $\mathbf{v} = c_1 v_1 + c_2 v_2 + c_3 v_3$ and $\mathbf{v} = (2c_2 - c_1)v_1 + (c_3 - c_2)v_2 + (c_2 - 1)v_3$.

24. Let $c_1 v_1 + c_2 v_2 + c_3 v_3 = d_1 v_1 + d_2 v_2 + d_3 v_3$ and $c_3 \neq d_3$. Show that $\{v_1, v_2, v_3\}$ is linearly dependent.

25. Prove that if $\{v_1, v_2, v_3\}$ is linearly independent, so is $\{v_1 - v_2, v_2 - v_3, v_3 + v_1\}$.

26. Prove the claims of Example 35.

27. Prove Theorem 11.

28. Prove Theorem 15.

29. Suppose the columns of the $m \times n$ matrix A are linearly independent. Show that for any m-vector $\mathbf{b}$, the system $[A : \mathbf{b}]$ has at most one solution.

30. Let A be an $n \times n$ matrix with linearly independent columns. Show that for any n-vector $\mathbf{b}$ the system $[A : \mathbf{b}]$ has exactly one solution.

31. Suppose that $S_1 = \{v_1, v_2\}$ and $S_2 = \{w_1, w_2\}$ are linearly independent subsets of $\mathbf{R}^3$. What geometric object is the intersection Span$(S_1) \cap$ Span(S_2)?

2.5   The Product A**x**

Reader's Goals for This Section

1. To know how to compute and interpret the matrix product A**x**.
2. To know how to write a linear system in terms of A**x**.

In this section we define the matrix-vector product A**x** and use it to introduce another notation for linear systems. Then we study the solution set of a homogeneous system and its properties.

The Product A**x**

Let us define the product between a matrix and a vector. If $A = \begin{bmatrix} -2 & 5 & -3 \\ 4 & 7 & 0 \end{bmatrix}$ and

$\mathbf{x} = \begin{bmatrix} -3 \\ 2 \\ 5 \end{bmatrix}$, then the product A**x** is the linear combination

$$\begin{bmatrix} -2 & 5 & -3 \\ 4 & 7 & 0 \end{bmatrix} \begin{bmatrix} -3 \\ 2 \\ 5 \end{bmatrix} = -3 \begin{bmatrix} -2 \\ 4 \end{bmatrix} + 2 \begin{bmatrix} 5 \\ 7 \end{bmatrix} + 5 \begin{bmatrix} -3 \\ 0 \end{bmatrix} = \begin{bmatrix} 1 \\ 2 \end{bmatrix}$$

In other words, we form the linear combination in the columns of A and coefficients the components of **x**. Note that this operation makes sense *only if the number of columns of the matrix equals the size of the vector.*

DEFINITION

Let A be an $m \times n$ matrix with columns $\mathbf{a}_1, \mathbf{a}_2, \ldots, \mathbf{a}_n$ and let $\mathbf{x}$ be an n-vector with components $x_1, x_2, \ldots, x_n$. The matrix product $A\mathbf{x}$ is the m-vector expressed as the linear combination

$$A\mathbf{x} = x_1\mathbf{a}_1 + x_2\mathbf{a}_2 + \cdots + x_n\mathbf{a}_n$$

■ EXAMPLE 41   Let

$$A = \begin{bmatrix} 2 & -2 \\ -1 & 6 \\ 4 & -3 \end{bmatrix}, \quad \mathbf{b} = \begin{bmatrix} -3 \\ 2 \end{bmatrix}, \quad \text{and} \quad \mathbf{c} = \begin{bmatrix} 10 \\ 20 \\ 30 \end{bmatrix}$$

Compute, if possible, the products $A\mathbf{b}$ and $A\mathbf{c}$.

SOLUTION

$$A\mathbf{b} = \begin{bmatrix} 2 & -2 \\ -1 & 6 \\ 4 & -3 \end{bmatrix} \begin{bmatrix} -3 \\ 2 \end{bmatrix} = -3 \begin{bmatrix} 2 \\ -1 \\ 4 \end{bmatrix} + 2 \begin{bmatrix} -2 \\ 6 \\ -3 \end{bmatrix} = \begin{bmatrix} -10 \\ 15 \\ -18 \end{bmatrix}$$

The product $A\mathbf{c}$ is undefined, because A is 3×2 and $\mathbf{c}$ is not a 2-vector. ■

■ EXAMPLE 42   Compute $A\mathbf{x}$, where

$$A = \begin{bmatrix} 2 & 3 & 1 \end{bmatrix} \quad \text{and} \quad \mathbf{x} = \begin{bmatrix} 2 \\ -1 \\ 4 \end{bmatrix}$$

SOLUTION

$$A\mathbf{x} = \begin{bmatrix} 2 & 3 & 1 \end{bmatrix} \begin{bmatrix} 2 \\ -1 \\ 4 \end{bmatrix} = 2 \cdot 2 + (-1) \cdot 3 + 4 \cdot 1 = 5$$

We see that this time $A\mathbf{x}$ is just the *dot product* $\begin{bmatrix} 2 \\ 3 \\ 1 \end{bmatrix} \cdot \begin{bmatrix} 2 \\ -1 \\ 4 \end{bmatrix}$. ■

Note that as **x** varies through $\mathbf{R}^n$, the product $A\mathbf{x}$ varies through $\mathbf{R}^m$. So, multiplication by A defines a **correspondence** between $\mathbf{R}^n$ and $\mathbf{R}^m$.

■ EXAMPLE 43   (Grade Recording)   A linear algebra instructor uses matrices
to keep statistics of her grades. Let M be the matrix defined by

Grade	Fall 1999	Spring 2000	Fall 2000
A	2	3	1
B	10	15	12
C	13	15	15
D	8	11	12

Compute and interpret the product $M \begin{bmatrix} 1 \\ 1 \\ 1 \end{bmatrix}$.

SOLUTION

$$\begin{bmatrix} 2 & 3 & 1 \\ 10 & 15 & 12 \\ 13 & 15 & 15 \\ 8 & 11 & 12 \end{bmatrix} \begin{bmatrix} 1 \\ 1 \\ 1 \end{bmatrix} = 1 \begin{bmatrix} 2 \\ 10 \\ 13 \\ 8 \end{bmatrix} + 1 \begin{bmatrix} 3 \\ 15 \\ 15 \\ 11 \end{bmatrix} + 1 \begin{bmatrix} 1 \\ 12 \\ 15 \\ 12 \end{bmatrix} = \begin{bmatrix} 6 \\ 37 \\ 43 \\ 31 \end{bmatrix}$$

The product is the vector with components the sums of the columns of M. We get the
total number of As, Bs, etc.

The $n \times n$ matrix with columns $\mathbf{e}_1, \mathbf{e}_2, \dots, \mathbf{e}_n$ is called the **identity matrix**, and it
is denoted by I_n or I.

■ EXAMPLE 44

$$I_2 = I = \begin{bmatrix} 1 & 0 \\ 0 & 1 \end{bmatrix}, \quad I_3 = I = \begin{bmatrix} 1 & 0 & 0 \\ 0 & 1 & 0 \\ 0 & 0 & 1 \end{bmatrix}, \quad I_4 = I = \begin{bmatrix} 1 & 0 & 0 & 0 \\ 0 & 1 & 0 & 0 \\ 0 & 0 & 1 & 0 \\ 0 & 0 & 0 & 1 \end{bmatrix}$$

The following theorem outlines the basic properties of the operation $A\mathbf{x}$.

THEOREM 17

Let A be an $m \times n$ matrix, let $\mathbf{x}, \mathbf{y}$ be n-vectors, and let c be any scalar. Then

1. $A(\mathbf{x} + \mathbf{y}) = A\mathbf{x} + A\mathbf{y}$
2. $A(c\mathbf{x}) = c(A\mathbf{x})$
3. $I_n\mathbf{x} = \mathbf{x}$

PROOF   We prove Part 1 and leave the remaining proofs as exercises. Let $\mathbf{a}_1, \dots, \mathbf{a}_n$
be the columns of A and let $x_1, \dots, x_n$ and $y_1, \dots, y_n$ be the components of $\mathbf{x}$ and $\mathbf{y}$,

respectively. Then

$$A(\mathbf{x} + \mathbf{y}) = A\left(\begin{bmatrix} x_1 \\ \vdots \\ x_n \end{bmatrix} + \begin{bmatrix} y_1 \\ \vdots \\ y_n \end{bmatrix} \right) = A \begin{bmatrix} x_1 + y_1 \\ \vdots \\ x_n + y_n \end{bmatrix}$$

By the definition of the product, this equals

$$(x_1 + y_1)\mathbf{a}_1 + (x_2 + y_2)\mathbf{a}_2 + \cdots + (x_n + y_n)\mathbf{a}_n$$

$$\begin{aligned}
&= x_1\mathbf{a}_1 + y_1\mathbf{a}_1 + x_2\mathbf{a}_2 + y_2\mathbf{a}_2 + \cdots + x_n\mathbf{a}_n + y_n\mathbf{a}_n && \text{by 6 of Theorem 1} \\
&= (x_1\mathbf{a}_1 + x_2\mathbf{a}_2 + \cdots + x_n\mathbf{a}_n) + (y_1\mathbf{a}_1 + y_2\mathbf{a}_2 + \cdots + y_n\mathbf{a}_n) && \text{by 1 and 2} \\
&= A\mathbf{x} + A\mathbf{y}
\end{aligned}$$

The Equation A**x** = **b**

The product A**x** offers a very elegant and useful way of representing a linear system. For example, consider the system

$$\begin{aligned} 2x - \ y &= 1 \\ -x + 4y &= 3 \end{aligned}$$

which we can rewrite in vector notation as

$$x \begin{bmatrix} 2 \\ -1 \end{bmatrix} + y \begin{bmatrix} -1 \\ 4 \end{bmatrix} = \begin{bmatrix} 1 \\ 3 \end{bmatrix}$$

This equation is equivalent to

$$\begin{bmatrix} 2 & -1 \\ -1 & 4 \end{bmatrix} \begin{bmatrix} x \\ y \end{bmatrix} = \begin{bmatrix} 1 \\ 3 \end{bmatrix}$$

by the definition of the matrix-vector product.

A linear system can be written as $A\mathbf{x} = \mathbf{b}$, where A is the coefficient matrix, **b** is the vector of constant terms, and **x** is the vector of unknowns. So we have the equivalent expressions

$$[A : \mathbf{b}] \Leftrightarrow A\mathbf{x} = \mathbf{b} \Leftrightarrow x_1\mathbf{a}_1 + \cdots + x_n\mathbf{a}_n = \mathbf{b}$$

where, $\mathbf{a}_1, \ldots, \mathbf{a}_n$ are the columns of A.

To say that the system $[A : \mathbf{b}]$ is consistent is equivalent to saying that there is a vector **x** such that **b** can be written as the product A**x**. So, the following statements are equivalent:

1. $[A : \mathbf{b}]$ is consistent.
2. There is a vector **x** such that $A\mathbf{x} = \mathbf{b}$.
3. **b** is a linear combination of the columns of A.
4. **b** is in the span of the columns of A.

By the equivalence of 2 and 4 we have

$$\text{Span}\{\mathbf{a}_1, \mathbf{a}_2, \ldots, \mathbf{a}_n\} = \{A\mathbf{x}, \text{ all } \mathbf{x} \text{ in } \mathbf{R}^n\} \tag{2.12}$$

▪ EXAMPLE 45   Find all possible vectors of the form

$$\begin{bmatrix} 1 & 0 \\ -1 & 1 \end{bmatrix} \begin{bmatrix} x \\ y \end{bmatrix}$$

as x and y take on any real values.

SOLUTION   By (2.12), $\left\{ \begin{bmatrix} 1 & 0 \\ -1 & 1 \end{bmatrix} \begin{bmatrix} x \\ y \end{bmatrix}, x, y \in \mathbf{R} \right\} = \text{Span} \left\{ \begin{bmatrix} 1 \\ -1 \end{bmatrix}, \begin{bmatrix} 0 \\ 1 \end{bmatrix} \right\}$. Be-

cause $\begin{bmatrix} 1 & 0 \\ -1 & 1 \end{bmatrix}$ has two pivots and two rows, the span—and hence the given set—is $\mathbf{R}^2$.

The Null Space

The **null space**, Null(A), of an $m \times n$ matrix A consists of all n-vectors $\mathbf{x}$ such that $A\mathbf{x} = \mathbf{0}$. This is the set of all solutions of the homogeneous system $A\mathbf{x} = \mathbf{0}$.

$$\text{Null}(A) = \{\mathbf{x} \text{ in } \mathbf{R}^n \text{ such that } A\mathbf{x} = \mathbf{0}\}$$

Note that $A\mathbf{x} = \mathbf{0}$ has only the trivial solution if and only if Null(A) = $\{\mathbf{0}\}$. Computing Null(A) amounts to finding all solutions of $A\mathbf{x} = \mathbf{0}$.

▪ EXAMPLE 46   Which of

$$\mathbf{u} = \begin{bmatrix} 1 \\ 2 \end{bmatrix} \quad \text{and} \quad \mathbf{v} = \begin{bmatrix} -1 \\ 2 \end{bmatrix}$$

is in the null space of $\begin{bmatrix} 2 & -1 \\ -4 & 2 \end{bmatrix}$?

SOLUTION   Since

$$\begin{bmatrix} 2 & -1 \\ -4 & 2 \end{bmatrix} \begin{bmatrix} 1 \\ 2 \end{bmatrix} = \begin{bmatrix} 0 \\ 0 \end{bmatrix} \quad \text{and} \quad \begin{bmatrix} 2 & -1 \\ -4 & 2 \end{bmatrix} \begin{bmatrix} -1 \\ 2 \end{bmatrix} = \begin{bmatrix} -4 \\ 8 \end{bmatrix} \neq \begin{bmatrix} 0 \\ 0 \end{bmatrix}$$

we see that $\mathbf{u}$ is in the null space and $\mathbf{v}$ is not.

One of the most important properties of the null space is described in the following theorem.

THEOREM 18   Let A be an $m \times n$ matrix. For any $\mathbf{x}_1$, $\mathbf{x}_2$ in Null(A) and any scalar c:

1. $\mathbf{x}_1 + \mathbf{x}_2$ is in Null(A);
2. $c\mathbf{x}_1$ is in Null(A).

PROOF   Because $A\mathbf{x}_1 = \mathbf{0}$ and $A\mathbf{x}_2 = \mathbf{0}$, we have

$$A(\mathbf{x}_1 + \mathbf{x}_2) = A\mathbf{x}_1 + A\mathbf{x}_2 = \mathbf{0} + \mathbf{0} = \mathbf{0}$$

$$A(c\mathbf{x}_1) = c(A\mathbf{x}_1) = c\mathbf{0} = \mathbf{0}$$

by Theorem 17. Therefore, $\mathbf{x}_1 + \mathbf{x}_2$, $c\mathbf{x}_1$ are in Null(A).

The Solutions of Ax = b and Ax = 0

In this paragraph we describe the relation between the solution sets of a linear system and its associated homogeneous system. We shall prove that if $\mathbf{x}_h$ is the general solution of $A\mathbf{x} = \mathbf{0}$ and $\mathbf{p}$ a particular solution of $A\mathbf{x} = \mathbf{b}$, then

$$\mathbf{x} = \mathbf{p} + \mathbf{x}_h$$

is the general solution of $A\mathbf{x} = \mathbf{b}$.

The relation between the two solution sets can be described elegantly in terms of the following notation: If $\mathbf{p}$ is an n-vector and S is a subset of $\mathbf{R}^n$, we denote by $\mathbf{p} + S$ the set of all sums $\mathbf{p} + \mathbf{x}$, where $\mathbf{x}$ is an element of S.

$$\mathbf{p} + S = \{\mathbf{p} + \mathbf{x}, \mathbf{x} \in S\}$$

THEOREM 19

Let S be the solution set of $A\mathbf{x} = \mathbf{b}$ and let $\mathbf{p}$ be in S. Then

$$S = \mathbf{p} + \text{Null}(A)$$

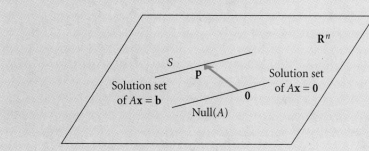

PROOF   If $\mathbf{p} + \mathbf{x}_h$ is any element of $\mathbf{p} + \text{Null}(A)$, then

$$A(\mathbf{x}_h + \mathbf{p}) = A\mathbf{x}_h + A\mathbf{p} = \mathbf{0} + \mathbf{b} = \mathbf{b}$$

So, $\mathbf{x}_h + \mathbf{p}$ is a solution of $A\mathbf{x} = \mathbf{b}$. Hence, $\mathbf{x}_h + \mathbf{p}$ is in S. Therefore, $\mathbf{p} + \text{Null}(A) \subseteq S$.

If $\mathbf{y}$ is any element of S, then $A\mathbf{y} = \mathbf{b}$. The difference $\mathbf{y} - \mathbf{p}$ is in Null(A), because

$$A(\mathbf{y} - \mathbf{p}) = A\mathbf{y} - A\mathbf{p} = \mathbf{b} - \mathbf{b} = \mathbf{0}$$

Therefore, $\mathbf{y} = \mathbf{p} + (\mathbf{y} - \mathbf{p})$ is an element of $\mathbf{p} + \text{Null}(A)$. Hence, $S = \mathbf{p} + \text{Null}(A)$, as asserted.

■ EXAMPLE 47   Verify Theorem 19 for the system

$$\begin{bmatrix} -2 & 4 & : & -6 \\ 1 & -2 & : & 3 \end{bmatrix}$$

SOLUTION   The solution set of the system is $S = \left\{ \begin{bmatrix} 3 + 2r \\ r \end{bmatrix}, r \in \mathbf{R} \right\}$ and the null

space of the coefficient matrix is $\text{Null}(A) = \left\{ \begin{bmatrix} 2r \\ r \end{bmatrix}, r \in \mathbf{R} \right\}$. (Check.) Theorem 19

holds in this case, because $\begin{bmatrix} 3 + 2r \\ r \end{bmatrix} = \begin{bmatrix} 3 \\ 0 \end{bmatrix} + \begin{bmatrix} 2r \\ r \end{bmatrix}$, and $\begin{bmatrix} 3 \\ 0 \end{bmatrix}$ is a particular solution

of the system.

The Row-Vector Computation of Ax

Sometimes we wish to compute only one component, say the ith, of the product $A\mathbf{x}$ without having to deal with the entire linear combination of the definition. This is easily done as follows: Take the ith row of A and $\mathbf{x}$. Multiply their corresponding entries together, and add all the products.

■ EXAMPLE 48   Compute components 1 and 3 of the product

$$\begin{bmatrix} 2 & 0 & 1 \\ 2 & 1 & 2 \\ 9 & 5 & 8 \\ 7 & 6 & 4 \end{bmatrix} \begin{bmatrix} 2 \\ 4 \\ 3 \end{bmatrix}$$

SOLUTION

$$\begin{bmatrix} 2 & 0 & 1 \\ 2 & 1 & 2 \\ 9 & 5 & 8 \\ 7 & 6 & 4 \end{bmatrix} \begin{bmatrix} 2 \\ 4 \\ 3 \end{bmatrix} = \begin{bmatrix} 7 \\ \cdot \\ \cdot \\ \cdot \end{bmatrix} \qquad \begin{bmatrix} 2 & 0 & 1 \\ 2 & 1 & 2 \\ 9 & 5 & 8 \\ 7 & 6 & 4 \end{bmatrix} \begin{bmatrix} 2 \\ 4 \\ 3 \end{bmatrix} = \begin{bmatrix} \cdot \\ \cdot \\ 62 \\ \cdot \end{bmatrix}$$

$$2 \cdot 2 + 0 \cdot 4 + 1 \cdot 3 = 7 \qquad\qquad 9 \cdot 2 + 5 \cdot 4 + 8 \cdot 3 = 62$$

Ax in Terms of the Dot Product

We see that $A\mathbf{x}$ is the vector with components the *dot product* of each row of A with $\mathbf{x}$. So, if $\mathbf{r}_1, \dots, \mathbf{r}_m$ are the rows of A, then

$$A\mathbf{x} = \begin{bmatrix} \mathbf{r}_1 \\ \vdots \\ \mathbf{r}_m \end{bmatrix} \mathbf{x} = \begin{bmatrix} \mathbf{r}_1 \cdot \mathbf{x} \\ \vdots \\ \mathbf{r}_m \cdot \mathbf{x} \end{bmatrix}$$

A**x** with CAS

Maple

```
> with(linalg) :
> A:=matrix([[2,-3],[1,4]]):x:=vector([5,-6]);
> evalm(A&*x);
```

$$[28, -19]$$

Mathematica

```
In[1]:=
     A={{2,-3},{1,4}};x={{5},{-6}};
In[2]:= A . x
Out[2]=
     {{28}, {-19}}
```

MATLAB

```
>> A = [2 -3; 1 4]; x = [5; -6];
>> A*x
ans =
   28
  -19
```

Exercises 2.5

Let

$$A = \begin{bmatrix} -3 & -2 \\ -1 & 0 \\ 5 & -3 \end{bmatrix}, \quad B = \begin{bmatrix} -3 & 7 & -3 & -2 \\ 4 & 6 & -1 & 0 \end{bmatrix}$$

$$\mathbf{u} = \begin{bmatrix} 4 \\ -1 \end{bmatrix}, \quad \mathbf{v} = \begin{bmatrix} 100 \\ 200 \\ 300 \end{bmatrix}, \quad \mathbf{w} = \begin{bmatrix} 1 \\ 2 \\ 3 \\ 4 \end{bmatrix}$$

1. Compute, if possible, $A\mathbf{u}, A\mathbf{v}, A\mathbf{w}$.
2. Compute, if possible, $B\mathbf{u}, B\mathbf{v}, B\mathbf{w}$.
3. Write $A\mathbf{x} = \mathbf{v}$ as a linear system.
4. Write $B\mathbf{z} = \mathbf{u}$ as a linear system.
5. Write the system in the form $A\mathbf{x} = \mathbf{b}$.

$$x - 7y = -5$$
$$-2x + 4y = 0$$

6. Write the system in the form $A\mathbf{x} = \mathbf{b}$.

$$x - 4y = -8$$
$$-2y = 7$$
$$x + y = 10$$

7. Find all possible n-vectors of the form $\begin{bmatrix} 1 & 2 \\ -1 & 4 \end{bmatrix} \begin{bmatrix} x \\ y \end{bmatrix}$. What is n?

8. Find all possible n-vectors of the form $\begin{bmatrix} -2 & 1 & -4 \\ 2 & -1 & 4 \end{bmatrix} \begin{bmatrix} x \\ y \\ z \end{bmatrix}$. What is n?

9. Find all possible n-vectors of the form

$$\begin{bmatrix} 1 & 0 & 1 & 2 \\ 2 & 0 & -1 & 4 \end{bmatrix} \begin{bmatrix} x \\ y \\ z \\ w \end{bmatrix}. \text{ What is } n?$$

10. Prove Parts 2 and 3 of Theorem 17.

11. Referring to Example 43 compute and interpret the following product.

$$\begin{bmatrix} 2 & 10 & 13 & 8 \\ 3 & 15 & 15 & 11 \\ 1 & 12 & 15 & 12 \end{bmatrix} \begin{bmatrix} 1 \\ 1 \\ 1 \\ 1 \end{bmatrix}$$

12. A sports company sells bicycles of types 1, 2, 3, 4 at three outlets. The outlets are supplied as follows:

	Outlet 1	Outlet 2	Outlet 3
Type			
Bike 1	25	15	35
Bike 2	20	25	25
Bike 3	15	35	20
Bike 4	20	30	10

If M is this matrix, compute and interpret the products

$$M\begin{bmatrix} 1 \\ 1 \\ 1 \end{bmatrix}, M\begin{bmatrix} 1 \\ 0 \\ 0 \end{bmatrix}, M\begin{bmatrix} 0 \\ 1 \\ 0 \end{bmatrix}, M\begin{bmatrix} 0 \\ 0 \\ 1 \end{bmatrix}.$$

13. Find the first and the last components of the product.

$$\begin{bmatrix} 2 & -2 & 2 \\ 3 & -3 & 3 \\ 4 & -4 & 4 \\ 5 & -5 & 5 \\ 6 & -6 & 6 \end{bmatrix} \begin{bmatrix} 1 \\ -1 \\ 1 \end{bmatrix}$$

Let $\mathbf{u} = \begin{bmatrix} 0 \\ 1 \\ 1 \end{bmatrix}$, $\mathbf{v} = \begin{bmatrix} 1 \\ 1 \\ 0 \end{bmatrix}$, and $\mathbf{w} = \begin{bmatrix} 1 \\ 1 \\ 1 \end{bmatrix}$.

14. Which of $\mathbf{u}$, $\mathbf{v}$, and $\mathbf{w}$ are in the null space of
$$A = \begin{bmatrix} 2 & -2 & 2 \\ 0 & -3 & 3 \end{bmatrix}?$$

15. Which of $\mathbf{u}$, $\mathbf{v}$, and $\mathbf{w}$ are in the null space of
$$A = \begin{bmatrix} 2 & -2 & 2 \\ 1 & -4 & 4 \\ 0 & 1 & -1 \end{bmatrix}?$$

16. Which of $\mathbf{u}$, $\mathbf{v}$, and $\mathbf{w}$ are in the null space of
$$A = \begin{bmatrix} 1 & -1 & 1 \\ 2 & -2 & 2 \\ 3 & -3 & 3 \\ 4 & -4 & 5 \end{bmatrix}?$$

17. Verify Theorem 19 for the system
$$\begin{bmatrix} 2 & -6 & : & 20 \\ -1 & 3 & : & -10 \end{bmatrix}.$$

18. Verify Theorem 19 for the system
$$\begin{bmatrix} 1 & -1 & 2 & : & 0 \\ 2 & -1 & 6 & : & 1 \\ -2 & 4 & 0 & : & 2 \end{bmatrix}.$$

19. Show that if $\mathbf{x}_1$ and $\mathbf{x}_2$ are solutions of the system $A\mathbf{x} = \mathbf{b}$, then $\mathbf{x}_2 - \mathbf{x}_1$ is a solution of the associated homogeneous system $A\mathbf{x} = \mathbf{0}$.

20. If the set of solutions of $A\mathbf{x} = \mathbf{b}$ satisfies the claims of Theorem 18, what can you say about $\mathbf{b}$?

2.6   Cross Product

Reader's Goals for This Section

1. To compute the cross product of two vectors and understand its geometry.
2. To understand and be able to use the basic properties of the cross product.

In this section we introduce the cross product. Although limited to 3-vectors, the cross product has many applications in engineering, physics, and mathematics.

CONVENTION

All vectors in this section are space vectors.

Right- and Left-Handed Systems

There are two kinds of coordinate systems in 3-space, the **right-handed** and the **left-handed** . A right-handed coordinate system is one with the positive semi-axes labeled as follows: When the fingers of a right hand placed along the positive x-direction curl toward the positive y-direction, the thumb should point toward the positive z-direction. For a left-handed system, the right hand is replaced with a left (Fig. 2.20).

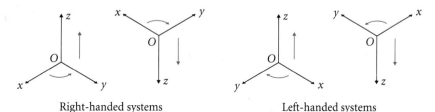

Right-handed systems   Left-handed systems

Figure 2.20

The Determinant Notation

The following *determinant* notation can be very handy in expressing some of the basic formulas in this section. First, we use the notation

$$\begin{vmatrix} a & b \\ c & d \end{vmatrix} = ad - bc$$

For example,

$$\begin{vmatrix} 1 & 2 \\ 3 & 4 \end{vmatrix} = 1 \cdot 4 - 2 \cdot 3 = -2$$

Second, we use the notation

$$\begin{vmatrix} a_1 & b_1 & c_1 \\ a_2 & b_2 & c_2 \\ a_3 & b_3 & c_3 \end{vmatrix} = a_1 \begin{vmatrix} b_2 & c_2 \\ b_3 & c_3 \end{vmatrix} - b_1 \begin{vmatrix} a_2 & c_2 \\ a_3 & c_3 \end{vmatrix} + c_1 \begin{vmatrix} a_2 & b_2 \\ a_3 & b_3 \end{vmatrix}$$

$$= a_1(b_2 c_3 - c_2 b_3) - b_1(a_2 c_3 - c_2 a_3) + c_1(a_2 b_3 - b_2 a_3)$$

For example,

$$\begin{vmatrix} 1 & 2 & 3 \\ 2 & -2 & 1 \\ 4 & -1 & 0 \end{vmatrix} = 1 \begin{vmatrix} -2 & 1 \\ -1 & 0 \end{vmatrix} - 2 \begin{vmatrix} 2 & 1 \\ 4 & 0 \end{vmatrix} + 3 \begin{vmatrix} 2 & -2 \\ 4 & -1 \end{vmatrix} = 27$$

Cross Product

DEFINITION

(Cross Product)

Let $\mathbf{u} = (u_1, u_2, u_3)$ and $\mathbf{v} = (v_1, v_2, v_3)$. The **cross product** $\mathbf{u} \times \mathbf{v}$ is the *vector* with components

$$\mathbf{u} \times \mathbf{v} = (u_2 v_3 - u_3 v_2, u_3 v_1 - u_1 v_3, u_1 v_2 - u_2 v_1)$$

This relationship may also be expressed in determinant notation:

$$\mathbf{u} \times \mathbf{v} = \begin{vmatrix} \mathbf{i} & \mathbf{j} & \mathbf{k} \\ u_1 & u_2 & u_3 \\ v_1 & v_2 & v_3 \end{vmatrix} = \begin{vmatrix} u_2 & u_3 \\ v_2 & v_3 \end{vmatrix} \mathbf{i} - \begin{vmatrix} u_1 & u_3 \\ v_1 & v_3 \end{vmatrix} \mathbf{j} + \begin{vmatrix} u_1 & u_2 \\ v_1 & v_2 \end{vmatrix} \mathbf{k}$$

which is the same as

$$\mathbf{u} \times \mathbf{v} = \left(\begin{vmatrix} u_2 & u_3 \\ v_2 & v_3 \end{vmatrix}, -\begin{vmatrix} u_1 & u_3 \\ v_1 & v_3 \end{vmatrix}, \begin{vmatrix} u_1 & u_2 \\ v_1 & v_2 \end{vmatrix} \right)$$

■ EXAMPLE 49   Let $\mathbf{u} = (2, -1, 3)$ and $\mathbf{v} = (1, -2, -1)$. Compute $\mathbf{u} \times \mathbf{v}$.

SOLUTION

$$\mathbf{u} \times \mathbf{v} = \begin{vmatrix} \mathbf{i} & \mathbf{j} & \mathbf{k} \\ 2 & -1 & 3 \\ 1 & -2 & -1 \end{vmatrix} = \begin{vmatrix} -1 & 3 \\ -2 & -1 \end{vmatrix} \mathbf{i} - \begin{vmatrix} 2 & 3 \\ 1 & -1 \end{vmatrix} \mathbf{j} + \begin{vmatrix} 2 & -1 \\ 1 & -2 \end{vmatrix} \mathbf{k}$$

$$= 7\mathbf{i} + 5\mathbf{j} - 3\mathbf{k} = (7, 5, -3)$$

■ EXAMPLE 50   Compute $\mathbf{i} \times \mathbf{j}$.

SOLUTION

$$\mathbf{i} \times \mathbf{j} = \begin{vmatrix} \mathbf{i} & \mathbf{j} & \mathbf{k} \\ 1 & 0 & 0 \\ 0 & 1 & 0 \end{vmatrix} = \begin{vmatrix} 0 & 0 \\ 1 & 0 \end{vmatrix} \mathbf{i} - \begin{vmatrix} 1 & 0 \\ 0 & 0 \end{vmatrix} \mathbf{j} + \begin{vmatrix} 1 & 0 \\ 0 & 1 \end{vmatrix} \mathbf{k} = 0\mathbf{i} - 0\mathbf{j} + 1\mathbf{k} = \mathbf{k}$$

Hence, $\mathbf{i} \times \mathbf{j} = \mathbf{k}$.

In fact, all possible cross products formed by $\mathbf{i}, \mathbf{j}$ and $\mathbf{k}$ satisfy

$$\mathbf{i} \times \mathbf{j} = \mathbf{k} \qquad \mathbf{j} \times \mathbf{i} = -\mathbf{k}$$
$$\mathbf{j} \times \mathbf{k} = \mathbf{i} \qquad \mathbf{k} \times \mathbf{j} = -\mathbf{i}$$
$$\mathbf{k} \times \mathbf{i} = \mathbf{j} \qquad \mathbf{i} \times \mathbf{k} = -\mathbf{j}$$

mnemonically, from Fig. 2.21: As we move clockwise, the cross product of two vectors gives the third. As we move counterclockwise, the cross product of two vectors gives the opposite of the third.

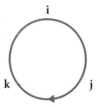

Figure 2.21

THEOREM 20   **(Properties of Cross Product)**

Let $\mathbf{u} = (u_1, u_2, u_3)$, $\mathbf{v} = (v_1, v_2, v_3)$ and $\mathbf{w} = (w_1, w_2, w_3)$ be space vectors and let c be any scalar. Then

1. $\mathbf{u} \times \mathbf{v} = -\mathbf{v} \times \mathbf{u}$;
2. $\mathbf{u} \times (\mathbf{v} + \mathbf{w}) = \mathbf{u} \times \mathbf{v} + \mathbf{u} \times \mathbf{w}$;
3. $(\mathbf{u} + \mathbf{v}) \times \mathbf{w} = \mathbf{u} \times \mathbf{w} + \mathbf{v} \times \mathbf{w}$;
4. $c(\mathbf{u} \times \mathbf{v}) = (c\mathbf{u}) \times \mathbf{v} = \mathbf{u} \times (c\mathbf{v})$;
5. $\mathbf{0} \times \mathbf{u} = \mathbf{u} \times \mathbf{0} = \mathbf{0}$;
6. $\mathbf{u} \times \mathbf{u} = \mathbf{0}$;
7. $\mathbf{u} \times (\mathbf{v} \times \mathbf{w}) = (\mathbf{u} \cdot \mathbf{w})\mathbf{v} - (\mathbf{u} \cdot \mathbf{v})\mathbf{w}$;
8. $\mathbf{u} \cdot (\mathbf{v} \times \mathbf{w}) = \begin{vmatrix} u_1 & u_2 & u_3 \\ v_1 & v_2 & v_3 \\ w_1 & w_2 & w_3 \end{vmatrix}$.

PROOF   We prove Parts 1, 7, and 8. The rest of the proof is left as an exercise.
For Part 1,

$$\mathbf{u} \times \mathbf{v} = (u_2 v_3 - u_3 v_2, u_3 v_1 - u_1 v_3, u_1 v_2 - u_2 v_1)$$
$$= -(v_2 u_3 - v_3 u_2, v_3 u_1 - v_1 u_3, v_1 u_2 - v_2 u_1) = -\mathbf{v} \times \mathbf{u}$$

For Part 7, because

$$\mathbf{v} \times \mathbf{w} = (v_2 w_3 - v_3 w_2, v_3 w_1 - v_1 w_3, v_1 w_2 - v_2 w_1)$$

the first component of $\mathbf{u} \times (\mathbf{v} \times \mathbf{w})$ is

$$u_2(v_1 w_2 - v_2 w_1) - u_3(v_3 w_1 - v_1 w_3) = u_2 v_1 w_2 - u_2 v_2 w_1 - u_3 v_3 w_1 + u_3 v_1 w_3$$

On the other hand, the first component of $(\mathbf{u} \cdot \mathbf{w})\mathbf{v} - (\mathbf{u} \cdot \mathbf{v})\mathbf{w}$ is

$$(u_1 w_1 + u_2 w_2 + u_3 w_3)v_1 - (u_1 v_1 + u_2 v_2 + u_3 v_3)w_1 = u_2 w_2 v_1 + u_3 w_3 v_1 - u_2 v_2 w_1 - u_3 v_3 w_1$$

Hence $\mathbf{u} \times (\mathbf{v} \times \mathbf{w})$ and $(\mathbf{u} \cdot \mathbf{w})\mathbf{v} - (\mathbf{u} \cdot \mathbf{v})\mathbf{w}$ have equal first components. The equality of the last two components is proved similarly.
For Part 8,

$$\mathbf{u} \cdot (\mathbf{v} \times \mathbf{w}) = (u_1, u_2, u_3) \cdot \left(\begin{vmatrix} v_2 & v_3 \\ w_2 & w_3 \end{vmatrix}, -\begin{vmatrix} v_1 & v_3 \\ w_1 & w_3 \end{vmatrix}, \begin{vmatrix} v_1 & v_2 \\ w_1 & w_2 \end{vmatrix} \right)$$

$$= u_1 \begin{vmatrix} v_2 & v_3 \\ w_2 & w_3 \end{vmatrix} - u_2 \begin{vmatrix} v_1 & v_3 \\ w_1 & w_3 \end{vmatrix} + u_3 \begin{vmatrix} v_1 & v_2 \\ w_1 & w_2 \end{vmatrix} = \begin{vmatrix} u_1 & u_2 & u_3 \\ v_1 & v_2 & v_3 \\ w_1 & w_2 & w_3 \end{vmatrix}$$

REMARK

1. According to Part 1 of Theorem 20, the cross product is **not** commutative but **anticommutative**.

2. The cross product is **not associative**.[5] This means that, in general, $(\mathbf{u} \times \mathbf{v}) \times \mathbf{w} \neq \mathbf{u} \times (\mathbf{v} \times \mathbf{w})$. For example,

$$(\mathbf{i} \times \mathbf{j}) \times \mathbf{j} = \mathbf{k} \times \mathbf{j} = -\mathbf{i}, \quad \text{whereas} \quad \mathbf{i} \times (\mathbf{j} \times \mathbf{j}) = \mathbf{i} \times \mathbf{0} = \mathbf{0}$$

Most of the identities involving cross products can be deduced from Theorem 20. For instance, it is easy to show that Property 8 implies

$$\mathbf{u} \cdot (\mathbf{u} \times \mathbf{v}) = 0 \quad \text{and} \quad \mathbf{v} \cdot (\mathbf{u} \times \mathbf{v}) = 0$$

Therefore, $\mathbf{u} \times \mathbf{v}$ is *orthogonal* to $\mathbf{u}$ and $\mathbf{v}$. So,

$$\mathbf{u} \perp (\mathbf{u} \times \mathbf{v}) \quad \text{and} \quad \mathbf{v} \perp (\mathbf{u} \times \mathbf{v})$$

If $\mathbf{u}$ and $\mathbf{v}$ are nonzero vectors, then the direction of $\mathbf{u} \times \mathbf{v}$ is perpendicular to the plane defined by $\mathbf{u}$ and $\mathbf{v}$. Furthermore, it can be shown that for a right-handed coordinate system, the vectors $\mathbf{u}, \mathbf{v}$ and $\mathbf{u} \times \mathbf{v}$ form also a right-handed system (Fig. 2.22). This determines the direction of the cross product. Next, we determine its length.

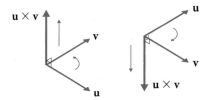

Figure 2.22   Direction of the cross product for a right-handed system.

By direct computation, Part 8 of Theorem 20 implies

$$\mathbf{u} \cdot (\mathbf{v} \times \mathbf{w}) = \mathbf{v} \cdot (\mathbf{w} \times \mathbf{u}) = \mathbf{w} \cdot (\mathbf{u} \times \mathbf{v}) \qquad (2.13)$$

(This property can also be memorized by a cyclic permutation diagram, as in Fig. 2.21, with $\mathbf{u}, \mathbf{v}, \mathbf{w}$ replacing $\mathbf{i}, \mathbf{j}, \mathbf{k}$.)

THEOREM 21

(Length of Cross Product)

1. The following identity holds:

$$\|\mathbf{u} \times \mathbf{v}\|^2 = \|\mathbf{u}\|^2 \|\mathbf{v}\|^2 - (\mathbf{u} \cdot \mathbf{v})^2 \qquad \textbf{Lagrange's identity}$$

2. If θ is the angle between $\mathbf{u}$ and $\mathbf{v}$, then

$$\|\mathbf{u} \times \mathbf{v}\| = \|\mathbf{u}\| \|\mathbf{v}\| \sin \theta \qquad (2.14)$$

---

[5] For a relationship between the nonassociativity of the cross product and the four-color problem, see Louis H. Kauffman's, "Map Coloring and the Vector Cross Product," *Journal of Combinatorial Theory*, series B, 48 (1990): 145–154.

PROOF

1. $\|\mathbf{u} \times \mathbf{v}\|^2 = (\mathbf{u} \times \mathbf{v}) \cdot (\mathbf{u} \times \mathbf{v})$

$= \mathbf{u} \cdot (\mathbf{v} \times (\mathbf{u} \times \mathbf{v}))$ by Equation (2.13)

$= \mathbf{u} \cdot ((\mathbf{v} \cdot \mathbf{v})\mathbf{u} - (\mathbf{v} \cdot \mathbf{u})\mathbf{v})$ by Theorem 20, Part 7

$= (\mathbf{v} \cdot \mathbf{v})(\mathbf{u} \cdot \mathbf{u}) - (\mathbf{v} \cdot \mathbf{u})(\mathbf{u} \cdot \mathbf{v})$

$= \|\mathbf{u}\|^2 \|\mathbf{v}\|^2 - (\mathbf{u} \cdot \mathbf{v})^2$

2. By Part 1 and Equation (2.7), Section 2.2, we have

$$\|\mathbf{u} \times \mathbf{v}\|^2 = \|\mathbf{u}\|^2 \|\mathbf{v}\|^2 - (\mathbf{u} \cdot \mathbf{v})^2 = \|\mathbf{u}\|^2 \|\mathbf{v}\|^2 - \|\mathbf{u}\|^2 \|\mathbf{v}\|^2 \cos^2 \theta$$

$$= \|\mathbf{u}\|^2 \|\mathbf{v}\|^2 (1 - \cos^2 \theta) = \|\mathbf{u}\|^2 \|\mathbf{v}\|^2 \sin^2 \theta$$

Therefore, $\|\mathbf{u} \times \mathbf{v}\| = \|\mathbf{u}\| \|\mathbf{v}\| \sin \theta$.

Part 2 of Theorem 21 determines the length of $\mathbf{u} \times \mathbf{v}$. Geometrically, this length is the area of the parallelogram defined by $\mathbf{u}$ and $\mathbf{v}$ (Fig. 2.23). Hence the area, A, of the parallelogram with adjacent sides $\mathbf{u}$ and $\mathbf{v}$ is

$$A = \|\mathbf{u} \times \mathbf{v}\|$$

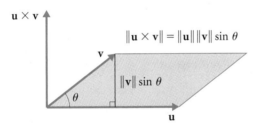

Figure 2.23

COROLLARY 22

$$\|\mathbf{u} \times \mathbf{v}\| \leq \|\mathbf{u}\| \|\mathbf{v}\|$$

PROOF   Because $0 \leq \theta \leq \pi, 0 \leq \sin \theta \leq 1$. Therefore, Equation (2.14) implies

$$\|\mathbf{u} \times \mathbf{v}\| = \|\mathbf{u}\| \|\mathbf{v}\| \sin \theta \leq \|\mathbf{u}\| \|\mathbf{v}\| \cdot 1 = \|\mathbf{u}\| \|\mathbf{v}\|$$

COROLLARY 23   (Criterion for Two Parallel Vectors)

Two nonzero vectors $\mathbf{u}$ and $\mathbf{v}$ are parallel if and only if $\mathbf{u} \times \mathbf{v} = \mathbf{0}$.

PROOF   By Equation (2.14),

$$\mathbf{u} \times \mathbf{v} = \mathbf{0} \Leftrightarrow \|\mathbf{u} \times \mathbf{v}\| = 0 \Leftrightarrow \|\mathbf{u}\| \|\mathbf{v}\| \sin \theta = 0$$

$$\Leftrightarrow \sin \theta = 0 \Leftrightarrow \theta = 0, \pi \Leftrightarrow \mathbf{u} \text{ and } \mathbf{v} \text{ are parallel}$$

Applications to Geometry

▪ EXAMPLE 51   (Area of a Parallelogram)   Compute the area of the parallelo-gram with adjacent sides PQ and PR, where $P(2, 1, 0)$, $Q(1, -2, 1)$, and $R(-2, 2, 4)$.

SOLUTION   By Part 2 of Theorem 21 the area is $\|\overrightarrow{PQ} \times \overrightarrow{PR}\|$. But

$$\|\overrightarrow{PQ} \times \overrightarrow{PR}\| = \|(-1, -3, 1) \times (-4, 1, 4)\| = \|(-13, 0, -13)\| = 13\sqrt{2}$$

Hence, the area is $13\sqrt{2}$ units.

▪ EXAMPLE 52   (Area of a Triangle)   Compute the area of the triangle with ver-tices the tips of $\mathbf{i}$, $\mathbf{j}$, and $\mathbf{k}$.

SOLUTION $\mathbf{j} - \mathbf{i}$ and $\mathbf{k} - \mathbf{i}$ are two sides of the triangle. Therefore, $\|(\mathbf{j} - \mathbf{i}) \times (\mathbf{k} - \mathbf{i})\|$ is the area of the parallelogram defined by these sides. One-half of that area is the area of the triangle.

$$\frac{1}{2}\|(\mathbf{j} - \mathbf{i}) \times (\mathbf{k} - \mathbf{i})\| = \frac{1}{2}\|(-1, 1, 0) \times (-1, 0, 1)\| = \frac{1}{2}\|(1, 1, 1)\| = \frac{1}{2}\sqrt{3}$$

THEOREM 24

(Volume of a Parallelepiped)

Show that the volume V of the parallelepiped with adjacent sides the position vectors $\mathbf{u}, \mathbf{v}$, and $\mathbf{w}$ (Fig. 2.24) is given by

$$V = |\mathbf{u} \cdot (\mathbf{v} \times \mathbf{w})| \tag{2.15}$$

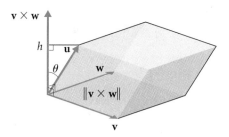

Figure 2.24

SOLUTION   Let A be the area of the base defined by $\mathbf{v}$ and $\mathbf{w}$. Let h be the height of the parallelepiped, and let θ be the angle between $\mathbf{u}$ and $\mathbf{v} \times \mathbf{w}$. Then $h = \|\mathbf{u}\| |\cos \theta|$ and $A = \|\mathbf{v} \times \mathbf{w}\|$. We have, by Equation (2.7), Section 2.2,

$$V = Ah = \|\mathbf{v} \times \mathbf{w}\| \|\mathbf{u}\| |\cos \theta| = |\mathbf{u} \cdot (\mathbf{v} \times \mathbf{w})|$$

■ EXAMPLE 53   Compute the volume of the parallelepiped with adjacent sides the position vectors $\mathbf{u} = (1, -1, 2)$, $\mathbf{v} = (0, 2, 1)$, and $\mathbf{w} = (3, -2, -1)$.

SOLUTION   By Part 8 of Theorem 20, we have

$$\mathbf{u} \cdot (\mathbf{v} \times \mathbf{w}) = \begin{vmatrix} 1 & -1 & 2 \\ 0 & 2 & 1 \\ 3 & -2 & -1 \end{vmatrix} = -15$$

Hence, the volume V of the parallelepiped is $|\mathbf{u} \cdot (\mathbf{v} \times \mathbf{w})| = |-15| = 15$.

Equation (2.15) implies an easy criterion for three points being coplanar. If $\mathbf{u}$, $\mathbf{v}$, and $\mathbf{w}$ are coplanar, then the volume of the parallelepiped with adjacent sides $\mathbf{u}$, $\mathbf{v}$, and $\mathbf{w}$ is zero (because the height is zero). Conversely, the only way this volume is zero is if $\mathbf{u}$, $\mathbf{v}$, and $\mathbf{w}$ are coplanar.

THEOREM 25   **(Criterion for Three Coplanar Vectors)**

The vectors $\mathbf{u}$, $\mathbf{v}$, and $\mathbf{w}$ are coplanar if and only if $\mathbf{u} \cdot (\mathbf{v} \times \mathbf{w}) = 0$.

Cross Product with CAS

Maple

```
> with(linalg):
> crossprod([1,2,3],[-3,2,-1]);

                    [-8  -8  8]
```

Mathematica

```
In[1]:=
      <<LinearAlgebra`CrossProduct`
In[2]:=
      Cross[{1,2,3},{-3,2,-1}]
Out[1]=
      {-8, -8, 8}
```

MATLAB

```
>> cross([1,2,3],[-3,2,-1])
ans =
    -8 -8 8
```

Exercises 2.6

1. Let $\mathbf{u} = (-1, 2, -2)$, $\mathbf{v} = (4, -3, 5)$, and $\mathbf{w} = (-4, -2, 0)$. Compute the expressions.

   $\mathbf{u} \times \mathbf{v}$ $\qquad$ $(\mathbf{u} \times \mathbf{v}) \times \mathbf{w}$

   $\mathbf{u} \times (\mathbf{v} \times \mathbf{w})$ $\qquad$ $(\mathbf{u} + \mathbf{v}) \times \mathbf{w}$

   $\mathbf{u} \times \mathbf{w} + \mathbf{v} \times \mathbf{w}$ $\qquad$ $\mathbf{u} + (\mathbf{v} \times \mathbf{w})$

2. Let $\mathbf{u} \times \mathbf{v} = (2, 1, -5)$. Find

   $\mathbf{v} \times \mathbf{u}$ $\qquad$ $-2(\mathbf{v} \times \mathbf{u})$

   $\mathbf{u} \times (10\mathbf{v})$ $\qquad$ $(-2\mathbf{u}) \times (10\mathbf{v})$

   $\|\mathbf{u} \times \mathbf{v}\|$ $\qquad$ $\|(2\mathbf{v}) \times \mathbf{u}\|$

3. Let $\mathbf{u} \cdot (\mathbf{v} \times \mathbf{w}) = -5$. Find

   $(\mathbf{v} \times \mathbf{w}) \cdot \mathbf{u}$ $\qquad$ $\mathbf{u} \cdot (\mathbf{w} \times \mathbf{v})$

   $\mathbf{v} \cdot (\mathbf{u} \times \mathbf{w})$ $\qquad$ $\mathbf{w} \cdot (\mathbf{u} \times \mathbf{v})$

Let $\mathbf{u}$, $\mathbf{v}$, and $\mathbf{w}$ be 3-vectors.

4. Which of the following expressions are *undefined* and why?

   $\mathbf{u} \times \mathbf{u} \times \mathbf{u}$ $\qquad$ $\mathbf{u} \times \mathbf{v} \times \mathbf{w}$

   $(\mathbf{u} \times \mathbf{u}) \times \mathbf{u}$ $\qquad$ $\mathbf{u} \times (\mathbf{v} \times \mathbf{w})$

   $\mathbf{u} \cdot (\mathbf{u} \times \mathbf{w})$ $\qquad$ $\mathbf{u} \times (\mathbf{u} \cdot \mathbf{w})$

   $(\mathbf{u} \times \mathbf{u}) \times (\mathbf{v} \times \mathbf{v})$ $\qquad$ $(\mathbf{u} \times \mathbf{w}) \cdot (\mathbf{v} \times \mathbf{w})$

5. Verify Lagrange's Identity for $\mathbf{u} = (-3, 4, 1)$ and $\mathbf{v} = (0, 5, -6)$.

6. Find the sine of the angle between $\mathbf{u}$ and $\mathbf{v}$ for the given values.

   a. $\mathbf{u} = (6, 1, -2)$, $\mathbf{v} = (7, 5, -1)$

   b. $\mathbf{u} = (9, -7, 4)$, $\mathbf{v} = (0, -4, 3)$

7. Find a unit vector perpendicular to the plane defined by $\mathbf{u} = (3, -4, 0)$ and $\mathbf{v} = (7, 5, -4)$.

8. Find a vector of length 4 that is perpendicular to the plane defined by $\mathbf{u} = (1, -1, 1)$ and $\mathbf{v} = (-1, 1, 0)$.

9. What is the area of the triangle with vertices $(1, 1, 1)$, $(1, -1, -1)$, and $(0, 1, -1)$?

10. What is the area of the parallelogram with adjacent sides $\overrightarrow{PQ}$ and $\overrightarrow{PR}$, where $P(1, 1, 1)$, $Q(1, -1, -1)$, and $R(0, 1, -1)$?

11. Find the volume of the parallelepiped with adjacent sides the position vectors $(1, -2, 3)$, $(2, 0, -5)$, and $(0, 4, -1)$.

12. Use the cross product to show that $(1, 2, -1)$ and $(-2, -4, 2)$ are parallel.

13. Use Theorem 25 to determine which of the vectors $\mathbf{u}$, $\mathbf{v}$, and $\mathbf{w}$ are coplanar.

    a. $\mathbf{u} = (-1, -1, 9)$, $\mathbf{v} = (0, 1, -3)$, $\mathbf{w} = (-1, 2, 0)$

    b. $\mathbf{u} = (1, -1, 1)$, $\mathbf{v} = (1, 0, 2)$, $\mathbf{w} = (1, -1, 0)$

14. Is it true that if $\mathbf{u} \times \mathbf{v} = \mathbf{0}$, then either $\mathbf{u} = \mathbf{0}$ or $\mathbf{v} = \mathbf{0}$?

15. Is it true that if $\mathbf{u} \times \mathbf{v} = \mathbf{u} \times \mathbf{w}$ and $\mathbf{u} \neq \mathbf{0}$, then $\mathbf{v} = \mathbf{w}$?

16. Complete the proof of Theorem 20.

17. Prove the identity

    $$(\mathbf{u} \times \mathbf{v}) \cdot \mathbf{w} = \mathbf{u} \cdot (\mathbf{v} \times \mathbf{w})$$

18. (**Jacobi's Identity**) Prove the identity

    $$(\mathbf{u} \times \mathbf{v}) \times \mathbf{w} + (\mathbf{v} \times \mathbf{w}) \times \mathbf{u} + (\mathbf{w} \times \mathbf{u}) \times \mathbf{v} = \mathbf{0}$$

19. (**Euler's Formula**) Let $\mathbf{u}$, $\mathbf{v}$, and $\mathbf{w}$ be three adjacent sides with common origin of a tetrahedron (Fig. 2.25). Show that the volume V is given by

    $$V = \frac{1}{6} |\mathbf{u} \cdot (\mathbf{v} \times \mathbf{w})|$$

    [*Hint:* We know from geometry that $V = \frac{1}{3}(\text{base area})(\text{height}).$]

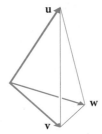

Figure 2.25

2.7   Lines, Planes, and Hyperplanes

Reader's Goals for This Section

1. To compute the equations of (a) a straight line, and (b) a plane and find the normal of a plane.
2. To realize that often questions about planes reduce to questions about their normals.
3. To compute the equation of (a) a straight line, and (b) a hyperplane in $\mathbf{R}^n$.

Lines

In this section we discuss the equation of a straight line l passing through a given point $P(x_0, y_0, z_0)$ parallel to a given *nonzero* vector $\mathbf{n} = (a, b, c)$. Let $X(x, y, z)$ be any point in l and let $\mathbf{p} = (x_0, y_0, z_0)$ and $\mathbf{x} = (x, y, z)$. The scalar multiples $t\mathbf{n}$ ($-\infty < t < \infty$) represent all possible vectors parallel to $\mathbf{n}$. Since $\mathbf{x} - \mathbf{p}$ is parallel to $\mathbf{n}$, we must have $\mathbf{x} - \mathbf{p} = t\mathbf{n}$ (Fig. 2.26) for some scalar t. Therefore,

$$\mathbf{x} = \mathbf{p} + t\mathbf{n} \tag{2.16}$$

This vector equation is called a **parametric equation of the straight line**, where t is the **parameter** of the equation. The parametric equation can also be expressed in terms of components; $(x, y, z) = (x_0, y_0, z_0) + t(a, b, c)$ is equivalent to

$$
\begin{aligned}
x &= x_0 + ta \\
y &= y_0 + tb \\
z &= z_0 + tc
\end{aligned}
\tag{2.17}
$$

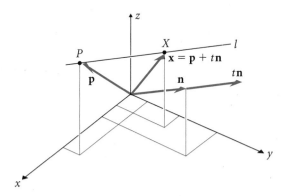

Figure 2.26

Equation (2.16) is also valid for plane lines. If $\mathbf{x} = (x, y)$, $\mathbf{p} = (x_0, y_0)$, and $\mathbf{n} = (a, b) \neq \mathbf{0}$, then $(x, y) = (x_0, y_0) + t(a, b)$, or

$$
\begin{aligned}
x &= x_0 + ta \\
y &= y_0 + tb
\end{aligned}
\tag{2.18}
$$

■ EXAMPLE 54   Let l be the line through $(1, -1, 2)$ in the direction of $(1, 1, 1)$. Find each of the following.

(a) A parametric equation of l
(b) Two points of l
(c) The intersection of l with the coordinate planes

SOLUTION

(a) Because $\mathbf{n} = (1, 1, 1)$ and $\mathbf{p} = (1, -1, 2)$ a parametric equation of the line is

$$\mathbf{x} = \begin{bmatrix} 1 \\ -1 \\ 2 \end{bmatrix} + t \begin{bmatrix} 1 \\ 1 \\ 1 \end{bmatrix}$$

or, in component form,

$$\begin{aligned} x &= 1 + t \\ y &= -1 + t \\ z &= 2 + t \end{aligned}$$

(b) To find points on l we need to evaluate the parameter t. For example, $t = 1$ and $t = -1$ yield $(1, -1, 2) + 1(1, 1, 1) = (2, 0, 3)$ and $(1, -1, 2) + (-1)(1, 1, 1) = (0, -2, 1)$.

(c) To find the intersection with the xy-plane we set $z = 0$. Hence $z = 2 + t = 0$, or $t = -2$. Substituting this into the first two parametric equations yields $x = -1$, $y = -3$. So, $(-1, -3, 0)$ is the intersection with the xy-plane. Similarly, we find that the intersections with the xz- and yz-planes are $(2, 0, 3)$ and $(0, -2, 1)$, respectively.

■ EXAMPLE 55   (Line Through Two Points)   Find a parametric equation of the line through the points $P(3, -1)$ and $Q(-1, 2)$.

SOLUTION   Because $\overrightarrow{PQ} = (-1, 2) - (3, -1) = (-4, 3)$ is parallel to the line, the direction vector is $\mathbf{n} = (-4, 3)$. Therefore, a parametric equation of the line is

$$\mathbf{x} = \begin{bmatrix} 3 \\ -1 \end{bmatrix} + t \begin{bmatrix} -4 \\ 3 \end{bmatrix}$$

■ EXAMPLE 56   (Parallel Lines)   Show that the lines with parametric equations

$$\begin{aligned} x &= 1 - 2t & x &= -t \\ y &= -1 + 4t \quad \text{and} \quad & y &= 2 + 2t \\ z &= 2 - 8t & z &= 7 - 4t \end{aligned}$$

are parallel.

SOLUTION   A direction vector for the first line is $(-2, 4, 8) = 2 \cdot (-1, 2, 4)$, which is a scalar multiple of the direction vector $(-1, 2, 4)$ of the second line. Both lines have the same direction, so they are parallel.

■ EXAMPLE 57   (Perpendicular Lines)   Show that the lines with parametric equations

$$
\begin{array}{ll}
x = 1 - 2t & x = -t \\
y = -1 + 4t \quad \text{and} \quad & y = 2 - 2t \\
z = 2 - 2t & z = 7 - 3t
\end{array}
$$

are perpendicular.

SOLUTION   This is true because the direction vectors $(-2, 4, -2)$ and $(-1, -2, -3)$ are orthogonal.

NOTE   The parametric equation of a line is not unique. We can use any point of the line or any vector parallel to the given direction vector.

Let us consider only plane lines for a moment. Because $\mathbf{n} \neq \mathbf{0}$, we can always eliminate t from (2.18) and get an equation of the straight line in the familiar form

$$
Ax + By = C \tag{2.19}
$$

as a relation between y and x.

For example, from the parametric set

$$
\begin{aligned}
x &= 2 - t \\
y &= 1 + 3t
\end{aligned} \tag{2.20}
$$

we can eliminate t to get

$$
y = -3x + 7 \tag{2.21}
$$

Conversely, (2.20) can be retrieved from (2.21) by letting $t = 2 - x$. Another (easier) set of parametric equations can be obtained by letting $x = t$ and $y = -3t + 7$.

$$
\begin{aligned}
x &= t \\
y &= -3t + 7
\end{aligned} \tag{2.22}
$$

Note that the two sets (2.20) and (2.22) are equivalent. Replacing t with $2 - t'$ in (2.20) yields (2.22).

In the case of space lines, eliminating t will not yield *one* equation as with plane lines. Thus, the *parametric equation is the only one available to describe a space line directly*. However, if $a \neq 0$, $b \neq 0$, and $c \neq 0$, we may still eliminate t from (2.17) to get *two* equations:

$$
\frac{x - x_0}{a} = \frac{y - y_0}{b} = \frac{z - z_0}{c}
$$

These are the **symmetric equations** of a line. They describe the line *indirectly*, as an intersection of two planes (discussed in the next subsection).

■ EXAMPLE 58   (Symmetric Equations of a Line)   Find the symmetric equations of the line passing through $(-2, 3, 1)$ in the direction of $(-1, -2, 1)$.

SOLUTION   Because $(x_0, y_0, z_0) = (-2, 3, 1)$ and $(a, b, c) = (-1, -2, 1)$, we have

$$\frac{x - (-2)}{-1} = \frac{y - 3}{-2} = \frac{z - 1}{1}$$

Planes

A nonzero vector $\mathbf{n} = (a, b, c)$ is called a **normal** to a plane $\mathcal{P}$ if it is perpendicular to $\mathcal{P}$ (Fig. 2.27). Let $P(x_0, y_0, z_0)$ be a given point in $\mathcal{P}$ and let $X(x, y, z)$ be any other point. If $\mathbf{p} = (x_0, y_0, z_0)$ and $\mathbf{x} = (x, y, z)$, then $\mathbf{x} - \mathbf{p}$ is parallel to $\mathcal{P}$ and, hence, orthogonal to the normal $\mathbf{n}$. Therefore, the dot product of $\mathbf{x} - \mathbf{p}$ and $\mathbf{n}$ is zero:

$$\mathbf{n} \cdot (\mathbf{x} - \mathbf{p}) = 0 \tag{2.23}$$

In terms of components, this equation can be rewritten in the form

$$a(x - x_0) + b(y - y_0) + c(z - z_0) = 0 \tag{2.24}$$

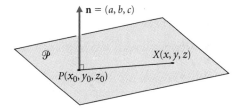

Figure 2.27

Equation (2.24) characterizes all the points $\mathbf{x}$ of $\mathcal{P}$ in terms of a normal vector $\mathbf{n}$ and a point $\mathbf{p}$ of $\mathcal{P}$. It is called a **point-normal** form of the equation of the plane $\mathcal{P}$.

This equation is analogous to the point-slope form of the equation of a straight line. Both point-slope and point-normal forms involve a given point and an inclination. In the case of the line, the inclination is expressed by the slope, whereas in the case of the plane, the inclination is expressed by a normal direction.

■ EXAMPLE 59   (Point-Normal Equation)   What is the equation of the plane passing through $(-1, 2, 3)$ and perpendicular to $(-2, 1, 4)$? Find another point on this plane.

SOLUTION   Because $\mathbf{p} = (-1, 2, 3)$ and $\mathbf{n} = (-2, 1, 4)$, Equation (2.24) yields

$$-2 \cdot (x + 1) + 1 \cdot (y - 2) + 4 \cdot (z - 3) = 0$$

A point on the plane can be obtained from any solution of this equation. For example, $x = 1, y = 2$ yields $4z - 16 = 0$, or $z = 4$. Hence $(1, 2, 4)$ is another point.

Equation (2.24) can be rewritten in the form

$$ax + by + cz + d = 0 \qquad (2.25)$$

with $d = -ax_0 - by_0 - cz_0$. This is the **general equation** of the plane. Even in this form, the coefficients of x, y, and z still give us a normal to the plane. Conversely, Equation (2.25) is the equation of a plane.

THEOREM 26

(Equation of a Plane)

If $(a, b, c) \neq \mathbf{0}$, then the graph of the equation

$$ax + by + cz + d = 0$$

is a plane with normal (a, b, c).

PROOF   If (x_0, y_0, z_0) satisfies the equation, then

$$ax + by + cz + d = 0 \quad \text{and} \quad ax_0 + by_0 + cz_0 + d = 0$$

imply

$$a(x - x_0) + b(y - y_0) + c(z - z_0) = 0 \quad \text{or} \quad (a, b, c) \cdot (x - x_0, y - y_0, z - z_0) = 0$$

Therefore, (a, b, c) is a normal to the plane containing the point (x_0, y_0, z_0) and the vector $(x - x_0, y - y_0, z - z_0)$.

■ EXAMPLE 60   (Parallel Planes)   Find the equation of the plane passing through $(1, -2, 4)$ and parallel to the plane $2x - 5y + 2z - 1 = 0$.

SOLUTION   Because the two planes are parallel, they have the same normals. The given plane has normal $(2, -5, 2)$. Therefore,

$$2 \cdot (x - 1) - 5 \cdot (y + 2) + 2 \cdot (z - 4) = 0$$

is the equation of the unknown plane.

■ EXAMPLE 61   (Plane Through Three Points)   Find the equation of the plane through the points $P(2, 0, 1), Q(1, 2, 0)$, and $R(-3, 2, 1)$.

SOLUTION   The cross product $\overrightarrow{PQ} \times \overrightarrow{PR} = (-1, 2, -1) \times (-5, 2, 0) = (2, 5, 8)$ is a normal to the plane. Hence, by the point-normal formula, $2(x - 2) + 5y + 8(z - 1) = 0$, or

$$2x + 5y + 8z - 12 = 0$$

using P as the point on the plane.

■ EXAMPLE 62   (Intersection of Two Planes)   Find the parametric equations for the line of intersection of the planes $x - y + z - 2 = 0$ and $2x + y + z + 1 = 0$.

SOLUTION   Let $z = t$. Then solving the system

$$x - y + t - 2 = 0$$
$$2x + y + t + 1 = 0$$

for x and y yields $x = -\frac{2}{3}t + \frac{1}{3}$ and $y = \frac{1}{3}t - \frac{5}{3}$. So, the parametric equations are $x = -\frac{2}{3}t + \frac{1}{3}, y = \frac{1}{3}t - \frac{5}{3}, z = t$.

DEFINITION

The **angle** between two planes is defined as the angle between two normals of the planes.

■ EXAMPLE 63   (Angle Between Two Planes)   Find the cosine of the angle between the planes $2x - y + z - 2 = 0$ and $x + 2y - z + 1 = 0$.

SOLUTION   Because $(2, -1, 1)$ and $(1, 2, -1)$ are the corresponding normals, the cosine of the angle of the planes is

$$\frac{(2, -1, 1) \cdot (1, 2, -1)}{\|(2, -1, 1)\| \, \|(1, 2, -1)\|} = -\frac{1}{6}$$

■ EXAMPLE 64   (Perpendicular Planes)   Show that the planes with equations $x + y + z = 0$ and $-x - y + 2z = 0$ are perpendicular.

SOLUTION   They are perpendicular because the normals $(1, 1, 1)$ and $(-1, -1, 2)$ are orthogonal.

Lines and Hyperplanes in $\mathbf{R}^n$

In $\mathbf{R}^n$ there are analogous equations for "lines" and for "planes," called *hyperplanes*. A vector from point $P(p_1, \ldots, p_n)$ to $Q(q_1, \ldots, q_n)$ has coordinates

$$\overrightarrow{PQ} = (q_1 - p_1, \ldots, q_n - p_n)$$

A **line** through the point $\mathbf{p} = (p_1, \ldots, p_n)$ in the direction $\mathbf{d} = (d_1, \ldots, d_n)$ is the set of points $\mathbf{x} = (x_1, \ldots, x_n)$ such that $\mathbf{x} - \mathbf{p}$ is parallel to $\mathbf{d}$. Therefore, $\mathbf{x} - \mathbf{p} = t\mathbf{d}$ for some scalar t. For $-\infty < t < \infty$ the equation

$$\mathbf{x} = \mathbf{p} + t\mathbf{d} \quad \text{or} \quad (x_1, \ldots, x_n) = (p_1, \ldots, p_n) + t(d_1, \ldots, d_n) \qquad (2.26)$$

is the **parametric equation of a line** in $\mathbf{R}^n$ with **parameter** t.

■ EXAMPLE 65   The parametric equation of the line in $\mathbf{R}^4$ passing through the point $(1, 2, 3, 4)$ and in the direction $(1, 1, 1, 1)$ is

$$(x_1, x_2, x_3, x_4) = (1, 1, 1, 1)t + (1, 2, 3, 4)$$

■ EXAMPLE 66 Determine the parametric equation of the line through the points $(-1, 1, -1, 1)$ and $(1, 2, 3, 4)$.

SOLUTION The direction of the line is $(1, 2, 3, 4) - (-1, 1, -1, 1) = (2, 1, 4, 3)$. Hence, the equation of the line is $(x_1, x_2, x_3, x_4) = (-1, 1, -1, 1) + t(2, 1, 4, 3)$.

We saw that the points $\mathbf{x}$ of a plane passing through a point $\mathbf{p}$ with normal $\mathbf{n}$ should satisfy the equation

$$\mathbf{n} \cdot (\mathbf{x} - \mathbf{p}) = 0 \tag{2.27}$$

because $\mathbf{n}$ and $\mathbf{x} - \mathbf{p}$ should be orthogonal. The analog of this equation in $\mathbf{R}^n$ is the **point-normal equation of a hyperplane**. If $\mathbf{n} = (a_1, \ldots, a_n)$, $\mathbf{x} = (x_1, \ldots, x_n)$, and $\mathbf{p} = (p_1, \ldots, p_n)$, then Equation (2.27) implies

$$a_1(x_1 - p_1) + \cdots + a_n(x_n - p_n) = 0 \tag{2.28}$$

Equation (2.28) can also be written in the form

$$a_1 x_1 + \cdots + a_n x_n + d = 0 \tag{2.29}$$

This is called the **general equation of a hyperplane**.

■ EXAMPLE 67 Find an equation of the hyperplane in $\mathbf{R}^4$ passing through the point $(1, 2, 3, 4)$ and normal to the direction $(-1, 2, -2, 1)$.

SOLUTION We have, by (2.28),

$$-1(x_1 - 1) + 2(x_2 - 2) - 2(x_3 - 3) + 1(x_4 - 4) = 0$$
$$\Leftrightarrow -x_1 + 2x_2 - 2x_3 + x_4 - 1 = 0$$

Exercises 2.7

Let l_1, l_2, l_3, and l_4 be the lines with respective parametric equations

$$l_1: \quad x = 5 - 3t \qquad l_2: \quad x = 1 + 6t$$
$$ \quad y = -4 + 2t \qquad \quad y = 6 - 4t$$
$$ \quad z = 2 - t \qquad \quad z = 8 + 2t$$

$$l_3: \quad x = 5 - s \qquad l_4: \quad x = 14 + s$$
$$ \quad y = -7 - s \qquad \quad y = -2 + 2s$$
$$ \quad z = 11 + s \qquad \quad z = 13 + 3s$$

and let P, Q, R, and S be the points

$$P(5, -4, 2) \qquad Q(2, -2, 1)$$
$$R(1, -2, 2) \qquad S(9, -12, -2)$$

1. Which of P, Q, R, and S are in l_1?

2. Which of P, Q, R, and S are in l_3?

3. Find three points in l_1.

4. Find the intersection of l_3 with the coordinate planes.

5. Find all pairs of parallel lines from l_1, l_2, l_3, and l_4.

6. Find all pairs of perpendicular lines from l_1, l_2, l_3, and l_4.

7. Show that l_1 and l_4 intersect. Find their point of intersection.

8. Show that l_1 and l_3 are *skew* lines. (That is, they are not parallel and they do not intersect.)

9. Find the symmetric equations of the lines l_1, l_2, l_3, and l_4.

10. For each of l_1, l_2, l_3, and l_4 find the equations of two planes whose intersection is the given line. (*Hint:* Use the symmetric equations.)

11. Find the parametric equations of the line through P parallel to $\mathbf{n} = (4, -3, 1)$.

12. Find the parametric equations of the line through $(0, 3, 2)$ parallel to $\mathbf{n} = (-1, 2, 4)$.

13. Find the parametric equations of the line through P and Q.

14. Find the parametric equations of the line through $(2, -3, 5)$ and $(-7, 4, 1)$.

15. Find the symmetric equations of the line through $(3, -1, -2)$ and $(-1, 2, 5)$.

16. Find the symmetric equations of the line passing through P in the direction of $\overrightarrow{SR}$.

17. Which of P, Q, R, and S belong to the plane $x + 3y + 3z + 1 = 0$?

In Exercises 18–19 find a point-normal form of the equation of the plane through the point X with normal $\mathbf{n}$.

18. $X = (-4, 2, 7)$ and $\mathbf{n} = (-3, 2, 1)$.

19. $X = P$ and $\mathbf{n} = (-6, 4, 5)$.

20. Find a general equation of the plane of Exercise 18.

21. Find a general equation of the plane of Exercise 19.

22. Find a point-normal form of the equation of the plane with each general equation.
   a. $2x - 3y + z - 9 = 0$
   b. $x - 7y + 3 = 0$

23. Find an equation of the plane through P, Q, and R.

24. Find an equation of the plane through $(2, -4, 1)$ and l_1.

25. Find an equation of the plane that contains the lines l_1 and l_4.

26. Find an equation of the plane passing through $(2, 3, -1)$ and perpendicular to $(-2, 4, 1)$.

27. Find the equation of the plane passing through $(-1, -2, 5)$ and parallel to the plane $x - 6y + 2z - 3 = 0$.

28. Find the parametric equations for the line of intersection of the planes $x - y + z - 3 = 0$ and $-x + 5y + 3z + 4 = 0$.

29. Find the cosine of the angle between the planes $6x + y + z - 1 = 0$ and $x + y - z + 1 = 0$.

30. Show that the planes with equations $x - y + 2z + 3 = 0$ and $-x + 2y + \frac{3}{2}z = 0$ are perpendicular.

31. Find an equation of the hyperplane in $\mathbf{R}^5$ passing through the point $(1, 2, 0, -1, 0)$ and normal $(-1, 3, -2, 8, 4)$.

32. Show that the lines $l_1: \mathbf{x} = \mathbf{x}_1 + t\mathbf{n}_1$ and $l_2: \mathbf{x} = \mathbf{x}_2 + s\mathbf{n}_2$ intersect if and only if $\mathbf{x}_2 - \mathbf{x}_1$ is in Span$\{\mathbf{n}_1, \mathbf{n}_2\}$.

2.8   Applications

Reader's Goal for This Section

To appreciate the many applications of vectors.

There are numerous applications of vectors to nearly all areas of mathematics, physics, and engineering. In this section we survey some of these important, yet elementary, applications. Although the emphasis is on the geometric applications of vectors to statics and engineering we also apply the product $A\mathbf{x}$ to the smoothing of data and, more importantly, to dynamical systems.

Smoothing of Data

In measuring various quantities that depend on time, we often collect data that include sudden disturbances. For example, suppose we measure wind velocities and record some high velocities of sudden gusts that only last a short time. We may want to minimize the impact of these brief gusts that could affect our interpretation of the data. One way of doing this is by *smoothing* the data. Such a smoothing scheme is *averaging*. If we have a sequence of numbers

$$a, b, c, d, e, \ldots$$

we transform it into the sequence of the successive *averages*

$$\frac{a}{2}, \frac{a+b}{2}, \frac{b+c}{2}, \frac{c+d}{2}, \frac{d+e}{2}, \ldots$$

starting with the average between a and 0 as the first new number. (Several other schemes are used in practice.)

Averaging is, in fact, multiplication by a finite-size square matrix of the form

$$A = \begin{bmatrix} \frac{1}{2} & 0 & 0 & 0 & 0 & \cdots \\ \frac{1}{2} & \frac{1}{2} & 0 & 0 & 0 & \cdots \\ 0 & \frac{1}{2} & \frac{1}{2} & 0 & 0 & \cdots \\ 0 & 0 & \frac{1}{2} & \frac{1}{2} & 0 & \cdots \\ 0 & 0 & 0 & \frac{1}{2} & \frac{1}{2} & \cdots \\ \vdots & \vdots & \vdots & \vdots & \vdots & \ddots \end{bmatrix}$$

Suppose, for example, that we record the following wind velocities in tens of miles per hour with measurements 1 h apart:

$$2 \quad 1 \quad 3 \quad 3 \quad 4 \quad 5 \quad 3 \quad 4 \quad 3 \quad 2 \quad 1 \quad 2$$

Plotting the data as a function of time yields the graph of Fig. 2.28.

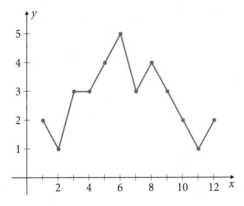

Figure 2.28 Data plotting.

Averaging transforms this sequence into

$$1 \quad \frac{3}{2} \quad 2 \quad 3 \quad \frac{7}{2} \quad \frac{9}{2} \quad 4 \quad \frac{7}{2} \quad \frac{7}{2} \quad \frac{5}{2} \quad \frac{3}{2} \quad \frac{3}{2}$$

Plotting the data yields a smoother graph (Fig. 2.29).

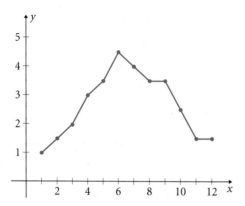

Figure 2.29   Data averaging.

If further smoothing by this technique is desirable, we average once more. This can also be done by multiplying the first sequence by a matrix of the form B (verify):

$$B = \begin{bmatrix} \frac{1}{4} & 0 & 0 & 0 & 0 & \cdots \\ \frac{1}{2} & \frac{1}{4} & 0 & 0 & 0 & \cdots \\ \frac{1}{4} & \frac{1}{2} & \frac{1}{4} & 0 & 0 & \cdots \\ 0 & \frac{1}{4} & \frac{1}{2} & \frac{1}{4} & 0 & \cdots \\ 0 & 0 & \frac{1}{4} & \frac{1}{2} & \frac{1}{4} & \cdots \\ \vdots & \vdots & \vdots & \vdots & \vdots & \ddots \end{bmatrix}$$

The new sequence is

$$\frac{1}{2} \quad \frac{5}{4} \quad \frac{7}{4} \quad \frac{5}{2} \quad \frac{13}{4} \quad 4 \quad \frac{17}{4} \quad \frac{15}{4} \quad \frac{7}{2} \quad 3 \quad 2 \quad \frac{3}{2}$$

Plotting now yields an even smoother graph (Fig. 2.30).

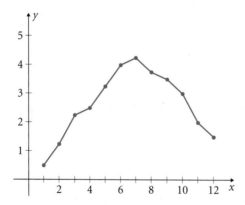

Figure 2.30   Data double averaging.

For this and other interesting examples of matrix transformations, see pp. 253–264 of Philip J. Davis' *The Mathematics of Matrices* (New York: Blaisdell Publishing Co., 1965).

Dynamical Systems

In this section we introduce discrete dynamical systems as applications of the product $A\mathbf{x}$. Because of its importance, we are going to revisit this topic several times as our linear algebra tools grow.

Roughly speaking, a dynamical system is an equation or system of equations used to study time-dependent quantities. A typical example is the equation involving the balance P_t of an interest earning account at time t. In a *discrete* dynamical system the time variable takes on *integer* values.[6] For instance, if an account earns 8% interest compounded annually, then P_0 represents the initial deposit, P_1, the balance at the end of the first year, P_2, the balance at the end of the second year, and so on. In this case, although noninteger t values such as $P_{1.5}$ are meaningful, it suffices to know P at integer values because $P_{1.5} = P_1$, $P_{2.7} = P_2$, etc. Let $P_0 = \$1000$. Then $P_1 = 0.08 \cdot P_0 + P_0 = 1080$, $P_2 = 0.08P_1 + P_1 = 1166.4,\ldots.$ At the end of the $(k+1)$st year, the balance is

$$P_{k+1} = 0.08P_k + P_k = 1.08P_k \tag{2.30}$$

Equation (2.30) is called a **discrete dynamical system**, or a **difference equation**, and it gives the next value of P in terms of the current value. We can compute P_k by repeated applications of (2.30):

$$P_k = 1.08P_{k-1} = 1.08 \cdot 1.08P_{k-2} = (1.08)^2 P_{k-2} = \cdots$$

Hence,

$$P_k = (1.08)^k P_0 \tag{2.31}$$

Equation (2.31) is called the **solution** of the dynamical system.

Sometimes the time-dependent quantity has several components and can be represented by a vector. We may then use matrix theory to study the corresponding dynamical system.

A Population Growth Model

Suppose we have a population of insects divided into three age groups, A, B, C. Group A consists of insects 0–1 wks old, group B consists of insects 1–2 wks old, and group C consists of insects 2–3 wks old. Suppose the groups have A_k, B_k, and C_k insects at the end of the kth week. We want to study how A, B, C change over time, given the following two conditions:

1. (**Survival Rate**) Only 10% of age group A survive a week. Hence,

$$B_{k+1} = \frac{1}{10}A_k \tag{2.32}$$

---

[6]For an excellent introduction see James T. Sandefur's *Discrete Dynamical Systems, Theory and Applications* (Oxford: Clarendon Press, 1990).

And only 40% of age group B survive a week. So,

$$C_{k+1} = \frac{2}{5}B_k \tag{2.33}$$

2. (**Birth Rate**) Each insect from group A has $\frac{2}{5}$ offspring, each insect from group B has 4 offspring, and each insect from group C has 5 offspring. In year $k + 1$ the insects of group A are offspring of insects in year k. Hence,

$$A_{k+1} = \frac{2}{5}A_k + 4B_k + 5C_k \tag{2.34}$$

▪ **PROBLEM**   If the insect population starts out with 1000 from each age group, how many insects are in each group at the end of the third week?

SOLUTION   Equations (2.32), (2.33), and (2.34) can be expressed in terms of vectors and matrices as follows:

$$\begin{bmatrix} A_{k+1} \\ B_{k+1} \\ C_{k+1} \end{bmatrix} = \begin{bmatrix} \frac{2}{5} & 4 & 5 \\ \frac{1}{10} & 0 & 0 \\ 0 & \frac{2}{5} & 0 \end{bmatrix} \begin{bmatrix} A_k \\ B_k \\ C_k \end{bmatrix}$$

This matrix equation is the *dynamical system* of the problem.

The condition on the initial population (**initial condition**) is

$$\begin{bmatrix} A_0 \\ B_0 \\ C_0 \end{bmatrix} = \begin{bmatrix} 1000 \\ 1000 \\ 1000 \end{bmatrix}$$

At the end of the first week we have:

$$\begin{bmatrix} A_1 \\ B_1 \\ C_1 \end{bmatrix} = \begin{bmatrix} \frac{2}{5} & 4 & 5 \\ \frac{1}{10} & 0 & 0 \\ 0 & \frac{2}{5} & 0 \end{bmatrix} \begin{bmatrix} 1000 \\ 1000 \\ 1000 \end{bmatrix} = \begin{bmatrix} 9400 \\ 100 \\ 400 \end{bmatrix}$$

At the end of the second week,

$$\begin{bmatrix} A_2 \\ B_2 \\ C_2 \end{bmatrix} = \begin{bmatrix} \frac{2}{5} & 4 & 5 \\ \frac{1}{10} & 0 & 0 \\ 0 & \frac{2}{5} & 0 \end{bmatrix} \begin{bmatrix} 9400 \\ 100 \\ 400 \end{bmatrix} = \begin{bmatrix} 6160 \\ 940 \\ 40 \end{bmatrix}$$

and at the end of the third week,

$$\begin{bmatrix} A_3 \\ B_3 \\ C_3 \end{bmatrix} = \begin{bmatrix} \frac{2}{5} & 4 & 5 \\ \frac{1}{10} & 0 & 0 \\ 0 & \frac{2}{5} & 0 \end{bmatrix} \begin{bmatrix} 6160 \\ 940 \\ 40 \end{bmatrix} = \begin{bmatrix} 6424 \\ 616 \\ 376 \end{bmatrix}$$

Therefore, after 3 wks age group A has 6424 insects, age group B has 616 insects, and age group C has 376 insects.

Analytical Geometry

■ EXAMPLE 68   (Distance from Point to Plane)   Find a formula for the shortest distance d from the point $P(x_0, y_0, z_0)$ to the plane $\mathcal{P}$ with equation $ax + by + cz + d = 0$.

SOLUTION   The equation for $\mathcal{P}$ implies that $\mathbf{n} = (a, b, c)$ is a normal vector. Let $Q(x, y, z)$ be any point in $\mathcal{P}$. Then d is the length of the orthogonal projection, $\mathbf{d}$, of $\overrightarrow{QP}$ along $\mathbf{n}$ (Fig. 2.31). Hence Equation (2.8) of Section 2.2 implies

$$d = \|\mathbf{d}\| = \left\| \frac{\overrightarrow{QP} \cdot \mathbf{n}}{\mathbf{n} \cdot \mathbf{n}} \mathbf{n} \right\| = |\overrightarrow{QP} \cdot \mathbf{n}| \frac{1}{\|\mathbf{n}\|} = \frac{|a(x_0 - x) + b(y_0 - y) + c(z_0 - z)|}{\sqrt{a^2 + b^2 + c^2}}$$

$$= \frac{|ax_0 + by_0 + cz_0 - (ax + by + cz)|}{\sqrt{a^2 + b^2 + c^2}}$$

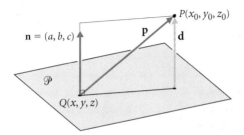

Figure 2.31   Distance from a point to a plane.

The coordinates (x, y, z) of Q satisfy the plane equation $ax + by + cz + d = 0$, because Q is in $\mathcal{P}$. Hence, $d = -ax - by - cz$. Therefore,

$$d = \frac{|ax_0 + by_0 + cz_0 + d|}{\sqrt{a^2 + b^2 + c^2}}$$

■ EXAMPLE 69   Find the distance from the point $P(-1, 3, -2)$ to the plane $2x - 3y + z - 1 = 0$.

SOLUTION   Referring to Example 68 we have

$$d = \frac{|2 \cdot (-1) + (-3) \cdot 3 + 1 \cdot (-2) - 1|}{\sqrt{2^2 + (-3)^2 + 1^2}} = \sqrt{14} \cong 3.74$$

Skew lines are nonparallel and nonintersecting lines.

■ EXAMPLE 70   (Distance Between Skew Lines)   Find the shortest distance d between two skew lines l_1 and l_2.

SOLUTION   Let P, Q and R, S be two pairs of points on l_1 and l_2, respectively (Fig. 2.32). The cross product $\mathbf{n} = \overrightarrow{PQ} \times \overrightarrow{RS}$ is orthogonal to $\overrightarrow{PQ}$ and $\overrightarrow{RS}$. Let $\mathcal{P}_1$ and $\mathcal{P}_2$ be the planes passing through P and R with normal $\mathbf{n}$. Then $\mathcal{P}_1$ and $\mathcal{P}_2$ are parallel, and they contain the lines l_1 and l_2. So, d is the distance between $\mathcal{P}_1$ and $\mathcal{P}_2$, and it can be computed as the length of the orthogonal projection of $\overrightarrow{PR}$ along the normal direction $\mathbf{n}$. Equation (2.8) implies

$$d = \left\| \frac{\overrightarrow{PR} \cdot \mathbf{n}}{\mathbf{n} \cdot \mathbf{n}} \mathbf{n} \right\| = |\overrightarrow{PR} \cdot \mathbf{n}| \frac{1}{\|\mathbf{n}\|}$$

So, in terms of the points P, Q, R, S, we have

$$d = |\overrightarrow{PR} \cdot (\overrightarrow{PQ} \times \overrightarrow{RS})| \frac{1}{\|\overrightarrow{PQ} \times \overrightarrow{RS}\|}$$

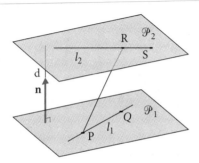

Figure 2.32   Distance between two skew lines.

▪ EXAMPLE 71   Find the shortest distance between the two skew lines l_1, passing through $P(1, -2, -1)$ and $Q(0, -2, 1)$, and l_2, passing through $R(-1, 2, 0)$ and $S(-1, 0, -2)$.

SOLUTION $\overrightarrow{PQ} = (-1, 0, 2)$, $\overrightarrow{RS} = (0, -2, -2)$, and $\overrightarrow{PR} = (-2, 4, 1)$. Hence, $\overrightarrow{PQ} \times \overrightarrow{RS} = (4, -2, 2)$, and

$$d = |(-2, 4, 1) \cdot (4, -2, 2)| \frac{1}{\|(4, -2, 2)\|} = \frac{7}{\sqrt{6}} \cong 2.86$$

Euclidean Geometry

Vectors can be used to prove theorems in Euclidean geometry, as the following three examples indicate.

▪ EXAMPLE 72   Prove that the line segment that bisects two sides of a triangle equals one-half of the third side.

SOLUTION   Referring to Fig. 2.33(a), we need to prove that if $AP = PB$ and $AQ = QC$, then $PQ = \frac{1}{2}BC$. We have

$$\overrightarrow{PQ} = \overrightarrow{PA} + \overrightarrow{AQ} = \frac{1}{2}\overrightarrow{BA} + \frac{1}{2}\overrightarrow{AC} = \frac{1}{2}\left(\overrightarrow{BA} + \overrightarrow{AC}\right) = \frac{1}{2}\overrightarrow{BC}$$

Therefore, the length PQ of $\overrightarrow{PQ}$ is one-half the length of BC.

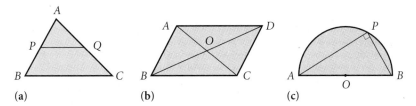

(a)   (b)   (c)

Figure 2.33

■ EXAMPLE 73   Show that the diagonals of a parallelogram bisect each other.

SOLUTION   Let $ABCD$ be the parallelogram and let O be the midpoint of the diagonal AC (Fig. 2.33(b)). Then $AO = OC$. It suffices to prove that BOD is a straight-line segment and that $BO = OD$. Because $\overrightarrow{BC} = \overrightarrow{AD}$, we have

$$\overrightarrow{BO} = \overrightarrow{BC} + \overrightarrow{CO} = \overrightarrow{AD} + \overrightarrow{OA} = \overrightarrow{OA} + \overrightarrow{AD} = \overrightarrow{OD}$$

Therefore, $\overrightarrow{BO}$ and $\overrightarrow{OD}$ have the same direction, a common point, and the same length. Hence $BO = OD$ and B, O, D are colinear, as asserted.

■ EXAMPLE 74   Show that any angle inscribed in a semicircle is a right angle.

SOLUTION   Referring to Fig. 2.33(c) it is enough to show that $\overrightarrow{AP}$ and $\overrightarrow{BP}$ are perpendicular, or that $\overrightarrow{AP} \cdot \overrightarrow{PB} = 0$. Let r be the radius. Then

$$\overrightarrow{AP} \cdot \overrightarrow{PB} = \left(\overrightarrow{AO} + \overrightarrow{OP}\right) \cdot \left(\overrightarrow{PO} + \overrightarrow{OB}\right)$$

$$= \left(\overrightarrow{OB} + \overrightarrow{OP}\right) \cdot \left(-\overrightarrow{OP} + \overrightarrow{OB}\right) = \overrightarrow{OB} \cdot \overrightarrow{OB} - \overrightarrow{OP} \cdot \overrightarrow{OP}$$

$$= \left\|\overrightarrow{OB}\right\|^2 - \left\|\overrightarrow{OP}\right\|^2 = r^2 - r^2 = 0$$

Physics and Engineering

Both the dot and the cross product have striking physical interpretations. The dot product can be interpreted as the work done by a constant force and the cross product, as the moment vector of a force.

If a constant force $\mathbf{F}$ moves an object distance $\mathbf{d}$ in the direction of $\mathbf{F}$ (Fig. 2.34(a)), then the work done is

$$W = \|\mathbf{F}\| \, \|\mathbf{d}\|$$

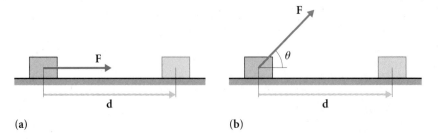

F

θ

d

d

(a)

(b)

Figure 2.34

If **F** and **d** are at an angle θ (Fig. 2.34(b)), then W is defined as the numerical component of **F** in the direction of **d**. In other words,

$$W = \|\mathbf{F}\| \, \|\mathbf{d}\| \cos \theta$$

Hence, Equation (2.7) implies that the **work** done by **F** is

$$W = \mathbf{F} \cdot \mathbf{d} \tag{2.35}$$

■ **EXAMPLE 75** (Work Done by a Constant Force)   Compute the work done by the constant force $\mathbf{F} = 4\mathbf{i} - 2\mathbf{j} + \mathbf{k}$ if its point of application moves from $P(0, 1, -2)$ to $Q(3, 0, 1)$.

SOLUTION

$$\mathbf{d} = \overrightarrow{PQ} = (3, 0, 1) - (0, 1, -2) = (3, -1, 3)$$

Therefore, by (2.35) we have

$$W = \mathbf{F} \cdot \mathbf{d} = (4, -2, 1) \cdot (3, -1, 3) = 17$$

If the force is in newtons and the distance in meters, W is in newton-meters (N·m).

Vectors can be used both graphically and algebraically to compute resultant forces or balance conditions on an object.

■ **EXAMPLE 76** (Inclined Plane)   Compute the force **F** that has to be exerted on the rope (Fig. 2.35) to balance the object weighing 500 lb if the inclined plane has angle of inclination 30°.

SOLUTION   The forces acting on the weight are its weight, **W**, the rope force, **F**, and the reaction of the inclined plane, **R**. Because the system is in balance, the vector sum of these forces should be zero. We know the direction of **F**, so we need only its magnitude. The weight **W** can be written as the sum of two component vectors, $\mathbf{W}_1$ and $\mathbf{W}_2$, in the directions of **F** and **R**, respectively. Because $\|\mathbf{F}\| = \|\mathbf{W}_1\|$, we have

$$\|\mathbf{F}\| = \|\mathbf{W}_1\| = 500 \cdot \cos 60° = 500 \cdot \frac{1}{2} = 250 \text{ lb}$$

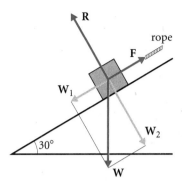

Figure 2.35

Let us now discuss the equilibrium of rigid bodies subjected to planar forces in more detail.

The **moment** m of a force $\mathbf{F}$ about a point P is the product $m = \|\mathbf{F}\|d$, where d is the shortest distance from P to the line l determined by the direction of $\mathbf{F}$ (Fig. 2.36). Let Q be any point on l and let $\mathbf{r} = \overrightarrow{PQ}$. Then $d = \|\mathbf{r}\| \sin \theta$, where θ is the angle between $\mathbf{r}$ and $\mathbf{F}$. Therefore,

$$m = \|\mathbf{r}\| \|\mathbf{F}\| \sin \theta$$

or, by Part 2 of Theorem 21,

$$m = \|\mathbf{r} \times \mathbf{F}\|$$

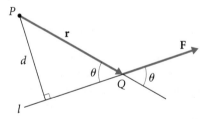

Figure 2.36

We define the **moment vector**, or **torque m**, as

$$\mathbf{m} = \mathbf{r} \times \mathbf{F}$$

The magnitude of $\mathbf{m}$ is the moment m and its direction is along the axis of rotation about P that $\mathbf{F}$ tends to cause.

■ EXAMPLE 77   (Moment of Force about a Point)   A force of 3 N at an angle of $60°$ with the positive x-axis is applied at the end of the position vector $\mathbf{r} = (\sqrt{3}, 1)$. Compute the torque of the force on $\mathbf{r}$. What is the moment about the origin?

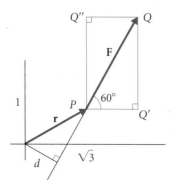

Figure 2.37

SOLUTION   Referring to Fig. 2.37, we have

$$PQ' = PQ\cos 60° = 3\cos 60° = \frac{3}{2}, \qquad PQ'' = PQ\sin 60° = 3\sin 60° = \frac{3\sqrt{3}}{2}$$

Hence, the force $\mathbf{F}$ is $(\frac{3}{2}, \frac{3\sqrt{3}}{2})$ as a position vector and $\mathbf{r} = (\sqrt{3}, 1)$. To compute the cross product $\mathbf{r} \times \mathbf{F}$, we need to add a third component of zero to these vectors. Therefore,

$$\mathbf{m} = \mathbf{r} \times \mathbf{F} = (\sqrt{3}, 1, 0) \times \left(\frac{3}{2}, \frac{3\sqrt{3}}{2}, 0 \right) = (0, 0, 3)$$

is the torque and $\|(0, 0, 3)\| = 3$ is the moment.

We can also compute the moment as $3d$ by finding the distance d from the origin to the line defined by $\mathbf{F}$ using triangles.

In practice, we often work with **signed** moments. The sign of m is positive if the force tends to produce counterclockwise rotation about the given point and negative otherwise. Note that for a right-handed system the (signed) moment is the third component of the torque if the xy-plane is the plane defined by $\mathbf{F}$ and $\mathbf{r}$.

We can now discuss the conditions under which a rigid body balances if coplanar forces act on it.

Equilibrium Conditions for Coplanar Forces

If coplanar forces act on a rigid body, then the body is in equilibrium if the following hold:

1. The vector sum of all forces is zero.
2. The algebraic sum of the signed moments of all forces about any point in the plane is zero.

The second condition is equivalent to: The vector sum of the torques of all forces about any point in the plane is zero.

■ EXAMPLE 78 (Balancing) The upper end of a uniform bar PQ, 5 ft long and weighing 50 lb, rests against a smooth vertical wall (Fig. 2.38(a)). The lower end rests against a smooth horizontal floor 3 ft away from the wall. A cord OR holds the system at equilibrium. If the distance RQ is 1 ft, what is the tension on the cord?

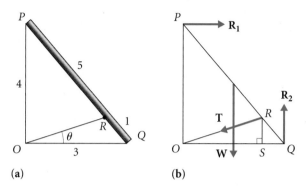

(a)   (b)

Figure 2.38

SOLUTION Let $\mathbf{R}_1$ and $\mathbf{R}_2$ be the reactions of the wall and the floor and let $\mathbf{T}$ be the tension on the cord (Fig. 2.38(b)). Let R_1, R_2, and T be the magnitudes of these vectors. The vector sum of $\mathbf{R}_1$, $\mathbf{R}_2$, $\mathbf{T}$, and the weight $\mathbf{W}$ should be zero. Hence, in terms of components we have

$$R_1 - T \cos \theta = 0$$

$$-50 + R_2 - T \sin \theta = 0$$

A third equation can be obtained from the second equilibrium condition. Setting the algebraic sum of the moments about O equal to zero yields

$$-4R_1 + 3R_2 - \frac{3}{2} \cdot 50 = 0$$

Next we solve the system for T to get

$$T = \frac{75}{4 \cos \theta - 3 \sin \theta}$$

Because,

$$\frac{RS}{4} = \frac{QS}{3} = \frac{1}{5}$$

we have $OS = 3 - QS = \frac{12}{5}$ and $OR = \sqrt{OS^2 + RS^2} = \sqrt{\left(\frac{12}{5}\right)^2 + \left(\frac{4}{5}\right)^2} = \frac{4}{5}\sqrt{10}$. Hence,

$$\cos \theta = \frac{OS}{OR} = \frac{3}{\sqrt{10}} \quad \text{and} \quad \sin \theta = \frac{RS}{OR} = \frac{1}{\sqrt{10}}$$

So,

$$T = \frac{75}{4(3/\sqrt{10}) - 3(1/\sqrt{10})} = \frac{25}{3}\sqrt{10} \cong 26.35 \text{ lb}$$

Exercises 2.8

Averaging

1. Plot the sequence and use matrices to average it twice. Plot each averaging.

$$
\begin{array}{cccccccc}
1 & 2 & 3 & 4 & 5 & 6 & 7 & 8 \\
\downarrow & \downarrow & \downarrow & \downarrow & \downarrow & \downarrow & \downarrow & \downarrow \\
2 & 3 & 7 & 2 & 3 & 9 & 1 & 10
\end{array}
$$

2. Plot the sequence and use matrices to average it twice. Plot each averaging.

$$
\begin{array}{cccccccc}
1 & 2 & 3 & 4 & 5 & 6 & 7 & 8 \\
\downarrow & \downarrow & \downarrow & \downarrow & \downarrow & \downarrow & \downarrow & \downarrow \\
25 & 15 & 45 & 15 & 20 & 30 & 20 & 50
\end{array}
$$

Discrete Dynamical Systems

Suppose a species consists of two age groups: the young and the adult. Let Y_k and A_k be the number of individuals after k time units. The young have survival rate s. The birth rate of the young is y (i.e., one young individual has y offspring) and the birth rate of the adult is a. In Exercises 3–8 write in matrix form the dynamical system that models this population. Find the number of individuals in each group after 3 time units for the given values of s, y, a, Y_0, and $A_0 = 100$.

3. $s = 4/5, y = 2, a = 10, Y_0 = 100$.

4. $s = 1/2, y = 2, a = 6, Y_0 = 100$.

5. $s = 1/2, y = 4, a = 10, Y_0 = 100$.

6. $s = 1/4, y = 2, a = 12, Y_0 = 100$.

7. $s = 1/3, y = 3, a = 12, Y_0 = 300$.

8. $s = 1/5, y = 5, a = 30, Y_0 = 100$.

9. A population of flies is divided into three age groups, A, B, and C. Group A consists of flies 0–2 wk old, group B consists of flies 2–4 wk old, and group C consists of flies 4–6 wk old. Suppose the groups have A_k, B_k, and C_k number of flies at the end of the $2k$th week. The survival rate of group A is 25%, whereas the survival rate of group B is 33.$\overline{33}$%. Each fly from group A has 0.25 offspring, each fly from group B has 2.5 offspring, and each fly from group C has 1.5 offspring. If the original population consists of 4800 flies in each age group, write in matrix form the dynamical system that models this population. Find the number of flies in each group after 6 wk.

Geometry and Physics

10. Find the distance from the point $P(8, 4, -5)$ to the plane $2x - 2y + z - 6 = 0$.

11. Find the distance from the point $P(1, 1, 6)$ to the plane $2x - 2y + z - 6 = 0$.

12. Find the distance from the point $P(1, 2, -4)$ to the plane through the points $(0, 0, 0)$, $(2, 1, 0)$ and $(3, 0, -1)$.

Let P, Q, R, and S be the points

$$
\begin{array}{ll}
P(0, -1, 2) & Q(2, -2, 1) \\
R(4, -2, 1) & S(0, 2, -3)
\end{array}
$$

In Exercises 13–15 find the shortest distance between the two skew lines determined by the given line segments.

13. $\overrightarrow{PQ}$ and $\overrightarrow{RS}$

14. $\overrightarrow{PS}$ and $\overrightarrow{RQ}$

15. $\overrightarrow{PR}$ and $\overrightarrow{QS}$

16. Find the shortest distance between the two lines with parametric equations

$$
\begin{array}{ll}
l_1: \quad x = 5 - 3t & l_2: \quad x = 5 - s \\
 \quad y = -4 + 2t & \quad y = -2 - 2s \\
 \quad z = 2 - t & \quad z = 1 + s
\end{array}
$$

17. Let AP be the line segment that bisects the hypotenuse BC of the right triangle ABC. Use dot products to prove that $AP = \frac{1}{2}BC$.

18. (**Centroid**) The centroid of n points (or of an n-gon with vertices) consisting of the tips of the vectors $\mathbf{v}_1, \ldots, \mathbf{v}_n$ is given by

$$
\frac{1}{n}(\mathbf{v}_1 + \cdots + \mathbf{v}_n)
$$

(Fig. 2.39). Find the centroid of the triangle PQR, where $P(1, 2)$, $Q(2, -4)$, and $R(-1, 7)$.

19. Referring to Exercise 18, find the centroid of $P_1(0, 0, 0)$, $P_2(1, 1, 1)$, $P_3(1, 1, -1)$, and $P_4(-2, 1, 0)$.

20. Show that the centroid of any triangle is the intersection of the three medians.

21. (**Center of Mass**) Let $m_1, \ldots, m_n$ be n masses located at the tips of the vectors $\mathbf{v}_1, \ldots, \mathbf{v}_n$ and let $M = m_1 + \cdots + m_n$ be the total mass. The center

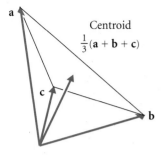

Centroid
$\frac{1}{3}(\mathbf{a} + \mathbf{b} + \mathbf{c})$

Figure 2.39

23. $\mathbf{F} = \mathbf{i} + \mathbf{j} + \mathbf{k}$, $X = (0, -2, 5)$, $Y = (1, 7, -2)$.

24. $\mathbf{F} = -2\mathbf{i} + 6\mathbf{j} + 8\mathbf{k}$, $X = (-2, 3, 0)$, $Y = (-1, 6, -4)$

25. Referring to Fig. 2.40, compute the rope force $\mathbf{F}$ that balances the 25-lb object.

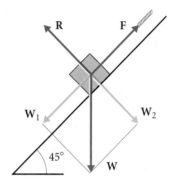

Figure 2.40

of mass of these systems is defined by

$$\frac{1}{M}(m_1\mathbf{v}_1 + \cdots + m_n\mathbf{v}_n)$$

Find the center of mass of the system with masses 1, 4, 5, and 2 kg located, respectively, at $P_1(-1, 2, 0)$, $P_2(0, 5, -1)$, $P_3(1, 1, -3)$, and $P_4(-6, 1, -3)$.

22. Referring to Exercises 18 and 21, show that if all masses are equal, then center of mass equals the centroid.

In Exercises 23–24 compute the work done by the constant force $\mathbf{F}$ if its point of application moves from X to Y.

26. Find the torque of $F = -2\mathbf{i} - 4\mathbf{j} + \mathbf{k}$ on $\mathbf{r} = (-2, 1, -3)$.

2.9   Miniprojects

1 ■ Long-Term Behavior of a Dynamical System

In this project we see that under special initial conditions we can predict some of the long-term behavior of a discrete dynamical system. We revisit this important topic in Chapter 7.

A Population Model

Suppose a species consists of two age groups: the young and the adult. Let A_k and B_k be the number of individuals after k time units. The young have survival rate $\frac{6}{7}$. The birth rate of the young is 3 (i.e., one young individual has 3 offspring), and the birth rate of the adult is 21.

Problem A

1. Write in matrix form the dynamical system that models this population.
2. Find a formula in terms of A_0 and B_0 that computes the number of individuals in each group after 3 time units.
3. Evaluate your formula for $A_0 = 700$ and $B_0 = 700$.

Problem B

Refer to Problem A.

1. Let $A_0 = 7$ and $B_0 = 1$ and let p_k be the ratio $A_k : B_k$. Find the long-term value p of p_k as k grows large. (That is, find $p = \lim_{k \to \infty} p_k$.) Justify your answer. (Knowledge of limits is not necessary.)
2. Now let $A_0 = 8$ and $B_0 = 2$, let q_k be the ratio $A_k : B_k$, and let q be the long-term value q_k. Is it easy to predict q this time? Why, or why not?
3. It is a fact that $p = q$. Find the first value of k in q_k such that q_k is within 0.5 of p.

Problem C

Consider the matrix form dynamical system:

$$\begin{bmatrix} A_{k+1} \\ B_{k+1} \\ C_{k+1} \end{bmatrix} = \begin{bmatrix} \frac{1}{4} & \frac{5}{2} & \frac{3}{2} \\ \frac{1}{4} & 0 & 0 \\ 0 & \frac{1}{3} & 0 \end{bmatrix} \begin{bmatrix} A_k \\ B_k \\ C_k \end{bmatrix}$$

1. Write a population growth problem that can be modeled by this system.
2. Find the long-term ratios $A_k : B_k : C_k$ if $\begin{bmatrix} A_0 \\ B_0 \\ C_0 \end{bmatrix}$ is (a) $\begin{bmatrix} 24 \\ 6 \\ 2 \end{bmatrix}$ and (b) $\begin{bmatrix} 24 \\ 5 \\ 1 \end{bmatrix}$.
3. Explain what these long-term ratios mean in your chosen problem.

2 ■ Plane Bisector of Two Planes; Line Bisector of Two Lines

In this project we are to accomplish the following:

1. Develop a method for computing the equation of the plane $\mathcal{P}$ that bisects two intersecting planes $\mathcal{P}_1$ and $\mathcal{P}_2$.
2. Find a formula for the distance from a point to a line.
3. Find a formula for the line internal and external bisector of two lines in the plane.

    With respect to the first question, if we know the equations for $\mathcal{P}_1$ and $\mathcal{P}_2$, we can find a point of the intersection and, hence, a point of $\mathcal{P}$. So by the point-normal formula, we need only a normal to $\mathcal{P}$. A normal should bisect the angles between the normals of $\mathcal{P}_1$ and $\mathcal{P}_2$ (Fig. 2.41(a)). Hence, it suffices to find a vector that bisects the angle between two given vectors. Out of the many such vectors let us compute one with unit length.

Problem A

Let $\mathbf{n}_1$ and $\mathbf{n}_2$ be two given vectors. Let $\mathbf{u} = (u_1, u_2, u_3)$ be a unit vector that bisects the angle between $\mathbf{n}_1$ and $\mathbf{n}_2$. Show that

$$\left(\frac{\mathbf{n}_1}{\|\mathbf{n}_1\|} - \frac{\mathbf{n}_2}{\|\mathbf{n}_2\|} \right) \cdot \mathbf{u} = 0 \tag{2.36}$$

$$\mathbf{u} \cdot (\mathbf{n}_1 \times \mathbf{n}_2) = 0 \tag{2.37}$$

$$u_1^2 + u_2^2 + u_3^2 = 1 \tag{2.38}$$

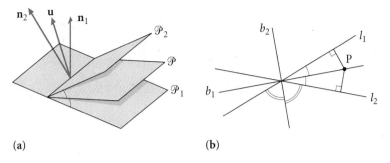

(a)   (b)

Figure 2.41

Problem B

Use Problem A to find a unit vector that bisects the angle between $\mathbf{n}_1 = (1, 1, 1)$ and $\mathbf{n}_2 = (1, 0, 0)$. Find an equation of the plane that bisects the planes $x + y + z = 1$ and the yz-plane.

The next two problems, which are independent of the first two, lead to a formula for the internal and external bisectors b_1 and b_2 of two plane lines l_1 and l_2 (Fig. 2.41(b)).

Problem C

Let $P(x_0, y_0)$ be a point and let $ax + by + c = 0$ be the equation of a line l in the xy-plane. Use the dot product to find a formula for the shortest distance from P to l. What is the shortest distance from $P(1, 1)$ to $3x - y + 2 = 0$?

Problem D

Let $a_1x + b_1y + c_1 = 0$ and $a_2x + b_2y + c_2 = 0$ be the equations of two lines l_1 and l_2 in the plane. Use Problem C to find a formula for the two lines that bisect the angles between l_1 and l_2. (*Hint:* Pick a point P on a bisector (Figure 2.41(b)) and draw the perpendicular distances from P to l_1 and l_2.) Find the two bisectors of the lines $3x - y + 2 = 0$ and $2x - 3y + 1 = 0$.

3 ■ Criterion for Collinearity and a Theorem of Pappus

In this project we are to use cross products to prove a simple necessary and sufficient condition for three coplanar vectors to be collinear. Then apply this criterion to solve a problem known as the *parallel case of a theorem of Pappus.*[7]

---

[7] **Pappus of Alexandria** was the last of the great Greek geometers. In about A.D. 320, he wrote the *Synagoge* (Mathematical Collection), a large treatise of 10 books in geometry. Besides the wealth of information on discoveries of Euclid, Apollonius, and Archimedes, the *Synagoge* includes names and important works of other Greek mathematicians that would otherwise be unknown. Many of the theorems in this treatise are due to Pappus himself.

THEOREM 27

(Criterion for Collinearity)

Three coplanar vectors $\mathbf{a} = (a_1, a_2, 0)$, $\mathbf{b} = (b_1, b_2, 0)$ and $\mathbf{c} = (c_1, c_2, 0)$ are colinear if and only if

$$\mathbf{a} \times \mathbf{b} + \mathbf{b} \times \mathbf{c} + \mathbf{c} \times \mathbf{a} = \mathbf{0}$$

Problem A

Prove the criterion for collinearity and show that it is equivalent to the formula

$$\begin{vmatrix} a_1 & a_2 & 1 \\ b_1 & b_2 & 1 \\ c_1 & c_2 & 1 \end{vmatrix} = 0$$

Problem B

Use the criterion for collinearity to prove the following.

THEOREM 28

(The Parallel Case of Pappus' Theorem)[8]

If P, Q, R and P', Q', R' are collinear triads of points, if QR' is parallel to $Q'R$, and if RP' is parallel to $R'P$, then the line PQ' is parallel to the line $P'Q$ (Fig. 2.42).

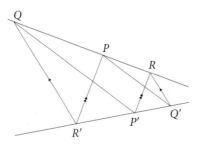

Figure 2.42

Hint: Choose an origin O and, using the notation $OP = \mathbf{p}$, $OP' = \mathbf{p}'$, etc., show that

$$(\mathbf{p} - \mathbf{q}') \times (\mathbf{p}' - \mathbf{q}) + (\mathbf{q} - \mathbf{r}') \times (\mathbf{q}' - \mathbf{r}) + (\mathbf{r} - \mathbf{p}') \times (\mathbf{r}' - \mathbf{p}) = \mathbf{0}$$

4 ■ Varignon's Theorem

Let $\mathbf{F}_1$ and $\mathbf{F}_2$ be two forces with the same origin P and let O be any point in the plane defined by $\mathbf{F}_1$ and $\mathbf{F}_2$ (Figure 2.43).

---

[8] See p. 47 of Dan Pedoe's *Geometry, a Comprehensive Course* (New York: Dover, 1988).

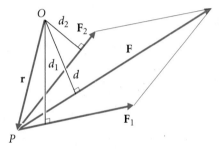

Figure 2.43

Problem A

1. Apply properties of the cross product to prove Varignon's theorem,[9] which states: The algebraic sum of the signed moments about O of $\mathbf{F}_1$ and $\mathbf{F}_2$ equals the signed moment of the resultant $\mathbf{F}_1 + \mathbf{F}_2$ about O.
2. Reprove Varignon's Theorem geometrically.

Problem B

Consider the uniform bar PQ of weight W touching the floor and a plane (Figure 2.44(a) and (b)). Prove that the horizontal force F applied at P that is required to balance the bar is

$$F = \frac{1}{2} W \cot \theta \quad \text{or} \quad F = \frac{1}{2} \frac{W}{\tan \theta + \cot \phi}$$

depending on whether the plane is vertical or at angle ϕ with the horizontal.

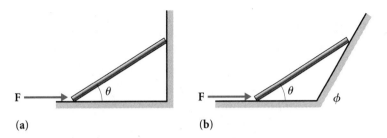

(a)   (b)

Figure 2.44

---

[9]**Pierre Varignon** (1654–1722), a Frenchman, was a contemporary of Newton. He discovered the principle of moments. Girwin, in his *A Historical Appraisal of Mechanics*, credits him as the first to deduce the equations of motion.

2.10   Computer Exercises

This computer section helps you learn how to manipulate vectors and matrices and perform basic vector operations with your mathematical software. It also helps review some of the basic material of this chapter.

Let

$$\mathbf{u} = \begin{bmatrix} 1 \\ 3 \\ 2 \end{bmatrix}, \qquad \mathbf{v} = \begin{bmatrix} -1 \\ 1 \\ 2 \end{bmatrix}, \qquad \mathbf{w} = \begin{bmatrix} 2 \\ 1 \\ -4 \end{bmatrix}, \qquad \mathbf{r} = \begin{bmatrix} 2 \\ -3 \\ 1 \\ -4 \end{bmatrix}$$

$$M = \begin{bmatrix} 1 & 3 & 5 \\ 7 & 9 & 2 \\ 4 & 6 & 8 \end{bmatrix}, \qquad N = \begin{bmatrix} 1 & 2 & 3 & 4 \\ 2 & 3 & 4 & 5 \\ 3 & 4 & 5 & 6 \end{bmatrix}$$

1. Compute

   a. $\mathbf{u} + \mathbf{v}$ **b.** $\mathbf{u} - \mathbf{v}$ **c.** $10\mathbf{u}$ **d.** $\mathbf{u} - 2\mathbf{v} + 3\mathbf{w}$

2. Check the identities

   a. $(\mathbf{u} + \mathbf{v}) + \mathbf{w} = \mathbf{u} + (\mathbf{v} + \mathbf{w})$ **b.** $10(\mathbf{u} + \mathbf{v}) = 10\mathbf{u} + 10\mathbf{v}$

3. If possible, write $\mathbf{v}$ as a linear combination of the columns of M.

4. If possible, write $\mathbf{w}$ as a linear combination of the columns of N.

5. Plot $\mathbf{u}$, $\mathbf{v}$, and $\mathbf{w}$ in separate graphs and in the same graph.

6. Compute

   a. $\|\mathbf{u}\| + \|\mathbf{v}\|$ **b.** $\|\mathbf{u} + \mathbf{v}\|$ **c.** $\mathbf{u} \cdot \mathbf{v} - \mathbf{u} \cdot \mathbf{w}$

7. Compute the angle between $\mathbf{u}$ and $\mathbf{v}$.

8. Compute the orthogonal projection of $\mathbf{v}$ onto $\mathbf{w}$. Verify your answer.

9. Which of $\mathbf{u}$, $\mathbf{v}$, and $\mathbf{w}$ are in the span of the columns of M?

10. Which of $\mathbf{u}$, $\mathbf{v}$, and $\mathbf{w}$ are in the span of the columns of N?

11. Is the set $\{\mathbf{u}, \mathbf{v}, \mathbf{w}\}$ linearly independent? Does it span $\mathbf{R}^3$?

12. Do the columns of M span $\mathbf{R}^3$? Are they linearly independent?

13. Do the columns of N span $\mathbf{R}^3$? Are they linearly independent?

14. True or false?

    a. $\{M\mathbf{x}, \mathbf{x} \in \mathbf{R}^3\} = \mathbf{R}^3$ **b.** $\{N\mathbf{x}, \mathbf{x} \in \mathbf{R}^4\} = \mathbf{R}^3$

15. Write the fourth column of N as a linear combination of the first three.

16. Show that the system is consistent for all values of b_1, b_2, and b_3.

$$x_1 + 2x_2 + x_3 + 2x_4 = b_1$$

$$x_1 + 2x_2 + 2x_3 + x_4 = b_2$$

$$x_1 + 2x_2 + x_3 + 2x_4 + x_5 = b_3$$

17. Find a solution for $b_1 = 1$, $b_2 = -1$, and $b_3 = 1$ and verify your answer by checking the corresponding vector equation.

18. Show that the columns of N are linearly independent and span $\mathbf{R}^3$. How is this double property affected if you add a column of your choice? What happens if you delete a column of your choice?

19. If possible, compute and interpret the products $M\mathbf{u}$, $M\mathbf{r}$, $N\mathbf{u}$, and $N\mathbf{r}$.

20. Verify that $\mathbf{u}$ and $\mathbf{u} \times \mathbf{v}$ are orthogonal.

21. Verify **Jacobi's identity**: $(\mathbf{u} \times \mathbf{v}) \times \mathbf{w} + (\mathbf{v} \times \mathbf{w}) \times \mathbf{u} + (\mathbf{w} \times \mathbf{u}) \times \mathbf{v} = \mathbf{0}$

22. Let l be the line with parametric equation

$$\mathbf{x} = (-1, 2, 1) + t(-4, 1, 5)$$

Which of $R(-9, 4, 11)$ and $S(7, 0, -10)$ are in l? Plot l from $t = -2$ to $t = 3$. Plot l from $P(-13, 5, 16)$ to $Q(-21, 7, 26)$.

23. Find a normal-point form equation for the plane through $P(1, 2, -3)$, $Q(-2, 4, 5)$, and $R(3, 3, 3)$. Plot this plane. Find two points one on and one off the plane.

24. Write and test the code for a function that computes the distance from a point to a plane.

25. Write and test the code for a function that computes the distance between two skew lines, given two points on each line.

Selected Solutions with Maple

```
# DATA and Remarks.
with(linalg):              # Loading the linalg package.
u := vector([1,3,2]);      # Defining u,v,w as vectors.
v := vector([-1,1,2]);     # Or u := array([1,2,3]);, etc..
w := vector([2,1,-4]);     # Vectors are not column matrices. A vector entry
                           # is determined by one number; a matrix enrty by two.
u;                         # Returns the name only and not the value.
evalm(u);                  # Now it returns the value.  Also eval(u); but limited.
M := matrix(3,3,[1,3,5, 7,9,2, 4,6,8]);          # M.
# Or M := matrix([[1,3,5],[7,9,2],[4,6,8]]);
# Or M := matrix(3,3,[[1,3,5],[7,9,2],[4,6,8]]);
M;                         # Just as with vectors the name of the matrix is returned.
eval(M);                   # Now M is evaluated.
N := matrix(3,4,[1,2,3,4,2,3,4,5,3,4,5,6]);
# Exercises 1,2.
evalm(u+v);                # Also add(u,v);
evalm(u-v);                # Also add(u,-v); But eval(u-v); will not do the arithmetic.
evalm(10*u);               # Also scalarmul(u,10);
evalm(u-2*v+3*w);          # Linear combination.
equal((u+v)+w,u+(v+w));    # Checking for equality.
equal(10*(u+v),10*u+10*v); # Same.
# Exercises 2,3.
am := augment(M, v);               # The augmented matrix [M:v].
rm := rref(am);                    # Reduction: The last column is non-pivot.
                                   # The system is consistent so v is a lin. comb.
                                   # in the columns of M. The coeffients of
```

```
evalm(rm[1,4]*col(M,1)+rm[2,4]*\    # the lin. com. are entries of the last column
col(M,2)+rm[3,4]*col(M,3));         # of rm. Indeed computing the lin. comb. yields v.
an := augment(N, w);                # [N:w].
rref(an);                           # The last column is pivot. No solutions.
                                    # w is not a lin. com. in the columns of N.
# Exercises 5-8.
with(plots):
o:=vector([0,0,0]):                 # The origin.
p1:=polygonplot3d([o,u]):           # Define and name the line segments, but do not
p2:=polygonplot3d([o,v]):           # display the graphs yet. Note that
p3:=polygonplot3d([o,w]):           # polygonplot3d([o,u],axes=boxed); does display u.
display(p1, axes=boxed);            # Display p1 in boxed axes. Repeat with p2 and p3.
display([p1,p2,p3],axes=boxed);         # Display all.
with(linalg):                           # The length of (1,2,3) is
norm(u,2)+norm(v,2);                    # norm([1,2,3],2); or norm(vector([1,2,3]),2);
evalf(");                               # or norm([1,2,3],frobenius);
norm(u+v,2); evalf(");                  # norm(vector[1,2,3],frobenius);
dotprod(u,v)-dotprod(u,w);              # dotprod(u,v) is the dot product of u and v.
angle(u,v); evalf(");
pr:=evalm(dotprod(v,w)/dotprod(w,w) * w); # The orthogonal projection of v on w.
vc:=evalm(v-pr);                        # The vector component of v orthogonal to w.
dotprod(pr,vc);                         # The dot product is zero and the
evalm(pr+vc);                           # sum is v as expected.
# Exercise 9 - Partial.
rref(augment(M, [1,3,2]));          # The last column is not pivot so u is in the span.
# Exercise 11.
concat(u,v,w); rref(");             # [u v w] has 3 pivots so the vecs. are independent.
# Exercise 15.
rref(N);                            # From the last column: (-2)xcol1+3xcol2 = col4.
# Exercise 16 - Hint:  Row reduce the coefficent matrix to see that its last
# column is pivot so the last column of the augmented matrix is not a pivot column.
# Exercise 19.
evalm(M &* vector([1,3,2]));                    # Etc.. Mr and Nu are undefined.
# Exercises 20, 21.
u:=[1,2,3]: v:=[-1,-1,1]: w:=[2,1,-4]:
with(linalg):
uv := crossprod(u,v);                       # Instead of lists [,,,] we could
dotprod(u,uv);                              # use vectors, vector([,,,]) .
evalm(crossprod(crossprod(u,v),w)+crossprod(crossprod(v,w),u)+
crossprod(crossprod(w,u),v));
# Exercise 22.
solve({-1-4*t=-9,2+t=4,1+5*t=11},t);    # t=2 for R so Q is in l.
solve({-1-3*t=7,-2+3*t=0,2-5*t=-10},t); # System has no solution so S is not in l.
with(plots):                            # Loading graphics package to use
spacecurve([-1-4*t,2+t,1+5*t],t=-2..3); # the command spacecurve. Etc..
```

```
# Exercise 25.
LineToLine := proc(p,q,r,s) local u,v,w,cr;              (* Code. *)
              u:=q-p; v:=s-r; w:=r-p; cr := linalg[crossprod](u,v);
              abs(linalg[dotprod](w,cr)) / linalg[norm](cr,2) end:
LineToLine([1,-2,-1],[0,-2,1],[-1,2,0],[-1,0,-2]);       (* Testing. *)
```

Selected Solutions with Mathematica

```
(* DATA and Remarks *)
<<LinearAlgebra'MatrixManipulation'; (*Loading package with matrix functions.*)
(* Problem A.*)                  (* Defining u,v,w as column vectors, or *)
u = {{1},{3},{2}}                (* column matrices. u = {1,2,3}, etc. would *)
v = {{-1},{1},{2}}               (* define them as "row vectors". *)
w = {{2},{1},{-4}}
M = {{1,3,5},{7,9,2},{4,6,8}}        (* M. *)
n = {{1,2,3,4},{2,3,4,5},{3,4,5,6}}
                                 (* N. Symbol N is reserved for numerical evaluation *)
(* To display a matrix m in row-column form use MatrixForm[m] .            *)
(* Exercises 1,2. *)
u + v                            (* Sum. *)
u - v                            (* Difference. *)
10 u                             (* Scalar product. *)
u - 2 v + 3 w                    (* Linear combination. *)
(u+v)+w===u+(v+w)        (* Checking for equality. Also, SameQ[(u+v)+w,u+(v+w)] *)
10 (u+v) === 10 u + 10 v         (* Same. *)
(* Exercises 2,3. *)
am = AppendRows[M,v]                (* The augmented matrix [M:v].              *)
rm = RowReduce[am]                  (* Reduction: The last column is non-pivot. *)
                                    (* The system is consistent so v is a lin. comb. *)
                                    (* in the columns of M. The coeffients of  *)
rm[[1,4]] TakeColumns[M,{1}]+       (* the lin. com. are entries of the last column *)
rm[[2,4]] TakeColumns[M,{2}]+       (* of rm. Indeed computing the lin. comb. yields v.*)
rm[[3,4]] TakeColumns[M,{3}]
an = AppendRows[n,w]                (* [N:w].                                   *)
RowReduce[an]                       (* The last column is pivot. No solutions.  *)
                                    (* w is not a lin. com. in the columns of N. *)
(* Exercises 5-8. *)
o={0,0,0}; u ={1,3,2}; v={-1,1,2}; w={2,1,-4};
p1=Line[{o,u}];             (* Define and name the line segments, but do not *)
p2=Line[{o,v}];             (* display the graphs yet. *)
p3=Line[{o,w}];                        (* Next display p1 with labeled axes.*)
Show[Graphics3D[p1,Axes->True]]        (* Then repeat with p2 and p3...        *)
Show[Graphics3D[{p1,p2,p3},Axes->True]] (* Display all.                         *)
(* Next we have to define our own norm-length-magnitude function. Starting *)
(* with a list-vector a, a^2 is list with elements the squares of the      *)
```

```
(* elements of a. Then we apply (@@ or Apply) Plus to a^2 to sum up all its  *)
(* components and finally take the square root of the sum.                    *)
Norm[a_]:=Sqrt[Plus@@(a^2)]
Norm[u] + Norm[v]
N[%]
Norm[u+v]
N[%]
u . v - u . w        (* The dot product of u and v is u.v or Dot[u,v].    *)
(* Define our own angle function.                                            *)
Angle[a_,b_]:=ArcCos[(a.b)/(Norm[a] Norm[b])]
N[Angle[u,v]]
pr = (v.w / w.w) w   (* Either use the formula for the orthogonal projection *)
(* or try Mathematica's built in Projection function in the                  *)
(*   LinearAlgebra'Orthogonalization' package.                               *)
<<LinearAlgebra'Orthogonalization';
pr=Projection[v,w]
vc = v - pr                  (* The vector component of v orthogonal to w. *)
pr . vc                      (* The dot product is zero and the            *)
pr + vc                      (* sum is v as expected.                      *)
(* Exercise 9 - Partial. *)
RowReduce[AppendRows[M,{{1},{3},{2}}]]    (* The last column is not pivot so u *)
                                          (* is in the span.*)
(* Exercise 11. *)
AppendRows[{{1},{3},{2}},{{-1},{1},{2}},{{2},{1},{-4}}]
RowReduce[%]              (* [u v w] has 3 pivots so the vecs. are independent. *)
(* Exercise 15. *)
RowReduce[n]                  (* From the last column: (-2)xcol1+3xcol2 = col4. *)
(* Exercise 16 - Hint:  Row reduce the coefficent matrix to see that its last  *)
(* column is pivot so the last column of the augmented matrix is not a pivot column. *)
(* Exercise 19. *)
M . {{1},{3},{2}}                        (* Etc.. Mr and Nu are undefined.*)
(* Exercises 20,21. *)
u={1,2,3}; v={-1,-1,1}; w={2,1,-4};
<<LinearAlgebra'CrossProduct';
uv = Cross[u,v]
u . uv
Cross[Cross[u,v],w]+Cross[Cross[v,w],u]+Cross[Cross[w,u],v]
(* Exercise 22. *)
Solve[{-1-4*t==-9,2+t==4,1+5*t==11},t]    (* t=2 so R is in l. *)
Solve[{-1-4*t==7,2+t==0,1+5*t==-11},t]   (* No solution. S is not in l.*)
ParametricPlot3D[{-1+3*t,-2+3*t,2-5*t},{t,-6,7}]
(* Exercise 25. *)
LineToLine[p_,q_,r_,s_] := Module[{u,v,w,cr},
                u=q-p; v=s-r; w=r-p; cr = Cross[u,v];
                Abs[w.cr] / Sqrt[cr[[1]]^2+cr[[2]]^2+cr[[3]]^2] ]
LineToLine[{1,-2,-1},{0,-2,1},{-1,2,0},{-1,0,-2}]          (* Testing *)
```

Selected Solutions with MATLAB

```
% DATA
u = [1; 3; 2]                          % Defining u,v,w.
v = [-1; 1; 2]
w = [2; 1; -4]
M = [1 3 5; 7 9 2; 4 6 8]              % M.
N = [1 2 3 4; 2 3 4 5; 3 4 5 6]        % N.
% Exercises 1,2.
u + v                                  % Sum.
u - v                                  % Difference.
10 * u                                 % Scalar product.
u-2*v+3*w                              % Linear combination.
(u+v)+w==u+(v+w)                       % Checking for equality. It returns
10*(u+v)==10*u+10*v                    % 1 (= TRUE) for each entry.
% Exercises 3,4.
am = [M v]                             % The augmented matrix [M:v].
rm = rref(am)                          % Reduction: The last column is non-pivot.
                                       % The system is consistent so v is a lin. comb.
                                       % in the columns of M. The coeffients of
rm(1,4)*M(:,1)+rm(2,4)*...             % the lin. com. are entries of the last column
M(:,2)+rm(3,4)*M(:,3)                  % of rm. Indeed computing the lin. comb. yields v.
an = [N w]                             % [N:w].
rref(an)                               % The last column is pivot. No solutions.
                                       % w is not a lin. com. in the columns of N.
% Exercises 5-8.
o=[0 0 0]; u=[1 3 2]; v=[-1 1 2]; w=[2 1 -4]
plot3([0 1], [0 3], [0 2])                       % Position vectors u,v,w.
plot3([0 -1], [0 1], [0 2])                      %
plot3([0 2], [0 1], [0 -4]),grid                 % grid adds a grid to the graph.
plot3([0 1 0 -1 0 2], [0 3 0 1 0 1], [0 2 0 2 0 -4])   % u,v,w together.
norm(u) + norm(v)
norm(u+v)
dot(u, v) - dot(u, w)                  % dot(a,b) is a.b .
acos(dot(u,v) / norm(u) / norm(v))     % Angle between u and v.
pr = (dot(v,w) / dot(w,w))*w           % The formula for the orthogonal projection.
vc = v - pr                            % The vector component of v orthogonal to w.
dot(pr, vc)                            % The dot product is (very close to) zero and
pr + vc                                %  the sum is v as expected.
% Exercise 9 - Partial.
rref([M [1;3;2]])                      % The last column is not pivot so u is in the span.
% Exercise 11.
rref([u; v; w])                        % [u v w] has 3 pivots so the vecs. are independent.
% Exercise 15.
rref(N)                                % From the last column: (-2)xcol1+3xcol2 = col4.
```

```
% Exercise 16 - Hint:  Row reduce the coefficent matrix to see that its last
% column is pivot so the last column of the augmented matrix is not a pivot column.
% Exercise 19.
M*[1;3;2]                        % Also: M*[2;-3;1;-4], etc.. Mr and Nu are undefined.
% Exercises 20,21.
u=[1 2 3]; v=[-1 -1 1]; w=[2 1 -4];
uv = cross(u,v)
cross(cross(u,v),w)+cross(cross(v,w),u)+cross(cross(w,u),v)
% Exercise 22.
% R is in l since the system -1-4*t=-9,2+t=4,1+5*t=11 is consistent because
[roots([-4 -1+9]) roots([1 2-4]) roots([5 1-11])]    % returns [2 2 2].
[roots([-4 -1-7]) roots([1 2-0]) roots([5 1+10])]    % [.2,.2,-2.2] so the
t = -2:.25:3;                         % system has no solution and S is not in l.
plot3(-1-4*t,2+t,1+5*t)               % Plotting the line.
% Exercise 25.
function [A] = LnToLn(p,q,r,s)                    % In a file.
           A = abs(dot(r-p,cross(q-p,s-r))) / norm(cross(q-p,s-r));
        end
LnToLn([1 -2 -1],[0 -2 1],[-1 2 0],[-1 0 -2])  % In session.
```

3

Matrices

For what I have accomplished I have to thank my industry much more than any outstanding talent.
—(Julius Wilhelm) Richard Dedekind (1831–1916)

Introduction

Sylvester and Cayley

Lord Cayley[1] was one of the founders of matrix theory, although it was his friend Sylvester[2] who first coined the term *matrix*. Both Sylvester and Cayley ranked among the best mathematicians of their times. Sylvester became the first professor of the mathematics department at The Johns Hopkins University and founded the prestigious *American Journal of Mathematics*.

Here is the type of example that Cayley considered. Three coordinate systems (x, y), (x', y'), and (x'', y'') are connected by the following transformations

$$x'' = x' - y' \qquad x' = x + 2y$$
$$y'' = x' + y' \quad \text{and} \quad y' = 2x - y$$

The relationship between (x, y) and (x'', y'') is given by the substitution

$$x'' = x' - y' = (x + 2y) - (2x - y) = -x + 3y$$
$$y'' = x' + y' = (x + 2y) + (2x - y) = 3x + y$$

---

[1](**Sir**) **Arthur Cayley** (1821–1895) was born in Surrey, England, and was educated at Cambridge University. He practiced law while writing papers in mathematics. A few years after meeting his colleague Sylvester, another lawyer-mathematician, he quit law to pursue mathematics full time.
[2]**James Joseph Sylvester** (1814–1897) was born in London of Jewish parents. He entered Cambridge University, but due to his religion, he was not awarded a diploma until several years after he had completed his studies. He studied for the bar while conducting mathematical research. He and Cayley had a long and fruitful collaboration on *invariant theory*, an area related to linear algebra.

This transformation can be also obtained as follows: If we abbreviate the three changes of coordinates by the *matrices* of the coefficients, we have

$$A = \begin{bmatrix} 1 & -1 \\ 1 & 1 \end{bmatrix} \qquad B = \begin{bmatrix} 1 & 2 \\ 2 & -1 \end{bmatrix} \qquad C = \begin{bmatrix} -1 & 3 \\ 3 & 1 \end{bmatrix}$$

Now C can be computed straight from A and B as $[\ A\mathbf{b}_1 \quad A\mathbf{b}_2\ ]$; i.e., C is the matrix with columns $A\mathbf{b}_1$ and $A\mathbf{b}_2$, where $\mathbf{b}_1$ and $\mathbf{b}_2$ are the columns of B. This procedure is called *matrix multiplication*, and it is discussed in Section 3.1.

3.1   Matrix Operations

Reader's Goals for This Section

1. To know how to perform addition, scalar multiplication, and matrix multiplication.
2. To understand the basic properties of these operations.
3. To be able to solve simple matrix equations.

In this section we introduce the basic matrix operations: addition, scalar multiplication, and matrix multiplication.

Recall that an $m \times n$ **matrix** A is a rectangular arrangement of $m \cdot n$ numbers into m horizontal **rows** and n vertical **columns**.

$$A = \begin{bmatrix} a_{11} & a_{12} & \cdots & a_{1n} \\ a_{21} & a_{22} & \cdots & a_{2n} \\ \vdots & \vdots & \ddots & \vdots \\ a_{m1} & a_{m2} & \cdots & a_{mn} \end{bmatrix}.$$

The number a_{ij} is the (i, j)**th entry** of A. The i**th row** and the j**th column** of A are

$$\begin{bmatrix} a_{i1} & a_{i2} & \cdots & a_{in} \end{bmatrix} \qquad \text{and} \qquad \begin{bmatrix} a_{1j} \\ a_{2j} \\ \vdots \\ a_{mj} \end{bmatrix}$$

respectively. The matrix A may also be viewed as a sequence of its columns $\mathbf{a}_1, \ldots, \mathbf{a}_n$:

$$A = \begin{bmatrix} \mathbf{a}_1 & \mathbf{a}_2 & \cdots & \mathbf{a}_n \end{bmatrix}$$

A $1 \times n$ matrix is called a **row matrix**. An $m \times 1$ matrix is called a **column matrix** or a **vector**. An $n \times n$ matrix is called a **square matrix** (also defined in Section 1.3). A square matrix has equal numbers of rows and columns. Corresponding examples are:

$$\begin{bmatrix} 1 & -2 & \sqrt{3} \end{bmatrix} \qquad \begin{bmatrix} 3.5 \\ 1 \end{bmatrix} \qquad \begin{bmatrix} 1 & 2 \\ 3 & 4 \end{bmatrix}$$

A matrix whose entries are all zero is called a **zero matrix**, and it is denoted by **0**.

$$\mathbf{0} = [0] \qquad \mathbf{0} = \begin{bmatrix} 0 \\ 0 \end{bmatrix} \qquad \mathbf{0} = \begin{bmatrix} 0 & 0 \end{bmatrix} \qquad \mathbf{0} = \begin{bmatrix} 0 & 0 \\ 0 & 0 \end{bmatrix}$$

$$\mathbf{0} = \begin{bmatrix} 0 & 0 \\ 0 & 0 \\ 0 & 0 \end{bmatrix} \qquad \mathbf{0} = \begin{bmatrix} 0 & 0 & 0 \\ 0 & 0 & 0 \end{bmatrix}$$

Let A be a *square* matrix with entries a_{ij}. The entries a_{ii} form the **main diagonal.** A is **upper triangular** if all entries below the main diagonal are zero. A is **lower triangular** if all entries above the main diagonal are zero. A is **diagonal** if all entries above and below the main diagonal are zero. A is **scalar** if it is diagonal with all diagonal entries equal.

▪ EXAMPLE 1   Let

$$A = \begin{bmatrix} 1 & 2 & -3 \\ 0 & 5 & -4 \\ 0 & 0 & 9 \end{bmatrix} \qquad B = \begin{bmatrix} 1 & 0 & 0 \\ 0 & 0 & 0 \\ 1 & 1 & 0 \end{bmatrix}$$

$$C = \begin{bmatrix} 1 & 0 & 0 \\ 0 & -2 & 0 \\ 0 & 0 & 2 \end{bmatrix} \qquad D = \begin{bmatrix} -2 & 0 & 0 \\ 0 & -2 & 0 \\ 0 & 0 & -2 \end{bmatrix}$$

A, C, D are upper triangular. B, C, D are lower triangular. C, D are diagonal. D is scalar. The main diagonal of A is 1, 5, 9, whereas that of C is 1, -2, 2.   ▪

Two matrices are **equal** if they have the same size and their corresponding entries are equal. This is the same as saying that the corresponding columns are equal as vectors. For example, $\begin{bmatrix} a & 5 \\ -1 & 0 \end{bmatrix} = \begin{bmatrix} 1 & 5 \\ -1 & b \end{bmatrix}$ only if $a = 1$ and $b = 0$. Matrices of *different sizes are never equal.*

Matrix Addition and Scalar Multiplication

The **sum**, $A + B$, of two matrices A and B of the same size is obtained by adding the corresponding entries of the matrices. Likewise, for the **difference**, $A - B$, we subtract the corresponding entries. Matrices of different sizes *cannot* be added or subtracted.

▪ EXAMPLE 2

$$\begin{bmatrix} -1 & 2 \\ 1 & 1 \\ 1 & 1 \end{bmatrix} + \begin{bmatrix} -1 & 0 \\ 1 & 2 \\ 4 & 1 \end{bmatrix} = \begin{bmatrix} -2 & 2 \\ 2 & 3 \\ 5 & 2 \end{bmatrix}$$

$$\begin{bmatrix} 1 & 1 & 2 \\ 2 & 0 & -3 \end{bmatrix} - \begin{bmatrix} 1 & -1 & 0 \\ -1 & 2 & -3 \end{bmatrix} = \begin{bmatrix} 0 & 2 & 2 \\ 3 & -2 & 0 \end{bmatrix}$$

Let A be any matrix and c be any scalar. The **scalar product**, cA, is the matrix obtained from multiplying each entry of A by c. If $c = -1$, $(-1)A$ is called the **opposite of A**, and it is denoted by $-A$.

■ EXAMPLE 3   Compute $2A$ and $(-1)A = -A$ if

$$A = \begin{bmatrix} 1 & -3 \\ 0 & 1 \\ -2 & 2 \end{bmatrix}$$

SOLUTION

$$2A = \begin{bmatrix} 2 & -6 \\ 0 & 2 \\ -4 & 4 \end{bmatrix}, \qquad (-1)A = -A = \begin{bmatrix} -1 & 3 \\ 0 & -1 \\ 2 & -2 \end{bmatrix}$$

Note that

$$A - B = A + (-B)$$

Matrix addition, subtraction, and scalar multiplication satisfy a few basic rules, summarized in the following theorem, which is the analogue of Theorem 1 of Section 2.1.

THEOREM 1

(Laws for Addition and Scalar Multiplication)

Let A, B, C be any $m \times n$ matrices and let a, b, c be any scalars. The following hold:

1. $(A + B) + C = A + (B + C)$ 6. $(a + b)C = aC + bC$
2. $A + B = B + A$ 7. $(ab)C = a(bC) = b(aC)$
3. $A + 0 = 0 + A = A$ 8. $1A = A$
4. $A + (-A) = (-A) + A = 0$ 9. $0A = 0$
5. $c(A + B) = cA + cB$

REMARKS

1. Property 1 is called the **associative law** for addition, whereas Property 2 is the **commutative law**. Properties 5 and 6 are known as **distributive laws**.

2. Because matrices may be viewed as sequences of vectors, Theorem 1 follows easily from Theorem 1 of Section 2.1. However, it is also useful to consider a matrix as a rectangular arrangement of its entries. Let us practice this by proving Property 5.

PROOF OF 5   Clearly, $c(A + B)$ and $cA + cB$ have the same size. Let p_{ij} be the (i, j) entry of $c(A + B)$ and q_{ij} be the (i, j) entry of $cA + cB$. We need to show that $p_{ij} = q_{ij}$. If a_{ij} and b_{ij} are the (i, j) entries of A and B, then $p_{ij} = c(a_{ij} + b_{ij})$ and $q_{ij} = ca_{ij} + cb_{ij}$. But $c(a_{ij} + b_{ij}) = ca_{ij} + cb_{ij}$. Therefore, $p_{ij} = q_{ij}$, so all the corresponding entries are equal. Hence, $c(A + B) = cA + cB$.

NOTE   The associative law allows us to drop the parentheses from multiple sums. So, all the equal expressions $(A + B) + (C + D), A + ((B + C) + D), A + (B + (C + D))$ can be simply written as $A + B + C + D$.

Simple Matrix Equations

Theorem 1 can be used in solving simple matrix equations. Let

$$A = \begin{bmatrix} 1 & 2 & 0 \\ 0 & 0 & -1 \end{bmatrix} \quad \text{and} \quad B = \begin{bmatrix} 0 & 1 & 3 \\ 1 & 0 & -1 \end{bmatrix}$$

■ EXAMPLE 4   Find the matrix X such that $2X - 4B = 3A$.

SOLUTION   Add $4B$ to both sides of the equation to get

$$(2X - 4B) + 4B = 3A + 4B$$
$$2X + (-4B + 4B) = 3A + 4B \quad \text{by 1 of Theorem 1}$$
$$2X + \mathbf{0} = 3A + 4B \quad \text{by 4}$$
$$2X = 3A + 4B \quad \text{by 3}$$

Multiply both sides of the last equation by $\frac{1}{2}$:

$$\frac{1}{2}(2X) = \frac{1}{2}(3A + 4B) \quad \text{or} \quad X = \frac{3}{2}A + 2B \quad \text{by 8, 5, and 7}$$

Therefore,

$$X = \frac{3}{2}\begin{bmatrix} 1 & 2 & 0 \\ 0 & 0 & -1 \end{bmatrix} + 2\begin{bmatrix} 0 & 1 & 3 \\ 1 & 0 & -1 \end{bmatrix} = \begin{bmatrix} \frac{3}{2} & 5 & 6 \\ 2 & 0 & -\frac{7}{2} \end{bmatrix}$$

Matrix Multiplication

Matrix multiplication is the most important matrix operation. Defining the product of two matrices as the matrix of the products of the corresponding entries is *not* very useful in applications. The following definition is far more useful.

DEFINITION

(Matrix Multiplication)

Let A be an $m \times k$ matrix and B be a $k \times n$ matrix. The **product** AB is the $m \times n$ matrix with columns $A\mathbf{b}_1, \ldots, A\mathbf{b}_n$, where $\mathbf{b}_1, \ldots, \mathbf{b}_n$ are the columns of B.

For example, if

$$A = \begin{bmatrix} 2 & 0 & 1 \\ 2 & 1 & 2 \end{bmatrix}, \qquad B = \begin{bmatrix} 3 & 2 & 4 \\ -2 & 4 & 5 \\ 0 & 3 & -2 \end{bmatrix}$$

then

$$\begin{bmatrix} 2 & 0 & 1 \\ 2 & 1 & 2 \end{bmatrix}\begin{bmatrix} 3 \\ -2 \\ 0 \end{bmatrix} = \begin{bmatrix} 6 \\ 4 \end{bmatrix}, \quad \begin{bmatrix} 2 & 0 & 1 \\ 2 & 1 & 2 \end{bmatrix}\begin{bmatrix} 2 \\ 4 \\ 3 \end{bmatrix} = \begin{bmatrix} 7 \\ 14 \end{bmatrix},$$

$$\begin{bmatrix} 2 & 0 & 1 \\ 2 & 1 & 2 \end{bmatrix}\begin{bmatrix} 4 \\ 5 \\ -2 \end{bmatrix} = \begin{bmatrix} 6 \\ 9 \end{bmatrix}$$

Hence,

$$AB = [\, A\mathbf{b}_1 \quad A\mathbf{b}_2 \quad A\mathbf{b}_3 \,] = \begin{bmatrix} 6 & 7 & 6 \\ 4 & 14 & 9 \end{bmatrix}$$

We also have

$$[\, 5 \quad 1 \quad -1 \,]\begin{bmatrix} 1 \\ -3 \\ 4 \end{bmatrix} = [-2], \qquad \begin{bmatrix} 1 \\ -3 \\ 4 \end{bmatrix}[\, 5 \quad 1 \quad -1 \,] = \begin{bmatrix} 5 & 1 & -1 \\ -15 & -3 & 3 \\ 20 & 4 & -4 \end{bmatrix}$$

CAUTION   Matrix multiplication is only possible if the number of columns of the first matrix equals the number of rows of the second matrix. (Otherwise, $A\mathbf{b}_1, \ldots$ would be undefined.)

Just as in the case of $A\mathbf{x}$, it is often useful to obtain AB one entry at a time. The (i, j) entry of AB can be computed as follows: Take the ith row of A and the jth column of B. Multiply their corresponding entries together and add all the products.

▪ **EXAMPLE 5**   Compute the $(1, 2)$ and the $(2, 3)$ entry of AB.

SOLUTION

$$\begin{bmatrix} \boxed{2 \quad 0 \quad 1} \\ 2 \quad 1 \quad 2 \end{bmatrix} \begin{bmatrix} 3 & \boxed{2} & 4 \\ -2 & \boxed{4} & 5 \\ 0 & \boxed{3} & -2 \end{bmatrix} = \begin{bmatrix} \cdot & 7 & \cdot \\ \cdot & \cdot & \cdot \end{bmatrix}$$

$$2 \cdot 2 + 0 \cdot 4 + 1 \cdot 3 = 7$$

$$\begin{bmatrix} 2 \quad 0 \quad 1 \\ \boxed{2 \quad 1 \quad 2} \end{bmatrix} \begin{bmatrix} 3 & 2 & \boxed{4} \\ -2 & 4 & \boxed{5} \\ 0 & 3 & \boxed{-2} \end{bmatrix} = \begin{bmatrix} \cdot & \cdot & \cdot \\ \cdot & \cdot & 9 \end{bmatrix}$$

Row-Column Computation of AB

In general, if $C = [c_{ij}] = AB$, the entries c_{ij} are given by

$$A = \begin{bmatrix} a_{11} & a_{12} & \cdots & a_{1k} \\ \vdots & \vdots & \vdots & \vdots \\ \boxed{a_{i1}} & \boxed{a_{i2}} & \cdots & \boxed{a_{ik}} \\ \vdots & \vdots & \vdots & \vdots \\ a_{m1} & a_{m2} & \cdots & a_{mk} \end{bmatrix}, \qquad B = \begin{bmatrix} b_{11} & \cdots & b_{1j} & \cdots & b_{1n} \\ b_{21} & \cdots & b_{2j} & \cdots & b_{2n} \\ \vdots & \vdots & \vdots & \vdots & \vdots \\ b_{k1} & \cdots & b_{kj} & \cdots & b_{kn} \end{bmatrix}$$

$$c_{ij} = a_{i1}b_{1j} + a_{i2}b_{2j} + \cdots + a_{ik}b_{kj}$$

In other words, the (i, j) entry of AB is the **dot product** of the ith row of A and the jth column of B.

We often need to compute only a certain row or column of a matrix product.

■ EXAMPLE 6   (Column of Product)   Compute the third column of AB.

SOLUTION   This is the same as the product of A with the third column of B.

$$\begin{bmatrix} 2 & 0 & 1 \\ 2 & 1 & 2 \end{bmatrix} \begin{bmatrix} 4 \\ 5 \\ -2 \end{bmatrix} = \begin{bmatrix} 6 \\ 9 \end{bmatrix}$$

■ EXAMPLE 7   (Row of Product)   Compute the second row of AB.

SOLUTION   This is the same as the product of the second row of A with all of B.

$$\begin{bmatrix} 2 & 1 & 2 \end{bmatrix} \begin{bmatrix} 3 & 2 & 4 \\ -2 & 4 & 5 \\ 0 & 3 & -2 \end{bmatrix} = \begin{bmatrix} 4 & 14 & 9 \end{bmatrix}$$

In general,

The ith row of AB is the product of the ith row of A with B.

The jth column of AB is the product of A with the jth column of B.

The basic properties of matrix multiplication can be summarized in the following.

THEOREM 2

(Laws of Matrix Multiplication)

Let A be an $m \times n$ matrix and B, C have sizes such that the operations below can be performed. Let a be any scalar.

1. $(AB)C = A(BC)$ **Associative law**
2. $A(B + C) = AB + AC$ **Left distributive law**
3. $(B + C)A = BA + CA$ **Right distributive law**
4. $a(BC) = (aB)C = B(aC)$
5. $I_m A = A I_n = A$ **Multiplicative identity**
6. $0A = 0$ and $A0 = 0$

PROOF OF 1   Let us now verify *associativity* and leave the remaining proofs as exercises. First we prove the special case where $C = \mathbf{v}$ is a vector (one-column matrix) with components $(v_1, \ldots, v_k)$. Then,

$$(AB)\mathbf{v} = [A\mathbf{b}_1 \cdots A\mathbf{b}_k]\mathbf{v} = v_1(A\mathbf{b}_1) + \cdots + v_k(A\mathbf{b}_k) = A(v_1\mathbf{b}_1 + \cdots + v_k\mathbf{b}_k)$$

$$= A(B\mathbf{v}) \tag{3.1}$$

Now let C have l columns. By (3.1),

$$(AB)C = [(AB)\mathbf{c}_1 \cdots (AB)\mathbf{c}_l] = [A(B\mathbf{c}_1) \cdots A(B\mathbf{c}_l)] = A[B\mathbf{c}_1 \cdots B\mathbf{c}_l] = A(BC)$$

REMARK   Just as with addition, this associative law allows us to drop parentheses from multiple products. So,

$$(AB)(CD) = A((BC)D) = A(B(CD)) = ABCD$$

■ EXAMPLE 8   Verify the left distributive law for matrix multiplication if

$$A = \begin{bmatrix} 1 & 1 \\ 0 & 1 \\ 1 & 0 \end{bmatrix}, \qquad B = \begin{bmatrix} 1 & 2 \\ 3 & 4 \end{bmatrix}, \qquad C = \begin{bmatrix} 1 & 0 \\ 1 & 1 \end{bmatrix}$$

SOLUTION

$$A(B + C) = \begin{bmatrix} 1 & 1 \\ 0 & 1 \\ 1 & 0 \end{bmatrix} \begin{bmatrix} 2 & 2 \\ 4 & 5 \end{bmatrix} = \begin{bmatrix} 6 & 7 \\ 4 & 5 \\ 2 & 2 \end{bmatrix}$$

$$AB + AC = \begin{bmatrix} 4 & 6 \\ 3 & 4 \\ 1 & 2 \end{bmatrix} + \begin{bmatrix} 2 & 1 \\ 1 & 1 \\ 1 & 0 \end{bmatrix} = \begin{bmatrix} 6 & 7 \\ 4 & 5 \\ 2 & 2 \end{bmatrix}$$

Powers of a Square Matrix

Let A be a square matrix. The product AA is also denoted by A^2. Likewise, $AAA = A^3$ and $AA \cdots A = A^n$ for n factors of A. We write A^1 for A. If A is nonzero, we also write A^0 for I.

$$A^n = \underbrace{AA \cdots A}_{n \text{ factors}}, \qquad A^1 = A, \qquad A^0 = I$$

■ **EXAMPLE 9** Let

$$A = \begin{bmatrix} 1 & -1 \\ -2 & 3 \end{bmatrix}, \quad B = \begin{bmatrix} 1 & 2 \\ 0 & 0 \end{bmatrix}, \quad \text{and} \quad C = \begin{bmatrix} 0 & 1 \\ 0 & 0 \end{bmatrix}$$

Then

$$A^1 = \begin{bmatrix} 1 & -1 \\ -2 & 3 \end{bmatrix}, \quad A^2 = \begin{bmatrix} 3 & -4 \\ -8 & 11 \end{bmatrix}, \quad A^3 = \begin{bmatrix} 11 & -15 \\ -30 & 41 \end{bmatrix} \cdots$$

$$B^1 = \begin{bmatrix} 1 & 2 \\ 0 & 0 \end{bmatrix}, \quad B^2 = \begin{bmatrix} 1 & 2 \\ 0 & 0 \end{bmatrix}, \quad B^3 = \begin{bmatrix} 1 & 2 \\ 0 & 0 \end{bmatrix} \cdots$$

$$C^1 = \begin{bmatrix} 0 & 1 \\ 0 & 0 \end{bmatrix}, \quad C^2 = \begin{bmatrix} 0 & 0 \\ 0 & 0 \end{bmatrix}, \quad C^3 = \begin{bmatrix} 0 & 0 \\ 0 & 0 \end{bmatrix} \cdots$$

THEOREM 3

The following relations hold for any positive integers n and m.

$$A^n A^m = A^{n+m}, \qquad (A^n)^m = A^{nm}, \qquad (cA)^n = c^n A^n$$

The Trouble with Matrix Multiplication

Theorem 2 describes properties shared by both matrix multiplication and ordinary multiplication (using I in place of 1). However, there are important differences that make matrix multiplication a little harder but far more interesting than ordinary multiplication. The most notable difference is that matrix multiplication is *noncommutative*, meaning that the property $ab = ba$ that holds for all numbers is no longer true. But let us be more specific.

1. If AB is defined, BA is not necessarily defined (for example, if A is 2×2 and B is 2×3).
2. If AB and BA are both defined, they do not have to be of the same size (for example, if A is 3×2 and B is 2×3).
3. If AB and BA are both defined and of the same size, they do not have to be equal. For example,

$$\begin{bmatrix} 1 & 1 \\ 1 & 1 \end{bmatrix} \begin{bmatrix} 0 & 1 \\ 0 & 0 \end{bmatrix} = \begin{bmatrix} 0 & 1 \\ 0 & 1 \end{bmatrix}, \quad \text{whereas} \quad \begin{bmatrix} 0 & 1 \\ 0 & 0 \end{bmatrix} \begin{bmatrix} 1 & 1 \\ 1 & 1 \end{bmatrix} = \begin{bmatrix} 1 & 1 \\ 0 & 0 \end{bmatrix}$$

4. $AB = \mathbf{0}$ does *not* always imply that either A or B is $\mathbf{0}$ (not even when $A = B$). For example,

$$\begin{bmatrix} 0 & 1 \\ 0 & 0 \end{bmatrix}^2 = \begin{bmatrix} 0 & 0 \\ 0 & 0 \end{bmatrix}$$

5. $CA = CB$ does *not* necessarily imply that $A = B$. For example,

$$\begin{bmatrix} 1 & 0 \\ 0 & 0 \end{bmatrix} \begin{bmatrix} 1 & 1 \\ 1 & 1 \end{bmatrix} = \begin{bmatrix} 1 & 1 \\ 0 & 0 \end{bmatrix} = \begin{bmatrix} 1 & 0 \\ 0 & 0 \end{bmatrix} \begin{bmatrix} 1 & 1 \\ 0 & 0 \end{bmatrix}$$

6. $AC = BC$ does *not* always imply that $A = B$. (Find an example.)
7. $A^2 = I$ does *not* necessarily imply that $A = \pm I$. (See Exercise 20.)
8. In general, $(AB)^n \neq A^n B^n$. (See Exercises 21 and 23.)

If two matrices A and B satisfy $AB = BA$, we say that they **commute**.

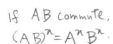

If AB commute,
$(AB)^n = A^n B^n$.

■ EXAMPLE 10

$$A = \begin{bmatrix} 0 & 0 \\ 1 & 1 \end{bmatrix} \text{ and } B = \begin{bmatrix} 1 & 0 \\ 2 & 3 \end{bmatrix} \text{ commute. (Verify.)}$$

Block Matrices

A **submatrix** of a matrix is the matrix obtained by deleting any rows and/or columns from the matrix. For example, $\begin{bmatrix} 1 & 2 \\ 4 & 5 \end{bmatrix}$, $\begin{bmatrix} 2 \\ 5 \end{bmatrix}$, $[\,4 \quad 5 \quad 6\,]$ are submatrices of $M = \begin{bmatrix} 1 & 2 & 3 \\ 4 & 5 & 6 \end{bmatrix}$.

We often partition a large matrix that cannot be easily manipulated into "blocks" of submatrices that are easier to handle. We partition a matrix by separating rows and columns using horizontal and vertical lines. The resulting matrix of submatrices is called a **block** matrix.

For example, let us partition a 3×6 matrix A as

$$A = \left[\begin{array}{ccc:ccc:c} 1 & 2 & 0 & 1 & -1 & & 1 \\ \hdashline 1 & 3 & 5 & 0 & 1 & & 2 \\ 2 & 4 & 6 & 1 & 0 & & 0 \end{array}\right]$$

We may view A as the 2×3 block matrix

$$\begin{bmatrix} A_{11} & A_{12} & A_{13} \\ A_{21} & A_{22} & A_{23} \end{bmatrix}$$

where $A_{11} = [\,1 \quad 2 \quad 0\,], \dots, A_{23} = \begin{bmatrix} 2 \\ 0 \end{bmatrix}$.

The main reason block matrices are used is that we can perform matrix operations on them just as if they were "plain" matrices, provided that the sizes of the blocks are compatible.[3] For example, if

$$A = \begin{bmatrix} 1 & 2 \\ 3 & 4 \\ \cdots & \cdots \\ 1 & 1 \\ 5 & 6 \end{bmatrix} = \begin{bmatrix} A_1 \\ A_2 \end{bmatrix}, \qquad B = \begin{bmatrix} -1 & 2 \\ -3 & 4 \\ \cdots & \cdots \\ -5 & 6 \\ -1 & 2 \end{bmatrix} = \begin{bmatrix} B_1 \\ B_2 \end{bmatrix}$$

then it is easy to check that

$$A + B = \begin{bmatrix} A_1 + B_1 \\ A_2 + B_2 \end{bmatrix}$$

Also, if

$$C = \begin{bmatrix} 1 & 3 & -4 & \vdots & 0 & 0 \\ 1 & -1 & 0 & \vdots & 0 & 0 \\ \cdots & \cdots & \cdots & \vdots & \cdots & \cdots \\ 0 & 2 & 4 & \vdots & 1 & 0 \\ 3 & 5 & 7 & \vdots & 0 & 1 \end{bmatrix} = \begin{bmatrix} C_{11} & C_{12} \\ C_{21} & C_{22} \end{bmatrix}$$

and

$$D = \begin{bmatrix} -1 & 0 \\ 3 & 1 \\ 2 & 5 \\ \cdots & \cdots \\ -4 & 0 \\ 0 & 2 \end{bmatrix} = \begin{bmatrix} D_1 \\ D_2 \end{bmatrix}$$

then we see that

$$CD = \begin{bmatrix} C_{11}D_1 + C_{12}D_2 \\ C_{21}D_1 + C_{22}D_2 \end{bmatrix}$$

Applications

Matrices and their operations can be used to record, update, and scale tabular data.

■ EXAMPLE 11 (Stock Update) A bookstore company has three outlets, each outlet carrying fiction, travel, and sports books, whose numbers are tabulated as follows:

Outlet	Fiction	Travel	Sports
1	300	300	100
2	300	100	240
3	50	150	200

[3]Block matrix operations are used mainly to manipulate very large matrices that are too big to fit into a computer's memory.

Suppose that deliveries D are made to each outlet. Find the updated stock.

$$D = \begin{bmatrix} 60 & 40 & 20 \\ 60 & 40 & 30 \\ 60 & 40 & 30 \end{bmatrix}$$

SOLUTION   To get the updated stock, we need to add the two matrices.

$$\begin{bmatrix} 360 & 340 & 120 \\ 360 & 140 & 270 \\ 110 & 190 & 230 \end{bmatrix}$$

Thus, the third outlet now carries 190 books on travel.

■ EXAMPLE 12   (Scaling)   Suppose the distances in miles between Annapolis, Baltimore, and Washington, D.C., are given by

City	Annapolis	Baltimore	Washington
Annapolis	0	30	25
Baltimore	30	0	18
Washington	25	18	0

If we want to draw a map scaled so that 1 in. of paper corresponds to 5 mi of real distance, what is the map's matrix of distances?

SOLUTION   A distance of 30 mi, for example, is scaled as $30 \text{ mi} \cdot \frac{1}{5} \frac{\text{in.}}{\text{mi}} = 6 \text{ in.}$ So we need to multiply all entries of the matrix by $\frac{1}{5} = 0.2$. This is the scalar product of the matrix and 0.2.

$$0.2 \begin{bmatrix} 0 & 30 & 25 \\ 30 & 0 & 18 \\ 25 & 18 & 0 \end{bmatrix} = \begin{bmatrix} 0 & 6 & 5 \\ 6 & 0 & 3.6 \\ 5 & 3.6 & 0 \end{bmatrix}$$

■ EXAMPLE 13   (Revenues from Several Sources)   Each of two appliance outlets o_1 and o_2 receives and sells daily from two factories, f_1 and f_2, TVs (t) and VCRs (v) according to

$$\begin{array}{cc} & \begin{array}{cc} t & v \end{array} \\ \begin{array}{c} f_1 \\ f_2 \end{array} & \begin{bmatrix} 40 & 50 \\ 70 & 80 \end{bmatrix} \end{array}$$

The charge per appliance, in dollars, for each outlet is as follows:

$$\begin{array}{cc} & \begin{array}{cc} o_1 & o_2 \end{array} \\ \begin{array}{c} t \\ v \end{array} & \begin{bmatrix} 200 & 250 \\ 300 & 280 \end{bmatrix} \end{array}$$

If A and B are the matrices of these tables, compute and interpret the product AB.

SOLUTION

$$AB = \begin{bmatrix} 40 & 50 \\ 70 & 80 \end{bmatrix} \begin{bmatrix} 200 & 250 \\ 300 & 280 \end{bmatrix} = \begin{bmatrix} 23{,}000 & 24{,}000 \\ 38{,}000 & 39{,}900 \end{bmatrix}$$

The $(1, 1)$st entry $40 \cdot 200 + 50 \cdot 300 = 23{,}000$ is the first outlet's revenue from selling all the appliances coming from the first factory. Similarly,

$$AB = \begin{bmatrix} \$ \text{ in } o_1 \text{ from } f_1 & \$ \text{ in } o_2 \text{ from } f_1 \\ \$ \text{ in } o_1 \text{ from } f_2 & \$ \text{ in } o_2 \text{ from } f_2 \end{bmatrix}$$

Some Useful Notes

Multiplication by a Scalar Matrix

Let A be an $m \times n$ matrix and let Q and R be scalar matrices of sizes $m \times m$ and $n \times n$, respectively, with the scalar c along their main diagonals. We can combine Properties 4 and 5 of Theorem 2 to compute $QA = (cI_m)A = c(I_mA) = cA$. Similarly, $AR = cA$. Hence,

$$QA = AR = cA$$

In other words, *multiplication by a scalar matrix is the same as multiplication by a scalar.* Moreover, the sum, difference, and product of scalar matrices of the same size is again a scalar matrix.

■ EXAMPLE 14

$$\begin{bmatrix} -2 & 0 & 0 \\ 0 & -2 & 0 \\ 0 & 0 & -2 \end{bmatrix} \begin{bmatrix} 1 & 2 \\ 3 & 2 \\ 1 & 1 \end{bmatrix} = \begin{bmatrix} -2 & -4 \\ -6 & -4 \\ -2 & -2 \end{bmatrix} = \begin{bmatrix} 1 & 2 \\ 3 & 2 \\ 1 & 1 \end{bmatrix} \begin{bmatrix} -2 & 0 \\ 0 & -2 \end{bmatrix} = -2 \begin{bmatrix} 1 & 2 \\ 3 & 2 \\ 1 & 1 \end{bmatrix}$$

Fast Calculation of A^k by Squaring

It is rather tedious to compute powers of a square matrix A. Say, for example, we need to compute A^8. If we try to use the definition

$$A^8 = (((((((AA)A)A)A)A)A)A$$

it would require 7 matrix multiplications. If, however, we compute A^2 first, then square it, $A^2A^2 = A^4$, and square that result, $A^4A^4 = A^8$, we need only 3 matrix multiplications.

$$A^8 = \left(\left(A^2 \right)^2 \right)^2$$

This method applies to any matrix power A^n. For example, we can compute A^{13} as follows: we find A^2, square it to get A^4, and square that result to get A^8. Now, $A^{13} = A^8A^4A$. This process takes only 5 matrix multiplications, instead of 12 if we choose the definition.

$$A^{13} = \left(\left(A^2 \right)^2 \right)^2 A^4A$$

Fast Calculation of $A^k\mathbf{x}$ without Matrix Powers

If we want to compute the product $A^k\mathbf{x}$ for an $n \times n$ matrix A and an n-vector $\mathbf{x}$, it is possible and smart to avoid the calculation of A^k. What we must do is first compute the n-vector $A\mathbf{x}$. Then compute the n-vector $A(A\mathbf{x})$ and continue this way until we are done. In other words, use

$$A(\dots(A(A\mathbf{x}))\dots) = A^k\mathbf{x}$$

where the left-hand side of the equation has $k-1$ pairs of parentheses. The advantage of this method is that we always compute the product of a matrix and a vector and never the product of two matrices.

To illustrate, let $n = 3$ and $k = 2$. Computing A^2 first requires 27 multiplications. For $A^2\mathbf{x}$ we need another 9 multiplications, a total of 36 multiplications. If, however, we compute $A\mathbf{x}$ first, we need 9 multiplications. Then for $A(A\mathbf{x})$, we need 9 more multiplications, a total of only 18 multiplications.

Matrix Operations with CAS

We can use the following to compute $AB - B^2$ for

$$A = \begin{bmatrix} 1 & -2 \\ 3 & 4 \end{bmatrix}, \qquad B = \begin{bmatrix} 0 & 7 \\ -5 & 2 \end{bmatrix}$$

Maple

```
> with(linalg):
> A:=matrix(2,2,[1,-2,3,4]):B:=matrix(2,2,[0,7,-5,2]):
> evalm(A&*B-B^2);
```

$$\begin{bmatrix} 45 & -11 \\ -10 & 60 \end{bmatrix}$$

Mathematica

```
In[1]:=
    A={{1,-2},{3,4}};B={{0,7},{-5,2}};
In[2]:=
    A . B-MatrixPower[B, 2]
Out[2]=
    {{45, -11}, {-10, 60}}
```

MATLAB

```
>> A=[1 -2; 3 4]; B=[0 7; -5 2];
>> A*B-B^2
ans =
       45      -11
      -10       60
```

Exercises 3.1

Matrices; Matrix Equality

1. Identify the rows, columns, sizes, and the $(2, 2)$nd entries of the matrices. Find the $(3, 1)$st entry of A and the $(2, 3)$rd entry of B.

$$A = \begin{bmatrix} -1 & 0 \\ 2 & 3 \\ -2 & 1 \end{bmatrix}, \qquad B = \begin{bmatrix} -1 & 0 & -2 \\ 2 & 2 & 1 \end{bmatrix}$$

2. Find the values of x, y, and z such that the following matrices are equal.

   a. $\begin{bmatrix} 1 & 0 & 1 \\ 0 & 2 & -3 \end{bmatrix} = \begin{bmatrix} x & 0 & 1 \\ x+y & 2 & x+z \end{bmatrix}$

   b. $\begin{bmatrix} 1 & 0 & 1 \\ 0 & 2 & -3 \end{bmatrix} = \begin{bmatrix} x+z & 0 & 1 \\ -y & 2 & -z \end{bmatrix}$

3. Show each of the following.

   a. $\begin{bmatrix} 1 & 0 \\ 0 & 2 \end{bmatrix} \neq \begin{bmatrix} x & y-2 \\ x-y & 2 \end{bmatrix}$

   b. $\begin{bmatrix} 1 & 1 \\ 1 & 3 \end{bmatrix} \neq \begin{bmatrix} x+y & x+y \\ -y+z & x+z \end{bmatrix}$

4. Which of the matrices

$$A = \begin{bmatrix} 1 & 1 \\ 1 & 0 \end{bmatrix}, \qquad B = \begin{bmatrix} 1 & 0 \\ 1 & 0 \end{bmatrix}$$

$$C = \begin{bmatrix} 1 & 0 \\ 1 & 1 \end{bmatrix}, \qquad D = \begin{bmatrix} 1 & 1 \\ 0 & 0 \end{bmatrix}$$

$$E = \begin{bmatrix} 0 & 1 \\ 1 & 0 \end{bmatrix}, \qquad F = \begin{bmatrix} 1 & 0 \\ 0 & 2 \end{bmatrix}$$

$$G = \begin{bmatrix} 3 & 0 \\ 0 & 3 \end{bmatrix}, \qquad H = \begin{bmatrix} 0 & 1 \\ 1 & 1 \end{bmatrix}$$

   are each of the following?

   a. Upper triangular

   b. Lower triangular

   c. Diagonal

   d. Scalar

   e. None of the above

Addition, Scalar Multiplication

5. Compute the following, if possible. If the operations cannot be performed, explain why.

   a. $\begin{bmatrix} 1 & -1 \\ 0 & 1 \end{bmatrix} + \begin{bmatrix} 0 & 1 \\ 1 & 2 \end{bmatrix}$

   b. $-3 \begin{bmatrix} 1 & 1 & -1 \\ 1 & 1 & -1 \end{bmatrix}$

   c. $\begin{bmatrix} 2 & 2 \\ -2 & -2 \end{bmatrix} + \begin{bmatrix} 0 \\ 0 \end{bmatrix}$

   d. $\begin{bmatrix} -2 & 3 \\ 4 & -5 \\ -6 & 7 \end{bmatrix} - \begin{bmatrix} 7 & -6 \\ -5 & 4 \\ 3 & -2 \end{bmatrix}$

   e. $-[1 \quad -2] + \begin{bmatrix} 1 & 2 \\ 4 & 3 \end{bmatrix}$

   f. $3 \begin{bmatrix} 0 & 2 \\ -4 & -6 \end{bmatrix} - 4 \begin{bmatrix} 3 & -5 \\ 7 & 0 \end{bmatrix}$

Let $A = \begin{bmatrix} 1 & -2 \\ -3 & 4 \end{bmatrix}$, $B = \begin{bmatrix} 0 & 1 \\ 5 & -2 \end{bmatrix}$, and

$C = \begin{bmatrix} 8 & -2 \\ -6 & 4 \end{bmatrix}$.

In Exercises 6–7 find the matrix X that satisfies the equations.

6. a. $2X + B = -3A + C$

   b. $\frac{1}{2}A - \frac{3}{2}X = B + 2C$

   c. $0C + X + 5B = 2A$

7. a. $2X + B + C = 0$

   b. $\frac{1}{2}A - X = B - C$

   c. $0C + \frac{1}{2}X = 2A - 4B$

8. Prove Parts 1–4 of Theorem 1.

9. Prove Parts 6–9 of Theorem 1.

Matrix Multiplication

10. Compute, if possible,

    a. $\begin{bmatrix} 3 \\ 4 \end{bmatrix} [1 \quad 2]$

    b. $[1 \quad 2] \begin{bmatrix} 3 \\ 4 \end{bmatrix}$

    c. $\begin{bmatrix} 1 & -2 \\ 4 & 0 \end{bmatrix} \begin{bmatrix} 1 & 2 & 3 & 4 \\ -2 & -4 & 3 & 0 \end{bmatrix}$

    d. $\begin{bmatrix} 1 & 2 & 3 & 4 \\ -2 & -4 & 3 & 0 \end{bmatrix} \begin{bmatrix} 1 & -2 \\ 4 & 0 \end{bmatrix}$

    e. $\begin{bmatrix} -3 & 0 \\ 2 & -5 \\ -7 & 4 \end{bmatrix} \begin{bmatrix} 1 & 0 & 1 \\ 0 & 1 & 1 \end{bmatrix}$

11. Compute the third row of AB if

$$A = \begin{bmatrix} 3 & 4 \\ 4 & 3 \\ 1 & 2 \end{bmatrix}, \qquad B = \begin{bmatrix} 1 & 2 & 5 & 6 \\ 6 & 5 & 2 & 1 \end{bmatrix}$$

12. Compute the second column of AB if

$$A = \begin{bmatrix} 1 & 2 & 5 \\ 6 & 5 & 2 \end{bmatrix}, \qquad B = \begin{bmatrix} 3 & 4 \\ 4 & 3 \\ 1 & 2 \end{bmatrix}$$

13. Compute the $(2, 2)$ entry of

$$\begin{bmatrix} 1 & 2 \\ 3 & 4 \end{bmatrix} \begin{bmatrix} 2 & -1 \\ -3 & 1 \end{bmatrix}$$

14. Find $(2A)^3$ if

$$A^3 = \begin{bmatrix} 1 & 1 \\ -5 & -2 \end{bmatrix}$$

15. Compute A^8 if

$$A = \begin{bmatrix} 1 & 1 \\ 0 & 1 \end{bmatrix}$$

Can you guess A^n?

16. Verify Theorem 2 for $A = \begin{bmatrix} -2 & 3 \\ 4 & -1 \end{bmatrix}$, $B = \begin{bmatrix} 2 & 5 \\ 0 & 3 \end{bmatrix}$, $C = \begin{bmatrix} 3 & 0 \\ 1 & -2 \end{bmatrix}$, and $a = -3$.

17. Prove Parts 2–6 of Theorem 2.

18. Find a 2×2 matrix A such that $A^2 = \mathbf{0}$.

19. Find a 2×2 matrix $A \neq I$ such that $A^2 = A$.

20. Recall that for real numbers the equation $a^2 = 1$ has only two solutions, namely, $a = \pm 1$. An analogous statement is no longer true for matrices. Find four 2×2 matrices A such that $A^2 = I$.

21. Find 2×2 matrices A and B such that $(AB)^2 \neq A^2 B^2$.

22. Find 2×2 matrices A and B that <u>commute</u>. $AB = BA$

23. Show that if A and B commute, then $(AB)^2 = A^2 B^2$.

24. a. Show that if A and B are square matrices of the same size, then

$$(A + B)^2 = A^2 + 2AB + B^2 \Leftrightarrow AB = BA$$

b. Find two 2×2 matrices A and B such that

$$(A + B)^2 \neq A^2 + 2AB + B^2$$

25. Verify Theorem 3 for $A = \begin{bmatrix} 1 & -2 \\ 3 & -1 \end{bmatrix}$, $n = 2, m = 1$, and $c = -3$.

26. Prove Theorem 3.

27. Find all the 2×2 submatrices of M.

$$M = \begin{bmatrix} 1 & 2 & 3 \\ 4 & 5 & 6 \end{bmatrix}$$

28. If

$$A = \begin{bmatrix} 3 & 2 \\ 1 & 5 \\ \cdots & \cdots \\ 1 & 1 \\ 5 & 6 \end{bmatrix} = \begin{bmatrix} A_1 \\ A_2 \end{bmatrix}$$

and

$$B = \begin{bmatrix} -1 & 2 \\ 8 & 9 \\ \cdots & \cdots \\ 5 & -6 \\ 1 & -2 \end{bmatrix} = \begin{bmatrix} B_1 \\ B_2 \end{bmatrix}$$

check that

$$A + B = \begin{bmatrix} A_1 + B_1 \\ A_2 + B_2 \end{bmatrix}$$

29. If

$$C = \begin{bmatrix} 7 & 3 & -4 & \vdots & 0 & 0 \\ 1 & 1 & 2 & \vdots & 0 & 0 \\ \cdots & \cdots & \cdots & \vdots & \cdots & \cdots \\ 0 & 2 & 3 & \vdots & 1 & 0 \\ 3 & 1 & -8 & \vdots & 0 & 1 \end{bmatrix} = \begin{bmatrix} C_{11} & C_{12} \\ C_{21} & C_{22} \end{bmatrix}$$

and

$$D = \begin{bmatrix} 1 & 0 \\ -3 & 1 \\ 5 & -5 \\ \cdots & \cdots \\ -3 & 0 \\ 2 & -2 \end{bmatrix} = \begin{bmatrix} D_1 \\ D_2 \end{bmatrix}$$

check that

$$CD = \begin{bmatrix} C_{11}D_1 + C_{12}D_2 \\ C_{21}D_1 + C_{22}D_2 \end{bmatrix}$$

Applications

30. A bottled water company delivers three kinds of bottled water, Spring, Mountain, and Polar, to four stores.

The stores are stocked, in hundreds of bottles, according to the following table.

Store	Spring	Mountain	Polar
1	8	4	3
2	10	5	4
3	6	3	2
4	6.5	3.5	2.5

Update the inventory if deliveries D are made.

$$D = \begin{bmatrix} 1 & 2 & 0.5 \\ 0.5 & 2 & 1 \\ 2 & 0.5 & 1 \\ 2.5 & 1.5 & 1.5 \end{bmatrix}$$

31. A department store company has three outlets. Each outlet carries jackets, slacks, and shoes, which are priced according to the following table. Find the values of these items if all three outlets have a 20% off sale.

Outlet	Jackets	Slacks	Shoes
1	80	40	30
2	100	50	40
3	60	30	20

32. Each of two department stores, d_1 and d_2, receive and sell weekly from two clothing factories, f_1 and f_2, slacks (s) and jackets (j) according to

$$\begin{array}{c} \\ f_1 \\ f_2 \end{array} \begin{array}{cc} s & j \\ \begin{bmatrix} 50 & 20 \\ 60 & 30 \end{bmatrix} \end{array}$$

The charge per item, in dollars, for each store is as follows:

$$\begin{array}{c} \\ s \\ j \end{array} \begin{array}{cc} d_1 & d_2 \\ \begin{bmatrix} 100 & 85 \\ 350 & 400 \end{bmatrix} \end{array}$$

If A and B are the matrices of these tables, compute and interpret the product AB.

3.2   Matrix Inverse

Reader's Goals for This Section

1. To know the definition and properties of the inverse.
2. To be able to compute the inverse of a matrix, if it exists.
3. To solve some square systems by inverting the coefficient matrix.

Matrix inversion is the last basic matrix operation. It applies only to *square* matrices. The inverse of a matrix, if it exists, is the analogue of the reciprocal of a nonzero number.

DEFINITION

(Inverse)

We say that the $n \times n$ matrix A is **invertible** or **nonsingular** if there exists a matrix B, called an **inverse** of A, such that

$$AB = I \quad \text{and} \quad BA = I$$

Note that the definition forces B to have size $n \times n$.

An invertible matrix has only one inverse, i.e., **the inverse is unique.** If C is another one, then

$$B = BI_n = B(AC) = (BA)C = I_nC = C$$

Therefore, $B = C$. The unique inverse of an invertible matrix A is denoted by A^{-1}. So,

$$AA^{-1} = I \quad \text{and} \quad A^{-1}A = I$$

A square matrix that has no inverse is called **noninvertible** or **singular.**

■ EXAMPLE 15   Show that

$$\begin{bmatrix} 2 & 7 \\ 1 & 4 \end{bmatrix} \text{ is the inverse of } \begin{bmatrix} 4 & -7 \\ -1 & 2 \end{bmatrix}.$$

SOLUTION

$$\begin{bmatrix} 2 & 7 \\ 1 & 4 \end{bmatrix} \begin{bmatrix} 4 & -7 \\ -1 & 2 \end{bmatrix} = \begin{bmatrix} 1 & 0 \\ 0 & 1 \end{bmatrix} \text{ and } \begin{bmatrix} 4 & -7 \\ -1 & 2 \end{bmatrix} \begin{bmatrix} 2 & 7 \\ 1 & 4 \end{bmatrix} = \begin{bmatrix} 1 & 0 \\ 0 & 1 \end{bmatrix}$$

In the case of 2×2 matrices, we can find exactly which matrices are invertible and give an explicit formula for the inverse (this is harder to do for sizes greater than 2).

THEOREM 4 $A = \begin{bmatrix} a & b \\ c & d \end{bmatrix}$ is invertible if and only if $ad - bc \neq 0$, in which case,

$$A^{-1} = \frac{1}{ad - bc} \begin{bmatrix} d & -b \\ -c & a \end{bmatrix}$$

PROOF   Exercise.

■ EXAMPLE 16   Show that

$$A = \begin{bmatrix} 1 & 2 \\ 3 & 4 \end{bmatrix}$$

is invertible and compute its inverse.

SOLUTION $1 \cdot 4 - 2 \cdot 3 \neq 0$. Hence, A is invertible by Theorem 4. Furthermore,

$$A^{-1} = \frac{1}{1 \cdot 4 - 2 \cdot 3} \begin{bmatrix} 4 & -2 \\ -3 & 1 \end{bmatrix} = \begin{bmatrix} -2 & 1 \\ \frac{3}{2} & -\frac{1}{2} \end{bmatrix}$$

■ EXAMPLE 17   Show that

$$A = \begin{bmatrix} 1 & 0 \\ 0 & 0 \end{bmatrix}$$

is noninvertible.

SOLUTION   This follows from Theorem 4, because $1 \cdot 0 - 0 \cdot 0 = 0$.

THEOREM 5   If the $n \times n$ matrix A is invertible, then the system $A\mathbf{x} = \mathbf{b}$ has exactly one solution for each n-vector $\mathbf{b}$. This unique solution is given by

$$\mathbf{x} = A^{-1}\mathbf{b}$$

$$\mathbf{x} = A^{-1}\mathbf{b}$$

$$A\mathbf{x} = A\,(A^{-1}\mathbf{b}) = (AA^{-1})\mathbf{b} = I\mathbf{b} = \mathbf{b}.$$

PROOF $\mathbf{x} = A^{-1}\mathbf{b}$ is a solution, because substitution into the system yields $A(A^{-1}\mathbf{b}) = \mathbf{b}$, or $(AA^{-1})\mathbf{b} = \mathbf{b}$, or $I\mathbf{b} = \mathbf{b}$, or $\mathbf{b} = \mathbf{b}$, which is true. Furthermore, the solution is unique, because if $\mathbf{y}$ is another one, then

$$A\mathbf{y} = \mathbf{b} \Rightarrow A^{-1}A\mathbf{y} = A^{-1}\mathbf{b} \Rightarrow \mathbf{y} = A^{-1}\mathbf{b}$$

Hence, all solutions are given by the same formula, $A^{-1}\mathbf{b}$.

A SPECIAL CASE   If A is invertible, then $A\mathbf{x} = \mathbf{0}$ has only the trivial solution.

Note that the systems discussed here have the same number of equations and unknowns, because A is a square matrix.

■ EXAMPLE 18   Use matrix inversion to solve the system

$$x - 4y = 2$$

$$x - 3y = 1$$

SOLUTION   Because the matrix form of the system is

$$\begin{bmatrix} 1 & -4 \\ 1 & -3 \end{bmatrix} \begin{bmatrix} x \\ y \end{bmatrix} = \begin{bmatrix} 2 \\ 1 \end{bmatrix}$$

we have

$$\begin{bmatrix} x \\ y \end{bmatrix} = \begin{bmatrix} 1 & -4 \\ 1 & -3 \end{bmatrix}^{-1} \begin{bmatrix} 2 \\ 1 \end{bmatrix} = \begin{bmatrix} -3 & 4 \\ -1 & 1 \end{bmatrix} \begin{bmatrix} 2 \\ 1 \end{bmatrix} = \begin{bmatrix} -2 \\ -1 \end{bmatrix}$$

Hence, $x = -2$ and $y = -1$.

The next theorem outlines the basic properties of inverses.

THEOREM 6

(Properties of Inverses)

1. The product of two invertible matrices is invertible. Its inverse is the product of the inverses of the factors in the reverse order. Hence, if A and B are $n \times n$ invertible, so is AB and

$$(AB)^{-1} = B^{-1}A^{-1}$$

2. The inverse of an invertible matrix is also invertible. Its inverse is the original matrix. Hence, if A is invertible, so is A^{-1} and

$$(A^{-1})^{-1} = A$$

3. Any nonzero scalar product of an invertible matrix is invertible. Its inverse is the scalar product of the reciprocal of the scalar and the inverse of the matrix. Hence, if A is invertible and c is a nonzero scalar, then cA is invertible and

$$(cA)^{-1} = \frac{1}{c}A^{-1}$$

$$(cA)^n = c^n A^n.$$

PROOF

1. We need to check that $(AB)(B^{-1}A^{-1}) = I = (B^{-1}A^{-1})(AB)$. We have

$$(AB)(B^{-1}A^{-1}) = A(BB^{-1})A^{-1} = AIA^{-1} = AA^{-1} = I$$
$$(B^{-1}A^{-1})(AB) = B^{-1}(A^{-1}A)B = B^{-1}IB = B^{-1}B = I$$

2. Because A is invertible, $AA^{-1} = I = A^{-1}A$. This shows that A^{-1} is invertible and $(A^{-1})^{-1} = A$.

3. The proof is left as an exercise.

REMARK   If $A_1, A_2, \ldots, A_{n-1}, A_n$ are invertible and of the same size, then so is the product $A_1 A_2 \ldots A_{n-1} A_n$. Its inverse is

$$(A_1 A_2 \ldots A_{n-1} A_n)^{-1} = A_n^{-1} A_{n-1}^{-1} \ldots A_2^{-1} A_1^{-1}$$

CAUTION   In general (see Exercises 25 and 26),

$$(A + B)^{-1} \neq A^{-1} + B^{-1} \quad \text{and} \quad (AB)^{-1} \neq A^{-1}B^{-1}$$

Matrix Powers with Negative Exponents

If A is invertible, we define powers of A with negative exponents as follows. For $n > 0$ and n an integer,

$$A^{-n} = (A^{-1})^n = \underbrace{A^{-1}A^{-1} \cdots A^{-1}}_{n \text{ factors}}$$

THEOREM 7

(Properties of Powers)

Let A be a $k \times k$ invertible matrix, let m and n be any two integers ($> 0, < 0,$ or $= 0$) and let c be a nonzero scalar. Then A^n is invertible and

1. $(A^n)^{-1} = (A^{-1})^n$ 3. $(A^n)^m = A^{nm}$
2. $A^n A^m = A^{n+m}$ 4. $(cA)^n = c^n A^n$

PROOF   The proof is left as exercise.

Recall from Section 3.1 that, in general, from a matrix equation $CA = CB$ we cannot cancel out the Cs. However, such a cancellation is possible if C is invertible.

THEOREM 8

(Cancellation Laws)

If C is invertible, then

1. $CA = CB \Rightarrow A = B$
2. $AC = BC \Rightarrow A = B$

PROOF   1. C^{-1} exists; hence,

$$CA = CB \Rightarrow C^{-1}(CA) = C^{-1}(CB) \Rightarrow (C^{-1}C)A = (C^{-1}C)B \Rightarrow IA = IB \Rightarrow A = B$$

Part 2 is proved similarly.

Simple Matrix Product Equations

The basic properties of matrix operations allow us to solve some matrix equations.

CAUTION   When we multiply both sides of a matrix equation by a matrix, we should use either left or right multiplication but not both. So,

$$A = B \Rightarrow CA = CB, \qquad A = B \Rightarrow AD = BD$$

whereas

$$A = B \text{ does not imply } CA = BC$$

■ EXAMPLE 19   Solve the matrix equation $CXA - B = 0$ for X, provided that A and C are invertible and that all sizes are compatible.

SOLUTION   By using properties of matrix operations, we can isolate X on the left-hand side of the equation.

$$AXC - B = 0$$
$$AXC = B$$
$$A^{-1}AXC = A^{-1}B$$
$$XC = A^{-1}B$$
$$XCC^{-1} = A^{-1}BC^{-1}$$
$$X = A^{-1}BC^{-1}$$

Steps 3 and 5 are possible because A^{-1} and C^{-1} exist. If we specify A, B, C as, for example,

$$A = \begin{bmatrix} 2 & 1 \\ 7 & 4 \end{bmatrix}, \qquad B = \begin{bmatrix} 1 & 1 \\ 2 & 2 \end{bmatrix}, \qquad C = \begin{bmatrix} 0 & -1 \\ -2 & -1 \end{bmatrix}$$

then

$$X = A^{-1}BC^{-1} = \begin{bmatrix} 4 & -1 \\ -7 & 2 \end{bmatrix}\begin{bmatrix} 1 & 1 \\ 2 & 2 \end{bmatrix}\begin{bmatrix} \frac{1}{2} & -\frac{1}{2} \\ -1 & 0 \end{bmatrix} = \begin{bmatrix} -1 & -1 \\ \frac{3}{2} & \frac{3}{2} \end{bmatrix}$$

A COMMON MISTAKE   The expression $A^{-1}BA$ does *not* usually simplify to B. (It does if A and B commute.)

Computation of A^{-1}

Let us now see how to use row reduction to compute the inverse. The next example illustrates and justifies the method (which was first introduced in Project 1 of Chapter 1).

Let $A = \begin{bmatrix} 2 & 3 \\ 1 & 2 \end{bmatrix}$ and let $A^{-1} = \begin{bmatrix} x & y \\ z & w \end{bmatrix}$. Because $AA^{-1} = I$,

$$\begin{bmatrix} 2 & 3 \\ 1 & 2 \end{bmatrix}\begin{bmatrix} x & y \\ z & w \end{bmatrix} = \begin{bmatrix} 1 & 0 \\ 0 & 1 \end{bmatrix} \quad \text{or} \quad \begin{bmatrix} 2x + 3z & 2y + 3w \\ x + 2z & y + 2w \end{bmatrix} = \begin{bmatrix} 1 & 0 \\ 0 & 1 \end{bmatrix}$$

By equating columns, we get two systems

$$\begin{array}{cc} 2x + 3z = 1 \\ x + 2z = 0 \end{array} \quad \text{and} \quad \begin{array}{cc} 2y + 3w = 0 \\ y + 2w = 1 \end{array}$$

with augmented matrices,

$$\begin{bmatrix} 2 & 3 & : & 1 \\ 1 & 2 & : & 0 \end{bmatrix} \quad \text{and} \quad \begin{bmatrix} 2 & 3 & : & 0 \\ 1 & 2 & : & 1 \end{bmatrix}$$

whose reduced row echelon forms are

$$\begin{bmatrix} 1 & 0 & : & 2 \\ 0 & 1 & : & -1 \end{bmatrix} \quad \text{and} \quad \begin{bmatrix} 1 & 0 & : & -3 \\ 0 & 1 & : & 2 \end{bmatrix}$$

Hence, $x = 2$, $z = -1$ from the first system and $y = -3$, $w = 2$ from the second. Therefore, $A^{-1} = \begin{bmatrix} 2 & -3 \\ -1 & 2 \end{bmatrix}$. Because the two systems have the same coefficient matrix, we can save in writing by combining them as

$$\begin{bmatrix} 2 & 3 & : & 1 & 0 \\ 1 & 2 & : & 0 & 1 \end{bmatrix} \quad \text{and then row-reduce to} \quad \begin{bmatrix} 1 & 0 & : & 2 & -3 \\ 0 & 1 & : & -1 & 2 \end{bmatrix}$$

Notice that we start out with $[A : I]$ and after reduction we have $[I : A^{-1}]$, from which we can read A^{-1}. This is the entire method. Let us practice.

■ EXAMPLE 20   Compute A^{-1} if

$$A = \begin{bmatrix} 1 & 0 & -1 \\ 3 & 4 & -2 \\ 3 & 5 & -2 \end{bmatrix}$$

SOLUTION   We row-reduce $[A : I]$,

$$\begin{bmatrix} 1 & 0 & -1 & : & 1 & 0 & 0 \\ 3 & 4 & -2 & : & 0 & 1 & 0 \\ 3 & 5 & -2 & : & 0 & 0 & 1 \end{bmatrix} \sim \begin{bmatrix} 1 & 0 & -1 & : & 1 & 0 & 0 \\ 0 & 4 & 1 & : & -3 & 1 & 0 \\ 0 & 5 & 1 & : & -3 & 0 & 1 \end{bmatrix}$$

$$\sim \begin{bmatrix} 1 & 0 & -1 & : & 1 & 0 & 0 \\ 0 & 4 & 1 & : & -3 & 1 & 0 \\ 0 & 0 & -\frac{1}{4} & : & \frac{3}{4} & -\frac{5}{4} & 1 \end{bmatrix} \sim \begin{bmatrix} 1 & 0 & -1 & : & 1 & 0 & 0 \\ 0 & 4 & 1 & : & -3 & 1 & 0 \\ 0 & 0 & 1 & : & -3 & 5 & -4 \end{bmatrix}$$

$$\sim \begin{bmatrix} 1 & 0 & 0 & : & -2 & 5 & -4 \\ 0 & 4 & 0 & : & 0 & -4 & 4 \\ 0 & 0 & 1 & : & -3 & 5 & -4 \end{bmatrix} \sim \begin{bmatrix} 1 & 0 & 0 & : & -2 & 5 & -4 \\ 0 & 1 & 0 & : & 0 & -1 & 1 \\ 0 & 0 & 1 & : & -3 & 5 & -4 \end{bmatrix}$$

Therefore,

$$A^{-1} = \begin{bmatrix} -2 & 5 & -4 \\ 0 & -1 & 1 \\ -3 & 5 & -4 \end{bmatrix}$$

This method not only computes inverses but also detects noninvertible matrices. For example, if $A = \begin{bmatrix} 1 & 2 \\ 2 & 4 \end{bmatrix}$, then

$$\begin{bmatrix} 1 & 2 & : & 1 & 0 \\ 2 & 4 & : & 0 & 1 \end{bmatrix} \sim \begin{bmatrix} 1 & 2 & : & 1 & 0 \\ 0 & 0 & : & -2 & 1 \end{bmatrix}$$

It is clear that we cannot obtain the identity matrix on the left, because the second entry of the second row is 0. Reinterpreting the answer in terms of systems, the second row implies the incorrect equations $0 = -2$ and $0 = 1$. So, the inverse does not exist.

The following algorithm tests a matrix for invertibility. If the answer is positive, it computes the inverse.

Algorithm 1

(Matrix Inversion Algorithm)

To find A^{-1}, if it exists, do the following:

1. Find the reduced row echelon form of the matrix $[A : I]$, say, $[B : C]$.
2. If B has a zero row, stop. A is noninvertible. Otherwise, go to Step 3.
3. The reduced matrix is now in the form $[I : A^{-1}]$. Read the inverse A^{-1}.

In Section 3.3 we analyze the matrix inversion algorithm and see why it works.

Transposition and Matrix Operations

The **transpose** A^T of an $m \times n$ matrix A is the $n \times m$ matrix whose columns are the rows of A in the same order. For example,

$$A = \begin{bmatrix} 1 & 2 & 3 \\ 4 & 5 & 6 \end{bmatrix}, \quad A^T = \begin{bmatrix} 1 & 4 \\ 2 & 5 \\ 3 & 6 \end{bmatrix}, \quad B = \begin{bmatrix} 1 & 2 \\ 2 & 3 \end{bmatrix}, \quad B^T = \begin{bmatrix} 1 & 2 \\ 2 & 3 \end{bmatrix}$$

The operation of taking the transpose of a matrix is called transposition. A matrix the same as its transpose is called **symmetric**. So, B is symmetric. Let us now see how transposition is affected by the basic matrix operations.

THEOREM 9

1. $(A^T)^T = A$.
2. $(A + B)^T = A^T + B^T$.
3. $(cA)^T = cA^T$.
4. $(AB)^T = B^T A^T$.
5. If A is invertible, so is A^T. In this case, $(A^T)^{-1} = (A^{-1})^T$.

Note that the matrices in Parts 1–4 are not necessarily square. Let us sketch the proof of Part 4 and leave the remaining proofs as exercises.

PARTIAL PROOF OF 4   First, the special case where A is $1 \times n$ and B is $n \times 1$ is implied by

$$[a_1 \ldots a_n] \begin{bmatrix} b_1 \\ \vdots \\ b_n \end{bmatrix} = a_1 b_1 + \cdots + a_n b_n = b_1 a_1 + \cdots + b_n a_n = [b_1 \ldots b_n] \begin{bmatrix} a_1 \\ \vdots \\ a_n \end{bmatrix}$$

Next we prove the special case, where A is $m \times n$ and B is an n-vector, say, $\mathbf{v}$. The general case, left as an exercise, is based on this one and is proved similarly. We have

$$j\text{th column of } (A\mathbf{v})^T = (j\text{th row of } A\mathbf{v})^T$$

$$= (j\text{th row of } A \text{ times } \mathbf{v})^T$$

$$= \mathbf{v}^T \text{ times the } (j\text{th row of } A)^T \qquad \text{(by the } 1 \times n \text{ by } n \times 1 \text{ case.)}$$

$$= \mathbf{v}^T \text{ times the } j\text{th column of } A^T$$

$$= j\text{th column of } \mathbf{v}^T A^T$$

■ EXAMPLE 21   Verify Part 5 of Theorem 9 for

$$A = \begin{bmatrix} 1 & -2 \\ 3 & -7 \end{bmatrix}$$

SOLUTION

$$(A^T)^{-1} = \begin{bmatrix} 1 & 3 \\ -2 & -7 \end{bmatrix}^{-1} = \begin{bmatrix} 7 & 3 \\ -2 & -1 \end{bmatrix}, \qquad (A^{-1})^T = \begin{bmatrix} 7 & -2 \\ 3 & -1 \end{bmatrix}^T = \begin{bmatrix} 7 & 3 \\ -2 & -1 \end{bmatrix}$$

Application to Engineering

As an example of matrix inversion, let us revisit the heat-transfer problem studied in Section 2.1. We were given a temperature distribution $b_1, b_2, \ldots, b_8$ for a square plate and asked to find the temperatures $x_1, \ldots, x_4$ in the interior (Fig. 2.8). The resulting linear system was

$$x_1 = \frac{1}{4}(x_2 + x_3 + b_1 + b_3)$$

$$x_2 = \frac{1}{4}(x_1 + x_4 + b_2 + b_4)$$

$$x_3 = \frac{1}{4}(x_1 + x_4 + b_5 + b_7)$$

$$x_4 = \frac{1}{4}(x_2 + x_3 + b_6 + b_8)$$

■ EXAMPLE 22 (Heat Conduction) Show that the heat-transfer problem with the preceding system has always a unique solution for any temperature values $b_1, b_2, \ldots, b_8$ of the boundary. Find this solution.

SOLUTION   To simplify the notation, we let $B_1 = b_1 + b_3, B_2 = b_2 + b_4, B_3 = b_5 + b_7,$ $B_4 = b_6 + b_8$. The system can now be written in the standard form $Ax = \mathbf{b}$ with

$$A = \begin{bmatrix} 4 & -1 & -1 & 0 \\ -1 & 4 & 0 & -1 \\ -1 & 0 & 4 & -1 \\ 0 & -1 & -1 & 4 \end{bmatrix}, \quad \mathbf{x} = \begin{bmatrix} x_1 \\ x_2 \\ x_3 \\ x_4 \end{bmatrix}, \quad \mathbf{b} = \begin{bmatrix} B_1 \\ B_2 \\ B_3 \\ B_4 \end{bmatrix}$$

A^{-1} (computed by the matrix inversion algorithm) is

$$\frac{1}{24} \begin{bmatrix} 7 & 2 & 2 & 1 \\ 2 & 7 & 1 & 2 \\ 2 & 1 & 7 & 2 \\ 1 & 2 & 2 & 7 \end{bmatrix}$$

Therefore, the system has a unique solution $\mathbf{x} = A^{-1}\mathbf{b}$ for any choice of constant terms, by Theorem 5. Hence,

$$\begin{bmatrix} x_1 \\ x_2 \\ x_3 \\ x_4 \end{bmatrix} = A^{-1} \begin{bmatrix} B_1 \\ B_2 \\ B_3 \\ B_4 \end{bmatrix} = \frac{1}{24} \begin{bmatrix} 7B_1 + 2B_2 + 2B_3 + B_4 \\ 2B_1 + 7B_2 + B_3 + 2B_4 \\ 2B_1 + B_2 + 7B_3 + 2B_4 \\ B_1 + 2B_2 + 2B_3 + 7B_4 \end{bmatrix}$$

Application to Economics

Economists often study conditions for *market equilibria*, i.e., conditions under which the prices of various commodities are related.

For instance, let us examine the related markets of pens and pencils. Let P_p and P_P be the dollar prices of a pencil and a pen, respectively. Some market conditions force the two prices to satisfy the relation $P_p + P_P = 1.5$, whereas other conditions require the relation $P_p - P_P = -0.5$. The *equilibrium price* for each market is the price that satisfies both conditions. This is the solution of the system of the two equations. So the equilibrium prices are $P_p = 0.5$ and $P_P = 1$.

■ EXAMPLE 23 (Equilibrium in Related Markets) The equilibrium conditions between three related markets (chicken, pork, and beef) are given by

$$\begin{aligned} 5P_c - P_p - 2P_b &= 1 \\ -2P_c + 6P_p - 3P_b &= 3 \\ -2P_c - P_p + 4P_b &= 10 \end{aligned}$$

Compute the equilibrium price, in dollars, for each market.

SOLUTION   The system $AP = B$, with

$$A = \begin{bmatrix} 5 & -1 & -2 \\ -2 & 6 & -3 \\ -2 & -1 & 4 \end{bmatrix}, \qquad P = \begin{bmatrix} P_c \\ P_p \\ P_b \end{bmatrix}, \qquad B = \begin{bmatrix} 1 \\ 3 \\ 10 \end{bmatrix}$$

can be solved by inverting A. Using the matrix inversion algorithm we get

$$P = \begin{bmatrix} P_c \\ P_p \\ P_b \end{bmatrix} = A^{-1}B = \frac{1}{63} \begin{bmatrix} 21 & 6 & 15 \\ 14 & 16 & 19 \\ 14 & 7 & 28 \end{bmatrix} \begin{bmatrix} 1 \\ 3 \\ 10 \end{bmatrix} = \begin{bmatrix} 3 \\ 4 \\ 5 \end{bmatrix}$$

Therefore, the equilibrium prices are \$3 for chicken, \$4 for pork, and \$5 for beef.

Inverse with CAS

Maple

```
> with(linalg):
> inverse(matrix([[1,1],[3,4]]));
```

$$\begin{bmatrix} 4 & -1 \\ -3 & 1 \end{bmatrix}$$

Mathematica

```
In[1]:=
       Inverse[{{1,1},{3,4}}]
Out[1]=
       {{4, -1}, {-3, 1}}
```

MATLAB

```
>> inv([1 1; 3 4])
ans =
     4.0000     -1.0000
    -3.0000      1.0000
```

Exercises 3.2

1. Use Theorem 4 to find the inverses of the following matrices.

$$\begin{bmatrix} 3 & 2 \\ 1 & 2 \end{bmatrix} \qquad \begin{bmatrix} 7 & 5 \\ 4 & 4 \end{bmatrix}$$

2. Use Theorem 4 to explain why the following matrices are noninvertible.

$$\begin{bmatrix} 1 & 1 \\ 2 & 2 \end{bmatrix} \qquad \begin{bmatrix} -10 & 20 \\ 20 & -40 \end{bmatrix}$$

3. Find a matrix whose inverse is

$$\begin{bmatrix} 1 & 3 \\ 2 & 8 \end{bmatrix}$$

4. Find the inverse of $10A$ if the inverse of A is

$$\begin{bmatrix} 4 & 4 \\ 8 & 6 \end{bmatrix}$$

5. If $a^2 + b^2 = 1$, use Theorem 4 to show that A is invertible. Compute A^{-1}.

$$A = \begin{bmatrix} a & b \\ -b & a \end{bmatrix}$$

6. Compute the inverse of

$$\begin{bmatrix} \cos\theta & \sin\theta \\ -\sin\theta & \cos\theta \end{bmatrix}$$

7. Find $(2A)^{-3}$ if

$$A^3 = \begin{bmatrix} 1 & 1 \\ -5 & -2 \end{bmatrix}$$

8. Compute $A^2, A^{-2}, A^3 - A, A^{-3}$ if

$$A = \begin{bmatrix} -1 & -1 \\ 2 & 1 \end{bmatrix}$$

9. Find the inverses of the matrices by inspection.

$$\begin{bmatrix} -1 & 0 & 0 \\ 0 & 1 & 0 \\ 0 & 0 & -1 \end{bmatrix}, \quad \begin{bmatrix} 5 & 0 & 0 \\ 0 & 5 & 0 \\ 0 & 0 & 5 \end{bmatrix}$$

10. Without actually computing, show that A is noninvertible.

$$A = \begin{bmatrix} -1 & 0 & 0 \\ 0 & 1 & 0 \\ 0 & 0 & 0 \end{bmatrix}$$

(*Hint:* If B is a hypothetical inverse, look at the $(3,3)$rd entry of AB.)

11. Without actually computing, show that A is noninvertible.

$$A = \begin{bmatrix} 0 & 0 & 0 \\ 0 & 1 & 0 \\ 0 & 0 & 1 \end{bmatrix}$$

(*Hint:* If B is a hypothetical inverse, look at the $(1,1)$st entry of AB.)

In Exercises 12–16 use the matrix inversion algorithm to compute, if possible, the inverse of the given matrix.

12. a. $\begin{bmatrix} -3 & -2 \\ -5 & -3 \end{bmatrix}$ **b.** $\begin{bmatrix} 1 & 2 \\ -2 & -4 \end{bmatrix}$

c. $\begin{bmatrix} 1 & 4 \\ 2 & 8 \end{bmatrix}$ **d.** $\begin{bmatrix} -\frac{1}{3} & \frac{1}{3} \\ -\frac{2}{3} & -\frac{1}{3} \end{bmatrix}$

13. a. $\begin{bmatrix} -1 & 1 & 0 \\ 0 & -1 & 0 \\ 0 & 0 & 1 \end{bmatrix}$ **b.** $\begin{bmatrix} -2 & 0 & -1 \\ 0 & -1 & -2 \\ 0 & 0 & -2 \end{bmatrix}$

c. $\begin{bmatrix} 1 & 2 & -1 \\ 1 & -2 & -1 \\ 1 & 6 & -1 \end{bmatrix}$ **d.** $\begin{bmatrix} -1 & 1 & 0 \\ 1 & 0 & -1 \\ 1 & 1 & -1 \end{bmatrix}$

14. a. $\begin{bmatrix} 1 & 3 & 2 \\ 3 & 2 & 1 \\ 3 & 3 & 1 \end{bmatrix}$ **b.** $\begin{bmatrix} -1 & 0 & 1 \\ 1 & \frac{2}{3} & -\frac{4}{3} \\ 0 & -1 & 1 \end{bmatrix}$

c. $\begin{bmatrix} 1 & 1 & -1 \\ 1 & 2 & -1 \\ 3 & 4 & -3 \end{bmatrix}$ **d.** $\begin{bmatrix} -1 & 1 & 1 \\ 1 & 0 & -2 \\ 2 & -1 & 0 \end{bmatrix}$

15. $\begin{bmatrix} -1 & 1 & 1 & -1 \\ -1 & 0 & 1 & 0 \\ 0 & 1 & -1 & 1 \\ 0 & 0 & 1 & -1 \end{bmatrix}$

16. $\begin{bmatrix} -1 & 0 & 1 & 2 \\ 0 & 0 & 1 & 1 \\ -1 & 1 & 1 & 2 \\ -1 & 1 & 1 & 1 \end{bmatrix}$

In Exercises 17–18 solve the systems by computing the inverse of the coefficient matrix first.

17. a. $\begin{aligned} x + y - z &= 1 \\ x \quad\quad - z &= 2 \\ -x + y \quad\quad &= 3 \end{aligned}$ **b.** $\begin{aligned} x - y - z &= 2 \\ x - y \quad\quad &= 4 \\ x - 2y - z &= -1 \end{aligned}$

18. a. $\begin{aligned} -2x + y \quad\quad &= a \\ x - y - z &= b \\ x + y - z &= c \end{aligned}$ **b.** $\begin{aligned} -x + y + z &= a \\ -x \quad\quad + z &= a^2 \\ y - z &= a^3 \end{aligned}$

19. Compute A if

$$A^{-1} = \begin{bmatrix} 1 & 1 & 0 \\ 1 & -1 & 1 \\ 0 & 0 & 1 \end{bmatrix}$$

20. Find all the values of a for which A^{-1} exists.

$$A = \begin{bmatrix} 1 & 1 & -1 \\ 0 & -1 & 1 \\ -1 & -2 & a \end{bmatrix}$$

21. Let c_1, c_2, c_3, c_4 be nonzero scalars. Compute A^{-1} and B^{-1}.

$$A = \begin{bmatrix} 0 & 0 & c_1 \\ 0 & c_2 & 0 \\ c_3 & 0 & 0 \end{bmatrix} \qquad B = \begin{bmatrix} 0 & 0 & 0 & c_1 \\ 0 & 0 & c_2 & 0 \\ 0 & c_3 & 0 & 0 \\ c_4 & 0 & 0 & 0 \end{bmatrix}$$

22. Find $(ABC)^{-1}$ if $A^{-1} = \begin{bmatrix} 1 & 1 \\ 0 & -1 \end{bmatrix}$, $B^{-1} = \begin{bmatrix} 1 & 0 \\ 1 & -1 \end{bmatrix}$, and $C^{-1} = \begin{bmatrix} 3 & 2 \\ -1 & -1 \end{bmatrix}$.

23. Compute $A^{-1}, A^{-2}, A^{-3}, A^{-24}, A^{-25}$, where

$$A = \begin{bmatrix} 1 & 0 & 0 \\ 0 & -1 & 1 \\ 0 & 0 & 1 \end{bmatrix}$$

24. Verify the identity $(ABC)^{-1} = C^{-1}B^{-1}A^{-1}$.

25. Find A, B such that $(A + B)^{-1} \neq A^{-1} + B^{-1}$.

26. Find A, B such that $(AB)^{-1} \neq A^{-1}B^{-1}$.

27. Show that a matrix with a row of zeros is noninvertible.

28. Show that a matrix with a column of zeros is non-invertible.

29. Show that a diagonal matrix A is invertible if and only if each element on the main diagonal is nonzero. What is A^{-1} in this case?

30. If A is invertible and $AB = \mathbf{0}$, show that $B = \mathbf{0}$.

31. Suppose $AB = \mathbf{0}$ and $B \neq \mathbf{0}$. Show that A is non-invertible.

32. If A is a square matrix with the property $A^2 = \mathbf{0}$, show that the inverse of $I - A$ exists and is equal to $A + I$.

33. If A is a square matrix with the property $A^3 = 0$, show that the inverse of $I - A$ exists and is equal to $A^2 + A + I$.

34. Suppose $A^2 + 2A - I = \mathbf{0}$. Show that $A^{-1} = A + 2I$.

35. Prove Theorem 4.

36. Prove Part 3 of Theorem 6.

37. Prove Theorem 7.

38. Prove Part 2 of Theorem 8.

39. For Theorem 9:

    a. Prove Parts 1–3 and 5.

    b. Complete the proof of Part 4.

40. Use matrix inversion to compute the equilibrium price in dollars for each of the related markets (cereal, crackers, and cookies) if the equilibrium conditions are given by

$$\begin{aligned} -P_{ce} + P_{cr} \qquad\quad &= -1 \\ P_{ce} \qquad\quad - P_{co} &= 1 \\ P_{ce} + P_{cr} - P_{co} &= 3 \end{aligned}$$

41. Prove that AA^T is always symmetric.

3.3   Elementary and Invertible Matrices

Reader's Goals for This Section

1. To understand the relation between elementary matrices and elementary row operations.
2. To know the different characterizations of invertible matrices.
3. To understand why the matrix inversion algorithm works.

In this section we study elementary matrices and use them to characterize invertible matrices in many interesting ways. Moreover, we justify the matrix inversion algorithm.

Elementary Matrices

DEFINITION   **(Elementary Matrix)**

An $n \times n$ matrix is called **elementary** if it can be obtained from the identity matrix I_n by using one and only one elementary row operation (elimination, scaling, or interchange). So, elementary matrices are always equivalent to I_n.

For example, the matrices

$$E_1 = \begin{bmatrix} 1 & -3 \\ 0 & 1 \end{bmatrix}, \qquad E_2 = \begin{bmatrix} 1 & 0 \\ 0 & 4 \end{bmatrix}, \qquad E_3 = \begin{bmatrix} 0 & 1 \\ 1 & 0 \end{bmatrix}$$

are elementary, because each is obtained from I_3 by applying $R_1 - 3R_2 \rightarrow R_1, 4R_2 \rightarrow R_2$, and $R_1 \leftrightarrow R_2$, respectively. Elementary matrices, such as E_3, which is obtained from I by interchanging two rows, are called **elementary permutation matrices**.

The key reason for studying elementary matrices is the observation that if we multiply a matrix A on the left by an elementary matrix E, then the product EA is the matrix obtained from A by using the same row operation that produced E from I_n. To illustrate:

$$\begin{bmatrix} 1 & -3 \\ 0 & 1 \end{bmatrix} \begin{bmatrix} a & b & c & d \\ e & f & g & h \end{bmatrix} = \begin{bmatrix} a-3e & b-3f & c-3g & d-3h \\ e & f & g & h \end{bmatrix}$$

$$\begin{bmatrix} 1 & 0 \\ 0 & 4 \end{bmatrix} \begin{bmatrix} a & b & c & d \\ e & f & g & h \end{bmatrix} = \begin{bmatrix} a & b & c & d \\ 4e & 4f & 4g & 4h \end{bmatrix}$$

$$\begin{bmatrix} 0 & 1 \\ 1 & 0 \end{bmatrix} \begin{bmatrix} a & b & c & d \\ e & f & g & h \end{bmatrix} = \begin{bmatrix} e & f & g & h \\ a & b & c & d \end{bmatrix}$$

Next, we note that elementary row operations are reversible, or **invertible**; i.e., for each such operation, there is another elementary row operation that reverses the effect of the first. (See also Exercise 6 of Section 1.2.) For example, $\frac{1}{4}R_2 \rightarrow R_2$ cancels out the effect of $4R_2 \rightarrow R_2$. In general,

Elementary Row Operation	Corresponding Inverse Operation
$R_i \leftrightarrow R_j$	$R_i \leftrightarrow R_j$
$cR_i \rightarrow R_i$	$(1/c)R_i \rightarrow R_i$
$R_i + cR_j \rightarrow R_i$	$R_i - cR_j \rightarrow R_i$

Because elementary row operations are invertible, we can recover I_n from an elementary matrix by performing the inverse operation. For example, I_3 is obtained from E_1, E_2, or E_3 by applying $R_1 + 3R_2 \rightarrow R_1, \frac{1}{4}R_2 \rightarrow R_2$, or $R_1 \leftrightarrow R_2$, respectively. This in turn implies that elementary matrices are invertible, because if E is an elementary matrix obtained from I_n by applying an elementary operation and E' is the elementary matrix obtained from I_n by applying the inverse operation, then $EE' = I_n$. Similarly, $E'E = I_n$. Hence, E is invertible and $E^{-1} = E'$. For example, E_1, E_2, and E_3 are invertible and

$$E_1^{-1} = \begin{bmatrix} 1 & 3 \\ 0 & 1 \end{bmatrix}, \qquad E_2^{-1} = \begin{bmatrix} 1 & 0 \\ 0 & \frac{1}{4} \end{bmatrix}, \qquad E_3^{-1} = \begin{bmatrix} 0 & 1 \\ 1 & 0 \end{bmatrix}$$

THEOREM 10

Every elementary matrix E has an inverse that is also an elementary matrix. E^{-1} is obtained from I by performing the inverse of the elementary row operation that produced E from I.

If matrices A and B are row-equivalent, then B can be obtained from A by a finite sequence of elementary row operations, say, $O_1, \ldots, O_k$. Let $E_1, \ldots, E_k$ be the elementary matrices corresponding to these operations. The effect of operation O_1 on A is the same as the product $E_1 A$. Likewise, the effect of O_2 is $E_2(E_1 A) = E_2 E_1 A$. Continuing in the same manner, we get $B = E_k \ldots E_2 E_1 A$.

THEOREM 11 $A \sim B$ if and only if there are elementary matrices $E_1, \ldots, E_k$ such that

$$B = E_k \cdots E_1 A$$

In particular, Theorem 11 applies to a matrix A and an echelon form U of A. If U is obtained from A by using row operations with corresponding elementary matrices $E_1, \ldots, E_k$, then

$$U = E_k \cdots E_1 A \tag{3.2}$$

Solving for A yields $A = (E_k \cdots E_1)^{-1} U$, or

$$A = E_1^{-1} \cdots E_k^{-1} U \tag{3.3}$$

Either of Equations (3.2) and (3.3) *records the row reduction of A*. The next example should be studied carefully to see how this works.

■ **EXAMPLE 24**   Row-reduce matrix A to U and write U as a product of elementary matrices and A. Then write A as a product of elementary matrices and U.

$$A = \begin{bmatrix} 1 & 3 & 7 \\ 2 & 6 & 8 \\ 0 & 4 & 3 \end{bmatrix}$$

SOLUTION   Let U be the echelon form matrix in the reduction

$$\begin{bmatrix} 1 & 3 & 7 \\ 2 & 6 & 8 \\ 0 & 4 & 3 \end{bmatrix} \sim \begin{bmatrix} 1 & 3 & 7 \\ 0 & 0 & -6 \\ 0 & 4 & 3 \end{bmatrix} \sim \begin{bmatrix} 1 & 3 & 7 \\ 0 & 4 & 3 \\ 0 & 0 & -6 \end{bmatrix} = U$$

The operations that yielded U are $-2R_1 + R_2 \rightarrow R_2$ and $R_2 \leftrightarrow R_3$. So the corresponding elementary matrices are

$$E_1 = \begin{bmatrix} 1 & 0 & 0 \\ -2 & 1 & 0 \\ 0 & 0 & 1 \end{bmatrix} \quad \text{and} \quad E_2 = \begin{bmatrix} 1 & 0 & 0 \\ 0 & 0 & 1 \\ 0 & 1 & 0 \end{bmatrix}$$

Hence, according to our analysis, $U = E_2 E_1 A$:

$$\begin{bmatrix} 1 & 3 & 7 \\ 0 & 4 & 3 \\ 0 & 0 & -6 \end{bmatrix} = \begin{bmatrix} 1 & 0 & 0 \\ 0 & 0 & 1 \\ 0 & 1 & 0 \end{bmatrix} \begin{bmatrix} 1 & 0 & 0 \\ -2 & 1 & 0 \\ 0 & 0 & 1 \end{bmatrix} \begin{bmatrix} 1 & 3 & 7 \\ 2 & 6 & 8 \\ 0 & 4 & 3 \end{bmatrix}$$

(Check.) So we *factored* U as the product $E_2 E_1 A$. Because $A = E_1^{-1} E_2^{-1} U$,

$$E_1^{-1} = \begin{bmatrix} 1 & 0 & 0 \\ 2 & 1 & 0 \\ 0 & 0 & 1 \end{bmatrix}, \quad \text{and} \quad E_2^{-1} = \begin{bmatrix} 1 & 0 & 0 \\ 0 & 0 & 1 \\ 0 & 1 & 0 \end{bmatrix}$$

we have

$$\begin{bmatrix} 1 & 3 & 7 \\ 2 & 6 & 8 \\ 0 & 4 & 3 \end{bmatrix} = \begin{bmatrix} 1 & 0 & 0 \\ 2 & 1 & 0 \\ 0 & 0 & 1 \end{bmatrix} \begin{bmatrix} 1 & 0 & 0 \\ 0 & 0 & 1 \\ 0 & 1 & 0 \end{bmatrix} \begin{bmatrix} 1 & 3 & 7 \\ 0 & 4 & 3 \\ 0 & 0 & -6 \end{bmatrix}$$

(Check.) This is a factorization of A in terms of one of its echelon forms and the elementary matrices of the inverse operations that produced it.

Characterization of Invertible Matrices

In this section we characterize an invertible matrix in several ways. First we observe the following.

THEOREM 12   The reduced row echelon form R of a square matrix A is either I or it has a row of zeros.

PROOF   If each row of A has a pivot, then $R = I$, because A is square. On the other hand, if some row of A has no pivot, the corresponding row of R is zero.

THEOREM 13   The following are equivalent:

1. A is invertible.
2. $A \sim I$.
3. A is a product of elementary matrices.

PROOF

$1 \Rightarrow 2$.   Let A be invertible. Then A is square, say, $n \times n$, and $A\mathbf{x} = \mathbf{b}$ is consistent for all n-vectors $\mathbf{b}$ by Theorem 5. So, each row of A has a pivot by Theorem 2 of Section 2.1. Hence, the reduced row echelon form of A is I_n (since A is square). Therefore, $A \sim I_n$.

$2 \Rightarrow 3$.   Let $A \sim I_n$ for some n. Then A is square $n \times n$, and there are elementary matrices $E_1, \ldots, E_k$ such that $A = E_k \cdots E_1 I_n = E_k \cdots E_1$, by Theorem 11. So A is a product of elementary matrices.

$3 \Rightarrow 1$.   Let A be a product of elementary matrices, say, $A = E_k \cdots E_1$. Then $E_1, \ldots, E_k$ are invertible by Theorem 10. So, A is invertible as a product of invertible matrices.

The Matrix Inversion Algorithm

Let us now use elementary matrices to explain why the matrix inversion algorithm works. If A is an $n \times n$ matrix with reduced row echelon form R, then there exist elementary matrices $E_1, \dots, E_k$ such that

$$A = E_k \cdots E_1 R$$

The row reduction of A using the operations that correspond to these elementary matrices can be described by

$$E_k^{-1} A = E_{k-1} \cdots E_1 R, \qquad E_{k-1}^{-1} E_k^{-1} A = E_{k-2} \cdots E_1 R, \quad \dots, \qquad E_1^{-1} \cdots E_k^{-1} A = R$$

So, in the reduction of $[A : I]$ the matrix obtained from placing $I_n = I$ next to A yields

$$[E_1^{-1} \cdots E_k^{-1} A : E_1^{-1} \cdots E_k^{-1}] \quad \text{or} \quad [R : E_1^{-1} \cdots E_k^{-1}]$$

If R has a row of zeros, then A is not invertible. Otherwise, R is I and A is invertible. Hence, $E_1^{-1} \cdots E_k^{-1} A = R = I$ which implies $A^{-1} = E_1^{-1} \cdots E_k^{-1}$. So, in this case

$$[R : E_1^{-1} \cdots E_k^{-1}] = [I : A^{-1}]$$

We conclude that the reduction of $[A : I]$, which is the matrix inversion algorithm, either detects a noninvertible matrix or computes its inverse, as claimed in Section 3.2.

■ EXAMPLE 25   Write

$$A = \begin{bmatrix} 1 & 1 \\ 1 & 2 \end{bmatrix}$$

and A^{-1} as products of elementary matrices.

SOLUTION   We compute A^{-1} as usual:

$$\begin{bmatrix} 1 & 1 & : & 1 & 0 \\ 1 & 2 & : & 0 & 1 \end{bmatrix} \sim \begin{bmatrix} 1 & 1 & : & 1 & 0 \\ 0 & 1 & : & -1 & 1 \end{bmatrix} \sim \begin{bmatrix} 1 & 0 & : & 2 & -1 \\ 0 & 1 & : & -1 & 1 \end{bmatrix}$$

The elementary matrices that correspond to the row operations $-R_1 + R_2 \rightarrow R_2$ and $R_1 - R_2 \rightarrow R_1$ are $E_1 = \begin{bmatrix} 1 & 0 \\ -1 & 1 \end{bmatrix}$ and $E_2 = \begin{bmatrix} 1 & -1 \\ 0 & 1 \end{bmatrix}$. Hence,

$$E_2 E_1 A = \begin{bmatrix} 1 & -1 \\ 0 & 1 \end{bmatrix} \begin{bmatrix} 1 & 0 \\ -1 & 1 \end{bmatrix} \begin{bmatrix} 1 & 1 \\ 1 & 2 \end{bmatrix} = I$$

Therefore,

$$A = (E_2 E_1)^{-1} = E_1^{-1} E_2^{-1} = \begin{bmatrix} 1 & 0 \\ 1 & 1 \end{bmatrix} \begin{bmatrix} 1 & 1 \\ 0 & 1 \end{bmatrix}$$

and

$$A^{-1} = E_2 E_1 = \begin{bmatrix} 1 & -1 \\ 0 & 1 \end{bmatrix} \begin{bmatrix} 1 & 0 \\ -1 & 1 \end{bmatrix}$$

are the required products.

An interesting implication of Theorems 12 and 13 is the next theorem, which shows that for square matrices of the same size *only one* of the two conditions $AB = I$ and $BA = I$ suffices to ensure that A (hence, B) is invertible.

THEOREM 14

Let A and B be $n \times n$ matrices. If $AB = I$, then A and B are invertible and $A^{-1} = B$, $B^{-1} = A$. In particular, $AB = I$ if and only if $BA = I$.

PROOF Let R be the reduced row echelon form of A. There are elementary matrices $E_1, \ldots, E_k$ such that $R = E_n \cdots E_1 A$, by Theorem 11. Therefore, $RB = E_n \cdots E_1 AB = E_n \cdots E_1$, because $AB = I$. Hence, RB is invertible as a product of invertible matrices. So, R cannot have a zero row. Therefore, $R = I$ by Theorem 12. Hence, A^{-1} exists and $AB = I$ implies $A^{-1}AB = A^{-1}I$ or $A^{-1} = B$. Hence, $B^{-1} = A$.

Let us now characterize invertible matrices in several basic ways.

THEOREM 15

Let A be an $n \times n$ matrix. The following are equivalent.

1. A is invertible.
2. $A \sim I_n$.
3. A is a product of elementary matrices.
4. There is an $n \times n$ matrix B such that $AB = I_n$.
5. There is an $n \times n$ matrix C such that $CA = I_n$.
6. Each column of A is a pivot column.
7. Each row of A has a pivot.
8. The columns of A are linearly independent.
9. The rows of A are linearly independent.
10. The columns of A span $\mathbf{R}^n$.
11. The rows of A span $\mathbf{R}^n$.
12. The system $A\mathbf{x} = \mathbf{b}$ has at least one solution for each n-vector $\mathbf{b}$.
13. The system $A\mathbf{x} = \mathbf{b}$ has exactly one solution for each n-vector $\mathbf{b}$.
14. The homogeneous system $A\mathbf{x} = \mathbf{0}$ has only the trivial solution.

Thm 2 → (handwritten annotation next to item 12)

PROOF $1 \Leftrightarrow 2 \Leftrightarrow 3$ by Theorem 13. That $2 \Rightarrow 4$ and $2 \Rightarrow 5$ is clear. Also $4 \Rightarrow 2$ and $5 \Rightarrow 2$ by Theorem 14. Therefore, 1–5 are all equivalent. Statements 6–12 and 14 are all equivalent by Theorems 2 and 12 (Sections 2.1 and 2.4) and the equivalent statements before Example 46, Section 2.5, for $m = n$. Also, $2 \Leftrightarrow 6$, because A is square. Clearly, $13 \Rightarrow 12$. Lastly, we prove that $12 \Rightarrow 13$. Let us assume 12. So, for each n-vector $\mathbf{b}$, the system $A\mathbf{x} = \mathbf{b}$ has at least one solution, say, $\mathbf{v}_1$. If $\mathbf{v}_2$ is another solution, then $A\mathbf{v}_1 = \mathbf{b} = A\mathbf{v}_2$. Hence, $A(\mathbf{v}_1 - \mathbf{v}_2) = \mathbf{0}$. Therefore, $\mathbf{v}_1 - \mathbf{v}_2 = \mathbf{0}$ or $\mathbf{v}_1 = \mathbf{v}_2$, since $12 \Leftrightarrow 14$. So, the solution of $A\mathbf{x} = \mathbf{b}$ is unique. This proves Statement 13.

A COMMON MISCONCEPTION Statements 12 and 13 of Theorem 15 may give the *wrong impression* that if a square system has a solution, it *should be unique*.

for one solution, all theorels fail. (handwritten annotation at bottom left)

This is not true. The system $\begin{bmatrix} 1 & 1 \\ 1 & 1 \end{bmatrix} \begin{bmatrix} x \\ y \end{bmatrix} = \mathbf{b}$ has infinitely many solutions for $\mathbf{b} = \begin{bmatrix} 1 \\ 1 \end{bmatrix}$ and no solutions for $\mathbf{b} = \begin{bmatrix} 1 \\ 2 \end{bmatrix}$. The key phrase is for each n-vector $\mathbf{b}$, which fails here.

Theorem 15 has an interesting consequence.

THEOREM 16   Let A and B be $n \times n$ matrices. If either A or B is noninvertible, so is AB.

PROOF

1. If B has no inverse, then $B\mathbf{x} = \mathbf{0}$ has a nontrivial solution by Theorem 15. Therefore, $AB\mathbf{x} = \mathbf{0}$ has a nontrivial solution, which in turn implies that AB is noninvertible.
2. If B is invertible, then A has no inverse, by assumption. So $A\mathbf{x} = \mathbf{0}$ has a nontrivial solution, say, $\mathbf{v} \neq \mathbf{0}$, by Theorem 15. Hence, $A\mathbf{v} = \mathbf{0}$. The vector $B^{-1}\mathbf{v}$ is not equal to $\mathbf{0}$ (why?) and $(AB)B^{-1}\mathbf{v} = \mathbf{0}$. So, $AB\mathbf{x} = \mathbf{0}$ has nontrivial solutions. Therefore, AB is not invertible.

Exercises 3.3

In Exercises 1–2 indicate which of the matrices are elementary. For each elementary matrix, identify the elementary row operation that yielded the matrix from the identity matrix.

1. $A = \begin{bmatrix} 1 & 1 \\ 0 & 1 \end{bmatrix}$, $B = \begin{bmatrix} -1 & -1 \\ 0 & 1 \end{bmatrix}$,

$C = \begin{bmatrix} 1 & 0 & -2 \\ 0 & 1 & 0 \\ 0 & 0 & 1 \end{bmatrix}$, $D = \begin{bmatrix} 1 & 0 & 0 & 0 \\ 0 & 0 & 0 & 1 \\ 0 & 0 & 1 & 0 \\ 0 & 1 & 0 & 0 \end{bmatrix}$

2. $E = \begin{bmatrix} 2 & 0 \\ 0 & 2 \end{bmatrix}$, $F = \begin{bmatrix} 1 & -1 \\ 0 & 1 \end{bmatrix}$,

$G = \begin{bmatrix} 1 & 0 & -1 \\ 0 & 0 & 1 \\ 0 & 1 & 0 \end{bmatrix}$, $H = \begin{bmatrix} 1 & 0 & 0 & 0 \\ 0 & 0 & 0 & 1 \\ 0 & 0 & 1 & 0 \\ 0 & -1 & 0 & 0 \end{bmatrix}$

3. What elementary row operation yields each of the elementary matrices from the identity matrix of the same size?

$J = \begin{bmatrix} 0 & 1 \\ 1 & 0 \end{bmatrix}$, $K = \begin{bmatrix} 1 & 0 & 0 \\ 0 & 2 & 0 \\ 0 & 0 & 1 \end{bmatrix}$

$L = \begin{bmatrix} 1 & 0 & -5 & 0 \\ 0 & 1 & 0 & 0 \\ 0 & 0 & 1 & 0 \\ 0 & 0 & 0 & 1 \end{bmatrix}$, $M = \begin{bmatrix} 1 & 0 & 0 \\ 0 & 1 & 0 \\ -1 & 0 & 1 \end{bmatrix}$

4. What row operation yields an identity matrix from each of J, K, L, M?

5. Write the inverses of the elementary matrix operations.

   a. $R_1 \leftrightarrow R_3$ **b.** $R_1 + 5R_3 \rightarrow R_1$

   c. $(1/2)R_4 \rightarrow R_4$ **d.** $10R_1 + R_3 \rightarrow R_3$

6. Multiply $A = \begin{bmatrix} 1 & 2 & 3 \\ -1 & -1 & -1 \\ 0 & 1 & 0 \end{bmatrix}$ on the left by a suitable elementary matrix to perform the following matrix operations.

   a. $R_1 \leftrightarrow R_3$ **b.** $R_1 - 2R_3 \rightarrow R_1$

   c. $-2R_2 \rightarrow R_2$ **d.** $5R_1 + R_3 \rightarrow R_3$

7. Write A and A^{-1} as products of elementary matrices.

$A = \begin{bmatrix} 2 & 1 \\ -1 & 0 \end{bmatrix}$

8. Show that the decomposition of an invertible matrix as a product of elementary matrices is *not unique* by finding a second factorization of matrix A in Exercise 7.

9. Express each of the following as products of elementary matrices.

$$A = \begin{bmatrix} 3 & -6 \\ 0 & 3 \end{bmatrix}, \qquad B = \begin{bmatrix} 2 & 0 \\ 1 & 1 \end{bmatrix}$$

$$C = \begin{bmatrix} 1 & 0 & 0 \\ 0 & 1 & 0 \\ 1 & 1 & 1 \end{bmatrix}$$

10. Matrices A and B are equivalent. Find elementary matrices E_1 and E_2 such that $A = E_2 E_1 B$.

$$A = \begin{bmatrix} 1 & 2 & 3 \\ -1 & -4 & -1 \\ 0 & 1 & 0 \end{bmatrix}, \qquad B = \begin{bmatrix} 1 & 3 & 3 \\ -1 & -4 & -1 \\ 1 & 2 & 3 \end{bmatrix}$$

11. Explain why the matrix cannot be written as a product of elementary matrices.

$$D = \begin{bmatrix} 1 & 1 & 1 \\ 0 & 1 & 0 \\ 1 & 0 & 1 \end{bmatrix}$$

12. Write matrix D of Exercise 11 as a product of two elementary matrices and a noninvertible matrix in reduced row echelon form.

13. Use Theorem 15 to show that the system has exactly one solution for any choices of b_1 and b_2.

$$x - y = b_1$$
$$x + 2y = b_2$$

14. Use Theorem 15 to show that the system has infinitely many solutions.

$$x + y + z = 0$$
$$y = 0$$
$$x + z = 0$$

15. Let c_1, c_2, c_3 be nonzero scalars. Write A and A^{-1} as products of elementary matrices.

$$A = \begin{bmatrix} 0 & 0 & c_1 \\ 0 & c_2 & 0 \\ c_3 & 0 & 0 \end{bmatrix}$$

16. Let c_1, c_2 be nonzero scalars. Write A as a product of elementary matrices and a noninvertible matrix in reduced row echelon form.

$$A = \begin{bmatrix} 0 & 0 & c_1 \\ 0 & c_2 & 0 \\ 0 & 0 & 0 \end{bmatrix}$$

17. Verify Statements 1–9 of Theorem 15 for

$$\begin{bmatrix} -1 & 0 & 2 \\ 0 & 1 & 0 \\ -2 & 0 & 3 \end{bmatrix}$$

18. Let A be an elementary permutation matrix. Show that $A^2 = I$. Deduce that $A^{-1} = A$.

19. True or false? (Justify your choices.) The matrix A is invertible if

a. All its columns are linearly independent.

b. Its reduced row echelon form is I.

c. There is a matrix B such that $AB = I$.

d. There is a matrix C such that $CA = I$.

e. Each row and each column of A has a pivot.

f. A^T is a product of elementary matrices.

20. Let A be a matrix and let U be one of its echelon forms. Can you determine whether A is invertible by looking at U?

Right and Left Inverses

Let A be an $m \times n$ matrix. A matrix B is called a **right inverse** of A if $AB = I$. Likewise, C is a **left inverse** of A if $CA = I$.

For example, if $P = \begin{bmatrix} 1 & 0 & 0 \\ 0 & 1 & 0 \end{bmatrix}$ and $Q = \begin{bmatrix} 1 & 0 \\ 0 & 1 \\ 0 & 0 \end{bmatrix}$, then

P is a left inverse of Q and Q is a right inverse of P, because

$$\begin{bmatrix} 1 & 0 & 0 \\ 0 & 1 & 0 \end{bmatrix} \begin{bmatrix} 1 & 0 \\ 0 & 1 \\ 0 & 0 \end{bmatrix} = \begin{bmatrix} 1 & 0 \\ 0 & 1 \end{bmatrix} \qquad (3.4)$$

21. If an $m \times n$ matrix A has a right inverse B, what is the size of B?

22. If an $m \times n$ matrix A has a left inverse C, what is the size of C?

23. Show that if A has a right inverse, then A^T has a left inverse.

24. Show that if A has a left inverse, then A^T has a right inverse.

25. Let A be an $m \times n$ matrix. Show that the following statements are equivalent:

a. A has a right inverse.

b. The system $A\mathbf{x} = \mathbf{b}$ is consistent for all m-vectors **b.**

c. Each row of A has a pivot.

d. The columns of A span $\mathbf{R}^m$.

(*Hint:* To prove (b) $\Rightarrow$ (a), consider the matrix $B = [\mathbf{b}_1 \cdots \mathbf{b}_n]$, where $\mathbf{b}_i$ is a solution of $A\mathbf{x} = \mathbf{e}_i$.)

26. Let A be an $m \times n$ matrix. Show that the following statements are equivalent:

a. A has a left inverse.

b. The system $A\mathbf{x} = \mathbf{0}$ has only the trivial solution.

c. Each column of A is a pivot column.

d. The columns of A are linearly independent.

27. Show that if an $m \times n$ matrix A has both a right inverse B and a left inverse C, then the following hold.

a. $m = n$

b. $B = C$

c. A is invertible.

3.4   LU Factorization

> ### Reader's Goals for This Section
>
> 1. To know how to factor a matrix as product of a lower and an upper triangular matrix.
> 2. To know how to use this factorization to solve linear systems.

In Section 3.3 we saw how to factor a matrix as a product of elementary matrices and one of its echelon forms. In general, a factorization of a matrix can be very useful in understanding the properties of the matrix. For instance, suppose we know how to factor an $m \times n$ matrix A as

$$A = LU$$

where L in $m \times m$ lower triangular and U is $m \times n$ row echelon form. Then the system

$$A\mathbf{x} = \mathbf{b} \tag{3.5}$$

can be solved in two easier steps. First we solve (3.6) for $\mathbf{y}$:

$$L\mathbf{y} = \mathbf{b} \tag{3.6}$$

Then we solve (3.7) for $\mathbf{x}$:

$$U\mathbf{x} = \mathbf{y} \tag{3.7}$$

Solving these two systems is, in fact, equivalent to solving the original system, because

$$LU\mathbf{x} = L(U\mathbf{x}) = L\mathbf{y} = \mathbf{b}$$

The advantage of not solving directly is that (3.6) is a lower triangular system and can be easily solved by a forward substitution and (3.7) is upper triangular and can be easily solved by a back substitution.

A factorization of A as shown—i.e., as a product of a lower triangular matrix L and an "upper triangular matrix" U, if it exists—is called an **LU factorization**, or **LU decomposition**.

■ EXAMPLE 26   Use the LU factorization of A,

$$A = \begin{bmatrix} 4 & -2 & 1 \\ 20 & -7 & 12 \\ -8 & 13 & 17 \end{bmatrix} = \begin{bmatrix} 1 & 0 & 0 \\ 5 & 1 & 0 \\ -2 & 3 & 1 \end{bmatrix} \begin{bmatrix} 4 & -2 & 1 \\ 0 & 3 & 7 \\ 0 & 0 & -2 \end{bmatrix} = LU \qquad (3.8)$$

to solve for $\mathbf{x} = (x_1, x_2, x_3)$ in the system

$$A\mathbf{x} = \begin{bmatrix} 11 \\ 70 \\ 17 \end{bmatrix} = \mathbf{b}$$

SOLUTION   Let $\mathbf{y} = (y_1, y_2, y_3)$ be a new vector of unknowns. We first solve the lower triangular system $L\mathbf{y} = \mathbf{b}$,

$$\begin{aligned} y_1 &= 11 \\ 5y_1 + y_2 &= 70 \\ -2y_1 + 3y_2 + y_3 &= 17 \end{aligned}$$

by forward elimination. Because $y_1 = 11$, the second equation yields $y_2 = 70 - 5y_1 = 15$, and the third yields $y_3 = 17 + 2y_1 - 3y_2 = -6$.

Then we solve the upper triangular system $U\mathbf{x} = \mathbf{y}$,

$$\begin{aligned} 4x_1 - 2x_2 + x_3 &= 11 \\ 3x_2 + 7x_3 &= 15 \\ -2x_3 &= -6 \end{aligned}$$

by back substitution to get $x_3 = 3$ from the third equation, $x_2 = -2$ from the second, and $x_1 = 1$ from the first. Hence, the solution of the original system is $(1, -2, 3)$.

It is clear from Example 26 that once we have an LU factorization of A, it is quite easy to solve the system $A\mathbf{x} = \mathbf{b}$. This is particularly useful when we have to solve *several* systems, all with the same coefficient matrix A. So, for instance, if we want to solve the system $A\mathbf{x} = (5, 21, -20)$, where A is from Example 26, we just solve

$$\begin{aligned} y_1 &= 5 \\ 5y_1 + y_2 &= 21 \\ -2y_1 + 3y_2 + y_3 &= -20 \end{aligned}$$

to get $y_1 = 5, y_2 = -4, y_3 = 2$. Then we solve

$$\begin{aligned} 4x_1 - 2x_2 + x_3 &= 5 \\ 3x_2 + 7x_3 &= -4 \\ -2x_3 &= 2 \end{aligned}$$

to get $x_1 = 2, x_2 = 1, x_3 = -1$.

If LU factorizations are so useful, how do we compute them? The answer is surprisingly simple. Recall from Section 3.3 that any matrix A can be factored as

$$A = E_1^{-1} \cdots E_k^{-1} U$$

where U is an echelon form of A and $E_1, \ldots, E_k$ are the elementary matrices corresponding to the elementary row operations used to reduce A to U.

Clearly any matrix A can be reduced to echelon form *without any scaling operations*, using only interchanges and eliminations. If our matrix A can be reduced using only eliminations (which is not always possible) then the matrices $E_1, \ldots, E_k$ are all *lower triangular* and have 1s on the main diagonal. Such matrices are called **unit** lower diagonal. It is easy to show that $E_1^{-1} \cdots E_k^{-1}$ is also unit lower triangular (see Exercises 25 and 28). If we let

$$L = E_1^{-1} \cdots E_k^{-1}$$

then $A = LU$ is an LU decomposition for A.

For example, the reduction

$$\begin{bmatrix} 4 & -2 & 1 \\ 20 & -7 & 12 \\ -8 & 13 & 17 \end{bmatrix} \sim \begin{bmatrix} 4 & -2 & 1 \\ 0 & 3 & 7 \\ 0 & 9 & 19 \end{bmatrix} \sim \begin{bmatrix} 4 & -2 & 1 \\ 0 & 3 & 7 \\ 0 & 0 & -2 \end{bmatrix}$$

yields matrix U of (3.8), and the elementary row operations yield

$$E_1 = \begin{bmatrix} 1 & 0 & 0 \\ -5 & 1 & 0 \\ 0 & 0 & 1 \end{bmatrix}, \qquad E_2 = \begin{bmatrix} 1 & 0 & 0 \\ 0 & 1 & 0 \\ 2 & 0 & 1 \end{bmatrix}, \qquad E_3 = \begin{bmatrix} 1 & 0 & 0 \\ 0 & 1 & 0 \\ 0 & -3 & 1 \end{bmatrix}$$

Hence,

$$E_1^{-1} = \begin{bmatrix} 1 & 0 & 0 \\ 5 & 1 & 0 \\ 0 & 0 & 1 \end{bmatrix}, \qquad E_2^{-1} = \begin{bmatrix} 1 & 0 & 0 \\ 0 & 1 & 0 \\ -2 & 0 & 1 \end{bmatrix}, \qquad E_3^{-1} = \begin{bmatrix} 1 & 0 & 0 \\ 0 & 1 & 0 \\ 0 & 3 & 1 \end{bmatrix}$$

So,

$$L = E_1^{-1} E_2^{-1} E_3^{-1} = \begin{bmatrix} 1 & 0 & 0 \\ 5 & 1 & 0 \\ -2 & 3 & 1 \end{bmatrix}$$

Therefore,

$$A = \begin{bmatrix} 4 & -2 & 1 \\ 20 & -7 & 12 \\ -8 & 13 & 17 \end{bmatrix} = \begin{bmatrix} 1 & 0 & 0 \\ 5 & 1 & 0 \\ -2 & 3 & 1 \end{bmatrix} \begin{bmatrix} 4 & -2 & 1 \\ 0 & 3 & 7 \\ 0 & 0 & -2 \end{bmatrix} = LU$$

A closer look at L shows that there is no need to compute inverses or products. In fact, L can be computed straight from the eliminations. First, it is lower triangular of size $m \times m$ if A is of size $m \times n$. Then its diagonal consists of 1s. The entry $(1, 2)$ is 5 and can be obtained from the operation $R_2 - 5R_1 \rightarrow R_2$ used to get a zero at $(1, 2)$. In L we used 5 instead of -5, because we had to invert E_1. Likewise, -2 is obtained from the operation $R_3 + 2R_1 \rightarrow R_3$ and 3 from $R_3 - 3R_1 \rightarrow R_3$ in the second stage of the reduction.

■ EXAMPLE 27   Find an LU factorization of A.

$$A = \begin{bmatrix} 2 & 3 & -1 & 4 & 1 \\ -6 & -6 & 5 & -11 & -4 \\ 4 & 18 & 6 & 14 & -1 \\ -2 & -9 & -3 & 4 & 9 \end{bmatrix}$$

SOLUTION L is size 4×4, and we have

$$A \sim \begin{bmatrix} 2 & 3 & -1 & 4 & 1 \\ 0 & 3 & 2 & 1 & -1 \\ 0 & 12 & 8 & 6 & -3 \\ 0 & -6 & -4 & 8 & 10 \end{bmatrix}, \quad \text{so } L = \begin{bmatrix} 1 & 0 & 0 & 0 \\ -3 & 1 & 0 & 0 \\ 2 & ? & 1 & 0 \\ -1 & ? & ? & 1 \end{bmatrix}$$

$$\sim \begin{bmatrix} 2 & 3 & -1 & 4 & 1 \\ 0 & 3 & 2 & 1 & -1 \\ 0 & 0 & 0 & 2 & 1 \\ 0 & 0 & 0 & 10 & 8 \end{bmatrix}, \quad \text{so } L = \begin{bmatrix} 1 & 0 & 0 & 0 \\ -3 & 1 & 0 & 0 \\ 2 & 4 & 1 & 0 \\ -1 & -2 & ? & 1 \end{bmatrix}$$

$$\sim \begin{bmatrix} 2 & 3 & -1 & 4 & 1 \\ 0 & 3 & 2 & 1 & -1 \\ 0 & 0 & 0 & 2 & 1 \\ 0 & 0 & 0 & 0 & 3 \end{bmatrix} = U, \quad \text{so } L = \begin{bmatrix} 1 & 0 & 0 & 0 \\ -3 & 1 & 0 & 0 \\ 2 & 4 & 1 & 0 \\ -1 & -2 & 5 & 1 \end{bmatrix}$$

■ EXAMPLE 28   Find an LU factorization of A.

$$A = \begin{bmatrix} 2 & 3 & -1 \\ -6 & -6 & 5 \\ 4 & 18 & 6 \\ -2 & -9 & -3 \end{bmatrix}$$

SOLUTION L is size 4×4 and

$$A \sim \begin{bmatrix} 2 & 3 & -1 \\ 0 & 3 & 2 \\ 0 & 12 & 8 \\ 0 & -6 & -4 \end{bmatrix}, \quad \text{so } L = \begin{bmatrix} 1 & 0 & 0 & 0 \\ -3 & 1 & 0 & 0 \\ 2 & ? & 1 & 0 \\ -1 & ? & ? & 1 \end{bmatrix}$$

$$\sim \begin{bmatrix} 2 & 3 & -1 \\ 0 & 3 & 2 \\ 0 & 0 & 0 \\ 0 & 0 & 0 \end{bmatrix} = U, \quad \text{so } L = \begin{bmatrix} 1 & 0 & 0 & 0 \\ -3 & 1 & 0 & 0 \\ 2 & 4 & 1 & 0 \\ -1 & -2 & ? & 1 \end{bmatrix}$$

In this case, there is no more elimination left, but because the $(4, 3)$ entry of L corresponds to the operation $R_4 - 0R_3 \to R_4$, we have

$$L = \begin{bmatrix} 1 & 0 & 0 & 0 \\ -3 & 1 & 0 & 0 \\ 2 & 4 & 1 & 0 \\ -1 & -2 & 0 & 1 \end{bmatrix}$$

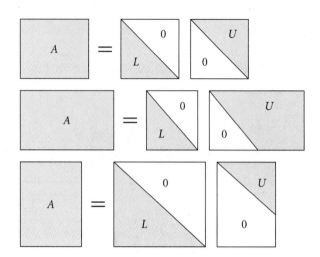

Our analysis so far yields the following.

THEOREM 17

Let A be an $m \times n$ matrix that can be reduced to the $m \times n$ echelon form U using only eliminations. Then A has an LU factorization. In particular, A can be factored as

$$A = LU$$

where L is $m \times m$ lower triangular with only 1s on the main diagonal. The (i, j)th entry l_{ij} $(i > j)$ of L comes from the operation $R_i - l_{ij}R_j \rightarrow R_i$ used to get 0 at this position during the reduction.

REMARKS

1. The entries of L below the main diagonal are sometimes called **Gauss multipliers**.
2. If A is square, the particular LU factorization we used is called **Doolittle**. There is another standard version, where the *upper* triangular matrix U has 1s on its main diagonal, called a **Crout factorization**.
3. Computer programs that find LU factorizations use **overwriting**. They compute L and U simultaneously and overwrite the original matrix, so that the part of A below the diagonal becomes L and on and above the diagonal becomes U. Overwriting for large matrices is very important, because it saves in memory storage. Additional saving is achieved by not explicitly storing the 1s on the main diagonal. Here is an example of LU reduction and gradual overwriting of the original matrix. The boxed numbers are the entries of L below the diagonal. The rest of the entries are those of U on and above the diagonal.

$$\begin{bmatrix} 4 & -2 & 1 \\ 20 & -7 & 12 \\ -8 & 13 & 17 \end{bmatrix} \rightarrow \begin{bmatrix} 4 & -2 & 1 \\ \boxed{5} & 3 & 7 \\ \boxed{-2} & 9 & 19 \end{bmatrix} \rightarrow \begin{bmatrix} 4 & -2 & 1 \\ \boxed{5} & 3 & 7 \\ \boxed{-2} & \boxed{3} & -2 \end{bmatrix}$$

Computational Efficiency with LU

If A is a square matrix, say, $n \times n$, it can be shown that to solve the linear system $A\mathbf{x} = \mathbf{b}$ using LU factorization it takes approximately $2n^3/3$ operations (for large n).[4] Of these operations $2n^2$ are performed during the forward and backward elimination. To get an idea of how useful LU factorization can be, suppose we need to solve two systems with 500 equations and 500 unknowns and the same coefficient matrix A. If we use Gauss elimination, it would take $2n^3/3 = 2 \cdot 500^3/3$ operations per system, a total of about 160 million operations. However, if we used an LU factorization of A to solve the first system ($2 \cdot 500^3/3$ operations), the second system would require only forward and backward elimination, an additional $2n^2 = 2 \cdot 500^2$ operations, a total of only about 83 million operations. Of course, if we have more or larger systems, the computational savings are truly enormous.

When Interchanges Are Necessary

So far our LU factorization exists only if the matrix A can be reduced to echelon form using only eliminations. What if this is not possible? For example, to row-reduce $\begin{bmatrix} 0 & 1 \\ 1 & 0 \end{bmatrix}$ we need an interchange. Since LU is too useful to ignore even in these cases, we offer a *brief* discussion to illustrate the *ideas* involved.[5]

Recall that the interchange of two rows of a matrix A can be expressed as $P_i A$, where P_i is the elementary permutation matrix that corresponds to the interchange. If during a row reduction of A we perform all the interchanges $P_1, \ldots, P_k$ first, then the matrix $P_k \cdots P_1 A$ can be row-reduced by using eliminations only. So, it has an LU factorization. The matrix $P = P_k \cdots P_1$, which is a product of elementary permutation matrices, is called a **permutation matrix**. What we do is compute P and then find an LU factorization for PA.

$$PA = LU$$

To illustrate, let us consider the following Gauss elimination, which requires two interchanges.

$$A = \begin{bmatrix} 0 & 0 & 4 \\ 1 & 2 & 3 \\ 1 & 4 & 1 \end{bmatrix} \sim \begin{bmatrix} 1 & 2 & 3 \\ 0 & 0 & 4 \\ 1 & 4 & 1 \end{bmatrix} \sim \begin{bmatrix} 1 & 2 & 3 \\ 0 & 0 & 4 \\ 0 & 2 & -2 \end{bmatrix} \sim \begin{bmatrix} 1 & 2 & 3 \\ 0 & 2 & -2 \\ 0 & 0 & 4 \end{bmatrix}$$

The elementary permutation matrices of the interchanges are

$$P_1 = \begin{bmatrix} 0 & 1 & 0 \\ 1 & 0 & 0 \\ 0 & 0 & 1 \end{bmatrix} \quad \text{and} \quad P_2 = \begin{bmatrix} 1 & 0 & 0 \\ 0 & 0 & 1 \\ 0 & 1 & 0 \end{bmatrix}$$

---

[4]This is exactly the number of operations in Gauss elimination.
[5]The theoretical details and efficient implementations can be found in more advanced texts on numerical linear algebra or on numerical analysis.

First we compute the permutation matrix P:

$$P = P_2 P_1 = \begin{bmatrix} 1 & 0 & 0 \\ 0 & 0 & 1 \\ 0 & 1 & 0 \end{bmatrix} \begin{bmatrix} 0 & 1 & 0 \\ 1 & 0 & 0 \\ 0 & 0 & 1 \end{bmatrix} = \begin{bmatrix} 0 & 1 & 0 \\ 0 & 0 & 1 \\ 1 & 0 & 0 \end{bmatrix}$$

Then we find PA:

$$PA = \begin{bmatrix} 0 & 1 & 0 \\ 0 & 0 & 1 \\ 1 & 0 & 0 \end{bmatrix} \begin{bmatrix} 0 & 0 & 4 \\ 1 & 2 & 3 \\ 1 & 4 & 1 \end{bmatrix} = \begin{bmatrix} 1 & 2 & 3 \\ 1 & 4 & 1 \\ 0 & 0 & 4 \end{bmatrix}$$

Finally, we find the LU decomposition of PA.

$$PA = \begin{bmatrix} 1 & 2 & 3 \\ 1 & 4 & 1 \\ 0 & 0 & 4 \end{bmatrix} = \begin{bmatrix} 1 & 0 & 0 \\ 1 & 1 & 0 \\ 0 & 0 & 1 \end{bmatrix} \begin{bmatrix} 1 & 2 & 3 \\ 0 & 2 & -2 \\ 0 & 0 & 4 \end{bmatrix} = LU$$

Even though A itself has no LU factorization, the LU factorization of PA is *just as useful.*

■ EXAMPLE 29   Use $PA = LU$ as before to solve the system.

$$\begin{bmatrix} 0 & 0 & 4 \\ 1 & 2 & 3 \\ 1 & 4 & 1 \end{bmatrix} \mathbf{x} = \begin{bmatrix} 12 \\ 14 \\ 12 \end{bmatrix} = \mathbf{b}$$

SOLUTION   First, we multiply $A\mathbf{x} = \mathbf{b}$ on the left by P.

$$PA\mathbf{x} = P\mathbf{b} = \begin{bmatrix} 0 & 1 & 0 \\ 0 & 0 & 1 \\ 1 & 0 & 0 \end{bmatrix} \begin{bmatrix} 12 \\ 14 \\ 12 \end{bmatrix} = \begin{bmatrix} 14 \\ 12 \\ 12 \end{bmatrix}$$

Then we use the LU factorization of PA to solve the *new* system, $PA\mathbf{x} = P\mathbf{b}$. We solve the lower triangular system $L\mathbf{y} = P\mathbf{b}$,

$$y_1 = 14$$

$$y_1 + y_2 = 12$$

$$y_3 = 12$$

to get $y_1 = 14$, $y_2 = -2$, and $y_3 = 12$. Then we solve the upper triangular system, $U\mathbf{x} = \mathbf{y}$,

$$x_1 + 2x_2 + 3x_3 = 14$$

$$2x_2 - 2x_3 = -2$$

$$4x_3 = 12$$

to get $x_1 = 1$, $x_2 = 2$, and $x_3 = 3$, which is the solution of the original system.

REMARK   Several mathematical packages have routines that find the LU factorization of a matrix A. Usually what is computed is a permutation matrix P, along with the factors L and U of PA. Note that if A can be reduced without interchanges, then P is just the identity matrix.

LU with CAS

Maple

```
> with(linalg):
> LUdecomp(matrix([[4,-2,1],[20,-7,12],
>          [-8,13,17]]),L='l',U='u'):
> evalm(l);evalm(u);
```

$$\begin{bmatrix} 1 & 0 & 0 \\ 5 & 1 & 0 \\ -2 & 3 & 1 \end{bmatrix} \begin{bmatrix} 4 & -2 & 1 \\ 0 & 3 & 7 \\ 0 & 0 & -2 \end{bmatrix}$$

Mathematica

```
In[1]:=
     LUDecomposition[{{4,-2,1},{20,-7,12},{-8,13,17}}]
Out[1]=    1    1           3             6
     {{{4, -, -(-)}, {20, 7, -(-)}, {-8, 19, -}}, {1, 3, 3}}
         2    4            7             7
```

MATLAB

```
>> [L,U] = lu([4 -2 1; 20 -7 12; -8 13 17])
 L =
     0.2000    -0.0588    1.0000
     1.0000         0         0
    -0.4000     1.0000         0
 U =
    20.0000    -7.0000    12.0000
         0     10.2000    21.8000
         0          0     -0.1176
```

Exercises 3.4

In Exercises 1–5 find the solution of the system $A\mathbf{x} = \mathbf{b}$, where A is already factored as LU. There is no need to compute A explicitly.

1. $\begin{bmatrix} 1 & 0 \\ -3 & 1 \end{bmatrix} \begin{bmatrix} 4 & 1 \\ 0 & -1 \end{bmatrix} \mathbf{x} = \begin{bmatrix} -11 \\ 32 \end{bmatrix}$

2. $\begin{bmatrix} 1 & 0 \\ 5 & 1 \end{bmatrix} \begin{bmatrix} 2 & 1 \\ 0 & -7 \end{bmatrix} \mathbf{x} = \begin{bmatrix} 12 \\ 46 \end{bmatrix}$

3. $\begin{bmatrix} 1 & 0 & 0 \\ 4 & 1 & 0 \\ -2 & 3 & 1 \end{bmatrix} \begin{bmatrix} 2 & -2 & 1 \\ 0 & 3 & -1 \\ 0 & 0 & -2 \end{bmatrix} \mathbf{x} = \begin{bmatrix} 2 \\ 7 \\ -3 \end{bmatrix}$

4. $\begin{bmatrix} 1 & 0 & 0 \\ 4 & 1 & 0 \\ -7 & 3 & 1 \end{bmatrix} \begin{bmatrix} -1 & 2 & 1 \\ 0 & 3 & -1 \\ 0 & 0 & -5 \end{bmatrix} \mathbf{x} = \begin{bmatrix} 0 \\ 3 \\ 9 \end{bmatrix}$

5. $\begin{bmatrix} 1 & 0 & 0 \\ 3 & 1 & 0 \\ -4 & 2 & 1 \end{bmatrix} \begin{bmatrix} 4 & 1 & 1 \\ 0 & 5 & -1 \\ 0 & 0 & 3 \end{bmatrix} \mathbf{x} = \begin{bmatrix} 6 \\ 22 \\ -13 \end{bmatrix}$

In Exercises 6–13 find an LU factorization of the matrix.

6. $\begin{bmatrix} 4 & 1 \\ 12 & 2 \end{bmatrix}$

7. $\begin{bmatrix} 2 & 1 \\ -10 & -12 \end{bmatrix}$

8. $\begin{bmatrix} 2 & -2 & 1 \\ -8 & 11 & -5 \\ 4 & -13 & 3 \end{bmatrix}$

9. $\begin{bmatrix} -1 & 2 & 1 \\ 4 & -5 & -5 \\ -7 & 5 & 5 \end{bmatrix}$

10. $\begin{bmatrix} 4 & 1 & 1 \\ -12 & 2 & -4 \\ 20 & -5 & 10 \end{bmatrix}$

11. $\begin{bmatrix} 4 & 1 & 1 & 2 \\ -12 & -1 & -4 & -4 \\ 0 & -4 & 5 & -2 \\ 20 & 3 & 6 & 7 \end{bmatrix}$

12. $\begin{bmatrix} 4 & 1 & 1 & 2 & 1 \\ -12 & -1 & -4 & -4 & -1 \\ 4 & -3 & 3 & 0 & -4 \end{bmatrix}$

13. $\begin{bmatrix} 4 & 1 & 1 \\ -12 & -1 & -4 \\ 0 & -4 & 5 \\ 20 & 3 & 6 \end{bmatrix}$

In Exercises 14–17 find the solution of the system $A\mathbf{x} = \mathbf{b}$, by using an LU factorization of the coefficient matrix A.

14. $\begin{bmatrix} 5 & 1 \\ -10 & -3 \end{bmatrix} \mathbf{x} = \begin{bmatrix} -2 \\ 1 \end{bmatrix}$

15. $\begin{bmatrix} 2 & 1 \\ 14 & 2 \end{bmatrix} \mathbf{x} = \begin{bmatrix} 6 \\ -8 \end{bmatrix}$

16. $\begin{bmatrix} 2 & 1 & 1 \\ 12 & 11 & 5 \\ -2 & 9 & 0 \end{bmatrix} \mathbf{x} = \begin{bmatrix} 1 \\ 17 \\ 18 \end{bmatrix}$

17. $\begin{bmatrix} 2 & 1 & 1 & 2 \\ 0 & 1 & -1 & 2 \\ -4 & 0 & -1 & 2 \\ 0 & 5 & -5 & 12 \end{bmatrix} \mathbf{x} = \begin{bmatrix} 3 \\ 1 \\ -4 \\ 5 \end{bmatrix}$

In Exercises 18–20 find a permutation matrix P and an LU factorization of PA.

18. $A = \begin{bmatrix} 0 & 3 \\ -5 & 4 \end{bmatrix}$

19. $A = \begin{bmatrix} 0 & 1 & 1 \\ -1 & 2 & -4 \\ 2 & -5 & 1 \end{bmatrix}$

20. $A = \begin{bmatrix} 0 & 0 & 2 \\ -1 & 5 & -2 \\ 3 & 6 & 7 \end{bmatrix}$

In Exercises 21–23 solve the system $A\mathbf{x} = \mathbf{b}$ by using a $PA = LU$ factorization.

21. $\begin{bmatrix} 0 & 3 & -1 \\ 2 & 0 & 1 \\ 2 & -6 & 1 \end{bmatrix} \mathbf{x} = \begin{bmatrix} -3 \\ -1 \\ -1 \end{bmatrix}$

22. $\begin{bmatrix} 0 & 3 & -1 \\ 0 & 0 & 1 \\ 2 & -6 & 1 \end{bmatrix} \mathbf{x} = \begin{bmatrix} 1 \\ 2 \\ -10 \end{bmatrix}$

23. $\begin{bmatrix} 0 & 1 & 1 \\ 0 & 2 & -4 \\ 2 & -5 & 1 \end{bmatrix} \mathbf{x} = \begin{bmatrix} 2 \\ 4 \\ -8 \end{bmatrix}$

24. Show that the product of two lower triangular matrices is lower triangular.

25. Show that the product of two unit lower triangular matrices is unit lower triangular.

26. Show that a lower triangular matrix is invertible if and only if all its diagonal entries are nonzero.

27. Show that the inverse of an invertible lower triangular matrix is also lower triangular.

28. Show that the inverse of a unit lower triangular matrix is also unit lower triangular.

29. (**Uniqueness**) Suppose A is invertible with two LU factorizations LU and $L'U'$ (L and L' are *unit* lower triangular). Show that $L = L'$ and $U = U'$.

3.5   Applications

Reader's Goals for This Section

1. To read and understand some of the applications.
2. To realize how powerful and useful matrices can be.

Graph Theory

Adjacency Matrices of Graphs and Digraphs

Matrix algebra has important applications to graph theory. Graphs are now major tools in operations research, electrical engineering, computer programming and networking, business administration, sociology, economics, marketing, and communications networks; the list can go on and on.

A **graph** is a set of points, called **vertices**, or **nodes**, together with a set of "lines," called **edges**, connecting some pairs of vertices. Two vertices P and Q connected by an edge e are called **adjacent**, or **neighbors**, and P and Q are said to be **incident** to e. An edge from a vertex to itself is called a **loop**. Vertices can be connected with more than one edge, in which case we say we have a **multiple edge**.

An example of a multiple-edged graph, also called a **multigraph**, is a communications network where two points can be connected with more than one phone line.

■ EXAMPLE 30   Figure 3.1 shows three graphs G_1, G_2, and G_3. G_1 is a multigraph with loops. Vertices A and B are adjacent, whereas A and E are not. Vertex D is no one's neighbor. Vertex P of G_2 is incident to edges e_1, e_5, e_4. Vertex Q is not incident to edge e_5. Vertex home is adjacent to the rest of the vertices of G_3.

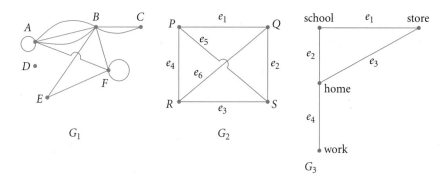

Figure 3.1   **Graphs.**

Matrices are used in the study of graphs. The easy computer programming of matrix operations enables us to study the behavior of very large graphs. (In applications to telephone networks, a graph may have tens of thousands of vertices.)[6]

The **matrix of a graph** is the $n \times n$ matrix whose (i, j) entry is the number of edges connecting the ith and the jth vertex.

■ EXAMPLE 31   Let us order the vertices of G_1, G_2, and G_3 of Fig. 3.1 as A, B, C, D, E, F, P, Q, R, S and school, store, home, work, respectively. Then the corresponding matrices are

$$
\begin{bmatrix}
1 & 3 & 0 & 0 & 0 & 1 \\
3 & 0 & 2 & 0 & 1 & 1 \\
0 & 2 & 0 & 0 & 0 & 0 \\
0 & 0 & 0 & 0 & 0 & 0 \\
0 & 1 & 0 & 0 & 0 & 1 \\
1 & 1 & 0 & 0 & 1 & 1
\end{bmatrix}
\qquad
\begin{bmatrix}
0 & 1 & 1 & 1 \\
1 & 0 & 1 & 1 \\
1 & 1 & 0 & 1 \\
1 & 1 & 1 & 0
\end{bmatrix}
\qquad
\begin{bmatrix}
0 & 1 & 1 & 0 \\
1 & 0 & 1 & 0 \\
1 & 1 & 0 & 1 \\
0 & 0 & 1 & 0
\end{bmatrix}
$$

The matrix of a graph is not used as often as the adjacency matrix of the graph, whose entries consist of only 1s and 0s. (Graph theorists often call such a matrix a $(0, 1)$-*matrix.*)

DEFINITION

The **adjacency matrix** $A(G)$ of a graph G is the matrix whose (i, j) entry is 1 if the ith and the jth vertex are adjacent and zero if they are not.

■ EXAMPLE 32   The adjacency matrices of the graphs G_1, G_2, and G_3 are

$$
A(G_1) =
\begin{bmatrix}
1 & 1 & 0 & 0 & 0 & 1 \\
1 & 0 & 1 & 0 & 1 & 1 \\
0 & 1 & 0 & 0 & 0 & 0 \\
0 & 0 & 0 & 0 & 0 & 0 \\
0 & 1 & 0 & 0 & 0 & 1 \\
1 & 1 & 0 & 0 & 1 & 1
\end{bmatrix}
\qquad
A(G_2) =
\begin{bmatrix}
0 & 1 & 1 & 1 \\
1 & 0 & 1 & 1 \\
1 & 1 & 0 & 1 \\
1 & 1 & 1 & 0
\end{bmatrix}
$$

$$
A(G_3) =
\begin{bmatrix}
0 & 1 & 1 & 0 \\
1 & 0 & 1 & 0 \\
1 & 1 & 0 & 1 \\
0 & 0 & 1 & 0
\end{bmatrix}
$$

For large graphs it is often easier to extract information from the adjacency matrix than from the graph itself, which may be visually complicated.

---

[6]For more information see Frederic Bien's article "Construction of Telephone Networks by Group Representations," *Notices of American Mathematical Society*, 36 (January 1989): 5–22.

Suppose we have a graph with no loops and no multiple edges, such as G_2 or G_3. Let us fix two vertices, P and Q. A **walk of length** m from P to Q is a sequence of vertices

$$P = V_1, V_2, \ldots, V_m, V_{m+1} = Q$$

such that V_i, V_{i+1} are adjacent for all i between 1 and m.

For example, P, Q, S, R, P is a walk of length 4 in G_2. Also home, school, store is a walk of length 2 in G_3.

The following result is useful in applications to physics and engineering, especially to networks.

THEOREM 18 The number of walks of length m from vertex i to vertex j in a graph G is equal to the (i, j)th entry of $A(G)^m$.

■ **EXAMPLE 33** The numbers of walks of length 2 and 3 in G_3 can be read from the entries of $A(G_3)^2$ and $A(G_3)^3$, respectively.

$$A(G_3)^2 = \begin{bmatrix} 2 & 1 & 1 & 1 \\ 1 & 2 & 1 & 1 \\ 1 & 1 & 3 & 0 \\ 1 & 1 & 0 & 1 \end{bmatrix}, \qquad A(G_3)^3 = \begin{bmatrix} 2 & 3 & 4 & 1 \\ 3 & 2 & 4 & 1 \\ 4 & 4 & 2 & 3 \\ 1 & 1 & 3 & 0 \end{bmatrix}$$

The number of walks of length 2 from home to the store is 1 because the only way to go is via school. This is also verified from the $(3, 2)$ entry of $A(G_3)^2$, which is 1. Likewise, the number of walks of length 2 from home back to home is 3, because we can go to the store and back, to school and back, or to work and back. This number is the $(3, 3)$ entry of $A(G_3)^2$.

The only walks of length 3 from school to the store (as awkward as they may be) are: school-store-school-store, school-home-school-store, and school-store-home-store, a total of three ways. This number is the $(1, 2)$ entry of $A(G_3)^3$.

COROLLARY 19 The number of walks of length 1, or 2, ..., or m from P to Q is equal to the (i, j) entry of the matrix $A(G) + A(G)^2 + \cdots + A(G)^m$. This is also interpreted as the number of accesses of the jth vertex from the ith in 1, or 2, ..., or m stages.

■ **EXAMPLE 34** The number of 1- or 2-stage accesses of the jth vertex from the ith for G_3 is the (i, j)th entry of $A(G_3) + A(G_3)^2$:

$$A(G_3) + A(G_3)^2 = \begin{bmatrix} 2 & 2 & 2 & 1 \\ 2 & 2 & 2 & 1 \\ 2 & 2 & 3 & 1 \\ 1 & 1 & 1 & 1 \end{bmatrix}$$

For instance, we can walk from school to home with length less than or equal to 2 in two ways, either straight or through the store. This number is the $(1, 3)$ entry of $A(G_3) + A(G_3)^2$.

DEFINITION

A **digraph**, or **directed graph**, is a graph whose edges are *directed* line segments.

■ EXAMPLE 35   Figure 3.2 shows three digraphs, D_1, D_2, and D_3. (Digraph D_3 could represent one- and two-way streets of a portion of the downtown of some city.)

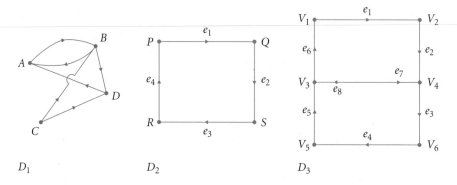

D_1 D_2 D_3

Figure 3.2   Digraphs.

DEFINITION

The **adjacency matrix** $A(D)$ of a digraph D is the matrix whose (i, j) entry is 1 if there is at least one directed edge connecting the ith to the jth vertex and zero otherwise.

■ EXAMPLE 36   Let us order the vertices of D_1, D_2, and D_3 of Fig. 3.2 as A, B, C, D, P, Q, R, S and V_1, V_2, V_3, V_4, V_5, V_6, respectively. Then the corresponding adjacency matrices are

$$A(D_1) = \begin{bmatrix} 0 & 1 & 0 & 0 \\ 1 & 0 & 0 & 1 \\ 0 & 1 & 0 & 1 \\ 1 & 0 & 0 & 0 \end{bmatrix}, \qquad A(D_2) = \begin{bmatrix} 0 & 1 & 0 & 0 \\ 0 & 0 & 0 & 1 \\ 1 & 0 & 0 & 0 \\ 0 & 0 & 1 & 0 \end{bmatrix},$$

$$A(D_3) = \begin{bmatrix} 0 & 1 & 0 & 0 & 0 & 0 \\ 0 & 0 & 0 & 1 & 0 & 0 \\ 1 & 0 & 0 & 1 & 0 & 0 \\ 0 & 0 & 1 & 0 & 0 & 1 \\ 0 & 0 & 1 & 0 & 0 & 0 \\ 0 & 0 & 0 & 0 & 1 & 0 \end{bmatrix}$$

Just as before, we can define a *walk* of length m from the ith to the jth vertex for digraphs. The only difference is that the edges are directed. We have, again, a theorem on the number of walks of length m.

THEOREM 20   The number of walks of length m from vertex i to vertex j in a digraph D is equal to the (i, j) entry of $A(D)^m$.

■ EXAMPLE 37   Show that the number of walks of length 4 from any vertex of D_2 back to itself is 1.

SOLUTION   Because of the direction of the edges, we need to walk over all four edges to go around once and return to the same vertex. This is also confirmed from the computation of $A(D_2)^4$, which is the identity matrix I_4 in this case:

$$A(D_2)^4 = \begin{bmatrix} 1 & 0 & 0 & 0 \\ 0 & 1 & 0 & 0 \\ 0 & 0 & 1 & 0 \\ 0 & 0 & 0 & 1 \end{bmatrix}$$

Incidence Matrices of Graphs and Digraphs; Line Graphs

Let G be a graph with vertex set $\{V_1, V_2, \ldots, V_m\}$ and edge set $\{e_1, e_2, \ldots, e_n\}$. The **incidence matrix** $I(G)$ of G is the $m \times n$ matrix $I(G) = [a_{ij}]$ with entries

$$a_{ij} = \begin{cases} 1, & \text{if } V_i \text{ and } e_j \text{ are incident} \\ 0, & \text{otherwise} \end{cases}$$

so $I(G)$ has the form

$$I(G) = \begin{array}{c} V_1 \\ V_2 \\ \vdots \\ V_m \end{array} \left\{ \begin{bmatrix} a_{11} & a_{12} & \cdots & a_{1n} \\ a_{21} & a_{22} & \cdots & a_{2n} \\ \vdots & & \ddots & \\ a_{m1} & a_{m2} & \cdots & a_{mn} \end{bmatrix} \right.$$

$$\underbrace{\qquad\qquad}_{e_1 \quad e_2 \quad \cdots \quad e_n}$$

with the entries a_{ij} being either 0 or 1.

■ EXAMPLE 38   For G_2 and G_3 of Fig. 3.1, we have

$$I(G_2) = \begin{bmatrix} 1 & 0 & 0 & 1 & 1 & 0 \\ 1 & 1 & 0 & 0 & 0 & 1 \\ 0 & 0 & 1 & 1 & 0 & 1 \\ 0 & 1 & 1 & 0 & 1 & 0 \end{bmatrix}, \quad I(G_3) = \begin{bmatrix} 1 & 1 & 0 & 0 \\ 1 & 0 & 1 & 0 \\ 0 & 1 & 1 & 1 \\ 0 & 0 & 0 & 1 \end{bmatrix}$$

DEFINITION

The *line graph* of a graph G is a graph $L(G)$ with vertices in one-to-one correspondence with the edges of G. Two vertices of $L(G)$ are adjacent only if the corresponding edges of G are incident to a common vertex. $L(G)$ is defined only if G has no loops or multiple edges.

■ EXAMPLE 39   Figure 3.3 shows three graphs and their line graphs.

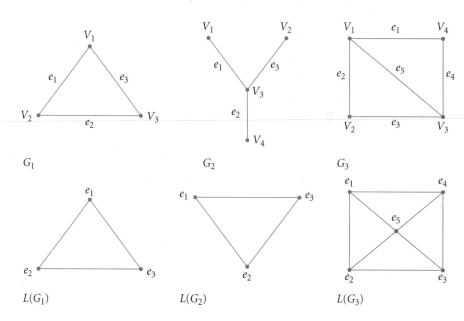

Figure 3.3   Line graphs of graphs.

There is an interesting connection between the incidence matrix of a graph G and the adjacency matrix of its line graph.

THEOREM 21

Let G be a graph with n edges. Then

$$A(L(G)) = I(G)^T I(G) - 2I_n$$

■ EXAMPLE 40   Verify the theorem for G_1 of Fig. 3.3.

SOLUTION   We have

$$\begin{bmatrix} 1 & 0 & 1 \\ 1 & 1 & 0 \\ 0 & 1 & 1 \end{bmatrix}^T \begin{bmatrix} 1 & 0 & 1 \\ 1 & 1 & 0 \\ 0 & 1 & 1 \end{bmatrix} - 2\begin{bmatrix} 1 & 0 & 0 \\ 0 & 1 & 0 \\ 0 & 0 & 1 \end{bmatrix} = \begin{bmatrix} 0 & 1 & 1 \\ 1 & 0 & 1 \\ 1 & 1 & 0 \end{bmatrix} = A(L(G_1))$$

We can also define incidence matrices of digraphs. Let D be a digraph with vertex set $\{V_1, V_2, \ldots, V_m\}$ and directed edge set $\{e_1, e_2, \ldots, e_n\}$. The **incidence matrix** $I(D)$ of D is the $m \times n$ matrix $I(D) = [d_{ij}]$ with entries:

$$d_{ij} = \begin{cases} 1, & \text{if } e_j \text{ points to } V_i \\ -1, & \text{if } e_j \text{ begins at } V_i \\ 0, & \text{otherwise} \end{cases}$$

so $I(D)$ has the form

$$I(D) = \begin{array}{c} V_1 \\ V_2 \\ \vdots \\ V_m \end{array} \left\{ \underbrace{\begin{bmatrix} d_{11} & d_{12} & \cdots & d_{1n} \\ d_{21} & d_{22} & \cdots & d_{2n} \\ \vdots & & \ddots & \\ d_{m1} & d_{m2} & \cdots & d_{mn} \end{bmatrix}}_{e_1 \quad e_2 \quad \cdots \quad e_n} \right.$$

with the entries d_{ij} being either 0, or 1, or -1.

■ EXAMPLE 41   The incidence matrix of the 3-node digraph D_4 (Fig. 3.4) is

$$I(D_4) = \begin{array}{c} V_1 \\ V_2 \\ V_3 \end{array} \underbrace{\begin{bmatrix} -1 & 0 & 1 \\ 1 & -1 & 0 \\ 0 & 1 & -1 \end{bmatrix}}_{e_1 \quad e_2 \quad e_3}$$

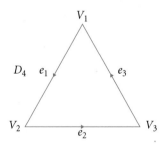

Figure 3.4   **A 3-node digraph.**

Sociology and Psychology

Dominance Graphs

Sociologists and psychologists use graphs to determine various kinds of relationships, such as influence, dominance, and communication, in groups. Suppose that in a group, for every pair of members V_i and V_j, either V_i influences (or dominates) V_j or V_j influences V_i, or there is no direct influence between V_i and V_j. This situation can be described by a digraph D that has *at most one directed edge* connecting any two vertices. Such a digraph is called a **dominance** digraph.

■ **EXAMPLE 42**   Figure 3.5 displays the dominance relationships among seven individuals, $V_1, \ldots, V_7$.

The adjacency matrix of a dominance digraph gives information about the influence relationships of a group. Rows with the most 1s represent group members with
greatest influence. Walks of length 1 represent direct influence, whereas walks of length
greater than 1 represent indirect influence. Hence the nth power of the adjacency matrix
gives the n-stage indirect influence of one member to another. By examining $A(D)$ of
the digraph in Fig. 3.5, we see that V_1 is the most influential member in direct influence,
because the first row has the most 1s. However, examination of $A(D)^2$ shows that V_5 has
more 2-stage influence than V_1. This is also clear from the graph, because V_5 influences
V_2, V_3 and V_7 in two stages, whereas V_1 only influences V_4 and V_6.

$$A(D) = \begin{bmatrix} 0 & 1 & 1 & 0 & 1 & 0 & 0 \\ 0 & 0 & 0 & 0 & 0 & 0 & 0 \\ 0 & 0 & 0 & 0 & 0 & 0 & 0 \\ 0 & 1 & 0 & 0 & 0 & 0 & 1 \\ 0 & 0 & 0 & 1 & 0 & 1 & 0 \\ 0 & 0 & 1 & 0 & 0 & 0 & 0 \\ 0 & 0 & 0 & 0 & 0 & 1 & 0 \end{bmatrix}, \qquad A(D)^2 = \begin{bmatrix} 0 & 0 & 0 & 1 & 0 & 1 & 0 \\ 0 & 0 & 0 & 0 & 0 & 0 & 0 \\ 0 & 0 & 0 & 0 & 0 & 0 & 0 \\ 0 & 0 & 0 & 0 & 0 & 1 & 0 \\ 0 & 1 & 1 & 0 & 0 & 0 & 1 \\ 0 & 0 & 0 & 0 & 0 & 0 & 0 \\ 0 & 0 & 1 & 0 & 0 & 0 & 0 \end{bmatrix}$$

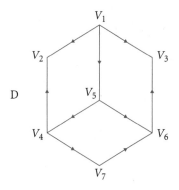

Figure 3.5   A dominance graph.

Stochastic and Doubly Stochastic Matrices; Markov Process

Stochastic matrices are special types of matrices that are often used in the study of
random phenomena in probability theory and statistics.

DEFINITION

A **stochastic matrix** is a square matrix with real nonnegative entries, for which each
column[7] adds up to 1. A stochastic matrix is **doubly stochastic** if each of its rows also
adds up to 1.

---

[7] Some authors want a stochastic matrix to have *rows* that add up to 1.

Note that a stochastic matrix has as entries only numbers between 0 and 1. (Why?) Also, the transpose of a doubly stochastic matrix is also doubly stochastic.

■ EXAMPLE 43   The following matrices are stochastic. In addition, matrices C and D are doubly stochastic.

$$A = \begin{bmatrix} 1 & \frac{1}{2} \\ 0 & \frac{1}{2} \end{bmatrix} \qquad B = \begin{bmatrix} \frac{3}{4} & \frac{1}{6} \\ \frac{1}{4} & \frac{5}{6} \end{bmatrix} \qquad C = \begin{bmatrix} 0.25 & 0.75 \\ 0.75 & 0.25 \end{bmatrix} \qquad D = \begin{bmatrix} 0 & \frac{3}{4} & \frac{1}{4} \\ \frac{1}{4} & \frac{1}{4} & \frac{1}{2} \\ \frac{3}{4} & 0 & \frac{1}{4} \end{bmatrix}$$

The matrix $\begin{bmatrix} 0 & \frac{1}{3} \\ 1 & \frac{1}{2} \end{bmatrix}$ is not stochastic, because for the second column $\frac{1}{3} + \frac{1}{2} \neq 1$.

The matrix $\begin{bmatrix} 2 & -1 \\ \frac{1}{2} & \frac{1}{2} \end{bmatrix}$ is not stochastic, because the entry 2 is greater than 1. Matrix A is not doubly stochastic, because for the first row $1 + \frac{1}{2} \neq 1$.

THEOREM 22

Let A and B be two $n \times n$ stochastic matrices. Then the product AB is stochastic.

PROOF   We prove the theorem only for $n = 2$. The idea of the proof generalizes easily to any size $n \times n$. Let

$$A = \begin{bmatrix} a & b \\ c & d \end{bmatrix}, \qquad B = \begin{bmatrix} a' & b' \\ c' & d' \end{bmatrix}$$

with $a + c = 1$, $b + d = 1$, $a' + c' = 1$, $b' + d' = 1$. Then

$$AB = \begin{bmatrix} a & b \\ c & d \end{bmatrix} \begin{bmatrix} a' & b' \\ c' & d' \end{bmatrix} = \begin{bmatrix} aa' + bc' & ab' + bd' \\ ca' + dc' & cb' + dd' \end{bmatrix}$$

But $(aa' + bc') + (ca' + dc') = a'(a + c) + c'(b + d) = a' + c' = 1$. Likewise, the second column adds to 1. Therefore, AB is stochastic.

COROLLARY 23

Let A be an $n \times n$ stochastic matrix. Then for any positive integer k the power A^k is stochastic.

NOTE   We can replace *stochastic* with *doubly stochastic* in the preceding theorem and corollary. The proof of this claim is left as an exercise.

Probability   Let us now introduce the notion of the *probability* of the occurrence of an event and see how this relates to stochastic matrices.

If an event is certain to occur, we say that its probability to occur is 1. If it will not occur, then its probability to occur is 0. Other values of probabilities are numbers between 0 and 1. For example, we say that the probability of rain is 70% = 0.7 for a given day. The larger the probability of occurrence of an event, the more likely the event will occur. If an event has n equally likely to occur outcomes, m of which interest us, the probability for one of our outcomes to occur is m/n.

For example, if we roll a die, all the possible outcomes are 6 (we can get a 1, or a 2, or a 3, or a 4, or a 5, or a 6, each with equal chance). The probability of getting 2 (one outcome) is 1 out of 6, or $\frac{1}{6}$. The probability of getting an even number is 3 (we can have a 2 or a 4 or a 6) out of 6, or $\frac{3}{6} = \frac{1}{2}$.

Because the elements of a stochastic matrix are numbers between 0 and 1, they can be viewed as probabilities of outcomes of events.

Let us look at the following study of the smoking habits of a group of people. Suppose that the probability of a smoker continuing to smoke a year later is 65% (hence there is a 35% probability of quitting), whereas the probability of a nonsmoker continuing not to smoke is 85% (hence there is a 15% probability of switching to smoking). This information can be tabulated by using the following stochastic matrix of probabilities, called a **matrix of transition probabilities**.

		Initial State	
		Smoker	Nonsmoker
Final State {	Smoker	0.65	0.15
	Nonsmoker	0.35	0.85

■ EXAMPLE 44   Suppose that when the study started in 1960, 70% of the group members were smokers and 30% were nonsmokers. Given that the preceding matrix of transition probabilities is valid for the next 10 y, what would the percentages of smokers and nonsmokers be in 1961? In 1962? In 1964?

SOLUTION   In 1961 the percentage of smokers consisted of those who were smokers in 1960, i.e., 70% · 65% = 0.455, plus those who picked up smoking after 1960, i.e., 30% · 15% = 0.045, a total of 0.5 = 50%. Likewise, the percentage of nonsmokers in 1961 was 0.7 · 0.35 + 0.3 · 0.85 = 0.5 = 50%. Both numbers can be computed as the matrix product

$$\begin{bmatrix} 0.65 & 0.15 \\ .035 & 0.85 \end{bmatrix} \begin{bmatrix} 0.7 \\ 0.3 \end{bmatrix} = \begin{bmatrix} 0.5 \\ 0.5 \end{bmatrix}$$

Similarly, for 1962, we use the same stochastic matrix, starting with the new vector (0.5, 0.5):

$$\begin{bmatrix} 0.65 & 0.15 \\ 0.35 & 0.85 \end{bmatrix} \begin{bmatrix} 0.5 \\ 0.5 \end{bmatrix} = \begin{bmatrix} 0.65 & 0.15 \\ 0.35 & 0.85 \end{bmatrix}^2 \begin{bmatrix} 0.7 \\ 0.3 \end{bmatrix} = \begin{bmatrix} 0.4 \\ 0.6 \end{bmatrix}$$

For 1964 we have

$$\begin{bmatrix} 0.65 & 0.15 \\ 0.35 & 0.85 \end{bmatrix}^4 \begin{bmatrix} 0.7 \\ 0.3 \end{bmatrix} = \begin{bmatrix} 0.325 \\ 0.675 \end{bmatrix}$$

Hence in 1964 there are 32.5% smokers and 67.5% nonsmokers. In general, in k years from 1960 the percentages can be computed as the initial vector $(0.7, 0.3)$ times the kth power of the transition of probabilities matrix:

$$\begin{bmatrix} 0.65 & 0.15 \\ 0.35 & 0.85 \end{bmatrix}^k \begin{bmatrix} 0.7 \\ 0.3 \end{bmatrix}$$

In this case $k \leq 10$.

As we substitute larger and larger k's, we see that although the percentages of nonsmokers grow, they never exceed $0.7 = 70\%$, even if k is allowed to be very large. We say that this process *tends* to the final vector $(0.3, 0.7)$. Hence, in the long run, 30% of people will be smokers, versus 70% for nonsmokers, provided that the current assumptions about the transition probabilities hold.

The process just described is an example of a **Markov process**, or **Markov chain**. In a Markov[8] process the next state of a system depends only on its current state. In our case the percentages of smokers and nonsmokers depended only on the percentages of the previous year. Each time the current percentage vector was multiplied by the fixed transition probabilities matrix.

As another example we have one of college football's best-known rivalries; the annual Army-Navy game. Suppose that the probability that Army wins one year and Navy wins the next year is 70%; hence the probability that Army wins one year and wins again the next year is 30%. Suppose that the probability that Navy wins one year and Army wins the next year is 30%; hence the probability that Navy wins one year and wins again the next year is 70%. This situation can be expressed by the following *doubly stochastic* matrix of transition probabilities:

		This Year	
		Navy wins	Army wins
Next Year	Navy wins	0.7	0.3
	Army wins	0.3	0.7

■ EXAMPLE 45 (Army-Navy Games)   Given that Navy won this year's game, what is the probability that it will win 2 y from now?

SOLUTION   Because Navy won this year, the probability of winning is 1, whereas the probability of Army winning is 0. Hence the initial state vector of probabilities is $(1, 0)$. According to our analysis of the last example, the probabilities of [Navy   Army]

---

[8]**Andrei Andreevitch Markov** (or Markoff) (1856–1922) was born in St. Petersburg (Leningrad), Russia. He was a student of Chebyshev, whose collected works he co-edited. He did important research in mathematical analysis and became a professor of mathematics at St. Petersburg University. He is best known for his contributions to the foundations of probability theory, particularly the study of the properties of processes, now known as Markov chains.

winning next year are given by the vector

$$\begin{bmatrix} 0.7 & 0.3 \\ 0.3 & 0.7 \end{bmatrix} \begin{bmatrix} 1 \\ 0 \end{bmatrix} = \begin{bmatrix} 0.7 \\ 0.3 \end{bmatrix}$$

which checks the assumptions. For 2 y from now the probabilities are

$$\begin{bmatrix} 0.7 & 0.3 \\ 0.3 & 0.7 \end{bmatrix}^2 \begin{bmatrix} 1 \\ 0 \end{bmatrix} = \begin{bmatrix} 0.58 & 0.42 \\ 0.42 & 0.58 \end{bmatrix} \begin{bmatrix} 1 \\ 0 \end{bmatrix} = \begin{bmatrix} 0.58 \\ 0.42 \end{bmatrix}$$

Hence the probability that Navy will win is 58%. In general, if Navy wins this year, the probability it will win k years from now is given by the first component of the vector $\begin{bmatrix} 0.7 & 0.3 \\ 0.3 & 0.7 \end{bmatrix}^k \begin{bmatrix} 1 \\ 0 \end{bmatrix}$. This can also be written as

$$\begin{bmatrix} 1 & 0 \end{bmatrix} \begin{bmatrix} 0.7 & 0.3 \\ 0.3 & 0.7 \end{bmatrix}^k \begin{bmatrix} 1 \\ 0 \end{bmatrix}$$

Inspection of this number for large k shows that it tends to $0.5 = 50\%$.

Economics: Leontief Input-Output Models

Let us consider an economy consisting of n industries, each producing only one commodity needed by the others and possibly by itself. For example, suppose we have coal, steel, and auto, which are interrelated in a way described by a 3×3 matrix as follows. Let c_{ij} be the dollar amount of the ith commodity needed to produce 1 dollar's worth of the jth commodity. Suppose it takes 0.30 dollars of coal to produce 1 dollar's worth of steel. The value 0.30 is the $(1, 2)$nd entry of the matrix:

	Coal	Steel	Auto
Coal	0.10	0.30	0.25
Steel	0.25	0.20	0.45
Auto	0.05	0.15	0.10

According to this matrix, it also takes 0.45 dollars of steel to produce 1 dollar's worth of automobile. Note that auto is the largest consumer of steel and steel is the largest consumer of coal. Steel is most dependent on auto to survive.

This matrix is an example of an **input-output**, or a **consumption** matrix describing the interdependency of the economic sectors. The entries of such a matrix are nonnegative less than 1. In addition, the sum of the entries of each column should be less than 1 if each sector is to produce more than it consumes. Consumption matrices were introduced and studied by the Harvard economist Wassily W. Leontief[9] in the 1930s.

Suppose n economic sectors are interrelated in a way described by a consumption matrix $C = [c_{ij}]$. Let x_i be the total amount of output needed to be produced by the ith sector to satisfy the demands of all sectors. Then $c_{ij}x_j$ is the amount needed from commodity i to produce x_j units of commodity j. Because the total output of sector i

---

[9]He received the 1973 Nobel Prize in Economics for this work.

equals the sum of the demands of all sectors, we have

$$x_1 = c_{11}x_1 + \cdots + c_{1n}x_n$$

$$\vdots$$

$$x_n = c_{n1}x_1 + \cdots + c_{nn}x_n$$

If $\mathbf{x}$ is the vector with components $x_1, \ldots, x_n$, then these relations can be expressed as the matrix equation

$$\mathbf{x} = C\mathbf{x}$$

Up to now we have considered demand for commodities only from producing economic sectors. This situation is known as a **Leontief closed model**. In reality, there is also demand from nonproducing sectors, such as consumers, governments, etc. For instance, the government may demand coal, steel, and automobiles in our example. All nonproducing sectors form the **open sector**. Suppose that d_i is the demand of the open sector from the ith producing sector. Then $x_i = c_{i1}x_1 + \cdots + c_{in}x_n + d_i$. If $\mathbf{d}$ is the vector with (the nonnegative) components $d_1, \ldots, d_n$, then

$$\mathbf{x} = C\mathbf{x} + \mathbf{d}$$

This matrix equation describes a **Leontief open model** (the open sector is taken into account); $\mathbf{x}$ is called the **output vector** and $\mathbf{d}$ is the **demand vector**. Economists are usually interested in computing the output vector $\mathbf{x}$ given the demand vector $\mathbf{d}$. This can be done by solving for $\mathbf{x}$ as follows:

$$\mathbf{x} = C\mathbf{x} + \mathbf{d} \Rightarrow (I - C)\mathbf{x} = \mathbf{d} \Rightarrow \mathbf{x} = (I - C)^{-1}\mathbf{d}$$

provided that the matrix $I - C$ is invertible. If, in addition, $(I - C)^{-1}$ has nonnegative entries, then the entries of $\mathbf{x}$ are nonnegative, and therefore they are acceptable as values of output. In general, a matrix C is called **productive** if $(I - C)^{-1}$ exists and has nonnegative entries.

■ EXAMPLE 46   Let C be the consumption matrix and $\mathbf{d}$ be the demand vector, in millions of dollars, for an open-sector economy with three interdependent industries. Compute the output demanded by the industries and the open sector.

$$C = \begin{bmatrix} \frac{1}{2} & 0 & \frac{1}{4} \\ \frac{1}{4} & \frac{1}{4} & 0 \\ 0 & \frac{1}{2} & \frac{1}{4} \end{bmatrix}, \qquad \mathbf{d} = \begin{bmatrix} 10 \\ 20 \\ 30 \end{bmatrix}$$

SOLUTION

$$I - C = \begin{bmatrix} 1 & 0 & 0 \\ 0 & 1 & 0 \\ 0 & 0 & 1 \end{bmatrix} - \begin{bmatrix} \frac{1}{2} & 0 & \frac{1}{4} \\ \frac{1}{4} & \frac{1}{4} & 0 \\ 0 & \frac{1}{2} & \frac{1}{4} \end{bmatrix} = \begin{bmatrix} \frac{1}{2} & 0 & -\frac{1}{4} \\ -\frac{1}{4} & \frac{3}{4} & 0 \\ 0 & -\frac{1}{2} & \frac{3}{4} \end{bmatrix}$$

Therefore,

$$\mathbf{x} = (I - C)^{-1}\mathbf{d} = \begin{bmatrix} \frac{9}{4} & \frac{1}{2} & \frac{3}{4} \\ \frac{3}{4} & \frac{3}{2} & \frac{1}{4} \\ \frac{1}{2} & 1 & \frac{3}{2} \end{bmatrix} \begin{bmatrix} 10 \\ 20 \\ 30 \end{bmatrix} = \begin{bmatrix} 55 \\ 45 \\ 70 \end{bmatrix}$$

We conclude that the output levels of the three industries should be $55, $45, and $70 million, to satisfy the demands.

Often analysts assume that the levels of production are known and want to compute the demand that can be placed upon the producing sectors. In such cases $\mathbf{x}$ is given and $\mathbf{d}$ is unknown; $\mathbf{d}$ is then computed very simply as

$$\mathbf{d} = \mathbf{x} - C\mathbf{x}$$

Input-output matrices are used to analyze the economy of a country or even an entire geographic region. The producing sectors are usually some key industries, such as agricultural goods, steel, chemicals, coal, livestock etc. For the U.S. national input-output matrix, the open sectors are the federal, state, and local governments.

Exercises 3.5

Graph Theory

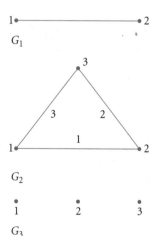

Figure 3.6

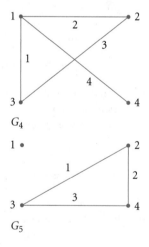

Figure 3.7

1. Find the matrices M_1, M_2, and M_3 of graphs G_1, G_2, and G_3, respectively (Fig. 3.6).

2. Find the matrices M_4, M_5, and M_6 of graphs G_4, G_5, and G_6, respectively (Figs. 3.7 and 3.8).

3. Find the adjacency matrices for graphs G_1, G_2, and G_3.

4. Find the adjacency matrices for graphs G_4, G_5, and G_6.

5. Compute $A(G_7)^2$. Count the number of walks of length 2 from (a) 1 to 1, (b) 1 to 2, and (c) 1 to 3. Do these answers agree with the $(1,1)$, $(1,2)$, $(1,3)$th entries of $A(G_7)^2$ (Fig. 3.8)?

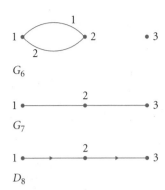

G_6

G_7

D_8

Figure 3.8

6. Answer the questions of Exercise 5 for the digraph D_8 (Fig. 3.8).

7. Compute the incidence matrices for graphs G_1, G_2, and G_3.

8. Compute the incidence matrices for graphs G_4, G_5, and G_6.

9. Draw the line graphs of G_1, G_2, and G_3.

10. Draw the line graphs of G_4, G_5, and G_6.

11. Let $L(G_7)$ be the line graph of G_7. Verify the relation $A(L(G_7)) = I(G_7)^T I(G_7) - 2I_2$.

Markov Processes

In Exercises 12–15 test the matrices for stochastic and doubly stochastic.

12. $A = \begin{bmatrix} \frac{1}{3} & 0 \\ \frac{2}{3} & 1 \end{bmatrix}$, $B = \begin{bmatrix} -0.1 & 1.1 \\ 1.1 & -0.1 \end{bmatrix}$

13. $C = \begin{bmatrix} \frac{1}{3} & 0 \\ \frac{2}{3} & 1 \end{bmatrix}$, $D = \begin{bmatrix} \frac{1}{2} & 0 \\ \frac{1}{3} & 1 \\ \frac{1}{6} & 0 \end{bmatrix}$

14. $E = \begin{bmatrix} \frac{1}{2} & \frac{1}{2} \\ \frac{1}{2} & \frac{1}{2} \end{bmatrix}$, $F = \begin{bmatrix} \frac{1}{4} & \frac{1}{2} \\ \frac{3}{4} & \frac{1}{2} \end{bmatrix}$

15. $G = \begin{bmatrix} 0 & \frac{1}{2} & \frac{1}{2} \\ 0 & \frac{1}{2} & \frac{1}{2} \\ 1 & 0 & 0 \end{bmatrix}$, $H = \begin{bmatrix} \frac{1}{2} & \frac{1}{3} & \frac{1}{6} \\ \frac{1}{3} & \frac{1}{6} & \frac{1}{2} \\ \frac{1}{6} & \frac{1}{2} & \frac{1}{3} \end{bmatrix}$

16. Is $\begin{bmatrix} \frac{1}{4} & \frac{1}{2} \\ \frac{3}{4} & \frac{1}{2} \end{bmatrix}^{100}$ stochastic?

17. Is $\begin{bmatrix} \frac{1}{2} & \frac{1}{2} \\ \frac{1}{2} & \frac{1}{2} \end{bmatrix}^{100}$ doubly stochastic?

18. Refer to Example 45 of the text.
   a. Given that Army won this year's game, what is the probability it will win 2 y from now?
   b. Given that Navy won this year's game, what is the probability it will win 3 y from now?

19. Find x, and y, if possible, such that the following matrices are doubly stochastic.

$$A = \begin{bmatrix} x & 0.2 \\ 0.2 & y \end{bmatrix}, \qquad B = \begin{bmatrix} \frac{1}{4} & x \\ \frac{3}{4} & y \end{bmatrix}$$

20. Find x, y, and z, if possible, such that the following matrix is stochastic.

$$A = \begin{bmatrix} x & 0 & 1 \\ \frac{1}{4} & y & 0 \\ \frac{1}{4} & \frac{1}{4} & z \end{bmatrix}$$

21. Let C denote a consumption matrix:

$$C = \begin{bmatrix} 0.5 & 0.4 \\ 0.1 & 0.6 \end{bmatrix}$$

   a. Is C productive?
   b. If so, let D denote a demand vector:

$$D = \begin{bmatrix} 10 \\ 20 \end{bmatrix}$$

   Find the output vector of production.

3.6   Miniprojects

1 ■ Codes

Governments, national security agencies, companies are often interested in the transmission of coded messages that are hard to be decoded by others, if intercepted, yet easily decoded by the receiving end. There are many interesting ways of coding a message, most of which use number theory or linear algebra. Let us discuss one that is effective, especially when a large-size invertible matrix is used.

Let us start out with an invertible matrix M that is known only to the transmitting and receiving ends. For example,

$$M = \begin{bmatrix} -3 & 4 \\ -1 & 2 \end{bmatrix}$$

Suppose we want to code the message

A T T A C K N O W

We replace each letter with the number that corresponds to the letter's position in the alphabet. We use 0 for an empty space.

A   T   T   A   C   K   N   O   W
↕   ↕   ↕   ↕   ↕   ↕   ↕   ↕   ↕   ↕
1 20 20 1 3 11 0 14 15 23

The message has now been converted into the sequence of numbers 1, 20, 20, 1, 3, 11, 0, 14, 15, 23, which we group as a sequence of column vectors,

$$\begin{bmatrix} 1 \\ 20 \end{bmatrix} \begin{bmatrix} 20 \\ 1 \end{bmatrix} \begin{bmatrix} 3 \\ 11 \end{bmatrix} \begin{bmatrix} 0 \\ 14 \end{bmatrix} \begin{bmatrix} 15 \\ 23 \end{bmatrix}$$

and multiply on the left by M:

$$M \begin{bmatrix} 1 \\ 20 \end{bmatrix} = \begin{bmatrix} 77 \\ 39 \end{bmatrix}, \qquad M \begin{bmatrix} 20 \\ 1 \end{bmatrix} = \begin{bmatrix} -56 \\ -18 \end{bmatrix}, \qquad M \begin{bmatrix} 3 \\ 11 \end{bmatrix} = \begin{bmatrix} 35 \\ 19 \end{bmatrix}$$

$$M \begin{bmatrix} 0 \\ 14 \end{bmatrix} = \begin{bmatrix} 56 \\ 28 \end{bmatrix}, \qquad M \begin{bmatrix} 15 \\ 23 \end{bmatrix} = \begin{bmatrix} 47 \\ 31 \end{bmatrix}$$

giving the sequence of numbers $77, 39, -56, -18, 35, 19, 56, 28, 47, 31$. This is the coded message. To decode it, the receiving end needs to compute M^{-1},

$$M^{-1} = \begin{bmatrix} -1 & 2 \\ -\frac{1}{2} & \frac{3}{2} \end{bmatrix}$$

and multiply it by the vectors $\begin{bmatrix} 77 \\ 39 \end{bmatrix}, \begin{bmatrix} -56 \\ -18 \end{bmatrix}, \begin{bmatrix} 35 \\ 19 \end{bmatrix}, \begin{bmatrix} 56 \\ 28 \end{bmatrix}, \begin{bmatrix} 47 \\ 31 \end{bmatrix}$ to get the original numbers back.

$$M^{-1} \begin{bmatrix} 77 \\ 39 \end{bmatrix} = \begin{bmatrix} 1 \\ 20 \end{bmatrix}, \qquad M^{-1} \begin{bmatrix} -56 \\ -18 \end{bmatrix} = \begin{bmatrix} 20 \\ 1 \end{bmatrix}, \dots$$

Problem A (Decoding Message)

Based on this approach, decode the message given by the numbers 17, 15, 29, 15, 17, 29, 16, 31, 47, 6, 19, 20, 35, 24, 39, 14, 19, 19 if

$$A = \begin{bmatrix} 1 & 0 & 1 \\ 0 & 1 & 1 \\ 0 & 1 & 2 \end{bmatrix}$$

Problem B (Code Breaking)

Suppose that you intercepted the following coded stock market message: 1156, −203, 624, −84, −228, 95, 1100, −165, 60, 19. Your sources inform you that the message was coded by using a 2 × 2 symmetric matrix. Your intuition tells you that the first word of the message is very likely to be either sell or buy. Can you break the code?

2 ■ The Fibonacci Numbers

In the early middle ages Fibonacci studied the sequence of numbers $f_0, f_1, f_2, \ldots$ that arises when one counts male-female pairs of rabbits that reproduce monthly and create another male-female pair. The process starts with one initial pair (so $f_0 = 1$), and it is assumed that the rabbits become reproductive from their second month on.

Starting with one pair, $f_0 = 1$, at the end of the first month we still have one pair, $f_1 = 1$. The pair reproduces, and at the end of the second month when we have $f_2 = 2$ pairs. At the end of the third month the old pair reproduces but the young pair is too young to reproduce. So at the end of the third month we have $f_3 = f_2 + f_1 = 2 + 1 = 3$ pairs. Likewise, at the end of the kth month we have the pairs we had at the end of the previous month, f_{k-1}, plus the number of the offspring of the reproducing pairs, f_{k-2}. So $f_k = f_{k-1} + f_{k-2}$. Hence the sequence that arises can be described by the recursive relations:

$$f_0 = 1, \qquad f_1 = 1, \qquad f_k = f_{k-1} + f_{k-2}, \qquad k \geq 2$$

The first few terms of the sequence are 1, 1, 2, 3, 5, 8, 13, 21, ...
In this project you are asked to do the following.

1. Compute $f_0, f_1, \ldots, f_{15}$.
2. Find a 2 × 2 matrix A with the following property:

$$A \begin{bmatrix} f_k \\ f_{k-1} \end{bmatrix} = \begin{bmatrix} f_{k+1} \\ f_k \end{bmatrix} \qquad \text{for } k = 1, 2, 3, \ldots$$

For example, if

$$A = \begin{bmatrix} a & b \\ c & d \end{bmatrix}$$

then

$$A \begin{bmatrix} 1 \\ 1 \end{bmatrix} = \begin{bmatrix} 2 \\ 1 \end{bmatrix}, \qquad A \begin{bmatrix} 2 \\ 1 \end{bmatrix} = \begin{bmatrix} 3 \\ 2 \end{bmatrix}, \qquad A \begin{bmatrix} 3 \\ 2 \end{bmatrix} = \begin{bmatrix} 5 \\ 3 \end{bmatrix}, \ldots$$

3. Compute

$$A^2 \begin{bmatrix} 1 \\ 1 \end{bmatrix}, \qquad A^3 \begin{bmatrix} 1 \\ 1 \end{bmatrix}, \qquad A^4 \begin{bmatrix} 1 \\ 1 \end{bmatrix}, \qquad A^5 \begin{bmatrix} 1 \\ 1 \end{bmatrix}$$

4. How does the product $A^k \begin{bmatrix} 1 \\ 1 \end{bmatrix}$ relate to the computation of the $(k + 1)$st Fibonacci number?

3 ■ Transition of Probabilities

Problem A

A group of people buys cars every 4 y from one of three automobile manufacturers, A, B, and C. The transition of probabilities of switching from one manufacturer to another is given by the following matrix:

$$\begin{bmatrix} 0.5 & 0.4 & 0.6 \\ 0.3 & 0.4 & 0.3 \\ 0.2 & 0.2 & 0.1 \end{bmatrix}$$

Suppose that in year 1995 manufacturer A sold 100 cars, B sold 80 cars, and C sold 40 cars.

1. How many cars were sold in 1999?
2. How many cars were sold in 1991?
3. Will one manufacturer eventually dominate the market over the others?

Problem B

Consider the stochastic matrix of transition probabilities

$$T = \begin{bmatrix} \frac{1}{3} & \frac{1}{2} \\ \frac{2}{3} & \frac{1}{2} \end{bmatrix}$$

expressing the flow of customers from and to markets A and B after one purchase. Suppose that the first time around $\frac{2}{3}$ of the customers buy from A and $\frac{1}{3}$ buy from B.

1. What are the market shares after the first purchase?
2. What are the market shares after the second purchase?
3. Is there a *market equilibrium* (i.e., a vector of shares (a, b) that remains the same from one purchase to the next)? If yes, compute it. (Keep in mind that the customers can buy only from A or from B. This means that $a + b = 1$.)

4 ■ Digraph Walks

Royal Ice Cream Company makes deliveries to four stores. The stores and the delivery routes (some one-way) form a digraph with adjacency matrix

$$A = \begin{bmatrix} 0 & 1 & 0 & 1 \\ 1 & 0 & 1 & 1 \\ 1 & 1 & 0 & 0 \\ 1 & 0 & 1 & 0 \end{bmatrix}$$

1. Draw the digraph D.
2. Compute the matrices representing the number of routes that can be traveled from one store to another so that a delivery truck would pass
   (a) By exactly one store;
   (b) By exactly two stores;

(c) By at most one store;

(d) By at most two stores.

(e) In how many ways can one go from store 3 to store 4 passing by exactly one other store?

3. Can A be the adjacency matrix of a graph (as opposed to a digraph)?

5 ■ A Theoretical Problem

Problem

Let A, B, C be $n \times n$ matrices and let r be any nonzero scalar.[10] If

$$A + B + rAB = 0$$

$$B + C + rBC = 0$$

$$C + A + rCA = 0$$

show that $A = B = C$. (*Hint:* Show that each of $I + rA, I + rB, I + rC$ is invertible by using Exercise 27, Section 3.3.)

3.7   Computer Exercises

This computer section will help you familiarize yourself with the basic matrix manipulation commands of your software. It is important to know how to access matrix entries, columns, rows, and submatrices and how to create new matrices out of old ones. You also need to know how to perform matrix operations and how to use them to illustrate known properties and perhaps explore new ones. All this is done with a simultaneous reviewing of some basic material. Note that an exercise designated as [**S**] requires symbolic manipulation.[11]

Let

$$M = \begin{bmatrix} 1 & 2 & 3 & 4 \\ -1 & -2 & -3 & -4 \\ 5 & 6 & 7 & 8 \\ -5 & -6 & -7 & -8 \end{bmatrix}, \quad A = \begin{bmatrix} 1 & -2 & 3 \\ 2 & -2 & 3 \\ 3 & -3 & 3 \end{bmatrix},$$

$$B = \begin{bmatrix} 2 & 3 & 6 \\ 3 & 3 & 6 \\ 6 & 6 & 6 \end{bmatrix}, \quad C = \begin{bmatrix} 2 & 4 \\ 0 & 4 \\ 1 & 4 \end{bmatrix}$$

First enter these matrices and name them as described. If a letter is already used by your program, change it. Use your program to solve the following exercises.

1. Display the fourth row, the third column and the $(2, 3)$rd entry of M.

2. Display the matrix obtained from M by using the first three columns.

---

[10]This interesting problem with $r = 1996$ was taken from the Summer 1996 National College Entrance Exams of Greece (the equivalent of SATs).

[11]Skip these exercises if symbolic manipulation is not available.

3. Display the matrix obtained from M by using the last two rows only.

4. Display the portion of M obtained by deleting the first row and the first two columns.

5. Display the matrix obtained from M by adding the numbers 4, 3, 2, 1 as a (a) first row, (b) last column.

6. Display the matrix obtained by writing A and B next to each other.

7. Display the matrix obtained by writing A above B.

8. Display a diagonal matrix with diagonal entries 1, 2, 3, 4.

Let T be the matrix obtained by reversing the rows of M. Hence, the last row becomes first, the fourth becomes second, and so on. Compute each value.

9. $M - T$

10. $15M - 35T$

11. Solve the matrix equation $17X - 51M = 62T$ for X.

12. Compute: (a) AB, (b) BA, (c) $(AB)C$, (d) $A(BC)$.

13. Compute: (a) $(AB)^2$, (b) A^2B^2, (c) $(A^3)^4$, (d) A^{12}.

14. Compute: (a) $(A + B)^2$, (b) $A^2 + 2AB + B^2$.

15. Compute A^{-1} by: (a) your program's command for inversion; (b) row-reducing $[A : I]$. Compare answers.

16. [S] Let $S = \begin{bmatrix} a & b \\ c & d \end{bmatrix}$. Compute S^3, S^{-3} and S^3S^{-3}.

17. Verify the identities: (a) $(A^{-1})^{-1} = A$, (b) $(10AB)^{-1} = \frac{1}{10}B^{-1}A^{-1}$, (c) $(ABA)^{-1} = A^{-1}B^{-1}A^{-1}$. (d) $A^5A^{-2} = A^3$.

18. Verify the identities: (a) $(A^T)^T = A$, (b) $(A^T)^{-1} = (A^{-1})^T$, (c) $(ABC)^T = C^TB^TA^T$, (d) $(A^{-2})^T = (A^T)^{-2}$.

19. Solve the matrix equations for the 3×3 matrix X: (a) $AX = B$, (b) $XA = B$.

Consider the sequence of matrices of size greater than or equal to 3 with 0s on the diagonal and 1s elsewhere:

$$A_3 = \begin{bmatrix} 0 & 1 & 1 \\ 1 & 0 & 1 \\ 1 & 1 & 0 \end{bmatrix}, \qquad A_4 = \begin{bmatrix} 0 & 1 & 1 & 1 \\ 1 & 0 & 1 & 1 \\ 1 & 1 & 0 & 1 \\ 1 & 1 & 1 & 0 \end{bmatrix}, \ldots$$

20. Write a one argument function, called `diagzero`, that displays these matrices according to size. So `diagzero(3)` is A_3, and so on.

21. Use `diagzero` to display A_3, A_4, and A_5, and compute A_3^{-1}, A_4^{-1}, and A_5^{-1}.

22. Guess the formula for A_n^{-1}.

23. Write and test the code of three functions that produce elementary matrices of a given size that are obtained from each of the elementary row operations.

24. Suppose a graph has adjacency matrix A_4, as given. Compute the matrix that yields: (a) the number of walks of length 4; (b) the number of walks of lengths 1, or 2, or 3, or 4.

25. Write the code for a function, called `sumpower`, that takes two arguments, a square matrix A and a positive integer n. The value of the function is the matrix

$$A + A^2 + \cdots + A^n$$

Apply sumpower with $A = A(G)$ and $n = 4$ to verify your answer from the second part of the last exercise.

Matrix Operations in Maple-Mathematica-MATLAB

For compatible-size matrices A and B we have

function	Maple	Mathematica	MATLAB
$A + B$	evalm($A + B$);	$A + B$	$A + B$
$2A - 3B$	evalm($2 * A - 3 * B$);	$2A - 3B$	$2 * A - 3 * B$
AB	evalm($A\& * B$);	$A.B$	$A * B$
Range $m, m + 1, \ldots n$	m..n	Range[m,n]	m:n
A^T	transpose(A);	Transpose[A]	A.' see also A'

Selected Solutions with Maple

```
with(linalg);                        #  Loading the packege linalg.
M := matrix(4,4,[1,2,3,4,-1,-2,-3,-4,5,6,7,8,-5,-6,-7,-8]); #Matrix definitions.
A := matrix(3,3,[1,-2,3,2,-2,3,3,-3,3]);
B := matrix(3,3,[2,3,6,3,3,6,6,6,6]);
C := matrix(3,2,[2,4,0,4,1,4]);
A := matrix(3,3,[[1,-2,3],[2,-2,3],[3,-3,3]]);       # Alternative way.
A := matrix([[1,-2,3],[2,-2,3],[3,-3,3]]);           # Yet another way.
# Exercises 1-5.
row(M,4);                # Fourth row.
submatrix(M,4..4,1..4);  # Fourth row again.
submatrix(M,1..4,3..3);  # Third column as a column matrix.
col(M,3);                # Third column as a vector.
matrix(4,1,col(M,3));    # Yet another way: Third column as a column matrix.
M[2,3];                  # (2,3)th entry.
submatrix(M,1..4,1..3);  # First three columns as a matrix.
col(M,1..3);             # First three columns as a sequence of vectors.
delcols(M,4..4);         # Yet another way: First three columns by deletion.
submatrix(M,3..4,1..4);  # Last two rows as a matrix.
row(M,3..4);             # Last two rows as a sequence of vectors.
delrows(M,1..2);         # Yet another way: Last two rows by deletion.
submatrix(M,2..4,3..4);  # Deletion of first row and first two columns.
delcols(delrows(M,1..1),1..2); # Another way.
v:=vector([4,3,2,1]);    # Define the vector (4,3,2,1)
stack(v,M);              # Add v to M as the first row.
augment(M,v);            # Add v to M as the last column.
concat(M,v);             # concat is the same as augment.
# Another useful command for enlarging a matrix is extend.
# Exercises 6,7.
concat(A,B);             # A and B next to each other.
```

```
stack(A,B);                    # A above B.
# Exercise 8.
diag(1,2,3,4);
# Exercises 9-11.
T:=matrix(4,4,(i,j)->M[5-i,j]);   # The row entries are flipped while
                                  # the column entries remain in tact.
evalm(M-T);
evalm(15*M-35*T);
evalm(1/17*(51*M+62*T));
# Exercises 12-14.
evalm(A&*B);                       # Matrix multiplication is denoted by &* .
evalm(B&*A);
evalm((A&*B)&*C);
evalm(A&*(B&*C));
evalm((A&*B)^2);                   # Matrix powers are denoted by ^ .
evalm((A&*B)&*(A&*B));             # Same as above.
evalm(A^2&*B^2);
evalm((A^3)^4);
evalm(A^12);
evalm((A+B)^2);
evalm(A^2+2*A&*B+B^2);
# Exercise 15.
inverse(A);                            # Computes A^(-1).
evalm(A^(-1));                         # Another way.
evalm(1/A);                            # Another way.
augment(A,[1,0,0],[0,1,0],[0,0,1]);    # [A:I]
rref(");                               # The right half is A^(-1).
# Exercise 16.
S := matrix(2,2,[a,b,c,d]);            # Symbolic matrix S.
evalm(S^3);                            # S^3,
s:=map(simplify,");                    # simplified. Need to "map"
evalm(S^(-3)):                         # the command simplify through
ss:=map(simplify,");                   # to the entries of the resulting
evalm(s &* ss);                        # matrix.
map(simplify,");
# Exercises 17-18.
inverse(inverse(A));
inverse(10*A&*B);
evalm((1/10)*inverse(B)&*inverse(A));
inverse(A&*B&*A);
evalm(inverse(A) &* inverse(B) &* inverse(A));
evalm(A^3);
evalm(A^5 &* A^(-2));
transpose(transpose(A));
transpose(inverse(A));
inverse(transpose(A));
```

```
transpose(A&*B&*C);
evalm(transpose(C) &* transpose(B) &* transpose(A));
transpose(A^(-2));
evalm(transpose(A)^(-2));
# Exercise 19.
evalm(inverse(A) &* B);
evalm(B &* inverse(A));
# Exercise 20-22.
diagzero := proc(n) local i; evalm(matrix(n,n,[seq(1,i=1..n^2)]) - &*()); end:
# Matrix of 1's minus I_n.
# It is also useful to know how to use fuller code such as:
diagzero := proc(n) local i,j,a;
                a := array(1..n,1..n):
                for i from 1 to n do
                   for j from 1 to n do
                      if i=j then a[i,j]:=0 else a[i,j]:=1 fi od: od:
                evalm(a);
                end:
diagzero(3);
diagzero(4);
diagzero(5);
inverse(diagzero(3));
inverse(diagzero(4));
inverse(diagzero(5));
# A_n^(-1) has -(n-2)/(n-1) on the main diagonal and 1/(n-1), elsewhere.
# Exercise 24.
AA := diagzero(4);       # A_4 using diagzero, or enter it directly.
evalm(AA^4);             # Matrix yielding the number of walks of length 4.
evalm(AA+AA^2+AA^3+AA^4);
# Exercise 25.
sumpower := proc(A,n) evalm(sum('A^i', 'i'=1..n)) end:
sumpower(AA,4);
```

Selected Solutions with Mathematica

```
M = {{1,2,3,4},{-1,-2,-3,-4},{5,6,7,8},{-5,-6,-7,-8}} (* Matrix      *)
A = {{1,-2,3},{2,-2,3},{3,-3,3}}                        (* definitions. *)
B = {{2,3,6},{3,3,6},{6,6,6}}
C1 = {{2,4},{0,4},{1,4}}        (* C is reserved for differential constants.*)
MatrixForm[A]                              (* Diplays A as a matrix. *)
(* Exercises 1-5.*)
(* First (and easy) approach. Using LinearAlgebra'MatrixManipulation'.     *)
<<LinearAlgebra'MatrixManipulation'      (* Loading the package.         *)
TakeRows[M,{4}]             (* The fourth row.                           *)
TakeColumns[M,{3}]          (* The third column.                         *)
```

```
TakeColumns[M,{1,3}]              (* The first three columns.               *)
TakeRows[M,{3,4}]                 (* The last two rows.                     *)
TakeMatrix[M,{2,3},{4,4}]   (* Deletion of first row and first two columns. *)
AppendColumns[{v},M]              (* Adds v to M as a first ROW.             *)
AppendRows[M,{{4},{2},{3},{1}}] (* Adds v to M as a last COLUMN.            *)
(* Second approach. Using list manipulations. More useful in the long run. *)
M[[4]]            (* The fourth row. Actually, the fourth element of list M. *)
M[[Range[4,4], Range[1,4]]] (* Another way by using column and row ranges.  *)
(* Caution: the two answers {13,14,15,16} and {{13,14,15,16}} are really    *)
(* different. The first list has 4 elements while the second has only one.  *)
M[[Range[1,4], Range[3,3]]](* Third column by using column and row ranges.  *)
Map[#[[3]]&, M]                   (* Another way: The third part of each row. *)
(* Again the numbers are the same but the lists are not.                    *)
#[[3]]& /@ M                      (* Same as last in different notation.     *)
Transpose[M][[3]]                 (* Yet another way.                        *)
M[[2,3]]                 (* (2,3)th entry. (The third part of the second row.) *)
M[[2]][[3]]               (* Same as above in different notation.            *)
M[[Range[1,4], Range[1,3]]] (*  First three columns.                        *)
Map[#[[Range[1,3]]]&, M]          (* Another way.                            *)
M[[Range[3,4], Range[1,4]]] (*  Last two rows.                              *)
M[[Range[3,4]]]                   (* Same as above.                          *)
M[[Range[2,4], Range[3,4]]] (*  Deleting first row, first two columns.       *)
v={4,3,2,1}                       (* Define the vector (4,3,2,1).            *)
Prepend[M,v]                      (* Add v to M as the first row.            *)
Join[{v}, M]                      (* Another way.                            *)
(* Append[M,v] would add v to M as a last row.                              *)
Transpose[Append[Transpose[M],v]]   (* Add v to M as the last column.        *)
(* This was done indirectly. First switch columns to rows and add v as      *)
(* last row then switch back.                                               *)
(* Exercises 6,7.*)
<<LinearAlgebra'MatrixManipulation'
AppendRows[A,B]                   (* A and B next to each other.   *)
AppendColumns[A,B]                (* A above B.                    *)
(*  Exercise 8. *)
DiagonalMatrix[{1,2,3,4}]
(* Exercises 9-11. *)
T=Reverse[M]  (* Reverses the order of the elements (=the rows) of the list. *)
M-T
15M-35T         (* Can use 15M instead of 15 M since the first factor        *)
                (*      is a number and the second a symbol.                 *)
1/17(51M+62T)
(* Exercises 12-14. *)
A.B                     (* Matrix multiplication is denoted by a dot . *)
B.A
(A.B).C1
A.(B.C1)
```

```
MatrixPower[A.B, 2]                (* A^n is denoted by MatrixPower[A,n] *)
MatrixPower[A,2].MatrixPower[B,2]
(A.A).(B.B)                                        (* Same as above. *)
(* Warning typing A^2 will yield the squares of the list elements
   of A and not the matrix A^2. *)
MatrixPower[MatrixPower[A,3],4]
MatrixPower[A,12]
(A+B).(A+B)
A.A+2A.B+B.B
(* Exercise 15. *)
Inverse[A]                          (* Computes A^(-1).              *)
<<LinearAlgebra`MatrixManipulation`;  (* Need AppendRows from the package. *)
AppendRows[A,{{1,0,0},{0,1,0},{0,0,1}}]
RowReduce[%]                        (* The right half is A^(-1).     *)
(* Exercise 16. *)
S = {{a,b},{c,d}}                   (* Symbolic matrix.              *)
MatrixPower[S,3]                    (* S^3.                          *)
s = Simplify[%]                     (* Needs simplification.         *)
MatrixPower[S,-3]                   (* S^(-3),                       *)
ss = Simplify[%]                    (* simplified.                   *)
Simplify[s.ss]                      (* Product is the identity.      *)
(* Exercises 17,18. *)
Inverse[Inverse[A]]
Inverse[10 A . B]
1/10 Inverse[B] . Inverse[A]
Inverse[A . B . A]
Inverse[A] . Inverse[B] . Inverse[A]
MatrixPower[A,3]
MatrixPower[A,5] . MatrixPower[A,-2]
Transpose[Transpose[A]]
Transpose[Inverse[A]]
Inverse[Transpose[A]]
Transpose[A.B.C1]
Transpose[C1].Transpose[B].Transpose[A]
Transpose[MatrixPower[A,-2]]
MatrixPower[Transpose[A],-2]
(* Exercise 19. *)
Inverse[A] . B
A . Inverse[B]
(* Exercises 20-22. *)
diagzero[n_] := Table[If[i==j,0,1], {i,1,n},{j,1,n}]
(* The 2 dimensional table = matrix with entries 0 on the         *)
(* diagonal and 1 elsewhere.                                      *)
(* Also we can use the nxn matrix of 1's minus I_n.               *)
diagzero[n_] := Table[1, {n},{n}] - IdentityMatrix[n]
diagzero[3]
```

```
diagzero[4]
diagzero[5]
Inverse[diagzero[3]]
Inverse[diagzero[4]]
Inverse[diagzero[5]]
(* A_n^(-1) has -(n-2)/(n-1) on the main diagonal and 1/(n-1), elsewhere.  *)
(* Exercise 24. *)
AA = diagzero[4]              (* Use diagzero or enter the matrix directly. *)
MatrixPower[AA,4]             (* Matrix yielding the number of walks of length 4.*)
AA+MatrixPower[AA,2]+MatrixPower[AA,3]+MatrixPower[AA,4]
(* Exercise 25. *)
sumpower[A_, n_] := Sum[MatrixPower[A,i], {i,1,n}]
sumpower[AA,4]
```

Selected Solutions with MATLAB

```
M = [1 2 3 4; -1 -2 -3 -4; 5 6 7 8; -5 -6 -7 -8]      % Matrix
A = [1 -2 3; 2 -2 3; 3 -3 3]                          % definitions.
B = [2 3 6; 3 3 6; 6 6 6]
C = [2 4; 0 4; 1 4]
% Exercises 1-5.
M(4,:)              %  Fourth row.
M(:,3)              %  Third column.
M(2,3)              %  (2,3)th entry.
M(:,1:3)            %  First three columns.
M(3:4,:)            %  Last two rows.
M(2:4,3:4)          %  Deleting first row, first two columns.
v = [4 3 2 1]       %  Define the vector (4,3,2,1).
[v ; M]             %  Adds v to M as a first row.
[M v.']             %  Adds v to M as a last column. v.' is v in column form.
[M [4;3;2;1]]       %  Same but entering v directly as a column vector.
% Exercises 6,7.
[A,B]                   % A and B next to each other.
[A;B]                   % A above B.
% Exercise 8.
diag([1,2,3,4])
% Exercises 9-11.
T=flipud(M)         %  flips M upside down. fliplr flips left to right.
M-T
15*M-35*T
1/17*(51*M+62*T)
% Exercises 12-14.
A*B                     % Matrix multiplication is denoted by * .
B*A
(A*B)*C
```

```
A*(B*C)
(A*B)^2                                 % Matrix powers are denoted by ^ .
(A*B)*(A*B)                             % Same as above.
A^2*B^2
(A^3)^4
A^12
(A+B)^2
A^2+2*A*B+B^2
% Exercise 15.
inv(A)                                  % Computes A^(-1).
A^(-1)                                  % Another way.
[A,[1 0,0; 0 1 0; 0 0 1]]               % [A:I]
rref(ans)                               % The right half is A^(-1).
% Exercise 16.    (ST) Requires the symbolic toolbox.
S = sym('[a b; c d]')                   % Symbolic matrix.
s=sympow(S,3)                           % S^3.
ss=sympow(S,-3)                         % S^(-1).
symmul(s,ss)                            % The product is the identity matrix.
% Exercises 17,18.
inv(inv(A))
inv(10*A*B)
(1/10)*inv(B)*inv(A)
inv(A*B*A)
inv(A) * inv(B) * inv(A)
A^3
A^5 * A^(-2)
A.'.'
inv(A).'
(inv(A)).'
(A*B*C).'
C.' * B.' * A.'
(A^(-2)).'
(A.')^(-2)
% Exercise 19.
inv(A)*B
B*inv(A)
% Exercises 20-22.
function [A] = diagzero(n)              % Write the code on the left in an  m-file.
            A = ones(n)-eye(n);         % Matrix with 1's minus I_n.
          end
% It is also useful to know how to use fuller code such as:
function [A] = diagzero(n)
                for i=1:n, for j=1:n,
  if i==j A(i,j)=0; else A(i,j)=1; end; end; end;
              A;
            end
```

```
% Then in MATLAB session in the same directory call these functions:
diagzero(3)
diagzero(4)
diagzero(5)
inv(diagzero(3))
inv(diagzero(4))
inv(diagzero(5))
% A_n^(-1) has -(n-2)/(n-1) on the main diagonal and 1/(n-1), elsewhere.
% Exercise 24.
AA = diagzero(4)          % Use diagzero or enter the matrix directly.
AA^4                      % Matrix yielding the number of walks of length 4.
AA+AA^2+AA^3+AA^4
% Exercise 25.
function [B] = sumpower(A,n)                    % Code in an file.
          B = A;
          for i=1:(n-1), B = B*A + A; end       % A->A^2+A->A^3+A^2+A->...
          end
sumpower(AA,4)                                  % Type in session
```

4

Vector Spaces

Mathematics—the unshaken Foundation of Science...
—Isaac Barrow (1630–1677)

Introduction

In this chapter we generalize the basic concepts of Chapter 2: vectors, span, and linear independence. The common features of matrix and vector arithmetic (Theorem 1 of Sections 2.1 and 3.1) become defining properties for a set of abstract, or generalized, vectors, called a *vector space*. The sets of matrices and ordinary vectors are examples of vector spaces. So are a wide variety of other sets.

The major advantage of such generalizations is enormous labor savings, because the properties of the abstract vectors apply automatically to all particular examples. Also, proofs become clear and easy, because they are free of the notation of any specific example.

Grassmann and Peano

The mathematician credited with the introduction and first use of these ideas is **Hermann Grassmann**.[1] According to historians (Bourbaki, van der Waerden) Grassmann seems to have been the first to define an n-dimensional vector space (he called it "system of hypercomplex numbers") and linear independence.

---

[1] **Hermann Günther Grassmann** (1809–1877) was born and died in Stettin, Germany. He taught at the gymnasium (high school) in his home town, and in 1844 he published a book containing a number of new ideas in n-dimensional geometry and vector spaces. Because Grassmann was not trained as a research mathematician, his book was difficult to read and did not receive the kind of recognition it does today.

The Italian mathematician **Giuseppe Peano**[2] clarified Grassmann's work. According to Bourbaki he is also responsible for the current (coordinate-free) definition of a linear transformation. Peano lived during the "axiomatic age" of mathematics and became one of the leading proponents of the axiomatic school. The definition of a vector space is based on Peano's reading of the works of Grassmann. Peano introduced some of the mathematical notation used today, such as the symbol $\in$, which is an abbreviation of the phrase, *belongs to*, or *is a member of*.

4.1   Subspaces of $\mathbf{R}^n$

Reader's Goals for This Section

1. To know what a subspace of $\mathbf{R}^n$ is.
2. To know what bases are and how to test for them.

In this section we discuss the fundamental concepts of *subspace* and *basis* in $\mathbf{R}^n$. This is preparation for the corresponding abstract concepts of Sections 4.2 and 4.3. We should master these topics through careful study and practice.

Subspaces of $\mathbf{R}^n$

DEFINITION

(Subspace of $\mathbf{R}^n$) $V \subseteq \mathbb{R}^n , V \neq \emptyset$

A nonempty subset V of $\mathbf{R}^n$ is called a (**vector** or **linear**) **subspace** of $\mathbf{R}^n$ if it satisfies the following properties.

1. If **u** and **v** are in V, then **u** + **v** is in V. $u, v \in V \Rightarrow u + v \in V$
2. If c is any scalar and **u** is in V, then c**u** is in V. $u \in V \Rightarrow cu \in V$

Properties 1 and 2 imply that any linear combination of elements of V is also in V. If a nonempty subset S of $\mathbf{R}^n$ satisfies Part 1 of the definition, we say that S is **closed under (vector) addition**. If S satisfies Part 2, we say that S is **closed under scalar multiplication**. Thus, a subspace of $\mathbf{R}^n$ is a subset that is closed under vector addition and scalar multiplication.

Any subspace V of $\mathbf{R}^n$ contains the zero vector **0**. (V is nonempty, so it has at least one element, say, **u**. But then $0\mathbf{u} = \mathbf{0}$ is in V by Part 2 of the definition.)

■ EXAMPLE 1 $\{\mathbf{0}\}$ and $\mathbf{R}^n$ are subspaces of $\mathbf{R}^n$.

[2]**Giuseppe Peano** (1858–1932) was raised from age 11 by his uncle in Turin, Italy. He graduated from the University of Turin, where he taught for the rest of his career. He also taught at a nearby military academy, from where he was forced to resign when he began teaching his "new symbolism." Peano is well known for his "space-filling" curve and his axiomatic definition of the natural numbers. Outside of mathematics, he was also very involved in improving secondary school education and in linguistics.

REASON $\{\mathbf{0}\}$ is a subspace of $\mathbf{R}^n$, because

$$\mathbf{0} + \mathbf{0} = \mathbf{0} \quad \text{and} \quad c\mathbf{0} = \mathbf{0} \qquad \text{for all } c \in \mathbf{R}$$

$\mathbf{R}^n$ is a subspace of $\mathbf{R}^n$, because the sum of any two n-vectors is an n-vector and any scalar multiple of an n-vector is again an n-vector.

$\{\mathbf{0}\}$ is called the **zero subspace** of $\mathbf{R}^n$. $\{\mathbf{0}\}$ and $\mathbf{R}^n$ are the **trivial subspaces** of $\mathbf{R}^n$.

■ EXAMPLE 2

$$V = \left\{ \begin{bmatrix} x \\ y \\ 0 \end{bmatrix}, \ x, y \in \mathbf{R} \right\}$$

is a subspace of $\mathbf{R}^3$.

REASON V is nonempty, because it contains the zero vector (take $x = y = 0$). The sum of two vectors in V,

$$\begin{bmatrix} x_1 \\ y_1 \\ 0 \end{bmatrix} + \begin{bmatrix} x_2 \\ y_2 \\ 0 \end{bmatrix} = \begin{bmatrix} x_1 + x_2 \\ y_1 + y_2 \\ 0 \end{bmatrix}$$

is again in V. So, Part 1 of the definition holds. Any scalar multiple of a vector in V

$$c \begin{bmatrix} x \\ y \\ 0 \end{bmatrix} = \begin{bmatrix} cx \\ cy \\ 0 \end{bmatrix}$$

is again in V. Hence, Part 2 of the definition also holds. Therefore, V is a subspace of $\mathbf{R}^3$.

■ EXAMPLE 3 $V = \{(x, y, x + y), \ x, y \in \mathbf{R}\}$ is a subspace of $\mathbf{R}^3$.

REASON V is nonempty. (Why?) Let $\mathbf{v}_1 = (x_1, y_1, x_1 + y_1)$ and $\mathbf{v}_2 = (x_2, y_2, x_2 + y_2)$ be any elements of V and let c be any scalar. Then

$$(x_1, y_1, x_1 + y_1) + (x_2, y_2, x_2 + y_2) = (x_1 + x_2, y_1 + y_2, (x_1 + x_2) + (y_1 + y_2))$$

and

$$c(x_1, y_1, x_1 + y_1) = (cx_1, cy_1, (cx_1) + (cy_1))$$

Hence, $\mathbf{v}_1 + \mathbf{v}_2$, $c\mathbf{v}_1 \in V$. Therefore, V is a subspace of $\mathbf{R}^3$.

■ EXAMPLE 4   Is the set

$$T = \left\{ \begin{bmatrix} x \\ x + 1 \end{bmatrix}, \ x \in \mathbf{R} \right\}$$

a subspace of $\mathbf{R}^2$?

ANSWER   No, the zero vector is not in T, because if $\begin{bmatrix} x \\ x+1 \end{bmatrix} = \begin{bmatrix} 0 \\ 0 \end{bmatrix}$, then $x = 0$ and $x + 1 = 0$, which is an inconsistent system. (We can also easily show that T is not closed under addition or scalar multiplication.)

■ EXAMPLE 5   Let $\mathbf{v}_1$ and $\mathbf{v}_2$ be in $\mathbf{R}^n$. Show that Span$\{\mathbf{v}_1, \mathbf{v}_2\}$ is a subspace of $\mathbf{R}^n$.

SOLUTION   Let $V = \text{Span}\{\mathbf{v}_1, \mathbf{v}_2\} = \{c_1\mathbf{v}_1 + c_2\mathbf{v}_2, \ c_1, c_2 \in \mathbf{R}\}$. Then V is nonempty, because it contains $\mathbf{v}_1$ and $\mathbf{v}_2$. Let $\mathbf{u}, \mathbf{v} \in V$. Then there are scalars c_1, c_2 and d_1, d_2 such that $\mathbf{u} = c_1\mathbf{v}_1 + c_2\mathbf{v}_2$ and $\mathbf{v} = d_1\mathbf{v}_1 + d_2\mathbf{v}_2$. We have

$$\mathbf{u} + \mathbf{v} = c_1\mathbf{v}_1 + c_2\mathbf{v}_2 + d_1\mathbf{v}_1 + d_2\mathbf{v}_2 = (c_1 + d_1)\mathbf{v}_1 + (c_2 + d_2)\mathbf{v}_2$$

Hence, $\mathbf{u} + \mathbf{v} \in V$. So V is closed under addition. Also, for any scalar c we have

$$c\mathbf{u} = c(c_1\mathbf{v}_1 + c_2\mathbf{v}_2) = (cc_1)\mathbf{v}_1 + (cc_2)\mathbf{v}_2$$

This shows that $c\mathbf{u} \in V$. Therefore, V is closed under scalar multiplication. Hence, V is a subspace of $\mathbf{R}^n$.

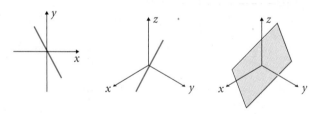

More generally we have the following, by Theorem 10, Section 2.3.

THEOREM 1   If $\mathbf{v}_1, \ldots, \mathbf{v}_k$ are n-vectors, then Span$\{\mathbf{v}_1, \ldots, \mathbf{v}_k\}$ is a subspace of $\mathbf{R}^n$.

■ EXAMPLE 6   Any line l through the origin in $\mathbf{R}^2$ ($\mathbf{R}^3$) is a subspace of $\mathbf{R}^2$ ($\mathbf{R}^3$) (Fig. 4.1).

Figure 4.1   Lines and planes through the origin are subspaces.

SOLUTION   Let $\mathbf{u}$ be a nonzero 2-vector in l. Then $l = \text{Span}\{\mathbf{u}\}$, as we saw in Section 2.3. But Span$\{\mathbf{u}\}$ is a subspace of $\mathbf{R}^2$, by Theorem 1. Similarly, we can show that a line l in $\mathbf{R}^3$ is a subspace of $\mathbf{R}^3$.

■ EXAMPLE 7   Any plane $\mathcal{P}$ through the origin is a subspace of $\mathbf{R}^3$ (Fig. 4.1).

SOLUTION   Let **u** and **v** be two vectors of $\mathcal{P}$ that are not collinear. Then $\mathcal{P} =$ Span$\{\mathbf{u}, \mathbf{v}\}$ by Section 2.3. Therefore, $\mathcal{P}$ is a subspace of $\mathbf{R}^3$, by Theorem 1.

Note that lines and planes not passing through the origin (Fig. 4.2) are **not** subspaces. (Why?)

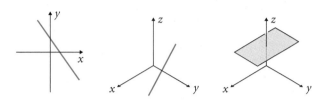

Figure 4.2   Lines and planes not through the origin are not subspaces.

We often prove that V is a subspace of $\mathbf{R}^n$ by writing it as the span of a set of vectors.

■ EXAMPLE 8   Is the set

$$V = \left\{ \begin{bmatrix} a - 3b \\ 2a + b \\ a \end{bmatrix}, \ a, b \in \mathbf{R} \right\}$$

a subspace of $\mathbf{R}^3$?

ANSWER   Yes. Because

$$\begin{bmatrix} a - 3b \\ 2a + b \\ a \end{bmatrix} = a \begin{bmatrix} 1 \\ 2 \\ 1 \end{bmatrix} + b \begin{bmatrix} -3 \\ 1 \\ 0 \end{bmatrix}$$

we see that $V = $ Span $\left\{ \begin{bmatrix} 1 \\ 2 \\ 1 \end{bmatrix}, \begin{bmatrix} -3 \\ 1 \\ 0 \end{bmatrix} \right\}$. Hence, V is a subspace of $\mathbf{R}^3$, by Theorem 1.

$$\begin{bmatrix} a-3b \\ 2a+b \\ a \end{bmatrix} = span \left\{ \begin{bmatrix} 1 \\ 2 \\ 1 \end{bmatrix}, \begin{bmatrix} -3 \\ 1 \\ 0 \end{bmatrix} \right\}$$

Since linear combo.

Basis of a Subspace of $\mathbf{R}^n$

In this paragraph we introduce the fundamental concept of a basis of a subspace of $\mathbf{R}^n$. Bases are very privileged sets of vectors. They are both linearly independent and spanning sets. Choosing and using a basis of a subspace is like choosing and using a coordinate frame in the plane or in space. Often, a smart choice of coordinates (or basis) simplifies an otherwise complex calculation.

DEFINITION

(Basis)

A nonempty subset $\mathcal{B}$ of a nonzero subspace V of $\mathbf{R}^n$ is a **basis** of V if

1. $\mathcal{B}$ is linearly independent;
2. $\mathcal{B}$ spans V.

See Figs. 4.3 and 4.4.

We also agree to say that the *empty set* is the only basis of the zero subspace $\{0\}$.

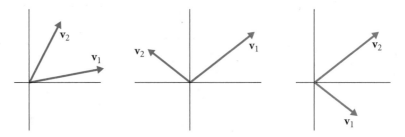

Figure 4.3　Some bases of the plane.

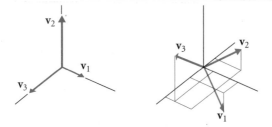

Figure 4.4　Bases of space.

■ **EXAMPLE 9**　Show $\mathcal{B} = \{e_1, \ldots, e_n\}$ is a basis of $\mathbf{R}^n$. It is called the **standard basis** of $\mathbf{R}^n$.

SOLUTION　By Examples 24, Section 2.3, and 35, Section 2.4, $\mathcal{B}$ spans $\mathbf{R}^n$ and is linearly independent. So, it is a basis of $\mathbf{R}^n$.

■ **EXAMPLE 10**　Show that

$$\mathcal{B} = \left\{ \begin{bmatrix} 1 \\ 1 \\ -1 \end{bmatrix}, \begin{bmatrix} 0 \\ 1 \\ 2 \end{bmatrix}, \begin{bmatrix} -2 \\ 1 \\ 0 \end{bmatrix} \right\}$$

is a basis of $\mathbf{R}^3$.

SOLUTION　We need to show that $\mathcal{B}$ is linearly independent and that it spans $\mathbf{R}^3$. Let A be the matrix with columns the vectors of $\mathcal{B}$. Because

$$A = \begin{bmatrix} 1 & 0 & -2 \\ 1 & 1 & 1 \\ -1 & 2 & 0 \end{bmatrix} \sim \begin{bmatrix} 1 & 0 & -2 \\ 0 & 1 & 3 \\ 0 & 0 & -8 \end{bmatrix}$$

we see that each column of A is a pivot column. Hence, the columns of A are linearly independent by Theorem 12, Section 2.4. Also each row of A has a pivot position. Therefore, the columns of A span $\mathbf{R}^3$ by Theorem 2, Section 2.1. ∎

Examples 9 and 10 show that a subspace can have several bases.

■ EXAMPLE 11   Is the set $S = \{(1, 0, 0), (0, 1, 0)\}$ a basis in $\mathbf{R}^3$? Is S a basis of the subspace $V = \{(x, y, 0), \ x, y \in \mathbf{R}\}$ of $\mathbf{R}^3$?

ANSWER S is linearly independent, because all columns of $A = \begin{bmatrix} 1 & 0 \\ 0 & 1 \\ 0 & 0 \end{bmatrix}$ are pivot columns. S is not a basis of $\mathbf{R}^3$ because A has 3 rows and only 2 pivots, so S does not span $\mathbf{R}^3$. On the other hand, S spans the smaller space V since

$$(x, y, 0) = x(1, 0, 0) + y(0, 1, 0)$$

Therefore, S *is* a basis of V. ∎

■ EXAMPLE 12   Is the set $T = \{(1, 1, 1), (2, 1, -1), (1, 0, -2)\}$ a basis of $\mathbf{R}^3$?

ANSWER   No. Because $\begin{bmatrix} 1 & 2 & 1 \\ 1 & 1 & 0 \\ 1 & -1 & -2 \end{bmatrix} \sim \begin{bmatrix} 1 & 2 & 1 \\ 0 & -1 & -1 \\ 0 & 0 & 0 \end{bmatrix}$, T is not linearly independent (3 columns, only 2 pivots). Actually, T does not even span $\mathbf{R}^3$ (3 rows, only 2 pivots). ∎

■ EXAMPLE 13   Is the set $S = \{(1, 0, 0), (0, 1, 0), (0, 0, 1), (1, 1, 1)\}$ a basis in $\mathbf{R}^3$?

ANSWER   No, because S is linearly dependent. ∎

■ EXAMPLE 14   Show that the set $\mathcal{B} = \{(2, 1, -1, 1), (1, 0, -2, 1), (0, 0, 0, 1)\}$ is a basis for the subspace $V = \{(2x + y, x, -x - 2y, x + y + z), \ x, y, z \in \mathbf{R}\}$ of $\mathbf{R}^4$.

SOLUTION   Because

$$(2x + y, x, -x - 2y, x + y + z) = x(2, 1, -1, 1) + y(1, 0, -2, 1) + z(0, 0, 0, 1)$$

we see that V is spanned by $\mathcal{B}$. We leave it to the reader to show that $\mathcal{B}$ is linearly independent. Therefore, $\mathcal{B}$ is a basis of V. ∎

Coordinates with Respect to Basis

In this paragraph V is a nonzero subspace of $\mathbf{R}^n$.

$$\mathcal{B} \le \mathbb{V}$$

THEOREM 2

(Uniqueness of Coordinates)

A subset $\mathcal{B} = \{v_1, \ldots, v_k\}$ of V is a basis of V if and only if for each vector $\mathbf{v}$ in V, there are **unique** scalars $c_1, \ldots, c_k$ such that

$$\mathbf{v} = c_1 \mathbf{v}_1 + \cdots + c_k \mathbf{v}_k$$

PROOF   Let $\mathcal{B}$ be a basis of V. Hence, $\mathcal{B}$ spans V, so each vector $\mathbf{v}$ is a linear combination of vectors in $\mathcal{B}$. Also $\mathcal{B}$ is linearly independent. Therefore, the representation of $\mathbf{v}$ as a linear combination of $\mathbf{v}_1, \ldots, \mathbf{v}_k$ is unique by Theorem 16, Section 2.4.
Conversely, let $\mathcal{B}$ be such that every vector $\mathbf{v}$ is uniquely written as $\mathbf{v} = c_1 \mathbf{v}_1 + \cdots + c_k \mathbf{v}_k$. Hence, $\mathcal{B}$ spans V. Let now $d_1 \mathbf{v}_1 + \cdots + d_k \mathbf{v}_k = \mathbf{0}$ for some scalars $d_1, \ldots, d_k$. Because $0\mathbf{v}_1 + \cdots + 0\mathbf{v}_k = \mathbf{0}$ the assumed uniqueness of the representation implies that $d_1 = \cdots = d_k = 0$. Therefore, $\mathcal{B}$ is linearly independent. Hence, $\mathcal{B}$ is a basis of V.

The unique scalars $c_1, \ldots, c_k$ that express the vector $\mathbf{v} \in V$ as a linear combination of a basis $\mathcal{B}$ of V are called the **coordinates of v with respect to** $\mathcal{B}$. The vector with components $c_1, \ldots, c_k$ is called the **coordinate vector of v with respect to** $\mathcal{B}$ and it is denoted by $[\mathbf{v}]_{\mathcal{B}}$.

$$[\mathbf{v}]_{\mathcal{B}} = \begin{bmatrix} c_1 \\ \vdots \\ c_k \end{bmatrix}$$

■ EXAMPLE 15   Find the coordinates and the coordinate vector of $\mathbf{v} = (4, 0, -4)$ with respect to the basis $\mathcal{B} = \{(1, 1, -1), (0, 1, 2), (-2, 1, 0)\}$ of $\mathbf{R}^3$.

SOLUTION   The fact that $\mathcal{B}$ is a basis of $\mathbf{R}^3$ was verified in Example 10. Now we need scalars $c_1, c_2,$ and c_3 such that $c_1 \begin{bmatrix} 1 \\ 1 \\ -1 \end{bmatrix} + c_2 \begin{bmatrix} 0 \\ 1 \\ 2 \end{bmatrix} + c_3 \begin{bmatrix} -2 \\ 1 \\ 0 \end{bmatrix} = \begin{bmatrix} 4 \\ 0 \\ -4 \end{bmatrix}$. Solving the

corresponding system yields $c_1 = 2, c_2 = -1,$ and $c_3 = -1$. Hence, $[\mathbf{v}]_{\mathcal{B}} = \begin{bmatrix} 2 \\ -1 \\ -1 \end{bmatrix}$.

■ EXAMPLE 16   Find the coordinates of $\mathbf{v} = (3, -2, 0)$ in the subspace $V = \{(x, y, 0), x, y \in \mathbf{R}\}$ with respect to the basis $\mathcal{B} = \{(1, 0, 0), (0, 1, 0)\}$ of V.

SOLUTION   Because

$$\mathbf{v} = \begin{bmatrix} 3 \\ -2 \\ 0 \end{bmatrix} = 3 \begin{bmatrix} 1 \\ 0 \\ 0 \end{bmatrix} + (-2) \begin{bmatrix} 0 \\ 1 \\ 0 \end{bmatrix}$$

we have

$$[\mathbf{v}]_{\mathcal{B}} = \begin{bmatrix} 3 \\ -2 \end{bmatrix}$$

■ **EXAMPLE 17**   If $\mathbf{v}$ is any n-vector and $\mathcal{B}$ is the standard basis of $\mathbf{R}^n$, show that

$$\mathbf{v} = [\mathbf{v}]_{\mathcal{B}}$$

SOLUTION   Exercise.

Exercises 4.1

Subspaces

In Exercises 1–10 show that the given sets of n-vectors are subspaces of $\mathbf{R}^n$.

1. $V = \left\{ \begin{bmatrix} a \\ 0 \end{bmatrix}, \, a \in \mathbf{R} \right\}$

2. $V = \left\{ \begin{bmatrix} a - b \\ 2a + b \end{bmatrix}, \, a, b \in \mathbf{R} \right\}$

3. $V = \left\{ \begin{bmatrix} a \\ 0 \\ -2a \end{bmatrix}, \, a \in \mathbf{R} \right\}$

4. $V = \left\{ \begin{bmatrix} a - c \\ b + c \\ 5c \end{bmatrix}, \, a, b, c \in \mathbf{R} \right\}$

5. $V = \left\{ \begin{bmatrix} a - b \\ b - c \\ c - d \\ d - a \end{bmatrix}, \, a, b, c, d \in \mathbf{R} \right\}$

6. The set of all 3-vectors with first and last components zero

7. The set of all 4-vectors with first three components zero

8. The set of all scalar multiples of $\begin{bmatrix} 1 \\ -3 \end{bmatrix}$

9. The set of all linear combinations in $\begin{bmatrix} 1 \\ -1 \\ 0 \end{bmatrix}$ and $\begin{bmatrix} 2 \\ 0 \\ 2 \end{bmatrix}$

10. Span $\left\{ \begin{bmatrix} 2 \\ -1 \end{bmatrix}, \begin{bmatrix} 7 \\ 0 \end{bmatrix} \right\}$

In Exercises 11–29 determine whether the given sets of n-vectors are subspaces of $\mathbf{R}^n$.

11. $\left\{ \begin{bmatrix} a \\ a \end{bmatrix}, \, a \in \mathbf{R} \right\}$

12. $\left\{ \begin{bmatrix} a + 1 \\ a \end{bmatrix}, \, a \in \mathbf{R} \right\}$

13. $\left\{ \begin{bmatrix} a \\ b \end{bmatrix}, \, a + b \leq 2, \, a, b \in \mathbf{R} \right\}$

14. $\left\{ \begin{bmatrix} a \\ b \end{bmatrix}, \, a^2 + b^2 = 1, \, a, b \in \mathbf{R} \right\}$

15. $\left\{ \begin{bmatrix} a \\ b \end{bmatrix}, \, a \leq 1, \, b \geq 1, \, a, b \in \mathbf{R} \right\}$

16. $\left\{ \begin{bmatrix} a \\ b \end{bmatrix}, \, a \leq b, \, a, b \in \mathbf{R} \right\}$

17. $\left\{ \begin{bmatrix} a \\ b \end{bmatrix}, \, a = -3b, \, a, b \in \mathbf{R} \right\}$

$c = 0$

$c\vec{w} = 0$ $a = (-3)(0) = 0$

18. $\left\{ \begin{bmatrix} a+1 \\ a+b \end{bmatrix}, \ a,b \in \mathbf{R} \right\}$

19. $\left\{ \begin{bmatrix} a-1 \\ 2a+1 \end{bmatrix}, \ a,b \in \mathbf{R} \right\}$

20. The set of all 3-vectors with equal first two components

21. The set of all 4-vectors with equal last two components

22. The set of all 3-vectors with first component twice the second

23. The set of all 2-vectors in the first quadrant

24. The set of all 2-vectors in the fourth quadrant

25. The set of all 3-vectors in the first octant

26. The line determined by the points with coordinates $(1,0), (0,1)$

27. The line determined by the points with coordinates $(-1,-1), (1,1)$

28. The plane determined by the points with coordinates $(1,0,0), (0,0,1), (0,0,0)$

29. The plane determined by the points with coordinates $(1,0,0), (0,1,0), (1,1,1)$

30. Let V_1 and V_2 be subspaces of $\mathbf{R}^n$. Show that the intersection $V_1 \cap V_2$ is also a subspace of $\mathbf{R}^n$.

31. Let V_1 and V_2 be two planes through the origin in $\mathbf{R}^3$. Explain geometrically why $V_1 \cap V_2$ is a subspace of $\mathbf{R}^3$.

32. Let $\mathbf{v}_1, \ldots, \mathbf{v}_k$ be n-vectors. Show that $\mathrm{Span}\{\mathbf{v}_1, \ldots, \mathbf{v}_k\}$ is the smallest subspace of $\mathbf{R}^n$ that contains the set $\{\mathbf{v}_1, \ldots, \mathbf{v}_k\}$. (*Hint:* If V is a subspace of $\mathbf{R}^n$ that contains $\mathbf{v}_1, \ldots, \mathbf{v}_k$, show that $\mathrm{Span}\{\mathbf{v}_1, \ldots, \mathbf{v}_k\} \subseteq V$.)

Bases

In Exercises 33–39 determine whether the given sets of n-vectors are bases of $\mathbf{R}^n$.

33. $\left\{ \begin{bmatrix} 0 \\ -4 \end{bmatrix}, \begin{bmatrix} -2 \\ 1 \end{bmatrix} \right\}$

34. $\left\{ \begin{bmatrix} 20 \\ 10 \end{bmatrix}, \begin{bmatrix} -2 \\ 1 \end{bmatrix} \right\}$

35. $\left\{ \begin{bmatrix} 1 \\ 0 \end{bmatrix}, \begin{bmatrix} 0 \\ 1 \end{bmatrix}, \begin{bmatrix} 0 \\ 0 \end{bmatrix} \right\}$

36. $\left\{ \begin{bmatrix} 2 \\ 0 \\ 0 \end{bmatrix}, \begin{bmatrix} 2 \\ 2 \\ 0 \end{bmatrix}, \begin{bmatrix} 2 \\ 2 \\ 2 \end{bmatrix} \right\}$

37. $\left\{ \begin{bmatrix} 1 \\ 0 \\ 0 \end{bmatrix}, \begin{bmatrix} 0 \\ 1 \\ 1 \end{bmatrix}, \begin{bmatrix} 1 \\ 0 \\ 1 \end{bmatrix} \right\}$

38. $\left\{ \begin{bmatrix} 1 \\ 0 \\ 0 \\ 1 \end{bmatrix}, \begin{bmatrix} 0 \\ 1 \\ 1 \\ 0 \end{bmatrix}, \begin{bmatrix} 1 \\ 0 \\ 1 \\ 0 \end{bmatrix}, \begin{bmatrix} 0 \\ 1 \\ 0 \\ 1 \end{bmatrix} \right\}$

39. $\left\{ \begin{bmatrix} 1 \\ 2 \\ 3 \\ 4 \end{bmatrix}, \begin{bmatrix} 2 \\ 2 \\ 3 \\ 4 \end{bmatrix}, \begin{bmatrix} 3 \\ 3 \\ 3 \\ 4 \end{bmatrix}, \begin{bmatrix} 4 \\ 4 \\ 4 \\ 4 \end{bmatrix} \right\}$

In Exercises 40–44 find a basis for the given subspaces.

40. The subspace V of $\mathbf{R}^2$ of Exercise 1

41. The subspace V of $\mathbf{R}^2$ of Exercise 2

42. The subspace V of $\mathbf{R}^3$ of Exercise 3

43. The subspace V of $\mathbf{R}^3$ of Exercise 4

44. The subspace V of $\mathbf{R}^4$ of Exercise 5

find a basis for V.

In Exercises 45–49 use *inspection* to determine whether the given sets of n-vectors are bases of $\mathbf{R}^n$.

45. $\left\{ \begin{bmatrix} 71 \\ 0 \end{bmatrix}, \begin{bmatrix} 25 \\ 34 \end{bmatrix} \right\}$

46. $\left\{ \begin{bmatrix} 75 \\ -45 \end{bmatrix}, \begin{bmatrix} -150 \\ 90 \end{bmatrix} \right\}$

47. $\left\{ \begin{bmatrix} 1 \\ 0 \end{bmatrix}, \begin{bmatrix} 0 \\ 1 \end{bmatrix}, \begin{bmatrix} 2 \\ 3 \end{bmatrix} \right\}$

48. $\left\{ \begin{bmatrix} 1 \\ 0 \\ 0 \end{bmatrix}, \begin{bmatrix} 3 \\ 8 \\ -1 \end{bmatrix}, \begin{bmatrix} 0 \\ -2 \\ 1 \end{bmatrix} \right\}$

49. $\left\{ \begin{bmatrix} -3 \\ 0 \\ 0 \\ 3 \end{bmatrix}, \begin{bmatrix} 0 \\ 1 \\ 1 \\ 0 \end{bmatrix}, \begin{bmatrix} 3 \\ 3 \\ 3 \\ 0 \end{bmatrix}, \begin{bmatrix} 3 \\ 4 \\ 4 \\ 0 \end{bmatrix} \right\}$

In Exercises 50–53 determine whether the columns of the $m \times n$ matrices form bases of $\mathbf{R}^m$.

50. a. $\begin{bmatrix} 1 & 2 \\ 0 & 5 \end{bmatrix}$ b. $\begin{bmatrix} 2 & 0 \\ -1 & 4 \end{bmatrix}$

51. a. $\begin{bmatrix} 1 & 3 & 8 \\ 0 & 1 & 7 \end{bmatrix}$ b. $\begin{bmatrix} 1 & 3 \\ -1 & 1 \\ 0 & 1 \end{bmatrix}$

52. a. $\begin{bmatrix} 1 & 3 & 8 \\ -5 & 1 & 7 \\ 0 & 0 & 4 \end{bmatrix}$ **b.** $\begin{bmatrix} 2 & 5 & 6 & 7 \\ 0 & 1 & 3 & 8 \\ 0 & 0 & 0 & 7 \\ 0 & 0 & 0 & 4 \end{bmatrix}$

53. a. $\begin{bmatrix} 2 & 2 & 2 & 2 \\ 0 & 3 & 3 & 3 \\ 0 & 0 & 4 & 4 \\ 0 & 0 & 5 & 5 \end{bmatrix}$ **b.** $\begin{bmatrix} 1 & 1 & 1 & 1 \\ 0 & 3 & 0 & 3 \\ 4 & 0 & 4 & 0 \\ 0 & 5 & 0 & 5 \end{bmatrix}$

54. Find a basis for $V = \text{Span}\left\{ \begin{bmatrix} 1 \\ -1 \end{bmatrix}, \begin{bmatrix} -3 \\ 8 \end{bmatrix} \right\}$.

55. Find a basis for $V = \text{Span}\left\{ \begin{bmatrix} 1 \\ -1 \end{bmatrix}, \begin{bmatrix} -2 \\ 0 \end{bmatrix}, \begin{bmatrix} 7 \\ 5 \end{bmatrix} \right\}$.

2.3 Thm 9.

56. Find a basis for

$$V = \text{Span}\left\{ \begin{bmatrix} 0 \\ -1 \\ 1 \end{bmatrix}, \begin{bmatrix} -2 \\ 1 \\ 0 \end{bmatrix} \right\}$$

57. Find a basis for

$$V = \text{Span}\left\{ \begin{bmatrix} -5 \\ -1 \\ 1 \end{bmatrix}, \begin{bmatrix} 4 \\ 1 \\ 7 \end{bmatrix}, \begin{bmatrix} -1 \\ 0 \\ 8 \end{bmatrix} \right\}$$

58. Find a basis for

$$V = \text{Span}\left\{ \begin{bmatrix} 1 \\ -1 \\ 0 \end{bmatrix}, \begin{bmatrix} 0 \\ 1 \\ -5 \end{bmatrix}, \begin{bmatrix} -1 \\ 0 \\ 0 \end{bmatrix} \right\}$$

59. Show that every nonzero subspace V of $\mathbf{R}^n$ has a basis $\mathcal{B}$ that consists of a finite number of vectors.

(*Hint:* Let $\mathbf{v}_1 \neq \mathbf{0}$ be in V. If $\text{Span}\{\mathbf{v}_1\} = V$, take $\mathcal{B} = \{\mathbf{v}_1\}$ and stop. Otherwise, pick a $\mathbf{v}_2 \in V$ that is not in $\text{Span}\{\mathbf{v}_1\}$. Then show that $\{\mathbf{v}_1, \mathbf{v}_2\}$ is linearly independent. Continue....)

Coordinates

In Exercises 60–62 find the n-vector $\mathbf{x}$, given a basis $\mathcal{B}$ of $\mathbf{R}^n$ and the coordinate n-vector $[\mathbf{x}]_{\mathcal{B}}$.

60. $\mathcal{B} = \left\{ \begin{bmatrix} 2 \\ 1 \end{bmatrix}, \begin{bmatrix} 5 \\ 0 \end{bmatrix} \right\}, [\mathbf{x}]_{\mathcal{B}} = \begin{bmatrix} 6 \\ -3 \end{bmatrix}$

61. $\mathcal{B} = \left\{ \begin{bmatrix} 2 \\ 1 \\ 1 \end{bmatrix}, \begin{bmatrix} -1 \\ 0 \\ 0 \end{bmatrix}, \begin{bmatrix} 0 \\ 2 \\ 1 \end{bmatrix} \right\}, [\mathbf{x}]_{\mathcal{B}} = \begin{bmatrix} -2 \\ 3 \\ 4 \end{bmatrix}$

62. $\mathcal{B} = \left\{ \begin{bmatrix} 2 \\ 2 \end{bmatrix}, \begin{bmatrix} 3 \\ -3 \end{bmatrix} \right\}, [\mathbf{x}]_{\mathcal{B}} = \begin{bmatrix} a \\ b \end{bmatrix}$

In Exercises 63–67 compute the coordinate n-vector $[\mathbf{x}]_{\mathcal{B}}$, given a basis $\mathcal{B}$ of $\mathbf{R}^n$ and $\mathbf{x}$.

63. $\mathcal{B} = \left\{ \begin{bmatrix} 2 \\ 1 \end{bmatrix}, \begin{bmatrix} -1 \\ 1 \end{bmatrix} \right\}, \mathbf{x} = \begin{bmatrix} 17 \\ 4 \end{bmatrix}$

64. $\mathcal{B} = \left\{ \begin{bmatrix} 4 \\ -7 \end{bmatrix}, \begin{bmatrix} -3 \\ 2 \end{bmatrix} \right\}, \mathbf{x} = \begin{bmatrix} -6 \\ 17 \end{bmatrix}$

65. $\mathcal{B} = \left\{ \begin{bmatrix} 2 \\ 2 \\ 1 \end{bmatrix}, \begin{bmatrix} -1 \\ 2 \\ 0 \end{bmatrix}, \begin{bmatrix} 0 \\ -2 \\ -1 \end{bmatrix} \right\}, \mathbf{x} = \begin{bmatrix} -8 \\ 6 \\ -1 \end{bmatrix}$

66. $\mathcal{B} = \left\{ \begin{bmatrix} 1 \\ 1 \\ 0 \\ 0 \end{bmatrix}, \begin{bmatrix} -1 \\ 0 \\ 0 \\ 0 \end{bmatrix}, \begin{bmatrix} 2 \\ 0 \\ -1 \\ 0 \end{bmatrix}, \begin{bmatrix} 1 \\ 2 \\ 3 \\ 4 \end{bmatrix} \right\}$,

$\mathbf{x} = \begin{bmatrix} 4 \\ -1 \\ -4 \\ -4 \end{bmatrix}$

67. $\mathcal{B} = \left\{ \begin{bmatrix} 2 \\ 1 \end{bmatrix}, \begin{bmatrix} 5 \\ 0 \end{bmatrix} \right\}, \mathbf{x} = \begin{bmatrix} 7a + 3b \\ a - b \end{bmatrix}$

68. Let $\mathcal{B}$ be a basis of a subspace V of $\mathbf{R}^n$ and let $\mathbf{v}_1, \ldots, \mathbf{v}_k$ be in V. Show each of the following.

a. $\mathbf{v}$ in V is a linear combination of $\mathbf{v}_1, \ldots, \mathbf{v}_k$ if and only if $[\mathbf{v}]_{\mathcal{B}}$ is linear combination of $[\mathbf{v}_1]_{\mathcal{B}}, \ldots, [\mathbf{v}_k]_{\mathcal{B}}$.

b. $\{\mathbf{v}_1, \ldots, \mathbf{v}_k\}$ is linearly dependent if and only if $\{[\mathbf{v}_1]_{\mathcal{B}}, \ldots, [\mathbf{v}_k]_{\mathcal{B}}\}$ is linearly dependent.

c. $\{\mathbf{v}_1, \ldots, \mathbf{v}_k\}$ is linearly independent if and only if $\{[\mathbf{v}_1]_{\mathcal{B}}, \ldots, [\mathbf{v}_k]_{\mathcal{B}}\}$ is linearly independent.

4.2   Vector Spaces

Reader's Goals for This Section

1. To study and practice with the definitions of vector space and subspace.
2. To work out the details of the examples of the section.

In this section we vastly generalize $\mathbf{R}^n$ and its operations. We consider general sets where abstract operations of addition and scalar multiplication can be defined not through any specific direct rules but by the requirement that they satisfy the basic properties of vector addition and scalar multiplication expressed in Theorems 1 of Sections 2.1 and 3.1.

DEFINITION

(Vector Space)

Let V be a set equipped with two operations named **addition** and **scalar multiplication**. Addition is a rule that associates any two elements $\mathbf{u}$ and $\mathbf{v}$ of V with a third one, the **sum** of $\mathbf{u}$ and $\mathbf{v}$, denoted by $\mathbf{u} + \mathbf{v}$. Scalar multiplication is a rule that associates any (real) scalar c and any element $\mathbf{u}$ of V with another element of V, the **scalar multiple** of $\mathbf{u}$ by c, denoted by $c\mathbf{u}$. Such a set V is called a (real) **vector space** if the two operations satisfy the following properties known as **axioms** for a vector space:

Addition:

(A1) $\mathbf{u} + \mathbf{v}$ belongs to V for all $\mathbf{u}, \mathbf{v} \in V$.
(A2) $\mathbf{u} + \mathbf{v} = \mathbf{v} + \mathbf{u}$ for all $\mathbf{u}, \mathbf{v} \in V$.
(A3) $(\mathbf{u} + \mathbf{v}) + \mathbf{w} = \mathbf{u} + (\mathbf{v} + \mathbf{w})$ for all $\mathbf{u}, \mathbf{v}, \mathbf{w} \in V$.
(A4) There exists a unique element $\mathbf{0} \in V$, called the **zero** of V, such that for all $\mathbf{u}$ in V,

$$\mathbf{u} + \mathbf{0} = \mathbf{0} + \mathbf{u} = \mathbf{u}$$

(A5) For each $\mathbf{u} \in V$ there exists a unique element $-\mathbf{u} \in V$, called the **negative** or **opposite** of $\mathbf{u}$, such that

$$\mathbf{u} + (-\mathbf{u}) = (-\mathbf{u}) + \mathbf{u} = \mathbf{0}$$

Scalar Multiplication:

(M1) $a\mathbf{u}$ belongs to V for all $\mathbf{u} \in V$ and all $a \in \mathbf{R}$.
(M2) $a(\mathbf{u} + \mathbf{v}) = a\mathbf{u} + a\mathbf{v}$ for all $\mathbf{u}, \mathbf{v} \in V$ and all $a \in \mathbf{R}$.
(M3) $(a + b)\mathbf{u} = a\mathbf{u} + b\mathbf{u}$ for all $\mathbf{u} \in V$ and all $a, b \in \mathbf{R}$.
(M4) $a(b\mathbf{u}) = (ab)\mathbf{u}$ for all $\mathbf{u} \in V$ and all $a, b \in \mathbf{R}$.
(M5) $1\mathbf{u} = \mathbf{u}$ for all $\mathbf{u} \in V$.

The elements of a vector space are called **vectors**. Axioms (A1) and (M1) are also expressed by saying that V **is closed under addition** and **scalar multiplication**. (A2) and (A3) are the **commutative** and **associative laws**, respectively, and (M2) and (M3) are the **distributive laws**. Note that *a vector space is a nonempty set*, because it has a zero by (A4).

REMARK   We should keep in mind that neither the vectors nor the operations in the definition of a vector space are specified. Acceptable operations are *any* ones that satisfy the axioms. Some authors use different notation for them, such as $\oplus$ and $\odot$, to distinguish from the ordinary addition and scalar multiplication of n-vectors. We do not do this here, since such notation becomes cumbersome after a while.

We usually write $\mathbf{u} - \mathbf{v}$ for the sum $\mathbf{u} + (-\mathbf{v})$.

$$\mathbf{u} - \mathbf{v} = \mathbf{u} + (-\mathbf{v})$$

(A1), (A2), (A3), and (M1) allow us to add several scalar multiples of vectors together without worrying about the order or grouping terms. In fact, we can add any *finite* set of scalar multiples of vectors together. If $\mathbf{v}_1, \ldots, \mathbf{v}_n$ are vectors and $c_1, \ldots, c_n$ are scalars, then the expression

$$c_1 \mathbf{v}_1 + \cdots + c_n \mathbf{v}_n$$

is well defined and is called a **linear combination** of $\mathbf{v}_1, \ldots, \mathbf{v}_n$. For example, $2\mathbf{u} - 4\mathbf{v} + 3\mathbf{w} + 0.5\mathbf{z}$ is a linear combination of $\mathbf{u}$, $\mathbf{v}$, $\mathbf{w}$, and $\mathbf{z}$.

THEOREM 3

Let V be a vector space. Let $\mathbf{u} \in V$ and $c \in \mathbf{R}$. Then

1. $0\mathbf{u} = \mathbf{0}$;
2. $c\mathbf{0} = \mathbf{0}$;
3. If $c\mathbf{u} = \mathbf{0}$, then $c = 0$ or $\mathbf{u} = \mathbf{0}$;
4. $(-c)\mathbf{u} = -(c\mathbf{u})$.

We prove Parts 1 and 4 and leave the remaining proofs as exercises.

PROOF OF 1

$$\begin{aligned}
& 0\mathbf{u} + 0\mathbf{u} = (0 + 0)\mathbf{u} = 0\mathbf{u} && \text{by (M3)} \\
\Rightarrow\quad & (0\mathbf{u} + 0\mathbf{u}) + (-0\mathbf{u}) = 0\mathbf{u} + (-0\mathbf{u}) && \text{by adding } -0\mathbf{u} \\
\Rightarrow\quad & 0\mathbf{u} + (0\mathbf{u} + (-0\mathbf{u})) = \mathbf{0} && \text{by (A3) and (A5)} \\
\Rightarrow\quad & 0\mathbf{u} + \mathbf{0} = \mathbf{0} && \text{by (A5)} \\
\Rightarrow\quad & 0\mathbf{u} = \mathbf{0} && \text{by (A4)}
\end{aligned}$$

PROOF OF 4 By (M3) and Part 1, we have

$$c\mathbf{u} + (-c)\mathbf{u} = (c + (-c))\mathbf{u} = 0\mathbf{u} = \mathbf{0}$$

So, $c\mathbf{u} + (-c)\mathbf{u} = \mathbf{0}$ and, hence, $(-c)\mathbf{u} + c\mathbf{u} = \mathbf{0}$, by (A2). Therefore, $(-c)\mathbf{u} = -c\mathbf{u}$, by (A5).

The axioms allow us to do n-vector-like arithmetic in a vector space. For example, to show that $\mathbf{u} + \mathbf{u} = 2\mathbf{u}$, we have

$$\mathbf{u} + \mathbf{u} = 1\mathbf{u} + 1\mathbf{u} = (1 + 1)\mathbf{u} = 2\mathbf{u}$$

Examples of Vector Spaces

To verify that a given set is a vector space we need to *define, or specify, explicitly*:

1. The two operations, addition and scalar multiplication;
2. The element that will serve as the zero;

3. The negative of each element;[3]
4. Then, verify the axioms.

The following are major examples of vector spaces. You should work out all details.

■ EXAMPLE 18 $\mathbf{R}^n$ (Special cases: $\mathbf{R}$, $\mathbf{R}^2$, $\mathbf{R}^3$):

1. *Operations:*   The usual vector addition and scalar multiplication (see Section 2.1).
2. *Zero:*   The zero n-vector **0**.
3. *Negative:*   The negative vector $-\mathbf{u}$ of each n-vector **u**.
4. *Axioms:*   All axioms are satisfied by Theorem 1, Section 2.1.

■ EXAMPLE 19   The set M_{mn} of all $m \times n$ matrices with real entries:

1. *Operations:*   The usual matrix addition and scalar multiplication (see Section 3.1).
2. *Zero:*   The $m \times n$ zero matrix **0**.
3. *Negative:*   The negative matrix $-A$ of each matrix A.
4. *Axioms:*   (A1) and (M1) are satisfied, because the sum of two $m \times n$ matrices is again an $m \times n$ matrix, and any scalar multiple of an $m \times n$ matrix is a matrix of size $m \times n$. The rest of the axioms follow from Theorem 1 of Section 3.1.

■ EXAMPLE 20   The set P of all polynomials with real coefficients:

1. *Operations:*   Let x be the indeterminant of the polynomials. (a) Addition: The sum of two polynomials is formed by adding the coefficients of the same powers of x of the polynomials. (b) Scalar multiplication is multiplication of a polynomial through by a constant.
2. *Zero:*   The zero polynomial, **0**, is the polynomial whose coefficients are all zero.
3. *Negative:*   The negative $-p$ of a polynomial p has as coefficients the opposites of the coefficients of p.
4. *Axioms:*   The verification of the axioms is left as exercise.

■ EXAMPLE 21   The set $F(\mathbf{R})$ of all real-valued functions defined on $\mathbf{R}$:

1. *Operations:*   Let f and g be two real-valued functions with domain $\mathbf{R}$ and let c be any scalar. (a) Addition: We define the sum $f + g$ of f and g as the function whose values are given by

$$(f + g)(x) = f(x) + g(x) \qquad \text{for all } x \in \mathbf{R}$$

Likewise, the scalar product cf is defined by

$$(c f)(x) = c f(x) \qquad \text{for all } x \in \mathbf{R}$$

(Figure 4.5).

---

[3]Once a scalar multiplication is defined, then the negative of **v** can be defined as $(-1)\,\mathbf{v}$.

2. *Zero:* The zero function **0** is the function whose values are all zero.

$$\mathbf{0}(x) = 0 \qquad \text{for all } x \in \mathbf{R}$$

3. *Negative:* The negative $-f$ of f is the function $(-1)f$.
4. *Axioms:* The verification of the axioms is left as an exercise.

More generally, the set $F(X)$ of all real-valued functions defined on a set X is a vector space. The operations, zero, and negative are defined the same way. The only difference is that x is in the set X instead of $\mathbf{R}$.

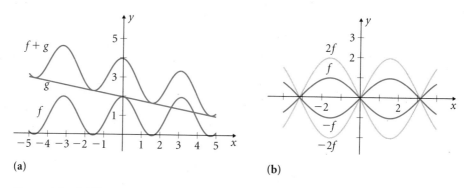

(a)   (b)

Figure 4.5   (a) The graphs of f, g and f + g, (b) f and some of its scalar multiples.

Subspaces

We now consider the important notion of a subspace. This generalizes our familiar notion of a subspace of $\mathbf{R}^n$.

DEFINITION

(Subspace)

A subset W of a vector space V is called a **subspace** of V if W itself is a vector space under the addition and scalar multiplication inherited from V.

In verifying that W is a subspace of V, there is no need to check all 10 axioms. W inherits its operations and their properties from V, for which most of the checking was already done. More precisely, we have the following theorem.

THEOREM 4

(Criterion for Subspace)

A nonempty subset W of a vector space V is a subspace if and only if

1. If $\mathbf{u}$ and $\mathbf{v}$ are in W, then $\mathbf{u} + \mathbf{v}$ is in W;
2. If c is any scalar and $\mathbf{u}$ is in W, then $c\mathbf{u}$ is in W.

PROOF   If W is a subspace of V then all axioms hold. In particular (A1) and (M1) hold. But these are exactly Conditions 1 and 2.

Conversely, let W be a subset that satisfies Conditions 1 and 2. Therefore, (A1) and (M1) hold. (A2), (A3), (M2), (M3), (M4), and (M5) are satisfied, because they are valid in V. We need to verify (A4) and (A5). Condition 2 implies that $0\mathbf{u} = \mathbf{0}$ is in W for $\mathbf{u}$ in W, by letting $c = 0$. Likewise, $(-1)\mathbf{u} = -\mathbf{u}$ is in W for any $\mathbf{u}$ in W, by choosing $c = -1$. The equations of (A4) and (A5) follow.

■ EXAMPLE 22   (The Trivial Subspaces) $\{\mathbf{0}\}$ and V are subspaces of V, called the **trivial subspaces** of V. $\{\mathbf{0}\}$ is the **zero subspace** of V.

SOLUTION V is clearly a subspace of itself. $\{\mathbf{0}\}$ is also a subspace, because Conditions 1 and 2 of Theorem 4 hold:

$$\mathbf{0} + \mathbf{0} = \mathbf{0} \quad \text{and} \quad c\mathbf{0} = \mathbf{0} \qquad \text{for all } c \in \mathbf{R}$$

The statement of Theorem 4 for $V = \mathbf{R}^n$ served as the definition of a subspace of $\mathbf{R}^n$ in Section 4.1.

■ EXAMPLE 23   Any subspace of $\mathbf{R}^n$ in the sense of Section 4.1 is a subspace of $\mathbf{R}^n$.

In particular, we have the following.

■ EXAMPLE 24   Any line through the origin in $\mathbf{R}^3$ ($\mathbf{R}^2$) is a subspace of $\mathbf{R}^3$ ($\mathbf{R}^2$).

■ EXAMPLE 25   Any plane through the origin is a subspace of $\mathbf{R}^3$.

NOTE   A subspace always contains the zero vector $\mathbf{0}$. A line not passing through the origin in $\mathbf{R}^3$ is not a subspace of $\mathbf{R}^3$. $\{(x, 1),\ x \in \mathbf{R}\}$ is not a subspace of $\mathbf{R}^2$, because $(0, 0)$ is not in the set.

Recall that the **degree** of a nonzero polynomial is the highest power of x with nonzero coefficient. For example, the degree of $1 - 2x^2 + x^6$ is 6. The degree of any nonzero constant polynomial is zero. The degree of the *zero polynomial is undefined.*

$f = 0$ undefined.

■ EXAMPLE 26   Show that the set P_n that consists of all polynomials of degree less than or equal n and the zero polynomial is a subspace of P.

SOLUTION   A polynomial in P_n is of the form

$$a_0 + a_1 x + a_2 x^2 + \cdots + a_n x^n$$

The sum of two such polynomials is a polynomial of degree less than or equal n, or zero. Also a constant multiple of such a polynomial is a polynomial of degree less than or equal n or zero. So, Conditions 1 and 2 of Theorem 4 are satisfied. Hence, P_n is a subspace of P.

■ **EXAMPLE 27**   Show that the set $W = \{c\mathbf{v}, c \in \mathbf{R}\}$ of all scalar multiples of the fixed vector $\mathbf{v}$ of a vector space V is a subspace of V.

SOLUTION W is nonempty. It contains $\mathbf{v}$. (Why?) Because

$$c_1\mathbf{v} + c_2\mathbf{v} = (c_1 + c_2)\mathbf{v} \quad \text{and} \quad r(c\mathbf{v}) = (rc)\mathbf{v}$$

W is closed under addition and scalar multiplication. So W is a subspace of V.

■ **EXAMPLE 28**   Let $\mathbf{a}$ be a fixed vector in $\mathbf{R}^3$ and let W be the set of all vectors orthogonal to $\mathbf{a}$. Show that W is a subspace of $\mathbf{R}^3$.

SOLUTION W is nonempty because it contains $\mathbf{0}$. Let $\mathbf{u}$ and $\mathbf{v}$ be vectors in W. Then $\mathbf{a} \cdot \mathbf{u} = 0$ and $\mathbf{a} \cdot \mathbf{v} = 0$. Hence,

$$\mathbf{a} \cdot (\mathbf{u} + \mathbf{v}) = \mathbf{a} \cdot \mathbf{u} + \mathbf{a} \cdot \mathbf{v} = 0 + 0 = 0$$

Therefore, $\mathbf{u} + \mathbf{v}$ is orthogonal to $\mathbf{a}$, so $\mathbf{u} + \mathbf{v}$ is an element of W. If c is any scalar, then

$$\mathbf{a} \cdot (c\mathbf{u}) = c(\mathbf{a} \cdot \mathbf{u}) = c0 = 0$$

implies that $c\mathbf{u}$ is in W. So W is closed under addition and scalar multiplication and, hence, is a subspace of $\mathbf{R}^3$.

■ **EXAMPLE 29**   (Requires Calculus)   The set $C(\mathbf{R})$ of all continuous real-valued functions defined on $\mathbf{R}$ is a subspace of $F(\mathbf{R})$. (Verify.)

■ **EXAMPLE 30**   Both P_n and P are subspaces of $F(\mathbf{R})$. (Verify.)

■ **EXAMPLE 31**   The set D_n of all diagonal matrices of size n is a subspace of M_{nn}. (Verify.)

Correspondence with $\mathbf{R}^n$

The elements of P_n can be viewed as $(n + 1)$-vectors if we decide to drop the variable and its powers and record only the coefficients in a definite order. (Consider the analogue between a linear system and its augmented matrix.)

For example, $p(x) = 1 + 4x - 2x^2$ and $q(x) = -3 + x^2$ in P_3 may be viewed as the 3-vectors

$$p' = \begin{bmatrix} 1 \\ 4 \\ -2 \end{bmatrix} \quad \text{and} \quad q' = \begin{bmatrix} -3 \\ 0 \\ 1 \end{bmatrix}$$

Note that adding p and q corresponds to adding p' and q'. Also any scalar multiple cp corresponds to cp'. So in a sense we may identify P_2 with $\mathbf{R}^3$ (and, in general, P_n with $\mathbf{R}^{n+1}$) as vector spaces and simplify the notation substantially. Although P_n can be viewed as $\mathbf{R}^{n+1}$, we should keep both notations in our toolkit. If we drop P_n altogether, we might lose insight. For instance, when we multiply polynomials, keeping the powers of x helps us compute the product correctly.

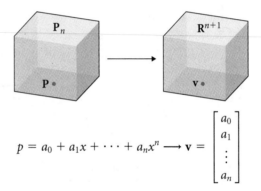

$$p = a_0 + a_1x + \cdots + a_nx^n \longrightarrow \mathbf{v} = \begin{bmatrix} a_0 \\ a_1 \\ \vdots \\ a_n \end{bmatrix}$$

Likewise, the elements of M_{mn} can be viewed as (mn)-vectors. For example, we may consider $\begin{bmatrix} 1 & 2 \\ 3 & 4 \end{bmatrix}$ as the 4-vector $\begin{bmatrix} 1 \\ 2 \\ 3 \\ 4 \end{bmatrix}$ and $\begin{bmatrix} 1 & 2 & 3 \\ 4 & 5 & 6 \end{bmatrix}$ as the 6-vector $\begin{bmatrix} 1 \\ 2 \\ 3 \\ 4 \\ 5 \\ 6 \end{bmatrix}$.

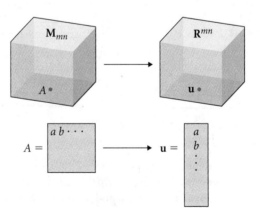

$$A = \begin{bmatrix} a & b & \cdots \\ & & \end{bmatrix} \longrightarrow \mathbf{u} = \begin{bmatrix} a \\ b \\ \vdots \end{bmatrix}$$

Exercises 4.2

NOTE Unless stated otherwise, all subsets of $\mathbf{R}^n$, P, M_{mn}, or $F(\mathbf{R})$ considered are equipped with the ordinary addition or scalar multiplication defined in Section 4.2.

1. Prove that P_n is a vector space.

2. Prove that P is a vector space.

3. Let p and q be in P. Prove that the set $\{ap + bq$, where $a, b \in \mathbf{R}\}$ is a vector space.

4. Is $\mathbf{R}^2$ with the following operations a vector space?

$$(x, y) + (x', y') = (x+x', y+y'), \qquad c(x, y) = (cx, y)$$

5. Is $\mathbf{R}^2$ with the following operations a vector space?

$$(x, y) + (x', y') = (x+x', y+y'), \qquad c(x, y) = (0, 0)$$

6. Show that the set of all real polynomials of degree n is *not* a vector space.

7. Show that the set of all invertible 2×2 matrices is *not* a vector space.

In Exercises 8–15 determine whether the given set is a subspace of P. Use the notation

$$p = a_0 + a_1 x + \cdots + a_n x^n$$

8. $\{p \in P, \; a_0 = 0\}$

9. $\{p \in P, \; a_n = 1\}$

10. $\{p \in P, \; a_0 > 0\}$

11. $\{p \in P, \; p(1) = 1\}$

12. $\{p \in P, \; p(1) = 0\}$

13. $\{p \in P, \; p(0) = 0\}$

14. $\{p \in P, \; a_0 + a_1 + \cdots + a_n = 0\}$

15. $\{p \in P, \; a_1 a_0 = 0, \; n > 1\}$

In Exercises 16–20 determine whether the given subset of M_{22} is a subspace of M_{22}.

16. The set of matrices of the form $\begin{bmatrix} 0 & -a \\ a & 0 \end{bmatrix}$

17. The set of matrices of the form $\begin{bmatrix} a & -b \\ b & a \end{bmatrix}$

18. All matrices $\begin{bmatrix} a & b \\ c & d \end{bmatrix}$ such that $a + d = 0$

19. All matrices $\begin{bmatrix} a & b \\ c & d \end{bmatrix}$ such that $a + d = 1$

20. The set of matrices $\begin{bmatrix} a & b \\ c & d \end{bmatrix}$ such that $ad = 0$

In Exercises 21–24 determine whether the given set can be made into a vector space using the usual rules of addition and scalar multiplication for matrices and polynomials. (x is the indeterminate of all polynomials considered.)

21. $\left\{ \begin{bmatrix} p & c \\ 0 & q \end{bmatrix}, \text{ where, } p, q \in P_1, \; c \in \mathbf{R} \right\}$

22. $\left\{ \begin{bmatrix} p & c \\ q & 0 \end{bmatrix}, \text{ where, } p \in P_1, \; q \in P_2, \; c \in \mathbf{R} \right\}$

23. $\{ \begin{bmatrix} xp & q \end{bmatrix}, \text{ where, } p, q \in P \}$

24. $\{ \begin{bmatrix} x & p \end{bmatrix}, \text{ where, } p \in P \}$

Consider the sets V_e and V_o of **even** and **odd** functions, respectively.

$$V_e = \{f \in F(\mathbf{R}) \text{ such that } f(x) = f(-x) \text{ for all } x \in \mathbf{R}\}$$

$$V_o = \{f \in F(\mathbf{R}) \text{ such that } f(x) = -f(-x) \text{ for all } x \in \mathbf{R}\}$$

25. Is V_e a subspace $F(\mathbf{R})$?

26. Is V_o a subspace $F(\mathbf{R})$?

27. Prove Parts 2 and 3 of Theorem 3.

28. Let V be a vector space and let $\mathbf{u}, \mathbf{v}, \mathbf{w} \in V$ be such that $\mathbf{u} + \mathbf{w} = \mathbf{v} + \mathbf{w}$. Use the axioms to show that $\mathbf{u} = \mathbf{v}$.

29. Let V be a vector space and let $\mathbf{u}, \mathbf{v} \in V$ be such that $r\mathbf{v} = r\mathbf{u}$, for some scalar r. Use the axioms to show that either $r = 0$ or $\mathbf{u} = \mathbf{v}$.

30. Show that the Cartesian product $V_1 \times V_2$ of two vector spaces V_1 and V_2

$$V_1 \times V_2 = \{(\mathbf{v}_1, \mathbf{v}_2), \text{ where } \mathbf{v}_1 \in V_1, \; \mathbf{v}_2 \in V_2\}$$

equipped with componentwise addition and scalar multiplication is a vector space.

$$(\mathbf{u}_1, \mathbf{u}_2) + (\mathbf{v}_1, \mathbf{v}_2) = (\mathbf{u}_1 + \mathbf{v}_1, \mathbf{u}_2 + \mathbf{v}_2)$$

$$c(\mathbf{v}_1, \mathbf{v}_2) = (c\mathbf{v}_1, c\mathbf{v}_2)$$

31. Let U and W be subspaces of a vector space V. Show that the intersection $U \cap W$ is a subspace of V.

4.3   Linear Independence and Bases

Reader's Goals for This Section

1. To review Sections 2.1, 2.3, 2.4, and 4.1.
2. To understand the concepts of span, linear independence, and basis.

The basic notions of spanning set, linear independence, and basis studied for n-vectors extend easily to abstract vectors. This section is rich in examples to help us understand these concepts thoroughly. The theorems generalize those of Sections 2.3, 2.4, and 4.1. And the proofs are essentially the same.

The Span of $v_1, \ldots, v_k$

DEFINITION   **(Span)**

Let V be a vector space and let $\mathbf{v}_1, \ldots, \mathbf{v}_k$ be vectors in V. The set of all linear combinations of $\mathbf{v}_1, \ldots, \mathbf{v}_k$ is called the **span** of $\mathbf{v}_1, \ldots, \mathbf{v}_k$ and is denoted by Span$\{\mathbf{v}_1, \ldots, \mathbf{v}_k\}$. If $V = \text{Span}\{\mathbf{v}_1, \ldots, \mathbf{v}_k\}$ we say that $\mathbf{v}_1, \ldots, \mathbf{v}_k$ **span** V and that $\{\mathbf{v}_1, \ldots, \mathbf{v}_k\}$ is a **spanning set** of V.

■ **EXAMPLE 32**   Show that the following vectors are in Span$\{\mathbf{v}_1, \mathbf{v}_2\}$.

$$\mathbf{0}, \qquad \mathbf{v}_1, \qquad \mathbf{v}_2, \qquad \mathbf{v}_1 + \mathbf{v}_2, \qquad -2\mathbf{v}_1, \qquad 3\mathbf{v}_1 - 2\mathbf{v}_2$$

SOLUTION   Each of these vectors is a linear combination of $\mathbf{v}_1$ and $\mathbf{v}_2$, because we may write

$$\mathbf{0} = 0\mathbf{v}_1 + 0\mathbf{v}_2 \qquad \mathbf{v}_1 = 1\mathbf{v}_1 + 0\mathbf{v}_2 \qquad \mathbf{v}_2 = 0\mathbf{v}_1 + 1\mathbf{v}_2$$
$$\mathbf{v}_1 + \mathbf{v}_2 = 1\mathbf{v}_1 + 1\mathbf{v}_2 \qquad -2\mathbf{v}_1 = -2\mathbf{v}_1 + 0\mathbf{v}_2 \qquad 3\mathbf{v}_1 - 2\mathbf{v}_2 = 3\mathbf{v}_1 + (-2)\mathbf{v}_2$$

■ **EXAMPLE 33**   Show that Span$\{\mathbf{v}\}$ is the set of all scalar multiples of $\mathbf{v}$.

$$\text{Span}\{\mathbf{v}\} = \{c\mathbf{v}, \ c \in \mathbf{R}\}$$

SOLUTION   Any linear combination of $\mathbf{v}$ is of the form $c\mathbf{v}$ for some scalar c. Conversely, any scalar multiple $c\mathbf{v}$ of $\mathbf{v}$ is a linear combination of $\mathbf{v}$.

For example, the span of $\{(1, -1)\}$ in $\mathbf{R}^2$ is the line through the origin in the direction of $(1, -1)$ (Fig. 4.6(a)). Likewise, the span of x^2 in P_2 consists of all the functions that are scalar multiples of x^2 (vectors $x^2, 2x^2, 4x^2, -1.5x^2, -3.5x^2$ of the span are pictured in Fig. 4.6(b)).

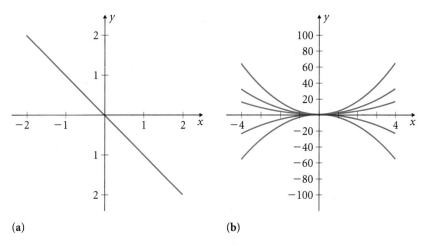

(a)   (b)

Figure 4.6   (a) Span$\{(1, -1)\}$; (b) Span$\{x^2\}$.

■ EXAMPLE 34   Is $-1 + x^2$ in the span of $p = 1 + x + x^3$ and $q = -x - x^2 - x^3$
in P_3?

SOLUTION   Let a and b be scalars such that $-1 + x^2 = a(1 + x + x^3) + b(-x - x^2 - x^3)$.
Then

$$-1 + x^2 = a + (a - b)x - bx^2 + (a - b)x^3$$

Equating coefficients of the same powers of x yields the system

$$a - b = 0, \qquad -b = 1, \qquad a - b = 0, \qquad a = -1$$

with solution $a = b = -1$. Hence, $-1 + x^2 = -p - q$. Because $-1 + x^2$ is a linear
combination of p and q it is in Span$\{p, q\}$.

■ EXAMPLE 35   Show that Span$(\{1, x, x^2, x^3\}) = P_3$.

SOLUTION   Any element $p = a_0 + a_1x + a_2x^2 + a_3x^3$ of P_3 is already a linear
combination of $\{1, x, x^2, x^3\}$. a_0, a_1, a_2, a_3 are the coefficients of $1, x, x^2, x^3$, respectively.

Let E_{ij} be the matrix in M_{mn} whose (i, j)th entry is 1 and the rest of the entries are
zero. For example, in M_{23}

$$E_{11} = \begin{bmatrix} 1 & 0 & 0 \\ 0 & 0 & 0 \end{bmatrix} \qquad E_{12} = \begin{bmatrix} 0 & 1 & 0 \\ 0 & 0 & 0 \end{bmatrix} \qquad E_{13} = \begin{bmatrix} 0 & 0 & 1 \\ 0 & 0 & 0 \end{bmatrix}$$

$$E_{21} = \begin{bmatrix} 0 & 0 & 0 \\ 1 & 0 & 0 \end{bmatrix} \qquad E_{22} = \begin{bmatrix} 0 & 0 & 0 \\ 0 & 1 & 0 \end{bmatrix} \qquad E_{23} = \begin{bmatrix} 0 & 0 & 0 \\ 0 & 0 & 1 \end{bmatrix}$$

■ EXAMPLE 36   Show that $\{E_{11}, E_{12}, E_{13}, E_{21}, E_{22}, E_{23}\}$ spans M_{23}.

SOLUTION   This is true because

$$\begin{bmatrix} a & b & c \\ d & e & f \end{bmatrix} = aE_{11} + bE_{12} + cE_{13} + dE_{21} + eE_{22} + fE_{23}$$

■ EXAMPLE 37   Show that $\{(1, 2, -1), (-1, 1, -2), (1, 1, 1)\}$ spans $\mathbf{R}^3$.

SOLUTION   Because

$$A = \begin{bmatrix} 1 & -1 & 1 \\ 2 & 1 & 1 \\ -1 & -2 & 1 \end{bmatrix} \sim \begin{bmatrix} 1 & -1 & 1 \\ 0 & 3 & -1 \\ 0 & 0 & 1 \end{bmatrix}$$

A has three pivots. Hence, the vectors span $\mathbf{R}^3$.

■ EXAMPLE 38   Compute the span of $\{A, B\}$ in M_{22}, where

$$A = \begin{bmatrix} 1 & 0 \\ 0 & 0 \end{bmatrix} \qquad B = \begin{bmatrix} 1 & 0 \\ 0 & -1 \end{bmatrix}$$

SOLUTION   Any linear combination of A and B is a diagonal matrix:

$$aA + bB = a\begin{bmatrix} 1 & 0 \\ 0 & 0 \end{bmatrix} + b\begin{bmatrix} 1 & 0 \\ 0 & -1 \end{bmatrix} = \begin{bmatrix} a + b & 0 \\ 0 & -b \end{bmatrix}$$

Conversely, any diagonal matrix can be written as a linear combination of A and B, because

$$\begin{bmatrix} a & 0 \\ 0 & b \end{bmatrix} = (a + b)\begin{bmatrix} 1 & 0 \\ 0 & 0 \end{bmatrix} - b\begin{bmatrix} 1 & 0 \\ 0 & -1 \end{bmatrix}$$

Therefore, $\text{Span}(\{A, B\}) = D_2$, the set of all 2×2 diagonal matrices.

REMARK   The following statements can be easily verified and are left as exercises.

1. $\{\mathbf{e}_1, \mathbf{e}_2, \ldots, \mathbf{e}_n\}$ spans $\mathbf{R}^n$.
2. $\{1, x, x^2, \ldots, x^n\}$ spans P_n.
3. $\{1, x, x^2, \ldots, x^n, \ldots\}$ spans P.
4. $\{E_{11}, E_{12}, E_{13}, \ldots, E_{mn}\}$ spans M_{mn}.
5. $\{E_{11}, E_{22}, E_{33}, \ldots, E_{nn}\}$ spans D_n.

THEOREM 5

Let S be a subset of a vector space V. Then

1. $\text{Span}(S)$ is a subspace of V;
2. $\text{Span}(S)$ is the smallest subspace of V that contains S.

PROOF

1. Let $\mathbf{u}_1,\ldots,\mathbf{u}_n$ and $\mathbf{v}_1,\ldots,\mathbf{v}_m$ be vectors in S and let $c_1,\ldots,c_n$ and $d_1,\ldots,d_m$ be scalars. Consider the two linear combinations of vectors in S:

$$c_1\,\mathbf{u}_1 + \cdots + c_n\,\mathbf{u}_n \quad \text{and} \quad d_1\,\mathbf{v}_1 + \cdots + d_m\,\mathbf{v}_m$$

The sum

$$c_1\,\mathbf{u}_1 + \cdots + c_n\,\mathbf{u}_n + d_1\,\mathbf{v}_1 + \cdots + d_m\,\mathbf{v}_m$$

is well defined in V and is again a linear combination of vectors in S. If c is any scalar, then

$$c\,(c_1\,\mathbf{u}_1 + \cdots + c_n\,\mathbf{u}_n) = c(c_1\,\mathbf{u}_1) + \cdots + c(c_n\,\mathbf{u}_n) = (cc_1)\,\mathbf{u}_1 + \cdots + (cc_n)\,\mathbf{u}_n$$

is also a linear combination of vectors in S. Therefore, Span(S) is closed under the addition and scalar multiplication of V. Hence, it is a subspace of V.

2. Let W be a subspace that contains S. As a subspace W contains all linear combinations of its elements. In particular, it contains all linear combinations of elements of S. But these are the elements of Span(S). Therefore, Span(S) $\subseteq W$. Hence, Span(S) is the subspace contained in any subspace W that contains S. This proves the second statement.

THEOREM 6

(Reduction of Spanning Set)

If one of the vectors $\mathbf{v}_1,\ldots,\mathbf{v}_k$ of the vector space V is a linear combination of the rest, then the span remains the same if we remove this vector.

PROOF   See the proof of Theorem 9, Section 2.3.

Linear Independence

DEFINITION

A set of vectors $\mathbf{v}_1,\ldots,\mathbf{v}_n$ from a vector space V is called **linearly dependent** if there are scalars $c_1,\ldots,c_n$ *not all zero* such that

$$c_1\,\mathbf{v}_1 + \cdots + c_n\,\mathbf{v}_n = \mathbf{0} \tag{4.1}$$

$\mathbf{v}_1,\ldots,\mathbf{v}_n$ is called **linearly independent** if it is not linearly dependent. In other words, (4.1) implies $c_1 = \cdots = c_n = 0$. Let S be any subset of V (possibly infinite). Then S is called linearly dependent if it contains a finite linearly dependent subset. Otherwise, S is linearly independent.

■ EXAMPLE 39   The set $\{2 - x + x^2, 2x + x^2, 4 - 4x + x^2\}$ is linearly dependent in P_3 because $4 - 4x + x^2 = 2(2 - x + x^2) - (2x + x^2)$.

▪ **EXAMPLE 40**   The set $\{A, B, C\}$ is linearly dependent in M_{22} because $A = B + C$.

$$A = \begin{bmatrix} 1 & -1 \\ 2 & 0 \end{bmatrix} \qquad B = \begin{bmatrix} 1 & 0 \\ 0 & -2 \end{bmatrix} \qquad C = \begin{bmatrix} 0 & -1 \\ 2 & 2 \end{bmatrix}$$

▪ **EXAMPLE 41**   Both $\{1, \cos 2x, \cos^2 x\}$ and $\{\sin x \cos x, \sin 2x\}$ are linearly dependent in $F(\mathbf{R})$, because

$$\cos^2 x = \frac{1}{2} \cdot 1 + \frac{1}{2} \cos 2x \quad \text{and} \quad \sin 2x = 2 \sin x \cos x \qquad \text{for all } x \in \mathbf{R}$$

▪ **EXAMPLE 42**   Show that the set $\{E_{11}, E_{12}, E_{21}, E_{22}\}$ is linearly independent in M_{22}.

SOLUTION   Let

$$c_1 \begin{bmatrix} 1 & 0 \\ 0 & 0 \end{bmatrix} + c_2 \begin{bmatrix} 0 & 1 \\ 0 & 0 \end{bmatrix} + c_3 \begin{bmatrix} 0 & 0 \\ 1 & 0 \end{bmatrix} + c_4 \begin{bmatrix} 0 & 0 \\ 0 & 1 \end{bmatrix} = \begin{bmatrix} 0 & 0 \\ 0 & 0 \end{bmatrix}$$

or

$$\begin{bmatrix} c_1 & c_2 \\ c_3 & c_4 \end{bmatrix} = \begin{bmatrix} 0 & 0 \\ 0 & 0 \end{bmatrix}$$

Hence, $c_1 = c_2 = c_3 = c_4 = 0$.

▪ **EXAMPLE 43**   Show that the set $\{1, x, \ldots, x^n\}$ is linearly independent in P_n.

SOLUTION   If a linear combination $p(x) = a_0 + a_1 x + \cdots + a_n x^n$ in $\{1, x, \ldots, x^n\}$ is the zero polynomial, $\mathbf{0}$, then

$$a_0 + a_1 r + \cdots + a_n r^n = 0 \qquad \text{for all } r \in \mathbf{R}$$

Recall now the basic fact from algebra that a *nonzero polynomial of degree n has at most n roots*. Since p has more than n roots it has to be the zero polynomial. This proves the assertion.

▪ **EXAMPLE 44**   Show that the set $\{x^2, 1 + x, -1 + x\}$ is linearly independent in P_3.

SOLUTION   If a linear combination $p(x) = ax^2 + b(1 + x) + c(-1 + x)$ is the zero polynomial, $\mathbf{0}$, then $p(x) = (b - c) + (b + c)x + ax^2 = 0$ for all $x \in \mathbf{R}$. Therefore, $b - c = 0, b + c = 0$ and $a = 0$. Hence, $a = b = c = 0$ and the set is linearly independent.

Matrix Operation Notation

In Example 44 we may exploit the identification of P_2 with $\mathbf{R}^3$ (discussed in Section 4.1) and show that $\left\{ \begin{bmatrix} 0 \\ 0 \\ 1 \end{bmatrix}, \begin{bmatrix} 1 \\ 1 \\ 0 \end{bmatrix}, \begin{bmatrix} -1 \\ 1 \\ 0 \end{bmatrix} \right\}$ is linearly independent instead of $\{x^2, 1 + x,$

$-1 + x$}. This is true because

$$\begin{bmatrix} 0 & 1 & -1 \\ 0 & 1 & 1 \\ 1 & 0 & 0 \end{bmatrix} \sim \begin{bmatrix} 1 & 0 & 0 \\ 0 & 1 & 1 \\ 0 & 0 & -2 \end{bmatrix} \qquad (4.2)$$

has three pivot columns.

■ EXAMPLE 45   Show that the set $\{\cos x, \sin x\}$ is linearly independent in $F(\mathbf{R})$.

SOLUTION   If a linear combination $a \cos x + b \sin x$ is the zero function, then

$$a \cos y + b \sin y = 0 \qquad \text{for all } y \in \mathbf{R}$$

By letting y be 0 and $\frac{\pi}{2}$, we get $a = 0$ and $b = 0$, respectively. Thus $\{\cos x, \sin x\}$ is linearly independent in $F(\mathbf{R})$.

REMARK   The following sets are linearly independent.

1. $\{\mathbf{e}_1, \mathbf{e}_2, \ldots, \mathbf{e}_n\} \subseteq \mathbf{R}^n$
2. $\{1, x, x^2, \ldots, x^n, \ldots\} \subseteq P$
3. $\{E_{11}, E_{12}, E_{13}, \ldots, E_{mn}\} \subseteq M_{mn}$
4. $\{1, \cos x, \cos 2x\} \subseteq F(\mathbf{R})$
5. $\{e^x, e^{2x}\} \subseteq F(\mathbf{R})$

THEOREM 7

(Test for Linear Dependence)

Let S be a subset of a vector space V.

1. If S consists of one vector $\mathbf{v}$, then S is linearly dependent if and only if $\mathbf{v} = \mathbf{0}$.
2. If S consists of two or more vectors $\mathbf{v}_1, \ldots, \mathbf{v}_k$, then S is linearly dependent if and only if at least one vector is a linear combination of the remaining vectors.
3. If S consists of two or more vectors $\mathbf{v}_1, \ldots, \mathbf{v}_k$, with $\mathbf{v}_1 \neq \mathbf{0}$, then S is linearly dependent if and only if at least one vector, say, $\mathbf{v}_i (i \geq 2)$, is a linear combination of the vectors that precede it, i.e., $\mathbf{v}_1, \ldots, \mathbf{v}_{i-1}$.

PROOF   See the proof of Theorem 14, Section 2.4.

The following parallels Theorem 15, Section 2.4. The proof is left as an exercise.

THEOREM 8

1. Any set of vectors that contains $\mathbf{0}$ is linearly dependent.
2. Two vectors are linearly dependent if and only if one is a scalar multiple of the other.
3. Any set of vectors that contains a linearly dependent set is itself linearly dependent.
4. Any subset of a linearly independent set is itself linearly independent.

THEOREM 9

Let $S = \{\mathbf{v}_1, \ldots, \mathbf{v}_n\}$ be a linearly independent set of vectors from a vector space V.

1. Any vector $\mathbf{v}$ in the span of S is uniquely expressible as a linear combination of vectors in S; i.e., the relations

$$\mathbf{v} = c_1 \mathbf{v}_1 + \cdots + c_n \mathbf{v}_n \quad \text{and} \quad \mathbf{v} = d_1 \mathbf{v}_1 + \cdots + d_n \mathbf{v}_n$$

imply

$$c_1 = d_1, \ldots, \, c_n = d_n$$

2. If $\mathbf{v}$ is not in the span of S, then the set $\{\mathbf{v}_1, \mathbf{v}_2, \ldots, \mathbf{v}_n, \mathbf{v}\}$ is linearly independent.

PROOF   See the proof of Theorem 16, Section 2.4.

Basis of a Vector Space

In this paragraph we discuss the fundamental concept of a basis of a vector space. Bases, just as in the case of subspaces of $\mathbf{R}^n$, are spanning sets that are also linearly independent. Knowing a basis of a vector space can be quite useful in understanding the space and its properties.

DEFINITION

(Basis)

A nonempty subset $\mathcal{B}$ of a nonzero vector space V is a basis of V if

1. $\mathcal{B}$ is linearly independent; and
2. $\mathcal{B}$ spans V.

The empty set is, by agreement, the only basis of the zero vector space $\{\mathbf{0}\}$.

The proof of the following main theorem requires set theory that is beyond the scope of this book and is omitted.[4]

THEOREM 10

(Existence of Basis)

Every vector space has a basis.

Here are some basic examples of bases. All sets were already seen to be linearly independent and spanning.

■ EXAMPLE 46 $\{\mathbf{e}_1, \mathbf{e}_2, \ldots, \mathbf{e}_n\}$ is a basis of $\mathbf{R}^n$ called the **standard basis** of $\mathbf{R}^n$.

---

[4]See Theorem 2, Section 7, Chapter 2 of N. Bourbaki's *Algebra I* (Paris: Hermann/Addison-Wesley, 1974).

■ **EXAMPLE 47** $\{1, x, x^2, \ldots, x^n\}$ is a basis of P_n called the **standard basis** of P_n (Fig. 4.7).

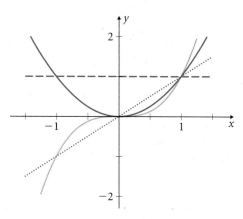

Figure 4.7   The standard basis $\{1, x, x^2, x^3\}$ of P_3.

■ **EXAMPLE 48** $\{1, x, x^2, \ldots, x^n, \ldots\}$ is a basis of P called the **standard basis** of P.

■ **EXAMPLE 49** $\{E_{11}, E_{12}, E_{13}, \ldots, E_{mn}\}$ is a basis of M_{mn} called the **standard basis** of M_{mn}.

In addition to the standard bases, $\mathbf{R}^n, P_n, P,$ and M_{mn} have plenty of other bases.

■ **EXAMPLE 50**   Show that $\mathcal{B} = \{1 + x, -1 + x, x^2\}$ is a basis of P_2.

SOLUTION   We need to show that $\mathcal{B}$ is linearly independent and that it spans P_2. In other words, we need to show that

$$a(1 + x) + b(-1 + x) + cx^2 = \mathbf{0} \quad \text{or} \quad (a - b) + (a + b)x + cx^2 = \mathbf{0}$$

implies $a = b = c = 0$ and that if $A + Bx + Cx^2$ is any polynomial in P_2, then there exist scalars a, b, c such that

$$a(1 + x) + b(-1 + x) + cx^2 = A + Bx + Cx^2 \quad \text{or} \quad (a-b) + (a+b)x + cx^2 = A + Bx + Cx^2$$

The two resulting systems

$$
\begin{array}{cc}
a - b = 0 & a - b = A \\
a + b = 0 & a + b = B \\
c = 0 & c = C
\end{array}
$$

have invertible coefficient matrices. Hence, the first system has only the trivial solution, and the second is consistent for all choices of A, B and C.

■ EXAMPLE 51   Show that neither of $X = \{1, 1+x, -1+x\}$ and $Y = \{1+x, -1+x\}$ is a basis of P_2.

SOLUTION X is not linearly independent, because $-2 \cdot 1 + (1 + x) - (-1 + x) = 0$ for all x in $\mathbf{R}$. Hence,

$$-2 \cdot 1 + (1 + x) - (-1 + x) = \mathbf{0}$$

is a nontrivial linear combination of the zero polynomial $\mathbf{0}$.

   Y, on the other hand, is linearly independent but does not span P_2. (Why?) Although Y is not a basis of P_2, it is a basis of P_1. (Verify.)

■ EXAMPLE 52   Consider the set $W = \{a \cos x + b \sin x, \ a, b \in \mathbf{R}\}$. Prove:

1. W is a vector space;
2. $\{\cos x, \sin x\}$ is a basis of W.

SOLUTION W is the set of all linear combinations of $\{\cos x, \sin x\}$ in $F(\mathbf{R})$; hence, it is a subspace of $F(\mathbf{R})$ by Theorem 5. In particular, W is a vector space. The set $\{\cos x, \sin x\}$ spans W by definition, and it was shown in Example 45 that it is linearly independent in $F(\mathbf{R})$. So, it is linearly independent in the smaller space W. Hence it forms a basis of W.

   The two basis vectors $\cos x, \sin x$ of W along with the vectors $2 \sin x + 3 \cos x$, $3 \cos x - 2 \sin x$ are depicted in Fig. 4.8.

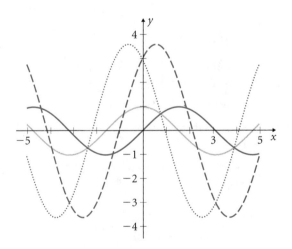

Figure 4.8   Some vectors of Span($\{\cos x, \sin x\}$).

One of the main characterizations of a basis is described in the following theorem.

THEOREM 11   (Uniqueness of Representation)

A subset $\mathcal{B} = \{v_1, \ldots, v_n\}$ of a vector space V is a basis of V if and only if for each vector v in V there are **unique** scalars $c_1, \ldots, c_n$ such that

$$v = c_1 v_1 + \cdots + c_n v_n$$

PROOF   See the proof of Theorem 2, Section 4.1.

Exercises 4.3

Span

Let p_1, p_2, p_3, and p_4 in P_2, where

$$p_1 = 1 + 3x - x^2 \qquad p_2 = -3x + 2x^2$$
$$p_3 = x^2 \qquad p_4 = 4 - x^2$$

1. a. Is p_4 in Span $\{p_1, p_2, p_3\}$?

  b. Is p_4 in Span $\{p_1, p_2\}$?

  c. Is p_4 in Span $\{p_2, p_3\}$?

2. a. Show that Span $\{p_1, p_2, p_3\} = P_2$.

  b. Show that Span $\{p_1, p_2, p_4\} = P_2$.

  c. Show that

$$\text{Span} \{p_1, p_3, p_4\} = \text{Span} \{p_2, p_3, p_4\}$$

In Exercises 3–7 determine whether the given set spans P_2.

3. $\{1 + x + x^2, \ 1 + 2x + x^2, \ x\}$

4. $\{1 - x + x^2, \ 1 + x - x^2, \ 1\}$

5. $\{1 + x, -1 + x, 2 + x + x^2\}$

6. $\{1 + x + x^2, \ 1 + x, \ 1\}$

7. $\{-1 + x + x^2, \ 1 - x + x^2, \ 1 + x - x^2\}$

8. Does $\{4, 1 + x, -1 + x^2, -1 + x^3\}$ span P_3?

In Exercises 9–13 determine whether the given set spans M_{22}.

9. $\left\{ \begin{bmatrix} 1 & 1 \\ 1 & 1 \end{bmatrix}, \begin{bmatrix} 0 & 3 \\ 0 & 3 \end{bmatrix}, \begin{bmatrix} 4 & 0 \\ 4 & 0 \end{bmatrix}, \begin{bmatrix} 0 & 5 \\ 0 & 5 \end{bmatrix} \right\}$

10. $\left\{ \begin{bmatrix} 2 & 2 \\ 2 & 2 \end{bmatrix}, \begin{bmatrix} 0 & 3 \\ 3 & 3 \end{bmatrix}, \begin{bmatrix} 0 & 0 \\ 4 & 4 \end{bmatrix}, \begin{bmatrix} 2 & 3 \\ 4 & 5 \end{bmatrix} \right\}$

11. $\left\{ \begin{bmatrix} 1 & 0 \\ 0 & 0 \end{bmatrix}, \begin{bmatrix} 1 & 1 \\ 1 & 0 \end{bmatrix}, \begin{bmatrix} 2 & 1 \\ 0 & 0 \end{bmatrix}, \begin{bmatrix} 3 & 0 \\ -1 & 1 \end{bmatrix} \right\}$

12. $\{E_{11}, E_{12}, -E_{21}, E_{11} + E_{12}\}$

13. $\{E_{11}, E_{11} + E_{12}, E_{11} + E_{12} + E_{21}, E_{11} + E_{12} + E_{21} + E_{22}\}$

14. Under what restriction(s) on a and b will $\{a + ax + ax^2, bx^2, 1\}$ span P_2?

15. True or false?

  a. P_9 can be spanned by exactly 9 polynomials in it.

  b. P_9 can be spanned by at least 9 polynomials in it.

  c. P_9 can be spanned by 10 polynomials from P_8.

  d. P_9 can be spanned by 10 polynomials in it.

  e. P_9 can be spanned by 11 polynomials in it.

  f. Any 20 polynomials of P_9 span P_9.

  g. Twenty polynomials of P_9 can span P_9.

16. Show that for any vectors u, v from a vector space V,

$$\text{Span} \{u, v\} = \text{Span} \{u + v, u - v\}$$

17. Show that for any vectors u, v from a vector space V,

$$\text{Span} \{u, v, w\} = \text{Span} \{u, u + v, u + v + w\}$$

Linear Independence

In Exercises 18–21 determine whether the set is linearly independent.

18. $\{-2x + x^2, 1 + x + x^2, 1 - x\}$

19. $\{1 + ax + ax^2, 1 + bx + bx^2, 1\}$

20. $\{1 + ax + ax^2, 1 + bx + bx^2, x^2\}$, for unequal constants a and b

21. $\left\{ \begin{bmatrix} 1 & 0 \\ 0 & 0 \end{bmatrix}, \begin{bmatrix} 1 & 1 \\ 1 & 0 \end{bmatrix}, \begin{bmatrix} 2 & 1 \\ 0 & 0 \end{bmatrix}, \begin{bmatrix} 3 & 0 \\ -1 & 1 \end{bmatrix} \right\}$

22. For which values of a is the set $\{1 + ax, a + (a+2)x\} \subseteq P_1$ linearly dependent?

23. True or false?

    a. Any two distinct vectors of a vector space are linearly independent.

    b. Any three distinct polynomials of P_1 span P_1.

    c. Any three linearly independent polynomials of P_2 span P_2.

    d. Any n linearly independent polynomials of P_n span P_n.

24. Let $\{v_1, v_2, v_3\}$ be a linearly independent set of vectors from a vector space V. Find $c_1, c_2,$ and c_3, if $v = c_1 v_1 + c_2 v_2 + c_3 v_3$ and $v = (2c_2 - c_1)v_1 + (c_3 - c_2)v_2 + (c_2 - 1)v_3$.

25. Let V be a vector space and let v_1, v_2, v_3 be in V. Prove that if $\{v_1, v_2, v_3\}$ is linearly independent so is $\{v_1 - v_2, v_2 - v_3, v_3 + v_1\}$.

26. Prove Theorem 8.

27. Let p, q, r be polynomials P_2. Suppose that $\{p, q\}$ and $\{q, r\}$ are linearly independent sets. Does this imply that $\{p, r\}$ is linearly independent? Explain.

Bases

In Exercises 28–30 determine whether the given sets are bases of M_{22}.

28. $\left\{ \begin{bmatrix} 1 & 2 \\ 1 & 2 \end{bmatrix}, \begin{bmatrix} 3 & 4 \\ 3 & 4 \end{bmatrix}, \begin{bmatrix} 5 & 6 \\ 5 & 6 \end{bmatrix}, \begin{bmatrix} 7 & 8 \\ 7 & 8 \end{bmatrix} \right\}$

29. $\left\{ \begin{bmatrix} 1 & 0 \\ 0 & 1 \end{bmatrix}, \begin{bmatrix} 0 & 1 \\ 1 & 0 \end{bmatrix}, \begin{bmatrix} 1 & 0 \\ 1 & 0 \end{bmatrix}, \begin{bmatrix} 0 & 1 \\ 0 & 1 \end{bmatrix} \right\}$

30. $\left\{ \begin{bmatrix} 1 & 2 \\ 3 & 4 \end{bmatrix}, \begin{bmatrix} 2 & 2 \\ 3 & 4 \end{bmatrix}, \begin{bmatrix} 3 & 3 \\ 3 & 4 \end{bmatrix}, \begin{bmatrix} 4 & 4 \\ 4 & 4 \end{bmatrix} \right\}$

31. Let $V \subseteq M_{22}$ denote the set of all matrices of the form $\begin{bmatrix} a & b \\ c & -a \end{bmatrix}$. Show that $\mathcal{B} = \{E_{11} - E_{22}, E_{12}, E_{21}\}$ is a basis for V.

32. Show that $\{x^2, (1+x)^2, (-1+x)^2\}$ is a basis of P_2.

33. Show that $\{1, x, 2x^2, 3 - 3x + x^3\}$ is a basis of P_3. (These are the first four **Chebyshev polynomials of the first kind**. They occur naturally in several areas of mathematics and physics.)

In Exercises 34–39:

(a) Find a basis for V.

(b) In each case check whether $V = P_2$.

34. $V = \text{Span}\left\{ 1 + x + x^2, 1, -1 - x^2, x^2 \right\}$

35. $V = \text{Span}\left\{ 2 + x + 2x^2, x^2, 1 - x - x^2, 1 \right\}$

36. $V = \text{Span}\left\{ x + x^2, 1 + x, -1 + x^2 \right\}$

37. $V = \text{Span}\left\{ x^2, 1 + x, -1 + x^2 \right\}$

38. $V = \text{Span}\left\{ 1 - x - 5x^2, 7 + x + 4x^2, 8 - x^2 \right\}$

39. $V = \text{Span}\left\{ -x + x^2, -5 + x, -x^2, 3 + x^2 \right\}$

In Exercises 40–45, extend the given linearly independent set to a basis of P_2.

40. $\{1 + x + x^2, 1\}$

41. $\{-x + x^2, x + x^2\}$

42. $\{x + x^2, 1 + x\}$

43. $\{1 + x, -1 + x^2\}$

44. $\{1 - x + x^2, 2 - x^2\}$

45. $\{-x + x^2, -5 + x\}$

4.4   Dimension

Reader's Goal for This Section

To understand the basic concept of the dimension of a vector space.

Up to this point we have freely said vectors in 2 or 3 dimensions, meaning 2- or 3-vectors. In this section we make the intuitive notion of dimension precise. We exploit this notion to classify all the subspaces of $\mathbf{R}^2$ and $\mathbf{R}^3$. To the applications-oriented mind, the material here seems pretty theoretical. However, it helps us understand the basic concepts so far as well as the important concept of *rank* discussed in Section 4.6.

The following theorem is crucial in proving that the dimension of a vector space is a well-defined number. It is due to **Steinitz**, and its proof is discussed at the end of the section.

THEOREM 12

(The Exchange Theorem)

If a vector space V is spanned by n vectors, then any subset of V with more than n vectors is linearly dependent. In other words, any linearly independent subset of V has at most n vectors.

As a consequence, we have Theorem 13.

THEOREM 13

If a vector space V has a basis with n elements, then every basis of V has n elements (Fig. 4.9).

PROOF   Let $\mathcal{B}$ be a basis with n vectors and let $\mathcal{B}'$ be another basis. If $\mathcal{B}'$ had more than n elements, it would be a linearly dependent set by Theorem 12, because $\mathcal{B}$ is a spanning set. Hence, $\mathcal{B}'$ is finite and if m is its number of elements, then $m \leq n$. By the same argument, with $\mathcal{B}$ and $\mathcal{B}'$ interchanged, we see that $n \leq m$. Therefore, $n = m$.

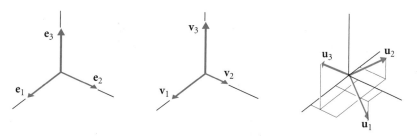

Figure 4.9   *All bases in a vector space have the same number of vectors.*

DEFINITION

(Dimension)

If a vector space V has a basis with n elements, then V is called **finite-dimensional** and n is the **dimension** of V. We write

$$\dim(V) = n$$

By Theorem 13, *the dimension is a well-defined number and does not depend on the choice of basis.* The dimension of the zero space $\{\mathbf{0}\}$ is *defined* to be zero. Hence, $\{\mathbf{0}\}$ is finite-dimensional. A vector space that has no finite basis is called **infinite-dimensional**.

A vector space that contains an infinite linearly independent set is infinite-dimensional, because if it had a finite basis it would be a finite spanning set. Hence, any set with more elements would be linearly dependent by Theorem 12.

By counting the number of elements of the standard bases, we conclude that $\mathbf{R}^n$, P_n, and M_{mn} are all finite-dimensional. Moreover,

1. $\dim(\mathbf{R}^n) = n$ (Fig. 4.10).
2. $\dim(P_n) = n + 1$.
3. $\dim(M_{mn}) = m \cdot n$.
4. P is *infinite*-dimensional.
5. $F(\mathbf{R})$ is *infinite*-dimensional.

Statements 4 and 5 hold because P and $F(\mathbf{R})$ both contain the infinite linearly independent set $\{1, x, x^2, \dots, x^n, \dots\}$.

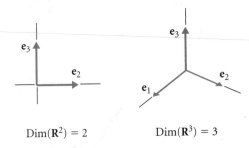

$$\text{Dim}(\mathbf{R}^2) = 2 \qquad \text{Dim}(\mathbf{R}^3) = 3$$

Figure 4.10   The dimensions of $\mathbf{R}^2$ and $\mathbf{R}^3$.

Note that because a subspace of a vector space is itself a vector space, it makes sense to talk about the *dimension of a subspace*.

■ EXAMPLE 53   Find the dimension of the subspace $V = \{(2x + y, x, -x - 2y, x + y + z),\ x, y, z \in \mathbf{R}\}$ of $\mathbf{R}^4$.

SOLUTION   In Example 14, Section 4.1 we found a basis of V with 3 elements. Hence, $\dim(V) = 3$.

■ EXAMPLE 54   Find the dimension of the subspace $V = \text{Span}\{(1, 1, 1), (2, 1, -1), (1, 0, -2)\}$ of $\mathbf{R}^3$.

SOLUTION   Because $(2, 1, -1) - (1, 1, 1) = (1, 0, -2)$, we have $V = \text{Span}\{(1, 1, 1), (2, 1, -1)\}$ by Theorem 6, Section 4.3. Hence, $\{(1, 1, 1), (2, 1, -1)\}$ spans V; because it is linearly independent (check), it is a basis. Therefore, $\dim(V) = 2$. Figure 4.11 shows the possible dimensions of the span of two vectors.

The next theorem tells us that any linearly independent set can have no more and a spanning set no fewer elements than the dimension.

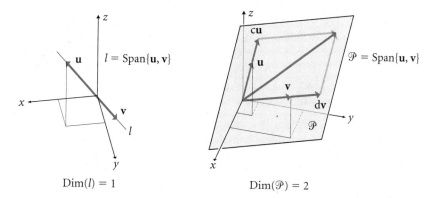

Figure 4.11   Dimensions of spans.

THEOREM 14

Let V be an n-dimensional vector space and let S be a set with m elements.

1. If S is linearly independent, then $m \le n$.
2. If S spans V, then $m \ge n$.

PROOF   Let $\mathcal{B}$ be a basis of V. Because $\dim(V) = n$, $\mathcal{B}$ is a spanning linearly independent set with n elements. Hence, by Theorem 12 any linearly independent set should have no more than n elements. This proves Part 1. By the same token, if a spanning set has less than n elements, then $\mathcal{B}$ would be linearly dependent, which is a contradiction. Hence, a spanning set should have n or more elements. This proves Part 2. ■

■ EXAMPLE 55   Let S be a set of 10 vectors in $\mathbf{R}^k$. What can we say about k if S (a) is linearly independent? (b) spans $\mathbf{R}^k$? (c) is a basis of $\mathbf{R}^k$?

SOLUTION   According to Theorem 14 we have (a) $k \ge 10$,(b) $k \le 10$, (c) $k = 10$. ■

The next theorem claims that a set with as many elements as the dimension of the vector space that either spans *or* is linearly independent is a basis. So we need not check both requirements. This cuts down on labor (it is often easy to show that a set is linearly independent). The cost is that we need to know the dimension of the space beforehand.

THEOREM 15

Let V be an n-dimensional vector space and let S be a set with n elements.

1. If S is linearly independent, then S is a basis.
2. If S spans V, then S is a basis.

PROOF

1. Let $S = \{\mathbf{v}_1, \ldots, \mathbf{v}_n\}$ be a linearly independent subset of V. If S does not span V, there is an element $\mathbf{v}$ of V that is not in Span(S). But then the set $S' = \{\mathbf{v}_1, \ldots, \mathbf{v}_n, \mathbf{v}\}$ would be linearly independent by Part 2 of Theorem 9, Section 4.3. This contradicts Part 1 of Theorem 14, because S' has $n + 1 > n$ elements. Hence S spans V, as asserted, and, because it is linearly independent, it is a basis.

2. Let S be a spanning set with n elements. If S is linearly dependent, then one element is a linear combination of the rest by Theorem 7, Section 4.3. If we delete this element the resulting set and S would have the same span by Theorem 6, Section 4.3. But then V would be spanned by fewer than n elements, which contradicts Part 2 of Theorem 14. We conclude that S is linearly independent. Hence, S is a basis, because it already spans V.

■ **EXAMPLE 56**   Show that $S = \{(1, -1), (0, 1)\}$ is a basis for $\mathbf{R}^2$.

SOLUTION S is linearly independent (why?) and has exactly two elements. Since the dimension of the $\mathbf{R}^2$ is also two, S is a basis, by Theorem 15.

The next theorem can be handy in practice. It tells us that we can obtain a basis by either adding elements to a linearly independent set or deleting elements from a spanning set, in an appropriate way.

THEOREM 16

Let V be an n-dimensional vector space and let S be a set with m elements.

1. If S is linearly independent and $m < n$, then S can be enlarged to a basis.
2. If S spans V, then S contains a basis. S can be trimmed to a basis.

PROOF

1. Let $S = \{\mathbf{v}_1, \ldots, \mathbf{v}_m\}$ be a linearly independent subset of V with $m < n$. By Theorem 13 S cannot span V. So there is an element $\mathbf{v}_{m+1}$ not in the span of S. Hence, the set $S' = \{\mathbf{v}_1, \ldots, \mathbf{v}_m, \mathbf{v}_{m+1}\}$ is linearly independent, by Part 2 of Theorem 9, Section 4.3. Now we repeat this process with S' in place of S and continue until the final set has n linearly independent elements. This set would be a basis by Theorem 15 and it contains S. So, S can be enlarged to a basis.

2. By Part 2 of Theorem 14, $m \geq n$, because S spans. If $m = n$, then S is a basis by Theorem 15. If $m > n$, then S is linearly dependent by Theorem 14. Let S' be the set obtained from deleting one element from S that is a linear combination of the remaining elements. Then S' has $m - 1$ elements and still spans V by Theorem 6, Section 4.3. We repeat this process with S'. We continue deleting elements from S in such a way that the remaining elements still span V. This process ends when we have a spanning subset with the smallest possible number of elements, i.e., exactly n elements, by Theorem 14. This final set is a basis by Theorem 15.

■ EXAMPLE 57   Extend the linearly independent set $S = \{-1 + x^2, 3 - 2x\}$ to a basis of P_3.

SOLUTION   First we enlarge S to S' that spans P_3 by adding the standard basis of P_3.

$$S' = \{-1 + x^2, 3 - 2x, 1, x, x^2, x^3\}$$

S' is linearly dependent by Theorem 7, Section 4.3, because the standard basis spans P_3. Hence, by the same theorem one element is a linear combination of the preceding. Because S is linearly independent we start with 1. 1 is not a linear combination in S. (Why?) However, both x and x^2 are linear combinations of $-1 + x^2$, $3 - 2x$, 1, so we drop them from S'. Lastly x^3 is not a linear combination of $-1 + x^2$, $3 - 2x$, 1 and we keep it. So $\{-1 + x^2, 3 - 2x, 1, x^3\}$ is linearly independent and still spans P_3. Hence, it is a basis that contains S.

The following theorem says that the dimension of a subspace cannot exceed the dimension of the vector space.

THEOREM 17

Let W be a subspace of an n-dimensional vector space V. Then

1. $\dim(W) \leq n$;
2. $\dim(W) = n$ if and only if $W = V$.

PROOF

1. Because any basis of W is in V and is linearly independent, it has at most n elements, by Theorem 12. Hence, $\dim(W) \leq n$.

2. Let $\dim(W) = n$. Then any basis $\mathcal{B}$ of W has n linearly independent elements. Hence, $\mathcal{B}$ is a basis of V, by Theorem 15. So, $V = \text{Span}(\mathcal{B}) = W$. The converse is trivial, of course.

The Subspaces of $\mathbf{R}^2$ and $\mathbf{R}^3$

Theorem 17 helps us classify all the subspaces of $\mathbf{R}^2$ and $\mathbf{R}^3$.

■ EXAMPLE 58   Find all the subspaces of $\mathbf{R}^2$.

SOLUTION   By Theorem 17 a subspace can be 0-, 1- or 2-dimensional. The zero subspace is the only 0-dimensional subspace. $\mathbf{R}^2$ is the only 2-dimensional subspace by Theorem 17. So we need to identify only the 1-dimensional subspaces. Let V be an 1-dimensional subspace and let $\{\mathbf{w}\}$ be a basis of V. Then $V = \text{Span}(\{\mathbf{w}\}) = \{r\mathbf{w}, r \in \mathbf{R}\}$. Hence, V is the set of all scalar multiples of $\mathbf{w}$, which is a line through the origin in the direction of $\mathbf{w}$. Conversely, any line through the origin is a subspace, because it is the span of any nonzero vector on the line. We have shown that the 1-dimensional subspaces comprise the lines through the origin. To summarize, we have the following.

The subspaces of $\mathbf{R}^2$ are

- 0-dimensional subspaces: $\{\mathbf{0}\}$;
- 1-dimensional subspaces: All lines through the origin;
- 2-dimensional subspaces: $\mathbf{R}^2$.

■ EXAMPLE 59   Find all the subspaces of $\mathbf{R}^3$.

SOLUTION   We have the following classification. The details are left as exercises.
The subspaces of $\mathbf{R}^3$ are (Fig. 4.12)

- 0-dimensional subspaces: $\{\mathbf{0}\}$;
- 1-dimensional subspaces: All lines through the origin;
- 2-dimensional subspaces: All planes through the origin;
- 3-dimensional subspaces: $\mathbf{R}^3$.

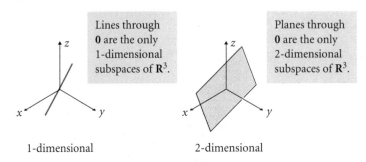

1-dimensional   2-dimensional

Figure 4.12   The 1- and 2-dimensional subspaces of $\mathbf{R}^3$.

Two Proofs of the Exchange Theorem

PROOF 1   Let $S = \{\mathbf{v}_1,\ldots,\mathbf{v}_n\}$ be the spanning set and let T be any set with at least $n + 1$ vectors. We shall prove that T is linearly dependent. Let T' be a subset of T consisting of exactly $n + 1$ vectors, say, $T' = \{\mathbf{u}_1,\ldots,\mathbf{u}_{n+1}\}$. If we prove that T' is linearly dependent, we are done by Part 3 of Theorem 8, Section 4.3.

Because S spans V, all the elements of T' can be written as linear combinations in S. Hence, there exist scalars a_{ij} such that

$$\mathbf{u}_1 = a_{11}\mathbf{v}_1 + \cdots + a_{1n}\mathbf{v}_n$$

$$\mathbf{u}_2 = a_{21}\mathbf{v}_1 + \cdots + a_{2n}\mathbf{v}_n$$

$$\vdots$$

$$\mathbf{u}_{n+1} = a_{n+1,1}\mathbf{v}_1 + \cdots + a_{n+1,n}\mathbf{v}_n$$

Consider the $n \times (n+1)$ matrix A' with (i, j) entry a_{ji} ($1 \le i \le n, 1 \le j \le n+1$). The homogeneous system $[A' : \mathbf{0}]$ has nontrivial solutions, because it has more unknowns than equations ($n+1$ versus n), by Theorem 5, Section 1.2. Therefore, there exists a nonzero $(n+1)$-vector $\mathbf{c} = (c_1, \ldots, c_{n+1})$ that solves the system. Now consider the sum

$$c_1 \mathbf{u}_1 + \cdots + c_{n+1} \mathbf{u}_{n+1} = c_1(a_{11}\mathbf{v}_1 + \cdots + a_{1n}\mathbf{v}_n) + c_2(a_{21}\mathbf{v}_1 + \cdots + a_{2n}\mathbf{v}_n)$$
$$+ \cdots + c_{n+1}(a_{n+1,1}\mathbf{v}_1 + \cdots + a_{n+1,n}\mathbf{v}_n)$$

If we multiply out and collect common terms, we see that the coefficient of each $\mathbf{v}_i$ is $c_1 a_{1i} + \cdots + c_{n+1} a_{n+1,i}$, which is zero, because $\mathbf{c}$ is a solution of the system. So, $c_1 \mathbf{u}_1 + \cdots + c_{n+1} \mathbf{u}_{n+1} = \mathbf{0}$. This is a nontrivial linear combination of the zero vector. Hence T' is linearly dependent, as asserted.

PROOF 2   Let $S = \{\mathbf{v}_1, \ldots, \mathbf{v}_n\}$ be the spanning set and let $T = \{\mathbf{u}_1, \ldots, \mathbf{u}_m\}$ be a linearly independent set. It suffices to prove that $m \le n$. The set

$$S' = \{\mathbf{u}_m, \mathbf{v}_1, \ldots, \mathbf{v}_n\}$$

is linearly dependent, by Theorem 7, Section 4.3, since S is a spanning set. Hence, by the same theorem one of the $\mathbf{v}$s, say, $\mathbf{v}_i$, is a linear combination of the preceding vectors. Thus the set S'' formed from S' by deleting $\mathbf{v}_i$ is still a spanning set.

$$S'' = \{\mathbf{u}_m, \mathbf{v}_1, \ldots, \mathbf{v}_{i-1}, \mathbf{v}_{i+1}, \ldots, \mathbf{v}_n\}$$

We now add $\mathbf{u}_{m-1}$ to S'' to get S''',

$$S''' = \{\mathbf{u}_{m-1}, \mathbf{u}_m, \mathbf{v}_1, \ldots, \mathbf{v}_{i-1}, \mathbf{v}_{i+1}, \ldots, \mathbf{v}_n\}$$

and use the same argument to show that S''' is linearly independent and spanning. None of the $\mathbf{u}$s is a linear combination of the preceding vectors, because the $\mathbf{u}$s are linearly independent, so one of the $\mathbf{v}$s can be deleted, as before. We continue this finite process and see that the $\mathbf{u}$s will be exhausted before the $\mathbf{v}$s. Otherwise, the remaining $\mathbf{u}$s would be linear combinations of the $\mathbf{u}$s already included in the set. This is impossible, because the $\mathbf{u}$s are linearly independent. Therefore, $m \le n$, as asserted.

REMARK   The name *exchange theorem* comes from Proof 2, where the spanning vectors were exchanged for linearly independent ones.

Exercises 4.4

In Exercises 1–8 find the dimension of V.

1. $V = \left\{ \begin{bmatrix} a \\ 0 \end{bmatrix}, a \in \mathbf{R} \right\}$

2. $V = \left\{ \begin{bmatrix} a - b \\ 2a + b \end{bmatrix}, a, b \in \mathbf{R} \right\}$

3. $V = \left\{ \begin{bmatrix} a \\ 0 \\ -2a \end{bmatrix}, a \in \mathbf{R} \right\}$

4. $V = \left\{ \begin{bmatrix} a - c \\ b + c \\ 5c \end{bmatrix}, a, b, c \in \mathbf{R} \right\}$

5. V is the set of all 3-vectors with first component zero.

6. V is the set of all 4-vectors with first and last components zero.

7. V is the set of all 4-vectors with first three components zero.

8. $V = \text{Span} \left\{ \begin{bmatrix} -5 \\ -1 \\ 1 \end{bmatrix}, \begin{bmatrix} 4 \\ 1 \\ 7 \end{bmatrix}, \begin{bmatrix} -1 \\ 0 \\ 8 \end{bmatrix} \right\}$

In Exercises 9–11 determine the dimension of the span of the given sets in M_{22}.

9. $\left\{ \begin{bmatrix} 1 & 2 \\ 1 & 2 \end{bmatrix}, \begin{bmatrix} 3 & 4 \\ 3 & 4 \end{bmatrix}, \begin{bmatrix} 5 & 6 \\ 5 & 6 \end{bmatrix} \right\}$

10. $\left\{ \begin{bmatrix} 1 & 0 \\ 0 & 1 \end{bmatrix}, \begin{bmatrix} 0 & 1 \\ 1 & 0 \end{bmatrix}, \begin{bmatrix} 1 & 1 \\ 1 & 1 \end{bmatrix} \right\}$

11. $\left\{ \begin{bmatrix} 1 & 2 \\ 3 & 4 \end{bmatrix}, \begin{bmatrix} 2 & 2 \\ 3 & 4 \end{bmatrix}, \begin{bmatrix} 3 & 3 \\ 3 & 4 \end{bmatrix} \right\}$

12. Find the dimension of the set V of all matrices of the form $\begin{bmatrix} a & b \\ c & -a \end{bmatrix}$.

In Exercises 13–18 find the dimension of $V \subseteq P_2$.

13. $V = \text{Span} \left\{ 1 + x + x^2, 1, -1 - x^2, x^2 \right\}$

14. $V = \text{Span} \left\{ 2 + x + x^2, x^2, 1 - x - x^2, 1 \right\}$

15. $V = \text{Span} \left\{ x + x^2, 1 + x, -1 + x^2 \right\}$

16. $V = \text{Span} \left\{ x^2, 1 + x, -1 + x^2 \right\}$

17. $V = \text{Span} \left\{ 1 - x - 5x^2, 7 + x + 4x^2, 8 - x^2 \right\}$

18. $V = \text{Span} \left\{ -x + x^2, -5 + x, -x^2, 3 + x^2 \right\}$

19. True or false?

  a. $\mathbf{R}^{10}$ has a basis with 11 elements.

  b. $\mathbf{R}^{10}$ has a basis with only 10 elements.

  c. $\mathbf{R}^{10}$ has only 10 elements.

d. $\mathbf{R}^{10}$ has a 9-dimensional subspace.

e. $\mathbf{R}^{10}$ has only one 9-dimensional subspace.

f. $\mathbf{R}^{10}$ is the only 10-dimensional subspace of $\mathbf{R}^{10}$.

g. $\mathbf{R}^2$ is a 2-dimensional subspace of $\mathbf{R}^{10}$.

20. True or false?

  a. A nonzero subspace V of $\mathbf{R}^{10}$ may have two distinct bases.

  b. A nonzero subspace V of $\mathbf{R}^{10}$ may have two bases with different number of elements.

  c. A nonzero subspace V of $\mathbf{R}^{10}$ may have a basis with 10 elements.

  d. A nonzero subspace V of $\mathbf{R}^{10}$ may have only 10 elements.

  e. A nonzero subspace V of $\mathbf{R}^{10}$ may have a basis with 11 elements.

  f. A nonzero subspace V of $\mathbf{R}^{10}$ may have a basis with 9 elements.

  g. The dimension of a subspace V of $\mathbf{R}^{10}$ is zero if and only if $V = \{\mathbf{0}\}$.

21. Find the dimension of Span $\{ e^x, e^{2x}, 2e^x \}$ in $F(R)$.

22. Find the dimension of Span $\{ \cos(x) \sin(x), \sin(2x) \}$ in $F(R)$.

23. Find the dimension of Span $\{ \cos^2(x), \sin^2(x), 1 \}$ in $F(R)$.

24. Show that for any vectors $\mathbf{u}, \mathbf{v}$ from a vector space V
$$\dim(\text{Span} \{\mathbf{u}, \mathbf{v}\}) = \dim(\text{Span} \{\mathbf{u} + \mathbf{v}, \mathbf{u} - \mathbf{v}\})$$

25. Describe geometrically all subspaces of $\mathbf{R}^4$.

4.5   Coordinate Vectors and Change of Basis

Reader's Goals for This Section

1. To compute the coordinate vector with respect to a given basis.
2. To find the transition matrix and use it to change from one basis to another.

Many physics and engineering problems can be greatly simplified by choosing the right coordinate system. Likewise, vector space problems can be simplified by choosing the right basis. First we study the coordinates of a general vector with respect to a fixed basis. Then we show how to change coordinates from an old basis to a new one. Because coordinates are numbers, many of the calculations can be performed by computer.

Coordinate Vectors

DEFINITION

Let V be a finite-dimensional vector space with basis $\mathcal{B} = \{v_1, \dots, v_n\}$. By Theorem 11, Section 4.3, for each $v \in V$, there exist unique scalars $c_1, \dots, c_n$ such that

$$v = c_1 v_1 + \cdots + c_n v_n$$

The vector with components the coefficients of v, written as $[v]_{\mathcal{B}}$, is called the **coordinate vector of v with respect to $\mathcal{B}$.**

$$[v]_{\mathcal{B}} = \begin{bmatrix} c_1 \\ \vdots \\ c_n \end{bmatrix}$$

$[v]_{\mathcal{B}}$ changes as the basis $\mathcal{B}$ changes (Fig. 4.13). Also, $[v]_{\mathcal{B}}$ depends on the *order* of elements of $\mathcal{B}$. We keep this order fixed by always using an **ordered basis.**

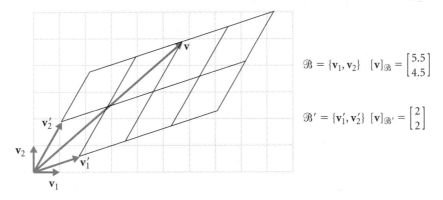

$$\mathcal{B} = \{v_1, v_2\} \quad [v]_{\mathcal{B}} = \begin{bmatrix} 5.5 \\ 4.5 \end{bmatrix}$$

$$\mathcal{B}' = \{v_1', v_2'\} \quad [v]_{\mathcal{B}'} = \begin{bmatrix} 2 \\ 2 \end{bmatrix}$$

Figure 4.13   Coordinates with respect to different bases.

REMARK   If $a = (a_1, \dots, a_n)$ is an n-vector and $\mathcal{B} = \{e_1, \dots, e_n\}$ is the standard basis of $\mathbf{R}^n$, then

$$[a]_{\mathcal{B}} = \begin{bmatrix} a_1 \\ \vdots \\ a_n \end{bmatrix} = a$$

because $a = a_1 e_1 + \cdots + a_n e_n$.

■ **EXAMPLE 60**   Consider the basis $\mathcal{B} = \{(1, 0, -1), (-1, 1, 0), (1, 1, 1)\}$ of $\mathbf{R}^3$ and the vector $v = (2, -3, 4)$.

(a) Find $[v]_{\mathcal{B}}$.

(b) Find the vector **w** if $[\mathbf{w}]_\mathcal{B} = \begin{bmatrix} 6 \\ -3 \\ 2 \end{bmatrix}$.

SOLUTION   (a) $[\mathbf{v}]_\mathcal{B}$ has as components the scalars c_1, c_2, c_3 such that

$$(2, -3, 4) = c_1(1, 0, -1) + c_2(-1, 1, 0) + c_3(1, 1, 1)$$

which implies $c_1 = -3, c_2 = -4, c_3 = 1$. Hence,

$$[\mathbf{v}]_\mathcal{B} = \begin{bmatrix} -3 \\ -4 \\ 1 \end{bmatrix}$$

(b) Because the components of $[\mathbf{w}]_\mathcal{B}$ are $6, -3, 2$, **w** is given by

$$\mathbf{w} = 6(1, 0, -1) - 3(-1, 1, 0) + 2(1, 1, 1) = (11, -1, -4)$$

■ EXAMPLE 61   Find the coordinate vector of $\mathbf{v} = (a, b, c)$ in $\mathbf{R}^3$ with respect to $\mathcal{B} = \{\mathbf{v}_1, \mathbf{v}_2, \mathbf{v}_3\}$, where

$$\mathbf{v}_1 = \mathbf{e}_3, \qquad \mathbf{v}_2 = \mathbf{e}_1, \qquad \mathbf{v}_3 = \mathbf{e}_2$$

SOLUTION   The components of $[\mathbf{v}]_{\mathcal{B}'}$ are scalars c_1, c_2, c_3 such that

$$(a, b, c) = c_1\mathbf{v}_1 + c_2\mathbf{v}_2 + c_3\mathbf{v}_3 = c_1\mathbf{e}_3 + c_2\mathbf{e}_1 + c_3\mathbf{e}_2 = (c_2, c_3, c_1)$$

Therefore, $c_1 = c, c_2 = a$, and $c_3 = c$. So,

$$[\mathbf{v}]_{\mathcal{B}'} = \begin{bmatrix} c \\ a \\ b \end{bmatrix}$$

■ EXAMPLE 62   Find the coordinate vector of $p = 1 + 2x + 3x^2$ in P_2 with respect to each of the following.

(a) The (standard) basis $\mathcal{B} = \{\mathbf{v}_1, \mathbf{v}_2, \mathbf{v}_3\}$, where

$$\mathbf{v}_1 = 1, \qquad \mathbf{v}_2 = x, \qquad \mathbf{v}_3 = x^2$$

(b) The basis $\mathcal{B}' = \{\mathbf{v}_1', \mathbf{v}_2', \mathbf{v}_3'\}$ where

$$\mathbf{v}_1' = 1 + x, \qquad \mathbf{v}_2' = 1 - x^2, \qquad \mathbf{v}_3' = 1 + x + x^2$$

SOLUTION   (a) Because $p = 1 \cdot 1 + 2 \cdot x + 3 \cdot x^2$, we have

$$[p]_\mathcal{B} = \begin{bmatrix} 1 \\ 2 \\ 3 \end{bmatrix}$$

(b) The components of $[p]_{\mathcal{B}'}$ are scalars c_1, c_2, c_3 such that

$$p = c_1\mathbf{v}_1' + c_2\mathbf{v}_2' + c_3\mathbf{v}_3' = c_1(1 + x) + c_2(1 - x^2) + c_3(1 + x + x^2)$$

$$\Rightarrow 1 + 2x + 3x^2 = (c_1 + c_2 + c_3) + (c_1 + c_3)x + (-c_2 + c_3)x^2$$

So, $1 = c_1 + c_2 + c_3$, $2 = c_1 + c_3$, $3 = -c_2 + c_3$ or, $c_1 = 0$, $c_2 = -1$, $c_3 = 2$. Therefore,

$$[p]_{\mathcal{B}'} = \begin{bmatrix} 0 \\ -1 \\ 2 \end{bmatrix}$$

■ EXAMPLE 63   Find the coordinate vector of

$$A = \begin{bmatrix} 2 & -3 \\ 1 & 4 \end{bmatrix}$$

in M_{22} with respect to each of the following.

(a) The (standard) basis $\mathcal{B} = \{\mathbf{v}_1, \mathbf{v}_2, \mathbf{v}_3, \mathbf{v}_4\}$, where

$$\mathbf{v}_1 = E_{11}, \qquad \mathbf{v}_2 = E_{12}, \qquad \mathbf{v}_3 = E_{21}, \qquad \mathbf{v}_4 = E_{22}$$

(b) The basis $\mathcal{B}' = \{\mathbf{v}_1', \mathbf{v}_2', \mathbf{v}_3', \mathbf{v}_4'\}$, where

$$\mathbf{v}_1' = -E_{21}, \qquad \mathbf{v}_2' = E_{22}, \qquad \mathbf{v}_3' = -E_{12}, \qquad \mathbf{v}_4' = E_{11}$$

SOLUTION   (a) The components of $[A]_{\mathcal{B}}$ are scalars c_1, c_2, c_3, c_4 such that

$$A = \begin{bmatrix} 2 & -3 \\ 1 & 4 \end{bmatrix} = c_1E_{11} + c_2E_{12} + c_3E_{21} + c_4E_{22} = \begin{bmatrix} c_1 & c_2 \\ c_3 & c_4 \end{bmatrix}$$

Therefore, $c_1 = 2$, $c_2 = -3$, $c_3 = 1$, and $c_4 = 4$. Hence,

$$[A]_{\mathcal{B}} = \begin{bmatrix} 2 \\ -3 \\ 1 \\ 4 \end{bmatrix}$$

(b) The components of $[A]_{\mathcal{B}'}$ are scalars c_1, c_2, c_3, c_4 such that

$$A = \begin{bmatrix} 2 & -3 \\ 1 & 4 \end{bmatrix} = -c_1E_{21} + c_2E_{22} - c_3E_{12} + c_4E_{11} = \begin{bmatrix} c_4 & -c_3 \\ -c_1 & c_2 \end{bmatrix}$$

Therefore, $c_1 = -1$, $c_2 = 4$, $c_3 = 3$, and $c_4 = 2$. Hence,

$$[A]_{\mathcal{B}'} = \begin{bmatrix} -1 \\ 4 \\ 3 \\ 2 \end{bmatrix}$$

REMARK   If $\mathcal{B} = \{\mathbf{v}_1, \ldots, \mathbf{v}_n\}$ is a basis of a finite-dimensional vector space V, then

$$[\mathbf{v}_i]_{\mathcal{B}} = \mathbf{e}_i, \qquad i = 1, 2, \ldots, n$$

THEOREM 18

Let $\mathcal{B} = \{\mathbf{v}_1, \ldots, \mathbf{v}_n\}$ be a basis of a finite-dimensional vector space V. Let $\mathbf{u}, \mathbf{u}_1, \ldots, \mathbf{u}_m$ be vectors in V. Then $\mathbf{u}$ is a linear combination of $\mathbf{u}_1, \ldots, \mathbf{u}_m$ in V if and only if $[\mathbf{u}]_{\mathcal{B}}$ is a linear combination of $[\mathbf{u}_1]_{\mathcal{B}}, \ldots, [\mathbf{u}_m]_{\mathcal{B}}$ in $\mathbf{R}^n$. Furthermore, for scalars, $c_1, \ldots, c_m$

$$\mathbf{u} = c_1 \mathbf{u}_1 + \cdots + c_m \mathbf{u}_m \tag{4.3}$$

if and only if

$$[\mathbf{u}]_{\mathcal{B}} = c_1 [\mathbf{u}_1]_{\mathcal{B}} + \cdots + c_m [\mathbf{u}_m]_{\mathcal{B}} \tag{4.4}$$

PROOF Let

$$[\mathbf{u}]_{\mathcal{B}} = \begin{bmatrix} u_1 \\ \vdots \\ u_n \end{bmatrix} \quad \text{and} \quad [\mathbf{u}_i]_{\mathcal{B}} = \begin{bmatrix} u_{i1} \\ \vdots \\ u_{in} \end{bmatrix}, \qquad i = 1, 2, \ldots, m$$

Assuming (4.3) we have

$$\mathbf{u} = c_1(u_{11}\mathbf{v}_1 + \cdots + u_{1n}\mathbf{v}_n) + \cdots + c_m(u_{m1}\mathbf{v}_1 + \cdots + u_{mn}\mathbf{v}_n)$$

$$= (c_1 u_{11} + \cdots + c_m u_{m1})\mathbf{v}_1 + \cdots + (c_1 u_{1n} + \cdots + c_m u_{mn})\mathbf{v}_n$$

Hence,

$$[\mathbf{u}]_{\mathcal{B}} = \begin{bmatrix} c_1 u_{11} + \cdots + c_m u_{m1} \\ \vdots \\ c_1 u_{1n} + \cdots + c_m u_{mn} \end{bmatrix} = c_1 \begin{bmatrix} u_{11} \\ \vdots \\ u_{1n} \end{bmatrix} + \cdots + c_m \begin{bmatrix} u_{m1} \\ \vdots \\ u_{mn} \end{bmatrix}$$

$$= c_1 [\mathbf{u}_1]_{\mathcal{B}} + \cdots + c_m [\mathbf{u}_m]_{\mathcal{B}}$$

which proves (4.4). All of the preceding steps can be reversed so the proof is complete.

Taking $\mathbf{u} = \mathbf{0}$, Theorem 18 has the following useful corollary:

THEOREM 19

Let $\mathcal{B}$ be a basis of an n-dimensional vector space V. Then $\{\mathbf{u}_1, \ldots, \mathbf{u}_m\}$ is linearly independent in V if and only if $\{[\mathbf{u}_1]_{\mathcal{B}}, \ldots, [\mathbf{u}_m]_{\mathcal{B}}\}$ is linearly independent in $\mathbf{R}^n$.

■ **EXAMPLE 64** Show that $p_1(x) = 1 - x^2$, $p_2(x) = -1 + x$, $p_3(x) = 1 + x + x^2$ are linearly independent in P_2.

SOLUTION By Theorem 19 it suffices to show independence of the coordinate vectors with respect to the standard basis $\mathcal{B}$,

$$[p_1]_{\mathcal{B}} = \begin{bmatrix} 1 \\ 0 \\ -1 \end{bmatrix}, \qquad [p_2]_{\mathcal{B}} = \begin{bmatrix} -1 \\ 1 \\ 0 \end{bmatrix}, \qquad [p_3]_{\mathcal{B}} = \begin{bmatrix} 1 \\ 1 \\ 1 \end{bmatrix}$$

which is easily verified.

Change of Basis

Let $\mathbf{v}$ be a vector in a finite-dimensional vector space V and let $\mathcal{B} = \{\mathbf{v}_1, \ldots, \mathbf{v}_n\}$ and $\mathcal{B}' = \{\mathbf{v}_1', \ldots, \mathbf{v}_n'\}$ be two bases. In this paragraph we find a relation between $[\mathbf{v}]_{\mathcal{B}}$ and $[\mathbf{v}]_{\mathcal{B}'}$.

Because $\mathcal{B}'$ is a basis, the elements of $\mathcal{B}$ are linear combinations of elements in $\mathcal{B}'$. So, there are scalars $a_{11}, a_{12}, \ldots, a_{nn}$ such that

$$\mathbf{v}_i = a_{1i}\mathbf{v}_1' + \cdots + a_{ni}\mathbf{v}_n' \qquad i = 1, 2, \ldots, n \qquad (4.5)$$

Let P be the matrix with (i, j) entry a_{ij}.

Because $\mathcal{B}$ spans V, there are scalars $c_1, \ldots, c_n$ such that $\mathbf{v} = c_1\mathbf{v}_1 + \cdots + c_n\mathbf{v}_n$. Hence,

$$[\mathbf{v}]_{\mathcal{B}} = \begin{bmatrix} c_1 \\ \vdots \\ c_n \end{bmatrix}$$

and by Theorem 18 we have

$$[\mathbf{v}]_{\mathcal{B}'} = c_1 [\mathbf{v}_1]_{\mathcal{B}'} + \cdots + c_n [\mathbf{v}_n]_{\mathcal{B}'}$$

By Equation (4.5) this can be rewritten as

$$[\mathbf{v}]_{\mathcal{B}'} = c_1 \begin{bmatrix} a_{11} \\ \vdots \\ a_{n1} \end{bmatrix} + c_2 \begin{bmatrix} a_{12} \\ \vdots \\ a_{n2} \end{bmatrix} + \cdots + c_n \begin{bmatrix} a_{1n} \\ \vdots \\ a_{nn} \end{bmatrix}$$

$$= P \begin{bmatrix} c_1 \\ \vdots \\ c_n \end{bmatrix} = P [\mathbf{v}]_{\mathcal{B}}$$

Therefore,

$$[\mathbf{v}]_{\mathcal{B}'} = P [\mathbf{v}]_{\mathcal{B}}$$

Hence $[\mathbf{v}]_{\mathcal{B}'}$ is the product of the matrix P with columns the coordinate vectors of the "old" basis $\mathcal{B}$ with respect to the "new" basis $\mathcal{B}'$ and $[\mathbf{v}]_{\mathcal{B}}$.

In addition, we can prove that the matrix P is *invertible* by showing that the system

$$P\mathbf{x} = \mathbf{b}$$

has a solution for every vector $\mathbf{b}$ in $\mathbf{R}^n$. Indeed, let

$$\mathbf{b} = \begin{bmatrix} b_1 \\ \vdots \\ b_n \end{bmatrix}$$

Consider the vector $\mathbf{v} = b_1\mathbf{v}_1' + \cdots + b_n\mathbf{v}_n'$. Then

$$\mathbf{b} = [\mathbf{v}]_{\mathcal{B}'} = P [\mathbf{v}]_{\mathcal{B}}$$

So we may take $\mathbf{x} = [\mathbf{v}]_{\mathcal{B}}$ to be the solution of $P\mathbf{x} = \mathbf{b}$ for the given $\mathbf{b}$. This completes the proof that P is invertible.

Lastly, we can show that P is the *only* matrix with the property that $[\mathbf{v}]_{\mathcal{B}'} = P[\mathbf{v}]_{\mathcal{B}}$. For if P' is another one, then $[\mathbf{v}]_{\mathcal{B}'} = P'[\mathbf{v}]_{\mathcal{B}}$. By letting $\mathbf{v} = \mathbf{v}_i$ we have

$$[\mathbf{v}_i]_{\mathcal{B}'} = P[\mathbf{v}_i]_{\mathcal{B}} = P\mathbf{e}_i \quad \text{and} \quad [\mathbf{v}_i]_{\mathcal{B}'} = P'[\mathbf{v}_i]_{\mathcal{B}} = P'\mathbf{e}_i$$

because $[\mathbf{v}_i]_{\mathcal{B}} = \mathbf{e}_i$ as vectors. Hence $P\mathbf{e}_i = P'\mathbf{e}_i$. Therefore, the ith columns of P and P' are equal for each $i = 1, 2, \ldots, n$. Thus, $P = P'$.

We have completely answered the question about the relation between $[\mathbf{v}]_{\mathcal{B}}$ and $[\mathbf{v}]_{\mathcal{B}'}$. Let us summarize and practice.

THEOREM 20 — (Change of Basis)

Let $\mathcal{B} = \{\mathbf{v}_1, \ldots, \mathbf{v}_n\}$ and $\mathcal{B}' = \{\mathbf{v}'_1, \ldots, \mathbf{v}'_n\}$ be two bases of a finite-dimensional vector space V. Let P be the $n \times n$ matrix with columns $[\mathbf{v}_1]_{\mathcal{B}'}, \ldots, [\mathbf{v}_n]_{\mathcal{B}'}$.

$$P = [[\mathbf{v}_1]_{\mathcal{B}'} \, [\mathbf{v}_2]_{\mathcal{B}'} \cdots [\mathbf{v}_n]_{\mathcal{B}'}]$$

Then P is invertible, and it is the only matrix such that for all $\mathbf{v} \in V$,

$$[\mathbf{v}]_{\mathcal{B}'} = P[\mathbf{v}]_{\mathcal{B}}$$

DEFINITION — (Transition Matrix)

The matrix P of Theorem 20 is called the **transition matrix** (or **change-of-basis matrix**) from $\mathcal{B}$ to $\mathcal{B}'$ (Fig. 4.14).

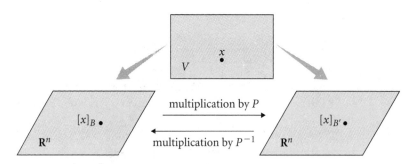

multiplication by P

multiplication by P^{-1}

Figure 4.14   The action of the transition matrix and its inverse.

COROLLARY 21

If P is the transition matrix from $\mathcal{B}$ to $\mathcal{B}'$, then P^{-1} is transition matrix from $\mathcal{B}'$ to $\mathcal{B}$ (Fig. 4.14).

PROOF   By Theorem 20 P^{-1} exists and $[\mathbf{v}]_{\mathcal{B}'} = P\,[\mathbf{v}]_{\mathcal{B}}$ for all $\mathbf{v} \in V$. Hence,

$$[\mathbf{v}]_{\mathcal{B}} = P^{-1}\,[\mathbf{v}]_{\mathcal{B}'} \quad \text{for all } \mathbf{v} \in V$$

Therefore, again, Theorem 20 implies that P^{-1} is the (only) transition matrix from $\mathcal{B}'$ to $\mathcal{B}$.

■ EXAMPLE 65   Let $\mathcal{B}$ be the standard basis of $\mathbf{R}^2$ and let $\mathcal{B}'$ be the basis $\mathcal{B}' = \{(1, 1), (-1, 1)\}$.

(a) Compute the transition matrix P from $\mathcal{B}$ to $\mathcal{B}'$.
(b) Compute the transition matrix from $\mathcal{B}'$ to $\mathcal{B}$.
(c) Verify the relation $[\mathbf{v}]_{\mathcal{B}'} = P\,[\mathbf{v}]_{\mathcal{B}}$ for $\mathbf{v} = (4, -2)$.

SOLUTION   (a) P is the matrix with columns $[\mathbf{e}_1]_{\mathcal{B}'}, [\mathbf{e}_2]_{\mathcal{B}'}$. For $[\mathbf{e}_1]_{\mathcal{B}'}$ we need scalars c_1, c_2 such that

$$\mathbf{e}_1 = \begin{bmatrix} 1 \\ 0 \end{bmatrix} = c_1 \begin{bmatrix} 1 \\ 1 \end{bmatrix} + c_2 \begin{bmatrix} -1 \\ 1 \end{bmatrix}$$

The solution of the resulting system is $c_1 = \frac{1}{2}, c_2 = -\frac{1}{2}$. Hence, $[\mathbf{e}_1]_{\mathcal{B}'} = \begin{bmatrix} \frac{1}{2} \\ -\frac{1}{2} \end{bmatrix}$.

Likewise, for $[\mathbf{e}_2]_{\mathcal{B}'}$ we need scalars c_1, c_2 such that

$$\mathbf{e}_2 = \begin{bmatrix} 0 \\ 1 \end{bmatrix} = c_1 \begin{bmatrix} 1 \\ 1 \end{bmatrix} + c_2 \begin{bmatrix} -1 \\ 1 \end{bmatrix}$$

Solving the resulting system yields $c_1 = \frac{1}{2}, c_2 = \frac{1}{2}$. Hence, $[\mathbf{e}_2]_{\mathcal{B}'} = \begin{bmatrix} \frac{1}{2} \\ \frac{1}{2} \end{bmatrix}$. Therefore, the transition matrix is

$$P = \begin{bmatrix} \frac{1}{2} & \frac{1}{2} \\ -\frac{1}{2} & \frac{1}{2} \end{bmatrix}$$

(b) The transition matrix from $\mathcal{B}'$ to $\mathcal{B}$ is P^{-1}, by Corollary 21. Hence,

$$P^{-1} = \begin{bmatrix} \frac{1}{2} & \frac{1}{2} \\ -\frac{1}{2} & \frac{1}{2} \end{bmatrix}^{-1} = \begin{bmatrix} 1 & -1 \\ 1 & 1 \end{bmatrix}$$

(c) The coordinate vector $[\mathbf{v}]_{\mathcal{B}'}$ can be computed in two different ways; by using P,

$$P\,[\mathbf{v}]_{\mathcal{B}} = \begin{bmatrix} \frac{1}{2} & \frac{1}{2} \\ -\frac{1}{2} & \frac{1}{2} \end{bmatrix} \begin{bmatrix} 4 \\ -2 \end{bmatrix} = \begin{bmatrix} 1 \\ -3 \end{bmatrix}$$

or directly from $\mathcal{B}'$, by finding c_1 and c_2 such that

$$\begin{bmatrix} 4 \\ -2 \end{bmatrix} = c_1 \begin{bmatrix} 1 \\ 1 \end{bmatrix} + c_2 \begin{bmatrix} -1 \\ 1 \end{bmatrix}$$

Solving the system gives us $c_1 = 1, c_2 = -3$. Hence, $[\mathbf{v}]_{\mathcal{B}'} = \begin{bmatrix} 1 \\ -3 \end{bmatrix}$ in each case.

■ EXAMPLE 66   Compute the transition matrix P from the standard basis $\mathcal{B}$ of $\mathbf{R}^4$ to the basis $\mathcal{B}' = \{\mathbf{v}_1', \mathbf{v}_2', \mathbf{v}_3', \mathbf{v}_4'\}$, where

$$\mathbf{v}_1' = \mathbf{e}_4, \quad \mathbf{v}_2' = \mathbf{e}_3, \quad \mathbf{v}_3' = \mathbf{e}_2, \quad \mathbf{v}_4' = \mathbf{e}_1$$

If $\mathbf{v} = (a, b, c, d)$ use P to find a formula for $[\mathbf{v}]_{\mathcal{B}'}$.

SOLUTION   Because

$$\mathbf{e}_1 = 0\mathbf{v}_1' + 0\mathbf{v}_2' + 0\mathbf{v}_3' + 1\mathbf{v}_4', \qquad \mathbf{e}_2 = 0\mathbf{v}_1' + 0\mathbf{v}_2' + 1\mathbf{v}_3' + 0\mathbf{v}_4'$$
$$\mathbf{e}_3 = 0\mathbf{v}_1' + 1\mathbf{v}_2' + 0\mathbf{v}_3' + 0\mathbf{v}_4', \qquad \mathbf{e}_4 = 1\mathbf{v}_1' + 0\mathbf{v}_2' + 0\mathbf{v}_3' + 0\mathbf{v}_4'$$

we have

$$[\mathbf{e}_1]_{\mathcal{B}'} = \mathbf{e}_4, \qquad [\mathbf{e}_2]_{\mathcal{B}'} = \mathbf{e}_3, \qquad [\mathbf{e}_3]_{\mathcal{B}'} = \mathbf{e}_2, \qquad [\mathbf{e}_4]_{\mathcal{B}'} = \mathbf{e}_1$$

Hence,

$$P = \begin{bmatrix} 0 & 0 & 0 & 1 \\ 0 & 0 & 1 & 0 \\ 0 & 1 & 0 & 0 \\ 1 & 0 & 0 & 0 \end{bmatrix}$$

Therefore,

$$[\mathbf{v}]_{\mathcal{B}'} = P[\mathbf{v}]_{\mathcal{B}} = \begin{bmatrix} 0 & 0 & 0 & 1 \\ 0 & 0 & 1 & 0 \\ 0 & 1 & 0 & 0 \\ 1 & 0 & 0 & 0 \end{bmatrix} \begin{bmatrix} a \\ b \\ c \\ d \end{bmatrix} = \begin{bmatrix} d \\ c \\ b \\ a \end{bmatrix}$$

■ EXAMPLE 67   Compute the transition matrix P from the standard basis $\mathcal{B}$ of $\mathbf{R}^2$ to the basis $\mathcal{B}'$ obtained by rotating $\mathcal{B}$ 45° counterclockwise about the origin. Find the new coordinates of the vector $(1, 1)$.

SOLUTION   Because $\sin 45° = \cos 45° = \frac{\sqrt{2}}{2}$, we have

$$\mathcal{B}' = \{\mathbf{e}_1', \mathbf{e}_2'\} = \left\{ \left(\frac{\sqrt{2}}{2}, \frac{\sqrt{2}}{2} \right), \left(-\frac{\sqrt{2}}{2}, \frac{\sqrt{2}}{2} \right) \right\}$$

(Fig. 4.15). Hence,

$$\mathbf{e}_1 = \frac{\sqrt{2}}{2}\mathbf{e}_1' - \frac{\sqrt{2}}{2}\mathbf{e}_2' \quad \text{and} \quad \mathbf{e}_2 = \frac{\sqrt{2}}{2}\mathbf{e}_1' + \frac{\sqrt{2}}{2}\mathbf{e}_2'$$

Thus,

$$P = \begin{bmatrix} \frac{\sqrt{2}}{2} & \frac{\sqrt{2}}{2} \\ -\frac{\sqrt{2}}{2} & \frac{\sqrt{2}}{2} \end{bmatrix}$$

The new coordinates of $(1, 1)$ are computed by

$$\begin{bmatrix} 1 \\ 1 \end{bmatrix}_{\mathcal{B}'} = \begin{bmatrix} \frac{\sqrt{2}}{2} & \frac{\sqrt{2}}{2} \\ -\frac{\sqrt{2}}{2} & \frac{\sqrt{2}}{2} \end{bmatrix} \begin{bmatrix} 1 \\ 1 \end{bmatrix} = \begin{bmatrix} \sqrt{2} \\ 0 \end{bmatrix}$$

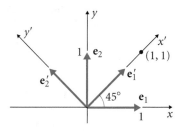

Figure 4.15   45° rotation about the origin.

Exercises 4.5

Coordinates

In Exercises 1–3 find the polynomial p, given a basis $\mathcal{B}$ of P_n and the coordinate vector $[p]_{\mathcal{B}}$.

1. $\mathcal{B} = \{1 + 2x, 5x\}$, $[p]_{\mathcal{B}} = \begin{bmatrix} -3 \\ 6 \end{bmatrix}$

2. $\mathcal{B} = \{1 + x + 2x^2, -x^2, 1 + 2x\}$, $[p]_{\mathcal{B}} = \begin{bmatrix} 4 \\ 3 \\ -2 \end{bmatrix}$

3. $\mathcal{B} = \{2 + 2x, -3 + 3x\}$, $[p]_{\mathcal{B}} = \begin{bmatrix} a \\ b \end{bmatrix}$

In Exercises 4–7 compute the coordinate vector $[p]_{\mathcal{B}}$, given a basis $\mathcal{B}$ of P_n and p.

4. $\mathcal{B} = \{1 + 2x, 1 - x\}$, $p = 4 + 17x$

5. $\mathcal{B} = \{-7 + 4x, 2 - 3x\}$, $p = 17 - 6x$

6. $\mathcal{B} = \{1 + 2x + 2x^2, 2x - x^2, -1 - 2x\}$, $p = -1 + 6x - 8x^2$

7. $\mathcal{B} = \{1 + 2x, 5x\}$, $p = (a - b) + (7a + 3b)x$

Let $\mathcal{B}$ be the following basis of M_{22}.

$$\mathcal{B} = \left\{ \begin{bmatrix} 1 & 1 \\ 0 & 0 \end{bmatrix}, \begin{bmatrix} -1 & 0 \\ 0 & 0 \end{bmatrix}, \begin{bmatrix} 2 & 0 \\ -1 & 0 \end{bmatrix}, \begin{bmatrix} 1 & 2 \\ 3 & 4 \end{bmatrix} \right\}$$

8. Find M if $[M]_{\mathcal{B}} = (4, -3, 8, 10)$.

9. Find the coordinate vector $\left[\begin{bmatrix} 4 & -1 \\ -4 & -4 \end{bmatrix} \right]_{\mathcal{B}}$.

Transition Matrix

10. Find the transition matrix from $\{v_1, v_2\}$ to $\{v_1', v_2'\}$, where

$$v_1 = (1, 1) \qquad v_2 = (1, 2)$$
$$v_1' = (1, 3) \qquad v_2' = (1, 4)$$

11. Find the transition matrix from $\{v_1, v_2\}$ to $\{v_1', v_2'\}$, where

$$v_1 = (1, 0) \qquad v_2 = (0, 1)$$
$$v_1' = (0, 1) \qquad v_2' = (1, 0)$$

12. Find the transition matrix from $\{v_1, v_2, v_3\}$ to $\{v_1', v_2', v_3'\}$, where

$$v_1 = e_3 \qquad v_2 = e_1 \qquad v_3 = e_2$$
$$v_1' = e_1 \qquad v_2' = e_2 \qquad v_3' = e_3$$

13. Find the transition matrix from $\{v_1, v_2, v_3\}$ to $\{v_1', v_2', v_3'\}$, where

$$v_1 = e_1 \qquad v_2 = e_2 \qquad v_3 = e_3$$
$$v_1' = e_3 \qquad v_2' = e_1 \qquad v_3' = e_2$$

In Exercises 14–19 find the transition matrix from the standard basis $\mathcal{B}_1 = \{1, x, x^2, x^3\}$ of P_3 to the given basis $\mathcal{B}_2$. (In each case the polynomials of $\mathcal{B}_2$ are eponymous, and they naturally arise in several areas of mathematics and physics).

14. (Chebyshev Polynomials, First Kind) $\mathcal{B}_2 = \{1, x, -1 + 2x^2, -3x + 4x^3\}$

15. (Chebyshev Polynomials, Second Kind) $\mathcal{B}_2 = \{1, 2x, -1 + 4x^2, -4x + 8x^3\}$

16. (Laguerre Polynomials) $\mathcal{B}_2 = \{1, 1 - x, 1 - 2x + (1/2)x^2, 1 - 3x + (3/2)x^2 - (1/6)x^3\}$

17. (Hermite Polynomials) $\mathcal{B}_2 = \{1, 2x, -2 + 4x^2, -12x + 8x^3\}$

18. (Legendre Polynomials) $\mathcal{B}_2 = \{1, x, -\frac{1}{2} + \frac{3}{2}x^2, -\frac{3}{2}x + \frac{5}{2}x^3\}$

19. (Euler Polynomials) $\mathcal{B}_2 = \{1, -\frac{1}{2} + x, -x + x^2, \frac{1}{4} - \frac{3}{2}x^2 + x^3\}$

20. Compute the transition matrix P from the standard basis $\mathcal{B}$ of $\mathbf{R}^2$ to the basis $\mathcal{B}'$ obtained by rotating $\mathcal{B}$ 45° clockwise about the origin. Find the new coordinates of the vector $(1, 1)$.

21. Compute the transition matrix P from the standard basis $\mathcal{B}$ of $\mathbf{R}^2$ to the basis $\mathcal{B}'$ obtained by reflecting $\mathcal{B}$ about the line $y = -x$. Find the new coordinates of the vector $(1, 1)$.

22. Compute the transition matrix P from the standard basis $\mathcal{B}$ of $\mathbf{R}^3$ to the basis $\mathcal{B}'$ obtained by rotating $\mathcal{B}$ about the z-axis counterclockwise by 90°. Find the new coordinates of the vector $(1, 1, 1)$.

4.6   Rank and Nullity

Reader's Goals for This Section

1. To know how to find bases for the null, row, and column spaces of a matrix.
2. To know how to find a basis for the span of vectors in $\mathbf{R}^n$.
3. To understand and be able to use the rank theorem.

In this section we study three important vector spaces associated with any matrix: the *null space*, the *column space*, and the *row space*. The null space was introduced in Section 2.5. The other two spaces are the spans of the columns and rows of the matrix. A by-product of this study yields two methods of computing bases for the span of a set of vectors.

Nullity

Recall from Section 2.5 that the **null space**, Null(A), of an $m \times n$ matrix A consists of all n-vectors $\mathbf{x}$ such $A\mathbf{x} = \mathbf{0}$. This is the set of all solutions of the homogeneous system $A\mathbf{x} = \mathbf{0}$.

$$\text{Null}(A) = \{\mathbf{x} \text{ in } \mathbf{R}^n \text{ such that } A\mathbf{x} = \mathbf{0}\}$$

By Theorem 18, Section 2.5 the null space of A is a *subspace* of $\mathbf{R}^n$. The dimension of Null(A) is called the **nullity** of A.

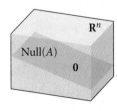

Because Null(A) is a subspace, we may ask how to find a basis for it. This is explained in the following algorithm and example.

Algorithm

(Computing a Basis for the Null Space)

To find a basis for Null(A):

1. Find the general solution vector of the system $Ax = 0$;
2. Write the solution vector as linear combination with coefficients the parameters (free variables);
3. The vectors of the linear combination form a basis for Null(A).

■ EXAMPLE 68   Find a basis for the null space of A. What is the nullity of A?

$$A = \begin{bmatrix} 1 & -1 & 2 & 3 & 0 \\ -1 & 0 & -4 & 3 & -1 \\ 2 & -1 & 6 & 0 & 1 \\ -1 & 2 & 0 & -1 & 1 \end{bmatrix}$$

SOLUTION   The augmented matrix $[A : 0]$ of the system $Ax = 0$ has reduced row echelon form

$$\begin{bmatrix} 1 & 0 & 4 & 0 & 1 & : & 0 \\ 0 & 1 & 2 & 0 & 1 & : & 0 \\ 0 & 0 & 0 & 1 & 0 & : & 0 \\ 0 & 0 & 0 & 0 & 0 & : & 0 \end{bmatrix}$$

Therefore, the original system is equivalent to

$$x_1 \quad + 4x_3 \quad + x_5 = 0$$
$$x_2 + 2x_3 \quad + x_5 = 0$$
$$x_4 \quad = 0$$

If $x_5 = r$ and $x_3 = s$ are any scalars, then the general solution is given by

$$x_1 = -4s - r$$
$$x_2 = -2s - r$$
$$x_3 = s$$
$$x_4 = 0$$
$$x_5 = r$$

Because

$$\begin{bmatrix} -4s - r \\ -2s - r \\ s \\ 0 \\ r \end{bmatrix} = r \begin{bmatrix} -1 \\ -1 \\ 0 \\ 0 \\ 1 \end{bmatrix} + s \begin{bmatrix} -4 \\ -2 \\ 1 \\ 0 \\ 0 \end{bmatrix}$$

the null space of A is spanned by the set

$$B = \left\{ \begin{bmatrix} -1 \\ -1 \\ 0 \\ 0 \\ 1 \end{bmatrix}, \begin{bmatrix} -4 \\ -2 \\ 1 \\ 0 \\ 0 \end{bmatrix} \right\}$$

which is readily seen to be linearly independent. Hence, B is a basis of Null(A). Since B has two elements, the nullity of A is 2.

NOTE   When we write the general solution of $A\mathbf{x} = \mathbf{0}$ as a linear combination with coefficients the parameters, the vectors of the linear combination not only span the null space, they are also *linearly independent*, because the parameters occur at different components in the combination. For example, let $r\mathbf{v}_1 + s\mathbf{v}_2 = \mathbf{0}$, where $\mathbf{v}_1$ and $\mathbf{v}_2$ are the two vectors of B in the last example. Then $r = 0$, because r corresponds to the free variable x_5, so the fifth component of the solution vector is $r \cdot 1$. Likewise, $s = 0$. Hence, $\mathbf{v}_1$ and $\mathbf{v}_2$ are automatically linearly independent. This is why our algorithm yields a *basis* for the null space.

Because the number of parameters determines the number of vectors in the basis of Null(A) we have the following theorem.

THEOREM 22   The nullity of a matrix A equals the number of free variables of $A\mathbf{x} = \mathbf{0}$.

The Column Space

Next we dissect a matrix into columns and rows and study the subspaces spanned by them.

The **column space**, Col(A), of a matrix A is the span of its columns. For example, if

$$A = \begin{bmatrix} 1 & -2 & 1 \\ 2 & 0 & 2 \end{bmatrix} \quad \text{then} \quad \text{Col}(A) = \text{Span} \left\{ \begin{bmatrix} 1 \\ 2 \end{bmatrix}, \begin{bmatrix} -2 \\ 0 \end{bmatrix} \right\}$$

The column space of an $m \times n$ matrix is a *subspace* of $\mathbf{R}^m$, because it is the span of m-vectors.

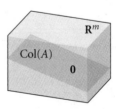

Because a linear system $A\mathbf{x} = \mathbf{b}$ is consistent if and only if $\mathbf{b}$ is in the span of the columns of A, we have the following.

4.6 ■ Rank and Nullity 275

THEOREM 23   A linear system $Ax = b$ is consistent if and only if b is in Col(A).

■ EXAMPLE 69   Which of

$$\mathbf{u} = \begin{bmatrix} -2 \\ 2 \end{bmatrix} \quad \text{and} \quad \mathbf{v} = \begin{bmatrix} 3 \\ 6 \end{bmatrix}$$

is in the column space of $A = \begin{bmatrix} 1 & -2 \\ 2 & -4 \end{bmatrix}$?

SOLUTION   Because

$$\begin{bmatrix} 1 & -2 & -2 \\ 2 & -4 & 2 \end{bmatrix} \sim \begin{bmatrix} 1 & -2 & -2 \\ 0 & 0 & 6 \end{bmatrix} \quad \text{and} \quad \begin{bmatrix} 1 & -2 & 3 \\ 2 & -4 & 6 \end{bmatrix} \sim \begin{bmatrix} 1 & -2 & 3 \\ 0 & 0 & 0 \end{bmatrix}$$

the system $Ax = \mathbf{u}$ is inconsistent, whereas $Ax = \mathbf{v}$ is consistent. Therefore, $\mathbf{v}$ is in Col(A) and $\mathbf{u}$ is not.

A Basis for the Column Space

Our next task is to show that the pivot columns of any matrix form a basis for its column space. This fact can be used to find a basis for the span of a finite set or sequence of vectors. Let us start with an example where the matrix is in echelon form.

■ EXAMPLE 70   Find a basis for Col(B).

$$B = \begin{bmatrix} 1 & -2 & 0 & -1 & 0 \\ 0 & 0 & 1 & 1 & 0 \\ 0 & 0 & 0 & 0 & 1 \\ 0 & 0 & 0 & 0 & 0 \end{bmatrix}$$

SOLUTION   Columns 1, 3, and 5 are pivot columns, and they are clearly linearly independent. The nonpivot columns can be written as linear combinations of the pivot columns. For example, it is easy to see that $\mathbf{b}_2 = -2\mathbf{b}_1$ and $\mathbf{b}_4 = -\mathbf{b}_1 + \mathbf{b}_3$. Hence,

$$\text{Col}(B) = \text{Span}\{\mathbf{b}_1, \mathbf{b}_2, \mathbf{b}_3, \mathbf{b}_4, \mathbf{b}_5\} = \text{Span}\{\mathbf{b}_1, \mathbf{b}_3, \mathbf{b}_5\}$$

by Theorem 9, Section 2.3 (or Theorem 6, Section 4.3). Therefore, the pivot columns $\{\mathbf{b}_1, \mathbf{b}_3, \mathbf{b}_5\}$ form a basis for Col(B).

■ EXAMPLE 71   Find a basis for Col(A).

$$A = \begin{bmatrix} 1 & -2 & 2 & 1 & 0 \\ -1 & 2 & -1 & 0 & 0 \\ 2 & -4 & 6 & 4 & 0 \\ 3 & -6 & 8 & 5 & 1 \end{bmatrix}$$

SOLUTION   It is not hard to show that A reduces to matrix B of Example 70. Therefore, the pivot columns of A are columns 1, 3, and 5. It takes a little more work to show that these are linearly independent. For the nonpivot columns we have $\mathbf{a}_2 = -2\mathbf{a}_1$ and $\mathbf{a}_4 = -\mathbf{a}_1 + \mathbf{a}_3$ (the same linear dependence relations as before). Hence, the pivot columns $\{\mathbf{a}_1, \mathbf{a}_3, \mathbf{a}_5\}$ form a basis for $\mathrm{Col}(A)$.

Examples 70 and 71 suggest that the columns of a matrix and those of an equivalent echelon form satisfy the same linear dependence relations.

THEOREM 24   If $A \sim B$, then the columns of A and B satisfy the same linear dependence relations, i.e.,

$$c_1\mathbf{a}_1 + \cdots + c_n\mathbf{a}_n = \mathbf{0} \Leftrightarrow c_1\mathbf{b}_1 + \cdots + c_n\mathbf{b}_n = \mathbf{0} \tag{4.6}$$

PROOF   Because $A \sim B$, the systems $A\mathbf{x} = \mathbf{0}$ and $B\mathbf{x} = \mathbf{0}$ have the same solutions. Hence

$$A\mathbf{c} = \mathbf{0} \Leftrightarrow B\mathbf{c} = \mathbf{0} \tag{4.7}$$

If $\mathbf{c}$ has components $c_1, \ldots, c_n$, we get (4.6), by the definition of $A\mathbf{c}$ and $B\mathbf{c}$.

Theorem 24 implies that any set of columns of A is linearly dependent (or independent) if and only if the corresponding set of columns of B is linearly dependent (or independent).

THEOREM 25   The pivot columns of any matrix form a basis for its column space.

PROOF   Let A be an $m \times n$ matrix and let B be its reduced row echelon form. We shall prove that the pivot columns of A are linearly independent and that the nonpivot columns are linear combinations of the pivot columns. By Theorem 24 it is sufficient to prove these claims for B.

Suppose B has k pivot columns, say, $\mathbf{b}_{i_1}, \ldots, \mathbf{b}_{i_k}$. Because B is in reduced row echelon form, $\mathbf{b}_{i_1} = \mathbf{e}_1, \ldots, \mathbf{b}_{i_k} = \mathbf{e}_k$, where each $\mathbf{e}_i$ is in $\mathbf{R}^m$. Hence, $\mathbf{b}_{i_1}, \ldots, \mathbf{b}_{i_k}$ are linearly independent. Because B is in echelon form, the last $m - k$ components of its columns are zero, so $\mathrm{Col}(B) \subseteq \mathrm{Span}\{\mathbf{e}_1, \ldots, \mathbf{e}_k\} \subseteq \mathbf{R}^m$. Therefore, $\dim \mathrm{Col}(B) \leq k$, by Theorem 17, Section 4.4. Because we already have k linearly independent pivot columns, $\dim \mathrm{Col}(B) \geq k$, by Theorem 14, Section 4.4. Hence, $\dim \mathrm{Col}(B) = k$, and the pivot columns form a basis by Theorem 15, Section 4.4.

WARNING   *Elementary row operations may change the column space of a matrix.* Referring to Examples 70 and 71, we see that because the last entry of the pivot columns of B is zero, the last column of A is not in $\mathrm{Col}(B)$. So although $A \sim B$, $\mathrm{Col}(A) \neq \mathrm{Col}(B)$ in this case. When computing a basis for $\mathrm{Col}(A)$, make sure to use the pivot columns of the *given* matrix A and *not* the columns of its echelon form B.

■ EXAMPLE 72 (Selecting a Basis from a Spanning Set) Find a basis from S for Span(S), where

$$S = \{(1, -1, 2, 3), (-2, 2, -4, -6), (2, -1, 6, 8), (1, 0, 4, 5), (0, 0, 0, 1)\}$$

SOLUTION It suffices to find a basis for the column space of the matrix with columns the vectors of S. This matrix is A of Example 71, whose pivot columns were columns 1, 3, and 5. Therefore,

$$\{(1, -1, 2, 3), (2, -1, 6, 8), (0, 0, 0, 1)\}$$

is a basis for Span(S), by Theorem 25.

Algorithm A

(Computation of Basis for Span(S))

Let $S = \{\mathbf{a}_1, \ldots, \mathbf{a}_n\} \subseteq \mathbf{R}^m$. A basis for Span($S$) can be found as follows:

1. Form the $m \times n$ matrix A with columns $\mathbf{a}_1, \ldots, \mathbf{a}_n$.
2. Row-reduce A to an echelon form B and identify the pivot columns of A.
3. A basis for Span(S) is the set of the pivot columns of A.

- **Advantage of Algorithm A:** The basis vectors are all in S.
- **Disadvantage:** The basis vectors may not have a lot of zero components.

Theorem 25 can be also used to extend a linearly independent set of $\mathbf{R}^n$ to a basis, as we see in the following example.

■ EXAMPLE 73 (Enlarging a Linearly Independent Set to a Basis) Extend the linearly independent set $S = \{(1, 0, -1, 0), (-1, 1, 0, 0)\}$ to a basis in $\mathbf{R}^4$.

SOLUTION We enlarge S to a spanning set S' by adding the standard basis of $\mathbf{R}^4$. In column notation we have

$$S' = \left\{ \begin{bmatrix} 1 \\ 0 \\ -1 \\ 0 \end{bmatrix}, \begin{bmatrix} -1 \\ 1 \\ 0 \\ 0 \end{bmatrix}, \begin{bmatrix} 1 \\ 0 \\ 0 \\ 0 \end{bmatrix}, \begin{bmatrix} 0 \\ 1 \\ 0 \\ 0 \end{bmatrix}, \begin{bmatrix} 0 \\ 0 \\ 1 \\ 0 \end{bmatrix}, \begin{bmatrix} 0 \\ 0 \\ 0 \\ 1 \end{bmatrix} \right\}$$

We then row-reduce the matrix with columns the elements of S' to get

$$\begin{bmatrix} 1 & 0 & 0 & 0 & -1 & 0 \\ 0 & 1 & 0 & 1 & 0 & 0 \\ 0 & 0 & 1 & 1 & 1 & 0 \\ 0 & 0 & 0 & 0 & 0 & 1 \end{bmatrix}$$

Therefore, columns 1, 2, 3, and 6 of S' form a basis for Span(S) = $\mathbf{R}^4$ by Theorem 25. Hence,

$$\{(1, 0, -1, 0), (-1, 1, 0, 0), (1, 0, 0, 0), (0, 0, 0, 1)\}$$

is a basis of $\mathbf{R}^4$ that extends the linearly independent set S.

NOTE　The method of Example 73 applies to *any*[5] linearly independent subset S of $\mathbf{R}^n$.

The Row Space

DEFINITION　The **row space**, Row(A), of a matrix A is the span of its rows.

$$\text{If } A = \begin{bmatrix} 2 & 0 \\ 1 & -2 \\ 2 & 0 \end{bmatrix}, \text{ then Row}(A) = \text{Span}\left\{ \begin{bmatrix} 2 \\ 0 \end{bmatrix}, \begin{bmatrix} 1 \\ -2 \end{bmatrix} \right\} = \text{Span}\{(2,0),(1,-2)\}.$$

The row space of an $m \times n$ matrix is a *subspace* of $\mathbf{R}^n$, because it is the span of n-vectors.

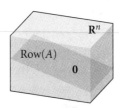

Unlike column spaces, *row spaces are not affected by elementary row operations.* Indeed, let B be obtained from A by one elementary row operation, $\mathcal{O}$. If $\mathcal{O}$ is $R_i \leftrightarrow R_j$, the set of rows remains the same. If $\mathcal{O}$ is either $cR_i \rightarrow R_i$ or $R_i + cR_j \rightarrow R_i$, then the new ith row is a linear combination of old ones. Hence, Row(B) $\subseteq$ Row(A). Because $\mathcal{O}$ is reversible, we also have Row(A) $\subseteq$ Row(B). Hence, in all cases Row(A) = Row(B).

THEOREM 26　If $A \sim B$, then Row(A) = Row(B).

PROOF　B is obtained from A by a finite set of elementary row operations, and all the row spaces of the intermediate matrices are equal.

THEOREM 27　The nonzero rows of a row echelon form matrix A are linearly independent.

PROOF　Let $c_1 \mathbf{r}_{i_1} + \cdots + c_k \mathbf{r}_{i_k} = \mathbf{0}$, where $\mathbf{r}_{i_1}, \ldots, \mathbf{r}_{i_k}$ are the nonzero rows of A. Because A is echelon form, all entries below the leading entry of $\mathbf{r}_{i_1}$ are 0. Hence, $c_1 = 0$. So we can drop the term $c_1 \mathbf{r}_{i_1}$ and repeat the argument. Eventually, all c_i will be zero. Hence, $\{\mathbf{r}_{i_1}, \ldots, \mathbf{r}_{i_k}\}$ is linearly independent.

---

[5]Let S have k elements, and let S' be obtained from S by appending to it the standard basis of $\mathbf{R}^n$. Because S is linearly independent, the first k columns of the reduced matrix of S' are pivot columns. Therefore, all vectors of S are picked by the row reduction as part of the basis.

Theorems 26 and 27 imply the following.

THEOREM 28

The nonzero rows of any echelon form of a matrix A form a basis for Row(A).

■ EXAMPLE 74   Find a basis for Row(A).

$$A = \begin{bmatrix} 1 & 2 & 2 & -1 \\ 1 & 3 & 1 & -2 \\ 1 & 1 & 3 & 0 \\ 0 & 1 & -1 & -1 \\ 1 & 2 & 2 & -1 \end{bmatrix}$$

SOLUTION A reduces to the row echelon form matrix

$$B = \begin{bmatrix} 1 & 2 & 2 & -1 \\ 0 & 1 & -1 & -1 \\ 0 & 0 & 0 & 0 \\ 0 & 0 & 0 & 0 \\ 0 & 0 & 0 & 0 \end{bmatrix}$$

The set $\{(1, 2, 2, -1), (0, 1, -1, -1)\}$ of nonzero rows of B forms a basis for Row(A) = Row(B).

Note that this basis does not consist exclusively of rows of A.

The method of Example 74 offers an alternative way of finding a basis for the span of a finite set of n-vectors. We form the matrix with rows are the given vectors and compute a basis for its row space. This time the basis may *not* consist entirely of the given vectors.

■ EXAMPLE 75   (Basis for the Span)   Find a basis for Span(S), where

$$S = \{(1, -1, 2, 3), (-2, 2, -4, -6), (2, -1, 6, 8), (1, 0, 4, 5), (0, 0, 0, 1)\}$$

SOLUTION   We answered this question in Example 72. Here is another way by computing the row space of the matrix with *rows* the elements of S. Because

$$A = \begin{bmatrix} 1 & -1 & 2 & 3 \\ -2 & 2 & -4 & -6 \\ 2 & -1 & 6 & 8 \\ 1 & 0 & 4 & 5 \\ 0 & 0 & 0 & 1 \end{bmatrix} \sim \begin{bmatrix} 1 & -1 & 2 & 3 \\ 0 & 1 & 2 & 2 \\ 0 & 0 & 0 & 1 \\ 0 & 0 & 0 & 0 \\ 0 & 0 & 0 & 0 \end{bmatrix}$$

$\{(1, -1, 2, 3), (0, 1, 2, 2), (0, 0, 0, 1)\}$ is a basis for Span(S). This basis is slightly different than the one found in Example 72. In particular, $(0, 1, 2, 2)$ is *not* in the original set S.

Another basis is $\{(1, 0, 4, 0), (0, 1, 2, 0), (0, 0, 0, 1)\}$. It is obtained from the reduced row echelon form of A. In general, such a basis has *more zero components*; hence it is often easier to use.

Algorithm B

(Computation of Basis for Span(S))

Let $S = \{\mathbf{r}_1, \ldots, \mathbf{r}_m\} \subseteq \mathbf{R}^n$. A basis for Span($S$) can be found as follows:

1. Form the $m \times n$ matrix A with rows $\mathbf{r}_1, \ldots, \mathbf{r}_m$.
2. Row-reduce A to an echelon form B.
3. A basis for Span(S) is the set of nonzero rows of B.

- **Advantage of Algorithm B:**   We get easy bases (several 0s in the vector components).
- **Disadvantage:**   The basis vectors may not be in S.

WARNING   *Elementary row operations do not preserve linear dependence relations among rows. For example, consider* $\begin{bmatrix} 1 & 2 \\ 2 & 4 \end{bmatrix} \sim \begin{bmatrix} 1 & 2 \\ 0 & 0 \end{bmatrix}$ *and* $\mathbf{r}_2 = 2\mathbf{r}_1$ *for the first matrix and* $\mathbf{r}_2 \neq 2\mathbf{r}_1$ *for the second.*

Rank

Because the dimension of Col(A) is the number of pivots of A, which is also the same as the number of nonzero rows of an echelon form of A, we have the following.

THEOREM 29

For any matrix A,

$$\dim \text{Col}(A) = \dim \text{Row}(A)$$

Graphically:

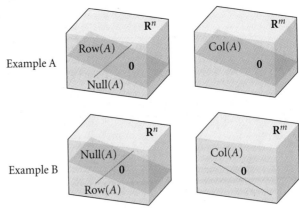

DEFINITION

(Rank)

The common dimension of the column and row spaces of A is called the **rank** of A and is denoted by Rank(A).

The rank is the number of the pivots of A. To compute it, we reduce A to echelon form and count the number of nonzero rows or the number of pivot columns.

■ EXAMPLE 76   The rank of A of Example 74 is 2, because the row echelon form B has two nonzero rows.

NOTE   The rank of an $m \times n$ matrix is less than or equal m, n. (Why?)

■ EXAMPLE 77   Can a 5×9 matrix have rank 6?

ANSWER   No; the rank cannot exceed 5.

An important consequence of Theorem 29 is the following.

COROLLARY 30 A and A^T have the same rank.

PROOF   The column space of A is the same as the row space of A^T.

The Rank Theorem

The next result is one of the *most important theorems* of linear algebra.

THEOREM 31   (The Rank Theorem)

For any matrix A,

$$\text{Rank}(A) + \text{Nullity}(A) = \text{number of columns of } A$$

PROOF   The rank of A is the number of pivot columns of A. On the other hand, the nullity of A is the number of free variables of $A\mathbf{x} = \mathbf{0}$, by Theorem 22. Because we have as many free variables as nonpivot columns, the nullity equals the number of nonpivot columns. The theorem now follows from

No. of pivot columns + no. of nonpivot columns = no. of columns

---

■ EXAMPLE 78   Verify the rank theorem for

$$A = \begin{bmatrix} 1 & -1 & 2 & 3 & 0 \\ -1 & 0 & -4 & 3 & -1 \\ 2 & -1 & 6 & 0 & 1 \\ -1 & 2 & 0 & -1 & 1 \end{bmatrix}$$

SOLUTION   In Example 68 we showed that the nullity of A is 2. On the other hand, the reduced row echelon form of A is

$$\begin{bmatrix} 1 & 0 & 4 & 0 & 1 \\ 0 & 1 & 2 & 0 & 1 \\ 0 & 0 & 0 & 1 & 0 \\ 0 & 0 & 0 & 0 & 0 \end{bmatrix}$$

Hence the rank of A is 3. Adding $2 + 3 = 5$ yields the number of columns of A, as predicted by the rank theorem.

■ EXAMPLE 79   Suppose the system $A\mathbf{x} = \mathbf{0}$ has 20 unknowns and its solution space is spanned by 6 linearly independent vectors.

(a) What is the rank of A?
(b) Can A have size 13×20?

SOLUTION   (a) The number of columns of A is 20, and the nullity is 6. Hence, the rank of A is $20 - 6 = 14$, by the rank theorem.
(b) No. The rank cannot exceed the number of rows, so A should have at least 14 rows.

■ EXAMPLE 80   Let $A\mathbf{x} = \mathbf{b}$ be a system with 20 equations and 24 unknowns. If the null space of A is spanned by four linearly independent vectors, can we be certain that the system is consistent for any choice of $\mathbf{b}$?

SOLUTION   Yes. A has 24 columns and nullity 4, so its rank is 20. (Why?) Hence, the column space has dimension 20 in $\mathbf{R}^{20}$. Therefore, the column space is all of $\mathbf{R}^{20}$ by Theorem 17, Section 4.4. So any vector $\mathbf{b}$ in $\mathbf{R}^{20}$ is spanned by the columns of A. Therefore, $A\mathbf{x} = \mathbf{b}$ is consistent for all $\mathbf{b}$ in $\mathbf{R}^{20}$.

Rank and Linear Systems

The theory and methods developed in this section are strongly related to linear systems.
    Because a linear system $A\mathbf{x} = \mathbf{b}$ is consistent if and only if $\mathbf{b}$ is in $\mathrm{Col}(A)$, by Theorem 23, we have the following.

THEOREM 32   The linear system $A\mathbf{x} = \mathbf{b}$ is consistent if and only if

$$\mathrm{Rank}(A) = \mathrm{Rank}([A : \mathbf{b}])$$

This theorem is mainly of theoretical interest. In practice, we have to reduce $[A : \mathbf{b}]$ anyway either to compute its rank or to see if the last column is a pivot column. So we do not gain any advantage.

Finally, let us see what happens in the extreme cases where the rank of an $m \times n$ matrix is m or n. The next two theorems summarize the main results of Chapter 2 to this point. The proofs have been discussed in pieces elsewhere and are left as exercises.

THEOREM 33

Let A be an $m \times n$ matrix. The following are equivalent.

1. A has rank m.
2. A has m pivots.
3. Each row of A has a pivot.
4. The system $Ax = b$ is consistent for all m-vectors b.
5. The columns of A span $\mathbf{R}^m$.
6. $Col(A) = \mathbf{R}^m$.
7. $\dim Col(A) = m$.
8. $\dim Row(A) = m$.
9. $Nullity(A) = n - m$.
10. A^T has rank m.

THEOREM 34

Let A be an $m \times n$ matrix. The following are equivalent.

1. A has rank n.
2. A has n pivots.
3. Each column of A is a pivot column.
4. The columns of A are linearly independent.
5. The homogeneous system $Ax = 0$ has only the trivial solution.
6. $Null(A) = \{0\}$.
7. $Nullity(A) = 0$.
8. $\dim Col(A) = n$.
9. $\dim Row(A) = n$.
10. A^T has rank n.

Uniqueness of Reduced Row Echelon Form (Optional)

As an application of the concepts developed in this section, we prove that the reduced row echelon form of any matrix is unique.

Theorem 1 of Section 1.2

Every matrix is row-equivalent to one and only one matrix in reduced row echelon form.

PROOF   Let A be any $m \times n$ matrix. That A has at least one reduced echelon form is guaranteed by the Gauss-Jordan elimination process, which computes one.

Let N be another reduced echelon form of A. We shall show that $M = N$. First $M \sim N$, because $M \sim A$ and $A \sim N$. So, the columns of M and N satisfy the same dependence relations, by Theorem 24. Let M have k pivot columns. These columns are precisely $\mathbf{e}_1, \ldots, \mathbf{e}_k$, with each $\mathbf{e}_i$ in $\mathbf{R}^m$, because M is in reduced echelon form. Moreover, a column of M (and of N) is a pivot column if and only it is *not* a linear combination of the columns to the left of it. Let $\mathbf{m}_i$ be the ith column of M.

Case 1: Let $\mathbf{m}_i$ be a pivot column. Then $\mathbf{m}_i = \mathbf{e}_j$, for some j, and $\mathbf{m}_i$ is not a linear combination of the preceding columns. Hence, the same is true for the ith column, $\mathbf{n}_i$, of N, because the columns of M and N satisfy the same dependence relations. So, $\mathbf{n}_i$ is a pivot column of N; because it is the jth pivot column, $\mathbf{n}_i = \mathbf{e}_j$. Hence, $\mathbf{m}_i = \mathbf{n}_i$.

Case 2: Let $\mathbf{m}_i$ be a nonpivot column. Then $\mathbf{m}_i$ is a linear combination of the preceding pivot columns by Theorem 25. So, the same is true for the ith column, $\mathbf{n}_i$, of N, because the columns of M and N satisfy the same dependence relations. But the pivot columns of M and N are the same; therefore, $\mathbf{m}_i = \mathbf{n}_i$.

We conclude that M and N have the same columns. So, $M = N$, as asserted.

Col(A) and Null(A) with CAS

Finding bases for $\mathrm{Col}(A)$ and $\mathrm{Null}(A)$:

Maple

```
> with(linalg):
> A:=matrix([[1,2,3,4],[2,3,4,5],[3,4,5,6]]);
```

$$A = \begin{bmatrix} 1 & 2 & 3 & 4 \\ 2 & 3 & 4 & 5 \\ 3 & 4 & 5 & 6 \end{bmatrix}$$

```
> colspace(A);nullspace(A);
```

$$\{[1, 0, -1], [0, 1, 2]\}$$
$$\{[1, -2, 1, 0], [2, -3, 0, 1]\}$$

Mathematica

```
    A={{1,2,3,4},{2,3,4,5},{3,4,5,6}};
In[2]:=
    RowReduce[A]
    NullSpace[A]
Out[2]=
    {{1, 0, -1, -2}, {0, 1, 2, 3}, {0, 0, 0, 0}}
Out[3]=
    {{2, -3, 0, 1}, {1, -2, 1, 0}}
```

Col(A): Indirect calculation

MATLAB

```
>> A=[1 2 3 4; 2 3 4 5; 3 4 5 6];
>> rref(A), null(A)
ans =                           ans =
      1    0   -1   -2             0.1507   -0.5266
      0    1    2    3             0.1916    0.8144
      0    0    0    0            -0.8352   -0.0491
                                   0.4929   -0.2388
```

Col(A): Indirect calculation

Exercises 4.6

Null Space

In Exercises 1–7 find a basis for the null space and the nullity of the given matrix. (Recall that the zero subspace has dimension 0 and basis the empty set.)

1. a. $\begin{bmatrix} -1 & 2 \\ 2 & -4 \end{bmatrix}$ **b.** $\begin{bmatrix} 2 & -2 & 2 \\ 3 & -3 & 3 \\ 4 & -4 & 5 \end{bmatrix}$

2. a. $\begin{bmatrix} 1 & 2 \\ 2 & 4 \\ 3 & 8 \end{bmatrix}$ **b.** $\begin{bmatrix} 1 & -1 & 1 \\ 2 & -2 & 2 \\ 3 & -3 & 3 \\ 4 & -4 & 4 \end{bmatrix}$

3. a. $\begin{bmatrix} 1 & 2 & -1 & -3 & 0 & 6 \\ 0 & 0 & 0 & 0 & 2 & 4 \\ 0 & 0 & 0 & 0 & 0 & 9 \end{bmatrix}$

b. $\begin{bmatrix} -1 & 1 & 1 \\ 0 & 2 & -2 \\ 0 & 0 & 3 \\ 0 & 0 & 1 \end{bmatrix}$

4. a. $\begin{bmatrix} -1 & 1 & 1 & 2 \\ 2 & 2 & 2 & 4 \\ 0 & -3 & 3 & 9 \end{bmatrix}$

b. $\begin{bmatrix} -1 & 1 & 1 & 2 \\ 2 & -2 & -2 & -4 \\ 0 & -3 & 3 & 9 \end{bmatrix}$

5. a. $\begin{bmatrix} 1 & -1 & 2 & -1 \\ -1 & 0 & -1 & 2 \\ 2 & -4 & 6 & 0 \end{bmatrix}$

b. $\begin{bmatrix} 1 & -1 \\ -1 & 1 \\ 1 & -1 \\ -4 & 4 \\ 0 & 0 \end{bmatrix}$

6. $\begin{bmatrix} 1 & -1 & 2 & -1 \\ -1 & 0 & -1 & 2 \\ 2 & -4 & 6 & 0 \\ 3 & 3 & 0 & -1 \\ 0 & -1 & 1 & 1 \end{bmatrix}$

7. $\begin{bmatrix} 1 & -1 & 2 & 3 & 0 \\ 2 & -1 & 6 & 0 & 1 \\ -1 & 2 & 0 & -1 & 1 \end{bmatrix}$

In Exercises 8–10 add the nullity to the number of pivot columns of the matrix. How does this sum relate to the size of the matrix?

8. a. $\begin{bmatrix} 1 & -1 \\ 2 & -2 \end{bmatrix}$ **b.** $\begin{bmatrix} 1 & -1 \\ 0 & 7 \end{bmatrix}$

9. a. $\begin{bmatrix} 1 & 0 & 1 & -1 \\ 0 & 2 & 2 & -2 \end{bmatrix}$ **b.** $\begin{bmatrix} 1 & 1 & 2 \\ 0 & 0 & 0 \\ 0 & 1 & 1 \\ 0 & -1 & -1 \end{bmatrix}$

10. $\begin{bmatrix} 1 & 2 & -1 & -3 & 0 \\ 0 & 0 & 0 & 0 & 2 \\ 0 & 0 & 0 & 0 & 0 \end{bmatrix}$

11. Let A be an $m \times n$ matrix. If the set of solutions of the system $A\mathbf{x} = \mathbf{b}$ forms a vector subspace of $\mathbf{R}^n$, what can you say about $\mathbf{b}$?

Column Space

Let

$$\mathbf{a} = \begin{bmatrix} -2 \\ 4 \end{bmatrix}, \quad \mathbf{b} = \begin{bmatrix} 3 \\ -6 \end{bmatrix}, \quad \mathbf{c} = \begin{bmatrix} -5 \\ 7 \end{bmatrix}$$

$$\mathbf{u} = \begin{bmatrix} 1 \\ 2 \\ 0 \end{bmatrix}, \quad \mathbf{v} = \begin{bmatrix} -4 \\ -8 \\ 1 \end{bmatrix}, \quad \mathbf{w} = \begin{bmatrix} 3 \\ -1 \\ 0 \end{bmatrix}$$

286   Chapter 4 ■ Vector Spaces

In Exercises 12–17 determine which of **a**, **b**, **c**, **u**, **v**, and **w** are in the column space of the given matrix.

12. $\begin{bmatrix} 1 & -5 \\ -2 & 10 \end{bmatrix}$

13. $\begin{bmatrix} 1 & 2 & 3 \\ 4 & 5 & 6 \end{bmatrix}$

14. $\begin{bmatrix} 1 & 2 & 3 \\ 3 & 6 & 9 \end{bmatrix}$

15. $\begin{bmatrix} 1 & -2 & 3 \\ 0 & -4 & 5 \\ 0 & 0 & 0 \end{bmatrix}$

16. $\begin{bmatrix} -2 & 1 & 4 & -6 \\ 0 & 4 & -5 & 0 \\ 0 & 0 & 1 & 1 \end{bmatrix}$

17. $\begin{bmatrix} -2 & 1 & 4 & -6 & 1 \\ 0 & 4 & -5 & 0 & 2 \\ 0 & -8 & 10 & 0 & -4 \end{bmatrix}$

In Exercises 18–24 find a basis for Col(A).

18. $A = \begin{bmatrix} 2 & 0 & 0 & -1 \\ 0 & 1 & 0 & 1 \\ 0 & 0 & 0 & 1 \end{bmatrix}$

19. $A = \begin{bmatrix} 2 & 0 & 0 & -1 \\ 0 & 1 & 1 & 1 \\ 0 & 0 & 1 & 1 \end{bmatrix}$

20. $A = \begin{bmatrix} 5 & 0 & 2 & -1 \\ 0 & 1 & 1 & 2 \\ 0 & 1 & 1 & 2 \end{bmatrix}$

21. $A = \begin{bmatrix} 1 & -2 & 1 & 0 & 0 & 0 & 0 \\ 0 & -2 & 2 & 2 & 4 & 6 & -5 \\ 0 & 2 & -4 & 0 & 2 & 1 & 2 \\ 0 & 0 & 0 & 0 & 2 & 1 & 2 \end{bmatrix}$

22. $A = \begin{bmatrix} 1 & -1 & 1 & 0 \\ 1 & -2 & 4 & 2 \\ 0 & 1 & -2 & -1 \\ 0 & 0 & 0 & 0 \\ 1 & 2 & 8 & 7 \\ 1 & -2 & -8 & -7 \end{bmatrix}$

23. $A = \begin{bmatrix} 1 & -2 & 1 & 0 & 0 & 0 & 0 \\ 0 & -2 & 2 & 2 & 4 & 6 & -5 \\ 0 & 0 & 0 & 0 & 2 & 1 & 2 \\ 0 & 0 & 0 & 0 & 2 & 1 & 2 \end{bmatrix}$

24. $A = \begin{bmatrix} 4 & 1 & 1 & 1 & 1 \\ 0 & 0 & 0 & 2 & -3 \\ 4 & 0 & 0 & -1 & 0 \\ 0 & 0 & 0 & 1 & 0 \\ 0 & 0 & 0 & 0 & 1 \end{bmatrix}$

25. Sketch Col(A) for $A = \begin{bmatrix} 2 & 1 & 0 & -1 \\ -2 & 1 & 1 & 1 \end{bmatrix}$

26. Sketch Col(A) for $A = \begin{bmatrix} 1 & 1 & -1 \\ 1 & 1 & 0 \\ -1 & 1 & 1 \end{bmatrix}$

27. Sketch Col(A) and Nul(A) for $A = \begin{bmatrix} 3 & 0 & 0 \\ 0 & 1 & -2 \\ 0 & -2 & 4 \end{bmatrix}$

28. Sketch Col(A) and Nul(A) for $A = \begin{bmatrix} 3 & -1 \\ -3 & 1 \\ 6 & -2 \end{bmatrix}$

29. Find matrices A and B such that $A \sim B$ and Col(A) $\neq$ Col(B).

In Exercises 30–33 find a basis for the span of the given set of vectors.

30. $\left\{ \begin{bmatrix} 1 \\ -3 \end{bmatrix}, \begin{bmatrix} -3 \\ 9 \end{bmatrix}, \begin{bmatrix} 0 \\ 4 \end{bmatrix}, \begin{bmatrix} 1 \\ -1 \end{bmatrix} \right\}$

31. $\left\{ \begin{bmatrix} -1 \\ 2 \\ 3 \end{bmatrix}, \begin{bmatrix} 3 \\ -6 \\ -9 \end{bmatrix}, \begin{bmatrix} 0 \\ 1 \\ 1 \end{bmatrix}, \begin{bmatrix} 3 \\ -5 \\ -8 \end{bmatrix} \right\}$

32. $\left\{ \begin{bmatrix} -1 \\ 0 \\ 2 \end{bmatrix}, \begin{bmatrix} 0 \\ 1 \\ -1 \end{bmatrix}, \begin{bmatrix} 2 \\ 0 \\ -4 \end{bmatrix}, \begin{bmatrix} -1 \\ 1 \\ 1 \end{bmatrix} \right\}$

33. $\left\{ \begin{bmatrix} 1 \\ -2 \\ -3 \\ 1 \end{bmatrix}, \begin{bmatrix} 0 \\ 1 \\ 2 \\ 0 \end{bmatrix}, \begin{bmatrix} 3 \\ -2 \\ -1 \\ 3 \end{bmatrix} \right\}$

In Exercises 34–38 enlarge the given linearly independent set of n-vectors to a basis of $\mathbf{R}^n$.

34. $\left\{ \begin{bmatrix} 1 \\ 1 \end{bmatrix} \right\}$

35. $\left\{ \begin{bmatrix} -1 \\ 0 \\ 1 \end{bmatrix} \right\}$

36. $\left\{ \begin{bmatrix} -1 \\ 0 \\ 1 \end{bmatrix}, \begin{bmatrix} 1 \\ -1 \\ 0 \end{bmatrix} \right\}$

37. $\left\{ \begin{bmatrix} -1 \\ 0 \\ 1 \\ 0 \end{bmatrix}, \begin{bmatrix} 1 \\ -1 \\ 0 \\ 0 \end{bmatrix} \right\}$

38. $\left\{ \begin{bmatrix} -1 \\ 0 \\ 1 \\ 0 \end{bmatrix}, \begin{bmatrix} 1 \\ -1 \\ 0 \\ 0 \end{bmatrix}, \begin{bmatrix} -1 \\ 1 \\ 1 \\ 0 \end{bmatrix} \right\}$

Row Space

In Exercises 39–42 find a basis for Row(A) and compute Rank(A).

39. $A = \begin{bmatrix} 1 & 2 & 2 & -1 \\ 0 & -1 & 2 & 3 \\ 1 & 1 & 4 & 2 \end{bmatrix}$

40. $A = \begin{bmatrix} 1 & 2 \\ 0 & -1 \\ 1 & 1 \\ 0 & 1 \\ 0 & 0 \end{bmatrix}$

41. $A = \begin{bmatrix} 1 & 2 & 2 \\ 0 & -1 & 2 \\ 1 & 1 & 4 \\ 0 & 1 & -1 \\ 0 & 0 & 2 \end{bmatrix}$

42. $A = \begin{bmatrix} 1 & 0 & 0 \\ 0 & -1 & 2 \\ 1 & 1 & -3 \\ 0 & 1 & -1 \\ 0 & 0 & 4 \\ 1 & 4 & -8 \end{bmatrix}$

In Exercises 43–45 use Algorithm B to find a basis for the span of the given set of vectors.

43. $\left\{ \begin{bmatrix} 1 \\ 1 \end{bmatrix}, \begin{bmatrix} 2 \\ 3 \end{bmatrix}, \begin{bmatrix} -1 \\ -2 \end{bmatrix} \right\}$

44. $\left\{ \begin{bmatrix} -1 \\ 1 \\ -2 \end{bmatrix}, \begin{bmatrix} 2 \\ -1 \\ 0 \end{bmatrix}, \begin{bmatrix} 1 \\ 0 \\ -2 \end{bmatrix} \right\}$

45. $\left\{ \begin{bmatrix} -1 \\ 1 \\ -2 \end{bmatrix}, \begin{bmatrix} 2 \\ -1 \\ 0 \end{bmatrix}, \begin{bmatrix} 1 \\ 0 \\ -2 \end{bmatrix}, \begin{bmatrix} 1 \\ -1 \\ 2 \end{bmatrix} \right\}$

Rank

Let

$$A = \begin{bmatrix} 1 & 2 & 2 \\ 0 & -1 & 2 \end{bmatrix}, \qquad B = \begin{bmatrix} 1 & 1 & 2 & 2 \\ 0 & 0 & -1 & 2 \\ 0 & 0 & 1 & -2 \end{bmatrix}$$

46. Verify Corollary 30 for A.

47. Verify Corollary 30 for B.

48. Verify the rank theorem for A.

49. Verify the rank theorem for B.

50. Verify the rank theorem for B^T.

51. Use Theorem 32 to show that the system $[B : \mathbf{b}]$, with $\mathbf{b} = \begin{bmatrix} 1 \\ 0 \\ 0 \end{bmatrix}$, is consistent.

52. Suppose the system $A\mathbf{x} = \mathbf{0}$ has 250 unknowns and its solution space is spanned by 50 linearly independent vectors.

   a. What is the rank of A?

   b. Can A have size 150×250?

   c. Can A have size 200×200?

   d. Can A have size 250×150?

   e. Can A have size 200×250?

   f. Can A have size 250×250?

53. Let $A\mathbf{x} = \mathbf{b}$ be a system with 400 equations and 450 unknowns. Suppose that the null space of A is spanned by 50 linearly independent vectors. Is the system consistent for all 400-vectors $\mathbf{b}$?

54. Prove Theorem 33.

55. Prove Theorem 34.

4.7 Applications to Coding Theory

Reader's Goal for This Section

To learn about the interesting application of vector spaces to coding theory.

Nearly all transmitted messages, from human speech to receiving data from a satellite, are subject to noise. It is important, therefore, to be able to encode a message such that after it gets scrambled by noise, it can be decoded to its original form (Fig. 4.16). This is done sometimes by repeating the message two or three times, something very common

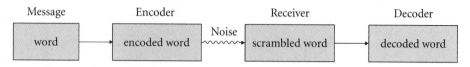

Figure 4.16　*The coding process.*

in human speech. However, repetition is not always very efficient: copying the data stored on a decaying floppy or hard disk once or twice requires a lot of extra space for storage.

In this paragraph we examine ways of encoding and decoding a message after it gets distorted by noise. This process is called **coding.** A code that detects errors in a scrambled message is called **error-detecting.** If, in addition, it can correct the error it is called **error-correcting**. It is much harder to find error-correcting than error-detecting codes. Let us now discuss a few examples.

Because most messages are *digital*—sequences of 0s and 1s, such as 10101 or 1010011—let us assume we want to send the message 1011. This binary "word" may stand for a real word, such as buy, or a sentence, such as buy stock in Beetles' songs. Encoding 1011 means to attach a binary tail to it so that if the message gets distorted to, say, 0011, we can detect the error. One simple thing to do is attach an 1 or 0, depending on whether we have an odd or an even number of 1s in the word. This way all encoded words will have an even number of 1s. So 1011 will be encoded as 10111. Now if this is distorted to 00111 we know that an error has occurred, because we only received an odd number of ones. This error-detecting code is called **parity check** (Fig. 4.17) and is too simple to be very useful. For example, if two digits were changed, our scheme will not detect the error. Even if there were only one error, we would not know where it was to fix it. This is definitely not an error-correcting code. Another approach would be to encode the message by repeating it twice, such as 10111011. Then if 00111011 were received, we know that one of the two equal halves was distorted. If only one error occurred, then it is clearly at position 1. This coding scheme also gives poor results and is not used often. We could get better results by repeating the message several times, but that takes space and time.

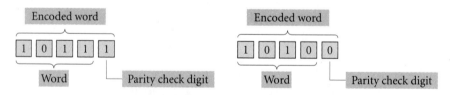

Figure 4.17　*Coding with one parity check.*

Hamming Code

Vector Spaces over Z_2

We are about to examine an interesting single error-correcting code introduced by R. H. Hamming in the 1950s. Before we plunge into the details we note that the definition of vector space can be extended so that scalars other than real numbers can be used—for example, rational numbers or even complex numbers. We are interested in the set of scalars $Z_2 = \{0, 1\}$, the *integers mod* 2. Addition and multiplication in Z_2 are defined by

$$0 + 0 = 0, \quad 1 + 0 = 1, \quad 0 + 1 = 1, \quad 1 + 1 = 0$$
$$0 \cdot 0 = 0, \quad 1 \cdot 0 = 0, \quad 0 \cdot 1 = 0, \quad 1 \cdot 1 = 1$$

Because $1 + 1 = 0$, the opposite of 1 is again 1, so $-1 = 1$. Thus, *subtraction is identical to addition*. These operations satisfy the usual properties of addition and multiplication. For example,

$$(1 + 1) + (1 \cdot 0 + 1) + 1 \cdot (0 + 1) = 0 + 1 + 1 = 0$$

Let Z_2^n be the set of n-vectors with components the elements of Z_2. If $n = 3$, Z_2^3 consists of the eight vectors

$$Z_2^3 = \{(0,0,0), (1,0,0), (0,1,0), (0,0,1), (1,1,0), (1,0,1), (0,1,1), (1,1,1)\}$$

In general, Z_2^n has 2^n elements.

Just as we did with $\mathbf{R}^n$, we equip Z_2^n with componentwise addition and scalar multiplication with Z_2 operations. In Z_2^4 we have

$$(1,1,0,1) + (0,1,1,0) = (1,0,1,1)$$
$$1(1,0,0,1) = (1,0,0,1)$$
$$0(1,1,1,0) = (0,0,0,0)$$

Under these operations Z_2^n satisfies all the axioms of a vector space, except that the scalars are from Z_2. We say that Z_2^n **is a vector space over** Z_2. All the basic concepts and properties, such as subspaces, bases, linearly independent vectors, spanning sets, row reduction of matrices, column space, row space, rank, and nullity, apply to vector spaces over Z_2 and to matrices with entries from Z_2.

THEOREM 35   If V is a vector subspace over Z_2 with dimension n, then V has 2^n elements.

PROOF   If $\{\mathbf{v}_1, \ldots, \mathbf{v}_n\}$ is a basis of V, then V consists of all different linear combinations

$$c_1\mathbf{v}_1 + \cdots + c_n\mathbf{v}_n \qquad \text{with } c_1, \ldots, c_n \text{ either 0 or 1}$$

For each coefficient there are two choices, so we have a total of 2^n different combinations.

■ **EXAMPLE 81**   Find bases and the number of elements of Col(A) and Null(A) over Z_2. Are the rank and nullity the same as in the case where A is viewed as a matrix with real entries?

$$A = \begin{bmatrix} 1 & 1 & 1 & 0 \\ 1 & 0 & 0 & 1 \\ 0 & 1 & 1 & 1 \end{bmatrix}$$

290 Chapter 4 ■ Vector Spaces

SOLUTION   Over Z_2 (keep in mind that the reduction is done with Z_2-arithmetic)

$$A = \begin{bmatrix} 1 & 1 & 1 & 0 \\ 1 & 0 & 0 & 1 \\ 0 & 1 & 1 & 1 \end{bmatrix} \sim \begin{bmatrix} 1 & 1 & 1 & 0 \\ 0 & 1 & 1 & 1 \\ 0 & 1 & 1 & 1 \end{bmatrix} \sim \begin{bmatrix} 1 & 1 & 1 & 0 \\ 0 & 1 & 1 & 1 \\ 0 & 0 & 0 & 0 \end{bmatrix} \sim \begin{bmatrix} 1 & 0 & 0 & 1 \\ 0 & 1 & 1 & 1 \\ 0 & 0 & 0 & 0 \end{bmatrix}$$

Hence, the rank is 2 and the first two columns,

$$\begin{bmatrix} 1 \\ 1 \\ 0 \end{bmatrix}, \quad \begin{bmatrix} 1 \\ 0 \\ 1 \end{bmatrix}$$

of A form a basis for its column space. The null space is obtained by setting the reduced row echelon equal to $(0, 0, 0)$ and solving for the leading variables. If $x_4 = r$ and $x_3 = s$, then $x_1 = -r = r$ and $x_2 = -r - s = r + s$, where $r, s \in \{0, 1\}$. So the null space over Z_2 is spanned by the vectors

$$\left\{ \begin{bmatrix} 1 \\ 1 \\ 0 \\ 1 \end{bmatrix}, \begin{bmatrix} 0 \\ 1 \\ 1 \\ 0 \end{bmatrix} \right\}$$

and the nullity is 2 over Z_2. Because $\text{Col}(A)$ has dimension 2, it has $2^2 = 4$ elements. Likewise, $\text{Null}(A)$ has 4 elements.

We do not get the same answers if we use **R**, however. A row echelon form of A is

$$\begin{bmatrix} 1 & 1 & 1 & 0 \\ 0 & -1 & -1 & 1 \\ 0 & 0 & 0 & 2 \end{bmatrix}$$

so the rank of A is 3 over **R**. Hence, the nullity over **R** is 1, by the rank theorem.

The Hamming $(7, 4)$ Code

We are now ready to define Hamming's interesting single-error-correcting code.[6] An (n, k) **linear code** is a subspace of Z_2^n of dimension k. All vectors of a linear code are called **codewords**, or **encoded words**.

Consider the matrix H over Z_2,

$$H = \begin{bmatrix} 0 & 0 & 0 & 1 & 1 & 1 & 1 \\ 0 & 1 & 1 & 0 & 0 & 1 & 1 \\ 1 & 0 & 1 & 0 & 1 & 0 & 1 \end{bmatrix}$$

Note that the columns $\mathbf{h}_1, \mathbf{h}_2, \ldots, \mathbf{h}_7$ of H are all the nonzero vectors of Z_2^3.

The null space of H is called a **Hamming $(7, 4)$ code**. Let $\text{Null}(H)$ be abbreviated as N_H. H is called a **parity check matrix** for the code N_H. Just as in Example 81, we may

---

[6]See (1) *Error Correcting Codes*, by Wesley Peterson (Cambridge, MA.: MIT Press, 1961); and (2) *Introduction to the Theory of Error Correcting Codes*, by Vera Pless (New York: Wiley, 1982).

easily compute a basis $\mathcal{B}$ for N_H,

$$\mathcal{B} = \{(1,0,0,0,0,1,1),(0,1,0,0,1,0,1),(0,0,1,0,1,1,0),(0,0,0,1,1,1,1)\}$$

So, N_H is a linear $(7,4)$ code and it has $2^4 = 16$ vectors. Because $H(\mathbf{e}_i) = \mathbf{h}_i$ for $i = 1,\dots,7$, we see that none of the standard basis vectors $\mathbf{e}_1,\dots,\mathbf{e}_7$ of Z_2^7 is in N_H.

The matrix G whose rows are the elements of $\mathcal{B}$,

$$G = \begin{bmatrix} 1 & 0 & 0 & 0 & 0 & 1 & 1 \\ 0 & 1 & 0 & 0 & 1 & 0 & 1 \\ 0 & 0 & 1 & 0 & 1 & 1 & 0 \\ 0 & 0 & 0 & 1 & 1 & 1 & 1 \end{bmatrix}$$

is called a **generator matrix** of the Hamming $(7,4)$ code.

THEOREM 36

Let $\mathbf{v} = (v_1,\dots,v_7)$ in Z_2^7.

1. If $\mathbf{v} \in N_H$, then $\mathbf{v} + \mathbf{e}_i \notin N_H$ for $i = 1,\dots,7$.
2. If $H\mathbf{v} = \mathbf{h}_j$, then $\mathbf{v} + \mathbf{e}_j \in N_H$. Furthermore, $\mathbf{v} + \mathbf{e}_i \notin N_H$ for $i \neq j$.

In other words, if any coordinate of a vector in N_H is changed, then the new vector is no longer in N_H. Also, if $H\mathbf{v}$ is the jth column of H, then changing the jth coordinate of $\mathbf{v}$ *only* will put the new vector in N_H.

PROOF

1. Because, $H\mathbf{v} = \mathbf{0}$,

$$H(\mathbf{v} + \mathbf{e}_i) = H(\mathbf{v}) + H(\mathbf{e}_i) = \mathbf{0} + \mathbf{h}_i = \mathbf{h}_i \neq \mathbf{0}$$

2.

$$H(\mathbf{v} + \mathbf{e}_j) = H(\mathbf{v}) + H(\mathbf{e}_j) = \mathbf{h}_j + \mathbf{h}_j = \mathbf{0}$$

so $\mathbf{v} + \mathbf{e}_i \in N_H$. Also,

$$H(\mathbf{v} + \mathbf{e}_i) = H(\mathbf{v}) + H(\mathbf{e}_i) = \mathbf{h}_j + \mathbf{h}_i \neq \mathbf{0}, \qquad i \neq j$$

Encoding and Decoding

Let us see now how to encode a message and decode the distorted reception of it. We are assuming that the word to be coded is binary of length 4, say, 1011, and that noise altered only one binary digit of the encoded word.

To encode 1011 we form the linear combination $\mathbf{v}$ in the basis $\mathcal{B}$ of the Hamming $(7,4)$ code with coefficients the digits $1,0,1,1$ of our message.

$$\mathbf{v} = 1(1,0,0,0,0,1,1) + 0(0,1,0,0,1,0,1) + 1(0,0,1,0,1,1,0) + 1(0,0,0,1,1,1,1)$$

$$= (1,0,1,1,0,1,0)$$

This is the same as right matrix multiplication by G over Z_2.

$$\mathbf{v}^T G = \begin{bmatrix} 1 & 0 & 1 & 1 \end{bmatrix} G = \begin{bmatrix} 1 & 0 & 1 & 1 & 0 & 1 & 0 \end{bmatrix}$$

The encoded word $\mathbf{v}$ is in N_H, by construction. It contains the original message in the first four components and adds a sort of parity check $0, 1, 0$ in the end. Suppose that the string 1011010 gets transmitted and received as 0011010. Let $\mathbf{u} = (0, 0, 1, 1, 0, 1, 0)$. To correct the received message, we compute the product $H\mathbf{u}$.

$$H\mathbf{u} = \begin{bmatrix} 0 & 0 & 0 & 1 & 1 & 1 & 1 \\ 0 & 1 & 1 & 0 & 0 & 1 & 1 \\ 1 & 0 & 1 & 0 & 1 & 0 & 1 \end{bmatrix} \begin{bmatrix} 0 \\ 0 \\ 1 \\ 1 \\ 0 \\ 1 \\ 0 \end{bmatrix} = \begin{bmatrix} 0 \\ 0 \\ 1 \end{bmatrix}$$

Because $H\mathbf{u}$ is the first column of H, Theorem 36, Part 2, implies that $\mathbf{u} + \mathbf{e}_1$, is in N_H and none of $\mathbf{u} + \mathbf{e}_i$, $i \neq 1$, is in N_H. Hence, $\mathbf{u} + \mathbf{e}_1 = \mathbf{v}$ is the only corrected coded message and the original message 1011 is recovered.

■ **EXAMPLE 82** Suppose we received the Hamming encoded messages 1010101 and 1100111. If there is at most one error in each transmission, what were the original messages?

SOLUTION Let $\mathbf{v}_1 = (1, 0, 1, 0, 1, 0, 1)$ and $\mathbf{v}_2 = (1, 1, 0, 0, 1, 1, 1)$.

1. $H\mathbf{v}_1 = (0, 0, 0)$. Hence, $\mathbf{v}_1 \in N_H$. Because the original encoded message was already in N_H, a single error would throw $\mathbf{v}_1$ out of N_H, by Theorem 36, Part 1. So there was no error in the transmission of the first message, which was 1010.

2. $H\mathbf{v}_2 = (1, 1, 1)$. So the seventh component of $\mathbf{v}_2$ needs to be corrected to 0, by Theorem 36, Part 2. Therefore, the original message was 1100. This time the noise affected the parity check part and the original message was never altered. ■

Let us summarize the method.

Algorithm for Error Correction with the Hamming (7, 4) Code

Suppose that a 4-binary-digit word $\mathbf{w}$ is coded as $\mathbf{u}$ so that $\mathbf{u}$ is in N_H. If $\mathbf{u}$ is distorted to $\mathbf{v} = (v_1, \ldots, v_7)$ by at most one component change, to recover the original message $\mathbf{w}$ do:

INPUT: $\mathbf{v}$

1. Compute $H\mathbf{v}$.
2. If $H\mathbf{v} = \mathbf{0}$ let $\mathbf{w} = v_1 v_2 v_3 v_4$. Stop.
3. If $H\mathbf{v} = \mathbf{h}_i$, change the ith component of $\mathbf{v}$ to get a new vector $\mathbf{v}' = (v'_1, \ldots, v'_7)$.
4. Let $\mathbf{w} = v'_1 v'_2 v'_3 v'_4$.

OUTPUT: $\mathbf{w}$

Other Types of Codes

Our study of the Hamming $(7, 4)$ code was intended as an illustration of some of the many fruitful ideas of C. E. Shannon, R. H. Hamming, and others in the late 1940s and early 1950s in the areas of electrical engineering and information theory. We did not attempt to be thorough. The Hamming code is good only for encoding binary words of length 4, of which there are only $2^4 = 16$. If we want a larger "alphabet" or if we want to correct at least two errors in a scrambled message, we need other types of codes.

Today in practice we use a wide variety of coding techniques that allow more words and thus longer messages to be coded. Also, many codes allow more noise errors than the Hamming code. Many of the interesting codes are **nonlinear**. Definitions and examples are included in standard texts on the subject.

In the study of codes mathematics is the main contributor, with linear algebra, number theory, and field theory in the front line.

Exercises 4.7

Z_2^n-Arithmetic

Let

$$\mathbf{u} = \begin{bmatrix} 1 \\ 0 \\ 1 \end{bmatrix}, \qquad \mathbf{v} = \begin{bmatrix} 0 \\ 1 \\ 1 \end{bmatrix}, \qquad \mathbf{w} = \begin{bmatrix} 1 \\ 1 \\ 0 \end{bmatrix}$$

$$A = [\mathbf{u}\,\mathbf{v}\,\mathbf{w}] = \begin{bmatrix} 1 & 0 & 1 \\ 0 & 1 & 1 \\ 1 & 1 & 0 \end{bmatrix}$$

1. Perform the indicated operations in Z_2^3.

   a. $\mathbf{u} + \mathbf{v}$ **b.** $-\mathbf{v}$

   c. $1\mathbf{u} + 0\mathbf{v} - 1\mathbf{w}$ **d.** $\mathbf{u} + \mathbf{v} + \mathbf{w}$

2. Solve the equation for $\mathbf{x}$ over Z_2.

$$\mathbf{x} - \mathbf{u} + \mathbf{v} = \mathbf{w} + \mathbf{u}$$

3. Compute A^2 and A^3 over Z_2.

4. Is $\{\mathbf{u}, \mathbf{v}\}$ linearly independent over Z_2? What about $\{\mathbf{u}, \mathbf{w}\}$?

5. Is $\{\mathbf{u}, \mathbf{v}, \mathbf{w}\}$ linearly independent over Z_2?

6. Add a vector to $\{\mathbf{u}, \mathbf{v}\}$ so that the resulting set is a basis of Z_2^3.

7. Find a basis and the vectors of the null space of A over Z_2. Repeat over **R**.

8. Find the inverse of $\begin{bmatrix} 1 & 0 & 0 \\ 0 & 1 & 1 \\ 1 & 1 & 0 \end{bmatrix}$ over Z_2 and verify your answer.

9. Let A and B be 2×2 binary commuting matrices. Prove that over Z_2:

$$(A + B)^2 = A^2 + B^2$$

Codes

10. Encode the message 1110.

11. Encode the message 0101.

12. Encode the message 0010.

In Exercises 13–17 suppose that a message word was encoded by the Hamming coding method. During transmission at most one coordinate was altered. Recover the original message from the received binary vector shown.

13. **a.** $(1, 1, 1, 1, 0, 1, 1)$ **b.** $(1, 1, 1, 1, 1, 0, 0)$

14. **a.** $(0, 1, 1, 1, 1, 0, 1)$ **b.** $(0, 1, 1, 1, 1, 0, 0)$

15. **a.** $(0, 1, 1, 0, 0, 0, 1)$ **b.** $(0, 1, 0, 0, 0, 1, 1)$

16. **a.** $(0, 1, 1, 0, 0, 1, 1)$ **b.** $(1, 1, 1, 1, 0, 0, 0)$

17. **a.** $(1, 1, 1, 0, 0, 1, 0)$ **b.** $(1, 1, 1, 0, 0, 0, 0)$

18. Write down all 16 elements of the Hamming $(7, 4)$ code N_H.

The **weight** $w(\mathbf{v})$ of a vector $\mathbf{v}$ in Z_2^n is the number of its nonzero entries. For example,

$$w(0, 1, 1, 0) = 2 \quad \text{and} \quad w(1, 0, 1, 1, 1) = 4$$

The **distance** $d(\mathbf{u}, \mathbf{v})$ between two vectors $\mathbf{u}$ and $\mathbf{v}$ in Z_2^n is the number of entries at which $\mathbf{u}$ and $\mathbf{v}$ differ. Hence,

$$d(\mathbf{u}, \mathbf{v}) = w(\mathbf{u} - \mathbf{v}) = w(\mathbf{u} + \mathbf{v})$$

19. Show that

$$d(\mathbf{u}, \mathbf{v}) = d(\mathbf{0}, \mathbf{u} - \mathbf{v})$$

20. Show that $w(\mathbf{v}) \geq 3$ for all nonzero vectors $\mathbf{v}$ in N_H. (*Hint:* Use Exercise 18.)

21. Show that $d(\mathbf{u} - \mathbf{v}) \geq 3$ for all distinct vectors $\mathbf{u}$ and $\mathbf{v}$ in N_H. (*Hint:* Use Exercise 20.)

Error-detecting codes can also be defined in terms of the distance function d. A linear code $V \subseteq Z_2^n$ is **single-error-detecting** if for any codeword $\mathbf{v} \in V$ and any vector $\mathbf{u}$ in Z_2^n, the relation $d(\mathbf{u}, \mathbf{v}) \leq 1$ implies that $\mathbf{u}$ is not a codeword unless $\mathbf{v} = \mathbf{u}$.

22. Use Exercise 21 to show that N_H is single-error-detecting according to this definition. Also show that this statement is equivalent to Part 1 of Theorem 36.

4.8 Miniprojects

The focus of this project section is to discuss an even further generalization of vector space. In Section 4.7 we defined vector spaces over Z_2. Now we allow more general types of scalars that are elements of a *field*.

1 ■ Fields

Definition

A **field** F is a set of elements called **scalars**, equipped with two operations, addition $(a + b)$ and multiplication (ab), that satisfy the following properties.

Addition:

(A1) $a + b$ belongs to F for all $a, b \in F$.
(A2) $a + b = b + a$ for all $a, b \in F$.
(A3) $(a + b) + c = a + (b + c)$ for all $a, b, c \in F$.
(A4) There exists a unique scalar $0 \in F$, called the **zero** of F, such that for all a in F,

$$a + 0 = a$$

(A5) For each $a \in F$, there exists a unique scalar $-a$, called the **negative**, or **opposite** of a, such that

$$a + (-a) = 0$$

Multiplication:

(M1) ab belongs to F for all $a, b \in F$.
(M2) $(a + b)c = ab + bc$ for all $a, b, c \in F$.
(M3) $ab = ba$ for all $a, b \in F$.
(M4) $(ab)c = a(bc)$ for all $a, b, c \in F$.
(M5) There exists a unique *nonzero* scalar $1 \in F$, called **one**, such that for all a in F,

$$a1 = a$$

(M6) For each $a \in F$, $a \neq 0$, there exists a unique scalar a^{-1} (or $\frac{1}{a}$), called the **inverse**, or **reciprocal**, of a, such that

$$aa^{-1} = 1$$

We usually write $a - b$ for the sum $a + (-b)$.

$$a - b = a + (-b)$$

Problem A

Show that in a field F,

$$\text{if } ab = 0, \quad \text{then} \quad a = 0 \quad \text{or} \quad b = 0$$

Problem B

Show that the following are fields. In each case use the usual addition, multiplication and reciprocation.

1. The set of real numbers $\mathbf{R}$
2. The set of rational numbers $\mathbf{Q}$
3. The set of complex numbers $\mathbf{C}$
4. The set of integers mod 2, Z_2
5. The set $\mathbf{Q}(\sqrt{2})$ of all numbers of the form $a + b\sqrt{2}$, where a and b are rational numbers

   Hint: For $\mathbf{Q}(\sqrt{2})$, the reciprocal of $a + b\sqrt{2}$ can be written in the form $A + B\sqrt{2}$ by multiplying and dividing $1/(a + b\sqrt{2})$ by the **conjugate** $a - b\sqrt{2}$. For example, the inverse of $1 - 3\sqrt{2}$ is

$$\frac{1}{1 - 3\sqrt{2}} = \frac{1 + 3\sqrt{2}}{(1 - 3\sqrt{2})(1 + 3\sqrt{2})} = \frac{1 + 3\sqrt{2}}{-17} = -\frac{1}{17} - \frac{3}{17}\sqrt{2}$$

Problem C

Explain why the following sets are **not** fields. In each case use the usual addition and multiplication.

1. The set of integers $\mathbf{Z}$
2. The set of positive integers $\mathbf{N}$
3. The set $\mathbf{R}^2$ with the usual componentwise addition, $(a, b) + (a', b') = (a + a', b + b')$, and componentwise multiplication, $(a, b)(a', b') = (aa', bb')$

2 ■ General Vector Spaces

A vector space V over a field F is a nonempty set equipped with two operations, addition and scalar multiplication, that satisfy all axioms of a vector space as defined in Section 4.2, except that all scalars come from the field F instead of from the real numbers $\mathbf{R}$. The elements are called **vectors** just as before. If the field F is **R**, we say that V is a **real** vector space. If $F = \mathbf{Q}$, the set of rational numbers, we say that V is a **rational** vector space. If $F = \mathbf{C}$, the set of complex numbers, we say that V is a **complex** vector space.

We denote by F^2 the set of all ordered pairs (a, b), where a and b are any elements of F. In general, we denote by F^n the set of all ordered n-tuples $(a_1, \ldots, a_n)$, where $a_1, \ldots, a_n$ are any elements of F. F^n is equipped with componentwise addition and scalar multiplication:

$$(a_1, \ldots, a_n) + (b_1, \ldots, b_n) = (a_1 + b_1, \ldots, a_n + b_n)$$

$$c(a_1, \ldots, a_n) = (ca_1, \ldots, ca_n)$$

Problem A

Show that the following sets are vector spaces over the specified field F.

1. $\mathbf{Q}$ over $\mathbf{Q}$
2. $\mathbf{Q}^n$ over $\mathbf{Q}$
3. $\mathbf{C}$ over $\mathbf{C}$
4. $\mathbf{C}^n$ over $\mathbf{C}$
5. Any field F over F
6. F^n over F

Problem B

Show that the following sets are vector spaces over the specified field F.

1. The real numbers **R** over the set of rational numbers **Q**. Addition is the usual $r_1 + r_2$, $r_1, r_2 \in \mathbf{R}$. Scalar multiplication is of the form qr, where q is a rational number and r is real.
2. The complex numbers **C** over the set of real numbers **R**. Addition is the usual $z_1 + z_2$, $z_1, z_2 \in \mathbf{C}$. Scalar multiplication is of the form rz, where r is a real number and z is complex.
3. The set $\mathbf{Q}(\sqrt{2})$ of all numbers of the form $a + b\sqrt{2}$, $a, b \in \mathbf{Q}$ over $\mathbf{Q}$.

Problem C

Find the *dimension* of the given vector spaces over the specified field F.

1. $\mathbf{C}$ over $\mathbf{C}$
2. $\mathbf{C}^2$ over $\mathbf{C}$
3. $\mathbf{C}$ over $\mathbf{R}$
4. F^n over F
5. $\mathbf{Q}(\sqrt{2})$ over $\mathbf{Q}$

3 ■ Vector Spaces over Finite Fields

In this paragraph we define some interesting fields that consist of finitely many elements and some vector spaces defined over them.

Z_p: The Integers Mod p

A prime number p is a positive integer whose only divisors are 1 and p. For example,

$$2, 3, 5, 7, 11, 13, 17, 19, 23, 29$$

are prime numbers. However,

$$4, 6, 8, 9, 10, 12, 14, 15, 16, 18$$

are not prime numbers, because each has divisors other than itself and 1.

Let p be a prime number. The set Z_p of integers mod p consists of the p elements $\{0, 1, \ldots, p - 1\}$. In Z_p we define addition and multiplication as follows.

If a and b are in Z_p, then $a + b$ is the smallest positive remainder that we get if we divide the integer $a + b$ by p. For example, if $p = 5$, $Z_5 = \{0, 1, 2, 3, 4\}$, then $2 + 3$ yields remainder 0 when divided by 5, so $2 + 3 = 0$ in Z_5. Also, $3 + 4 = 2$ in Z_5, because $7 = 5 \cdot 1 + 2$ has remainder 2 when divided by 5.

If a and b are in Z_p, then ab is defined in the same way, i.e., the smallest positive remainder that we get if we divide the integer ab by p. For example, in Z_5, $2 \cdot 3 = 1$, because $6 = 5 \cdot 1 + 1$. Likewise, $3 \cdot 3 \cdot 3 \cdot 3 = 1$, because $81 = 16 \cdot 5 + 1$.

The operations we just defined are called **mod p** operations of the integers.

Let us now practice with Z_p arithmetic by finding the opposites of all elements of Z_5 and the reciprocals of all its nonzero elements. Clearly the opposite of 0, -0, is 0, because $0 + 0 = 0$. The opposite of -1 is 4, because $1 + 4 = 0$ in Z_5. So we may write $-1 = 4$. Likewise, $-2 = 3$, $-3 = 2$, $-4 = 1$. The reciprocal of 1 is 1, because $1 \cdot 1 = 1$. The reciprocal of 2 is 3, because $2 \cdot 3 = 1$ in Z_5. We may write $\frac{1}{2} = 3$. Likewise, $\frac{1}{3} = 2$ and $\frac{1}{4} = 4$.

Problem A

1. Find -1 in Z_7.
2. Find -10 in Z_{17}.
3. Find $\frac{1}{3}$ in Z_{17}.
4. Find $\frac{1}{6}$ in Z_7.
5. Find $\frac{1}{10}$ in Z_{11}.
6. Find $\frac{1}{p-1}$ in Z_p (p is prime).

If m is any positive integer, $Z_m = \{0, \ldots, m - 1\}$—the integers mod m—is defined just as Z_p and is given the same mod m operations.

Problem B

With respect to the mod p operations show each of the following.

1. Z_3 is a field.
2. Z_7 is a field.
3. Z_p is a field (p is prime).
4. Z_4 is **not** a field. (*Hint:* Does $\frac{1}{2}$ exist?)
5. If m is not a prime integer, then Z_m is not a field.

Because Z_p is a field for any prime p, we may talk about vector spaces over Z_p. Because F^n is a vector space over F by Project 2, Z_p^n, the set of n-tuples $(a_1, \ldots, a_n)$ with $a_i \in Z_p$, is a vector space over Z_p of dimension n.

Problem C

Consider the vector space Z_p^n over Z_p.

1. Prove that Z_p^n has p^n elements.
2. If V is a subspace of Z_p^n of dimension m, then V has p^m elements.
3. Find a basis for Z_3^2.
4. Let

$$A = \begin{bmatrix} 4 & 1 & 0 \\ 4 & 3 & 2 \\ 3 & 3 & 1 \end{bmatrix}$$

be a matrix with entries in Z_5. Row-reduce A using only mod 5 elementary row operations.
5. Find bases for the null and column spaces of A over Z_5. Verify the rank theorem.

4.9   Computer Exercises

This computer session will help you master the commands of your software related to the topics of this chapter. In addition it will help review several of the basic concepts.

$$M = [\mathbf{v}_1 \ \mathbf{v}_2 \ \mathbf{v}_3 \ \mathbf{v}_4 \ \mathbf{v}_5 \ \mathbf{v}_6] = \begin{bmatrix} 1 & 2 & 3 & 4 & 5 & 6 \\ 2 & 3 & 4 & 5 & 6 & 7 \\ 3 & 4 & 5 & 6 & 7 & 8 \end{bmatrix},$$

$$N = [\mathbf{u}_1 \ \mathbf{u}_2 \ \mathbf{u}_3 \ \mathbf{u}_4 \ \mathbf{u}_5] = \begin{bmatrix} 1 & 2 & 1 & 1 & 1 \\ 2 & 4 & 3 & 4 & 5 \\ 3 & 6 & 5 & 7 & 9 \\ 4 & 8 & 7 & 10 & 13 \end{bmatrix}$$

and let

$$S = \{\mathbf{v}_1, \mathbf{v}_2, \mathbf{v}_3, \mathbf{v}_4, \mathbf{v}_5, \mathbf{v}_6\}, \qquad T = \{\mathbf{u}_1, \mathbf{u}_2, \mathbf{u}_3, \mathbf{u}_4, \mathbf{u}_5\}, \qquad \mathcal{B} = \{\mathbf{e}_1, \mathbf{e}_2, \mathbf{u}_3, \mathbf{u}_4\}$$

1. Is S a basis of $\mathbf{R}^3$?
2. Is T a basis of $\mathbf{R}^4$?
3. Show that $\mathcal{B}$ is a basis of $\mathbf{R}^4$.
4. Find $[\mathbf{u}_1]_{\mathcal{B}}$ and $[\mathbf{u}_2]_{\mathcal{B}}$.
5. If $[\mathbf{x}]_{\mathcal{B}} = \mathbf{u}_5$ what is $\mathbf{x}$?
6. Find a basis for Span$\{\mathbf{v}_1, \mathbf{v}_2, \mathbf{v}_3\}$.
7. Find a basis for Span$\{\mathbf{u}_1, \mathbf{u}_2, \mathbf{u}_3, \mathbf{u}_4\}$.
8. Find bases for Col(M), Row(M), and Null(M).
9. Compute the rank and nullity of M. Verify the rank theorem.

10. Verify that $\text{Rank}(M) = \text{Rank}(M^T)$.

11. Find bases for $\text{Col}(N)$, $\text{Row}(N)$, and $\text{Null}(N)$.

12. Compute the rank and nullity of N. Verify the rank theorem.

13. Find two bases for $\text{Span}\{\mathbf{u}_1, \mathbf{u}_2, \mathbf{e}_1, \mathbf{e}_2\}$ by using Algorithms A and B, Section 4.6.

14. Find the pivot columns of N and show that they are linearly independent. Show that the fourth column is a linear combination of the preceding pivot columns.

15. Enlarge the linearly independent set $\{\mathbf{u}_2, \mathbf{u}_3\}$ to a basis of $\mathbf{R}^4$.

16. Let

$$ a = -7, \qquad b = 2, \qquad \mathbf{u} = \begin{bmatrix} 1 & 3 \\ 0 & 2 \end{bmatrix}, $$

$$ \mathbf{v} = \begin{bmatrix} -1 & 1 \\ 0 & 8 \end{bmatrix}, \qquad \mathbf{w} = \begin{bmatrix} 2 & 1 \\ 0 & -9 \end{bmatrix} $$

Verify Axioms (A2), (A3), (M2), (M3), and (M4) for a, b, $\mathbf{u}$, $\mathbf{v}$, and $\mathbf{w}$. Why is the set of all matrices of the form $\begin{bmatrix} c_1 & c_2 \\ 0 & c_3 \end{bmatrix}$ a subspace of M_{22}?

17. Define the three-variable function $f(a, b, x) = a\cos(3x) + b\sin(2x)$ that represents the linear combinations of $\cos(3x)$ and $\sin(2x)$ in $F(\mathbf{R})$. Use f to plot in one graph the linear combinations with $\{a = 1, b = 1\}$, $\{a = 3, b = 0\}$, $\{a = 0, b = -3\}$, $\{a = 3, b = -4\}$, and $\{a = -3, b = 4\}$.

18. Show that the set V of polynomials of the form $ax^3 + bx^2$, $a, b \in \mathbf{R}$ is a subspace of P_3.

19. Show that the following sets of polynomials form bases of P_3.

$$ \mathcal{B}_1 = \{1, -x + 1, x^2 - x, -x^3 + x^2 - 1\} $$
$$ \mathcal{B}_2 = \{x + 2, 4x^2 - x, x^3 - x, x^2 + 1\} $$

20. Find the transition matrix P from $\mathcal{B}_1$ to $\mathcal{B}_2$.

21. Find the transition matrix Q from $\mathcal{B}_2$ to $\mathcal{B}_1$.

22. Verify that $P = Q^{-1}$.

23. Verify that $[-x^3 + 5x + 1]_{\mathcal{B}_2} = P[-x^3 + 5x + 1]_{\mathcal{B}_1}$.

24. Find a basis for the set V of polynomials of the form $ax^3 + bx^2 + cx + d$ such that

$$ a + 2c = 0, \qquad 3b - d = 0 $$

25. Find a such that the polynomials form a basis of P_2.

$$ x + a, \qquad ax^2 + ax + 1, \qquad ax^2 + x + 1 - a $$

Selected Solutions with Maple

Commands augment, col, collect, colspace, evalm, expand, gausselim, matrix, nullspace, plot, proc, rowspace, rref, stack, submatrix, transpose, vector.

```
# Data.
with(linalg):
v1 := vector([1,2,3]); v2 := vector([2,3,4]); v3 := vector([3,4,5]);
v4 := vector([4,5,6]); v5 := vector([5,6,7]); v6 := vector([6,7,8]);
```

```
u1 := vector([1,2,3,4]); u2 := vector([2,4,6,8]);
u3 := vector([1,3,5,7]); u4 := vector([1,4,7,10]);
u5 := vector([1,5,9,13]);
e1 := vector([1,0,0,0]); e2 := vector([0,1,0,0]); # Standard basis vectors.
e3 := vector([0,0,1,0]); e4 := vector([0,0,0,1]);
M := augment(v1,v2,v3,v4,v5,v6);
N := augment(u1,u2,u3,u4,u5);
B := augment(e1,e2,u3,u4);   # The matrix with columns the vectors of B.
# Exercises 1-7.
rref(M);        # 2 pivots, 3 rows: does not span R^3. Not a basis.
rref(N);        # 2 pivots, 4 rows: does not span R^4. Not a basis.
rref(B);        # 4 pivots, 4 rows, 4 columns: spanning and
                # linearly independent: Basis of R^4.
rref(augment(B,u1));  # We solve the system [B:u1] by reduction. The last
col(",5);             # column gives the coordinates of u1. Repeat with u2.
evalm(B &* u5);       # Exercise 5: x is just Bu5.
rref(augment(v1,v2,v3));    # Pivots at (1,1),(2,2). So {v1,v2} is a basis.
rref(augment(u1,u2,u3,u4)); # Pivots at (1,1),(2,3). So {u1,u3} is a basis.
# Exercises 8-12.
colspace(M);            # Basis for column space. 2 vectors so rank=2.
colspace(M, 'r'); r;    # Another way that yields the rank too.
rowspace(M);            # Basis for row space. Also rowspace(M,'r'); r;
nullspace(M);           # Basis for null space. 4 vectors so nullity=4.
nullspace(M, 'n'); n;   # Another way that yields the nullity too.
n + r;                  # Equals the number of columns. Rank Theorem.
colspace(transpose(M), 'r1'); r1; # The rank of the transpose is also 2.
# Repeat with N.
# Exercise 13.
augment(u1,u2,e1,e2);   # Algorithm A: The matrix [u1 u2 e1 e2].
rref(");                # Pivots at (1,1), (2,3), (3,3). {u1,e1,e2} is a basis.
stack(u1,u2,e1,e2);     # Algorithm B.  The vectors as rows of a matrix.
gausselim(");           # The nonzero rows (the first 3) form a basis.
rref(");                # Complete reduction yields even more zeros.
# Exercise 14.
rref(N);                        # The pivot columns are 1 and 3.
pcol := augment(col(N,1),col(N,3)); # Reduction 2 pivots 2 columns,
rref(pcol);                     # so the pivot columns are independent.
rref(augment(pcol, col(N,4)));  # Reduce[col1,col3,col4] to get (-1,2,0,0)
evalm(-1*col(N,1)+2*col(N,3)-col(N,4)); # as last column so -1*col1+2*col3=col4.
# Exercise 15.
m := augment(u1,u2,e1,e2,e3,e4);  # Pivot columns 1,3,4,5. So basis:
rref(m);                          # {u1,e1,e2,e3}.
# Exercise 16.
a:=-7: b:=2: u:=matrix(2,2,[1,3,0,2]):          # Entering scalars
v:=matrix(2,2,[-1,1,0,8]): w:=matrix(2,2,[2,1,0,-9]): # and matrices.
evalm((u+v) - (v+u));                      # A(2),
```

```
evalm((u+v)+w - (u+(v+w)));                    # A(3),
evalm(a*(u+v) - (a*u+a*v));                    # M(2),
evalm((a+b)*u - (a*u+b*u));                    # M(3), and
evalm((a*b)*u - a*(b*u));                      # M(4)  hold.
# Exercise 17.
f := proc(a,b,x) a*cos(3*x)+b*sin(2*x) end:         # Defining f.
plot({f(1,1,x),f(3,0,x),f(0,-3,x),f(3,-4,x),f(4,-3,x)},x=0..Pi);  # Plotting.
# Exercise 18.
p1 := a1*x^3 + b1*x^2:  p2 := a2*x^3 + b2*x:       # General p
collect(p1+p2,x);           # The sum p1 + p2 is in V.
collect(expand(c*p1),x);   # c*p1 is in V. So V is a subspace
# Exercise 19.
B1 := matrix([[0,0,0,1],[0,0,-1,1],[0,1,-1,0],[-1,1,0,-1]]);
B2 := matrix([[0,0,1,2],[0,4,-1,0],[1,0,-1,0],[0,1,0,1]]);
rref(B1);   # All matrices have 4 pivot rows, so the polynomials are linearly
rref(B2);   # independent, hence they form a basis, since dim(P_3)=4.
# Exercises 20-23.
P1:=rref(augment(B2,B1));            # Reduce [B1:B2] and keep the last 4
P:=submatrix(P1,1..4,5..8);          # columns to get P.
Q1:=rref(augment(B1,B2));            # Repeat with Q.
Q:=submatrix(Q1,1..4,5..8);
evalm(P &* Q);                       # PQ=I and same size so P^(-1)=Q.
p:=vector([-1,0,5,1]);               # x^3+5x+1 in vector form.
pb1:=col(rref(augment(B1,p)),5);     # [p]_B1.
pb2:=col(rref(augment(B2,p)),5);     # [p]_B2.
evalm(P &* pb1);                     # P[p]_B2 yields [p]_B1 as expected.
```

Selected Solutions with Mathematica

Commands AppendRows, Clear, Collect, Expand, Flatten, NullSpace, Plot, RowReduce, TakeColumns.

```
(* Data. *)
<<LinearAlgebra`MatrixManipulation`;
v1 = {{1},{2},{3}}; v2 = {{2},{3},{4}}; v3 = {{3},{4},{5}};
v4 = {{4},{5},{6}}; v5 = {{5},{6},{7}}; v6 = {{6},{7},{8}};
u1 = {{1},{2},{3},{4}}; u2 = {{2},{4},{6},{8}};
u3 = {{1},{3},{5},{7}}; u4 = {{1},{4},{7},{10}};
u5 = {{1},{5},{9},{13}};
e1 = {{1},{0},{0},{0}}; e2 = {{0},{1},{0},{0}}; (* Standard basis vectors. *)
e3 = {{0},{0},{1},{0}}; e4 = {{0},{0},{0},{1}};
M = AppendRows[v1,v2,v3,v4,v5,v6]
n = AppendRows[u1,u2,u3,u4,u5]          (* N is already used by the program. *)
B = AppendRows[e1,e2,u3,u4]    (* The matrix with columns the vectors of B. *)
(* Exercises 1-7. *)
RowReduce[M]         (* 2 pivots, 3 rows: does not span R^3. Not a basis.  *)
```

```
RowReduce[n]              (* 2 pivots, 4 rows: does not span R^4. Not a basis.  *)
RowReduce[B]              (* 4 pivots, 4 rows, 4 columns: spanning and           *)
                          (* linearly independent: Basis of R^4.                *)
RowReduce[AppendRows[B,u1]]   (* We solve the system {B:u1} by reduction.  *)
TakeColumns[%,{5}]        (* The last col. gives the coords. of u1. Repeat with u2. *)
B . u5                    (* Exercise 5: x is just Bu5. *)
RowReduce[AppendRows[v1,v2,v3]]    (* Pivots at (1,1),(2,2). So {v1,v2} basis. *)
RowReduce[AppendRows[u1,u2,u3,u4]] (* Pivots at (1,1),(2,3). So {u1,u3} basis. *)
(* Exercises 8-12. *)
RowReduce[M]              (* Pivot cols. 1,2 so basis {v1,v2}. Rank=2.   *)
RowReduce[Transpose[M]]   (* The first two rows form a basis for row space. *)
NullSpace[M]              (* Basis for null space. 4 vectors so nullity=4. *)
(* Nulity + rank = 4 +2 equals the number of columns, 6. Rank Theorem is OK.*)
(* For Exer. 11 again: RowReduce[Transpose[M]] .*)
(* Repeat with N. *)
(* Exercise 13. *)
AppendRows[u1,u2,e1,e2]   (* Algorithm A: The matrix {u1 u2 e1 e2}. *)
RowReduce[%]              (* Pivots at (1,1), (2,3), (3,3). {u1,e1,e2} is a basis. *)
{Flatten[u1],Flatten[u2],Flatten[e1],Flatten[e2]} (* Algorithm B.  The vectors *)
RowReduce[%]  (* as rows of a matrix. The nonzero rows (the first 3) form a basis. *)
(* Exercise 14. *)
RowReduce[n]                       (* The pivot columns are 1 and 3. *)
pcol = AppendRows[TakeColumns[n,{1}],TakeColumns[n,{3}]] (* Reduction 2 pivots *)
RowReduce[pcol]          (* 2 columns,  so the pivot columns are independent. *)
RowReduce[AppendRows[TakeColumns[n,{1}],\
                            (* Reduce [col1,col3,col4] to get (-1,2,0,0) *)
TakeColumns[n,{3}],TakeColumns[n,{4}]]]   (* as last column. So -1*col1+2*col3=col4. *)
-1 TakeColumns[n,{1}]+2 TakeColumns[n,{3}]-TakeColumns[n,{4}]   (* Verification.*)
(* Exercise 15. *)
m = AppendRows[u1,u2,e1,e2,e3,e4]        (* Pivot columns 1,3,4,5. So basis: *)
RowReduce[m]                             (* {u1,e1,e2,e3}.                    *)
(* Exercise 16. *)
a=-7; b=2; u={{1,3},{0,2}};              (* Entering scalars           *)
v={{-1,1},{0,8}}; w={{2,1},{0,-9}};      (* and matrices.              *)
(u+v) - (v+u)                            (* A(2),                      *)
(u+v)+w - (u+(v+w))                      (* A(3),                      *)
a (u+v) - (a u+a v)                      (* M(2),                      *)
(a+b) u - (a u+b u)                      (* M(3), and                  *)
(a b) u - a (b u)                        (* M(4)  hold.                *)
(* Exercise 17. *)
Clear[a,b]                               (* Clear the values of a and b *)
f[a_,b_,x_] = a Cos[3 x] + b Sin[2 x]    (* Defining f.                *)
Plot[{f[1,1,x],f[3,0,x],f[0,-3,x],f[3,-4,x],f[4,-3,x]},{x,0,Pi}](* Plotting.*)
(* Exercise 18. *)
p1 = a1 x^3 + b1 x^2;   p2 = a2 x^3 + b2 x^2;  (* General polynomials of V. *)
Collect[p1+p2,x]        (* The sum p1 + p2 is in V.                         *)
```

```
Collect[Expand[c p1],x]  (*  c*p1 is in V. So V is a subspace.           *)
(* Exercise 19. *)
B1 = {{0,0,0,1},{0,0,-1,1},{0,1,-1,0},{-1,1,0,-1}}
B2 = {{0,0,1,2},{0,4,-1,0},{1,0,-1,0},{0,1,0,1}}
RowReduce[B1]   (* All matrices have 4 pivot rows, so the polynomials are linearly *)
RowReduce[B2]   (* independent, hence they form a basis, since dim(P_3)=4. *)
(* Exercises 20-23. *)
P1=RowReduce[AppendRows[B2,B1]]      (* Reduce [B1:B2] and keep the *)
P=TakeColumns[P1,{5,8}]              (* last 4 columns to get P.    *)
Q1=RowReduce[AppendRows[B1,B2]]      (* Repeat with Q.              *)
Q=TakeColumns[Q1,{5,8}]
P . Q                         (* PQ=I and same size so P^(-1)=Q. *)
p={{-1},{0},{5},{1}}                  (* x^3+5x+1 in vector form.      *)
pb1=TakeColumns[RowReduce[AppendRows[B1,p]],{5}]    (* [p]_B1.       *)
pb2=TakeColumns[RowReduce[AppendRows[B2,p]],{5}]    (* [p]_B2.       *)
P . pb1              (* P[p}_B2 yields [p]_B1 as expected.     *)
```

Selected Solutions with MATLAB

NOTE The indication (ST) means that the command is from the symbolic toolbox.

Commands colspace, fplot, function, null, nullspace, rref.

```
% Data.
v1 = [1; 2; 3]; v2 = [2; 3; 4]; v3 = [3; 4; 5];
v4 = [4; 5; 6]; v5 = [5; 6; 7]; v6 = [6; 7; 8];
u1 = [1; 2; 3; 4]; u2 = [2; 4; 6; 8];
u3 = [1; 3; 5; 7]; u4 = [1; 4; 7; 10];
u5 = [1; 5; 9; 13];
e1 = [1; 0; 0; 0]; e2 = [0; 1; 0; 0];      % Standard basis vectors.
e3 = [0; 0; 1; 0]; e4 = [0; 0; 0; 1];
M = [v1 v2 v3 v4 v5 v6]
N = [u1 u2 u3 u4 u5]
B = [e1 e2 u3 u4]    % The matrix with columns the vectors of B.
% Exercises 1-7.
rref(M)              % 2 pivots, 3 rows: does not span R^3. Not a basis.
rref(N)              % 2 pivots, 4 rows: does not span R^4. Not a basis.
rref(B)              % 4 pivots, 4 rows, 4 columns: spanning and
                     % linearly independent: Basis of R^4.
rref([B u1])         % We solve the system [B:u1] by reduction. The last
ans(:,5)             % column gives the coordinates of u1. Repeat with u2.
B * u5               % Exercise 5: x is just Bu5.
rref([v1 v2 v3])     % Pivots at (1,1),(2,2). So {v1,v2} is a basis.
rref([u1 u2 u3 u4])  % Pivots at (1,1),(2,3). So {u1,u3} is a basis.
% Exercises 8-12.
rref(M)              % Pivot cols. 1,2 so basis {v1,v2}. Rank=2.
colspace(M)          % (ST) Another way. Basis of column space with ST.
```

```
rank(M)                        % The rank is 2.
rref(M')                       % The first two rows form a basis for row space.
null(M)                        % Basis for null space. 4 vectors so nullity=4.
nullspace(M)                   % (ST) Another basis using ST.
% Nulity + rank = 4 +2 equals the number of columns, 6. Rank Theorem is OK.
rank(M')                       % The rank of the transpose is also 2.
% Repeat with N.
% Exercise 13.
[u1 u2 e1 e2]                  % Algorithm A: The matrix [u1 u2 e1 e2].
rref(ans)                      % Pivots at (1,1), (2,3), (3,3). {u1,e1,e2} is a basis.
[u1'; u2'; e1'; e2']           % Algorithm B.  The vectors as rows of a matrix.
rref(ans)                      % The nonzero rows (the first 3) form a basis.
% Exercise 14.
rref(N)                            % The pivot columns are 1 and 3.
pcol = [N(:,1),N(:,3)]             % Reduction 2 pivots 2 columns,
rref(pcol)                         % so the pivot columns are independent.
rref([pcol N(:,4)] )               % Reduce[col1,col3,col4] to get (-1,2,0,0)
-N(:,1)+2*N(:,3)-N(:,4)            % as last column so -1*col1+2*col3=col4.
% Exercise 15.
m = [u1 u2 e1 e2 e3 e4]            % Pivot columns 1,3,4,5. So basis:
rref(m)                            % {u1,e1,e2,e3}.
% Exercise 16.
a=-7; b=2; u=[1 3; 0 2];                  % Entering scalars
v=[-1 1; 0 8]; w=[2 1; 0 -9];             % and matrices.
(u+v) - (v+u)                             % A(2),
(u+v)+w - (u+(v+w))                       % A(3),
a*(u+v) - (a*u+a*v)                       % M(2),
(a+b)*u - (a*u+b*u)                       % M(3), and
(a*b)*u - a*(b*u)                         % M(4)  hold.
% Exercise 17.
function [A] = f(a,b,x)           % Defining f in an m-file. Type the
    A=a*cos(3*x)+b*sin(2*x);      % code on the left in a file named f.m .
    end                          % Then in MATLAB session type:
fplot('[f(1,1,x) f(3,0,x) f(0,-3,x) f(4,-3,x) f(-3,4,x)]',[0 pi]) % Plotting.
% Exercise 19.
B1 = [0 0 0 1; 0 0 -1 1; 0 1 -1 0; -1 1 0 -1]
B2 = [0 0 1 2; 0 4 -1 0; 1 0 -1 0; 0 1 0 1]
rref(B1)       % All matrices have 4 pivot rows, so the polynomials are linearly
rref(B2)       % independent, hence they form a basis, since dim(P_3)=4.
% Exercises 20-23.
P1=rref([B2 B1])               % Reduce [B1:B2] and keep the last 4
P=P1(:,5:8)                    % columns to get P.
Q1=rref([B1 B2])               % Repeat with Q.
Q=Q1(:,5:8)
P * Q                          % PQ=I and same size so P^(-1)=Q.
p=[-1;0;5;1]                   % x^3+5x+1 in vector form.
```

```
rref([B1,p])
pb1=ans(:,5)              % [p]_B1.
rref([B2,p])
pb2=ans(:,5)              % [p]_B2.
P * pb1                   % P[p]_B2 yields [p]_B1 as expected.
```

Linear Transformations

I have no special gift. I am only passionately curious.
—Albert Einstein (1879–1955)

Introduction

Vectors and matrices are intimately connected via matrix multiplication. For a fixed $m \times n$ matrix A, any given n-vector $\mathbf{x}$ corresponds to the m-vector $A\mathbf{x}$. This correspondence defined by the matrix product $A\mathbf{x}$ is the main example of a linear transformation, whose current definition is due to Peano. Linear transformations play a very important role in mathematics, physics, engineering, image processing, computer graphics, and many other areas of science and everyday life. Let us now focus our attention on a contemporary type of application.

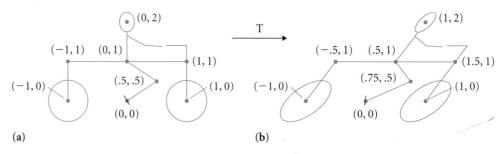

Figure 5.1 *Linear transformation of an image.*

A Cartoonist's Question

A modern cartoonist uses computers and linear algebra to transform images she draws. Suppose she is interested in conveying the sensation of motion by gradually tilting and stretching (horizontally) the image of Fig. 5.1(a) to get that of Fig. 5.1(b). If the necessary gradual stretching, say, along the x-axis, is 50%, how can she model this situation mathematically and have a computer draw the tilted image? The method should be independent of the initial image (frame) so it can be applied to other frames. As we see in Section 5.1 the answer involves a simple matrix-vector multiplication. In fact, what we need to do is multiply the coordinate vector of any plane point that we want to transform on the left by the matrix $\begin{bmatrix} 1 & 0.5 \\ 0 & 1 \end{bmatrix}$.

5.1 Matrix Transformations

Reader's Goals for This Section

1. To know what a matrix transformation is and how to compute its range.
2. To understand basic matrix transformations of $\mathbf{R}^2$ geometrically.

In this section we introduce matrix transformations and study some plane matrix transformations that play an important role in computer graphics. We also discuss an example from physics, the Galilean transformations.

General Transformations

We often need to know how the elements of one set relate to those of another. Sometimes a defining *rule* is used to describe this relation. Let us consider some examples of such rules.

(R_1) To each 2-vector (x, y) we assign the 3-vector $(x - y, 0, y)$.

(R_2) To each 2-vector $\begin{bmatrix} x \\ y \end{bmatrix}$ we assign the 3-vector defined by $\begin{bmatrix} 1 & -1 \\ 0 & 0 \\ 0 & 1 \end{bmatrix} \begin{bmatrix} x \\ y \end{bmatrix}$.

(R_3) To each $x > 0$ we assign the real solution for y of $y^2 - x = 0$.

(R_4) To each real x we assign the real solution of $x^2 + 1 = 0$.

There are some sharp differences between some of these rules. Rule (R_4) is meaningless, because $x^2 + 1 = 0$ has no real solution. (R_3) is ambiguous, because $y^2 - x = 0$ implies $\pm\sqrt{x}$. So, to each x there correspond *two* numbers not one. On the other hand, Rules (R_1) and (R_2) are free of these problems. Each 2-vector is assigned exactly one 3-vector defined by the corresponding rule. (R_1) and (R_2) are *well defined*. They are examples of *transformations*.

A **transformation** T (or **map** or **function**) from a set A to a set B, denoted by $T : A \rightarrow B$, is a rule that associates with *each* element a of A a *unique* element b of B, called the **image** of a under T. We write $T(a) = b$ and say that a **maps to** $T(a)$. A is called the **domain** of T. B is the **codomain** of T. The subset of B that consists of all images of elements of A is called the **range** of T, and it is denoted by $R(T)$ or by $T(A)$. It is possible for two or more elements of A to have the same image (Fig. 5.2). Two transformations $T_1, T_2 : A \rightarrow B$ are **equal** (we write $T_1 = T_2$) if their corresponding images are equal, i.e., if

$$T_1(a) = T_2(a) \qquad \text{for all } a \text{ in } A$$

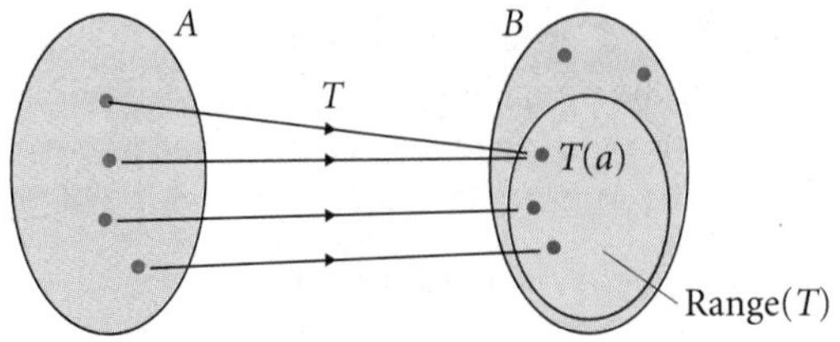

Figure 5.2   Domain, codomain, and range.

(R_1) and (R_2) define equal transformations, because

$$\begin{bmatrix} 1 & -1 \\ 0 & 0 \\ 0 & 1 \end{bmatrix} \begin{bmatrix} x \\ y \end{bmatrix} = \begin{bmatrix} x - y \\ 0 \\ y \end{bmatrix}$$

■ EXAMPLE 1   Let $T : \mathbf{R}^3 \to \mathbf{R}^2$ be the transformation given by

$$T(x, y, z) = (x - y + z, x + y - z)$$

(a) Why is T a transformation? What is its domain? What is its codomain?
(b) Which of the vectors $(1, -2, 3), (1, 2, -3)$, and $(1, 0, 5)$ have the same image under T?
(c) Find all 3-vectors that map to $(0, 0)$.
(d) Describe the range of T.

SOLUTION

(a) T is a transformation because *each* 3-vector, say, (x, y, z), corresponds to *exactly* one 2-vector, namely, $(x - y + z, x + y - z)$. The domain of T is $\mathbf{R}^3$. The codomain is $\mathbf{R}^2$.
(b) $T(1, -2, 3) = (1 - (-2) + 3, 1 + (-2) - 3) = (6, -4)$. Likewise, $T(1, 2, -3) = (-4, 6)$ and $T(1, 0, 5) = (6, -4)$. Hence, $(1, -2, 3)$ and $(1, 0, 5)$ have the same image.
(c) We need all vectors (x, y, z) such that $T(x, y, z) = (x - y + z, x + y - z) = (0, 0)$. Hence,

$$x - y + z = 0$$
$$x + y - z = 0$$

with general solution $(0, r, r), r \in \mathbf{R}$. These are all the 3-vectors that map to $(0, 0)$.
(d) To find the range we need all 2-vectors (a, b) for which there exist numbers x, y, z such that $T(x, y, z) = (a, b)$. So we need all (a, b) that make the system

$$x - y + z = a$$
$$x + y - z = b$$

consistent. Because the coefficient matrix has two pivots (check), the system is consistent for all a, b. Therefore, the range of T is $\mathbf{R}^2$.

Matrix Transformations

Let us fix a matrix, say, $A = \begin{bmatrix} 1 & -1 & 1 \\ 1 & 1 & -1 \end{bmatrix}$. If we take 3-vectors $\mathbf{x}$ and form the products $A\mathbf{x}$, we get unique 2-vectors. For example,

$$\begin{bmatrix} 1 & -1 & 1 \\ 1 & 1 & -1 \end{bmatrix} \begin{bmatrix} 1 \\ 0 \\ 1 \end{bmatrix} = \begin{bmatrix} 2 \\ 0 \end{bmatrix}, \quad \begin{bmatrix} 1 & -1 & 1 \\ 1 & 1 & -1 \end{bmatrix} \begin{bmatrix} 1 \\ -2 \\ 3 \end{bmatrix} = \begin{bmatrix} 6 \\ -4 \end{bmatrix}$$

and, in general,

$$\begin{bmatrix} 1 & -1 & 1 \\ 1 & 1 & -1 \end{bmatrix} \begin{bmatrix} x \\ y \\ z \end{bmatrix} = \begin{bmatrix} x - y + z \\ x + y - z \end{bmatrix}$$

It is clear that we may define a transformation $T : \mathbf{R}^3 \rightarrow \mathbf{R}^2$ by the rule $T(\mathbf{x}) = A\mathbf{x}$. In fact, T is the map of Example 1. This an example of a *matrix transformation*. Matrix transformations are the most important transformations of linear algebra.

DEFINITION

(Matrix Transformation)

A matrix transformation T is a transformation $T : \mathbf{R}^n \rightarrow \mathbf{R}^m$ for which there is an $m \times n$ matrix A such that

$$T(\mathbf{x}) = A\mathbf{x}$$

for all $\mathbf{x} \in \mathbf{R}^n$. A is called the **(standard) matrix** of T.

For example, (R_1) and (R_2) define the matrix transformation $T : \mathbf{R}^2 \rightarrow \mathbf{R}^3$, $T(\mathbf{x}) = A\mathbf{x}$, with $A = \begin{bmatrix} 1 & -1 \\ 0 & 0 \\ 0 & 1 \end{bmatrix}$.

■ **EXAMPLE 2** For the preceding A and T compare $R(T)$ and $\text{Col}(A)$. Find an explicit description for $R(T)$.

SOLUTION A 3-vector $\mathbf{w} = \begin{bmatrix} a \\ b \\ c \end{bmatrix}$ is in $R(T)$ if and only if there is a 2-vector $\begin{bmatrix} x \\ y \end{bmatrix}$

such that $T\left(\begin{bmatrix} x \\ y \end{bmatrix}\right) = \begin{bmatrix} 1 & -1 \\ 0 & 0 \\ 0 & 1 \end{bmatrix} \begin{bmatrix} x \\ y \end{bmatrix} = \begin{bmatrix} x - y \\ 0 \\ y \end{bmatrix} = \begin{bmatrix} a \\ b \\ c \end{bmatrix}$. This is equivalent to

saying that the system with augmented matrix $[A : \mathbf{w}]$ is consistent, or equivalently, that $\mathbf{w}$ is in $\text{Col}(A)$. Therefore, $R(T) = \text{Col}(A)$.

Because $T(\mathbf{x}) = \mathbf{w}$ implies $a = x - y$, $b = 0$, and $c = y$, we have $x = a + c$, $b = 0$ and $y = c$. So a and c can be any scalars and $b = 0$. Therefore, R(T) = $$\left\{ \begin{bmatrix} a \\ 0 \\ c \end{bmatrix}, \quad a, c \in \mathbf{R} \right\}.$$

THEOREM 1

If $T(\mathbf{x}) = A\mathbf{x}$ is any matrix transformation, then

$$R(T) = \text{Col}(A)$$

PROOF

$$\mathbf{w} \in R(T) \Leftrightarrow T(\mathbf{x}) = \mathbf{w} \text{ for some } \mathbf{x}$$

$$\Leftrightarrow A\mathbf{x} = \mathbf{w} \text{ for some } \mathbf{x}$$

$$\Leftrightarrow [A : \mathbf{w}] \text{ is consistent}$$

$$\Leftrightarrow \mathbf{w} \in \text{Col}(A)$$

The next theorem outlines the two most important properties of a matrix transformation. Its proof follows from Theorem 17, Section 2.5.

THEOREM 2

Any matrix transformation $T : \mathbf{R}^n \rightarrow \mathbf{R}^m$, $T(\mathbf{x}) = A\mathbf{x}$ satisfies

1. $T(\mathbf{x} + \mathbf{y}) = T(\mathbf{x}) + T(\mathbf{y})$ for all $\mathbf{x}, \mathbf{y}$ in $\mathbf{R}^n$
2. $T(c\mathbf{x}) = cT(\mathbf{x})$ for all $\mathbf{x}$ in $\mathbf{R}^n$ and all scalars c

Some Matrix Transformations of the Plane

Let us now study some interesting geometric matrix transformations of the plane ($\mathbf{R}^2 \rightarrow \mathbf{R}^2$): reflections, compressions–expansions, shears, rotations, and projections.

Reflections

Reflections are defined about any line in the plane. We are interested in reflections about a line through the origin, especially about the coordinate axes (R_x and R_y) and the diagonal line $y = x$ (R_d). See Fig. 5.3. These reflections are defined by the formulas

$$R_y(x, y) = (-x, y) \qquad R_x(x, y) = (x, -y) \qquad R_d(x, y) = (y, x)$$

and they are all matrix transformations with corresponding matrices

$$\overset{R_y}{\begin{bmatrix} -1 & 0 \\ 0 & 1 \end{bmatrix}} \qquad \overset{R_x}{\begin{bmatrix} 1 & 0 \\ 0 & -1 \end{bmatrix}} \qquad \overset{y=x}{\begin{bmatrix} 0 & 1 \\ 1 & 0 \end{bmatrix}}$$

For example, $R_y \left(\begin{bmatrix} x \\ y \end{bmatrix} \right) = \begin{bmatrix} -1 & 0 \\ 0 & 1 \end{bmatrix} \begin{bmatrix} x \\ y \end{bmatrix} = \begin{bmatrix} -x \\ y \end{bmatrix}$, and so on.

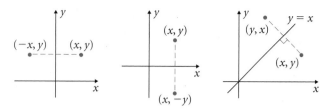

Figure 5.3   Basic reflections.

We also have the basic reflection about the origin with formula and matrix:

$$R_O(x, y) = (-x, -y) \qquad \begin{bmatrix} -1 & 0 \\ 0 & -1 \end{bmatrix}$$

This may also be viewed as rotation by 180° about the origin.

Compressions–Expansions

Compressions and expansions are scalings along the coordinate axes. More precisely, for $c > 0$ the transformation $C_x(x, y) = (cx, y)$ scales the x-coordinates by a factor of c while leaving the y-coordinates unchanged. If $0 < c < 1$, we have **compression** in the direction of the positive x-axis. If $c > 1$, we have **expansion** (Fig. 5.4). We also have compressions and expansions along the y-axis given by $C_y(x, y) = (x, cy)$ for $c > 0$.

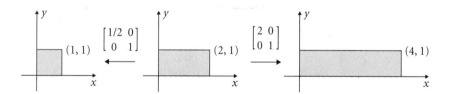

Figure 5.4   Compression and expansion along the x-axis.

We may also have simultaneous scalings along the x- and the y-axes, i.e., $C_{xy}(x, y) = (cx, dy)$ with scale factors $c > 0$ and $d > 0$ along the x- and y-directions (Fig. 5.5).

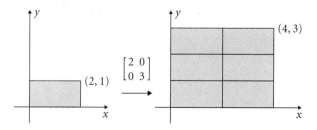

Figure 5.5   Scaling along the x- and y-axes.

C_x, C_y, and C_{xy} are all matrix transformations with respective matrices

$$C_x \qquad\qquad C_y \qquad\qquad C_{xy}$$
$$\begin{bmatrix} c & 0 \\ 0 & 1 \end{bmatrix} \qquad \begin{bmatrix} 1 & 0 \\ 0 & c \end{bmatrix} \qquad \begin{bmatrix} c & 0 \\ 0 & d \end{bmatrix}$$

Shears

A **shear along the x-axis** is a transformation of the form

$$S_x(x, y) = (x + cy, y)$$

In other words, each point is moved along the x-direction by an amount proportional to the distance from the x-axis (Fig. 5.6). We also have shears along the y-axis:

$$S_y(x, y) = (x, cx + y)$$

S_x and S_y are matrix transformations with matrices

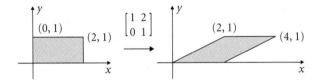

$$S_x \qquad\qquad S_y$$
$$\begin{bmatrix} 1 & c \\ 0 & 1 \end{bmatrix} \qquad \begin{bmatrix} 1 & 0 \\ c & 1 \end{bmatrix}$$

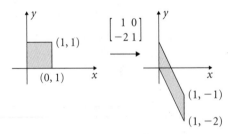

Figure 5.6   Shear along the x-axis.

Note that the constant c in the formula for a shear can be negative. Figure 5.7 illustrates this for $S_y(x, y) = (x, -2x + y)$.

Figure 5.7   Shear along the negative y-direction.

Computer graphists use shears and other types of matrix transformations to transform images. Matrix transformations are well suited for computer calculations because products Ax are easily implemented.

■ EXAMPLE 3 (Application to Computer Graphics) Find the transformation S_x that shears along the positive x-direction by a factor of 0.5. Referring to Fig. 5.8, find the images of the points $(0, 2)$, $(0, 1)$, $(0.5, 0.5)$, $(0, 0)$, $(1, 0)$, $(1, 1)$, $(-1, 1)$, $(-1, 0)$.[1]

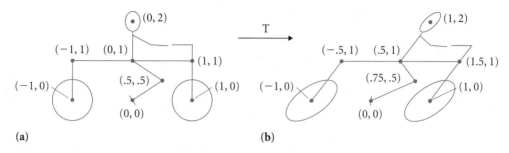

Figure 5.8 *Shearing in image transformations.*

SOLUTION S_x is given by $S_x(x, y) = (x + 0.5y, y)$. Hence,

$$S_x\left(\begin{bmatrix} x \\ y \end{bmatrix}\right) = \begin{bmatrix} 1 & 0.5 \\ 0 & 1 \end{bmatrix} \begin{bmatrix} x \\ y \end{bmatrix} \tag{5.1}$$

The images of the labeled points can be computed by substituting their coordinates into (5.1). We could save space, however, by writing the matrix-vector products in one matrix product:

$$\begin{bmatrix} 1 & 0.5 \\ 0 & 1 \end{bmatrix} \begin{bmatrix} 0 & 0 & 0.5 & 0 & 1 & 1 & -1 & -1 \\ 2 & 1 & 0.5 & 0 & 0 & 1 & 1 & 0 \end{bmatrix}$$

$$= \begin{bmatrix} 1 & 0.5 & 0.75 & 0 & 1 & 1.5 & -0.5 & -1 \\ 2 & 1 & 0.5 & 0 & 0 & 1 & 1 & 0 \end{bmatrix}$$

The coordinates of the images are $(1, 2)$, $(0.5, 1)$, $(0.75, 0.5)$, $(0, 0)$, $(1, 0)$, $(1.5, 1)$, $(-0.5, 1)$, $(-1, 0)$.

Rotations

Another common type of a plane transformation is rotation about any point in the plane. We are interested primarily in rotations about the origin.

■ EXAMPLE 4 (Plane Rotation) The transformation $R_\theta : \mathbf{R}^2 \to \mathbf{R}^2$ defined by

$$R_\theta\left(\begin{bmatrix} x \\ y \end{bmatrix}\right) = \begin{bmatrix} \cos\theta & -\sin\theta \\ \sin\theta & \cos\theta \end{bmatrix} \begin{bmatrix} x \\ y \end{bmatrix}$$

rotates each vector counterclockwise θ rad about the origin.

---

[1]To generate the picture on the right in Fig. 5.8, we would have to compute the images of several points of the picture on the left.

SOLUTION   Referring to Fig. 5.9(a), let $\overrightarrow{OB}$ be the rotation of $\overrightarrow{OA}$ by θ. Then

$$x = r\cos\phi \qquad\qquad y = r\sin\phi$$
$$x' = r\cos(\phi + \theta) \qquad y' = r\sin(\phi + \theta)$$

Using trigonometric identities we have

$$x' = r\cos\phi\cos\theta - r\sin\phi\sin\theta \qquad y' = r\sin\phi\cos\theta + r\cos\phi\sin\theta$$
$$x' = x\cos\theta - y\sin\theta \qquad\qquad y' = y\cos\theta + x\sin\theta$$

Hence,

$$\begin{bmatrix} x' \\ y' \end{bmatrix} = \begin{bmatrix} x\cos\theta - y\sin\theta \\ x\sin\theta + y\cos\theta \end{bmatrix} = \begin{bmatrix} \cos\theta & -\sin\theta \\ \sin\theta & \cos\theta \end{bmatrix} \begin{bmatrix} x \\ y \end{bmatrix} = R_\theta\left(\begin{bmatrix} x \\ y \end{bmatrix}\right)$$

which shows that R_θ is a rotation of θ rad about the origin.

As an example, let us compute the image of $(1, 1)$ for $\theta = \frac{\pi}{2}$ (Fig. 5.9(b)).

$$R_{\pi/2}\left(\begin{bmatrix} 1 \\ 1 \end{bmatrix}\right) = \begin{bmatrix} \cos\frac{\pi}{2} & -\sin\frac{\pi}{2} \\ \sin\frac{\pi}{2} & \cos\frac{\pi}{2} \end{bmatrix} \begin{bmatrix} 1 \\ 1 \end{bmatrix} = \begin{bmatrix} 0 & -1 \\ 1 & 0 \end{bmatrix} \begin{bmatrix} 1 \\ 1 \end{bmatrix} = \begin{bmatrix} -1 \\ 1 \end{bmatrix}$$

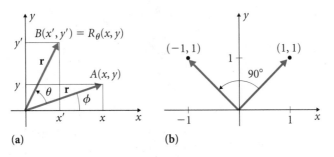

(a)   (b)

Figure 5.9   Rotation about the origin.

NOTE   In justifying the claim of Example 4, we actually showed that *every* plane rotation about the origin is given by the preceding matrix transformation for some θ.

Projections

Projections of the plane onto a line are also transformations of the plane. We are interested in orthogonal projections onto lines through the origin, especially onto the axes.

The **projections** onto the x-axis P_x and the y-axis P_y (Fig. 5.10) are given by

$$P_x(x, y) = (x, 0) \qquad P_y(x, y) = (0, y)$$

and they are matrix transformations with matrices

$$\underset{P_x}{\begin{bmatrix} 1 & 0 \\ 0 & 0 \end{bmatrix}} \qquad \underset{P_y}{\begin{bmatrix} 0 & 0 \\ 0 & 1 \end{bmatrix}}$$

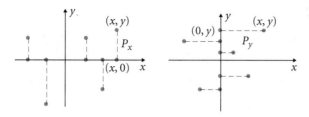

Figure 5.10 **Orthogonal projections along the axes.**

An Example from Physics (Galilean Transformations)

Let F and F' be two frames of reference with parallel coordinate axes, x, y, z and x', y', z'. Let us assume that frame F' is moving away from frame F at a constant relative velocity v in a direction along the x, x'-axes (Fig. 5.11). Furthermore, suppose that we have two observers, say, a mathematician and a physicist, on F and F', respectively, both contacting measurements (such as measuring distances, velocities, etc.).[2]

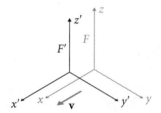

Figure 5.11 **Frames moving at a constant relative speed.**

If (x, y, z, t) are the space-time coordinates that the mathematician is using to carry out his measurements and (x', y', z', t') are the space-time coordinates that the physicist is using to carry out hers, what is the connection between the two-coordinate frames?

Because there is no motion along the y, y'- and z, z'-axes, we must have $y = y'$ and $z = z'$. Also in Newtonian mechanics, $t = t'$ for both frames. For measurements along the x, x'-axes we have: an x-measurement made at time t in frame F by the mathematician exceeds the corresponding measurement x' made by the physicist at the same time by an amount vt. Therefore, $x' = x - vt$. So we have

$$x' = x - vt$$

$$y' = y$$

$$z' = z$$

$$t' = t$$

---

[2] An example of such a situation would be when the mathematician and the physicist were riding on two different trains from New York to Baltimore, with both trains moving at constant speeds along a straight-line stretch, but the physicist is using the faster MetroLiner.

or, in matrix notation,

$$
\begin{bmatrix} x' \\ y' \\ z' \\ t' \end{bmatrix} = \begin{bmatrix} 1 & 0 & 0 & -v \\ 0 & 1 & 0 & 0 \\ 0 & 0 & 1 & 0 \\ 0 & 0 & 0 & 1 \end{bmatrix} \begin{bmatrix} x \\ y \\ z \\ t \end{bmatrix}
$$

We see that system $F'(x', y', z', t')$ is given by a *matrix transformation* of system $F(x, y, z, t)$. This is known as the **Galilean transformation** of Newtonian mechanics. It is named after Galileo,[3] who first expressed the principle that systems in uniform motion with respect to each other are equivalent in describing the laws of kinematics.

Notational Simplification

When we study transformations $\mathbf{R}^n \rightarrow \mathbf{R}^m$, we often skip the parentheses in $T(\mathbf{u})$ when the vector $\mathbf{u}$ is given in component form. For instance, we write $T \begin{bmatrix} 1 \\ 2 \end{bmatrix}$ and $T(1, 2)$, instead of the more proper $T \left(\begin{bmatrix} 1 \\ 2 \end{bmatrix} \right)$ and $T((1, 2))$.

Defining Transformations with CAS

Define $T(x, y, z) = (x - 2y, y - 3z, z - 4x)$ and compute $T(1, 2, 3)$.

Maple

```
> T := (x,y,z,) -> [x-2*y,y-3*z,z-4*x];
            T:=(x,y,z)->[x-2y,y-3z,z-4x]
> T(1,2,3);

                        [-3, -7, -1]
```

Mathematica

```
In[1]:=
      T[x_,y_,z_]:={x-2y,y-3z,z-4x}
In[2]:=
      T[1,2,3]
Out[2]=
      {-3, -7, -1}
```

---

[3] **Galileo Galilei** (1564–1642), was born in Pisa, Italy, and is regarded as one of the greatest physicists of all time. He studied medicine and mathematics at the University of Pisa. He taught at Pisa and Padua and is famous for his fundamental work in physics and astronomy. Some of his work offended the church and caused him to be persecuted and imprisoned.

MATLAB

```
% Function file:
function [A] = T(x,y,z)
               A=[x-2*y,y-3*z,z-4*x];
               end
% In session:
T(1,2,3)
ans =
       -3      -7      -1
```

Exercises 5.1

General Transformations

In Exercises 1–5 determine which of the rules are transformations. If a rule is a transformation, identify the domain and codomain.

To each 2-vector (x, y) we assign

1. The 2-vector $(x - y, y)$
2. The 2-vector $(x - y, y^2)$
3. The 2-vector $(x - y, \sqrt{y})$
4. $\begin{bmatrix} 1 & -1 \\ 0 & 2 \end{bmatrix} \begin{bmatrix} x \\ y \end{bmatrix}$
5. $A \begin{bmatrix} x \\ y \end{bmatrix}$, where A is a 3×3 matrix

Let

$$T_1 : \mathbf{R} \to \mathbf{R}, \quad T_1(x) = x^2$$

In Exercises 6–8 determine whether transformations T_1 and T_2 are equal.

6. $T_2 : \mathbf{R} \to \mathbf{R}$, $T_2(x)$ is the solution for y of $\sqrt{y} - x = 0$
7. $T_2 : \mathbf{R} \to \mathbf{R}^+$, $T_2(x) = x^2$ ($\mathbf{R}^+$ is the set of positive real numbers.)
8. $T_2 : \mathbf{R}^+ \to \mathbf{R}$, $T_2(x) = x^2$

In Exercises 9–15 for the given transformations $T : \mathbf{R}^n \to \mathbf{R}^m$ find each.

(a) n and m
(b) The domain and codomain of T
(c) All vectors of the domain whose image is the zero m-vector
(d) The range of T

9. $T(x, y) = (x + y, x - y)$
10. $T(x, y) = (x - y, 0)$
11. $T(x, y, z) = (x + y, x - z)$
12. $T(x, y) = (x - y, x - y, 0)$
13. $T(x, y, z) = (x - z, -x + z, x - z)$
14. $T \begin{bmatrix} x \\ y \end{bmatrix} = \begin{bmatrix} 1 & -2 \\ 1 & 0 \end{bmatrix} \begin{bmatrix} x \\ y \end{bmatrix}$
15. $T(\mathbf{x}) = \begin{bmatrix} 1 & 0 & 1 & -2 \\ 0 & 1 & 1 & 0 \end{bmatrix} \mathbf{x}$

Matrix Transformations

16. Prove Theorem 2.

In Exercises 17–22 find a basis for $R(T) = \text{Col}(A)$.

17. $T : \mathbf{R}^2 \to \mathbf{R}^2$, $T(\mathbf{x}) = \begin{bmatrix} 1 & 2 \\ 0 & 1 \end{bmatrix} \mathbf{x}$
18. $T : \mathbf{R}^2 \to \mathbf{R}^2$, $T(\mathbf{x}) = \begin{bmatrix} 1 & 2 \\ -3 & -6 \end{bmatrix} \mathbf{x}$
19. $T : \mathbf{R}^4 \to \mathbf{R}^2$, $T(\mathbf{x}) = \begin{bmatrix} 1 & 2 & 7 & -5 \\ 0 & 1 & 6 & 6 \end{bmatrix} \mathbf{x}$
20. $T : \mathbf{R}^3 \to \mathbf{R}^2$, $T(\mathbf{x}) = \begin{bmatrix} 1 & 0 & -8 \\ -2 & 0 & 16 \end{bmatrix} \mathbf{x}$
21. $T : \mathbf{R}^3 \to \mathbf{R}^2$, $T(\mathbf{x}) = \begin{bmatrix} 1 & 0 & -2 \\ -2 & 0 & 1 \\ 0 & -1 & 4 \end{bmatrix} \mathbf{x}$

22. $T : \mathbf{R}^2 \rightarrow \mathbf{R}^4, \quad T(\mathbf{x}) = \begin{bmatrix} 1 & 0 \\ 2 & 1 \\ 1 & 1 \\ 0 & 0 \end{bmatrix} \mathbf{x}$

Plane Matrix Transformations

For Exercises 23–30 consider the matrix plane transformation $T : \mathbf{R}^2 \rightarrow \mathbf{R}^2$, $T(x) = Ax$, with the given matrix A.

(i) Compute and draw the images of

$$\begin{bmatrix} 1 \\ 0 \end{bmatrix}, \quad \begin{bmatrix} 0 \\ 1 \end{bmatrix}, \quad \begin{bmatrix} -1 \\ 2 \end{bmatrix}$$

(ii) Identify T as one of the following.

1. Reflection
2. Compression-expansion
3. Shear
4. Rotation
5. Projection
6. None of the above

23. a. $\begin{bmatrix} -1 & 0 \\ 0 & -1 \end{bmatrix}$ **b.** $\begin{bmatrix} -1 & 0 \\ 0 & 3 \end{bmatrix}$

24. a. $\begin{bmatrix} 1 & 0 \\ 0 & 3 \end{bmatrix}$ **b.** $\begin{bmatrix} 1 & 3 \\ 0 & 1 \end{bmatrix}$

25. a. $\begin{bmatrix} 1 & -3 \\ 0 & 1 \end{bmatrix}$ **b.** $\begin{bmatrix} 1 & -3 \\ -3 & 1 \end{bmatrix}$

26. a. $\begin{bmatrix} 1 & 0 \\ 3 & 1 \end{bmatrix}$ **b.** $\begin{bmatrix} 1 & 0 \\ -3 & 1 \end{bmatrix}$

27. a. $\begin{bmatrix} 1 & 2 \\ 3 & 4 \end{bmatrix}$ **b.** $\begin{bmatrix} 3 & 3 \\ 3 & 3 \end{bmatrix}$

28. a. $\begin{bmatrix} 3 & 0 \\ 0 & 3 \end{bmatrix}$ **b.** $\begin{bmatrix} 3 & 0 \\ 0 & -3 \end{bmatrix}$

29. a. $\begin{bmatrix} 1 & 0 \\ 0 & 0 \end{bmatrix}$ **b.** $\begin{bmatrix} 3 & 0 \\ 0 & 0 \end{bmatrix}$

30. a. $\begin{bmatrix} 0 & 3 \\ 0 & 0 \end{bmatrix}$ **b.** $\begin{bmatrix} 0 & 0 \\ 0 & 3 \end{bmatrix}$

31. Find a formula and the matrix for the reflection about the line $y = -x$.

32. Find a formula and the matrix for the reflection about the line $y = 2x$.

33. Find a formula and the matrix for the clockwise rotation θ radians about the origin.

34. Sketch the image of the square with vertices $(0,0)$, $(1,0)$, $(1,1)$, and $(0,1)$ under the shear

$$S_x \begin{bmatrix} x \\ y \end{bmatrix} = \begin{bmatrix} 1 & -3 \\ 0 & 1 \end{bmatrix} \begin{bmatrix} x \\ y \end{bmatrix}$$

35. Show that any matrix transformation $T : \mathbf{R}^2 \rightarrow \mathbf{R}^2$ maps straight lines to straight lines or to points.

Space Matrix Transformations

Reflections: In space the main reflections are about the *origin, coordinate planes, coordinate axes, bisectors of the coordinate axes,* and *bisector planes.*

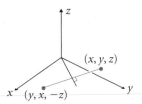

Figure 5.12   Reflection about the xy-bisector.

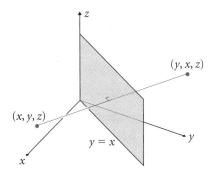

Figure 5.13   Reflection about the plane $y = x$.

36. In the following table fill in the question marks.

Reflections about the:	$T(\mathbf{x})$	Matrix A
?	?	$\begin{bmatrix} -1 & 0 & 0 \\ 0 & -1 & 0 \\ 0 & 0 & -1 \end{bmatrix}$
?-plane	?	$\begin{bmatrix} 1 & 0 & 0 \\ 0 & 1 & 0 \\ 0 & 0 & -1 \end{bmatrix}$
?-axis	?	$\begin{bmatrix} 1 & 0 & 0 \\ 0 & -1 & 0 \\ 0 & 0 & -1 \end{bmatrix}$
First quadrant line bisecting the ?-plane (Fig. 5.12)	?	$\begin{bmatrix} 0 & 1 & 0 \\ 1 & 0 & 0 \\ 0 & 0 & -1 \end{bmatrix}$

Reflections about the:	$T(\mathbf{x})$	Matrix A
Plane $y =$? (Fig. 5.13)	?	$\begin{bmatrix} 0 & 1 & 0 \\ 1 & 0 & 0 \\ 0 & 0 & 1 \end{bmatrix}$

Rotations: The rotation transformation R_θ^z that rotates any 3-vector θ radians about the z-axis in the positive direction[4] (Fig. 5.14) is given by

$$R_\theta^z \begin{bmatrix} x \\ y \\ z \end{bmatrix} = \begin{bmatrix} \cos\theta & -\sin\theta & 0 \\ \sin\theta & \cos\theta & 0 \\ 0 & 0 & 1 \end{bmatrix} \begin{bmatrix} x \\ y \\ z \end{bmatrix}$$

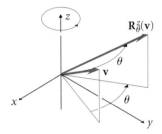

Figure 5.14   Rotation by θ about the z-axis.

37. Find formulas for the corresponding rotations about the x- and y-axes.

Projections: There are several orthogonal projections, mainly onto the coordinate planes and the coordinate axes.

38. In the following table fill in the question marks.

Projection onto the:	$T(\mathbf{x})$	Matrix A
?-plane (Fig. 5.15)	?	$\begin{bmatrix} 1 & 0 & 0 \\ 0 & 1 & 0 \\ 0 & 0 & 0 \end{bmatrix}$
?-axis (Fig. 5.16)	?	$\begin{bmatrix} 1 & 0 & 0 \\ 0 & 0 & 0 \\ 0 & 0 & 0 \end{bmatrix}$

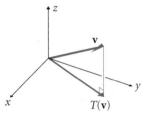

Figure 5.15   Projection onto the xy-plane.

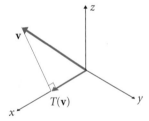

Figure 5.16   Projection onto the x-axis.

5.2   Linear Transformations

Reader's Goal for This Section

To really understand the definition and how to test for a linear transformation.

Linear transformations are maps of fundamental importance in linear algebra and its applications. They are transformations between vector spaces that preserve vector addition and scalar multiplication. In this section we discuss the definition of a linear transformation and offer several examples that should be studied carefully.

[4]Determined by the **right-hand rule**. As the figures of a right hand curl from the positive x-axis to the positive y-axis, the thumb points to the positive z-axis.

Definition and Examples

DEFINITION

(Linear Transformation)

Let V, W be two vector spaces. A **linear transformation** (or **linear map**) from V to W is a transformation $T : V \rightarrow W$ such that for all vectors **u** and **v** of V and any scalar c,

1. $T(\mathbf{u} + \mathbf{v}) = T(\mathbf{u}) + T(\mathbf{v})$;
2. $T(c\mathbf{u}) = cT(\mathbf{u})$.

The $+$ in $\mathbf{u} + \mathbf{v}$ is addition in V, whereas the $+$ in $T(\mathbf{u}) + T(\mathbf{v})$ is addition in W. Likewise, scalar multiplications $c\mathbf{u}$ and $cT(\mathbf{u})$ occur in V and W, respectively. In the special case where $V = W$, the linear transformation $T : V \rightarrow V$ is called a **linear operator** of V.

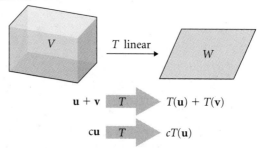

■ EXAMPLE 5 Show that the transformation $T : \mathbf{R}^2 \rightarrow \mathbf{R}^2$ defined by

$$T\begin{bmatrix} x \\ y \end{bmatrix} = \begin{bmatrix} 2x - 3y \\ x + 4y \end{bmatrix}$$

is linear.

SOLUTION Let $\mathbf{u} = \begin{bmatrix} x_1 \\ y_1 \end{bmatrix}$ and $\mathbf{v} = \begin{bmatrix} x_2 \\ y_2 \end{bmatrix}$. Then

$$T(\mathbf{u} + \mathbf{v}) = T\left(\begin{bmatrix} x_1 \\ y_1 \end{bmatrix} + \begin{bmatrix} x_2 \\ y_2 \end{bmatrix} \right)$$

$$= T\begin{bmatrix} x_1 + x_2 \\ y_1 + y_2 \end{bmatrix}$$

$$= \begin{bmatrix} 2(x_1 + x_2) - 3(y_1 + y_2) \\ (x_1 + x_2) + 4(y_1 + y_2) \end{bmatrix}$$

$$= \begin{bmatrix} 2x_1 - 3y_1 \\ x_1 + 4y_1 \end{bmatrix} + \begin{bmatrix} 2x_2 - 3y_2 \\ x_2 + 4y_2 \end{bmatrix}$$

$$= T\begin{bmatrix} x_1 \\ y_1 \end{bmatrix} + T\begin{bmatrix} x_2 \\ y_2 \end{bmatrix}$$

$$= T(\mathbf{u}) + T(\mathbf{v})$$

For any scalar c,

$$T(c\mathbf{u}) = T \begin{bmatrix} cx_1 \\ cy_1 \end{bmatrix}$$

$$= \begin{bmatrix} 2cx_1 - 3cy_1 \\ cx_1 + 4cy_1 \end{bmatrix}$$

$$= c \begin{bmatrix} 2x_1 - 3y_1 \\ x_1 + 4y_1 \end{bmatrix}$$

$$= cT \begin{bmatrix} x_1 \\ y_1 \end{bmatrix}$$

$$= cT(\mathbf{u})$$

Both parts of the definition are satisfied, so T is linear.

■ EXAMPLE 6   Show that the transformation $T : \mathbf{R}^3 \to \mathbf{R}^2$ is linear:

$$T(x, y, z) = (x - z, y + z)$$

SOLUTION   Let $\mathbf{u} = (x_1, y_1, z_1)$ and $\mathbf{v} = (x_2, y_2, z_2)$. Then

$$\begin{aligned} T(\mathbf{u} + \mathbf{v}) &= T(x_1 + x_2, y_1 + y_2, z_1 + z_2) \\ &= ((x_1 + x_2) - (z_1 + z_2), (y_1 + y_2) + (z_1 + z_2)) \\ &= (x_1 - z_1, y_1 + z_1) + (x_2 - z_2, y_2 + z_2) \\ &= T(\mathbf{u}) + T(\mathbf{v}) \end{aligned}$$

proves Part 1 of the definition. The verification of Part 2 is left for practice.

The most important examples of linear transformations are matrix transformations. In fact, as we shall see, matrix transformations are the *only* linear maps from $\mathbf{R}^n$ to $\mathbf{R}^m$.

■ EXAMPLE 7   (Matrix Transformations)   Show that every matrix transformation is linear.

SOLUTION   The two properties of the definition hold by Theorem 2, Section 5.1. Let us practice by reworking the proof. If $T : \mathbf{R}^n \to \mathbf{R}^m$, $T(\mathbf{x}) = A\mathbf{x}$ is a matrix transformation, then

$$T(\mathbf{x} + \mathbf{y}) = A(\mathbf{x} + \mathbf{y}) = A\mathbf{x} + A\mathbf{y} = T(\mathbf{x}) + T(\mathbf{y})$$

$$T(c\mathbf{x}) = A(c\mathbf{x}) = c(A\mathbf{x}) = cT(\mathbf{x})$$

Hence, T is linear.

■ EXAMPLE 8 (Geometric Linear Transformations)   Show that reflections, shears, compressions-expansions (all) about the coordinate axes, rotations about the origin, and projections onto the coordinate axes are linear transformations.

SOLUTION   All these maps[5] are matrix transformations by Section 5.1. Hence, they are linear by Example 7.

Properties of Linear Transformations

The definition of a linear transformation may also be rephrased as follows.

THEOREM 3

$T : V \longrightarrow W$ is a linear transformation if and only if for all vectors v_1 and $v_2 \in V$ and all scalars c_1 and c_2, we have

$$T(c_1 v_1 + c_2 v_2) = c_1 T(v_1) + c_2 T(v_2)$$

PROOF   If T is linear, then

$$T(c_1 v_1 + c_2 v_2) = T(c_1 v_1) + T(c_2 v_2)$$
$$= c_1 T(v_1) + c_2 T(v_2)$$

by Parts 1 and 2 of the definition.

Conversely, if T is a transformation such that $T(c_1 v_1 + c_2 v_2) = c_1 T(v_1) + c_2 T(v_2)$ for all $v_1, v_2 \in R^n$ and all $c_1, c_2 \in R$, then letting $c_1 = c_2 = 1$ yields Part 1 of the definition. Letting $c_2 = 0$ yields Part 2.

More generally, if $v_1, \ldots, v_n$ are any vectors in V and $c_1, \ldots, c_n$ are any scalars, then

$$T(c_1 v_1 + \cdots + c_n v_n) = c_1 T(v_1) + \cdots + c_n T(v_n)$$

So, linear transformations map a linear combination of vectors to the same linear combination of the images of these vectors.

THEOREM 4

$T : V \longrightarrow W$ is a linear transformation. Then

1. $T(0) = 0$;
2. $T(u - v) = T(u) - T(v)$.

PROOF

1. We have, by Part 2 of the definition,

$$T(0) = T(0 v) = 0 T(v) = 0$$

[5]Reflections about *any* line through the origin and projections onto *any* line through the origin are also matrix (thus, linear) transformations.

2. By Theorem 3 with $c_1 = 1, c_2 = -1$, we have

$$T(\mathbf{u} - \mathbf{v}) = T(1\mathbf{u} + (-1)\mathbf{v}) = 1T(\mathbf{u}) + (-1)T(\mathbf{v}) = T(\mathbf{u}) - T(\mathbf{v})$$

■ EXAMPLE 9   Is $f : \mathbf{R}^2 \rightarrow \mathbf{R}^2$ defined by $f(x, y) = (x, 1)$ a linear transformation?

ANSWER   If f were linear, then $f(0, 0)$ would be $(0, 0)$ by Part 1 of Theorem 4. However, $f(0, 0) = (0, 1)$, so f is nonlinear.

More Examples

The transformation $\mathbf{0} : V \rightarrow W$ that maps all vectors of V to $\mathbf{0}$ in W is called the **zero transformation**.

$$\mathbf{0}(\mathbf{v}) = \mathbf{0} \qquad \text{for all } \mathbf{v} \in V$$

■ EXAMPLE 10   Show that the zero transformation $\mathbf{0} : V \rightarrow W$ is linear.

SOLUTION   If $\mathbf{v}$ and $\mathbf{u}$ are vectors of V and c is a scalar, then

$$\mathbf{0}(\mathbf{v} + \mathbf{u}) = \mathbf{0} = \mathbf{0} + \mathbf{0} = \mathbf{0}(\mathbf{v}) + \mathbf{0}(\mathbf{u})$$

and

$$\mathbf{0}(c\mathbf{v}) = \mathbf{0} = c\mathbf{0} = c\,\mathbf{0}(\mathbf{v})$$

The transformation $I : V \rightarrow V$ that maps each vector of V to itself is called the **identity transformation of V**.

$$I(\mathbf{v}) = \mathbf{v} \qquad \text{for all } \mathbf{v} \in V$$

■ EXAMPLE 11   Show that the identity transformation $I : V \rightarrow V$ is linear.

SOLUTION   Exercise.

■ EXAMPLE 12   (Homothety)   For a fixed scalar c, show that $T : V \rightarrow V$ is linear.

$$T(\mathbf{v}) = c\,\mathbf{v}$$

SOLUTION   Let $\mathbf{u}, \mathbf{w} \in V$ and $r \in \mathbf{R}$. T is linear, because

$$T(\mathbf{u} + \mathbf{w}) = c(\mathbf{u} + \mathbf{w}) = c\mathbf{u} + c\mathbf{w} = T(\mathbf{u}) + T(\mathbf{w})$$

$$T(r\mathbf{u}) = c(r\mathbf{u}) = r(c\mathbf{u}) = rT(\mathbf{u})$$

The transformation defined in Example 12 is called a **homothety**. If $c > 1$, the homothety is a **dilation**, and its effect on $\mathbf{v}$ is to stretch $\mathbf{v}$ by a factor of c. If $0 < c < 1$, the homothety is a **contraction**, and its effect on $\mathbf{v}$ is to shrink $\mathbf{v}$ by a factor of c (Fig. 5.17). If $c < 0$, then this transformation reverses the direction of $\mathbf{v}$.

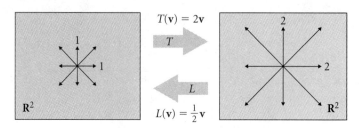

Figure 5.17   Dilation and contraction by a factor of 2.

■ EXAMPLE 13   (Multiplication by Fixed Matrix)   Let A be a fixed $m \times n$ matrix. Show that the transformation $T : M_{nk} \to M_{mk}$ is linear.

$$T(B) = AB$$

SOLUTION

$$T(B + C) = A(B + C) = AB + AC = T(B) + T(C)$$

$$T(cB) = A(cB) = c(AB) = cT(B)$$

■ EXAMPLE 14   Show that the transformation $T : P_2 \to P_1$ is linear.

$$T(a + bx + cx^2) = b + 2cx$$

SOLUTION   Let $p_1 = a_1 + b_1 x + c_1 x^2$ and $p_2 = a_2 + b_2 x + c_2 x^2$. Then

$$T(p_1 + p_2) = T\left((a_1 + a_2) + (b_1 + b_2)x + (c_1 + c_2)x^2\right)$$
$$= (b_1 + b_2) + 2(c_1 + c_2)x$$
$$= (b_1 + 2c_1 x) + (b_2 + 2c_2 x)$$
$$= T(p_1) + T(p_2)$$

The verification of Part 2 of the definition is left as an exercise.

■ EXAMPLE 15 $T : M_{22} \to \mathbf{R}^2$ is linear. (Why?)

$$T\begin{bmatrix} a & b \\ c & d \end{bmatrix} = \begin{bmatrix} b - a \\ c + d \end{bmatrix}$$

■ EXAMPLE 16   (Dotting by Fixed Vector)   Let $\mathbf{u}$ be a fixed vector in $\mathbf{R}^n$. Show that the transformation $T : \mathbf{R}^n \to \mathbf{R}$ is linear.

$$T(\mathbf{v}) = \mathbf{u} \cdot \mathbf{v}$$

SOLUTION   Let $v_1, v_2 \in \mathbf{R}^n$ and $c_1, c_2 \in \mathbf{R}$. We have

$$T(c_1 v_1 + c_2 v_2) = \mathbf{u} \cdot (c_1 v_1 + c_2 v_2) = \mathbf{u} \cdot (c_1 v_1) + \mathbf{u} \cdot (c_2 v_2)$$
$$= c_1(\mathbf{u} \cdot v_1) + c_2(\mathbf{u} \cdot v_2) = c_1 T(v_1) + c_2 T(v_2)$$

Hence, T is linear by Theorem 3.

■ **EXAMPLE 17**   (Crossing by Fixed Vector)   Show that the transformation $T :$ $\mathbf{R}^3 \to \mathbf{R}^3$ is linear.

$$T(\mathbf{v}) = \mathbf{u} \times \mathbf{v}$$

SOLUTION   The verification is similar to that of Example 16.

■ **EXAMPLE 18**   (Projection to Line through the Origin)   Let $\mathbf{u}$ be a fixed nonzero vector in $\mathbf{R}^3$. The transformation $T : \mathbf{R}^3 \to \mathbf{R}^3$ defined by taking the orthogonal projection of each $\mathbf{v} \in \mathbf{R}^3$ on $\mathbf{u}$ (see Section 2.2) is linear (Fig. 5.18).

$$T(\mathbf{v}) = \frac{\mathbf{v} \cdot \mathbf{u}}{\mathbf{u} \cdot \mathbf{u}} \mathbf{u}$$

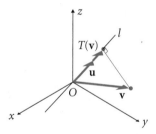

Figure 5.18   Orthogonal projection onto the line l.

SOLUTION   Because

$$T(c_1 v_1 + c_2 v_2) = \frac{(c_1 v_1 + c_2 v_2) \cdot \mathbf{u}}{\mathbf{u} \cdot \mathbf{u}} \mathbf{u} = \frac{c_1 v_1 \cdot \mathbf{u} + c_2 v_2 \cdot \mathbf{u}}{\mathbf{u} \cdot \mathbf{u}} \mathbf{u}$$
$$= c_1 \frac{v_1 \cdot \mathbf{u}}{\mathbf{u} \cdot \mathbf{u}} \mathbf{u} + c_2 \frac{v_2 \cdot \mathbf{u}}{\mathbf{u} \cdot \mathbf{u}} \mathbf{u} = c_1 T(v_1) + c_2 T(v_2)$$

we are done by Theorem 3.

■ **EXAMPLE 19**   (Requires Calculus)   Let V be the vector space of all differentiable real valued functions defined on $\mathbf{R}$. Show that the transformation $T : V \to V$

defined by differentiating each $f \in V$

$$T(f) = f'$$

is linear.

SOLUTION   If $f, g \in V$ and $c \in \mathbf{R}$, then basic properties of derivatives yield

$$T(f + g) = (f + g)' = f' + g' = T(f) + T(g)$$
$$T(cf) = (cf)' = cf' = cT(f)$$

So, T is linear.

■ EXAMPLE 20   (Requires Calculus)   Let $C[0, 1]$ be the vector space of all continuous real-valued differentiable functions defined on the interval $[0, 1]$. Show that the transformation $T : C[0, 1] \rightarrow C[0, 1]$ defined by (Riemann) integration

$$T(f) = \int_0^1 f(x)\, dx$$

is linear.

SOLUTION   If $f, g \in V$ and $c \in \mathbf{R}$, then basic properties of the integral yield

$$T(f + g) = \int_0^1 (f(x) + g(x))\, dx = \int_0^1 f(x)\, dx + \int_0^1 g(x)\, dx = T(f) + T(g)$$

$$T(cf) = \int_0^1 cf(x)\, dx = c \int_0^1 f(x)\, dx = cT(f)$$

Hence, T is linear.

■ EXAMPLE 21   Is the transformation $f : \mathbf{R} \rightarrow \mathbf{R}$ defined by $f(x) = x^2$ linear?

ANSWER   No, Part 1 of the definition fails, since

$$f(x + y) = (x + y)^2 = x^2 + 2xy + y^2$$

and

$$f(x) + f(y) = x^2 + y^2$$

are not equal if $xy \neq 0$. (Also, Part 2 fails. Why?)

■ EXAMPLE 22   Show that the transformation $f : \mathbf{R}^2 \rightarrow \mathbf{R}$ defined by $f(x, y) = xy$ is not linear.

SOLUTION   Verify.

Determination of a Linear Map from Its Values on a Basis

One of the most important properties of a linear transformation is that it can be determined *uniquely* given only its values on a basis. Let us explain this by an example.

■ **EXAMPLE 23**   Let $T : \mathbf{R}^2 \to \mathbf{R}^3$ be a linear transformation such that

$$T\begin{bmatrix} 1 \\ 1 \end{bmatrix} = \begin{bmatrix} -1 \\ 3 \\ 1 \end{bmatrix}, \quad T\begin{bmatrix} -1 \\ 2 \end{bmatrix} = \begin{bmatrix} -8 \\ -6 \\ 5 \end{bmatrix}$$

Compute $T\begin{bmatrix} -9 \\ 6 \end{bmatrix}$ and $T\begin{bmatrix} x \\ y \end{bmatrix}$.

$V = \{ \underline{v} \mid \underline{v} = c_1 \begin{bmatrix} 1 \\ 1 \end{bmatrix} + c_2 \begin{bmatrix} -1 \\ 2 \end{bmatrix} , \; c_1, c_2 \in \mathbf{R} \}$
$= \mathbf{R}^2$.

SOLUTION   Because

$B = \{ \begin{bmatrix} 1 \\ 1 \end{bmatrix}, \begin{bmatrix} -1 \\ 2 \end{bmatrix} \}$

$$\begin{bmatrix} -9 \\ 6 \end{bmatrix} = -4 \begin{bmatrix} 1 \\ 1 \end{bmatrix} + 5 \begin{bmatrix} -1 \\ 2 \end{bmatrix}$$

$T(\underline{v}) = T(c_1 \begin{bmatrix} 1 \\ 1 \end{bmatrix} + c_2 \begin{bmatrix} -1 \\ 2 \end{bmatrix})$
$= c_1 T(\begin{bmatrix} 1 \\ 1 \end{bmatrix}) + c_2 T(\begin{bmatrix} -1 \\ 2 \end{bmatrix})$
$= c_1 \begin{bmatrix} -1 \\ 3 \\ 1 \end{bmatrix} + c_2 \begin{bmatrix} -8 \\ -6 \\ 5 \end{bmatrix}$

if we apply T to both sides of the equation, we get

$$T\begin{bmatrix} -9 \\ 6 \end{bmatrix} = T\left(-4 \begin{bmatrix} 1 \\ 1 \end{bmatrix} + 5 \begin{bmatrix} -1 \\ 2 \end{bmatrix} \right) = -4T\begin{bmatrix} 1 \\ 1 \end{bmatrix} + 5T\begin{bmatrix} -1 \\ 2 \end{bmatrix}$$

$$= -4 \begin{bmatrix} -1 \\ 3 \\ 1 \end{bmatrix} + 5 \begin{bmatrix} -8 \\ -6 \\ 5 \end{bmatrix} = \begin{bmatrix} -36 \\ -42 \\ 21 \end{bmatrix}$$

The second equality holds because T is linear.
    It is easy to see that

$$\begin{bmatrix} x \\ y \end{bmatrix} = \left(\frac{2}{3}x + \frac{1}{3}y \right) \begin{bmatrix} 1 \\ 1 \end{bmatrix} + \left(-\frac{1}{3}x + \frac{1}{3}y \right) \begin{bmatrix} -1 \\ 2 \end{bmatrix}$$

Therefore,

$$T\begin{bmatrix} x \\ y \end{bmatrix} = \left(\frac{2}{3}x + \frac{1}{3}y \right) \begin{bmatrix} -1 \\ 3 \\ 1 \end{bmatrix} + \left(-\frac{1}{3}x + \frac{1}{3}y \right) \begin{bmatrix} -8 \\ -6 \\ 5 \end{bmatrix} = \begin{bmatrix} 2x - 3y \\ 4x - y \\ -x + 2y \end{bmatrix}$$

Matrix Operation Notation

All we did in Example 23 was to perform the matrix operations

$$\begin{bmatrix} 1 & -1 & : & x \\ 1 & 2 & : & y \end{bmatrix} \sim \begin{bmatrix} 1 & 0 & : & \frac{2}{3}x + \frac{1}{3}y \\ 0 & 1 & : & \frac{1}{3}y - \frac{1}{3}x \end{bmatrix}$$

and

$$\begin{bmatrix} -1 & -8 \\ 3 & -6 \\ 1 & 5 \end{bmatrix} \begin{bmatrix} \frac{2}{3}x + \frac{1}{3}y \\ \frac{1}{3}y - \frac{1}{3}x \end{bmatrix} = \begin{bmatrix} 2x - 3y \\ 4x - y \\ -x + 2y \end{bmatrix}$$

NOTE   We have seen how to compute the values of a linear transformation if we are given its values on a basis. If T is not given on an entire basis, we cannot find the image of *every* element. Referring to Example 23, if we were given only the images of $(1, 1)$ and $(-1, -1)$, then we would not be able to find $T(-9, 6)$, because $(-9, 6)$ is not in Span $\{(1, 1), (-1, -1)\}$ = Span $\{(1, 1)\}$.

■ **EXAMPLE 24**   Let $T : P_1 \rightarrow P_1$ be a linear transformation such that

$$T(-1 + x) = -7 + 2x, \qquad T(1 + x) = 4 + x$$

Compute $T(2 + 6x)$.

SOLUTION $\mathcal{B} = \{-1 + x, 1 + x\}$ is a basis of P_2. First we write $2 + 6x$ as a linear combination in $\mathcal{B}$:

$$2 + 6x = 2(-1 + x) + 4(1 + x)$$

Then we apply T.

$$T(2 + 6x) = 2T(-1 + x) + 4T(1 + x) = 2(-7 + 2x) + 4(4 + x) = 2 + 8x$$

In other words,

$$\begin{bmatrix} -1 & 1 & : & 2 \\ 1 & 1 & : & 6 \end{bmatrix} \sim \begin{bmatrix} 1 & 0 & : & 2 \\ 0 & 1 & : & 4 \end{bmatrix} \quad \text{and} \quad \begin{bmatrix} -7 & 4 \\ 2 & 1 \end{bmatrix} \begin{bmatrix} 2 \\ 4 \end{bmatrix} = \begin{bmatrix} 2 \\ 8 \end{bmatrix}$$

In general we have the following.

THEOREM 5   Let $T : V \rightarrow W$ be a linear transformation and let $\mathcal{B} = \{v_1, \ldots, v_n\}$ span V. Then the set $T(\mathcal{B}) = \{T(v_1), \ldots, T(v_n)\}$ spans the range of T.

PROOF   Let $w \in R(T)$. Then there exists a $v \in V$ such that $T(v) = w$. Because $\mathcal{B}$ spans V, there are scalars $c_1, \ldots, c_n$ such that $v = c_1 v_1 + \cdots + c_n v_n$. Then

$$w = T(v) = T(c_1 v_1 + \cdots + c_n v_n) = c_1 T(v_1) + \cdots + c_n T(v_n)$$

Hence w is a linear combination of $T(\mathcal{B})$.

Values of Linear Transformations with CAS

To redo Example 24 with CAS:

```
Maple

> with(linalg):
> col(rref([[-1,1,2],[1,1,6]]),3);

                          [2, 4]

> evalm([[-7,4],[2,1]] &* ");

                          [2, 8]
```

Mathematica

```
In[1]:=
    Map[Last,RowReduce[{{-1,1,2},{1,1,6}}]]
Out[1]=
    {2, 4}
In[2]:=
    {{-7,4},{2,1}} . %
Out[2]=
    {2, 8}
```

MATLAB

```
>> A=rref([-1 1 2; 1 1 6]); A(:,3)
ans =
          2
          4
>> [-7 4; 2 1] * ans
ans =
          2
          8
```

Exercises 5.2

In Exercises 1–12 find the domain and codomain of the transformation T and determine whether T is linear.

1. $T\begin{bmatrix} x \\ y \end{bmatrix} = \begin{bmatrix} 3x + y \\ x - 2y \end{bmatrix}$

2. $T\begin{bmatrix} x \\ y \end{bmatrix} = x\begin{bmatrix} 3 \\ 1 \end{bmatrix} + y\begin{bmatrix} 1 \\ -2 \end{bmatrix}$

3. $T\begin{bmatrix} x \\ y \end{bmatrix} = \begin{bmatrix} x - 1 \\ x - y \end{bmatrix}$

4. $T\begin{bmatrix} x \\ y \end{bmatrix} = x\begin{bmatrix} 3 \\ 1 \end{bmatrix} + y\begin{bmatrix} 1 \\ -2 \end{bmatrix} - \begin{bmatrix} 0 \\ 1 \end{bmatrix}$

5. $T\begin{bmatrix} x \\ y \end{bmatrix} = \begin{bmatrix} x + y \\ y \end{bmatrix}$

6. $T\begin{bmatrix} x \\ y \end{bmatrix} = \begin{bmatrix} xy \\ -y \end{bmatrix}$

7. $T(\mathbf{x}) = 3\mathbf{x}, \mathbf{x} \in \mathbf{R}^2$

8. $T(\mathbf{x}) = 3\mathbf{x} - \begin{bmatrix} 1 \\ 2 \end{bmatrix}, \mathbf{x} \in \mathbf{R}^2$

9. $T\begin{bmatrix} x \\ y \end{bmatrix} = \begin{bmatrix} x + y \\ y \\ 0 \end{bmatrix}$

10. $T\begin{bmatrix} x \\ y \\ z \end{bmatrix} = \begin{bmatrix} x - y \\ y + z \end{bmatrix}$

11. $T\begin{bmatrix} x \\ y \\ z \end{bmatrix} = \begin{bmatrix} 1 & -1 & 0 \\ 0 & 1 & 1 \end{bmatrix}\begin{bmatrix} x \\ y \\ z \end{bmatrix}$

12. $T\begin{bmatrix} x \\ y \\ z \end{bmatrix} = \begin{bmatrix} x - y \\ y + z \\ -7z \end{bmatrix}$

Use Theorem 3 to show that the following two transformations are linear.

13. $T : \mathbf{R}^3 \rightarrow \mathbf{R}^2, T\begin{bmatrix} x \\ y \\ z \end{bmatrix} = \begin{bmatrix} x + y + z \\ x - y - z \end{bmatrix}$

14. $T : \mathbf{R}^3 \rightarrow \mathbf{R}^2, T(\mathbf{x}) = \begin{bmatrix} 1 & 1 & 1 \\ 1 & -1 & -1 \end{bmatrix}\mathbf{x}$

In Exercises 15–20 determine whether $T : P_1 \rightarrow P_1$ is linear.

15. $T(a + bx) = (3a + b) + (a - 2b)x$

16. $T(a + bx) = (a - b) + (a + b + 1)x$

17. $T(a + bx) = (a - 5b) + abx$

18. $T(p) = 10p$

19. $T(p) = 10p - 2$

20. $T(p) = 10p^2$

21. Is $T : P_2 \rightarrow P_1$ linear?

$$T(a + bx + cx^2) = (a - b) + (b + c)x$$

22. Is $T : M_{22} \rightarrow \mathbf{R}^2$ linear?

$$T\begin{bmatrix} a & b \\ c & d \end{bmatrix} = \begin{bmatrix} a \\ c \end{bmatrix}$$

23. Is $T : \mathbf{R}^2 \rightarrow M_{22}$ linear?

$$T\begin{bmatrix} a \\ b \end{bmatrix} = \begin{bmatrix} a & b \\ 0 & 2 \end{bmatrix}$$

24. Let T be the linear transformation such that

$$T\begin{bmatrix} 1 \\ -1 \end{bmatrix} = \begin{bmatrix} 5 \\ -8 \end{bmatrix}, \qquad T\begin{bmatrix} 2 \\ 2 \end{bmatrix} = \begin{bmatrix} -9 \\ 4 \end{bmatrix}$$

Find $T\begin{bmatrix} x \\ y \end{bmatrix}, T\begin{bmatrix} 10 \\ -15 \end{bmatrix}$, and $T\begin{bmatrix} y \\ x \end{bmatrix}$.

25. Let $T : \mathbf{R}^3 \rightarrow \mathbf{R}^2$ be the linear transformation such that

$$T(\mathbf{e}_1 + \mathbf{e}_2 + \mathbf{e}_3) = \begin{bmatrix} 3 \\ -1 \end{bmatrix}$$

$$T(-\mathbf{e}_1 + \mathbf{e}_2 + \mathbf{e}_3) = \begin{bmatrix} 2 \\ -3 \end{bmatrix}$$

$$T(\mathbf{e}_1 - \mathbf{e}_2 + \mathbf{e}_3) = \begin{bmatrix} 2 \\ 1 \end{bmatrix}$$

Find $T(\mathbf{x})$ and $T\begin{bmatrix} -10 \\ 15 \\ -25 \end{bmatrix}$.

26. Let $T : P_1 \rightarrow P_1$ be the linear transformation such that

$$T(-1 + x) = -8 + 5x, \qquad T(2 + 2x) = 4 - 9x$$

Find $T(a + bx)$, $T(-15 + 10x)$, and $T(b + ax)$.

27. Let $T : P_2 \rightarrow P_1$ be the linear transformation such that

$$T(1 + x + x^2) = -1 + 3x$$

$$T(1 + x - x^2) = -3 + 2x$$

$$T(1 - x + x^2) = 1 + 2x$$

Find $T(a + bx + cx^2)$ and $T(-25 + 15x - 10x^2)$.

28. Let $T : P \rightarrow P$ be the linear transformation that satisfies

$$T(x^n) = \frac{1}{n + 1}x^{n+1}, \qquad n \geq 0$$

Compute: $T(x + x^2)$, $T(-1 + x^3)$, and $T((1 + x^2)^2)$.

29. Let C be an invertible $n \times n$ matrix. Show that $T : M_{nn} \rightarrow M_{nn}$ is linear.

$$T(X) = C^{-1}XC$$

30. Let C be an invertible $n \times n$ matrix. Show that $T : M_{nn} \rightarrow M_{nn}$ is linear.

$$T(X) = C^{-1}XC - X$$

31. Explain why a linear transformation T such that

$$T\begin{bmatrix} 1 \\ -1 \end{bmatrix} = \begin{bmatrix} 3 \\ -1 \end{bmatrix}, \qquad T\begin{bmatrix} -2 \\ 2 \end{bmatrix} = \begin{bmatrix} -6 \\ 2 \end{bmatrix}$$

cannot be uniquely determined. Find at least two such linear transformations.

32. Prove that the identity transformation is linear.

33. Give an example of a nonlinear transformation T with the property $T(\mathbf{0}) = \mathbf{0}$.

34. Let $T : \mathbf{R}^2 \rightarrow \mathbf{R}^2$ be a linear transformation. Show that if $\{\mathbf{v}_1, \mathbf{v}_2\} \subseteq \mathbf{R}^2$ is linearly dependent, then $\{T(\mathbf{v}_1), T(\mathbf{v}_2)\}$ is linearly dependent.

35. Give an example of a linear transformation $T : \mathbf{R}^2 \rightarrow \mathbf{R}^2$ and linearly independent 2-vectors $\mathbf{v}_1, \mathbf{v}_2$ such that $\{T(\mathbf{v}_1), T(\mathbf{v}_2)\}$ is linearly dependent.

36. Let $T : V \rightarrow W$ be a linear transformation of the vector spaces V and W. Show that if $\{\mathbf{v}_1, \ldots, \mathbf{v}_k\} \subseteq V$ is linearly dependent, then $\{T(\mathbf{v}_1), \ldots, T(\mathbf{v}_k)\} \subseteq W$ is linearly dependent.

37. Show that any linear transformation $T : \mathbf{R}^2 \rightarrow \mathbf{R}^2$ maps straight lines to straight lines or to points.

38. Show that

$$T : \mathbf{R}^m \rightarrow \mathbf{R}^m, \qquad T(\mathbf{x}) = \mathbf{x} + \mathbf{b}, \qquad \mathbf{b} \neq \mathbf{0}$$

is nonlinear. Such a transformation is called a **translation** of $\mathbf{R}^m$.

39. Find the sizes of A and $\mathbf{b}$ and show that

$$T : \mathbf{R}^n \rightarrow \mathbf{R}^m, \qquad T(\mathbf{x}) = A\mathbf{x} + \mathbf{b}, \qquad \mathbf{b} \neq \mathbf{0}$$

is nonlinear. Such a transformation is called an **affine transformation**.

40. Show that a translation (defined in Exercise 38) is a special case of an affine transformation (defined in Exercise 39).

In Exercises 41–47 show that the linear transformations are matrix transformations and find their standard matrices.

41. $T\begin{bmatrix} x \\ y \end{bmatrix} = \begin{bmatrix} x - y \\ y \\ x \end{bmatrix}$

42. $T\begin{bmatrix} x \\ y \\ z \end{bmatrix} = \begin{bmatrix} x - y \\ y - 4z \end{bmatrix}$

43. $\mathbf{0} : \mathbf{R}^2 \rightarrow \mathbf{R}^3$

44. $T : \mathbf{R}^3 \rightarrow \mathbf{R}^3, \ T(\mathbf{v}) = -10\mathbf{v}$

45. The reflection about the y-axis in the plane

46. The reflection about the line $y = -x$ in the plane

47. The counterclockwise rotation θ rad about the origin in the plane

48. Let $\mathcal{B}$ be a basis of $\mathbf{R}^n$ and let T be the transformation

$$T : \mathbf{R}^n \rightarrow \mathbf{R}^n, \qquad T(\mathbf{v}) = [\mathbf{v}]_{\mathcal{B}}$$

  a. Show that T is linear.

  b. Let $n = 2$ and $[\mathbf{e}_1]_{\mathcal{B}} = \begin{bmatrix} -2 \\ 6 \end{bmatrix}$, $[\mathbf{e}_2]_{\mathcal{B}} = \begin{bmatrix} 1 \\ -5 \end{bmatrix}$.

  Compute $T\begin{bmatrix} -9 \\ 6 \end{bmatrix}$.

49. Find all the linear transformations from $\mathbf{R}$ to $\mathbf{R}$.

In Example 16 we showed that dotting by a fixed vector in $\mathbf{R}^n$ is a linear transformation. In the next exercise we show that these are the *only* linear transformations from $\mathbf{R}^n$ to $\mathbf{R}$.

50. Let $T : \mathbf{R}^n \rightarrow \mathbf{R}$ be linear. Show that there exists an n-vector $\mathbf{u}$ such that

$$T(\mathbf{v}) = \mathbf{u} \cdot \mathbf{v} \qquad \text{for all } \mathbf{v} \in \mathbf{R}^n$$

(*Hint:* Let $\mathbf{u} = (T(\mathbf{e}_1), \ldots, T(\mathbf{e}_n))$.)

5.3 Kernel and Range

Reader's Goals for This Section

1. To know how to compute the kernel and range of a linear transformation.
2. To understand and know how to use the dimension theorem.
3. To understand the concepts one-to-one, onto, and isomorphism.

In this section we focus our attention on two basic subspaces that accompany every linear transformation, the *kernel* and the *range*. In addition, we discuss how we can distinguish between two different vector spaces and when we consider them to be the same, or *isomorphic*.

DEFINITION

(Kernel)

Let $T : V \rightarrow W$ be a linear transformation. The **kernel**, Ker(T), of T consists of all vectors in V that map to zero in W.

$$\text{Ker}(T) = \{\mathbf{v} \in V, \quad T(\mathbf{v}) = \mathbf{0} \in W\}$$

Recall that the **range** R(T), of T is the set of all images of T in W.

$$R(T) = \{\mathbf{w} \in W, \quad \mathbf{w} = T(\mathbf{v}) \text{ for some } \mathbf{v} \in V\}$$

Note that both Ker(T) and R(T) of a linear transformation T are nonempty sets: $T(\mathbf{0}) = \mathbf{0}$ implies that $\mathbf{0} \in V$ is in Ker(T) and that $\mathbf{0} \in W$ is in R(T).

■ EXAMPLE 25   Compute the kernel and range of

(a) The zero linear transformation $\mathbf{0} : V \rightarrow W$;
(b) The identity linear transformation $I : V \rightarrow V$;
(c) The projection $p : \mathbf{R}^2 \rightarrow \mathbf{R}^2$, $p(x, y) = (x, 0)$.

SOLUTION

(a) $\mathbf{0}(\mathbf{v}) = \mathbf{0}$ for all $\mathbf{v} \in V$, so the kernel is V. Because $\mathbf{0}$ is the only image, the range is $\{\mathbf{0}\}$.

$$\mathrm{Ker}(\mathbf{0}) = V, \qquad \mathrm{R}(\mathbf{0}) = \{\mathbf{0}\}$$

(b) Because $I(\mathbf{v}) = \mathbf{v}$ for all $\mathbf{v} \in V$, every nonzero vector maps to nonzero. So the kernel is $\{\mathbf{0}\}$. Since every $\mathbf{v}$ is its own image, the range is V.

$$\mathrm{Ker}(I) = \{\mathbf{0}\}, \qquad \mathrm{R}(I) = V$$

(c) (x, y) is in $\mathrm{Ker}(p)$ if and only if $p(x, y) = (x, 0) = (0, 0)$. So, $x = 0$ and the kernel consists of the points $(0, y)$. Also, (z, w) is in the range if and only if there is (x, y) such that $p(x, y) = (x, 0) = (z, w)$. Hence, $w = 0$, so the range consists of the points $(x, 0)$ (Fig. 5.19):

$$\mathrm{Ker}(p) = \{(0, y), \quad y \in \mathbf{R}\}, \qquad \mathrm{R}(p) = \{(x, 0), \quad x \in \mathbf{R}\}$$

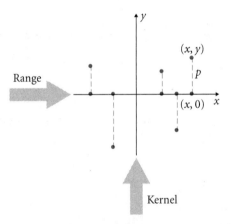

Figure 5.19   The kernel and range of the projection onto the x-axis.

■ EXAMPLE 26   Find the kernel of $T : \mathbf{R}^3 \rightarrow \mathbf{R}^2$.

$$T \begin{bmatrix} x \\ y \\ z \end{bmatrix} = \begin{bmatrix} x - z \\ y + z \end{bmatrix}$$

SOLUTION   Ker(T) is the set of all vectors $\begin{bmatrix} x \\ y \\ z \end{bmatrix}$ such that $\begin{bmatrix} x - z \\ y + z \end{bmatrix} = \begin{bmatrix} 0 \\ 0 \end{bmatrix}$. Solving

the system $x - z = 0, y + z = 0$, we get $(r, -r, r), r \in \mathbf{R}$. Hence,

$$\text{Ker}(T) = \left\{ \begin{bmatrix} r \\ -r \\ r \end{bmatrix}, r \in \mathbf{R} \right\} = \text{Span} \left\{ \begin{bmatrix} 1 \\ -1 \\ 1 \end{bmatrix} \right\}$$

The most basic property of the kernel and the range is that each is a vector space.

THEOREM 6

Let $T : V \rightarrow W$ be a linear transformation. Then

1. Ker(T) is a subspace of V;
2. R(T) is a subspace of W.

PROOF

1. Let $\mathbf{u}, \mathbf{v} \in \text{Ker}(T)$ and let $c \in \mathbf{R}$. Because Ker(T) is nonempty, it suffices to show that $\mathbf{u} + \mathbf{v}, c\mathbf{u} \in \text{Ker}(T)$. T is linear, so

$$T(\mathbf{u} + \mathbf{v}) = T(\mathbf{u}) + T(\mathbf{v}) = \mathbf{0} + \mathbf{0} = \mathbf{0}$$

$$T(c\mathbf{u}) = c\,T(\mathbf{u}) = c\mathbf{0} = \mathbf{0}$$

Therefore, $\mathbf{u} + \mathbf{v}, c\mathbf{u} \in \text{Ker}(T)$. Hence, Ker(T) is a subspace of V.
2. Let $\mathbf{u}', \mathbf{v}' \in \text{R}(T)$ and let $c \in \mathbf{R}$. Then there are vectors $\mathbf{u}$ and $\mathbf{v}$ of V such that $\mathbf{u}' = T(\mathbf{u})$ and $\mathbf{v}' = T(\mathbf{v})$. Because T is linear, we have

$$\mathbf{u}' + \mathbf{v}' = T(\mathbf{u}) + T(\mathbf{v}) = T(\mathbf{u} + \mathbf{v})$$

$$c\mathbf{u}' = c\,T(\mathbf{u}) = T(c\mathbf{u})$$

We found vectors $\mathbf{u} + \mathbf{v}$ and $c\mathbf{u}$ that map to $\mathbf{u}' + \mathbf{v}'$ and $c\mathbf{u}'$, respectively. Therefore, $\mathbf{u}' + \mathbf{v}'$ and $c\mathbf{u}'$ are in R(T). Hence, R(T) is a subspace of W.

The dimension of the kernel is called the **nullity** of T. The dimension of the range is called the **rank** of T.

■ EXAMPLE 27   Compute the nullity and rank of T in Example 26.

SOLUTION   Because $\text{Ker}(T) = \{(r, -r, r), r \in \mathbf{R}\} = \text{Span}\{(1, -1, 1)\}$ the nullity is 1. Because $\text{R}(T) = \mathbf{R}^2$, the rank is 2 (Fig. 5.20).

■ EXAMPLE 28   Compute the kernel and range of T of Example 16, Section 5.2.

SOLUTION   The kernel consists of all vectors $\mathbf{v}$ such that $\mathbf{u} \cdot \mathbf{v} = 0$, i.e., all n-vectors *orthogonal* to $\mathbf{u}$. This is the hyperplane through the origin with normal $\mathbf{u}$. To compute

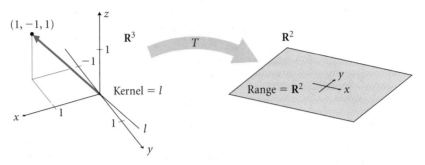

Figure 5.20   The kernel and range of $T(x, y, z) = (x - z, y + z)$.

the range we observe that because $\mathbf{u}$ is nonzero,

$$T(\mathbf{u}) = \mathbf{u} \cdot \mathbf{u} = \|\mathbf{u}\|^2 > 0$$

Hence, the nonzero number $\|\mathbf{u}\|^2$ is in the range of T. So, the range contains the span of $\|\mathbf{u}\|^2$, which is $\mathbf{R}$. Therefore, $R(T) = \mathbf{R}$ (Fig. 5.21).

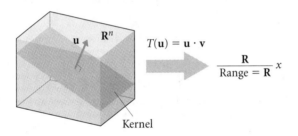

Figure 5.21   Kernel and image of dotting by fixed vector.

■ EXAMPLE 29   Compute the kernel, range, nullity, and rank of T of Example 17, Section 5.2.

SOLUTION   We compute the kernel and range *geometrically*, which is simpler than an algebraic calculation. The kernel consists of all vectors $\mathbf{v}$ such that $\mathbf{u} \times \mathbf{v} = \mathbf{0}$. These are vectors parallel to $\mathbf{u}$. Hence, the kernel is the line through the origin in the direction of $\mathbf{u}$. The nullity of T is 1.

The range consists of all vectors $\mathbf{b}$ for which there is a vector $\mathbf{v}$ such that $\mathbf{u} \times \mathbf{v} = \mathbf{b}$. Hence, $\mathbf{u}$ and $\mathbf{b}$ are orthogonal. So, the range is contained in the set of vectors orthogonal to $\mathbf{u}$. Now let $\mathbf{b}$ be a nonzero vector orthogonal to $\mathbf{u}$. We show that $\mathbf{b}$ is in the range of T. The vectors $\mathbf{b}$ and $(\mathbf{u} \times \mathbf{b}) \times \mathbf{u}$ are parallel, because each is perpendicular to both $\mathbf{u}$ and $\mathbf{u} \times \mathbf{b}$. So, $(\mathbf{u} \times \mathbf{b}) \times \mathbf{u} = c\mathbf{b}$ for some nonzero scalar c. Therefore,

$$\mathbf{u} \times \left(-\frac{1}{c}\mathbf{u} \times \mathbf{b} \right) = \mathbf{b}$$

So the vector $-\frac{1}{c}\mathbf{u} \times \mathbf{b}$ maps to $\mathbf{b}$; hence $\mathbf{b}$ is in the range of T. Thus we proved that the range is the plane through the origin with normal $\mathbf{u}$. The rank of T is 2 (Fig. 5.22).

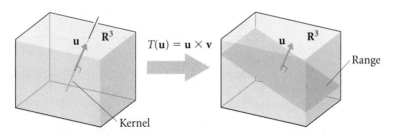

Figure 5.22   The kernel and image of crossing by fixed vector.

In the special case when our linear transformation is a matrix transformation, we have the following.

THEOREM 7

Let $T : \mathbf{R}^n \to \mathbf{R}^m$ be a matrix transformation with standard matrix A. Then

1. $\text{Ker}(T) = \text{Null}(A)$;
2. $\text{R}(T) = \text{Col}(A)$;
3. $\text{Nullity}(T) = \text{Nullity}(A)$;
4. $\text{Rank}(T) = \text{Rank}(A)$.

PROOF   Because $T(\mathbf{x}) = A\mathbf{x}$, $T(\mathbf{x}) = \mathbf{0}$ if and only if $A\mathbf{x} = \mathbf{0}$. Hence, $\text{Ker}(T) = \text{Null}(A)$. Likewise, if $\mathbf{b}$ is in $\mathbf{R}^m$, then there is an $\mathbf{x}$ in $\mathbf{R}^n$ such that $T(\mathbf{x}) = \mathbf{b}$ if and only if $A\mathbf{x} = \mathbf{b}$. Hence, $\text{R}(T) = \text{Col}(A)$. The claims on the nullities and ranks follow.

■ EXAMPLE 30   Find bases for the kernel and range and compute the nullity and rank of

$$T : \mathbf{R}^4 \to \mathbf{R}^3, \qquad T(x, y, z, w) = (x + 3z, y - 2z, w)$$

SOLUTION   By Theorem 7 it suffices to write T as a matrix transformation and find bases for the column and null space of its standard matrix A. A is given by

$$A = \begin{bmatrix} 1 & 0 & 3 & 0 \\ 0 & 1 & -2 & 0 \\ 0 & 0 & 0 & 1 \end{bmatrix}$$

and is in reduced row echelon form. Hence, the vectors $\{(1, 0, 0), (0, 1, 0), (0, 0, 1)\}$ form a basis for $\text{Col}(A) = \text{R}(T) = \mathbf{R}^3$. On the other hand, $\{(-3, 2, 1, 0)\}$ is a basis for $\text{Null}(A) = \text{Ker}(T)$. Hence, the rank of T is 3 and the nullity is 1.

Theorem 7 implies that

$$\text{Ker}(T) = \{\mathbf{0}\} \Leftrightarrow A\mathbf{x} = \mathbf{0} \text{ has only the trivial solution}$$

and

$$\text{R}(T) = \mathbf{R}^m \Leftrightarrow \text{the columns of } A \text{ span } \mathbf{R}^m$$

The next theorem is one of the cornerstones of linear algebra. It generalizes the rank theorem and it is proved in Section 5.4.

THEOREM 8

(The Dimension Theorem)

If $T : V \rightarrow W$ is a linear transformation from a finite-dimensional vector space V into a vector space W. Then

$$\text{Nullity}(T) + \text{Rank}(T) = \dim(V)$$

■ **EXAMPLE 31**   Verify the dimension theorem for T of Example 14, Section 5.2.

SOLUTION   We leave as exercise the verification of

$$\text{Ker}(T) = P_0 = \{a_0,\ a_0 \in \mathbf{R}\} \subseteq P_2$$

and

$$\text{R}(T) = P_1 = \{a_0 + a_1 x,\ a_0, a_1 \in \mathbf{R}\} \subseteq P_2$$

Hence,

$$\dim(P_2) = 3 = 1 + 2 = \dim(P_0) + \dim(P_1)$$

Theorem 8 is very useful, because it is often much easier to compute the null space and the nullity of a linear transformation than its range and rank, as the work in Examples 28 and 29 suggests.

■ **EXAMPLE 32**   Determine the range of the linear transformation $T : \mathbf{R}^4 \rightarrow P_2$.

$$T(a, b, c, d) = (a - b) + (c + d)x + (2a + b)x^2$$

SOLUTION   The null space of T is spanned by $(0, 0, -1, 1)$, which is obtained immediately from solving the system $a - b = 0, c + d = 0, 2a + b = 0$. Therefore, the

---

[5] Not necessarily finite-dimensional.

nullity of T is 1. Hence, by the dimension theorem,

$$\text{Rank}(T) = \dim(\mathbf{R}^4) - \text{Nullity}(T) = 4 - 1 = 3$$

Thus the range is a three-dimensional subspace of P_2, so it is all of P_2.

■ EXAMPLE 33   Suppose that a linear transformation $T : \mathbf{R}^4 \rightarrow \mathbf{R}^3$ has kernel spanned by one nonzero vector. What is the range of T?

SOLUTION   The nullity of T is 1. By the dimension theorem,

$$\text{Rank}(T) = 4 - \text{Nullity}(T) = 4 - 1 = 3$$

Hence, the range is a three-dimensional subspace of $\mathbf{R}^3$. Therefore, $\text{R}(T) = \mathbf{R}^3$.

One-to-one, Onto, and Isomorphisms

We know already that there are many vector spaces that mathematicians and scientists are interested in. However, if we set the different notations of the individual examples aside, we may find that many of these spaces are "essentially the same." In this paragraph we analyze the notion that two vector spaces are the same. We call such spaces isomorphic. But let us start from the beginning.

The definition of a transformation $T : A \rightarrow B$ between two sets allows for

1. Two or more elements of A to have the same image;
2. The range of T to be strictly contained in the codomain B.

If (1) does not occur, T is one-to-one.

DEFINITION   **(One-to-One)**

A transformation $T : A \rightarrow B$ is called **one-to-one** if for each element b of the range, there is *exactly one* element a with image $b = T(a)$. This can be rephrased as

$$T(a_1) = T(a_2) \Rightarrow a_1 = a_2 \tag{5.2}$$

or, equivalently,

$$a_1 \neq a_2 \Rightarrow T(a_1) \neq T(a_2) \tag{5.3}$$

(See Fig. 5.23.)

If (2) is false, T is onto.

DEFINITION   **(Onto)**

A transformation $T : A \rightarrow B$ is called **onto** if its range equals its codomain, i.e.,

$$\text{R}(T) = B \tag{5.4}$$

(See Fig. 5.23.)

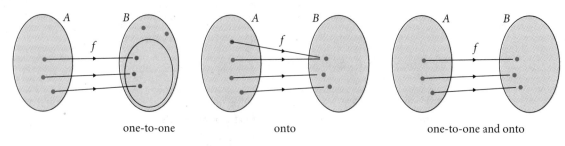

one-to-one   onto   one-to-one and onto

Figure 5.23   One-to-one and onto.

■ EXAMPLE 34   Which of the transformations are one-to-one? Onto?

(a) $T : \mathbf{R}^2 \to \mathbf{R}^3$, $T(x, y) = (x + y, y, 0)$
(b) $T : \mathbf{R}^3 \to \mathbf{R}^2$, $T(x, y, z) = (x, z)$

SOLUTION

(a) If $T(x_1, y_1) = T(x_2, y_2)$, then $(x_1 + y_1, y_1, 0) = (x_2 + y_2, y_2, 0)$. Therefore, $y_1 = y_2$. Hence, $x_1 = x_2$. So $(x_1, y_1) = (x_2, y_2)$ and T is one-to-one. T is not onto, because $(0, 0, 1)$ is not the image of any 2-vector.
(b) T is not one-to-one, because $T(0, 0, 0) = (0, 0) = T(0, 1, 0)$. T is onto, because for any 2-vector (a, b), there is at least one 3-vector that maps to it. For example, $T(a, 0, b) = (a, b)$.

The next theorem tells us that in order to show that a linear transformation is one-to-one, it suffices to check that only zero maps to zero, and ignore all other images. This reduces the amount of labor involved in checking a *linear* transformation for one-to-one.

I is onto   ⇔   Rank (T) = dim (W)

THEOREM 9   Let $T : V \to W$ be a linear transformation. Then:

$$T \text{ is one-to-one} \quad \Leftrightarrow \quad \text{Ker}(T) = \{\mathbf{0}\}$$

T is not onto ⇔ Rank (T) < dim (W)

PROOF   Suppose T is one-to-one. If $\mathbf{v}$ in the kernel of T, then $T(\mathbf{v}) = \mathbf{0}$. But we know that $T(\mathbf{0}) = \mathbf{0}$. Hence $T(\mathbf{v}) = T(\mathbf{0})$, which implies $\mathbf{v} = \mathbf{0}$, because T is one-to-one. This proves that any element of the kernel is the zero vector. Hence, $\text{Ker}(T) = \{\mathbf{0}\}$.
    Conversely, suppose $\text{Ker}(T) = \{\mathbf{0}\}$. We shall prove that it is one-to-one. Let $\mathbf{u}$ and $\mathbf{v}$ be vectors of V such that $T(\mathbf{u}) = T(\mathbf{v})$. We must show that $\mathbf{u} = \mathbf{v}$. Because T is linear,

$$T(\mathbf{u}) = T(\mathbf{v}) \Rightarrow T(\mathbf{u}) - T(\mathbf{v}) = \mathbf{0} \Rightarrow T(\mathbf{u} - \mathbf{v}) = \mathbf{0}$$

Hence, $\mathbf{u} - \mathbf{v}$ is in the kernel of T, which was assumed to be $\{\mathbf{0}\}$. Hence $\mathbf{u} - \mathbf{v} = \mathbf{0}$ or $\mathbf{u} = \mathbf{v}$. Therefore, T is one-to-one (Fig. 5.24).

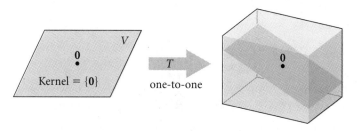

Figure 5.24 The kernel of a one-to-one linear transformation is zero.

■ EXAMPLE 35   Show that $T : \mathbf{R}^2 \to \mathbf{R}^2$, $T(x, y) = (x - y, x + 2y)$ is one-to-one and onto.

SOLUTION   Let $T(x, y) = (0, 0)$. Then $(x - y, x + 2y) = (0, 0)$ implies that $x = 0, y = 0$, or $(x, y) = (0, 0)$. Hence, T is one-to-one by Theorem 9 and its nullity is zero. The rank is 2. So, the range is all of $\mathbf{R}^2$, and the transformation is onto.

■ EXAMPLE 36   Is the transformation of Example 33 one-to-one? Is it onto?

SOLUTION   It is not one-to-one, because the kernel is nonzero. It is onto, because its range was found to be $\mathbf{R}^3$.

■ EXAMPLE 37   Show that $T : P_1 \to P_1$, $T(a + bx) = (a - b) + 2ax$ is one-to-one and onto.

SOLUTION   That T is linear is left as an exercise. Let $p = a + bx$ be in the kernel of T. Then $T(p) = \mathbf{0}$. So, $(a - b) + 2ax$ is the zero polynomial, which implies that $a - b = 0$ and $2a = 0$, or $a = b = 0$. Hence, $p = \mathbf{0}$ and the kernel is $\{\mathbf{0}\}$. So, T is one-to-one by Theorem 9, and its nullity is 0. Hence, the rank is 2, by the dimension theorem. Therefore, the range is P_1, and the transformation is onto.

THEOREM 10

A one-to-one linear transformation maps linearly independent sets to linearly independent sets. In other words, if $T : V \to W$ is linear and one-to-one and $\{\mathbf{v}_1, \ldots, \mathbf{v}_k\}$ is a linearly independent subset of V, then

$$\{T(\mathbf{v}_1), \ldots, T(\mathbf{v}_k)\}$$

is a linearly independent subset of W.

PROOF   Let

$$c_1 T(\mathbf{v}_1) + \cdots + c_k T(\mathbf{v}_k) = \mathbf{0}$$

then

$$T(c_1\mathbf{v}_1 + \cdots + c_k\mathbf{v}_k) = \mathbf{0}$$

since T is linear. Hence,

$$c_1\mathbf{v}_1 + \cdots + c_k\mathbf{v}_k = \mathbf{0}$$

by Theorem 9. But because $\{\mathbf{v}_1, \ldots, \mathbf{v}_k\}$ is linearly independent, $c_1 = \cdots = c_k = 0$. Therefore, $\{T(\mathbf{v}_1), \ldots, T(\mathbf{v}_k)\}$ is also linearly independent.

Our next theorem is a labor-saving one. It says that a one-to-one linear transformation between spaces of the same dimension is automatically onto, and vice versa.

THEOREM 11   Let $T : V \rightarrow W$ be a linear transformation between two finite-dimensional vector spaces V, W with $\dim(V) = \dim(W)$. Then T is one-to-one if and only if it is onto.

PROOF   Exercise. (*Hint:*   Use Theorem 9 and the dimension theorem.)

DEFINITION

(Isomorphism)

A linear transformation between two vector spaces that is one-to-one and onto is called an **isomorphism**. Two vector spaces are called **isomorphic** if there is an isomorphism between them. We consider isomorphic spaces to be the same because their elements correspond one for one and the structure of the vector space operations is preserved through linearity.

■ EXAMPLE 38   The transformation T of Example 37 is an isomorphism.

■ EXAMPLE 39   Show that $\mathbf{R}^n$ and P_{n-1} are isomorphic.

SOLUTION   It suffices to find an isomorphism between the two spaces. Consider $T : \mathbf{R}^n \rightarrow P_{n-1}$ defined by

$$T(a_0, \ldots, a_{n-1}) = a_0 + a_1 x + \cdots + a_{n-1}x^{n-1}$$

$\mathbf{R}^3$ and P_2 are isomorphic.

We leave as an exercise the proof that T is linear, one-to-one, and onto and, hence, an isomorphism.

■ EXAMPLE 40   Show that $\mathbf{R}^6$ and M_{23} are isomorphic.

SOLUTION   The map $T : \mathbf{R}^6 \rightarrow M_{23}$ defined by

$$T(a_1, \ldots, a_6) = \begin{bmatrix} a_1 & a_2 & a_3 \\ a_4 & a_5 & a_6 \end{bmatrix}$$

is easily seen to be linear, one-to-one, and onto.

■ EXAMPLE 41   Show that $\mathbf{R}^{mn}$ and M_{mn} are isomorphic.

SOLUTION   Exercise.

■ EXAMPLE 42   Show that M_{32} and M_{23} are isomorphic.

SOLUTION   The map $T : M_{32} \rightarrow M_{23}$ defined by

$$T \begin{bmatrix} a_1 & a_2 \\ a_3 & a_4 \\ a_5 & a_6 \end{bmatrix} = \begin{bmatrix} a_1 & a_2 & a_3 \\ a_4 & a_5 & a_6 \end{bmatrix}$$

is easily seen to be linear, one-to-one, and onto.

The most important criterion for determining whether two vector spaces are isomorphic is also the simplest one:

THEOREM 12

Let V and W be finite-dimensional vector spaces. Then

$$V \text{ and } W \text{ are isomorphic} \quad \Leftrightarrow \quad \dim(V) = \dim(W)$$

PROOF   If V and W are isomorphic, then there is an isomorphism $T : V \rightarrow W$. If $\dim(V) = n$ and $\mathcal{B} = \{\mathbf{v}_1, \ldots, \mathbf{v}_n\}$ is a basis of V, then the set of the images $T(\mathcal{B}) = \{T(\mathbf{v}_1), \ldots, T(\mathbf{v}_n)\}$ is linearly independent by Theorem 10. Because T is onto the span of $T(\mathcal{B})$ is all of W. Therefore, $T(\mathcal{B})$ is a basis of W; because it has n elements, $\dim(W) = n$.

Conversely, suppose V and W have the same dimension, n. If $\mathcal{B} = \{\mathbf{v}_1, \ldots, \mathbf{v}_n\}$ and $\mathcal{B}' = \{\mathbf{w}_1, \ldots, \mathbf{w}_n\}$ are bases of V and W, respectively, we establish an isomorphism $T : V \rightarrow W$ as follows: Let $\mathbf{v}$ be in V. Because $\mathcal{B}$ spans V, there are scalars c_i such that

$$\mathbf{v} = c_1 \mathbf{v}_1 + \cdots + c_n \mathbf{v}_n$$

We define T by

$$T(\mathbf{v}) = c_1\mathbf{w}_1 + \cdots + c_n\mathbf{w}_n$$

The map T is well defined; the coefficients c_i are uniquely determined because $\mathcal{B}$ is a basis. We let the reader prove that T is linear. It is also one-to-one, because if $\mathbf{v}$ is in $\text{Ker}(T)$, then

$$\mathbf{0} = T(\mathbf{v}) = c_1\mathbf{w}_1 + \cdots + c_n\mathbf{w}_n$$

implies $c_1 = \cdots = c_n = 0$, because $\mathcal{B}'$ is linearly independent. Therefore, $\mathbf{v} = \mathbf{0}$, by Theorem 9. So, T is onto by Theorem 11 and, hence, is an isomorphism.

■ EXAMPLE 43   Show that $\mathbf{R}^2$ and $\mathbf{R}^3$ are not isomorphic.

SOLUTION   They cannot be isomorphic, because they do not have the same dimension.

■ EXAMPLE 44   Show that $\mathbf{R}^n$ and P_n are not isomorphic.

SOLUTION   This is true because

$$n = \dim(\mathbf{R}^n) \neq \dim(P_n) = n + 1$$

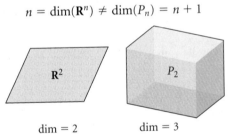

dim = 2   dim = 3

$\mathbf{R}^2$ and P_2 are *not* isomorphic.

Kernel and Range with CAS

To find bases for the kernel and range of $T(x, y, z) = (x + 2y + 3z, x + 3y + 2z)$:

```
Maple

> with(linalg):
> A:=matrix([[1,2,3],[1,3,2]]):
> nullspace(A);colspace(A);

          {[-5, 1, 1]}     {[0, 1], [1, 0]}
```

Mathematica

```
In[1]:= A={{1,2,3},{1,3,2}};
In[2]:= NullSpace[A]
        RowReduce[A]
Out[2]= {{-5, 1, 1}}
Out[3]= {{1, 0, 5}, {0, 1, -1}}
```

> The range is computed indirectly

MATLAB

```
>> A=[1 2 3; 1 3 2];
>> null(A), rref(A)
ans =            ans =
  -0.9623
   0.1925         1   0    5
   0.1925         0   1   -1
```

> The range is computed indirectly

Exercises 5.3

Kernel and Range

In Exercises 1–10 find bases for the kernel and range, and compute the nullity and rank of T. In each case verify the dimension theorem.

1. $T \begin{bmatrix} x \\ y \\ z \end{bmatrix} = \begin{bmatrix} x - y \\ y - z \\ -x + z \end{bmatrix}$

2. $T \begin{bmatrix} x \\ y \\ z \\ w \end{bmatrix} = \begin{bmatrix} x - y \\ y - w \\ z + w \end{bmatrix}$

3. $T \begin{bmatrix} x \\ y \end{bmatrix} = \begin{bmatrix} x - 2y \\ x - 2y \\ 2x - 4y \\ 0 \end{bmatrix}$ **4.** $I : \mathbf{R}^2 \rightarrow \mathbf{R}^2$

5. $T : \mathbf{R}^3 \rightarrow \mathbf{R}^3, T(\mathbf{v}) = -2\mathbf{v}$

6. $T : \mathbf{R}^2 \rightarrow \mathbf{R}^2$ is the projection onto the y-axis.

7. $T : \mathbf{R}^2 \rightarrow \mathbf{R}^2$ is the reflection about the line $y = -x$.

8. $T : P_2 \rightarrow P_2$ such that

$T(a + bx + cx^2) = (a - b) + (b - c)x + (-a + c)x^2$

9. $T : P_3 \rightarrow P_2$ such that

$T(a + bx + cx^2 + dx^3) = (a-b) + (b-d)x + (c+d)x^2$

10. $T : P_1 \rightarrow P_3$ such that

$T(a + bx) = (a - 2b)x + (a - 2b)x^2 + (2a - 4b)x^3$

In Exercises 11–13 find a basis for the kernel and the range of $T : P_2 \rightarrow P_2$ if T satisfies the given equations.

11. $T(1) = x, T(x) = x^2, T(x^2) = -1$

12. $T(p(x)) = p(1 + x)$

13. $T(1) = 0, T(x) = 0, T(x^2) = 1$

In Exercises 14–16 for the given A find the dimension of the kernel of the linear transformation $T : M_{33} \rightarrow M_{33}$ defined by

$$T(X) = AX$$

14. $A = \begin{bmatrix} 1 & 0 & 0 \\ 0 & 1 & 0 \\ 0 & 0 & 0 \end{bmatrix}$ **15.** $A = \begin{bmatrix} 1 & 1 & 1 \\ 1 & 0 & 1 \\ 0 & 0 & 0 \end{bmatrix}$

16. $A = \begin{bmatrix} 1 & -1 & 0 \\ -2 & 2 & 0 \\ 0 & 0 & 0 \end{bmatrix}$

In Exercises 17–19 for the given n and A, find the nullity and rank of the linear transformation $T : M_{22} \rightarrow M_{22}$ defined by

$$T(X) = AX - XA$$

17. $A = \begin{bmatrix} 2 & 0 \\ 0 & 3 \end{bmatrix}$ **18.** $A = E_{12}$ **19.** $A = E_{22}$

20. Referring to Exercise 50, Section 5.2, suppose that $\mathbf{u} \neq \mathbf{0}$. Find the kernel and range of T. If $n = 3$ describe the kernel and range geometrically and verify the dimension theorem.

One-to-one, Onto, and Isomorphisms

In Exercises 21–26 use Theorem 9 to show that the linear transformations are one-to-one.

21. $T \begin{bmatrix} x \\ y \end{bmatrix} = \begin{bmatrix} 2x + y \\ -3x + 4y \end{bmatrix}$

22. $T \begin{bmatrix} x \\ y \\ z \end{bmatrix} = \begin{bmatrix} 2x - y \\ -y + z \\ -3x + z \end{bmatrix}$

23. $T \begin{bmatrix} x \\ y \end{bmatrix} = \begin{bmatrix} 2x - y \\ x - y \\ -x + y \\ x - 2y \end{bmatrix}$

24. $T : P_1 \rightarrow P_1$ such that

$$T(a + bx) = (2a + b) + (-3a + 4b)x$$

25. $T : P_2 \rightarrow P_2$ such that

$$T(a + bx + cx^2) = (2a - b) + (-b + c)x + (-3a + c)x^2$$

26. $T : P_1 \rightarrow P_4$ such that

$$T(a + bx) = (2a - b) + (a - b)x + (-a + b)x^2$$
$$+ (a - 2b)x^3$$

In Exercises 27–32 show that the linear transformations are onto.

27. $T \begin{bmatrix} x \\ y \end{bmatrix} = \begin{bmatrix} x - y \\ -x + 2y \end{bmatrix}$

28. $T \begin{bmatrix} x \\ y \\ z \end{bmatrix} = \begin{bmatrix} x \\ z \end{bmatrix}$

29. $T \begin{bmatrix} x \\ y \\ z \end{bmatrix} = \begin{bmatrix} x - y + z \\ -x + y + z \end{bmatrix}$

30. $T(\mathbf{x}) = \begin{bmatrix} 1 & 2 & 3 \\ 2 & 2 & 3 \\ 3 & 3 & 3 \end{bmatrix} \mathbf{x}, \; \mathbf{x} \in \mathbf{R}^3$

31. $T : P_2 \rightarrow P_1$ such that

$$T(a + bx + cx^2) = (a + b) + (a + c)x$$

32. $T : P_2 \rightarrow P_2$ such that

$$T(a + bx + cx^2) = c + bx + ax^2$$

In Exercises 33–37 show that the linear transformations are isomorphisms.

33. $T : P_1 \rightarrow P_1$ such that

$$T(a + bx) = (a - 2b) + (-2a + b)x$$

34. $T \begin{bmatrix} x \\ y \end{bmatrix} = \begin{bmatrix} x - 2y \\ -2x + y \end{bmatrix}$

35. $T \begin{bmatrix} x \\ y \\ z \end{bmatrix} = \begin{bmatrix} x - y + z \\ -x + y + z \\ -y + z \end{bmatrix}$

36. $T(\mathbf{x}) = \begin{bmatrix} 1 & 0 & 1 \\ 1 & 0 & -1 \\ 0 & -2 & 6 \end{bmatrix} \mathbf{x}, \; \mathbf{x} \in \mathbf{R}^3$

37. $T : P_2 \rightarrow P_2$ such that

$$T(a + bx + cx^2) = c + bx + (a - b)x^2$$

Which of the following maps are isomorphisms?

38. $T \begin{bmatrix} x \\ y \end{bmatrix} = \begin{bmatrix} x \\ 0 \\ y \end{bmatrix}$ **39.** $T \begin{bmatrix} x \\ y \\ z \end{bmatrix} = \begin{bmatrix} x \\ y \end{bmatrix}$

40. $I : \mathbf{R}^n \rightarrow \mathbf{R}^n, I(\mathbf{x}) = \mathbf{x}$

41. $T : \mathbf{R}^n \rightarrow \mathbf{R}^n, T(\mathbf{x}) = -10\mathbf{x}$

42. $T : \mathbf{R}^n \rightarrow \mathbf{R}^n, T(\mathbf{x}) = \mathbf{x} + \mathbf{b}, \mathbf{b} \neq \mathbf{0}$

43. $T : \mathbf{R}^n \rightarrow \mathbf{R}^m, T(\mathbf{x}) = A\mathbf{x}, m \neq n$

44. Show that any isomorphism $T : \mathbf{R}^2 \rightarrow \mathbf{R}^2$ maps straight lines to straight lines.

45. Let $\mathcal{B}$ be a basis of $\mathbf{R}^n$. Show that the coordinate transformation is an isomorphism.

$$T : \mathbf{R}^n \rightarrow \mathbf{R}^n, \quad T(\mathbf{v}) = [\mathbf{v}]_{\mathcal{B}}$$

46. Let A be an $n \times n$ matrix of rank n and let $\mathbf{b}$ be an n-vector. Show that the affine transformation (defined in Exercise 39, Section 5.2)

$$T : \mathbf{R}^n \rightarrow \mathbf{R}^n, \quad T(\mathbf{x}) = A\mathbf{x} + \mathbf{b}, \quad \mathbf{b} \neq \mathbf{0}$$

is one-to-one and onto. Is it an isomorphism?

Some Theoretical Results

Let V and W be finite-dimensional vector spaces. Prove the following statements:

47. If T is one-to-one, then $\dim(V) \leq \dim(W)$.

48. If T is onto and a set S spans V, then $T(S)$ spans W.

49. If T is onto, then $\dim(V) \geq \dim(W)$.

50. If T is an isomorphism and a set $\mathcal{B}$ is a basis of V, then $T(\mathcal{B})$ is a basis of W.

5.4 The Matrix of a Linear Transformation

Reader's Goals for This Section

1. To know how to compute the matrix of a linear transformation.
2. To be able to evaluate a linear transformation from its matrix.
3. To know how to compute the matrix of a linear transformation with respect to a new basis.

In this section we generalize the concept of the standard matrix of a matrix transformation. We show that *any* linear transformation between finite-dimensional vector spaces can be represented by a matrix transformation. This useful result allows us to evaluate linear transformations by using matrix multiplication, which can be done by computer.

THEOREM 13

(Matrix of Linear Transformation)[6]

Let $T : V \to W$ be a linear transformation between two finite-dimensional vector spaces V and W. Let $\mathcal{B} = \{\mathbf{v}_1, \ldots, \mathbf{v}_n\}$ be a basis of V and let $\mathcal{B}' = \{\mathbf{v}_1', \ldots, \mathbf{v}_m'\}$ be a basis of W. The $m \times n$ matrix A with columns

$$A = \left[\, [T(\mathbf{v}_1)]_{\mathcal{B}'}, \ldots, [T(\mathbf{v}_n)]_{\mathcal{B}'} \,\right]$$

is the only matrix that satisfies

$$[T(\mathbf{v})]_{\mathcal{B}'} = A\,[\mathbf{v}]_{\mathcal{B}}$$

for all $\mathbf{v} \in V$.

PROOF Because $\mathcal{B}$ spans V, there are scalars $c_1, \ldots, c_n$ such that $\mathbf{v} = c_1\mathbf{v}_1 + \cdots + c_n\mathbf{v}_n$. So

$$T(\mathbf{v}) = c_1 T(\mathbf{v}_1) + \cdots + c_n T(\mathbf{v}_n)$$

because T is linear. Hence, by Theorem 18, Section 4.5, we have

$$[T(\mathbf{v})]_{\mathcal{B}'} = c_1\,[T(\mathbf{v}_1)]_{\mathcal{B}'} + \cdots + c_n\,[T(\mathbf{v}_n)]_{\mathcal{B}'}$$

$$= A \begin{bmatrix} c_1 \\ \vdots \\ c_n \end{bmatrix} = A\,[\mathbf{v}]_{\mathcal{B}}$$

The proof of the fact that A is the *only* matrix with the property $[T(\mathbf{v})]_{\mathcal{B}'} = A[\mathbf{v}]_{\mathcal{B}}$ for all $\mathbf{v} \in V$ is left as an exercise.

DEFINITION

The matrix A of Theorem 13 is called the **matrix of T with respect to $\mathcal{B}$ and $\mathcal{B}'$**. If $V = W$ and $\mathcal{B} = \mathcal{B}'$, then A is called the **matrix of T with respect to $\mathcal{B}$** (Fig. 5.25).

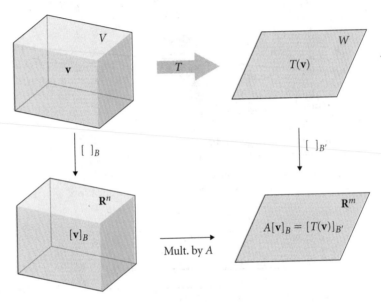

Figure 5.25   Matrix of linear transformation.

REMARKS

1. Theorem 13 is very useful. If we know A, we can evaluate $T(\mathbf{v})$ by computing $[T(\mathbf{v})]_{\mathcal{B}'}$ as $A[\mathbf{v}]_{\mathcal{B}}$, which is just matrix multiplication.
2. The matrix of T depends on T, $\mathcal{B}$, and $\mathcal{B}'$. Even if the order of the vectors in one of the basis changes, the matrix of T changes (see Example 48).

Theorem 13 has the important consequence that the only linear transformations from $\mathbf{R}^n$ to $\mathbf{R}^m$ are the matrix transformations.

THEOREM 14   Every linear transformation $T : \mathbf{R}^n \rightarrow \mathbf{R}^m$ is a matrix transformation.

PROOF   Let $\mathcal{B}$ and $\mathcal{B}'$ be the standard bases of $\mathbf{R}^n$ and $\mathbf{R}^m$, respectively. Then by Theorem 13 there is a matrix A such that

$$[T(\mathbf{v})]_{\mathcal{B}'} = A[\mathbf{v}]_{\mathcal{B}}$$

for all $\mathbf{v} \in \mathbf{R}^n$. But since $T(\mathbf{v}) = [T(\mathbf{v})]_{\mathcal{B}'}$ and $A[\mathbf{v}]_{\mathcal{B}} = A\mathbf{v}$ for the standard bases we have

$$T(\mathbf{v}) = A\mathbf{v}$$

Therefore, T is a matrix transformation with standard matrix A.

■ EXAMPLE 45   Let V be an n-dimensional vector space and let $\mathcal{B} = \{\mathbf{v}_1, \ldots, \mathbf{v}_n\}$ be any basis. Show that the matrix of the identity $I : V \rightarrow V$ with respect to $\mathcal{B}$ is I_n.

SOLUTION   Since $I(\mathbf{v}_i) = \mathbf{v}_i$ and $[\mathbf{v}_i]_{\mathcal{B}} = \mathbf{e}_i$, we have $[I(\mathbf{v}_i)]_{\mathcal{B}} = \mathbf{e}_i$ for $i = 1, \ldots, n$. Therefore, the matrix of I with respect to $\mathcal{B}$ has columns $\mathbf{e}_1, \ldots, \mathbf{e}_n$; hence, it is I_n.

■ EXAMPLE 46   Let $T : \mathbf{R}^2 \rightarrow \mathbf{R}^3$ be the linear transformation defined by

$$T\begin{bmatrix} x \\ y \end{bmatrix} = \begin{bmatrix} 2x + y \\ x - y \\ x + 4y \end{bmatrix}$$

and let $\mathcal{B} = \{\mathbf{v}_1, \mathbf{v}_2\}$ and $\mathcal{B}' = \{\mathbf{v}_1', \mathbf{v}_2', \mathbf{v}_3'\}$ be the bases of $\mathbf{R}^2$ and $\mathbf{R}^3$, with

$$\mathbf{v}_1 = \mathbf{e}_2, \quad \mathbf{v}_2 = \mathbf{e}_1 \quad \text{and} \quad \mathbf{v}_1' = \mathbf{e}_3, \quad \mathbf{v}_2' = \mathbf{e}_2, \quad \mathbf{v}_3' = \mathbf{e}_1$$

respectively.

(a) Compute the matrix A of T with respect to the bases $\mathcal{B}$ and $\mathcal{B}'$.
(b) Is the standard matrix of T the same as A in Part (a)?
(c) Evaluate $T\begin{bmatrix} -4 \\ 6 \end{bmatrix}$ directly and by using Part (a).

SOLUTION

(a) We have

$$T(\mathbf{e}_2) = T\begin{bmatrix} 0 \\ 1 \end{bmatrix} = \begin{bmatrix} 1 \\ -1 \\ 4 \end{bmatrix} \quad \text{and} \quad T(\mathbf{e}_1) = T\begin{bmatrix} 1 \\ 0 \end{bmatrix} = \begin{bmatrix} 2 \\ 1 \\ 1 \end{bmatrix}$$

Next, we need $[T(\mathbf{e}_2)]_{\mathcal{B}'}$ and $[T(\mathbf{e}_1)]_{\mathcal{B}'}$. It is easy to check that

$$\begin{bmatrix} 1 \\ -1 \\ 4 \end{bmatrix}_{\mathcal{B}'} = \begin{bmatrix} 4 \\ -1 \\ 1 \end{bmatrix} \quad \text{and} \quad \begin{bmatrix} 2 \\ 1 \\ 1 \end{bmatrix}_{\mathcal{B}'} = \begin{bmatrix} 1 \\ 1 \\ 2 \end{bmatrix}$$

Hence,

$$A = \begin{bmatrix} 4 & 1 \\ -1 & 1 \\ 1 & 2 \end{bmatrix}$$

(b) The standard matrix of T, which is also the matrix of T with respect to the standard basis, is

$$\begin{bmatrix} 2 & 1 \\ 1 & -1 \\ 1 & 4 \end{bmatrix}$$

which is not the same as A.

(c) By substitution into the formula for T, we get

$$T \begin{bmatrix} -4 \\ 6 \end{bmatrix} = \begin{bmatrix} -2 \\ -10 \\ 20 \end{bmatrix}$$

On the other hand, to use A we need $\begin{bmatrix} -4 \\ 6 \end{bmatrix}_B$, which is $\begin{bmatrix} 6 \\ -4 \end{bmatrix}$. Hence,

$$\left[T \begin{bmatrix} -4 \\ 6 \end{bmatrix} \right]_{B'} = \begin{bmatrix} 4 & 1 \\ -1 & 1 \\ 1 & 2 \end{bmatrix} \begin{bmatrix} 6 \\ -4 \end{bmatrix} = \begin{bmatrix} 20 \\ -10 \\ -2 \end{bmatrix}$$

So, by the definition of a coordinate vector with respect to B',

$$T \begin{bmatrix} -4 \\ 6 \end{bmatrix} = 20 \begin{bmatrix} 0 \\ 0 \\ 1 \end{bmatrix} - 10 \begin{bmatrix} 0 \\ 1 \\ 0 \end{bmatrix} - 2 \begin{bmatrix} 1 \\ 0 \\ 0 \end{bmatrix} = \begin{bmatrix} -2 \\ -10 \\ 20 \end{bmatrix}$$

■ EXAMPLE 47   Let $T : P_1 \rightarrow P_2$ be the linear transformation defined by $T(a + bx) = ax + bx^2$.

(a) Find the matrix A of T with respect to the bases $B = \{v_1, v_2\}$ and $B' = \{v_1', v_2', v_3'\}$, where

$$v_1 = x, \quad v_2 = 1 \quad \text{and} \quad v_1' = 1, \quad v_2' = x, \quad v_3' = x^2$$

(b) Evaluate $T(-3 + 4x)$ directly and by using A, B, B'.
(c) Recover the general formula for T by using A, B, B'.

SOLUTION

(a) Since $T(x) = x^2$ and $T(1) = x$, we have

$$[T(x)]_{B'} = [x^2]_{B'} = \begin{bmatrix} 0 \\ 0 \\ 1 \end{bmatrix}$$

and

$$[T(1)]_{B'} = [x]_{B'} = \begin{bmatrix} 0 \\ 1 \\ 0 \end{bmatrix}$$

Hence,

$$A = \begin{bmatrix} 0 & 0 \\ 0 & 1 \\ 1 & 0 \end{bmatrix}$$

(b) Direct evaluation yields $T(-3 + 4x) = -3x + 4x^2$. On the other hand, because

$$[-3 + 4x]_{\mathcal{B}} = \begin{bmatrix} 4 \\ -3 \end{bmatrix}$$

we have

$$[T(-3 + 4x)]_{\mathcal{B}'} = \begin{bmatrix} 0 & 0 \\ 0 & 1 \\ 1 & 0 \end{bmatrix} \begin{bmatrix} 4 \\ -3 \end{bmatrix} = \begin{bmatrix} 0 \\ -3 \\ 4 \end{bmatrix}$$

Hence,

$$T(-3 + 4x) = 0 \cdot 1 - 3 \cdot x + 4 \cdot x^2$$

yielding the same answer.

(c) Because

$$[a + bx]_{\mathcal{B}} = \begin{bmatrix} b \\ a \end{bmatrix}$$

we have

$$[T(a + bx)]_{\mathcal{B}'} = \begin{bmatrix} 0 & 0 \\ 0 & 1 \\ 1 & 0 \end{bmatrix} \begin{bmatrix} b \\ a \end{bmatrix} = \begin{bmatrix} 0 \\ a \\ b \end{bmatrix}$$

Hence,

$$T(a + bx) = 0 \cdot 1 + a \cdot x + b \cdot x^2$$

■ EXAMPLE 48   Let $T : P_2 \rightarrow P_2$ be the linear transformation defined by $T(a + bx + cx^2) = b + 2cx$.

(a) Find the matrix A of T with respect to the basis $\mathcal{B} = \{v_1, v_2, v_3\}$, where $v_1 = x^2$, $v_2 = x$, $v_3 = 1$.
(b) Find the matrix A' of T with respect to the basis of $\mathcal{B}' = \{v_1', v_2', v_3'\}$, where $v_1' = 1$, $v_2' = x$, $v_3' = x^2$.
(c) Evaluate $T(3x - 4x^2)$: (i) directly, (ii) by using A, and (iii) by using A'.

SOLUTION

(a) Because $T(x^2) = 2x$, $T(x) = 1$, and $T(1) = 0$, we have

$$\left[T(x^2)\right]_{\mathcal{B}} = [2x]_{\mathcal{B}} = \begin{bmatrix} 0 \\ 2 \\ 0 \end{bmatrix}, \qquad [T(x)]_{\mathcal{B}} = [1]_{\mathcal{B}} = \begin{bmatrix} 0 \\ 0 \\ 1 \end{bmatrix},$$

$$[T(1)]_{\mathcal{B}} = [0]_{\mathcal{B}} = \begin{bmatrix} 0 \\ 0 \\ 0 \end{bmatrix}$$

Hence,

$$A = \begin{bmatrix} 0 & 0 & 0 \\ 2 & 0 & 0 \\ 0 & 1 & 0 \end{bmatrix}$$

(b) We have

$$[T(1)]_{\mathcal{B}'} = [0]_{\mathcal{B}'} = \begin{bmatrix} 0 \\ 0 \\ 0 \end{bmatrix}, \qquad [T(x)]_{\mathcal{B}'} = [1]_{\mathcal{B}'} = \begin{bmatrix} 1 \\ 0 \\ 0 \end{bmatrix},$$

$$\left[T(x^2)\right]_{\mathcal{B}'} = [2x]_{\mathcal{B}'} = \begin{bmatrix} 0 \\ 2 \\ 0 \end{bmatrix}$$

Hence,

$$A' = \begin{bmatrix} 0 & 1 & 0 \\ 0 & 0 & 2 \\ 0 & 0 & 0 \end{bmatrix}$$

(c) (i) Direct evaluation yields $T(3x - 4x^2) = 3 - 8x$.
    (ii) Because

$$\left[3x - 4x^2\right]_{\mathcal{B}} = \begin{bmatrix} -4 \\ 3 \\ 0 \end{bmatrix}$$

we have

$$\left[T(3x - 4x^2)\right]_{\mathcal{B}} = \begin{bmatrix} 0 & 0 & 0 \\ 2 & 0 & 0 \\ 0 & 1 & 0 \end{bmatrix} \begin{bmatrix} -4 \\ 3 \\ 0 \end{bmatrix} = \begin{bmatrix} 0 \\ -8 \\ 3 \end{bmatrix}$$

Hence,

$$T(3x - 4x^2) = 0 \cdot x^2 - 8 \cdot x + 3 \cdot 1 = 3 - 8x$$

giving the same answer.
    (iii) Likewise, since

$$\left[3x - 4x^2\right]_{\mathcal{B}'} = \begin{bmatrix} 0 \\ 3 \\ -4 \end{bmatrix}$$

we have

$$\left[T(3x - 4x^2)\right]_{\mathcal{B}'} = \begin{bmatrix} 0 & 1 & 0 \\ 0 & 0 & 2 \\ 0 & 0 & 0 \end{bmatrix} \begin{bmatrix} 0 \\ 3 \\ -4 \end{bmatrix} = \begin{bmatrix} 3 \\ -8 \\ 0 \end{bmatrix}$$

Hence,

$$T(3x - 4x^2) = 3 \cdot 1 - 8 \cdot x + 0 \cdot x^2 = 3 - 8x$$

Note that $A \neq A'$ in Example 48.

THEOREM 15

Let $T : V \to W$ be a linear transformation between two finite-dimensional vector spaces V and W. Let A be the matrix of T with respect to bases $\mathcal{B} = \{v_1, \ldots, v_n\} \subseteq V$ and $\mathcal{B}' = \{v_1', \ldots, v_m'\} \subseteq W$. Then

1. v is in the kernel of T if and only if $[v]_{\mathcal{B}}$ is in the null space of A;
2. w is in the range of T if and only if $[w]_{\mathcal{B}'}$ is in the column space of A.

PROOF

1. The equivalence holds, because

$$v \in \text{Ker}(T) \Leftrightarrow T(v) = 0 \in W$$
$$\Leftrightarrow [T(v)]_{\mathcal{B}'} = 0 \in \mathbf{R}^m \qquad \text{by Theorem 18, Section 4.5}$$
$$\Leftrightarrow A[v]_{\mathcal{B}} = 0 \in \mathbf{R}^m \qquad \text{by Theorem 13}$$
$$\Leftrightarrow [v]_{\mathcal{B}} \in \text{Null}(A)$$

2. The proof of this part is left as an exercise.

Theorem 15 implies that $\text{Ker}(T) = \{0\}$ if and only if $\text{Null}(A) = \{0\}$ and that $R(T) = W$ if and only if $\text{Col}(A) = \mathbf{R}^m$.

THEOREM 16

In the notation of Theorem 15, we have

1. T is one-to-one if and only if A has n pivots;
2. T is onto if and only if A has m pivots;
3. T is an isomorphism if and only if A is invertible.

Change of Basis and the Matrix of a Linear Transformation

In this paragraph we study how the matrix of a linear transformation $T : V \to V$ is affected when we change bases in V. In general, a linear transformation has different matrices with respect to different bases (see Example 48). Sometimes there are bases that yield a very simple matrix for T, for instance, a *diagonal matrix*. Evaluation of T then becomes very easy. The next theorem tells us how to find a new (potentially easy) matrix representation of T from an old one.

THEOREM 17

Let $T : V \to V$ be a linear transformation from a finite-dimensional vector space V into itself. Let $\mathcal{B}$ and $\mathcal{B}'$ be two bases of V and let P be the transition matrix from $\mathcal{B}'$ to $\mathcal{B}$. If A is the matrix of T with respect to $\mathcal{B}$ and A' is the matrix of T with respect to $\mathcal{B}'$, then

$$A' = P^{-1}AP$$

PROOF   Because P is the transition matrix from $\mathcal{B}'$ to $\mathcal{B}$, P^{-1} is the transition matrix from $\mathcal{B}$ to $\mathcal{B}'$ (by Corollary 21, Section 4.5). Hence, $[\mathbf{w}]_{\mathcal{B}'} = P^{-1}[\mathbf{w}]_{\mathcal{B}}$ for all $\mathbf{w}$ in V. In particular, $[T(\mathbf{v})]_{\mathcal{B}'} = P^{-1}[T(\mathbf{v})]_{\mathcal{B}}$ for all $\mathbf{v}$ in V. Therefore,

$$[T(\mathbf{v})]_{\mathcal{B}'} = P^{-1}[T(\mathbf{v})]_{\mathcal{B}} = P^{-1}(A[\mathbf{v}]_{\mathcal{B}})$$
$$= (P^{-1}A)[\mathbf{v}]_{\mathcal{B}} = P^{-1}A(P[\mathbf{v}]_{\mathcal{B}'}) = (P^{-1}AP)[\mathbf{v}]_{\mathcal{B}'}$$

So the matrix $P^{-1}AP$ satisfies $[T(\mathbf{v})]_{\mathcal{B}'} = (P^{-1}AP)[\mathbf{v}]_{\mathcal{B}'}$ for all $\mathbf{v}$ in V. Hence, it has to be the transition matrix of T with respect to $\mathcal{B}'$ (Fig. 5.26).

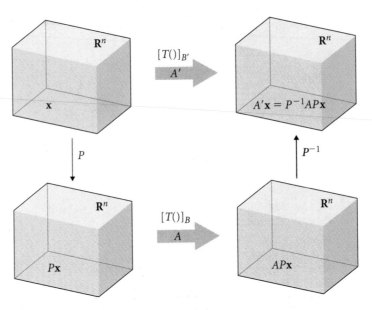

Figure 5.26   The effect of change of basis.

Let us illustrate Theorem 17.

■ EXAMPLE 49   Let $T : \mathbf{R}^2 \rightarrow \mathbf{R}^2$ be the linear transformation given by

$$T\begin{bmatrix} x \\ y \end{bmatrix} = \begin{bmatrix} -5x + 6y \\ -3x + 4y \end{bmatrix}$$

and let $\mathcal{B}$ and $\mathcal{B}'$ be the bases

$$\mathcal{B} = \left\{ \begin{bmatrix} 1 \\ 0 \end{bmatrix}, \begin{bmatrix} 0 \\ 1 \end{bmatrix} \right\} \qquad \mathcal{B}' = \left\{ \begin{bmatrix} 1 \\ 1 \end{bmatrix}, \begin{bmatrix} 2 \\ 1 \end{bmatrix} \right\}$$

(a) Compute the matrix A of T with respect to $\mathcal{B}$. (This is the standard matrix.)
(b) Compute the transition matrix P from $\mathcal{B}'$ to $\mathcal{B}$.
(c) Use Theorem 17 to find the matrix of T with respect to $\mathcal{B}'$.
(d) Compute the matrix A' of T with respect to $\mathcal{B}'$ directly from $\mathcal{B}'$.

SOLUTION

(a) Because

$$T\begin{bmatrix} 1 \\ 0 \end{bmatrix} = \begin{bmatrix} -5 \\ -3 \end{bmatrix}, \qquad T\begin{bmatrix} 0 \\ 1 \end{bmatrix} = \begin{bmatrix} 6 \\ 4 \end{bmatrix}$$

the standard matrix A of T is

$$A = \begin{bmatrix} -5 & 6 \\ -3 & 4 \end{bmatrix}$$

(b) To find P we need the coordinate vectors of the vectors of $\mathcal{B}'$ with respect to $\mathcal{B}$. We have

$$\begin{bmatrix} 1 \\ 1 \end{bmatrix}_{\mathcal{B}} = \begin{bmatrix} 1 \\ 1 \end{bmatrix}, \qquad \begin{bmatrix} 2 \\ 1 \end{bmatrix}_{\mathcal{B}} = \begin{bmatrix} 2 \\ 1 \end{bmatrix}$$

Hence,

$$P = \begin{bmatrix} 1 & 2 \\ 1 & 1 \end{bmatrix}$$

(c) It is easy to see that

$$P^{-1} = \begin{bmatrix} -1 & 2 \\ 1 & -1 \end{bmatrix}$$

By Theorem 17

$$A' = P^{-1}AP = \begin{bmatrix} -1 & 2 \\ 1 & -1 \end{bmatrix}\begin{bmatrix} -5 & 6 \\ -3 & 4 \end{bmatrix}\begin{bmatrix} 1 & 2 \\ 1 & 1 \end{bmatrix} = \begin{bmatrix} 1 & 0 \\ 0 & -2 \end{bmatrix}$$

(d) Let us compute A' straight from $\mathcal{B}'$ and T. Evaluating T at $\mathcal{B}'$ we get

$$T\begin{bmatrix} 1 \\ 1 \end{bmatrix} = \begin{bmatrix} 1 \\ 1 \end{bmatrix}, \qquad T\begin{bmatrix} 2 \\ 1 \end{bmatrix} = \begin{bmatrix} -4 \\ -2 \end{bmatrix}$$

Because

$$\begin{bmatrix} 1 \\ 1 \end{bmatrix}_{\mathcal{B}'} = \begin{bmatrix} 1 \\ 0 \end{bmatrix}, \qquad \begin{bmatrix} -4 \\ -2 \end{bmatrix}_{\mathcal{B}'} = \begin{bmatrix} 0 \\ -2 \end{bmatrix}$$

we have

$$A' = \begin{bmatrix} 1 & 0 \\ 0 & -2 \end{bmatrix}$$

yielding the same answer as in Part (c).

Note that A' in Example 49 is a diagonal matrix.

DEFINITION

(Similar Matrices)

Let A and B be two $n \times n$ (square) matrices. We say that B **is similar to** A if there exists an invertible matrix P such that $B = P^{-1}AP$.

Let us point out the following basic facts (whose proofs are left as exercises):

1. A is similar to itself.
2. If B is similar to A, then A is similar to B.
3. If B is similar to A and C is similar to B, then C is similar to A.

Theorem 17 can be rephrased by saying that

The matrices of a linear transformation with respect to two bases are similar.

In fact, as we show in the exercises, two similar matrices give rise to the *same* linear transformation with respect to different bases.

Proof of the Dimension Theorem

Let us now prove the dimension theorem (Section 5.3), which states that

If $T : V \rightarrow W$ is a linear transformation from a finite-dimensional vector space V into a vector space W, then

$$\text{Nullity}(T) + \text{Rank}(T) = \dim(V)$$

PROOF   Because V is finite-dimensional, Theorem 5, Section 5.2, implies that $R(T)$ is finite-dimensional. Hence, T may be viewed as a linear transformation between the two finite-dimensional vector spaces V and $R(T)$. Let A be the matrix of T with respect to bases $\mathcal{B}$ and $\mathcal{B}'$ of V and $R(T)$, respectively. Then the number of columns of A is $\dim(V)$. Theorems 18 and 19, Section 4.5, along with Theorem 15 imply that

$$\text{Nullity}(T) = \text{Nullity}(A) \quad \text{and} \quad \text{Rank}(T) = \text{Rank}(A)$$

The theorem now follows from the Rank Theorem (Section 4.6).   ■

Exercises 5.4

Let $\mathcal{B}$, $\mathcal{B}'$, and $\mathcal{B}''$ be the following bases of P_1, P_2, and P_3, respectively.

$$\mathcal{B} = \{1 + x, -1 + x\} \subseteq P_1$$

$$\mathcal{B}' = \{-x + x^2, 1 + x, x\} \subseteq P_2$$

$$\mathcal{B}'' = \{-x + x^3, 1 + x^2, x, -1 + x\} \subseteq P_3$$

Let p, p', p'' be any polynomials of P_1, P_2, and P_3, respectively.

$$p = a + bx \quad \text{in} \quad P_1$$

$$p' = a + bx + cx^2 \quad \text{in} \quad P_2$$

$$p'' = a + bx + cx^2 + dx^3 \quad \text{in} \quad P_3$$

In Exercises 1–3 find the matrix of T with respect to each of the following.

1. $\mathcal{B}$ and $\mathcal{B}'$, if $T(p) = (a - b) + bx + ax^2$
2. $\mathcal{B}'$ and $\mathcal{B}$, if $T(p') = (a - b) + (b - 4c)x$
3. $\mathcal{B}'$, if $T(p') = (a - b) + (b - c)x + (-a + 3c)x^2$
4. Let $T : P_1 \rightarrow P_1$, $T(p) = (-a + b) + (2a - 3b)x$, and let $S = \{2 + x, 1\}$.

   a. Compute the matrix A of T with respect to $\mathcal{B}$.

   b. Compute the transition matrix P from S to $\mathcal{B}$.

   c. Use (a) and (b) to find the matrix of T with respect to S.

   d. Compute the matrix A' of T with respect to S directly from S.

5. Let $T : P_2 \rightarrow P_2$ be the linear transformation defined by $T(p') = -2c + bx$.

a. Find the matrix A of T with respect to the standard basis $\{1, x, x^2\}$.

b. Find the matrix A' of T with respect to the basis of $\mathcal{B}'$.

c. Evaluate $T(6x - 2x^2)$ (a) directly, (b) by using A, and (c) by using A'.

Let $T : P_1 \to P_2$ be a linear transformation. In Exercises 6–7 for the given T:

(a) Find the matrix A of T with respect to $\mathcal{B}$ and $\mathcal{B}'$.

(b) Evaluate $T(5 - 2x)$ directly and by using A, $\mathcal{B}$, $\mathcal{B}'$.

6. $T(p) = b + ax - ax^2$

7. $T(p) = a - ax - bx^2$

Let $\mathcal{R}$ be the following basis of M_{22}:

$$\mathcal{R} = \{E_{11} - E_{12}, E_{12} - E_{21}, E_{21} - E_{22}, E_{22} + E_{11}\}$$

8. Find the matrix of $T : M_{22} \to M_{22}$ with respect to $\mathcal{R}$, where

$$T \begin{bmatrix} a & b \\ c & d \end{bmatrix} = \begin{bmatrix} -b & a \\ -d & c \end{bmatrix}$$

9. Find $T \begin{bmatrix} 1 & -2 \\ -4 & 3 \end{bmatrix}$ if the matrix of the linear transformation $T : M_{22} \to M_{22}$ with respect to $\mathcal{R}$ is

$$A = \begin{bmatrix} 1 & 3 & 0 & 2 \\ 0 & 1 & 0 & 0 \\ 0 & -2 & 1 & -1 \\ 1 & 0 & 0 & 0 \end{bmatrix}$$

10. Prove Part 2 of Theorem 15.

11. Prove the uniqueness of the transition matrix as claimed by Theorem 13.

In Exercises 12–14 A is the matrix of a linear transformation $T : P_n \to P_m$. Find n and m and a formula for $T(q)$, $q \in P_n$ with respect to each.

12. $\mathcal{B}'$, if $A = \begin{bmatrix} 1 & 0 & 0 \\ 0 & 2 & 0 \\ 0 & 0 & 3 \end{bmatrix}$

13. $\mathcal{B}$ and $\mathcal{B}''$, if $A = \begin{bmatrix} 2 & 0 & 1 & 0 \\ -4 & 1 & 2 & -8 \end{bmatrix}$

14. $\mathcal{B}''$ and $\mathcal{B}$, if $A = \begin{bmatrix} -4 & 2 \\ 0 & 9 \\ 1 & -1 \\ 2 & -3 \end{bmatrix}$

15. Compute the matrix A of $T : M_{22} \to M_{22}$,

$$T \begin{bmatrix} a & b \\ c & d \end{bmatrix} = \begin{bmatrix} -b & d \\ c & -a \end{bmatrix}$$

with respect to the basis $\mathcal{R}$. Prove that T is an isomorphism by row-reducing A. Which theorem did you use?

16. Prove that A is similar to itself.

17. Prove that if B is similar to A, then A is similar to B.

18. Prove that if B is similar to A and C is similar to B, then C is similar to A.

19. For two $n \times n$ matrices A and B with at least one of them invertible, show that AB is similar to BA.

5.5   The Algebra of Linear Transformations

Reader's Goals for This Section

1. To know the operations of linear transformations and how they relate to matrix operations.
2. To know the definition and properties of an invertible linear transformation (isomorphism).

We define the basic operations of linear transformations, addition, scalar multiplication, and inversion, and relate them to the basic matrix operations. Also, we revisit and further explore the concept of isomorphism.

NOTE   Throughout this section V, W, and U are finite-dimensional vector spaces.

Sums and Scalar Products

Let $f, g : V \to W$ be linear transformations. The **sum** $f + g$ of f and g is the map $f + g : V \to W$ defined by

$$(f + g)(\mathbf{v}) = f(\mathbf{v}) + g(\mathbf{v})$$

for all $\mathbf{v} \in V$. Let c be any scalar. The **scalar multiple** cf of f by c is the map $cf : V \to W$ defined by

$$(cf)(\mathbf{v}) = cf(\mathbf{v})$$

for all $\mathbf{v} \in V$.

■ **EXAMPLE 50**   Evaluate $f + g$ and $5f$ at $a + bx + cx^2$ if

$$f, g : P_2 \to P_1$$

$$f(a + bx + cx^2) = b + cx \qquad \text{and} \qquad g(a + bx + cx^2) = c - ax$$

SOLUTION   The sum $f + g$ is given by

$$\begin{aligned}(f + g)(a + bx + cx^2) &= f(a + bx + cx^2) + g(a + bx + cx^2) \\ &= (b + cx) + (c - ax) \\ &= (b + c) + (c - a)x\end{aligned}$$

The scalar multiple $5f$ is the map with values

$$(5f)(a + bx + cx^2) = 5f(a + bx + cx^2) = 5(b + cx) = 5b + 5cx$$

THEOREM 18 $f + g$ and cf are linear transformations.

PROOF   Let $\mathbf{v}_1, \mathbf{v}_2 \in V$ and let $c_1, c_2 \in \mathbf{R}$. Then

$$\begin{aligned}(f + g)(c_1 \mathbf{v}_1 + c_2 \mathbf{v}_2) &= f(c_1 \mathbf{v}_1 + c_2 \mathbf{v}_2) + g(c_1 \mathbf{v}_1 + c_2 \mathbf{v}_2) \\ &= c_1 f(\mathbf{v}_1) + c_2 f(\mathbf{v}_2) + c_1 g(\mathbf{v}_1) + c_2 g(\mathbf{v}_2) \\ &= c_1 (f(\mathbf{v}_1) + g(\mathbf{v}_1)) + c_2 (f(\mathbf{v}_2) + g(\mathbf{v}_2)) \\ &= c_1 (f + g)(\mathbf{v}_1) + c_2 (f + g)(\mathbf{v}_2)\end{aligned}$$

Hence, $f + g$ is linear. The verification that cf is linear is left as an exercise.

Note that we can form linear combinations

$$c_1 f_1 + \cdots + c_n f_n$$

for scalars $c_1, \ldots, c_n$ and linear transformations $f_1, \ldots, f_n$.

Addition and scalar multiplication satisfy properties identical to those of matrix addition and scalar multiplication, stated in Theorem 1, Section 3.1. Recall that I is the identity transformation $I : V \to V$, $I(\mathbf{v}) = \mathbf{v}$ and $\mathbf{0}$ is the zero transformation $\mathbf{0} : V \to W$, $\mathbf{0}(\mathbf{v}) = \mathbf{0}$.

THEOREM 19 **(Laws for Addition and Scalar Multiplication)**

Let $f, g,$ and h be linear transformations such that the operations below can be performed. Let c be any scalar. Then the following hold.

1. $(f + g) + h = f + (g + h)$
2. $f + g = g + f$
3. $f + \mathbf{0} = \mathbf{0} + f = f$
4. $f + (-f) = (-f) + f = \mathbf{0}$
5. $c(f + g) = cf + cg$
6. $(a + b)f = af + bf$
7. $(ab)f = a(bf) = b(af)$
8. $1f = f$
9. $0f = \mathbf{0}$

PROOF   Exercise.

Composition of Linear Transformations

Composition of two transformations $g : A \to B$ and $f : B \to C$ is the new transformation $f \circ g : A \to C$ obtained by evaluating the second transformation f at the values of the first g. Hence, $f \circ g(a) = f(g(a))$ for all $a \in A$. This operation is very important in all of mathematics and its applications. Let us now study the case where f and g are linear transformations.

DEFINITION **(Composition)**

Let $g : U \to V$ and $f : V \to W$ be linear transformations. The **composition of f with g** is the map $f \circ g : U \to W$ defined by

$$f \circ g(\mathbf{v}) = f(g(\mathbf{v}))$$

for all $\mathbf{v} \in U$ (Fig. 5.27).

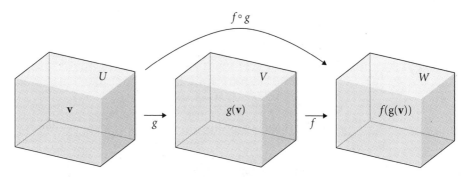

Figure 5.27   Composition of transformations.

■ EXAMPLE 51   Let $g : \mathbf{R}^2 \rightarrow \mathbf{R}^3$ and $f : \mathbf{R}^3 \rightarrow \mathbf{R}^4$ be the linear transformations defined by

$$g(x, y) = (x + y, x - y, 2x) \quad \text{and} \quad f(x, y, z) = (x - y, x + y, x + z, 2z)$$

Find (a) $f \circ g(1, 2)$ and (b) $f \circ g(x, y)$.

SOLUTION

(a) The composition is a map from $\mathbf{R}^2$ to $\mathbf{R}^4$, $f \circ g : \mathbf{R}^2 \rightarrow \mathbf{R}^4$. Because $g(1, 2) = (3, -1, 2)$, we have

$$f \circ g(1, 2) = f(g(1, 2)) = f(3, -1, 2) = (4, 2, 5, 4)$$

(b) Likewise,

$$f \circ g(x, y) = f(g(x, y)) = f(x + y, x - y, 2x) = (2y, 2x, 3x + y, 4x)$$

THEOREM 20 $f \circ g : U \rightarrow W$ is a linear transformation.

PROOF   Let $\mathbf{v}_1, \mathbf{v}_2 \in V$ and let $c_1, c_2 \in \mathbf{R}$. Then

$$
\begin{aligned}
f \circ g(c_1 \mathbf{v}_1 + c_2 \mathbf{v}_2) &= f(g(c_1 \mathbf{v}_1 + c_2 \mathbf{v}_2)) \\
&= f(c_1 g(\mathbf{v}_1) + c_2 g(\mathbf{v}_2)) \\
&= c_1 f(g(\mathbf{v}_1)) + c_2 f(g(\mathbf{v}_2)) \\
&= c_1 f \circ g(\mathbf{v}_1) + c_2 f \circ g(\mathbf{v}_2)
\end{aligned}
$$

Hence, $f \circ g$ is linear.

Compositions of linear transformations satisfy the following properties. (Notice the similarity to Theorem 2, Section 3.1, with composition replacing matrix multiplication.)

THEOREM 21   **(Laws of Composition)**

Let f, g, and h be linear transformations such that the following operations can be performed. Let c be any scalar. Then the following hold.

1. $(f \circ g) \circ h = f \circ (g \circ h)$
2. $f \circ (g + h) = f \circ g + f \circ h$
3. $(g + h) \circ f = g \circ f + h \circ f$
4. $c(g \circ h) = (cg) \circ h = g \circ (ch)$
5. $I \circ f = f \circ I = f$
6. $\mathbf{0} \circ f = \mathbf{0}, \quad f \circ \mathbf{0} = \mathbf{0}$

We use the same names as in Theorem 2, Section 3.1, i.e., associative law, left distributive law, etc.

PROOF   We prove Parts 1 and 2 and leave the remaining proofs as exercises.

1. We have, by the definition of composition,

$$(f \circ g) \circ h(\mathbf{v}) = f \circ g(h(\mathbf{v}))$$
$$= f(g(h(\mathbf{v}))$$
$$= f(g \circ h(\mathbf{v}))$$
$$= f \circ (g \circ h)(\mathbf{v})$$

for all $\mathbf{v}$.[7]

2. For each $\mathbf{v}$ we have

$$f \circ (g + h)(\mathbf{v}) = f((g + h)(\mathbf{v}))$$
$$= f((g(\mathbf{v}) + h(\mathbf{v}))$$
$$= f(g(\mathbf{v})) + f(h(\mathbf{v})) \qquad \text{because } f \text{ is linear}$$
$$= f \circ g(\mathbf{v}) + f \circ h(\mathbf{v})$$
$$= (f \circ g + f \circ h)(\mathbf{v})$$

As in the case of matrix multiplication, composition is not commutative. So, in general,

$$f \circ g \neq g \circ f$$

Powers of a Linear Transformation

Let $f : V \to V$ be a linear transformation. The composition $f \circ f$ is usually written as f^2. Likewise, we write f^3 for $(f^2) \circ f$, etc. We also define f^1 to be f and f^0 to be I, the identity map. We refer to these compositions as **powers** of f.

$$f^0 = I, \quad f^1 = f, \quad f^2 = f \circ f, \quad \ldots, \quad f^k = f \circ f \circ \cdots \circ f \qquad (k \text{ "factors"})$$

Linear Transformation and Matrix Operations

We know from Section 5.4 that there is a very close relationship between matrices and linear transformations; namely, any linear transformation $f : V \to W$ can be represented by a matrix transformation via

$$\left[f(\mathbf{v}) \right]_{\mathcal{B}'} = A \left[\mathbf{v} \right]_{\mathcal{B}} \qquad \text{for all } \mathbf{x} \in V \tag{5.5}$$

where $\mathcal{B}$ and $\mathcal{B}'$ are fixed bases of V and W, respectively. A is the matrix of f with respect to $\mathcal{B}$ and $\mathcal{B}'$. Recall that A is the only matrix that satisfies (5.5) and is given by

$$A = \left[[f(\mathbf{v}_1)]_{\mathcal{B}'}, \ldots, [f(\mathbf{v}_n)]_{\mathcal{B}'} \right]$$

The next theorem tells us how the linear transformation operations correspond to matrix operations.

---

[7]It is worth noticing that the assumption that the transformations are linear was never used in the proof of 1, because associativity is true in general as long as the composite maps are defined.

THEOREM 22

Let f and g be linear transformations between finite-dimensional vector spaces with matrices A and B with respect to fixed bases. Then the matrix of the linear transformation

1. $f + g$ is $A + B$;
2. $f - g$ is $A - B$;
3. $-f$ is $-A$;
4. cf is cA;
5. $f \circ g$ is AB.

PROOF   We prove Part 1 and leave the remaining proofs as exercises. For all $\mathbf{v} \in V$,

$$[(f + g)(\mathbf{v})]_{\mathcal{B}'} = [f(\mathbf{v}) + g(\mathbf{v})]_{\mathcal{B}'}$$
$$= [f(\mathbf{v})]_{\mathcal{B}'} + [g(\mathbf{v})]_{\mathcal{B}'}$$
$$= A[\mathbf{v}]_{\mathcal{B}} + B[\mathbf{v}]_{\mathcal{B}}$$
$$= (A + B)[\mathbf{v}]_{\mathcal{B}}$$

so $A + B$ is the matrix of $f + g$ with respect to $\mathcal{B}$ and $\mathcal{B}'$.

■ EXAMPLE 52   Verify Part 5 of Theorem 22 for the transformations of Example 51 using the standard bases.

SOLUTION   We have

$$g(1, 0) = (1, 1, 2), \qquad g(0, 1) = (1, -1, 0)$$
$$f(1, 0, 0) = (1, 1, 1, 0), \qquad f(0, 1, 0) = (-1, 1, 0, 0), \qquad f(0, 0, 1) = (0, 0, 1, 2)$$

Hence the standard matrices of f and g are, respectively,

$$\begin{bmatrix} 1 & -1 & 0 \\ 1 & 1 & 0 \\ 1 & 0 & 1 \\ 0 & 0 & 2 \end{bmatrix} \quad \text{and} \quad \begin{bmatrix} 1 & 1 \\ 1 & -1 \\ 2 & 0 \end{bmatrix}$$

On the other hand, the composition $f \circ g$ is given by $(f \circ g)(x, y) = (2y, 2x, 3x + y, 4x)$. Hence,

$$(f \circ g)(1, 0) = (0, 2, 3, 4) \quad \text{and} \quad (f \circ g)(0, 1) = (2, 0, 1, 0)$$

So, the standard matrix of $f \circ g$ is

$$\begin{bmatrix} 0 & 2 \\ 2 & 0 \\ 3 & 1 \\ 4 & 0 \end{bmatrix} = \begin{bmatrix} 1 & -1 & 0 \\ 1 & 1 & 0 \\ 1 & 0 & 1 \\ 0 & 0 & 2 \end{bmatrix} \begin{bmatrix} 1 & 1 \\ 1 & -1 \\ 2 & 0 \end{bmatrix}$$

Invertible Linear Transformations

DEFINITION

A linear transformation $f : V \to V$ is **invertible** if there is a map $g : V \to V$ with the property

$$f \circ g = I \quad \text{and} \quad g \circ f = I$$

The map g is called an **inverse** of f. If an inverse exists, it is unique (the proof is identical to that of the uniqueness of the inverse of a matrix). This unique inverse is denoted by f^{-1}. Hence,

$$f \circ f^{-1} = I \quad \text{and} \quad f^{-1} \circ f = I$$

Note that if f is invertible, then

$$f(\mathbf{v}) = \mathbf{w} \Leftrightarrow f^{-1}(\mathbf{w}) = \mathbf{v}$$

Indeed, $f(\mathbf{v}) = \mathbf{w}$ implies $f^{-1}(f(\mathbf{v})) = f^{-1}(\mathbf{w})$, which implies $f^{-1} \circ f(\mathbf{v}) = f^{-1}(\mathbf{w})$ or $\mathbf{v} = f^{-1}(\mathbf{w})$. Since these steps are reversible the equivalence follows.

The inverse of a map, if it exists, reverses the effect of the map (Fig. 5.28).

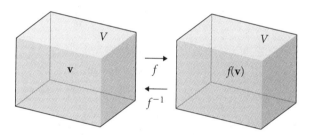

Figure 5.28 A linear transformation and its inverse.

The next theorem identifies the invertible linear transformations with isomorphisms, i.e., linear transformations that are one-to-one and onto, studied in Section 5.3.

THEOREM 23

Let $f : V \to V$ be a linear transformation.

1. f is invertible if and only if it is an isomorphism.
2. If f is invertible then f^{-1} is linear.

PROOF

1. Let f be invertible and let g be its inverse. We prove that f is one-to-one and onto. For $\mathbf{v}_1, \mathbf{v}_2 \in V$ let us assume that $f(\mathbf{v}_1) = f(\mathbf{v}_2)$. Then

$$f(\mathbf{v}_1) = f(\mathbf{v}_2)$$

$$\Rightarrow g(f(\mathbf{v}_1)) = g(f(\mathbf{v}_2))$$

$$\Rightarrow g \circ f(\mathbf{v}_1) = g \circ f(\mathbf{v}_2)$$

$$\Rightarrow \mathbf{v}_1 = \mathbf{v}_2$$

Therefore, f is one-to-one. Let $\mathbf{w} \in V$ and let $\mathbf{v} = g(\mathbf{w})$. Then

$$f(\mathbf{v}) = f(g(\mathbf{w})) = f \circ g(\mathbf{w}) = \mathbf{w}$$

Hence for each element $\mathbf{w}$ there exists an element $\mathbf{v}$ that maps to $\mathbf{w}$ under f. So f is onto.

Conversely, suppose that f is one-to-one and onto. We define the inverse g of f. Let $\mathbf{v} \in V$. There exists a unique $\mathbf{w}$ such that $f(\mathbf{w}) = \mathbf{v}$. We define $g(\mathbf{v}) = \mathbf{w}$. It follows that g is well defined and is the inverse of f. (Check.)

2. Let f be invertible. We prove that f^{-1} is linear. Let $\mathbf{v}_1, \mathbf{v}_2 \in V$. Then there are unique vectors $\mathbf{w}_1, \mathbf{w}_2 \in V$ such that $f(\mathbf{w}_1) = \mathbf{v}_1$ and $f(\mathbf{w}_2) = \mathbf{v}_2$, because f is one-to-one and onto by Part 1. Therefore, $\mathbf{w}_1 = f^{-1}(\mathbf{v}_1)$ and $\mathbf{w}_2 = f^{-1}(\mathbf{v}_2)$. We have

$$\begin{aligned} f^{-1}(\mathbf{v}_1 + \mathbf{v}_2) &= f^{-1}(f(\mathbf{w}_1) + f(\mathbf{w}_2)) \\ &= f^{-1}(f(\mathbf{w}_1 + \mathbf{w}_2)) \\ &= f^{-1} \circ f(\mathbf{w}_1 + \mathbf{w}_2) \\ &= I(\mathbf{w}_1 + \mathbf{w}_2) \\ &= \mathbf{w}_1 + \mathbf{w}_2 \\ &= f^{-1}(\mathbf{v}_1) + f^{-1}(\mathbf{v}_2) \end{aligned}$$

We proved that $f^{-1}(\mathbf{v}_1 + \mathbf{v}_2) = f^{-1}(\mathbf{v}_1) + f^{-1}(\mathbf{v}_2)$. We leave it to the reader to show that for each $\mathbf{v} \in V$ and each scalar c,

$$f^{-1}(c\mathbf{v}) = c f^{-1}(\mathbf{v})$$

Hence, f^{-1} is a linear transformation.

THEOREM 24   Let $f : V \to V$ be a linear transformation with matrix A with respect to bases $\mathcal{B}$ and $\mathcal{B}'$ of V. Then

1. f is invertible if and only if A is invertible;
2. If f is invertible, then A^{-1} is the matrix of f^{-1} with respect to $\mathcal{B}'$ and $\mathcal{B}$.

PROOF   Exercise.

■ **EXAMPLE 53**   Show that the transformation $T : \mathbf{R}^3 \to \mathbf{R}^3$ is invertible and compute its inverse:

$$T\begin{bmatrix} x \\ y \\ z \end{bmatrix} = \begin{bmatrix} x+y \\ y+z \\ z \end{bmatrix}$$

SOLUTION The standard matrix A of T is invertible. More precisely,

$$A = \begin{bmatrix} 1 & 1 & 0 \\ 0 & 1 & 1 \\ 0 & 0 & 1 \end{bmatrix}, \qquad A^{-1} = \begin{bmatrix} 1 & -1 & 1 \\ 0 & 1 & -1 \\ 0 & 0 & 1 \end{bmatrix}$$

Therefore, T is invertible, by Theorem 24, and the standard matrix of T^{-1} is A^{-1}. Hence,

$$T^{-1}\begin{bmatrix} x \\ y \\ z \end{bmatrix} = \begin{bmatrix} 1 & -1 & 1 \\ 0 & 1 & -1 \\ 0 & 0 & 1 \end{bmatrix}\begin{bmatrix} x \\ y \\ z \end{bmatrix} = \begin{bmatrix} x-y+z \\ y-z \\ z \end{bmatrix}$$

Exercises 5.5

In Exercises 1–3 evaluate $f + g$ and $-4f$ at (x, y, z) and at $(-1, 2, 0)$.

1. $f\begin{bmatrix} x \\ y \\ z \end{bmatrix} = \begin{bmatrix} x - y + z \\ x + y \end{bmatrix}$

$g\begin{bmatrix} x \\ y \\ z \end{bmatrix} = \begin{bmatrix} -x + y \\ 2x - y - z \end{bmatrix}$

2. $f\begin{bmatrix} x \\ y \\ z \end{bmatrix} = \begin{bmatrix} x \\ x + 2y \\ z \end{bmatrix}$

$g\begin{bmatrix} x \\ y \\ z \end{bmatrix} = \begin{bmatrix} x + y - z \\ 2x - 2y \\ x - z \end{bmatrix}$

3. $f\begin{bmatrix} x \\ y \\ z \end{bmatrix} = \begin{bmatrix} -x - 2y \\ x - z \end{bmatrix}$

$g\begin{bmatrix} x \\ y \\ z \end{bmatrix} = \begin{bmatrix} y + z \\ x - 3y \end{bmatrix}$

4. Evaluate $f + g$ and $-3f$ at $a + bx$ and at $-6 + 7x$ if

$$f(a + bx) = b - ax, \quad g(a + bx) = (3a + b) - bx$$

In Exercises 5–7 verify Parts 1 and 4 of Theorem 22 for the given f and g. Use $c = -2$.

5. f and g as in Exercise 1
6. f and g as in Exercise 2

7. f and g as in Exercise 3

In Exercises 8–10 find $f \circ g(x, y)$ and $f \circ g(-1, -3)$.

8. $f\begin{bmatrix} x \\ y \end{bmatrix} = \begin{bmatrix} x - y \\ x + y \end{bmatrix}$

$g\begin{bmatrix} x \\ y \end{bmatrix} = \begin{bmatrix} -5x + y \\ x + 3y \end{bmatrix}$

9. $f\begin{bmatrix} x \\ y \\ z \end{bmatrix} = \begin{bmatrix} x - y + z \\ x + 2z \end{bmatrix}$

$g\begin{bmatrix} x \\ y \end{bmatrix} = \begin{bmatrix} x + y \\ x - 4y \\ x - y \end{bmatrix}$

10. $f\begin{bmatrix} x \\ y \\ z \\ w \end{bmatrix} = \begin{bmatrix} -x - 2y + z \\ x - w \end{bmatrix}$

$g\begin{bmatrix} x \\ y \end{bmatrix} = \begin{bmatrix} y \\ x - 3y \\ x - y \\ x \end{bmatrix}$

In Exercises 11–13 verify Part 5 of Theorem 22 for the given f and g.

11. f and g as in Exercise 8
12. f and g as in Exercise 9
13. f and g as in Exercise 10

14. Referring to Exercise 4, find $f \circ g$ and $g \circ f$.

15. Explain why $f \circ g$ is undefined. Is $g \circ f$ defined?

    a. $f(x, y) = (x - y, x + y, y)$
        $g(x, y) = (-x + y, x, y)$

    b. $f(x, y, z) = (x + z, x + y)$
        $g(x, y) = (-x + y, x + 2y)$

    c. $f(x, y) = (x - y, x + y)$
        $g(x, y) = (-x + y, x + y, x)$

16. Find two linear transformations f and g such that $f \circ g \neq g \circ f$.

17. Let $f : \mathbf{R}^2 \to \mathbf{R}^2$, $f(x, y) = (x - y, x + y)$. Find $f^3(x, y)$ and $f^3(1, -1)$.

18. Show that f and g are inverse to each other.

$$f \begin{bmatrix} x \\ y \\ z \end{bmatrix} = \begin{bmatrix} -x + y \\ x - z \\ x + y - z \end{bmatrix}$$

$$g \begin{bmatrix} x \\ y \\ z \end{bmatrix} = \begin{bmatrix} -x - y + z \\ -y + z \\ -x - 2y + z \end{bmatrix}$$

19. Show that f is invertible. (Do not compute the inverse.)

$$f \begin{bmatrix} x \\ y \\ z \end{bmatrix} = \begin{bmatrix} -x + y + z \\ x - 2z \\ 2x - y \end{bmatrix}$$

20. Show that f is invertible. (Do not compute the inverse.)

$$f \begin{bmatrix} x \\ y \\ z \end{bmatrix} = \begin{bmatrix} x + 3y + 2z \\ 3x + 2y + z \\ 3x + 3y + z \end{bmatrix}$$

In Exercises 21–25 show that the transformation is invertible and compute its inverse.

21. $f(x, y, z) = (-2x - z, -y - 2z, -2z)$

22. $f(x, y, z) = (x + 2y - z, x - 2y - z, x + 6y + z)$

23. $f(x, y, z) = (x + y - z, x + 2y - z, 3x + 4y + 3z)$

24. $f \begin{bmatrix} x \\ y \\ z \end{bmatrix} = \begin{bmatrix} -x + z \\ x + \frac{2}{3}y - \frac{4}{3}z \\ -y + z \end{bmatrix}$

25. $f : \mathbf{R}^4 \to \mathbf{R}^4$, given by

$$f(\mathbf{x}) = \begin{bmatrix} -1 & 1 & 1 & -1 \\ -1 & 0 & 1 & 0 \\ 0 & 1 & -1 & 1 \\ 0 & 0 & 1 & -1 \end{bmatrix} \mathbf{x}$$

26. Complete the proof of Theorem 18.

27. Prove Theorem 19.

28. Prove Parts 3–6 of Theorem 21.

29. Prove Parts 2–5 of Theorem 22.

30. Prove Theorem 24.

Right and Left Inverses

In this paragraph we introduce right and left inverses of linear transformations, as we did for matrices in the exercises of Section 3.3.

Let $f : V \to W$ be a linear transformation between finite-dimensional vector spaces. We say that the linear transformation g is a **right inverse** of f, if $f \circ g = I$. Likewise, h is a **left inverse** of f if $h \circ f = I$. For example, if $p(x, y, z) = (x, y)$, and $q(x, y) = (x, y, 0)$, then p is a left inverse of q and q is a right inverse of p, because

$$p \circ q(x, y) = (x, y) \tag{5.6}$$

Let A be the matrix of f with respect to fixed bases of V and W. Recall the definitions of left and right inverses of matrices in the exercises of Section 3.3.

31. Show that the statements are equivalent:

    a. f has a right inverse.

    b. f is onto.

    c. A has a right inverse.

32. Show that the statements are equivalent:

    a. f has a left inverse.

    b. f is one-to-one.

    c. A has a left inverse.

33. Show that if f has both a right inverse g and a left inverse h, then the following hold.

    a. $g = h$.

    b. f is one-to-one and onto.

    c. f is an isomorphism.

    d. A has a left and a right inverse that coincide.

    e. A is invertible.

Negative Powers of Invertible Transformations

Just as we did with invertible matrices, we may define negative powers of invertible transformations. If f is invertible and n is a positive integer, we define

$$f^{-n} = (f^{-1})^n = \underbrace{f^{-1} \circ f^{-1} \circ \cdots \circ f^{-1}}_{n \text{ factors}}$$

34. Find $f^{-3}(x, y)$ and $f^{-3}(-1, 2)$ if

$$f\begin{bmatrix} x \\ y \end{bmatrix} = \begin{bmatrix} x - y \\ x + y \end{bmatrix}$$

35. Compare $f^{-2} \circ f^{-1}(x, y)$ and $f^{-3}(x, y)$ if

$$f\begin{bmatrix} x \\ y \end{bmatrix} = \begin{bmatrix} x - 2y \\ 2x + y \end{bmatrix}$$

5.6 Applications

> ### Reader's Goal for This Section
>
> To get an idea of the applications related to linear transformations.

In this section we offer some important applications of the material discussed in this chapter, emphasizing the relations of linear transformations to computer graphics and fractals.

Affine Transformations and Computer Graphics

DEFINITION

> Let A be an $m \times n$ matrix. An **affine transformation** $T : \mathbf{R}^n \to \mathbf{R}^m$ is a transformation of the form
>
> $$T(\mathbf{x}) = A\mathbf{x} + \mathbf{b}$$
>
> for some fixed m-vector $\mathbf{b}$. This transformation is *nonlinear* if $\mathbf{b} \neq \mathbf{0}$. Since, $T(\mathbf{0}) \neq \mathbf{0}$. In the special case, where $m = n$ and A is the $n \times n$ identity matrix I, we have
>
> $$T(\mathbf{x}) = I\mathbf{x} + \mathbf{b} = \mathbf{x} + \mathbf{b}$$
>
> Such a T is called a **translation** by $\mathbf{b}$.

A translation by a vector $\mathbf{b} \neq \mathbf{0}$ translates a figure by adding $\mathbf{b}$ to all its points. An affine transformation is a linear transformation followed by a translation. Figure 5.29(a)

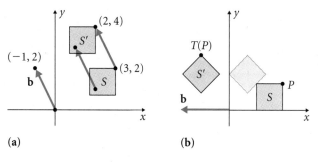

(a)   **(b)**

Figure 5.29 (a) Translation, (b) affine transformation: rotation then translation.

shows the image S' of the square S after translation by $(2, -1)$. Figure 5.29(b) shows the image S' of the square S under the affine transformation

$$T(\mathbf{x}) = \begin{bmatrix} \frac{\sqrt{2}}{2} & -\frac{\sqrt{2}}{2} \\ \frac{\sqrt{2}}{2} & \frac{\sqrt{2}}{2} \end{bmatrix} \mathbf{x} + \begin{bmatrix} -2 \\ 0 \end{bmatrix}$$

T consists of a rotation by $45°$ followed by a translation by $(-2, 0)$.

Affine transformations with $n = m = 2$ and $n = m = 3$ are very useful in computer graphics.

■ EXAMPLE 54   Find the affine transformation T that transformed the left image of Fig. 5.30 to the right image, given that the following points were used:

$$(1, 0), (0.7, 0.7), (0, 1), (-0.7, 0.7), (-1, 0), (-0.7, -0.7), (0, -1), (0.7, -0.7), (1, 0)$$

with respective images

$$(2, -1), (2.05, -0.3), (1.5, 0), (0.65, -0.3), (0, -1), (-0.05, -1.7), (0.5, -2),$$
$$(1.35, -1.7), (2, -1)$$

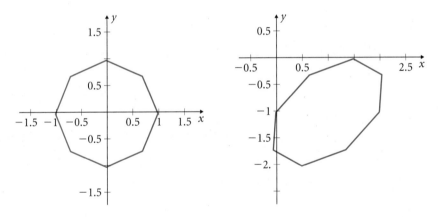

Figure 5.30   Sheared translated polygon.

SOLUTION   Let $T(\mathbf{x}) = A\mathbf{x} + \mathbf{b}$ with $\mathbf{b} = \begin{bmatrix} b_1 \\ b_2 \end{bmatrix}$ and $A = \begin{bmatrix} a & b \\ c & d \end{bmatrix}$. Then

$$T \begin{bmatrix} x_1 \\ x_2 \end{bmatrix} = \begin{bmatrix} a & b \\ c & d \end{bmatrix} \begin{bmatrix} x_1 \\ x_2 \end{bmatrix} + \begin{bmatrix} b_1 \\ b_2 \end{bmatrix} = \begin{bmatrix} ax_1 + bx_2 + b_1 \\ cx_1 + dx_2 + b_2 \end{bmatrix}$$

Because

$$T \begin{bmatrix} 1 \\ 0 \end{bmatrix} = \begin{bmatrix} a + b_1 \\ c + b_2 \end{bmatrix} = \begin{bmatrix} 2 \\ -1 \end{bmatrix}$$

$$T \begin{bmatrix} 0 \\ 1 \end{bmatrix} = \begin{bmatrix} b + b_1 \\ d + b_2 \end{bmatrix} = \begin{bmatrix} 1.5 \\ 0 \end{bmatrix}$$

$$T \begin{bmatrix} -1 \\ 0 \end{bmatrix} = \begin{bmatrix} -a + b_1 \\ -c + b_2 \end{bmatrix} = \begin{bmatrix} 0 \\ -1 \end{bmatrix}$$

we get the system

$$a + b_1 = 2$$
$$c + b_2 = -1$$
$$b + b_1 = 1.5$$
$$d + b_2 = 0$$
$$-a + b_1 = 0$$
$$-c + b_2 = -1$$

with solution $a = 1, b = 0.5, c = 0, d = 1$ and $b_1 = 1, b_2 = -1$. Therefore,

$$T(\mathbf{x}) = A\mathbf{x} + \mathbf{b} = \begin{bmatrix} 1 & 0.5 \\ 0 & 1 \end{bmatrix} \mathbf{x} + \begin{bmatrix} 1 \\ -1 \end{bmatrix}$$

provided that this equation is valid for the rest of the points used in the graph. (It is; check.)

Hence, T is the shear by 0.5 along the x-axis followed by the translation by $(1, -1)$.

Referring to Fig. 5.31, the affine transformation T given by the rotation of 45° about the z-axis in the positive direction followed by a translation by $(1, 1, 1)$ has been applied to the tetrahedron on the left and produced the tetrahedron on the right.

$$T(\mathbf{x}) = \begin{bmatrix} \frac{\sqrt{2}}{2} & -\frac{\sqrt{2}}{2} & 0 \\ \frac{\sqrt{2}}{2} & \frac{\sqrt{2}}{2} & 0 \\ 0 & 0 & 1 \end{bmatrix} \mathbf{x} + \begin{bmatrix} 1 \\ 1 \\ 1 \end{bmatrix}$$

The images of a rather lengthy list of points were computed in order to produce the transformed image.

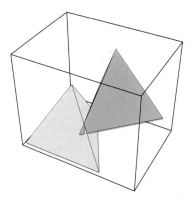

Figure 5.31   Rotated translated tetrahedron.

Affine Transformations and Fractals

In recent years a new area of mathematics, called *fractal geometry*, has emerged. Although fractal geometry has its roots in important works by Cantor, Sierpinski, von Koch, Peano, and other mathematicians of the nineteenth century, it is only since the late 1960s that it has become a "new field." This was due to the pioneering work of Benoit Mandelbrot of IBM corporation and to the emergence of fast computers. The word *fractal*, introduced by Mandelbrot, is used to describe figures with "infinite repetition of the same shape" (Figs. 5.32 and 5.33).[8] In this paragraph we describe two fractals, the Sierpinski triangle and a fractal that looks like a fir tree. (It is an analogue of M. Barnsley's fern.)

It was observed by M. Barnsley that many "fractallike" objects can be obtained by plotting *iterations* of certain affine transformations.

The Sierpinski Triangle

Let f_1, f_2, f_3 be the three affine transformations from $\mathbf{R}^2$ to $\mathbf{R}^2$ given by

$$f_1(\mathbf{x}) = \begin{bmatrix} \frac{1}{2} & 0 \\ 0 & \frac{1}{2} \end{bmatrix} \mathbf{x}$$

$$f_2(\mathbf{x}) = \begin{bmatrix} \frac{1}{2} & 0 \\ 0 & \frac{1}{2} \end{bmatrix} \mathbf{x} + \begin{bmatrix} \frac{1}{2} \\ 0 \end{bmatrix}$$

$$f_3(\mathbf{x}) = \begin{bmatrix} \frac{1}{2} & 0 \\ 0 & \frac{1}{2} \end{bmatrix} \mathbf{x} + \begin{bmatrix} 0 \\ \frac{1}{2} \end{bmatrix}$$

The Sierpinski triangle can be generated as follows. Starting with a triangle, say, the triangle with vertices $(0,0)$, $(1,0)$, $(0,1)$, we pick a point inside it and plot it; say, the point $\left(\frac{1}{2}, \frac{1}{2}\right)$. Then we randomly select one of f_1, f_2, f_3, say, f_i, and compute and

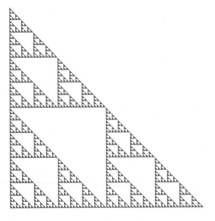

Figure 5.32   A Sierpinski triangle.

[8]For more information on fractals, we recommend M. Barnsley's *Fractals Everywhere*, 2nd ed. (San Diego, CA: Academic Press, 1993), and R. L. Devaney's *An Introduction to Chaotic Dynamical Systems*, 2nd ed. (Reading, MA: Addison-Wesley, 1989).

plot $f_i \left(\frac{1}{2}, \frac{1}{2}\right)$. Making this point our new starting point, we repeat the process as often as desired. The resulting picture is a "fractal object" that looks like a triangle with triangular holes in it (if enough points are plotted). See Fig. 5.32.

A Fir Tree Like Barnsley's Fern

Let f_1, f_2, f_3, f_4 be the four affine transformations from $\mathbf{R}^2$ to $\mathbf{R}^2$ given by

$$f_1(\mathbf{x}) = \begin{bmatrix} 0.75 & 0.03 \\ -0.07 & 0.7 \end{bmatrix} \mathbf{x} + \begin{bmatrix} 10 \\ 150 \end{bmatrix}$$

$$f_2(\mathbf{x}) = \begin{bmatrix} -0.15 & 0.51 \\ 0.5 & 0.15 \end{bmatrix} \mathbf{x} + \begin{bmatrix} 10 \\ 40 \end{bmatrix}$$

$$f_3(\mathbf{x}) = \begin{bmatrix} 0.2 & -0.25 \\ 0.21 & 0.4 \end{bmatrix} \mathbf{x} + \begin{bmatrix} 30 \\ 150 \end{bmatrix}$$

$$f_4(\mathbf{x}) = \begin{bmatrix} 0.02 & -0.05 \\ 0.03 & 0.2 \end{bmatrix} \mathbf{x} + \begin{bmatrix} 10 \\ 1 \end{bmatrix}$$

The fir tree–like fractal of Fig. 5.33 can be generated as follows: We pick and plot any point, say, $(5, 5)$. Then we randomly select one of f_1, f_2, f_3, f_4, say, f_i, and compute and plot $f_i(5, 5)$. Making this point the new starting point, we repeat the process. The resulting picture looks like a piece of a fir tree.

Figure 5.33   A fir tree.

Let us outline the procedure that generated both fractals. *The following procedure produces a fractal image for* some *sets of affine transformations.*[9]

---

[9]It is not true that iterations of any set of affine transformations will produce fractallike images.

Algorithm (Generator of Fractal Image)

1. Start with an *appropriate* set of affine transformations $S = \{f_1, f_2, \ldots, f_n\}$, and initial point (x_k, y_k).
2. Randomly choose an affine transformation of S, say f_i.
3. Compute and plot the point $f_i(x_k, y_k)$. Set $(x_k, y_k) = f_i(x_k, y_k)$.
4. Go to Step 2. Repeat as often as desired.

Exercises 5.6

In Exercises 1–4 find the images of the zero vector and the images of the standard basis vectors for the given affine transformations.

1. $T(\mathbf{x}) = \begin{bmatrix} 1 & 0 \\ 0 & 5 \end{bmatrix} \mathbf{x} + \begin{bmatrix} 1 \\ -1 \end{bmatrix}$

2. $T(\mathbf{x}) = \begin{bmatrix} 4 & -3 \\ 2 & 5 \end{bmatrix} \mathbf{x} + \begin{bmatrix} 1 \\ -1 \end{bmatrix}$

3. $T(\mathbf{x}) = \begin{bmatrix} -1 & 2 & 0 \\ 1 & 2 & -1 \\ 0 & -4 & 1 \end{bmatrix} \mathbf{x} + \begin{bmatrix} 1 \\ -1 \\ 0 \end{bmatrix}$

4. $T(\mathbf{x}) = \begin{bmatrix} 1 & 2 & -1 & -3 \\ 0 & -4 & 1 & 0 \end{bmatrix} \mathbf{x} + \begin{bmatrix} 1 \\ -1 \end{bmatrix}$

In Exercises 5–8 write the given affine transformations in the form $T(\mathbf{x}) = A\mathbf{x} + \mathbf{b}$.

5. $T \begin{bmatrix} x \\ y \end{bmatrix} = \begin{bmatrix} x - y \\ -x + y - 1 \end{bmatrix}$

6. $T \begin{bmatrix} x \\ y \end{bmatrix} = \begin{bmatrix} -2y - 1 \\ 2x + 7y - 1 \end{bmatrix}$

7. $T \begin{bmatrix} x \\ y \\ z \end{bmatrix} = \begin{bmatrix} -x + 3y + 1 \\ x - z \\ x - 5y + z - 1 \end{bmatrix}$

8. $T \begin{bmatrix} x \\ y \\ z \\ w \end{bmatrix} = \begin{bmatrix} x - z - 9w + 1 \\ -3z - 6w \end{bmatrix}$

In Exercises 9–13 find A and $\mathbf{b}$ for the given affine transformation $T(\mathbf{x}) = A\mathbf{x} + \mathbf{b}$ if

9. $T : \mathbf{R}^2 \rightarrow \mathbf{R}^2, T(\mathbf{e}_1) = \begin{bmatrix} -1 \\ 3 \end{bmatrix}$,

$T(\mathbf{e}_2) = \begin{bmatrix} 4 \\ -7 \end{bmatrix}, T(\mathbf{0}) = \begin{bmatrix} -1 \\ 1 \end{bmatrix}$

10. $T : \mathbf{R}^2 \rightarrow \mathbf{R}^2, T(\mathbf{e}_1) = \begin{bmatrix} 4 \\ -2 \end{bmatrix}$,

$T(\mathbf{e}_2) = \begin{bmatrix} -5 \\ 9 \end{bmatrix}, T(\mathbf{0}) = \begin{bmatrix} -1 \\ 3 \end{bmatrix}$

11. $T : \mathbf{R}^2 \rightarrow \mathbf{R}^2, T(\mathbf{e}_1) = \begin{bmatrix} -2 \\ 5 \end{bmatrix}$,

$T(\mathbf{e}_2) = \begin{bmatrix} -5 \\ 6 \end{bmatrix}, T(\mathbf{e}_1 + \mathbf{e}_2) = \begin{bmatrix} -4 \\ 9 \end{bmatrix}$

12. $T : \mathbf{R}^3 \rightarrow \mathbf{R}^3$ and

$T(\mathbf{e}_1) = \begin{bmatrix} 0 \\ -1 \\ 1 \end{bmatrix}, \quad T(\mathbf{e}_2) = \begin{bmatrix} 1 \\ 0 \\ -2 \end{bmatrix}$

$T(\mathbf{e}_3) = \begin{bmatrix} 2 \\ -2 \\ 0 \end{bmatrix}, \quad T(\mathbf{0}) = \begin{bmatrix} 1 \\ -1 \\ 0 \end{bmatrix}$

13. $T : \mathbf{R}^3 \rightarrow \mathbf{R}^2$ and

$T(\mathbf{e}_1) = \begin{bmatrix} 2 \\ -5 \end{bmatrix}, \quad T(\mathbf{e}_2) = \begin{bmatrix} -1 \\ 4 \end{bmatrix}$

$T(\mathbf{e}_3) = \begin{bmatrix} 4 \\ -7 \end{bmatrix}, \quad T(\mathbf{e}_1 + \mathbf{e}_2 + \mathbf{e}_3) = \begin{bmatrix} 3 \\ -6 \end{bmatrix}$

14. Show that if T is an affine transformation with $T(\mathbf{0}) = \mathbf{b}$, then $L(\mathbf{x}) = T(\mathbf{x}) - \mathbf{b}$ is a linear transformation.

15. Let $T : \mathbf{R}^n \rightarrow \mathbf{R}^m, T(\mathbf{x}) = A\mathbf{x} + \mathbf{b}$ be any affine transformation. Show that T is uniquely determined by the values

$$T(\mathbf{e}_1), T(\mathbf{e}_2), \ldots, T(\mathbf{e}_n) \quad \text{and} \quad T(\mathbf{0})$$

(*Hint:* Note that $T(\mathbf{0}) = \mathbf{b}$ and that $L(\mathbf{x}) = T(\mathbf{x}) - \mathbf{b}$ is a linear (matrix) transformation.)

16. Show that any affine transformation $T : \mathbf{R}^2 \to \mathbf{R}^2$, $T(\mathbf{x}) = A\mathbf{x} + \mathbf{b}$ maps straight lines to straight lines or to points.

17. Let $T : \mathbf{R}^n \to \mathbf{R}^m$, $T(\mathbf{x}) = A\mathbf{x} + \mathbf{b}$ be an affine transformation. What is the relation between: the set

$\{\mathbf{x} \in \mathbf{R}^n, \quad T(\mathbf{x}) = \mathbf{0}\}$ and the set of solutions of the system $[A : -\mathbf{b}]$?

18. Draw the image of the straight line $x - y = -1$ under

$$T(\mathbf{x}) = \begin{bmatrix} 1 & -3 \\ 0 & 5 \end{bmatrix} \mathbf{x} + \begin{bmatrix} -1 \\ 0 \end{bmatrix}.$$

5.7 Miniprojects

1 ■ Some Special Affine Transformations

In this project you are to teach yourselves the basics of some special affine transformations, the similitudes. Similitudes are extensively used in computer graphics, dynamical systems, and fractals.

A **similarity transformation** or **similitude** $T : \mathbf{R}^2 \to \mathbf{R}^2$ is a special affine transformation of one of the forms

$$T(\mathbf{x}) = \begin{bmatrix} r\cos\theta & -r\sin\theta \\ r\sin\theta & r\cos\theta \end{bmatrix} \mathbf{x} + \begin{bmatrix} b_1 \\ b_2 \end{bmatrix}$$

$$T(\mathbf{x}) = \begin{bmatrix} r\cos\theta & r\sin\theta \\ r\sin\theta & -r\cos\theta \end{bmatrix} \mathbf{x} + \begin{bmatrix} b_1 \\ b_2 \end{bmatrix}$$

for some scalar $r \neq 0$, some angle θ, $0 \leq \theta < 2\pi$, and some scalars b_1 and b_2. Similitudes are scaled rotations followed by translations or scaled reflected rotations followed by translations. As such, they preserve angles, as we shall see later on.

Problem A

Show that the following transformations are similitudes:

1. Any rotation about the origin.
2. Reflections about the axes, the diagonal, or the origin.
3. Any homothety of $\mathbf{R}^2$.

Problem B

1. Are shears in general similitudes? Check the shears with standard matrices:

$$\begin{bmatrix} 1 & 0.5 \\ 0 & 1 \end{bmatrix} \qquad \begin{bmatrix} 1 & 0.5 \\ -0.5 & 0 \end{bmatrix}$$

2. Are projections onto the axes similitudes?
3. Are translations similitudes?

Problem C

1. Find a formula for the similitude T that maps the triangle $(0,0), (1,0), (0,1)$ to the triangle $(1,1), (-1,1), (1,-1)$.
2. Let S_1 be the image under T of the rectangle S with vertices $(0,0), (2,0), (2,1), (0,1)$ and let S_2 be the image of S_1. Compute that areas $(S), (S_1)$, and (S_2) and compare the ratios of the areas $(S_2) : (S_1)$ and $(S_1) : (S)$.
3. Find the formula of the similitude R that rotates any point $45°$ about the origin, then scales it by a factor of 2, and, finally, translates it by $(1,1)$.
4. Find the image L_1 under R of the triangle L with vertices $(0,0), (1,0), (0,1)$ and find the image L_2 of L_1. Compute the area ratios $(L_2) : (L_1)$ and $(L_1) : (L)$. What did you observe?

2 ■ Another Fractal

Usually, fractal images cannot be plotted without the help of a computer. In this project we study a fractal that, to some degree, can be visualized by hand plotting.

Consider the rectangles with the given vertices:

$$A: \quad (1,-1), (1,1), (-1,1), (-1,-1)$$
$$B: \quad (1,0), (1,1), (-1,1), (-1,0)$$
$$C: \quad (2,0), (2,1), (-2,1), (-2,0)$$

Also consider the affine transformations

$$R(\mathbf{x}) = \begin{bmatrix} 0 & -\frac{1}{2} \\ \frac{1}{2} & 0 \end{bmatrix} \mathbf{x}$$

$$T(\mathbf{x}) = \begin{bmatrix} 0 & -\frac{1}{2} \\ \frac{1}{2} & 0 \end{bmatrix} \mathbf{x} + \begin{bmatrix} \frac{1}{4} \\ \frac{1}{2} \end{bmatrix}$$

Let $A_1^R = R(A)$ (the image of rectangle A under R), $A_2^R = R(A_1^R), A_3^R = R(A_2^R), A_4^R = R(A_3^R)$. Likewise, let A_1^T, A_2^T, A_3^T, and A_4^T the corresponding images under T. Also, we have the consecutive images of B, $B_1^R, B_2^R, B_3^R, B_4^R$ under R and $B_1^T, B_2^T, B_3^T, B_4^T$ under T. The images of C are defined the same way.

Problem A is designed to show you the effects of R and T on A, B, C and their iterated images.

Problem A

1. Plot $A, A_1^R, A_2^R, A_3^R, A_4^R$ on one graph and $A, A_1^T, A_2^T, A_3^T, A_4^T$ on another.
2. Plot $B, B_1^R, B_2^R, B_3^R, B_4^R$ on one graph and $B, B_1^T, B_2^T, B_3^T, B_4^T$ on another.
3. Plot $C, C_1^R, C_2^R, C_3^R, C_4^R$ on one graph and $C, C_1^T, C_2^T, C_3^T, C_4^T$ on another.

Problem B is designed to show you the fractal image generated by applying R and T at the origin and iterating.

Problem B

Let $P(0,0)$. Find the two images P_1 and P_2 of P under R and T. Then find the images P_3, P_4 of P_1 under R and T and the images P_5, P_6 under T. Continue this process for as long as you please. Then plot *all* points found. It takes about 5 to 6 iterations to see a descent formation of a fractal object.

Problem C is designed to show you how the fractal image is affected if you start at a different point.

Problem C

Answer the questions of Problem B starting with the point $Q(0.5, 0.5)$.

5.8 Computer Exercises

This computer session will help you master the commands of your software related to the topics of this chapter. In addition it will help review several of the basic concepts.

Let

$$T_1 : \mathbf{R}^3 \rightarrow \mathbf{R}^3, \qquad T_1(x, y, z) = (2x - y + z, x + y, 2y - 3z)$$
$$T_2 : \mathbf{R}^2 \rightarrow \mathbf{R}^3, \qquad T_2(x, y) = (3x - 4y, x + 3y, -y)$$
$$T_3 : \mathbf{R}^3 \rightarrow \mathbf{R}^2, \qquad T_3(x, y, z) = (2x - y + z, x - z)$$
$$T_4 : \mathbf{R}^3 \rightarrow \mathbf{R}^3, \qquad T_4(x, y, z) = (-y + z, x + 1, 2y)$$

1. Define T_1, T_2, T_3 and T_4.
2. Compute $T_1(1, 1, 1), T_2(1, 1), T_3(1, 1, 1), T_4(1, 1, 1)$.
3. Show that T_1, T_2, T_3 are linear.
4. Show that T_4 is nonlinear.
5. Find the standard matrices of T_1, T_2, T_3.

A linear transformation T is such that:

$$T(1, 2, 3, 4) = (1, 0, -1, 1)$$
$$T(1, 3, 5, 7) = (0, 1, 0, -1)$$
$$T(3, 3, 4, 4) = (1, 1, 1, -1)$$
$$T(4, 4, 4, 5) = (1, 1, -1, 1)$$

6. Find the standard matrix of T.
7. Compute $T(2, 2, -2, -2)$.

Let

$$T_1(x, y, z, w) = (x + 2y + 3z + 4w, 2x + 3y + 4z + 5w, 3x + 4y + 5z + 6w)$$
$$T_2(x, y, z) = (x + 2y + 3z + 4w, 2x + 2y + 3z + 4w, 3x + 3y + 3z + 4w)$$
$$T_3(x, y, z, w, t) = (x + 2y + 3z + 4w, 2x + 2y + 3z + 4w, 3x + 3y + 3z + 4w,$$
$$4x + 4y + 4z + 4t)$$

8. Which of $v_1 = (-1, 0, 3, -2)$, $v_2 = (18, -31, 8, 5)$, and $v_3 = (1, -1, 8, 7)$ are in $\text{Ker}(T_1)$?

9. Which of $w_1 = (2, 7, 12)$, $w_2 = (42, 59, 76)$, $w_3 = (42, 59, 77)$ are in $R(T_1)$?

For each of T_1, T_2, and T_3:

10. Find the standard matrix.

11. Find a basis for the kernel. What is the nullity?

12. Find a basis for the range. What is the rank?

13. Verify the dimension theorem.

14. Which of the transformations T_1, T_2, and T_3 is: one-to-one, onto, an isomorphism, none of these choices?

15. True or false?
    a. $R(T_1) = \mathbf{R}^3$
    b. $R(T_2) = \mathbf{R}^3$
    c. $R(T_3) = \mathbf{R}^4$

16. Define T and evaluate $T(x + 1)$. Show that T is linear.
    $$T : P_1 \longrightarrow P_2, \qquad T(ax + b) = (3a - 4b)x^2 + (a + 3b)x - b$$

17. Compute $L(2x^3 + 2x^2 - 2x - 2)$, if L is linear such that
    $$L(x^3 + 2x^2 + 3x + 4) = x^3 - x + 1$$
    $$L(x^3 + 3x^2 + 5x + 7) = x^2 - 1$$
    $$L(3x^3 + 3x^2 + 4x + 4) = x^3 + x^2 + x - 1$$
    $$L(4x^3 + 4x^2 + 4x + 5) = x^3 + x^2 - x + 1$$

18. Find a basis for the null space of F, where
    $$F(ax^3 + bx^2 + cx + d) = (a + 2b + 3c + 4d)x^2$$
    $$+ (2a + 3b + 4c + 5d)x$$
    $$+ (3a + 4b + 5c + 6d)$$

19. Find the matrix M of T with respect to $\mathcal{B} = \{x - 1, x + 1\}$ and $\mathcal{B}' = \{x^2 - 1, x + 1, x - 1\}$.
    $$T : P_1 \longrightarrow P_2, \qquad T(ax + b) = (3a - 4b)x^2 + (a + 3b)x - b$$

Selected Solutions with Maple

```
with(linalg):     # We load the package for the entire set.
# Exercises 1-5.
# Instead of T1:= (x,y,z) -> vector([2*x-y+z, x+y, 2*y-3*z]); etc.,
# we may also use the simpler:
T1 := (x,y,z) -> [2*x-y+z, x+y, 2*y-3*z];
T2 := (x,y) -> [3*x-4*y, x+3*y, -y];
T3 := (x,y,z) -> [2*x-y+z, x-z];
T4 := (x,y,z) -> [-y+z, x+1, 2*y];
```

```
# Also T1 := proc(x,y,z) [2*x-y+z, x+y, 2*y-3*z] end, etc,.
T1(1,1,1);                    # Also T2(1,1); T3(1,1,1); T4(1,1,1);
# Next, we need the equal command, loaded from the linalg
# package. equal tests vectors and matrices for equality.
# First check Part 1 of the definition:
equal(T1(x1+x2,y1+y2,z1+z2),T1(x1,y1,z1)+T1(x2,y2,z2));
equal(T2(x1+x2,y1+y2),T2(x1,y1)+T2(x2,y2));
equal(T3(x1+x2,y1+y2,z1+z2),T3(x1,y1,z1)+T3(x2,y2,z2));
equal(T4(x1+x2,y1+y2,z1+z2),T4(x1,y1,z1)+T4(x2,y2,z2));
# For the second part be careful. A product like c*[x-y,x+y]
# does not automatically simplify to [c*x-c*y,c*x+c*y].
# First we use evalm(c*[x-y,x+y]); to pass the scalar inside:
# [c*(x-y),c*(x+y)]. Now expand([c*(x-y),c*(x+y)]); will not
# do any good. We need to go pass the vector or list and expand
# each component separately, by map(expand, [c*(x-y),c*(x+y)]); .
equal(T1(c*x,c*y,c*z),map(expand, evalm(c*T1(x,y,z))));
equal(T2(c*x,c*y),map(expand, evalm(c*T2(x,y))));
equal(T3(c*x,c*y,c*z),map(expand, evalm(c*T3(x,y,z))));
equal(T4(c*x,c*y,c*z),map(expand, evalm(c*T4(x,y,z))));
MT1 := augment(T1(1,0,0),T1(0,1,0),T1(0,0,1));  # Evaluation at the
MT2 := augment(T2(1,0),T2(0,1));                 # basis vectors to get
MT3 := augment(T3(1,0,0),T3(0,1,0),T3(0,0,1));   # the standard matrix.
# Exercises 6,7.
M := matrix([[1,1,3,4],[2,3,3,4],[3,5,4,4],[4,7,4,5]]);   # The domain vectors.
Id := matrix([[1,0,0,0],[0,1,0,0],[0,0,1,0],[0,0,0,1]]);  # I_4.
N := matrix([[1,0,1,1],[0,1,1,1],[-1,0,1,-1],[1,-1,-1,1]]);# The values.
M1 := rref(augment(M, Id));
M2 := delcols(M1,1..4);                          # Reduction of [M:I_4].
STM := evalm(N &* M2);
evalm(STM &* vector([2,2,-2,-2]));               # The standard matrix.
# Exercises 8-15.                                # T(2,2,-2,-2) .
T1:=(x,y,z,w)->[x+2*y+3*z+4*w, 2*x+3*y+4*z+5*w, 3*x+4*y+5*z+6*w];
T2:=(x,y,z,w)->[x+2*y+3*z+4*w, 2*x+2*y+3*z+4*w, 3*x+3*y+3*z+4*w];
T3:=(x,y,z,w)->[x+2*y+3*z+4*w,2*x+2*y+3*z+4*w,3*x+3*y+3*z+4*w,4*x+4*y+4*z+4*w];
T1(-1,0,3,-2);        # v1 in the kernel.
T1(18,-31,8,5);       # v2 in the kernel.
T1(1,-1,8,7);         # v3 not in the kernel.
M1:=augment(T1(1,0,0,0),T1(0,1,0,0),T1(0,0,1,0),T1(0,0,0,1)); # Standard matrix.
Mw1:=rref(augment(M1,vector([2,7,12])));  # In Range(T1). Last column is pivot.
Mw2:=rref(augment(M1,vector([42,59,76])));# In Range(T1). Last column is pivot.
Mw3:=rref(augment(M1,vector([42,59,77])));# Not in Range(T1). Last column is non-pivot.
evalm(M1);        # Standard matrix of T1 already found. The remaining are:
M2:=augment(T2(1,0,0,0),T2(0,1,0,0),T2(0,0,1,0),T2(0,0,0,1));
M3:=augment(T3(1,0,0,0),T3(0,1,0,0),T3(0,0,1,0),T3(0,0,0,1));
              # Or, kernel(M2) etc..
              # k1 has two vectors, so nullity 2.
```

```
k1:=nullspace(M1);   # kernel non-zero, T1 is not one-to-one.
                     # Hence, not an isomorphism.
# Also,  k1:=kernel(T1);
                     # k2 has one vector, so nullity 2.
k2:=nullspace(M2);   # kernel non-zero, T2 is not one-to-one.
                     # Hence, not an isomorphism.
# Also,  k2:=kernel(T2);
k3:=nullspace(M3);   # k3 has no vectors, so nullity 0. T3 is one-to-one.
# Also,  k3:=kernel(T3);
                  # The first 2 columns form a basis for the range.
r1:=rref(M1);     # Rank is 2. 2 + 2 = number of columns.
                  # The third row has no pivot, so not onto.
# Also,  r1:=range(T1);
                  # The first 3 columns form a basis for the range.
r2:=rref(M2);     # Rank is 3. 3 + 1 = number of columns.
                  # Each row has a pivot, so onto.
# Also,  r2:=range(T2);
                  # All columns form a basis for the range.
r3:=rref(M3);     # Rank is 4. 4 + 0 = number of columns.
                  # T3 is one-to-one and onto, hence, an isomorphism.
# Also,  r3:=range(T3);
# False. T1 is not onto.
# True. T2 is onto.
# True. T3 is onto.
# Exercise 16.
T := (a,b) -> [3*a-4*b, a+3*b, -b];   # We use [a,b,c] for ax^2+bx+c.
T(1,1);                               # T(x+1).
equal(T(a1+a2,b1+b2),T(a1,b1)+T(a2,b2));#linalg[equal] tests for matrix equality.
# To simplify c*[a-b,a+b] to [c*a-c*b,c*a+c*b], we use evalm(c*[a-b,a+b]);
# to pass the scalar inside then "map(expand())" to expand each component.
equal(T(c*a,c*b),map(expand, evalm(c*T(a,b))));
# Exercise 17.
# First we form a matrix with the coefficients of the given polynomials,
M := matrix([[1,1,3,4],[2,3,3,4],[3,5,4,4],[4,7,4,5]]);
# then a matrix with the coefficients of their values.
N := matrix([[1,0,1,1],[0,1,1,1],[-1,0,1,-1],[1,-1,-1,1]]);
M1 := rref(augment(M, vector([2,2,-2,-2])));
# The last column of M1 has entries the coefficients of 2x^3+2x^2-2x-2
delcols(M1,1..4);            # in terms of the given polynomials.
evalm(N &* ("));             # Multiplication by N evaluates T(2,2,-2,-2).
# Exercise 18.
# Set F=0. We need to solve a+2b+3c+4d=0,2a+3b+4c+5d=0,3a+4b+5c+6d=0.
M := matrix(3,4,[1,2,3,4, 2,3,4,5, 3,4,5,6]);  # We compute a basis for
NM := nullspace(M);          # the nullspace of the coefficient matrix.
evalm(matrix([NM[1],NM[2]])&*vector([x^3,x^2,x,1])); # In polynomial form.
# Exercise 19.
```

```
T := (a,b) -> matrix(3,1,[3*a-4*b, a+3*b, -b]); # The transformation.
b2:=matrix([[1,0,0],[0,1,1],[-1,1,-1]]); # The coefficients of B'.
aug := augment(b2,T(1,-1),T(1,1)); # The coefficients of T(x-1),T(x+1)
rref(aug);                    # in terms of B' are computed by rref(aug);.
delcols(",1..3);              # The last 2 columns form the matrix of T.
```

Selected Solutions with Mathematica

```
<<LinearAlgebra`MatrixManipulation`; (* We load the package once for all.*)
(* Exercises 1-5. *)
T1[x_,y_,z_] := {2 x-y+z, x+y, 2 y-3 z}  (* Definitions. Underscore _ *)
T2[x_,y_] := {3 x-4 y, x+3 y, -y}         (* declares a position for a *)
T3[x_,y_,z_] := {2 x-y+z, x-z}            (* variable. x_ is declaration of *)
T4[x_,y_,z_] := {-y+z, x+1, 2 y}          (* a variable named x.   *)
(* Colon equal := is the delayed assigment operator. It was used so that *)
(* the funtion gets evaluated when it is called and not during definition.*)
T1[1,1,1]          (* Also T2[1,1]   T3[1,1,1] T4[1,1,1]          *)
(* To test for equality we use SameQ or its synonym === . For example, *)
(* SameQ[a, b] or a===b both test the equality a = b. *)
(* Before we test for equality we must expand the two sides. *)
SameQ[Expand[T1[x1+x2,y1+y2,z1+z2]],Expand[T1[x1,y1,z1]+T1[x2,y2,z2]]]
SameQ[Expand[c T1[x,y,z]],Expand[T1[c x, c y, c z]]]
SameQ[Expand[T2[x1+x2,y1+y2]],Expand[T2[x1,y1]+T2[x2,y2]]]
SameQ[Expand[c T2[x,y]],Expand[T2[c x, c y]]]
SameQ[Expand[T3[x1+x2,y1+y2,z1+z2]],Expand[T3[x1,y1,z1]+T3[x2,y2,z2]]]
SameQ[Expand[c T3[x,y,z]],Expand[T3[c x, c y, c z]]]
SameQ[Expand[T4[x1+x2,y1+y2,z1+z2]],Expand[T4[x1,y1,z1]+T4[x2,y2,z2]]]
SameQ[Expand[c T4[x,y,z]],Expand[T4[c x, c y, c z]]]
Transpose[{T1[1,0,0],T1[0,1,0],T1[0,0,1]}]  (* Evaluation at the basis *)
Transpose[{T2[1,0],T2[0,1]}]                 (* vectors to get the standard *)
Transpose[{T3[1,0,0],T3[0,1,0],T3[0,0,1]}]   (* matrix. *)
(* Exercises 6,7. *)
M = {{1,1,3,4},{2,3,3,4},{3,5,4,4},{4,7,4,5}}        (*  The domain vectors.  *)
n = {{1,0,1,1},{0,1,1,1},{-1,0,1,-1},{1,-1,-1,1}}   (* The values.           *)
M1 = RowReduce[AppendRows[M, IdentityMatrix[4]]]     (* Reduction of [M:I_4]. *)
M2 = TakeColumns[M1,{5,8}]
STM = n . M2                                  (*  The standard matrix.  *)
STM . {{2},{2},{-2},{-2}}                     (*  T(2,2,-2,-2) .        *)
(*  Exercises 8-15. *)
T1[x_,y_,z_,w_]:={x+2y+3z+4w, 2x+3y+4z+5w, 3x+4y+5z+6w}
T2[x_,y_,z_,w_]:={x+2y+3z+4w, 2x+2y+3z+4w, 3x+3y+3z+4w}
T3[x_,y_,z_,w_]:={x+2y+3z+4w, 2x+2y+3z+4w, 3x+3y+3z+4w,4x+4y+4z+4w}
T1[-1,0,3,-2]            (* v1 in the kernel.     *)
T1[18,-31,8,5]          (* v2 in the kernel.     *)
T1[1,-1,8,7]            (* v3 not in the kernel. *)
(* Standard matrix of T1: *)
```

```
M1=Transpose[{T1[1,0,0,0],T1[0,1,0,0],T1[0,0,1,0],T1[0,0,0,1]}]
Mw1=RowReduce[AppendRows[M1,{{2},{7},{12}}]]
                            (* In Range(T1). Last column is pivot. *)
Mw2=RowReduce[AppendRows[M1,{{42},{59},{76}}]]
                            (* In Range(T1). Last column is pivot. *)
Mw3=RowReduce[AppendRows[M1,{{42},{59},{77}}]]
                        (* Not in Range(T1). Last column is non-pivot.*)
M1          (* Standard matrix of T1 already found. The remaining are:*)
M2=Transpose[{T2[1,0,0,0],T2[0,1,0,0],T2[0,0,1,0],T2[0,0,0,1]}]
M3=Transpose[{T3[1,0,0,0],T3[0,1,0,0],T3[0,0,1,0],T3[0,0,0,1]}]
                    (* k1 has two vectors, so nullity 2.       *)
k1=NullSpace[M1]    (* kernel non-zero, T1 is not one-to-one. *)
                    (* Hence, not an isomorphism.             *)

                    (* k2 has one vector, so nullity 2.       *)
k2=NullSpace[M2]    (* kernel non-zero, T2 is not one-to-one. *)
                    (* Hence, not an isomorphism.             *)

k3=NullSpace[M3]    (* k3 has no vectors, so nullity 0. T3 is one-to-one. *)

                    (* The first 2 columns form a basis for the range.   *)
r1=RowReduce[M1]    (* Rank is 2. 2 + 2 = number of columns.             *)
                    (* The third row has no pivot, so not onto.          *)

                    (* The first 3 columns form a basis for the range.   *)
r2=RowReduce[M2]    (* Rank is 3. 3 + 1 = number of columns.             *)
                    (* Each row has a pivot, so onto.                    *)

                    (* All columns form a basis for the range.           *)
r3=RowReduce[M3]    (* Rank is 4. 4 + 0 = number of columns.             *)
                    (* T3 is one-to-one and onto, hence, an isomorphism. *)
(* False. T1 is not onto.  *)
(* True. T2 is onto.       *)
(* True. T3 is onto.       *)
(* Exercise 16. *)
T[a_,b_] := {3 a-4 b, a+3 b, -b}    (* We used {a,b,c} for ax^2+bx+c. *)
T[1,1]                              (* T(1,1) .                       *)
(* Before we test for equality we must expand the two sides.         *)
SameQ[Expand[T[a1+a2,b1+b2]],Expand[T[a1,b1]+T[a2,b2]]]
SameQ[Expand[c T[a,b]],Expand[T[c a, c b]]]
(* Exercise 17. *)
(* First we form a matrix with the coefficients of the given polynomials,*)
M = {{1,1,3,4},{2,3,3,4},{3,5,4,4},{4,7,4,5}}
n = {{1,0,1,1},{0,1,1,1},{-1,0,1,-1},{1,-1,-1,1}}
(* then a matrix with the coefficients of their values.              *)
M1 = RowReduce[AppendRows[M, {{2},{2},{-2},{-2}}]]
```

```
(* The last column of M1 has entries the coefficients of 2x^3+2x^2-2x-2*)
TakeColumns[M1,{5}]        (* in terms of the given polynomials.        *)
n . %                     (* Multiplication by N evaluates T(2,2,-2,-2). *)
(* Exercise 18. *)
(* Set F=0. We need to solve a+2b+3c+4d=0,2a+3b+4c+5d=0,3a+4b+5c+6d=0. *)
M = {{1,2,3,4}, {2,3,4,5}, {3,4,5,6}} (*  We compute a basis for the   *)
MN = NullSpace[M]              (* nullspace of the coefficient matrix.   *)
MN . {{x^3}, {x^2},{x},{1}} (* The answer in polynomial form.           *)
(* Exercise 19. *)
T[a_,b_] := {3 a-4 b, a+3 b, -b}
b2={{1,0,0},{0,1,1},{-1,1,-1}} (* The coefficients of B'and the coefficients *)
aug = AppendRows[b2,Transpose[{T[1,-1],T[1,1]}]]      (* of T(x-1),T(x+1) *)
RowReduce[aug]                      (* in terms of B' are computed by rref(aug) *)
TakeColumns[%,{4,5}]          (* The last 2 columns form the matrix of T. *)
```

Selected Solutions with MATLAB

```
% Exercises 1-5.
function [A] = T1(x,y,z)          % Define function T1 by editing and saving
    A=[2*x-y+z; x+y; 2*y-3*z];    % a file named T1.m in the current working
    end                          % directory and type the 3 lines on the left.
function [A] = T2(x,y)            % Repeat with T2.
    A = [3*x-4*y; x+3*y; -y];
    end
function [A] = T3(x,y,z)          % T3.
    A = [2*x-y+z; x-z];
    end
function [A] = T4(x,y,z)          % T4.
    A = [-y+z; x+1; 2*y];
    end                          % Then in MATLAB session type:
T1(1,1,1)                        % Also T2(1,1) T3(1,1,1) T4(1,1,1)
T4(0,0,0)                        % Get non-zero, so T4 is non-linear.
T4(0,0,0) == [0; 0; 0]           % Also we may test for equality.
[T1(1,0,0) T1(0,1,0) T1(0,0,1)]  % Evaluation at the basis vectors
[T2(1,0) T2(0,1)]                % to get the standard matrix.
[T3(1,0,0) T3(0,1,0) T3(0,0,1)]
% Exercises 6,7.
M = [1 1 3 4; 2 3 3 4; 3 5 4 4; 4 7 4 5]    %  The domain vectors.
N = [1,0,1,1; 0,1,1,1; -1,0,1,-1; 1,-1,-1,1] % The values.
M1 = rref([M eye(4)])            % Reduction of [M:I_4].
M2 = M1(:,5:8)
STM = N * M2
STM * [2;2;-2;-2]                % The standard matrix
% Exercises 8-15.                % T(2,2,-2,-2)
```

```
% As usual, define the functions by editing and saving files named T1.m, T2.m,
% and T3.m in the current working. (The code follows.) When in MATLAB session
% eveluate each function as needed.
function [A] = T1(x,y,z,w)
              A = [x+2*y+3*z+4*w; 2*x+3*y+4*z+5*w; 3*x+4*y+5*z+6*w]; end
function [A] = T2(x,y,z,w)
              A = [x+2*y+3*z+4*w; 2*x+2*y+3*z+4*w; 3*x+3*y+3*z+4*w]; end
function [A] = T3(x,y,z,w)
         A = [x+2*y+3*z+4*w;2*x+2*y+3*z+4*w;3*x+3*y+3*z+4*w;4*x+4*y+4*z+4*w]; end
T1(-1,0,3,-2)               % v1 in the kernel.
T1(18,-31,8,5)              % v2 in the kernel.
T1(1,-1,8,7)               % v3 not in the kernel.
M1=[T1(1,0,0,0) T1(0,1,0,0) T1(0,0,1,0) T1(0,0,0,1)]   % Standard matrix.
Mw1=rref([M1 [2;7;12]])       % In Range(T1). Last column is pivot.
Mw2=rref([M1 [42;59;76]])     % In Range(T1). Last column is pivot.
Mw3=rref([M1 [42;59;77]])     % Not in Range(T1). Last column is non-pivot.
M1                % Standard matrix of T1 already found. The remaining are:
M2=[T2(1,0,0,0) T2(0,1,0,0) T2(0,0,1,0) T2(0,0,0,1)]
M3=[T3(1,0,0,0) T3(0,1,0,0) T3(0,0,1,0) T3(0,0,0,1)]
                  % k1 has two vectors, so nullity 2.
k1=null(M1)       % kernel non-zero, T1 is not one-to-one.
                  % Hence, not an isomorphism.
                  % k2 has one vector, so nullity 2.
k2=null(M2)       % kernel non-zero, T2 is not one-to-one.
                  % Hence, not an isomorphism.
k3=null(M3)       % k3 has no vectors, so nullity 0. T3 is one-to-one.
                  % The first 2 columns form a basis for the range.
r1=rref(M1)       % Rank is 2. 2 + 2 = number of columns.
                  % The third row has no pivot, so not onto.
                  % The first 3 columns form a basis for the range.
r2=rref(M2)       % Rank is 3. 3 + 1 = number of columns.
                  % Each row has a pivot, so onto.
                  % All columns form a basis for the range.
r3=rref(M3)       % Rank is 4. 4 + 0 = number of columns.
                  % T3 is one-to-one and onto, hence, an isomorphism.
% False. T1 is not onto.
% True. T2 is onto.
% True. T3 is onto.
%  Exercise 16.
function [A] = T(a,b)              % In a function file we type the code:
    A = [3*a-4*b; a+3*b; -b];
    end
T(1,1)
% Exercise 17.
% First we form a matrix with the coefficients of the given polynomials,
M = [1 1 3 4; 2 3 3 4; 3 5 4 4; 4 7 4 5]
```

```
% then a matrix with the coefficients of their values.
N = [1 0 1 1; 0 1 1 1; -1 0 1 -1; 1 -1 -1 1]
% The last column of M1 has entries the coefficients of 2x^3+2x^2-2x-2
M1 = rref([M [2;2;-2;-2]])   % in terms of the given polynomials.
M1(:,5), N * ans             % Multiplication by N evaluates T(2,2,-2,-2).
% Exercise 18.
% Set F=0. We need to solve a+2b+3c+4d=0,2a+3b+4c+5d=0,3a+4b+5c+6d=0.
M = [1 2 3 4; 2 3 4 5; 3 4 5 6]         % We compute a basis for
null(M)                      % the nullspace of the coefficient matrix.
% WARNING: The answer is way off the correct N/norm(N) where N=[1 2;-2 -3;1 0;0 1].
% In polynomial form this is {0.4082x^3-0.8165x+0.4082x,0.5345x^3-0.8018x^2+0.2673}
% Exercise 19.
function [A] = T(a,b)        % In a function file type the transformation.
    A = [3*a-4*b; a+3*b; -b];
b2=[1,0,0; 0,1,1; -1,1,-1]   % The coefficients of B'.
aug = [b2,T(1,-1),T(1,1)]    % The coefficients of T(x-1),T(x+1)
rref(aug)                    % in terms of B' are computed by rref(aug).
ans(:,4:5)                   % The last 2 columns form the matrix of T.
```

6

Determinants

Algebra is generous; she often gives more than is asked of her.

—Jean Le Rond D'Alembert (1717–1783)

Introduction

Determinants are among the most useful topics of linear algebra, with numerous applications in engineering, physics, economics, mathematics, and other sciences. In geometry they offer a natural setting for writing very elegant formulas that compute areas and volumes, as well as equations of geometric objects such as lines, circles, planes, spheres, etc.

Dirichlet informs us that determinants were introduced by Leibniz in a letter to L'Hôpital, dated April 28, 1693. There is also evidence that the Japanese mathematician Seki Takakazu was already using determinants by 1683. Main contributors in this area were Laplace, Cauchy, Jacobi, Bezout, Sylvester, and Cayley.

Newton's Cows and Fields Problem

In 1707, Sir Isaac Newton[1] posed the following problem. Suppose

a_1 cows graze b_1 fields bare in c_1 days
a_2 cows graze b_2 fields bare in c_2 days
a_3 cows graze b_3 fields bare in c_3 days

Assuming that all the fields provide the same amount of grass, that the daily growth of the fields remains constant, and that the cows eat the same amount each day, what relation exists between the nine numbers $a_1, \ldots, c_3$?

---

[1] **Sir Isaac Newton** (1642–1727) was born on Christmas Day, the year Galileo died, in Woolsthorpe-by-Colsterworth, England. He revolutionized both mathematics and physics with the discovery of calculus and the theory of gravitation in mechanics. His work on gravity explains the planetary motions outlined by Kepler. About his astonishing discoveries he said, "If I have seen further than others it is only because I have stood on the shoulders of giants." He died on March 20, 1727, and was buried in Westminster Abbey. Newton ranks with Archimedes and Gauss as one of the three greatest mathematicians of all time.

What may seem surprising at first is that any relation exists at all. However, as we explain in Section 6.5, these numbers satisfy the following condition. The determinant of the matrix

$$\begin{bmatrix} a_1c_1 & b_1c_1 & b_1 \\ a_2c_2 & b_2c_2 & b_2 \\ a_3c_3 & b_3c_3 & b_3 \end{bmatrix}$$

is zero.

Although determinants appeared in the literature in the late 1600s (well before matrices[2]), the first work that systematically studied them for their own sake was written in 1772, by Vandermonde.[3]

CONVENTION

Unless stated otherwise, all matrices in this chapter are square.

6.1 Determinants and Cofactor Expansion

Reader's Goal for This Section

To compute determinants by the cofactor expansion.

Let $A = \begin{bmatrix} a_{11} & a_{12} \\ a_{21} & a_{22} \end{bmatrix}$. The determinant of A is the number

$$\det(A) = a_{11}a_{22} - a_{12}a_{21}$$

■ EXAMPLE 1

$$\det\begin{bmatrix} 1 & 2 \\ 3 & 4 \end{bmatrix} = 1 \cdot 4 - 2 \cdot 3 = -2, \quad \det\begin{bmatrix} 2 & -1 \\ 4 & -2 \end{bmatrix} = 2 \cdot (-2) - (-1) \cdot 4 = 0$$

Let

$$B = \begin{bmatrix} a_{11} & a_{12} & a_{13} \\ a_{21} & a_{22} & a_{23} \\ a_{31} & a_{32} & a_{33} \end{bmatrix}$$

---

[2] For a brief history of the subject, see *Lessons Introductory to the Higher Modern Algebra* by George Salmon, D.D., 5th ed. (Chelsea Publishing Company, 1885), pp. 338–339.
[3] **A. T. Vandermonde** (1735–1796), a Frenchman, made some of the earliest contributions to the theory of determinants. He also worked in geometry, and his work was known to Gauss. (He should not be confused with C. A. Vandermonde, a mathematician in the same period who worked on finding a formula for the roots of polynomial equations.)

The determinant of B can be written in terms of 2×2 determinants.

$$\det(B) = a_{11} \det \begin{bmatrix} a_{22} & a_{23} \\ a_{32} & a_{33} \end{bmatrix} - a_{12} \det \begin{bmatrix} a_{21} & a_{23} \\ a_{31} & a_{33} \end{bmatrix} + a_{13} \det \begin{bmatrix} a_{21} & a_{22} \\ a_{31} & a_{32} \end{bmatrix}$$

or explicitly as

$$\det(B) = a_{11}(a_{22}a_{33} - a_{23}a_{32}) - a_{12}(a_{21}a_{33} - a_{23}a_{31}) + a_{13}(a_{21}a_{32} - a_{22}a_{31})$$

There is also a mnemonic device for memorizing this formula, called the **Sarrus scheme**. We add the first two columns to the right of B and form the products of the entries covered by the arrows. The products of the arrows going from upper left to lower right are taken with a plus sign and the others with a minus. Then all signed products are added.

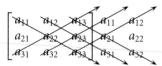

$$\det(B) = a_{11}a_{22}a_{33} + a_{12}a_{23}a_{31} + a_{13}a_{21}a_{32} - a_{13}a_{22}a_{31} - a_{11}a_{23}a_{32} - a_{12}a_{21}a_{33}$$

WARNING   The Sarrus scheme *does not apply* to $4 \times 4, 5 \times 5, \ldots$ determinants.

Let

$$C = \begin{bmatrix} 1 & 2 & 0 \\ 1 & 0 & -2 \\ 0 & 2 & -1 \end{bmatrix}, \qquad D = \begin{bmatrix} 1 & 2 & 0 & 1 \\ -1 & 1 & 2 & 0 \\ -2 & 1 & 0 & -2 \\ 1 & 0 & 2 & -1 \end{bmatrix}$$

■ EXAMPLE 2

$$\det(C) = 1 \det \begin{bmatrix} 0 & -2 \\ 2 & -1 \end{bmatrix} - 2 \det \begin{bmatrix} 1 & -2 \\ 0 & -1 \end{bmatrix} + 0 \det \begin{bmatrix} 1 & 0 \\ 0 & 2 \end{bmatrix} = 1 \cdot 4 - 2 \cdot (-1) + 0 \cdot 2 = 6$$

In the same manner we can define 4×4 determinants.

$$\det \begin{bmatrix} a_{11} & a_{12} & a_{13} & a_{14} \\ a_{21} & a_{22} & a_{23} & a_{24} \\ a_{31} & a_{32} & a_{33} & a_{34} \\ a_{41} & a_{42} & a_{43} & a_{44} \end{bmatrix} = a_{11} \det \begin{bmatrix} a_{22} & a_{23} & a_{24} \\ a_{32} & a_{33} & a_{34} \\ a_{42} & a_{43} & a_{44} \end{bmatrix} - a_{12} \det \begin{bmatrix} a_{21} & a_{23} & a_{24} \\ a_{31} & a_{33} & a_{34} \\ a_{41} & a_{43} & a_{44} \end{bmatrix}$$

$$+ a_{13} \det \begin{bmatrix} a_{21} & a_{22} & a_{24} \\ a_{31} & a_{32} & a_{34} \\ a_{41} & a_{42} & a_{44} \end{bmatrix} - a_{14} \det \begin{bmatrix} a_{21} & a_{22} & a_{23} \\ a_{31} & a_{32} & a_{33} \\ a_{41} & a_{42} & a_{43} \end{bmatrix}$$

■ EXAMPLE 3

$$\det(D) = 1 \det \begin{bmatrix} 1 & 2 & 0 \\ 1 & 0 & -2 \\ 0 & 2 & -1 \end{bmatrix} - 2 \det \begin{bmatrix} -1 & 2 & 0 \\ -2 & 0 & -2 \\ 1 & 2 & -1 \end{bmatrix} + 0 \det \begin{bmatrix} -1 & 1 & 0 \\ -2 & 1 & -2 \\ 1 & 0 & -1 \end{bmatrix}$$

$$- 1 \det \begin{bmatrix} -1 & 1 & 2 \\ -2 & 1 & 0 \\ 1 & 0 & 2 \end{bmatrix} = 1 \cdot 6 - 2 \cdot (-12) + 0 \cdot (-3) - 1 \cdot 0 = 30$$

We can continue similarly and define $n \times n$ determinants in terms of $(n-1) \times (n-1)$ determinants called *minors*. The (i, j) **minor**, M_{ij}, of a matrix A is the determinant obtained by deleting the ith row and the jth column.

We have introduced what is known as the *cofactor expansion of a determinant about its first row*. Each entry of the first row was multiplied by the corresponding minor. Each such product was multiplied by ± 1, depending on the position of the entry. The signed products were added together. In fact, there is nothing special about the choice of the first row in the computation of the determinant. We could have used *any other row or column*. Here is how.

Let A be a square matrix. First we assign a sign to each entry of A according to a checkerboard pattern of pluses and minuses.

$$\begin{bmatrix} + & - & + & \cdots \\ - & + & - & \cdots \\ + & - & + & \cdots \\ \vdots & \vdots & \vdots & \ddots \end{bmatrix}.$$

Then we pick any row or column and multiply each signed (according to the chart) entry by the corresponding minor. Finally, we add all these products. Note that the sign of the (i, j) position in the checkerboard pattern is given by $(-1)^{i+j}$.

■ EXAMPLE 4 $\det(C)$ expanded about the third row is

$$\det(C) = 0 \det \begin{bmatrix} 2 & 0 \\ 0 & -2 \end{bmatrix} - 2 \det \begin{bmatrix} 1 & 0 \\ 1 & -2 \end{bmatrix} - 1 \det \begin{bmatrix} 1 & 2 \\ 1 & 0 \end{bmatrix} = 6$$

■ EXAMPLE 5 $\det(D)$ expanded about the second column is

$$\det(D) = -2 \det \begin{bmatrix} -1 & 2 & 0 \\ -2 & 0 & -2 \\ 1 & 2 & -1 \end{bmatrix} + 1 \det \begin{bmatrix} 1 & 0 & 1 \\ -2 & 0 & -2 \\ 1 & 2 & -1 \end{bmatrix} - 1 \det \begin{bmatrix} 1 & 0 & 1 \\ -1 & 2 & 0 \\ 1 & 2 & -1 \end{bmatrix}$$

$$+ 0 \det \begin{bmatrix} 1 & 0 & 1 \\ -1 & 2 & 0 \\ -2 & 0 & -2 \end{bmatrix} = -2 \cdot (-12) + 1 \cdot 0 - 1 \cdot (-6) + 0 \cdot 10 = 30$$

NOTE   *We usually try to expand a determinant about the row or column with the most zeros.* This avoids the computation of some of the minors.

More generally, let

$$A = \begin{bmatrix} a_{11} & a_{12} & \cdots & a_{1n} \\ a_{21} & a_{22} & \cdots & a_{2n} \\ \vdots & \vdots & \ddots & \vdots \\ a_{n1} & a_{n2} & \cdots & a_{nn} \end{bmatrix}$$

The (i, j) **cofactor**, C_{ij}, of A is the signed (i, j) minor

$$C_{ij} = (-1)^{i+j} M_{ij}$$

Cofactor Expansion about the ith Row

The determinant of A can be expanded about the ith row in terms of the cofactors as

$$\det A = a_{i1} C_{i1} + a_{i2} C_{i2} + \cdots + a_{in} C_{in}$$

Cofactor Expansion about the jth Column

The determinant of A can be expanded about the jth column in terms of the cofactors as

$$\det A = a_{1j} C_{1j} + a_{2j} C_{2j} + \cdots + a_{nj} C_{nj}$$

This method of computing determinants by using cofactors is called the **cofactor**, or **Laplace**, **expansion**, and it is attributed to Vandermonde and Laplace.[4]

■ EXAMPLE 6   Let

$$A = \begin{bmatrix} -1 & 2 & 2 \\ 4 & 3 & -2 \\ -5 & 0 & 3 \end{bmatrix}$$

Then

$$M_{11} = \det \begin{bmatrix} 3 & -2 \\ 0 & 3 \end{bmatrix} = 9 \qquad C_{11} = (-1)^{1+1} M_{11} = 9$$

$$M_{12} = \det \begin{bmatrix} 4 & -2 \\ -5 & 3 \end{bmatrix} = 2 \qquad C_{12} = (-1)^{1+2} M_{12} = -1 \cdot 2 = -2$$

$$M_{13} = \det \begin{bmatrix} 4 & 3 \\ -5 & 0 \end{bmatrix} = 15 \qquad C_{13} = (-1)^{1+3} M_{13} = 15$$

---

[4]**(Marquis de) Pierre Simon Laplace** (1749–1827) was born in Beumont-en-Auge, Normandy, France. He wrote his first published paper, on the calculus of finite differences, at the age of 16. He made important contributions to calculus, celestial mechanics, and probability theory. He served briefly as minister of interior under Napoleon, became president of the senate, and was later made a count.

$$M_{21} = \det \begin{bmatrix} 2 & 2 \\ 0 & 3 \end{bmatrix} = 6 \qquad C_{21} = (-1)^{2+1}M_{21} = -1 \cdot 6 = -6$$

$$M_{22} = \det \begin{bmatrix} -1 & 2 \\ -5 & 3 \end{bmatrix} = 7 \qquad C_{22} = (-1)^{2+2}M_{22} = 7$$

$$M_{23} = \det \begin{bmatrix} -1 & 2 \\ -5 & 0 \end{bmatrix} = 10 \qquad C_{23} = (-1)^{2+3}M_{23} = -1 \cdot 10 = -10$$

$$M_{31} = \det \begin{bmatrix} 2 & 2 \\ 3 & -2 \end{bmatrix} = -10 \qquad C_{31} = (-1)^{3+1}M_{31} = -10$$

$$M_{32} = \det \begin{bmatrix} -1 & 2 \\ 4 & -2 \end{bmatrix} = -6 \qquad C_{32} = (-1)^{3+2}M_{32} = (-1)(-6) = 6$$

$$M_{33} = \det \begin{bmatrix} -1 & 2 \\ 4 & 3 \end{bmatrix} = -11 \qquad C_{33} = (-1)^{3+3}M_{33} = -11$$

$$\det A = a_{11}C_{11} + a_{12}C_{12} + a_{13}C_{13} = (-1)9 + 2(-2) + 2 \cdot 15 = 17$$
$$\det A = a_{21}C_{21} + a_{22}C_{22} + a_{23}C_{23} = 4(-6) + 3 \cdot 7 + (-2)(-10) = 17$$
$$\det A = a_{31}C_{31} + a_{32}C_{32} + a_{33}C_{33} = (-5)(-10) + 0 \cdot 6 + 3(-11) = 17$$
$$\det A = a_{11}C_{11} + a_{21}C_{21} + a_{31}C_{31} = (-1)9 + 4(-6) + (-5)(-10) = 17$$
$$\det A = a_{12}C_{12} + a_{22}C_{22} + a_{32}C_{32} = 2(-2) + 3 \cdot 7 + 0 \cdot 6 = 17$$
$$\det A = a_{13}C_{13} + a_{23}C_{23} + a_{33}C_{33} = 2 \cdot 15 + (-2)(-10) + 3(-11) = 17$$

NOTE   If we multiply the entries of a row (or column) by the corresponding cofactors from *another* row (or column) the answer is always *zero*. (See proof of Theorem 10, Section 6.3). For example,

$$a_{11}C_{21} + a_{12}C_{22} + a_{13}C_{23} = (-1)(-6) + 2 \cdot 7 + 2(-10) = 0$$
$$a_{11}C_{12} + a_{21}C_{22} + a_{31}C_{32} = (-1)(-2) + 4 \cdot 7 + (-5)6 = 0$$

REMARK   The cofactor expansion implies that the determinant of any upper or lower triangular matrix is the product of its main diagonal entries.

For example, repeated expansion about the first column yields

$$\det \begin{bmatrix} 4 & 5 & 6 \\ 0 & 7 & 8 \\ 0 & 0 & 9 \end{bmatrix} = 4 \cdot \det \begin{bmatrix} 7 & 8 \\ 0 & 9 \end{bmatrix} = 4 \cdot 7 \cdot 9 = 252$$

The || Notation

We often write $|A|$ instead of $\det(A)$ (not to be confused with the absolute value). For example,

$$\det \begin{bmatrix} 1 & 2 \\ 3 & 4 \end{bmatrix} = \begin{vmatrix} 1 & 2 \\ 3 & 4 \end{vmatrix} = -2$$

It is occasionally convenient to talk about 1×1 determinants. The determinant of a 1×1 matrix $A = [a]$ is simply the single entry a. For example, $\det[-2] = -2$, $\det[3] = 3$.

The Amazing Geometry of the Determinant

In this section we briefly explore one of the most striking properties of the determinant. In general, if we apply a linear transformation to a region in the plane, the area of its image changes. The question that arises then is, given the matrix of the transformation, how can we predict the area of the image? Here we demonstrate that the area of the unit square is scaled by a factor equal to the absolute value of the determinant of the matrix of the transformation.

For example, consider the effect of

$$T(\mathbf{x}) = \begin{bmatrix} 3 & 1 \\ 0 & 2 \end{bmatrix} \mathbf{x}$$

on the unit square. The image is the rectangle with vertices $(0,0)$, $(3,0)$, $(1,2)$, and $(4,2)$. The area of the image is 6, which happens to be the determinant of the matrix.

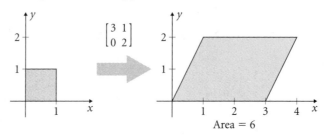

Likewise, consider the effect of

$$T(\mathbf{x}) = \begin{bmatrix} 2 & 0 \\ 1 & 2 \end{bmatrix} \mathbf{x}$$

on the unit square. The image is the rectangle with vertices $(0,0)$, $(2,1)$, $(0,2)$, and $(2,3)$. Again the area of the image, 4, is equal to the determinant of the matrix.

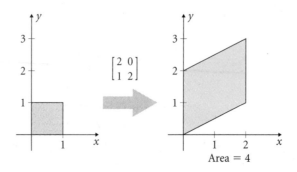

In general, if we apply

$$T(\mathbf{x}) = \begin{bmatrix} a & b \\ c & d \end{bmatrix} \mathbf{x}$$

to the unit square, the images of $(0,0)$, $(1,0)$, $(0,1)$, $(1,1)$ are $(0,0)$, (a,c), (b,d), $(a+b, c+d)$, respectively. These define a parallelogram if (a,c) is not proportional to (b,d)— i.e., if the matrix is invertible (which also means that the determinant $ad - bc \neq 0$) (why?). In this case we can compute the area of the parallelogram quickly utilizing cross products. Recall from Chapter 2, Section 2.6, that the area of the parallelogram with adjacent sides two given 3-vectors $\mathbf{u}$ and $\mathbf{v}$ equals the length of $\mathbf{u} \times \mathbf{v}$. In this case, we let $\mathbf{u} = (a, c, 0)$ and $\mathbf{v} = (b, d, 0)$ (converting the 2-vectors to 3-vectors allows us to use the cross product). Thus the area of the image of T is

$$\|\mathbf{u} \times \mathbf{v}\| = |ad - bd|$$

which is the absolute value of the determinant of T.

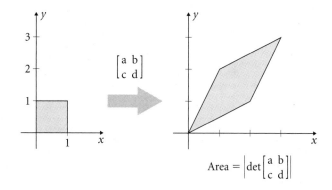

$$\text{Area} = \left| \det \begin{bmatrix} a & b \\ c & d \end{bmatrix} \right|$$

Determinants with CAS

Maple

```
> with(linalg):
> det(matrix([[1,7,-8], [5,2,-3], [1,-3,-2]]));
```
$$172$$

Mathematica

```
In[1]:=
    Det[{{1,7,-8},{5,2,-3},{1,-3,-2}}]
Out[1]=
    172
```

MATLAB
```
>> det([1 7 -8; 5 2 -3; 1 -3 -2])
ans =
    172
```

Exercises 6.1

In Exercises 1–11 compute the determinants.

1. a. $\begin{vmatrix} 3 & 6 \\ 4 & 7 \end{vmatrix}$ **b.** $\begin{vmatrix} 3 & 4 \\ 6 & 7 \end{vmatrix}$

2. a. $\begin{vmatrix} 0 & -1 \\ -1 & 100 \end{vmatrix}$ **b.** $\begin{vmatrix} \sqrt{\frac{1}{2}} & -\sqrt{\frac{1}{2}} \\ \sqrt{\frac{1}{2}} & \sqrt{\frac{1}{2}} \end{vmatrix}$

3. a. $\begin{vmatrix} 1 & 0 & 2 \\ 0 & 5 & 0 \\ 3 & 0 & 4 \end{vmatrix}$ **b.** $\begin{vmatrix} 1 & 2 & 0 \\ 3 & 4 & 0 \\ 0 & 0 & 5 \end{vmatrix}$

4. a. $\begin{vmatrix} 0 & 0 & 3 \\ 0 & 2 & 0 \\ 1 & 0 & 0 \end{vmatrix}$ **b.** $\begin{vmatrix} 0 & 7 & -2 \\ 1 & 3 & 8 \\ 0 & 1 & 1 \end{vmatrix}$

5. a. $\begin{vmatrix} 11 & 1 & 0 \\ -2 & 7 & 0 \\ 3 & 0 & 2 \end{vmatrix}$ **b.** $\begin{vmatrix} -3 & 5 & 1 \\ 1 & 0 & -2 \\ 0 & 3 & 0 \end{vmatrix}$

6. a. $\begin{vmatrix} 1 & 2 & 3 \\ 4 & 5 & 6 \\ 7 & 8 & 9 \end{vmatrix}$ **b.** $\begin{vmatrix} 1 & 2 & 3 \\ 2 & 2 & 3 \\ 3 & 3 & 3 \end{vmatrix}$

7. a. $\begin{vmatrix} 1 & 0 & 0 & 0 \\ 0 & 0 & 0 & 1 \\ 0 & 1 & 0 & 0 \\ 0 & 0 & 1 & 0 \end{vmatrix}$ **b.** $\begin{vmatrix} 1 & 0 & 0 & 0 \\ 5 & 2 & 0 & 0 \\ 7 & 0 & 3 & 0 \\ 9 & -1 & 4 & 0 \end{vmatrix}$

8. a. $\begin{vmatrix} 1 & 2 & -3 & -4 \\ 0 & 2 & 5 & 6 \\ 0 & 0 & 3 & 9 \\ 0 & 0 & 0 & 4 \end{vmatrix}$ **b.** $\begin{vmatrix} 1 & 0 & 1 & 0 \\ 2 & 1 & 0 & -1 \\ 3 & 0 & 1 & 1 \\ 4 & 0 & 1 & 0 \end{vmatrix}$

9. $\begin{vmatrix} 1 & -1 & 2 & 0 & 0 \\ 0 & 4 & 2 & 0 & 0 \\ 0 & 5 & 0 & 0 & 0 \\ 4 & 6 & 7 & 8 & 0 \\ -4 & 7 & 0 & 1 & 1 \end{vmatrix}$

10. $\begin{vmatrix} 2 & 0 & 0 & 0 & 0 \\ 1 & 1 & 0 & 0 & 0 \\ 1 & 2 & 1 & 0 & -1 \\ 1 & 3 & 0 & 1 & 1 \\ 1 & 4 & 0 & 1 & 0 \end{vmatrix}$

11. $\begin{vmatrix} a & a & a \\ a & a & a \\ b & c & d \end{vmatrix}$

12. Compute the determinant of the matrix from Newton's cows and fields problem.

$$\begin{vmatrix} a_1 c_1 & b_1 c_1 & b_1 \\ a_2 c_2 & b_2 c_2 & b_2 \\ a_3 c_3 & b_3 c_3 & b_3 \end{vmatrix}$$

In Exercises 13–14 compute the determinant of the coefficient matrix of the system.

13. $5x - 2y = 1$
    $-x + y + 2z = -3$
        $-7y + 2z = -2$

14. $y - z = 0$
    $x \quad - z = 1$
    $-x + y \quad = -1$

15. Let

$$A = \begin{bmatrix} 1 & 2 \\ 3 & 4 \end{bmatrix} \quad B = \begin{bmatrix} 4 & -2 \\ -3 & 1 \end{bmatrix}$$

a. Show that $\det(A) = \det(A^T)$.

b. Show that $\det(AB) = \det(A) \det(B)$.

c. Show that $\det(A^{-1}) = 1/\det(A)$.

16. Write all minors M_{ij} and all cofactors C_{ij} of A.

$$A = \begin{bmatrix} a & b \\ c & d \end{bmatrix}$$

17. **a.** Find all the minors and all the cofactors of A.

$$A = \begin{bmatrix} 1 & -2 & 2 \\ 3 & 5 & -4 \\ 7 & 0 & -6 \end{bmatrix}$$

b. Compute $\det(A)$ by using cofactor expansion about

(b_1) the first row
(b_2) the second row
(b_3) the third row
(b_4) the first column
(b_5) the second column
(b_6) the third column

Let

$$A = \begin{bmatrix} 0 & 0 & 2 & 0 \\ 0 & 1 & 0 & 0 \\ 1 & 0 & 0 & 0 \\ 0 & 0 & 0 & -1 \end{bmatrix}$$

$$B = \begin{bmatrix} 1 & 0 & 0 & 0 \\ 0 & 0 & 2 & 0 \\ 0 & -5 & 0 & 0 \\ 0 & 0 & 0 & 1 \end{bmatrix}$$

18. Verify that $\det(AB) = \det(A)\det(B)$.

19. Verify that $\det(A^{-1}) = 1/\det(A)$.

20. Prove

$$\left| \begin{bmatrix} a & b \\ c & d \end{bmatrix} \begin{bmatrix} e & f \\ g & h \end{bmatrix} \right| = \begin{vmatrix} a & b \\ c & d \end{vmatrix} \begin{vmatrix} e & f \\ g & h \end{vmatrix}$$

21. Let $A = \begin{bmatrix} a & b \\ c & d \end{bmatrix}$ with $ad - bc \neq 0$. Prove

$$\det(A^{-1}) = \frac{1}{\det(A)}$$

Determinants of Elementary Matrices

In Exercises 22–24 compute the determinants of the elementary matrices.

22. $E_1 = \begin{bmatrix} 1 & 0 & 0 \\ r & 1 & 0 \\ 0 & 0 & 1 \end{bmatrix}$, $E_2 = \begin{bmatrix} 1 & 0 & 0 \\ 0 & 1 & 0 \\ 0 & r & 1 \end{bmatrix}$

23. $E_3 = \begin{bmatrix} r & 0 & 0 \\ 0 & 1 & 0 \\ 0 & 0 & 1 \end{bmatrix}$, $E_4 = \begin{bmatrix} 1 & 0 & 0 \\ 0 & 1 & 0 \\ 0 & 0 & r \end{bmatrix}$

24. $E_5 = \begin{bmatrix} 1 & 0 & 0 \\ 0 & 0 & 1 \\ 0 & 1 & 0 \end{bmatrix}$, $E_6 = \begin{bmatrix} 0 & 0 & 1 \\ 0 & 1 & 0 \\ 1 & 0 & 0 \end{bmatrix}$

25. Based on your calculations from Exercises 22–24, form a statement about the determinants of the three kinds of elementary matrices: those obtained from I by (a) elimination, (b) scaling, and (c) interchange.

Let

$$A = \begin{bmatrix} a & b & c \\ d & e & f \\ g & h & i \end{bmatrix}$$

and let

Be obtained from A by
A_1 $R_2 + rR_1 \rightarrow R_2$
A_2 $R_3 + rR_2 \rightarrow R_3$
A_3 $rR_1 \rightarrow R_1$
A_4 $rR_3 \rightarrow R_3$
A_5 $R_2 \leftrightarrow R_3$
A_6 $R_1 \leftrightarrow R_3$

In Exercises 26–28 prove the identities.

26. **a.** $\det(E_1 A) = \det(E_1)\det(A) = \det(A_1)$
    b. $\det(E_2 A) = \det(E_2)\det(A) = \det(A_2)$

27. **a.** $\det(E_3 A) = \det(E_3)\det(A) = \det(A_3)$
    b. $\det(E_4 A) = \det(E_4)\det(A) = \det(A_4)$

28. **a.** $\det(E_5 A) = \det(E_5)\det(A) = \det(A_5)$
    b. $\det(E_6 A) = \det(E_6)\det(A) = \det(A_5)$

Equations with Determinants

In Exercises 29–31 compute all values of λ (real or complex) such that the determinants are zero.

29. **a.** $\begin{vmatrix} 1-\lambda & 2 \\ 1 & 3-\lambda \end{vmatrix}$

    b. $\begin{vmatrix} 1-\lambda & 5 \\ 2 & 10-\lambda \end{vmatrix}$

30. **a.** $\begin{vmatrix} \lambda & 1 \\ 4 & \lambda \end{vmatrix}$

    b. $\begin{vmatrix} 1-\lambda & -1 \\ 1 & 1-\lambda \end{vmatrix}$

31. $\begin{vmatrix} 1 & 0 & 0 \\ 0 & \lambda-2 & 1 \\ 0 & 2 & \lambda-1 \end{vmatrix}$

32. $\begin{vmatrix} 2-\lambda & 0 & 0 \\ 1 & 3-\lambda & 0 \\ 0 & 1 & 1-\lambda \end{vmatrix}$

33. Solve the equation for x.

$$\begin{vmatrix} x & x \\ 1 & 2 \end{vmatrix} = \begin{vmatrix} x-1 & 1 \\ -1 & x-1 \end{vmatrix}$$

34. Solve the equation for x.

$$\begin{vmatrix} x & 0 & 2 \\ 0 & x-3 & -1 \\ 2 & 0 & x-6 \end{vmatrix} = \begin{vmatrix} x-1 & 1 & 0 \\ -2 & x-7 & 0 \\ 3 & 0 & x-2 \end{vmatrix}$$

35. Find a, b such that

$$\begin{vmatrix} a & b \\ 1 & 2 \end{vmatrix} = 0 = \begin{vmatrix} a & 4 \\ 1 & a \end{vmatrix}$$

Volumes and Determinants

Let R be the unit cube in space. This is the cube of side 1 in the first octant with adjacent sides the coordinate planes.

36. Let T be the linear transformation with matrix

$$A = \begin{bmatrix} 2 & 1 & 0 \\ 0 & 3 & 0 \\ 0 & 0 & -4 \end{bmatrix}$$

Compute the volume of the image of R under T and relate it to the determinant of A.

37. Let T be the general linear transformation with matrix

$$A = \begin{bmatrix} a & b & c \\ d & e & f \\ g & h & i \end{bmatrix}$$

Prove that the volume of the image of R under T equals $|\det(A)|$. (*Hint:* If T is invertible use the formula from Section 2.6 for the volume of the parallelepiped determined by three vectors.)

6.2   Properties of Determinants

Reader's Goals for This Section

1. To understand and apply the basic properties of determinants.
2. To simplify determinants by a correct row or column reduction.

Cofactor expansion of determinants is quite tedious unless the matrix is small or it has many zeros. A better method is based on Gauss elimination. First we need to study the effects of the elementary row operations on determinants.

Elementary Operations and Determinants

Proofs of special cases of the next theorem are discussed in exercises. A complete proof is outlined in the exercise set of Section 6.4.

THEOREM 1

(Basic Properties)

Let A be an $n \times n$ matrix. (To illustrate we let $n = 3$.)

1. A and its transpose have the same determinant: $\det(A) = \det(A^T)$.

$$\begin{vmatrix} a_1 & a_2 & a_3 \\ b_1 & b_2 & b_3 \\ c_1 & c_2 & c_3 \end{vmatrix} = \begin{vmatrix} a_1 & b_1 & c_1 \\ a_2 & b_2 & c_2 \\ a_3 & b_3 & c_3 \end{vmatrix}$$

2. Let B be obtained from A by multiplying one of its rows (or columns) by a nonzero constant. Then $\det(B) = k\det(A)$.

$$\begin{vmatrix} a_1 & a_2 & a_3 \\ kb_1 & kb_2 & kb_3 \\ c_1 & c_2 & c_3 \end{vmatrix} = k\begin{vmatrix} a_1 & a_2 & a_3 \\ b_1 & b_2 & b_3 \\ c_1 & c_2 & c_3 \end{vmatrix} \qquad \begin{vmatrix} a_1 & a_2 & ka_3 \\ b_1 & b_2 & kb_3 \\ c_1 & c_2 & kc_3 \end{vmatrix} = k\begin{vmatrix} a_1 & a_2 & a_3 \\ b_1 & b_2 & b_3 \\ c_1 & c_2 & c_3 \end{vmatrix}$$

3. Let B be obtained from A by interchanging any two rows (or columns). Then $\det(B) = -\det(A)$.

$$\begin{vmatrix} a_1 & a_2 & a_3 \\ b_1 & b_2 & b_3 \\ c_1 & c_2 & c_3 \end{vmatrix} = -\begin{vmatrix} b_1 & b_2 & b_3 \\ a_1 & a_2 & a_3 \\ c_1 & c_2 & c_3 \end{vmatrix} \qquad \begin{vmatrix} a_1 & a_2 & a_3 \\ b_1 & b_2 & b_3 \\ c_1 & c_2 & c_3 \end{vmatrix} = -\begin{vmatrix} a_3 & a_2 & a_1 \\ b_3 & b_2 & b_1 \\ c_3 & c_2 & c_1 \end{vmatrix}$$

4. Let B be obtained from A by adding a multiple of one row (or column) to another. Then $\det(B) = \det(A)$.

$$\begin{vmatrix} a_1 & a_2 & a_3 \\ ka_1 + b_1 & ka_2 + b_2 & ka_3 + b_3 \\ c_1 & c_2 & c_3 \end{vmatrix} = \begin{vmatrix} a_1 & a_2 & a_3 \\ b_1 & b_2 & b_3 \\ c_1 & c_2 & c_3 \end{vmatrix}$$

$$\begin{vmatrix} a_1 & a_2 & ka_2 + a_3 \\ b_1 & b_2 & kb_2 + b_3 \\ c_1 & c_2 & kc_2 + c_3 \end{vmatrix} = \begin{vmatrix} a_1 & a_2 & a_3 \\ b_1 & b_2 & b_3 \\ c_1 & c_2 & c_3 \end{vmatrix}$$

We see that elimination, $R_i + kR_j \rightarrow R_i$, does not change the determinant; scaling, $kR_i \rightarrow R_i$, scales the determinant by k; and interchange, $R_i \leftrightarrow R_j$, changes the sign of the determinant.

■ EXAMPLE 7

$$\begin{vmatrix} 1 & 3 \\ 2 & 4 \end{vmatrix} = \begin{vmatrix} 1 & 2 \\ 3 & 4 \end{vmatrix}, \qquad \begin{vmatrix} 1 & 1 \\ 3 & 2 \end{vmatrix} = \frac{1}{2}\begin{vmatrix} 1 & 2 \\ 3 & 4 \end{vmatrix}, \qquad \begin{vmatrix} 1 & 3 \\ 2 & 4 \end{vmatrix} = -\begin{vmatrix} 3 & 1 \\ 4 & 2 \end{vmatrix}$$

$$\begin{vmatrix} 1 & 2 \\ 3 & 4 \end{vmatrix} = \begin{vmatrix} 1 & 2 \\ 0 & -2 \end{vmatrix} \quad \text{by} \quad \boxed{-3R_1 + R_2 \rightarrow R_2}$$

$$\begin{vmatrix} 1 & 2 \\ 3 & 4 \end{vmatrix} = \begin{vmatrix} 1 & 1 \\ 3 & 1 \end{vmatrix} \quad \text{by} \quad \boxed{-C_1 + C_2 \rightarrow C_2}$$

A COMMON MISTAKE   Property 4 of Theorem 1 is sometimes misused. A row (or column) is replaced by a multiple of another added to the first, *not to a multiple of it*. If the original row (column) is scaled, so is the determinant. For example,

$$\begin{vmatrix} 3 & 1 \\ 1 & 2 \end{vmatrix} = \begin{vmatrix} 3 & 1 \\ 0 & \frac{5}{3} \end{vmatrix} = 5 \quad \text{by} \quad \boxed{-\tfrac{1}{3}R_1 + R_2 \rightarrow R_2}$$

whereas $\boxed{R_1 - 3R_2 \rightarrow R_2}$ implies $\begin{vmatrix} 3 & 1 \\ 0 & -5 \end{vmatrix} = -15$

The properties discussed in Theorem 1 can be used to write a determinant in triangular form, by Gauss elimination, and then compute the product of the diagonal entries.

■ EXAMPLE 8

$$
\begin{vmatrix}
2 & 4 & 6 & -2 & 16 \\
0 & 0 & 4 & 2 & -1 \\
0 & -5 & 5 & 3 & 7 \\
0 & 0 & 0 & 1 & 6 \\
1 & 2 & 3 & -2 & -9
\end{vmatrix}
= 2
\begin{vmatrix}
1 & 2 & 3 & -1 & 8 \\
0 & 0 & 4 & 2 & -1 \\
0 & -5 & 5 & 3 & 7 \\
0 & 0 & 0 & 1 & 6 \\
1 & 2 & 3 & -2 & -9
\end{vmatrix}
\quad \text{by} \quad \boxed{\tfrac{1}{2}R_1 \rightarrow R_1}
$$

$$
= 2
\begin{vmatrix}
1 & 2 & 3 & -1 & 8 \\
0 & 0 & 4 & 2 & -1 \\
0 & -5 & 5 & 3 & 7 \\
0 & 0 & 0 & 1 & 6 \\
0 & 0 & 0 & -1 & -17
\end{vmatrix}
\quad \text{by} \quad \boxed{-R_1 + R_5 \rightarrow R_5}
$$

$$
= -2
\begin{vmatrix}
1 & 2 & 3 & -1 & 8 \\
0 & -5 & 5 & 3 & 7 \\
0 & 0 & 4 & 2 & -1 \\
0 & 0 & 0 & 1 & 6 \\
0 & 0 & 0 & -1 & -17
\end{vmatrix}
\quad \text{by} \quad \boxed{R_2 \leftrightarrow R_3}
$$

$$
= -2
\begin{vmatrix}
1 & 2 & 3 & -1 & 8 \\
0 & -5 & 5 & 3 & 7 \\
0 & 0 & 4 & 2 & -1 \\
0 & 0 & 0 & 1 & 6 \\
0 & 0 & 0 & 0 & -11
\end{vmatrix}
\quad \text{by} \quad \boxed{R_4 + R_5 \rightarrow R_5}
$$

$$
= -2 \cdot (-5) \cdot 4 \cdot 1(-11) = -440
$$

The method of Example 8 also yields a formula for the determinant: first note that we can always reduce any matrix to echelon form *without using any scaling operations*. (In Example 8 that would mean to not perform $\tfrac{1}{2}R_1 \rightarrow R_1$ and not put the extra factor of 2 out.)

Let A be an $n \times n$ matrix that row reduces without scaling to the upper triangular matrix B. The only operations that alter $\det(A)$ (and only by a sign) are interchanges. Hence,

$$
\det(A) = (-1)^k \det(B)
$$

where k is the number of interchanges in the reduction process. If A is invertible, then B has n pivots, say, $p_1, \ldots, p_n$, all on the main diagonal, because $A \sim B \sim I$. Hence, $\det(A) = (-1)^k \det(B) = (-1)^k p_1 p_2 \cdots p_n$. If A is noninvertible, then B has at least one row of zeros, so $\det(A) = \det(B) = 0$. Thus we have proved Theorem 2.

THEOREM 2

In the preceding notation,

$$\det(A) = \begin{cases} (-1)^k p_1 p_2 \cdots p_n & \text{if } A \text{ is invertible} \\ 0 & \text{if } A \text{ is not invertible} \end{cases}$$

Referring to Example 8, if we reduce without scaling, the pivots of the reduced matrix are $-2, -5, 4, 1, -11$. Their product is the determinant. Since pivots are always nonzero, Theorem 2 implies the following basic theorem.

THEOREM 3

An $n \times n$ matrix A is invertible if and only if $\det(A) \neq 0$.

■ EXAMPLE 9 A, B, and C are noninvertible, because their determinants are 0.

$$A = \begin{bmatrix} 1 & 3 \\ 2 & 6 \end{bmatrix}, \qquad B = \begin{bmatrix} 2a & a \\ 2 & 1 \end{bmatrix}, \qquad C = \begin{bmatrix} 1 & 2 & 3 \\ 4 & 5 & 6 \\ 7 & 8 & 9 \end{bmatrix}$$

■ EXAMPLE 10   Are the rows of D linearly independent? What about the columns?

$$D = \begin{bmatrix} 2 & -4 & 5 \\ 8 & 0 & -3 \\ 5 & -2 & 1 \end{bmatrix}$$

SOLUTION   No, $\det(D) = 0$. Hence, the rows and columns of D are linearly dependent by Theorem 3 and Theorem 15, Section 3.3.

■ EXAMPLE 11   Are the following vectors linearly independent?

$$\begin{bmatrix} 2 \\ 8 \\ 5 \end{bmatrix}, \qquad \begin{bmatrix} -4 \\ -7 \\ -2 \end{bmatrix}, \qquad \begin{bmatrix} 5 \\ -3 \\ 1 \end{bmatrix}$$

ANSWER   Yes, because $\begin{vmatrix} 2 & -4 & 5 \\ 8 & -7 & -3 \\ 5 & -2 & 1 \end{vmatrix} = 161 \neq 0.$

As a consequence of Theorem 3 and Theorem 15, Section 3.3, we have the following very useful theorem.

THEOREM 4

The square homogeneous system $A\mathbf{x} = \mathbf{0}$ has nontrivial solutions if and only if $\det(A) = 0$.

■ EXAMPLE 12   Does the system have nontrivial solutions?

$$2x + 3y + z = 0$$
$$x - y + 2z = 0$$
$$x + 4y - z = 0$$

ANSWER   Yes, by Theorem 4, because

$$\begin{vmatrix} 2 & 3 & 1 \\ 1 & -1 & 2 \\ 1 & 4 & -1 \end{vmatrix} = 0$$

Let us now draw some easy but important conclusions from Theorem 1 and cofactor expansions.

THEOREM 5

1. If A has a row (or column) of zeros, then $\det(A) = 0$.

$$\begin{vmatrix} a_1 & a_2 & a_3 \\ 0 & 0 & 0 \\ c_1 & c_2 & c_3 \end{vmatrix} = 0, \qquad \begin{vmatrix} a_1 & a_2 & 0 \\ b_1 & b_2 & 0 \\ c_1 & c_2 & 0 \end{vmatrix} = 0$$

2. If A has two rows (or columns) that are equal, then $\det(A) = 0$.

$$\begin{vmatrix} a_1 & a_2 & a_3 \\ a_1 & a_2 & a_3 \\ c_1 & c_2 & c_3 \end{vmatrix} = 0, \qquad \begin{vmatrix} a_1 & a_2 & a_1 \\ b_1 & b_2 & b_1 \\ c_1 & c_2 & c_1 \end{vmatrix} = 0$$

3. If A has two rows (or columns) that are multiples of each other, then $\det(A) = 0$.

$$\begin{vmatrix} a_1 & a_2 & a_3 \\ ka_1 & ka_2 & ka_3 \\ c_1 & c_2 & c_3 \end{vmatrix} = 0, \qquad \begin{vmatrix} a_1 & a_2 & ka_1 \\ b_1 & b_2 & kb_1 \\ c_1 & c_2 & kc_1 \end{vmatrix} = 0$$

4. If a row (or column) of A is the sum of multiples of two other rows (or columns), then $\det(A) = 0$.

$$\begin{vmatrix} a_1 & a_2 & a_3 \\ ka_1 + lc_1 & ka_2 + lc_2 & ka_3 + lc_3 \\ c_1 & c_2 & c_3 \end{vmatrix} = 0, \qquad \begin{vmatrix} a_1 & a_2 & ka_1 + la_2 \\ b_1 & b_2 & kb_1 + lb_2 \\ c_1 & c_2 & kc_1 + lc_2 \end{vmatrix} = 0$$

PROOF   Part 1 is clear (use cofactor expansion about the zero row or column). It suffices to prove Part 4, because Parts 2 and 3 follow from Part 4 by letting $k = 1, l = 0$, and $l = 0$, respectively. To illustrate, we use a 3×3 determinant.

$$\begin{vmatrix} a_1 & a_2 & a_3 \\ ka_1 + lc_1 & ka_2 + lc_2 & ka_3 + lc_3 \\ c_1 & c_2 & c_3 \end{vmatrix} = \begin{vmatrix} a_1 & a_2 & a_3 \\ lc_1 & lc_2 & lc_3 \\ c_1 & c_2 & c_3 \end{vmatrix} \qquad \text{by 4 of Theorem 1}$$

$$= 1 \begin{vmatrix} a_1 & a_2 & a_3 \\ c_1 & c_2 & c_3 \\ c_1 & c_2 & c_3 \end{vmatrix} \qquad \text{by 2 of Theorem 1}$$

$$= \begin{vmatrix} a_1 & a_2 & a_3 \\ c_1 & c_2 & c_3 \\ 0 & 0 & 0 \end{vmatrix} \qquad \text{by 4 of Theorem 1}$$

The last determinant is zero by Part 1.

■ EXAMPLE 13

$$\begin{vmatrix} 10 & -11 & 3 \\ 20 & 50 & 40 \\ 0 & 0 & 0 \end{vmatrix} = 0 \qquad \begin{vmatrix} 1 & 2 & 4 & 4 \\ 2 & 2 & 2 & 2 \\ 3 & 3 & 4 & 4 \\ 1 & -1 & -4 & -4 \end{vmatrix} = 0$$

$$\begin{vmatrix} 10 & 15 & -5 \\ 12 & 21 & -6 \\ -120 & 26 & 60 \end{vmatrix} = \begin{vmatrix} (-2) \cdot (-5) & 15 & -5 \\ (-2) \cdot (-6) & 21 & -6 \\ (-2) \cdot 60 & 26 & 60 \end{vmatrix} = 0$$

$$\begin{vmatrix} 1 & 2 & 3 \\ 3 & 3 & 3 \\ 1 & -1 & -3 \end{vmatrix} = \begin{vmatrix} 1 & 2 & 3 \\ 3 & 3 & 3 \\ 3 - 2 \cdot 1 & 3 - 2 \cdot 2 & 3 - 2 \cdot 3 \end{vmatrix} = 0$$

Matrix Operations and Determinants

Let us see how determinants are affected by the matrix operations $A + B$, kA, and AB.

Unfortunately, there is no easy formula for the determinant of the sum of two matrices, $\det(A + B)$. In general,

$$\det(A + B) \neq \det(A) + \det(B)$$

■ EXAMPLE 14

$$\begin{vmatrix} 1 & 2 \\ 2 & 3 \end{vmatrix} + \begin{vmatrix} 1 & 1 \\ 1 & 2 \end{vmatrix} = -1 + 1 = 0, \quad \text{but} \quad \det\left(\begin{bmatrix} 1 & 2 \\ 2 & 3 \end{bmatrix} + \begin{bmatrix} 1 & 1 \\ 1 & 2 \end{bmatrix} \right) = \begin{vmatrix} 2 & 3 \\ 3 & 5 \end{vmatrix} = 1$$

On the positive side, we have Theorem 6.

THEOREM 6

(Sum of Rows)

If every entry in any row (or column) of a determinant is the sum of two others, then the determinant is the sum of two others.

$$\begin{vmatrix} a_1 & a_2 & a_3 \\ b_1 & b_2 & b_3 \\ c_1 + d_1 & c_2 + d_2 & c_3 + d_3 \end{vmatrix} = \begin{vmatrix} a_1 & a_2 & a_3 \\ b_1 & b_2 & b_3 \\ c_1 & c_2 & c_3 \end{vmatrix} + \begin{vmatrix} a_1 & a_2 & a_3 \\ b_1 & b_2 & b_3 \\ d_1 & d_2 & d_3 \end{vmatrix}$$

PROOF   Exercise. (*Hint:*   If the *i*th row is $[c_{i1} + d_{i1}, \ldots, c_{in} + d_{in}]$, then cofactor expand about it.)

The determinant of a scalar product $\det(kA)$ can be computed from the next theorem.

THEOREM 7

(Determinant of Scalar Product)

Let A be an $n \times n$ matrix, and let k be any scalar. Then

$$\det(kA) = k^n \det(A)$$

PROOF   By repeated application of Property 2 of Theorem 1, we factor out of each row one k at a time. If $A = [\mathbf{a}_1 \mathbf{a}_2 \cdots \mathbf{a}_n]$, then

$$\det(kA) = \det[k\mathbf{a}_1 k\mathbf{a}_2 \cdots k\mathbf{a}_n] = k \det[\mathbf{a}_1 k\mathbf{a}_2 \cdots k\mathbf{a}_n]$$

$$= k^2 \det[\mathbf{a}_1 \mathbf{a}_2 \cdots k\mathbf{a}_n] = \cdots = k^n \det[\mathbf{a}_1 \mathbf{a}_2 \cdots \mathbf{a}_n] = k^n \det(A)$$

Next, we turn to $\det(AB)$. This time we get a nice and useful formula, known as **Cauchy's theorem.**

THEOREM 8

(Determinant of a Product)

The determinant of a product of matrices is the product of the determinants of the factors.

$$\det(AB) = \det(A) \det(B)$$

$$\det(A_1 A_2 \cdots A_m) = \det(A_1) \det(A_2) \cdots \det(A_m)$$

The proof, due to Cauchy,[5] is outlined in the exercises.
Let

$$A = \begin{bmatrix} 0 & 1 & 0 \\ 1 & 1 & 0 \\ 1 & 0 & 3 \end{bmatrix}, \qquad B = \begin{bmatrix} 0 & 2 & 0 \\ -5 & 0 & 0 \\ 0 & 0 & 1 \end{bmatrix}$$

■ EXAMPLE 15   Verify that $\det(AB) = \det(A) \det(B)$.

SOLUTION
$$\det(A) = -3 \qquad \det(B) = 10 \qquad \det(A) \det(B) = -30$$

$$AB = \begin{bmatrix} -5 & 0 & 0 \\ -5 & 2 & 0 \\ 0 & 2 & 3 \end{bmatrix} \qquad \det(AB) = -30$$

---

[5]This fact was also discovered by Gauss in the cases of 2×2 and 3×3 matrices.

Theorem 8 has an important implication.

THEOREM 9

If A is invertible, then

$$\det(A^{-1}) = \frac{1}{\det(A)}$$

PROOF A is invertible, so $AA^{-1} = I$. Thus, $\det(AA^{-1}) = \det(A)\det(A^{-1}) = \det(I) = 1$, by Theorem 8. Hence, $\det(A) \neq 0$. Division of $\det(A)\det(A^{-1}) = 1$ by $\det(A)$ yields the formula.

■ EXAMPLE 16   Show that A, B are invertible. Verify that $\det(B^{-1}) = 1/\det(B)$.

SOLUTION $\det(A) = -3$, $\det(B) = 10$. The determinants are nonzero; hence, the matrices are invertible by Theorem 9.

$$B^{-1} = \begin{bmatrix} 0 & -\frac{1}{5} & 0 \\ \frac{1}{2} & 0 & 0 \\ 0 & 0 & 1 \end{bmatrix}, \qquad \det(B^{-1}) = \frac{1}{10} = \frac{1}{\det(B)}$$

Exercises 6.2

In Exercises 1–7 evaluate the determinants by inspection.

1. a. $\begin{vmatrix} 5 & 1 & 14 \\ 0 & 10 & 2 \\ 0 & 0 & -1 \end{vmatrix}$
   b. $\begin{vmatrix} 1 & -1 & 1 \\ 2 & -1 & 2 \\ 4 & -2 & 4 \end{vmatrix}$

2. a. $\begin{vmatrix} 1 & 1 & 2 \\ 2 & 2 & 3 \\ 4 & 4 & 8 \end{vmatrix}$
   b. $\begin{vmatrix} 1 & 1 & 1 & 0 \\ 1 & 1 & 1 & 0 \\ 0 & 1 & 1 & 1 \\ 0 & 0 & 1 & 1 \end{vmatrix}$

3. a. $\begin{vmatrix} 0 & 0 & 0 & 1 \\ 0 & 1 & 0 & 0 \\ 0 & 0 & 1 & 0 \\ 1 & 0 & 0 & 0 \end{vmatrix}$
   b. $\begin{vmatrix} 1 & 0 & 0 & 0 \\ 0 & 0 & 0 & 1 \\ 0 & 0 & 1 & 0 \\ 0 & 1 & 0 & 0 \end{vmatrix}$

4. a. $\begin{vmatrix} 0 & 0 & 1 & 0 \\ 0 & 1 & 0 & 0 \\ 1 & 0 & 0 & 0 \\ 0 & 0 & 0 & 1 \end{vmatrix}$
   b. $\begin{vmatrix} 0 & 1 & 0 & 0 \\ 0 & 0 & 1 & 0 \\ 0 & 0 & 0 & 1 \\ 1 & 0 & 0 & 0 \end{vmatrix}$

5. a. $\begin{vmatrix} 0 & 1 & 0 & 0 \\ 0 & 0 & 0 & 1 \\ 0 & 0 & 1 & 0 \\ 1 & 0 & 0 & 0 \end{vmatrix}$
   b. $\begin{vmatrix} 1 & 0 & 0 & 0 \\ 0 & 0 & 1 & 0 \\ 0 & 0 & 0 & 1 \\ 0 & 1 & 0 & 0 \end{vmatrix}$

6. $\begin{vmatrix} 1 & 0 & 0 & 0 & 0 \\ 2 & 2 & 0 & 0 & 0 \\ 3 & 6 & 3 & 0 & 0 \\ 4 & 6 & 6 & 4 & 0 \\ 5 & 5 & 5 & 5 & -1 \end{vmatrix}$

7. $\begin{vmatrix} 0 & 0 & 0 & 0 & 1 \\ 0 & 2 & 0 & 0 & 2 \\ 0 & 6 & 3 & 0 & 3 \\ 0 & 6 & 6 & 4 & 5 \\ -1 & 5 & 5 & 5 & 5 \end{vmatrix}$

Let

$$\begin{vmatrix} a & b & c \\ d & e & f \\ g & h & i \end{vmatrix} = 3$$

In Exercises 8–15 explain the identities. (Do not compute.)

8. $\begin{vmatrix} a & b & c \\ d & e & f \\ 2g & 2h & 2i \end{vmatrix} = 6$

9. $\begin{vmatrix} g & h & i \\ d & e & f \\ a & b & c \end{vmatrix} = -3$

10. $\begin{vmatrix} a - 4c & b & c \\ d - 4f & e & f \\ g - 4i & h & i \end{vmatrix} = 3$

11. $\begin{vmatrix} 2b - 4c & b & c \\ 2e - 4f & e & f \\ 2h - 4i & h & i \end{vmatrix} = 0$

12. $\begin{vmatrix} c & b & -a \\ f & e & -d \\ i & h & -g \end{vmatrix} = 3$

13. $\begin{vmatrix} a & b & c \\ d - a & e - b & f - c \\ 3g & 3h & 3i \end{vmatrix} = 9$

14. $\begin{vmatrix} -1 & a & e & i \\ 0 & a & b & c \\ 0 & d & e & f \\ 0 & g & h & i \end{vmatrix} = -3$

15. $\begin{vmatrix} 2a & 2b & 2c \\ 2d & 2e & 2f \\ 2g & 2h & 2i \end{vmatrix} = 24$

In Exercises 16–17 explain without computing why the substitutions $x = 0, 2$ make the determinants zero.

16. $\begin{vmatrix} x & x & 2x \\ 0 & 1 & 0 \\ 2 & 2 & 4 \end{vmatrix}$

17. $\begin{vmatrix} x & x & 2 \\ 0 & x^2 & 0 \\ 8 & 8 & x^3 \end{vmatrix}$

18. Explain without computing why the following determinants are equal.

$$\begin{vmatrix} a & b & c \\ d & e & f \\ g & h & i \end{vmatrix} = \begin{vmatrix} a & -b & c \\ -d & e & -f \\ g & -h & i \end{vmatrix}$$

In Exercises 19–24 evaluate the determinants by row reduction.

19. a. $\begin{vmatrix} 2 & 2 & 4 \\ -1 & -1 & -1 \\ 0 & 2 & 0 \end{vmatrix}$ b. $\begin{vmatrix} 1 & 1 & 0 \\ 1 & 1 & 1 \\ 0 & 1 & 1 \end{vmatrix}$

20. a. $\begin{vmatrix} 2 & -2 & 4 \\ 0 & 1 & 7 \\ -2 & 9 & 0 \end{vmatrix}$ b. $\begin{vmatrix} 1 & -2 & 5 \\ -2 & 6 & -4 \\ 3 & -5 & 0 \end{vmatrix}$

21. $\begin{vmatrix} 2 & -4 & 2 & 8 \\ -2 & 3 & 0 & -7 \\ 0 & 1 & 5 & -1 \\ 1 & 0 & 1 & 0 \end{vmatrix}$

22. $\begin{vmatrix} 1 & -1 & 2 & 0 & 0 \\ 0 & 1 & 2 & -2 & 7 \\ 0 & 0 & 1 & -1 & 2 \\ 0 & 0 & 4 & 0 & 3 \\ 0 & 3 & 0 & 0 & 1 \end{vmatrix}$

23. $\begin{vmatrix} 1 & 0 & -1 & 2 & 1 \\ 2 & 2 & 4 & 2 & 8 \\ 0 & 0 & 3 & 0 & 3 \\ 0 & 0 & 1 & 5 & 0 \\ 0 & 1 & 0 & 1 & 0 \end{vmatrix}$

24. $\begin{vmatrix} 2 & 1 & 1 & 1 & -1 & 0 \\ 0 & 1 & -1 & 2 & 0 & 0 \\ 0 & 0 & 1 & 2 & -2 & 3 \\ 0 & 0 & 0 & 1 & -1 & 0 \\ 0 & 0 & 0 & 4 & 2 & 3 \\ 0 & 0 & 0 & 0 & 1 & 1 \end{vmatrix}$

25. Prove the following properties of determinants.

a. $\begin{vmatrix} a_1 & a_2 & ka_3 \\ b_1 & b_2 & kb_3 \\ c_1 & c_2 & kc_3 \end{vmatrix} = k \begin{vmatrix} a_1 & a_2 & a_3 \\ b_1 & b_2 & b_3 \\ c_1 & c_2 & c_3 \end{vmatrix}$

b. $\begin{vmatrix} a_1 & a_2 & a_3 \\ b_1 & b_2 & b_3 \\ c_1 & c_2 & c_3 \end{vmatrix} = - \begin{vmatrix} c_1 & c_2 & c_3 \\ b_1 & b_2 & b_3 \\ a_1 & a_2 & a_3 \end{vmatrix}$

26. Prove the following properties of determinants.

a. $\begin{vmatrix} a_1 & a_2 & a_3 \\ b_1 & b_2 & b_3 \\ c_1 & c_2 & c_3 \end{vmatrix} = \begin{vmatrix} a_1 & b_1 & c_1 \\ a_2 & b_2 & c_2 \\ a_3 & b_3 & c_3 \end{vmatrix}$

b. $\begin{vmatrix} a_1 + kb_1 & a_2 + kb_2 & a_3 + kb_3 \\ b_1 & b_2 & b_3 \\ c_1 & c_2 & c_3 \end{vmatrix}$

$$= \begin{vmatrix} a_1 & a_2 & a_3 \\ b_1 & b_2 & b_3 \\ c_1 & c_2 & c_3 \end{vmatrix}$$

27. Prove Theorem 6.

In Exercises 28–31 use Theorem 3 to find which matrices are invertible.

28. $A = \begin{bmatrix} 1 & -1 & 1 \\ 0 & 2 & 4 \\ 0 & 0 & 0 \end{bmatrix}$ $B = \begin{bmatrix} 1 & -1 & 1 \\ 0 & 2 & 4 \\ 0 & 0 & 3 \end{bmatrix}$

29. $C = \begin{bmatrix} 1 & 1 & 0 \\ 1 & 1 & 1 \\ 0 & 1 & 1 \end{bmatrix}$ $D = \begin{bmatrix} 1 & 1 & 0 & 0 \\ 1 & 1 & 1 & 0 \\ 0 & 1 & 1 & 1 \\ 0 & 0 & 1 & 1 \end{bmatrix}$

30. $E = \begin{bmatrix} 1 & -1 & -2 \\ 0 & 1 & 0 \\ 2 & 3 & -4 \end{bmatrix}$ $F = \begin{bmatrix} 1 & 2 & 3 \\ 4 & 5 & 6 \\ 7 & 8 & 9 \end{bmatrix}$

31. $G = \begin{bmatrix} 1 & 2 & 3 \\ 2 & 2 & 3 \\ 3 & 3 & 3 \end{bmatrix}$ $H = \begin{bmatrix} 0 & 1 & 1 \\ 1 & 0 & 1 \\ 1 & 1 & 0 \end{bmatrix}$

In Exercises 32–33 find all values of k such that the matrices are noninvertible.

32. $\begin{vmatrix} k & k-1 & 1 \\ 0 & k+1 & 4 \\ k & 0 & k \end{vmatrix}$

33. $\begin{vmatrix} k & k^2 & 0 \\ 0 & k^3 & 4 \\ -k & 0 & k \end{vmatrix}$

Let A and B be $n \times n$ matrices.

34. Prove the identity

$$\det(AB) = \det(BA)$$

(This is true even if $AB \neq BA$.)

35. If B is invertible, show that

$$\det(B^{-1}AB) = \det(A)$$

36. Let A be a 3×3 matrix with $\det(A) = -2$. Compute each value.

   a. $\det(A^3)$ b. $\det(A^{-1})$
   c. $\det(A^{-3})$ d. $\det(A^T)$
   e. $\det(AA^T)$ f. $\det(-7A)$

37. Let A and B be 4×4 matrices, such that $\det(A) = -2$ and $\det(B) = 7$. Compute each value.

   a. $\det(AB)$ b. $\det(A^3B)$
   c. $\det(ABA^{-1})$ d. $\det(BAB^{-1})$
   e. $\det(AB^T)$ f. $\det(3AB)$

38. Prove that the square of any determinant, $\det(A)^2$, can be expressed as the determinant of a symmetric matrix.[6] (*Hint*: Consider the *symmetric* matrix AA^T.)

In Exercises 39–42 use row reduction to prove the identities.

39. $\begin{vmatrix} 1 & 1 & 1 \\ a & b & c \\ a^2 & b^2 & c^2 \end{vmatrix} = (b-a)(c-b)(c-a)$

40. $\begin{vmatrix} 1 & a & b \\ 1 & a^2 & b^2 \\ 1 & a^3 & b^3 \end{vmatrix} = ab(b-a)(a-1)(b-1)$

41. $\begin{vmatrix} 1 & 1 & 1 \\ a & b & c \\ a^3 & b^3 & c^3 \end{vmatrix} = (b-a)(c-b)(c-a)(a+b+c)$

42. $\begin{vmatrix} 1 & a & b \\ 1 & a^2 & b^2 \\ 1 & a^4 & b^4 \end{vmatrix} = ab(b-a)(a-1)(b-1)(a+b+1)$

A Proof of Cauchy's Theorem

The following exercises outline a proof of the basic formula $\det(AB) = \det(A)\det(B)$ using Theorem 1.

43. Let E be an elementary matrix. Use Theorem 1 to show

	If E is obtained from I by
$\det(E) = 1$	$R_i + cR_j \longrightarrow R_i$
$\det(E) = c$	$cR_i \longrightarrow R_i$
$\det(E) = -1$	$R_i \longleftrightarrow R_j$

44. Let A, E be $n \times n$ matrices, with E elementary. Use Exercise 43 to show that

$$\det(EA) = \det(E)\det(A)$$

45. Let A and B be $n \times n$ matrices with A noninvertible. Prove that

$$\det(AB) = \det(A)\det(B)$$

(*Hint*: Use Theorem 16, Section 3.3, and Theorem 3.)

46. Let A and B be $n \times n$ matrices with A invertible. Prove that

$$\det(AB) = \det(A)\det(B)$$

(*Hint*: Write A as a product of elementary matrices. Then use Exercise 44.)

---

[6](a) Recall that B is symmetric if $B^T = B$. (b) The claim of this exercise is due to Lagrange.

6.3   The Adjoint; Cramer's Rule

Reader's Goals for This Section

1. To compute matrix inverses by using the adjoint matrix.
2. To solve $n \times n$ linear systems by using Cramer's Rule.

Adjoint and Inverse

DEFINITION   Let A be an $n \times n$ matrix. The matrix whose (i, j) entry is the cofactor C_{ij} of A is the **matrix of cofactors** of A. Its transpose is the **adjoint of A,** and it is denoted by $\mathrm{Adj}(A)$.

$$\mathrm{Adj}(A) = \begin{bmatrix} C_{11} & C_{21} & \cdots & C_{n1} \\ C_{12} & C_{22} & \cdots & C_{n2} \\ \vdots & \vdots & \ddots & \vdots \\ C_{1n} & C_{2n} & \cdots & C_{nn} \end{bmatrix}$$

In Section 6.1 we found the cofactors of

$$A = \begin{bmatrix} -1 & 2 & 2 \\ 4 & 3 & -2 \\ -5 & 0 & 3 \end{bmatrix}$$

to be

$$\begin{array}{lll} C_{11} = 9 & C_{12} = -2 & C_{13} = 15 \\ C_{21} = -6 & C_{22} = 7 & C_{23} = -10 \\ C_{31} = -10 & C_{32} = 6 & C_{33} = -11 \end{array}$$

■ EXAMPLE 17   The matrix of cofactors of A is

$$\begin{bmatrix} 9 & -2 & 15 \\ -6 & 7 & -10 \\ -10 & 6 & -11 \end{bmatrix}$$

The adjoint of A is

$$\mathrm{Adj}(A) = \begin{bmatrix} 9 & -6 & -10 \\ -2 & 7 & 6 \\ 15 & -10 & -11 \end{bmatrix}$$

In Chapter 3 we discussed an algorithm—but gave *no formula*—for computing A^{-1} by row reducing $[A : I]$. Now with determinants in our toolkit, we can give an *explicit formula* for A^{-1}.

THEOREM 10

Let A be an $n \times n$ matrix. Then

$$A \, \text{Adj}(A) = \det(A)I_n = \text{Adj}(A)A$$

PROOF Consider the product $\text{Adj}(A)A$.

$$\text{Adj}(A)A = \begin{bmatrix} C_{11} & \cdots & C_{j1} & \cdots & C_{n1} \\ \vdots & & \vdots & & \vdots \\ C_{1i} & \cdots & C_{ji} & \cdots & C_{ni} \\ \vdots & & \vdots & & \vdots \\ C_{1n} & \cdots & C_{jn} & \cdots & C_{nn} \end{bmatrix} \begin{bmatrix} a_{11} & \cdots & a_{1j} & \cdots & a_{1n} \\ \vdots & & \vdots & & \vdots \\ a_{i1} & \cdots & a_{ij} & \cdots & a_{in} \\ \vdots & & \vdots & & \vdots \\ a_{n1} & \cdots & a_{nj} & \cdots & a_{nn} \end{bmatrix}$$

The (i, j) entry is

$$C_{1i}a_{1j} + C_{2i}a_{2j} + \cdots + C_{ni}a_{nj}$$

This sum can be viewed as the determinant cofactor expansion about the jth column of the matrix A' obtained from A by replacing the jth column with the ith. If $i = j$, the sum is $\det(A)$, because $A = A'$. If $i \neq j$, the sum is 0, because A has a repeated column, by Theorem 5. Therefore,

$$\text{Adj}(A)A = \begin{bmatrix} \det(A) & \cdots & 0 \\ \vdots & \ddots & \vdots \\ 0 & \cdots & \det(A) \end{bmatrix} = \det(A)I_n$$

A similar argument shows that $A \, \text{Adj}(A) = \det(A)I_n$.

THEOREM 11

(Inverse via Adjoint)

Let A be an invertible matrix. Then

$$A^{-1} = \frac{1}{\det(A)} \, \text{Adj}(A)$$

PROOF If we multiply $A \, \text{Adj}(A) = \det(A)I_n$ of Theorem 10 on the left by A^{-1}, we get

$$\text{Adj}(A) = A^{-1} \det(A)$$

Because A is invertible, $\det(A) \neq 0$ by Theorem 3. Hence,

$$\frac{1}{\det(A)} \, \text{Adj}(A) = A^{-1}$$

■ EXAMPLE 18   Let A be as in Example 17. Compute A^{-1} by applying Theorem 11.

SOLUTION $\det(A) = 17$. Hence,

$$A^{-1} = \frac{1}{\det(A)} \, \text{Adj}(A) = \frac{1}{17} \begin{bmatrix} 9 & -6 & -10 \\ -2 & 7 & 6 \\ 15 & -10 & -11 \end{bmatrix} = \begin{bmatrix} \frac{9}{17} & -\frac{6}{17} & -\frac{10}{17} \\ -\frac{2}{17} & \frac{7}{17} & \frac{6}{17} \\ \frac{15}{17} & -\frac{10}{17} & -\frac{11}{17} \end{bmatrix}$$

■ EXAMPLE 19   Let

$$A = \begin{bmatrix} a & b \\ c & d \end{bmatrix}$$

Then

$$\det(A) = ad - bc \quad \text{and} \quad \begin{matrix} C_{11} = d & C_{12} = -c \\ C_{21} = -b & C_{22} = a \end{matrix}$$

Therefore,

$$\text{Adj}(A) = \begin{bmatrix} d & -b \\ -c & a \end{bmatrix}; \quad \text{hence,} \quad A^{-1} = \frac{1}{ad - bc} \begin{bmatrix} d & -b \\ -c & a \end{bmatrix} \quad \text{by Theorem 11.}$$

Thus, we obtained the familiar formula for the general 2×2 inverse.

NOTE   Computing $\text{Adj}(A)$ involves the calculation of n^2 determinants of size $(n-1) \times (n-1)$. If $n = 10$, we would need one hundred 9×9 determinants. Because of this type of computational intensity, Theorem 11 is rarely used to find A^{-1}. Row reduction of $[A : I]$ *is still the method of choice*. However, the theorem can be very handy in proving theoretical properties of inverses. We can also use it in our next topic to prove a formula for the solution of a square linear system.

Cramer's Rule

Let $A\mathbf{x} = \mathbf{b}$ be a square system, with

$$A = \begin{bmatrix} a_{11} & \cdots & a_{1n} \\ \vdots & \ddots & \vdots \\ a_{n1} & \cdots & a_{nn} \end{bmatrix} \qquad \mathbf{x} = \begin{bmatrix} x_1 \\ \vdots \\ x_n \end{bmatrix} \qquad \mathbf{b} = \begin{bmatrix} b_1 \\ \vdots \\ b_n \end{bmatrix}$$

Let A_i denote the matrix obtained from A by replacing the ith column with $\mathbf{b}$.

$$A_i = \begin{bmatrix} a_{11} & \cdots & a_{1,i-1} & b_1 & a_{1,i+1} & \cdots & a_{1n} \\ \vdots & \vdots & \vdots & \vdots & \vdots & \vdots & \vdots \\ a_{n1} & \cdots & a_{n,i-1} & b_n & a_{n,i+1} & \cdots & a_{nn} \end{bmatrix}$$

Cramer's rule is an explicit formula for the solution of any consistent square system.

THEOREM 12   (Cramer's Rule)

If $\det(A) \neq 0$, the system $A\mathbf{x} = \mathbf{b}$ has a unique solution, given by

$$x_1 = \frac{\det(A_1)}{\det(A)}, x_2 = \frac{\det(A_2)}{\det(A)}, \ldots, x_n = \frac{\det(A_n)}{\det(A)}$$

PROOF   Because $\det(A) \neq 0$, A is invertible by Theorem 3. Hence, $A\mathbf{x} = \mathbf{b}$ has the unique solution $\mathbf{x} = A^{-1}\mathbf{b}$. But A^{-1} can be computed by Theorem 11,

$$\mathbf{x} = A^{-1}\mathbf{b} = \frac{1}{\det(A)} \text{Adj}(A)\mathbf{b} = \frac{1}{\det(A)} \begin{bmatrix} C_{11}b_1 + C_{21}b_2 + \cdots + C_{n1}b_n \\ \vdots \\ C_{1n}b_1 + C_{2n}b_2 + \cdots + C_{nn}b_n \end{bmatrix}$$

So, the ith component x_i of $\mathbf{x}$ equals the ith component of the right-hand side.

$$x_i = \frac{1}{\det(A)}(C_{1i}b_1 + C_{2i}b_2 + \cdots + C_{ni}b_n)$$

Because A and A_i differ only by the ith column, the cofactors of that column are the same. Hence, $C_{1i}b_1 + C_{2i}b_2 + \cdots + C_{ni}b_n$ is $\det(A_i)$ by cofactor expansion about its ith column. Therefore,

$$x_i = \frac{\det(A_i)}{\det(A)}$$

for $i = 1, \ldots, n$ as asserted.

■ EXAMPLE 20   Use Cramer's rule to find the solution to the system

$$x + y - z = 2$$
$$x - y + z = 3$$
$$-x + y + z = 4$$

SOLUTION   First we compute the determinant of the coefficient matrix

$$A = \begin{bmatrix} 1 & 1 & -1 \\ 1 & -1 & 1 \\ -1 & 1 & 1 \end{bmatrix}$$

as well as the determinants of the matrices

$$A_1 = \begin{bmatrix} 2 & 1 & -1 \\ 3 & -1 & 1 \\ 4 & 1 & 1 \end{bmatrix}, \quad A_2 = \begin{bmatrix} 1 & 2 & -1 \\ 1 & 3 & 1 \\ -1 & 4 & 1 \end{bmatrix}, \quad A_3 = \begin{bmatrix} 1 & 1 & 2 \\ 1 & -1 & 3 \\ -1 & 1 & 4 \end{bmatrix}$$

We get

$$\det(A) = -4 \qquad \det(A_1) = -10 \qquad \det(A_2) = -12 \qquad \det(A_3) = -14$$

which yields

$$x = \frac{\det(A_1)}{\det(A)} = \frac{5}{2}, \qquad y = \frac{\det(A_2)}{\det(A)} = 3, \qquad z = \frac{\det(A_3)}{\det(A)} = \frac{7}{2}$$

■ EXAMPLE 21   Suppose $ad - bc \neq 0$. Use Cramer's rule to find the solution to the general linear system

$$ax + by = e$$
$$cx + dy = f$$

SOLUTION

$$A = \begin{bmatrix} a & b \\ c & d \end{bmatrix}, \qquad A_1 = \begin{bmatrix} e & b \\ f & d \end{bmatrix}, \qquad A_2 = \begin{bmatrix} a & e \\ c & f \end{bmatrix}$$

$$\det(A) = ad - bc \qquad \det(A_1) = de - bf \qquad \det(A_2) = af - ce$$

Because $\det(A) = ad - bc \neq 0$, we can apply Cramer's rule to get

$$x = \frac{de - bf}{ad - bc}, \qquad y = \frac{af - ce}{ad - bc}$$

Exercises 6.3

In Exercises 1–6 use Theorem 11 to find the inverses of the given matrices.

1. a. $\begin{bmatrix} 2 & 4 \\ 3 & 5 \end{bmatrix}$ **b.** $\begin{bmatrix} -1 & 2 \\ -3 & 4 \end{bmatrix}$

2. a. $\begin{bmatrix} -3 & 4 \\ 8 & -11 \end{bmatrix}$ **b.** $\begin{bmatrix} 2.2 & 1.0 \\ 3.2 & 1.5 \end{bmatrix}$

3. $\begin{bmatrix} 1 & 1 & 1 \\ 1 & -1 & 1 \\ 1 & 1 & -1 \end{bmatrix}$

4. $\begin{bmatrix} 1 & 2 & 3 \\ 0 & 1 & 4 \\ 0 & 0 & 1 \end{bmatrix}$

5. $\begin{bmatrix} 1 & 0 & 0 \\ 2 & 1 & 0 \\ 3 & 4 & 1 \end{bmatrix}$

6. $\begin{bmatrix} 1 & 2 & 3 \\ 2 & 2 & 3 \\ 3 & 3 & 3 \end{bmatrix}$

7. Let

$$A = \begin{bmatrix} 1 & 2 & 1 & 0 \\ 2 & 2 & 1 & 1 \\ 2 & 1 & 1 & 2 \\ 1 & 2 & 1 & 1 \end{bmatrix}$$

a. Compute A^{-1} by row reduction.

b. Compute A^{-1} by using Theorem 11.

c. Which method is more efficient?

8. Let A be a square matrix with *integer* entries. Show the following.

a. $\det(A)$ is an integer.

b. $\text{Adj}(A)$ has integer entries.

c. If $\det(A)$ divides exactly all the entries of $\text{Adj}(A)$, then A^{-1} has integer entries.

d. If $\det(A) = \pm 1$, then A^{-1} has integer entries.

In Exercises 9–11 use Cramer's rule to solve the systems.

9. a. $x + y = 1$ **b.** $x + 2z = 1$
 $x - y = 1$ $3x + 4z = 1$

10. $x + y + z = 1$
  $x - y + z = 1$
  $x + y - z = 1$

11. $x + 2y + 3z = 1$
        $y + 4z = 1$
              $z = 1$

12. Use Cramer's rule to solve the system for x and y.

$$(\cos \theta)x - (\sin \theta)y = \cos \theta - 2 \sin \theta$$

$$(\sin \theta)x + (\cos \theta)y = \sin \theta + 2 \cos \theta$$

In Exercises 13–14 use Cramer's rule to solve the systems for z only.

13. $3x - 3y - 2z = 3$
  $-x - 4y + 2z = 2$
  $5x + 4y + \ \ z = 1$

14. $x - y - z - w = 0$
  $-x - y + z + w = 2$
  $x + y - z + w = 1$
  $x + y + z + w = 1$

15. Let A be any $n \times n$ matrix. Prove that

$$\det(\mathrm{Adj}(A)) = (\det(A))^{n-1}$$

(*Hint:* Use the equation $A \, \mathrm{Adj}(A) = \det(A)I_n$ from the proof of Theorem 11.)

16. Let A be a 4×4 with $\det(A) = 3$. Find $\det(\mathrm{Adj}(A))$.

17. Prove that an $n \times n$ matrix A is invertible if and only if $\mathrm{Adj}(A)$ is invertible. (*Hint:* Use Exercise 15.)

6.4 Determinants with Permutations

Reader's Goals for This Section

1. To know what a permutation is and how to compute its sign.
2. To understand the relation between permutations and determinants.
3. To use complete expansion to compute a determinant.

In Section 6.1 we learned that

$$\begin{vmatrix} a_{11} & a_{12} \\ a_{21} & a_{22} \end{vmatrix} = a_{11}a_{22} - a_{12}a_{21}$$

$$\begin{vmatrix} a_{11} & a_{12} & a_{13} \\ a_{21} & a_{22} & a_{23} \\ a_{31} & a_{32} & a_{33} \end{vmatrix} = \begin{aligned} & a_{11}a_{22}a_{33} + a_{12}a_{23}a_{31} + a_{13}a_{21}a_{32} \\ & - a_{13}a_{22}a_{31} - a_{11}a_{23}a_{32} - a_{12}a_{21}a_{33} \end{aligned}$$

These are examples of what is called the **complete expansion**, a formula that gives the determinant in terms of the *entries* of the matrix;[7] for an $n \times n$ matrix A this formula is obtained as follows:

1. Form all products each consisting of n entries of A coming from *different* rows and columns.
2. Assign a sign to each product. Add all signed products.

All terms of the two preceding determinants are as in Step 1. For example, $-a_{13}a_{22}a_{31}$ in the 3×3 determinant is the signed product formed by picking a_{13}, a_{22}, a_{31} from the first row and third column, the second row and second column, and

---

[7]Contrast with the cofactor expansion, which computes a determinant in terms of smaller size determinants.

the third row and first column, respectively. To find all such products, called **elementary products**, we use the concept of permutation.

DEFINITION

A **permutation** of the set of integers $\{1, 2, \ldots, n\}$ is a rearrangement of these integers.

■ EXAMPLE 22   The permutations of $\{1, 2\}$, $\{1, 2, 3\}$, $\{1, 2, 3, 4\}$ are
$\{1, 2\}$:

$$(1, 2) \qquad (2, 1)$$

$\{1, 2, 3\}$:

$$(1, 2, 3) \qquad (2, 1, 3) \qquad (3, 1, 2)$$
$$(1, 3, 2) \qquad (2, 3, 1) \qquad (3, 2, 1)$$

$\{1, 2, 3, 4\}$:

$(1, 2, 3, 4)$	$(1, 4, 2, 3)$	$(2, 3, 1, 4)$	$(3, 1, 2, 4)$	$(3, 4, 1, 2)$	$(4, 2, 1, 3)$
$(1, 2, 4, 3)$	$(1, 4, 3, 2)$	$(2, 3, 4, 1)$	$(3, 1, 4, 2)$	$(3, 4, 2, 1)$	$(4, 2, 3, 1)$
$(1, 3, 2, 4)$	$(2, 1, 3, 4)$	$(2, 4, 1, 3)$	$(3, 2, 1, 4)$	$(4, 1, 2, 3)$	$(4, 3, 1, 2)$
$(1, 3, 4, 2)$	$(2, 1, 4, 3)$	$(2, 4, 3, 1)$	$(3, 2, 4, 1)$	$(4, 1, 3, 2)$	$(4, 3, 2, 1)$

There are $6 = 1 \cdot 2 \cdot 3 = 3!$ permutations of $\{1, 2, 3\}$ and $24 = 4!$ permutations of $\{1, 2, 3, 4\}$. In general, the number of permutations of $\{1, 2, \ldots, n\}$ is $n!$. This can be seen as follows: To fill the first position, there are n choices, because any one of the numbers can be used. For the second position, there are $n - 1$ choices, because one number has been already used in the first position. So, to fill the first two positions, there are $n \cdot (n - 1)$ choices. We continue in the same manner with the rest of the positions to get a total of $n \cdot (n - 1) \cdots 2 \cdot 1 = n!$ choices. Note that $n!$ grows very rapidly with n.

$$0! = 1 \qquad\qquad 6! = 720$$
$$1! = 1 \qquad\qquad 7! = 5040$$
$$2! = 2 \qquad\qquad 8! = 40{,}320$$
$$3! = 6 \qquad\qquad 9! = 362{,}880$$
$$4! = 24 \qquad\qquad 10! = 3{,}628{,}800$$
$$5! = 120 \qquad\qquad 11! = 39{,}916{,}800$$

So, it is hard to write all permutations even for relatively small n.

DEFINITION

Let $p = (j_1, j_2, \ldots, j_n)$ be any permutation of $\{1, 2, \ldots, n\}$. We say that p has an **inversion** (j_i, j_k) if a larger integer, j_i, precedes a smaller one, j_k. Also, p is called **even** or **odd** if it has an even or an odd total number of inversions. The **sign** of p, denoted by sign(p), is 1 if p is even and -1 if it is odd. $(1, 2, \ldots, n)$ is considered as even with sign 1.

For example, $(1, 3, 2, 4)$ has one inversion: $(3, 2)$. So it is odd with sign -1. $(4, 2, 1, 3)$ has four inversions: $(4, 2)$, $(4, 1)$, $(4, 3)$, $(2, 1)$. So it is even with sign 1.

■ EXAMPLE 23

Permutation	Inversions	Even/Odd	Sign
$(1, 2, 3)$	None	Even	1
$(1, 3, 2)$	$(3, 2)$	Odd	-1
$(2, 1, 3)$	$(2, 1)$	Odd	-1
$(2, 3, 1)$	$(2, 1), (3, 1)$	Even	1
$(3, 1, 2)$	$(3, 1), (3, 2)$	Even	1
$(3, 2, 1)$	$(3, 2), (3, 1), (2, 1)$	Odd	-1

NOTE   If r is the number of inversions of a permutation p, then $\text{sign}(p) = (-1)^r$.

Let us see now how permutations are involved in the computation of determinants.[8] For the 2×2 determinant, to get all elementary products we form

$$a_{1\_}a_{2\_}$$

with each row number represented in the first indices. The blanks in the second indices are to be filled with all the choices for the column numbers. In this case they are either 1, 2 or 2, 1. These are just the permutations of $\{1, 2\}$. By following this process, we ensure that no two entries come from the same row or column. In addition, the sign of each elementary product is the sign of the current permutation of the column indices. The sign of $a_{11}a_{12}$ is 1, because $(1, 2)$ is even, whereas the sign of $a_{12}a_{21}$ is -1, because $(2, 1)$ is odd.

Likewise, in the 3×3 determinant all elementary products are obtained from

$$a_{1\_}a_{2\_}a_{3\_}$$

and the blanks are filled with the permutations of $\{1, 2, 3\}$. For example $a_{13}a_{21}a_{32}$ corresponds to permutation $(3, 1, 2)$. The sign of each term is the sign of the corresponding permutation. The sign of $a_{13}a_{21}a_{32}$ is 1 since $(3, 1, 2)$ is even. The sign of $a_{12}a_{21}a_{33}$ is -1 since $(2, 1, 3)$ is odd. In general, we may **define** an $n \times n$ determinant using permutations as follows:

DEFINITION

If A is an $n \times n$ matrix with entries a_{ij} the **complete expansion** of the determinant of A is

$$\det(A) = \sum \text{sign}(j_1, j_2, \ldots, j_n)a_{1j_1}a_{2j_2} \cdots a_{nj_n} \qquad (6.1)$$

where the sum is over all permutations $(j_1, j_2, \ldots, j_n)$ of $\{1, 2, \ldots, n\}$ and $\text{sign}(j_1, j_2, \ldots, j_n)$ denotes the sign of $(j_1, j_2, \ldots, j_n)$.

REMARKS   This definition enables us to prove *all the properties of determinants, including cofactor expansion and Theorem 1* (see the exercise set). It is also very useful

---

[8]This method of studying determinants is due to Bezout and Laplace.

in proving theoretical properties of determinants. However, computing determinants by using Equation (6.1) is not very practical. For example, for an 11×11 determinant we would need 39,916,800 terms, each consisting of 11 factors. This requires a total of 439,084,800 operations, provided we have a good method for writing all possible elementary products. At the rate of 10,000 operations a minute, it would take more than 30 years of calculation. We have already encountered a far more practical method: the cofactor expansion preceded by the (ever present) Gaussian elimination. Using (6.1) can be handy for matrices with only few nonzero elementary products.

■ EXAMPLE 24   Compute the determinant

$$\begin{vmatrix} 3 & 0 & 4 \\ 0 & 5 & 0 \\ 6 & 0 & 7 \end{vmatrix}$$

SOLUTION   Because each elementary product should have no factors from the same column or row, the only nonzero terms are

$$3 \cdot 5 \cdot 7, \quad \text{and} \quad 4 \cdot 5 \cdot 6$$

corresponding to

$$a_{11}a_{22}a_{33} \quad \text{and} \quad a_{13}a_{22}a_{31}$$

and hence to the permutations

$$(1, 2, 3) \quad \text{and} \quad (3, 2, 1)$$

respectively. The first permutation is even and the second is odd; hence the signs are 1 and -1, respectively. Therefore, the determinant is

$$3 \cdot 5 \cdot 7 - 4 \cdot 5 \cdot 6 = 105 - 120 = -15$$

We conclude this section by discussing an interesting way of computing the sign of a permutation. This is done by using crossing diagrams. (Yet another way, the method of *permutation matrices*, is offered in the exercise section.)

Crossing Diagrams (Optional)

DEFINITION   A **crossing diagram** of a permutation $(j_1, j_2, \ldots, j_n)$ is a diagram consisting of two columns of vertices, each column labeled $1, 2, \ldots, n$, and for which a line segment (or a curve) has been drawn from each node, say, i, to the ith coordinate j_i of the permutation. The **crossing number** of a crossing diagram is its number of crossings. This number is odd if the permutation is odd and even if the permutation is even. The case of 0 crossings corresponds to $(1, 2, \ldots, n)$, which is an even permutation.

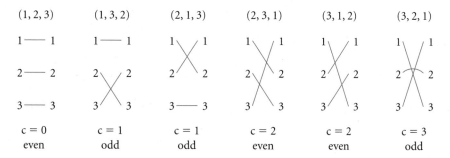

Figure 6.1   *The crossing diagrams of the permutations of* {1,2,3}.

Consider the permutation $(3, 1, 2)$. We write two columns, each labeled 1, 2, 3. Because $j_1 = 3$, $j_2 = 1$, $j_3 = 2$, we draw a line connecting 1 of the first column with 3 of the second; another line connecting 2 of the first column with 1 of the second, and a third line connecting 3 of the first column with 2 of the second (Fig. 6.1). The number of crossings in the diagram is 2, so the crossing number c is 2, which is an even number. Therefore, $(3, 1, 2)$ is an even permutation. The crossing diagrams of all permutations of $\{1, 2, 3\}$ are shown in Fig. 6.1.

■ EXAMPLE 25   Use Equation (6.1) and crossing diagrams to compute the determinant

$$\begin{vmatrix} 1 & 0 & 0 & 0 & 2 \\ 0 & 3 & 0 & 4 & 0 \\ 0 & 0 & 5 & 0 & 0 \\ 0 & 6 & 0 & 7 & 0 \\ 8 & 0 & 0 & 0 & 9 \end{vmatrix}$$

SOLUTION   A nonzero elementary product can have a factor of 1 or 2 from the first row. If it starts with 1, then the last factor is 9 and not 8, because 1 and 8 are in the same column. Likewise, if a product starts with 2, the last factor is 8. So we have only products of the form

$$1 \cdot \_ \cdot \_ \cdot \_ \cdot 9 \quad \text{and} \quad 2 \cdot \_ \cdot \_ \cdot \_ \cdot 8$$

The rest of the factors come from the submatrix

$$\begin{bmatrix} 3 & 0 & 4 \\ 0 & 5 & 0 \\ 6 & 0 & 7 \end{bmatrix}$$

used in Example 24. The possible products here are $3 \cdot 5 \cdot 7$ and $4 \cdot 5 \cdot 6$. Hence, we get a total of four nonzero products, namely, $1 \cdot 3 \cdot 5 \cdot 7 \cdot 9$, $1 \cdot 4 \cdot 5 \cdot 6 \cdot 9$, $2 \cdot 3 \cdot 5 \cdot 7 \cdot 8$, and $2 \cdot 4 \cdot 5 \cdot 6 \cdot 8$. The signs of the corresponding permutations are computed from the crossing diagrams of Fig. 6.2.

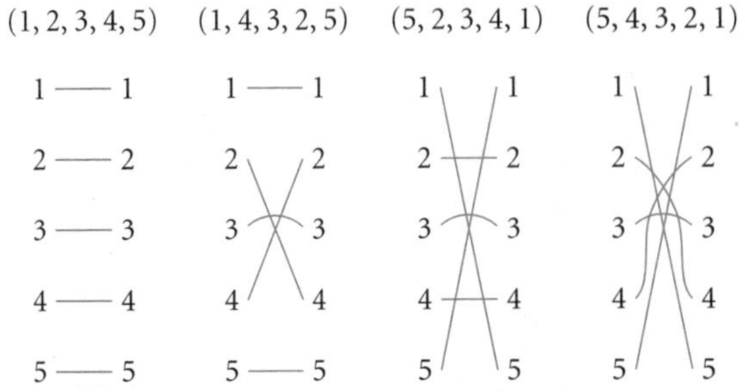

$(1, 2, 3, 4, 5)$ $(1, 4, 3, 2, 5)$ $(5, 2, 3, 4, 1)$ $(5, 4, 3, 2, 1)$

Figure 6.2   Crossing diagrams of some permutations of $\{1, 2, 3, 4, 5\}$.

We have:

Elementary Product	Permutation	Even/Odd	Sign	Value
$1 \cdot 3 \cdot 5 \cdot 7 \cdot 9 = 945$	$(1, 2, 3, 4, 5)$	Even	$+$	945
$1 \cdot 4 \cdot 5 \cdot 6 \cdot 9 = 1080$	$(1, 4, 3, 2, 5)$	Odd	$-$	-1080
$2 \cdot 3 \cdot 5 \cdot 7 \cdot 8 = 1680$	$(5, 2, 3, 4, 1)$	Odd	$-$	-1680
$2 \cdot 4 \cdot 5 \cdot 6 \cdot 8 = 1920$	$(5, 4, 3, 2, 1)$	Even	$+$	1920

The determinant is $945 - 1080 - 1680 + 1920 = 105$.

Permutations with CAS

Maple

```
# Permutations of {1,2,3}
> with(combinat) :
> permute(3);
```

$$[[1, 2, 3], [1, 3, 2], [2, 1, 3], [2, 3, 1], [3, 1, 2], [3, 2, 1]]$$

Mathematica

```
(* Permutations of {1,2,3}*)
In[1]:=
    Permutations[{1,2,3}]
Out[1]= {{1, 2, 3}, {1, 3, 2}, {2, 1, 3}, {2, 3, 1},
        {3, 1, 2}, {3, 2, 1}}
```

MATLAB

```
% A random permutation
>> randperm(3)
ans =
        2    1    3
```

Exercises 6.4

In Exercises 1–3 determine the sign and classify the permutations as odd and even.

1. $(2, 1, 3, 4)$, $(1, 4, 2, 3)$, $(1, 5, 2, 4, 3)$, $(1, 4, 3, 5, 2)$

2. $(2, 3, 1, 4)$, $(3, 1, 4, 2)$, $(2, 3, 4, 1, 5)$, $(5, 3, 1, 4, 2)$

3. $(3, 1, 4, 2)$, $(4, 2, 1, 3)$, $(3, 4, 2, 1, 5)$, $(4, 2, 5, 1, 3)$

In Exercises 4–9 compute the determinants of the matrices by using the complete expansion.

4. a. $\begin{bmatrix} 2 & 0 & 0 \\ 0 & 3 & 0 \\ 0 & 0 & 4 \end{bmatrix}$ **b.** $\begin{bmatrix} 2 & 0 & 0 \\ 0 & 0 & 3 \\ 0 & 4 & 0 \end{bmatrix}$

5. a. $\begin{bmatrix} 0 & 2 & 0 \\ 3 & 0 & 0 \\ 0 & 0 & 4 \end{bmatrix}$ **b.** $\begin{bmatrix} 0 & 2 & 0 \\ 0 & 0 & 3 \\ 4 & 0 & 0 \end{bmatrix}$

6. a. $\begin{bmatrix} 0 & 0 & 2 \\ 3 & 0 & 0 \\ 0 & 4 & 0 \end{bmatrix}$ **b.** $\begin{bmatrix} 0 & 0 & 2 \\ 0 & 3 & 0 \\ 4 & 0 & 0 \end{bmatrix}$

7. a. $\begin{bmatrix} 2 & 0 & 5 \\ 0 & 3 & 0 \\ 6 & 0 & 4 \end{bmatrix}$ **b.** $\begin{bmatrix} -2 & 0 & 5 \\ 0 & 3 & 0 \\ -6 & 0 & -4 \end{bmatrix}$

8. $\begin{bmatrix} 1 & 0 & 0 & 5 \\ 0 & 2 & 6 & 0 \\ 0 & 7 & 3 & 0 \\ 8 & 0 & 0 & 4 \end{bmatrix}$

9. $\begin{bmatrix} 1 & 0 & 0 & 0 & 0 \\ 0 & 0 & 0 & 0 & 2 \\ 0 & 3 & 0 & 0 & 0 \\ 0 & 0 & 0 & 4 & 0 \\ 0 & 0 & 5 & 0 & 0 \end{bmatrix}$

10. Compute the signs of the permutations by using crossing diagrams.

$$(3, 1, 2, 4) \qquad (4, 2, 3, 1)$$
$$(1, 4, 2, 3, 5) \qquad (5, 1, 2, 4, 3)$$

Permutation Matrices

A *permutation matrix* is a square matrix consisting of 1s and 0s such that there is exactly one 1 in each row and each column (see also Section 3.4). Matrices A, B, C, D are

permutation matrices:

$$A = \begin{bmatrix} 1 & 0 & 0 \\ 0 & 1 & 0 \\ 0 & 0 & 1 \end{bmatrix}, \qquad B = \begin{bmatrix} 1 & 0 & 0 \\ 0 & 0 & 1 \\ 0 & 1 & 0 \end{bmatrix}$$

$$C = \begin{bmatrix} 1 & 0 & 0 & 0 \\ 0 & 0 & 1 & 0 \\ 0 & 1 & 0 & 0 \\ 0 & 0 & 0 & 1 \end{bmatrix}$$

$$D = \begin{bmatrix} 0 & 1 & 0 & 0 & 0 \\ 1 & 0 & 0 & 0 & 0 \\ 0 & 0 & 0 & 1 & 0 \\ 0 & 0 & 1 & 0 & 0 \\ 0 & 0 & 0 & 0 & 1 \end{bmatrix}$$

A permutation matrix gives rise to one and only one permutation, as follows: For each row of the matrix, write the column number of the entry with 1. All these numbers form the entries of the corresponding permutation. For example, the permutations corresponding to A, B, C, D are $(1, 2, 3)$, $(1, 3, 2)$, $(1, 3, 2, 4)$, and $(2, 1, 4, 3, 5)$, respectively.

11. Write all 3×3 permutation matrices. For each such matrix write its corresponding permutation.

12. Write the permutation matrices corresponding to the following permutations.

$$(4, 1, 2, 3) \qquad (3, 2, 4, 1)$$
$$(4, 2, 3, 1, 5) \qquad (3, 5, 4, 2, 1)$$

13. Prove that the sign of a permutation equals the determinant of the corresponding permutation matrix.

14. Determine the signs of the following permutations by computing the determinant of the corresponding permutation matrix.

$$p = (1, 4, 2, 3) \qquad q = (3, 2, 1, 4)$$
$$s = (2, 1, 3, 4, 5) \qquad t = (1, 5, 4, 2, 3)$$

In the next two paragraphs we outline the proofs of Theorem 1 and the cofactor expansion by using permutations. For this material A and B are two $n \times n$ matrices with respective entries a_{ij} and b_{ij}.

Proof of Theorem 1

15. Let B be obtained from A by multiplying one of its rows by a nonzero constant k. Prove that $\det(B) = k \det(A)$. (*Hint:* If $kR_i \rightarrow R_i$ yields B from A, then $\det(B) = \sum \pm a_{1j_1} \ldots (ka_{ij_i}) \ldots a_{nj_n}$. Factor out k.)

16. Show that if we interchange any two *consecutive* entries in a permutation, then the number of inversions increases or decreases by 1.

17. Use Exercise 16 to show that if we interchange *any* two entries in a permutation the number of inversions changes by an *odd* integer.

18. Use Exercise 17 to show that if we interchange *any* two entries in a permutation, the new and the old permutations have opposite signs.

19. Let B be obtained from A by interchanging any two rows. Prove that $\det(B) = -\det(A)$. (*Hint:* If $R_i \leftrightarrow R_l$ ($i < l$) yields B from A, then $\det(B) = \sum \pm a_{1j_1} \ldots a_{lj_i} \ldots a_{ij_l} \ldots a_{nj_n}$. Use Exercise 18.)

20. If two rows of A are equal, show that $\det(A) = 0$. (*Hint:* Interchange the two equal rows and use Exercise 19.)

21. Let B be obtained from A by adding a multiple of one row to another. Then $\det(B) = \det(A)$. (*Hint:* If $R_i + kR_l \rightarrow R_i$ yields B from A, then $\det(B) = \sum \pm a_{1j_1} \ldots (a_{ij_i} + ka_{lj_i}) \ldots a_{nj_n}$. Separate into two sums. The second sum is zero by Exercise 20 after factoring out k.)

22. A and its transpose have the same determinant, $\det(A) = \det(A^T)$. (*Hint:* $\det(A^T) = \sum \pm a_{j_1 1} \ldots a_{j_n n}$. Rearrange $a_{j_1 1} \ldots a_{j_n n}$ in the form $a_{1l_1} \ldots a_{nl_n}$ and compare the signs of $(j_1, \ldots, j_n)$ and $(l_1, \ldots, l_n)$.)

23. Prove Theorem 1.

Proof of the Cofactor Expansion

Recall that the terms of the sum

$$\det(A) = \sum \pm a_{1j_1} \ldots a_{ij_i} \ldots a_{nj_n}$$

contain exactly one entry from each row and each column. Hence, the factor a_{i1} of the ith row occurs in exactly $(n-1)!$ terms, whereas a_{i2} occurs in $(n-1)!$ terms, distinct from the first ones, and finally a_{in} occurs in $(n-1)!$ terms, distinct from the preceding ones. Because the sum of all these terms is $\det(A)$, we may write

$$\det(A) = a_{i1}D_{i1} + a_{i2}D_{i2} + \cdots + a_{in}D_{in}$$

where D_{ij} is the sum in $\det(A)$ that is left after we factor a_{ij} out of all terms that contain it. In the next two exercises we prove that $D_{ij} = C_{ij}$, the (i, j) cofactor of A, thus proving the cofactor expansion of $\det(A)$ about the ith row.

$$\det(A) = a_{i1}C_{i1} + a_{i2}C_{i2} + \cdots + a_{in}C_{in} \qquad (6.2)$$

NOTATION   We denote by $A(i, j)$ the matrix obtained from A by deleting the ith row and the jth column.

24. Show that $D_{11} = C_{11}$. (*Hint:* $D_{11} = \sum \pm a_{2j_2} \ldots a_{nj_n}$, where the sum is over all permutations of the form $(j_2, \ldots, j_n)$, because $j_1 = 1$. But this is the determinant of $A(1, 1)$.)

25. Show that $D_{ij} = C_{ij}$. (*Hint:* Let A' be the matrix obtained from A by $i - 1$ successive interchanges of adjacent rows and $j - 1$ successive interchanges of adjacent columns that may bring a_{ij} into the top left position maintaining the relative orders of the other elements. Then $\det(A) = (-1)^{i+j} \det(A')$, by several applications of Part 3 of Theorem 1. Note that $a_{ij} = a'_{11}$ and that $\det(A(i, j)) = \det(A'(1, 1))$. Now use Exercise 24.)

26. Prove the cofactor expansion of $\det(A)$ about the ith row. (This is Formula (6.2).)

27. Prove the formula for the cofactor expansion of $\det(A)$ about the jth column.

$$\det(A) = a_{1j}C_{1j} + a_{2j}C_{2j} + \cdots + a_{nj}C_{nj}$$

(*Hint:* Use Exercises 22 and 26.)

6.5 Applications

Reader's Goal for This Section

To realize the wide range of applicability of determinants.

There are numerous applications of determinants to engineering, physics, mathematics and other sciences. In this section we discuss a few. The backbone of all these applications is Theorem 4.

Geometry: Equations of Geometric Objects

In analytic geometry determinants play a major role in computations of areas and volumes and in finding equations of geometric objects such as straight lines, circles, ellipses, parabolas, planes, and spheres. In this section we concentrate on the equations of some geometric objects.

Line Through Two Points

We can compute the equation of the straight line passing through two distinct points with coordinates (x_1, y_1), (x_2, y_2).

Let $ax + by + c = 0$ be the equation of the line. Because both points lie on the line, (x_1, y_1) and (x_2, y_2) should satisfy this equation. Hence,

$$ax + by + c = 0$$

$$ax_1 + by_1 + c = 0$$

$$ax_2 + by_2 + c = 0$$

This is a homogeneous system with unknowns a, b, c. By Theorem 4 it has a nontrivial solution if and only if

$$\begin{vmatrix} x & y & 1 \\ x_1 & y_1 & 1 \\ x_2 & y_2 & 1 \end{vmatrix} = 0 \tag{6.3}$$

This relation is, upon expansion of the determinant, the equation of the line.

■ **EXAMPLE 26**   Compute the equation of the line passing through the points $(1, 2)$, $(-2, 0)$.

SOLUTION   Substitution of the two points into Equation (6.3) yields

$$\begin{vmatrix} x & y & 1 \\ 1 & 2 & 1 \\ -2 & 0 & 1 \end{vmatrix} = 0$$

Expansion of the determinant gives $2x - 3y + 4 = 0$.

Three Points on the Same Line

We can show that a necessary and sufficient condition for three points with coordinates (x_1, y_1), (x_2, y_2), (x_3, y_3) to lie on the same line is

$$\begin{vmatrix} x_1 & y_1 & 1 \\ x_2 & y_2 & 1 \\ x_3 & y_3 & 1 \end{vmatrix} = 0 \tag{6.4}$$

According to (6.3), the line defined by (x_2, y_2), (x_3, y_3) has equation

$$\begin{vmatrix} x & y & 1 \\ x_2 & y_2 & 1 \\ x_3 & y_3 & 1 \end{vmatrix} = 0 \tag{6.5}$$

If the point with coordinates (x_1, y_1) is also on the line, then the substitution $x = x_1$, $y = y_1$ yields Equation (6.4).

Conversely, Equation (6.4) implies that the equation of the line through (x_2, y_2) and (x_3, y_3), (6.5), is satisfied if we let $x = x_1$ and $y = y_1$. Therefore, the point (x_1, y_1) is on the line defined by (x_2, y_2) and (x_3, y_3).

Circle through Three Points

As another application we compute the equation of the circle passing through the noncollinear points (x_1, y_1), (x_2, y_2) and (x_3, y_3).

Let $(x - a)^2 + (y - b)^2 = r^2$ be the equation of the circle of radius r centered at (a, b). This equation expanded can be written in the form $A(x^2 + y^2) + Bx + Cy + D = 0$. (Note that $A = 1$). Because the points (x_1, y_1), (x_2, y_2), (x_3, y_3) satisfy the equation, we have

$$\begin{aligned} A(x^2 + y^2) + Bx\ + Cy\ + D &= 0 \\ A(x_1^2 + y_1^2) + Bx_1 + Cy_1 + D &= 0 \\ A(x_2^2 + y_2^2) + Bx_2 + Cy_2 + D &= 0 \\ A(x_3^2 + y_3^2) + Bx_3 + Cy_3 + D &= 0 \end{aligned}$$

This can be viewed as a homogeneous system of four equations in four unknowns A, B, C, D (one of which is already known, namely, $A = 1$). By Theorem 4 a nontrivial solution should force the coefficient determinant to be zero. Therefore,

$$\begin{vmatrix} x^2 + y^2 & x & y & 1 \\ x_1^2 + y_1^2 & x_1 & y_1 & 1 \\ x_2^2 + y_2^2 & x_2 & y_2 & 1 \\ x_3^2 + y_3^2 & x_3 & y_3 & 1 \end{vmatrix} = 0 \tag{6.6}$$

This relation yields the equation of the circle.

■ EXAMPLE 27   Find the equation of the circle passing through $(1, 4)$, $(3, 2)$, and $(-1, 2)$.

SOLUTION   Substitution of the three points into Equation (6.6) gives

$$\begin{vmatrix} x^2 + y^2 & x & y & 1 \\ 17 & 1 & 4 & 1 \\ 13 & 3 & 2 & 1 \\ 5 & -1 & 2 & 1 \end{vmatrix} = 0$$

This equation reduces to

$$-8x^2 - 8y^2 + 32y + 16x - 8 = 0$$

Division by -8 and completion of the two squares results to an equation of a circle of radius 2 centered at $(1, 2)$.

$$(x - 1)^2 + (y - 2)^2 = 4$$

Plane Through Three Points

We can use determinants to compute the equation of the plane passing through the noncollinear points (x_1, y_1, z_1), (x_2, y_2, z_2), and (x_3, y_3, z_3).

Let $ax + by + cz + d = 0$ be the equation of the plane. Substitution of the three points yields

$$\begin{aligned} ax + by + cz + d &= 0 \\ ax_1 + by_1 + cz_1 + d &= 0 \\ ax_2 + by_2 + cz_2 + d &= 0 \\ ax_3 + by_3 + cz_3 + d &= 0 \end{aligned}$$

Once more we get a homogeneous system of four equations in four unknowns a, b, c, d. By Theorem 4 a nontrivial solution should make the coefficient determinant zero. Hence,

$$\begin{vmatrix} x & y & z & 1 \\ x_1 & y_1 & z_1 & 1 \\ x_2 & y_2 & z_2 & 1 \\ x_3 & y_3 & z_3 & 1 \end{vmatrix} = 0 \qquad (6.7)$$

This relation reduces to the equation of the plane.

■ EXAMPLE 28 Find the equation of the plane through the points $(1, 1, 7)$, $(3, 2, 6)$, and $(-2, -2, 4)$.

SOLUTION Substitution of the three points into Equation (6.7) results in

$$\begin{vmatrix} x & y & z & 1 \\ 1 & 1 & 7 & 1 \\ 3 & 2 & 6 & 1 \\ -2 & -2 & 4 & 1 \end{vmatrix} = 0$$

This equation reduces to

$$-6x + 9y + -3z + 18 = 0$$

or to

$$2x - 3y + z - 6 = 0$$

Parabola Through Three Points

Let us finally compute the equation of a parabola of the form $ay + bx^2 + cx + d = 0$ passing through three noncollinear points (x_1, y_1), (x_2, y_2), (x_3, y_3).

Substitution of the three points yields

$$ay + bx^2 + cx + d = 0$$
$$ay_1 + bx_1^2 + cx_1 + d = 0$$
$$ay_2 + bx_2^2 + cx_2 + d = 0$$
$$ay_3 + bx_3^2 + cx_3 + d = 0$$

Just as before, we get a homogeneous system of four equations in four unknowns, a, b, c, d. By Theorem 4 a nontrivial solution implies that the coefficient determinant is zero. Hence,

$$\begin{vmatrix} y & x^2 & x & 1 \\ y_1 & x_1^2 & x_1 & 1 \\ y_2 & x_2^2 & x_2 & 1 \\ y_3 & x_3^2 & x_3 & 1 \end{vmatrix} = 0 \qquad (6.8)$$

reduces to the equation of the parabola.

■ **EXAMPLE 29**   Find the equation of the parabola of the form $ay + bx^2 + cx + d = 0$ passing through the points with coordinates $(1, 0)$, $(2, 1)$ and $(0, 3)$.

SOLUTION   Substitution of the three points into Equation (6.8) gives

$$\begin{vmatrix} y & x^2 & x & 1 \\ 0 & 1 & 1 & 1 \\ 1 & 4 & 2 & 1 \\ 3 & 0 & 0 & 1 \end{vmatrix} = 0$$

Hence,

$$-2y + 4x^2 - 10x + 6 = 0 \quad \Rightarrow \quad y = 2x^2 - 5x + 3$$

Algebra: Elimination Theory and Resultants

Elimination theory is about eliminating a number of unknowns from a system of polynomial equations in one or more variables. Usually, the goal is to find the common solutions of these equations.[9]

Elimination theory flourished around the end of the nineteenth and the beginning of the twentieth century with Sylvester, Cayley, Dixon, and Macaulay (early contributors include Euler and Bezout). Though it is among the most elegant and useful mathematical theories, it went out of fashion for the best part of the twentieth century.[10] Now

---

[9] The special case when the polynomials are linear belongs, as we have seen, to the domain of linear algebra.

[10] See S. S. Abyankar's 1976 paper titled "Historical Ramblings in Algebraic Geometry and Related Algebra," *American Mathematical Monthly*, 83 (6): 409–448.

there is increasingly renewed interest, due partly to symbolic packages such as Maple and Mathematica and their needs to solve systems of polynomial equations efficiently. More importantly, because of the existence of such packages, exploration and further development of this theory is more feasible than ever before.

Suppose we have a system of two polynomial[11] equations in one variable, x. If we are interested in the common solutions, a crude approach would be to solve each equation separately (usually a difficult task) and then find the common roots of the solution sets. If we are interested only in the *existence* of solutions, then this method is impractical. This is also the case if we have to solve a system of two polynomial equations in two variables, x and y.

Let

$$a_1 x^2 + b_1 x + c_1 = 0$$
$$a_2 x^2 + b_2 x + c_2 = 0$$

be a system of two quadratics p_1 and p_2 in x. Let us find a necessary and sufficient condition for the existence of a common solution. If p_1 and p_2 have a common solution, they must have a common linear factor, say, Q. Let $q_1 = p_1/Q$ and $q_2 = p_2/Q$ be the two linear quotients and let $q_1 = A_1 x + B_1$ and $q_2 = -A_2 x - B_2$ (the peculiar choice of signs will make sense in just a moment). Then $Q = p_1/q_1 = p_2/q_2$; hence, $p_1 q_2 = p_2 q_1$. Explicitly,

$$(a_1 x^2 + b_1 x + c_1)(-A_2 x - B_2) = (a_2 x^2 + b_2 x + c_2)(A_1 x + B_1)$$

Expansion and collection of terms in powers of x yields

$$(a_1 A_2 + a_2 A_1)x^3 + (b_2 A_1 + b_1 A_2 + a_1 B_2 + a_2 B_1)x^2$$
$$+ (c_1 A_2 + c_2 A_1 + b_2 B_1 + b_1 B_2)x + c_1 B_2 + c_2 B_1 = 0$$

Because this polynomial equation is valid for all x, the coefficients of x^3, x^2, x^1, and x^0 should be zero. Hence,

$$a_1 A_2 + a_2 A_1 = 0$$
$$b_2 A_1 + b_1 A_2 + a_1 B_2 + a_2 B_1 = 0$$
$$c_1 A_2 + c_2 A_1 + b_2 B_1 + b_1 B_2 = 0$$
$$c_1 B_2 + c_2 B_1 = 0$$

This is a homogeneous system in A_2, B_2, A_1, B_1. By Theorem 4 it has nontrivial solutions if and only if the coefficient determinant is zero. So,

$$\begin{vmatrix} a_1 & 0 & a_2 & 0 \\ b_1 & a_1 & b_2 & a_2 \\ c_1 & b_1 & c_2 & b_2 \\ 0 & c_1 & 0 & c_2 \end{vmatrix} = 0$$

---

[11] In this section we write our polynomials in descending powers of x so that the end formulas match those in other books.

Most authors use the determinant of the transpose of this matrix. So, a necessary and sufficient condition for the existence of a common root is that

$$\begin{vmatrix} a_1 & b_1 & c_1 & 0 \\ 0 & a_1 & b_1 & c_1 \\ a_2 & b_2 & c_2 & 0 \\ 0 & a_2 & b_2 & c_2 \end{vmatrix} = 0$$

This determinant is called the **Sylvester resultant**[12] of p_1 and p_2. It has size 4, and it consists of the coefficients of the two polynomials padded by zeros. In general, the Sylvester resultant of two polynomials of degrees m and n has size $m + n$. For example, consider

$$a_1 x^2 + b_1 x + c_1 = 0$$

$$a_2 x^3 + b_2 x^2 + c_2 x + d_2 = 0$$

Then the Sylvester resultant of this system is of size $2 + 3 = 5$ and is given by

$$\begin{vmatrix} a_1 & b_1 & c_1 & 0 & 0 \\ 0 & a_1 & b_1 & c_1 & 0 \\ 0 & 0 & a_1 & b_1 & c_1 \\ a_2 & b_2 & c_2 & d_2 & 0 \\ 0 & a_2 & b_2 & c_2 & d_2 \end{vmatrix}$$

This determinant is zero if and only if the system has a common solution.

THEOREM 13

(Vanishing of the Sylvester Resultant)

Let f and g be two polynomials in x. The system $f = 0, g = 0$ has a solution if and only if the Sylvester resultant is zero, provided that not both coefficients of the highest powers of x are zero.

■ EXAMPLE 30   Without solving the equations, show that the following system has a solution.

$$x^2 - 5x + 6 = 0$$

$$x^2 + 2x - 8 = 0$$

SOLUTION   The Sylvester resultant of the system is

$$\begin{vmatrix} 1 & -5 & 6 & 0 \\ 0 & 1 & -5 & 6 \\ 1 & 2 & -8 & 0 \\ 0 & 1 & 2 & -8 \end{vmatrix} = 0$$

Hence, the system has a common root, by Theorem 13.

---

[12]Originally due to Euler, the current formulation is due to Sylvester.

NOTE   Example 30 served only as an illustration of the Sylvester resultant method and not as a demonstration of its power. We could certainly solve each of the quadratics and see that the common root is 2. What makes the method powerful, however, is that it is very general, and it imposes almost no restrictions on the coefficients of the polynomials. The coefficients themselves can be polynomials in another variable. This observation can be used to solve multivariate polynomial systems.

■ EXAMPLE 31   Solve the system

$$x^2 + y^2 - 1 = 0$$
$$x^2 - 2x + y^2 - 2y + 1 = 0$$

SOLUTION   Let us view this system as a system in y with coefficients polynomials in x.

$$y^2 + (x^2 - 1) = 0$$
$$y^2 - 2y + (x^2 - 2x + 1) = 0$$

By Theorem 13, if there is a solution, then the Sylvester resultant is zero.

$$\begin{vmatrix} 1 & 0 & x^2 - 1 & 0 \\ 0 & 1 & 0 & x^2 - 1 \\ 1 & -2 & x^2 - 2x + 1 & 0 \\ 0 & 1 & -2 & x^2 - 2x + 1 \end{vmatrix} = 0$$

Expansion of the determinant and simplification yields

$$8x^2 - 8x = 0$$

Therefore, $x = 0, 1$. If $x = 0$, substitution into the first equation implies $y = -1, 1$. However, substitution into the second equation implies that y can only be 1. Likewise, if $x = 1$, substitution into the system implies that $y = 0$. We conclude that there are two common solutions $(1, 0)$ and $(0, 1)$.

Note that the system can be rewritten as $x^2 + y^2 = 1$ and $(x-1)^2 + (y-1)^2 = 1$. So, we have the equations of the circles of radius 1 centered at $(0, 0)$ and $(1, 1)$, respectively. Their intersection is geometrically obvious, namely the points $(1, 0)$, and $(0, 1)$ (Fig. 6.3).

Electrical Engineering and Graph Theory: Spanning Trees

In 1847 in a famous paper G. R. Kirchhoff laid the foundations for the study of electrical circuits, known since as **Kirchhoff's laws**. Several of the many formulas presented there depend on only the geometry of the circuit and not on the resistors, inductors, or voltage sources present. To study geometric properties, Kirchhoff replaced the electrical circuit with the underlying graph (Fig. 6.4).

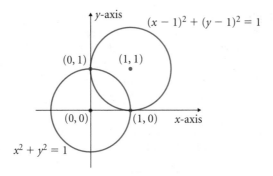

Figure 6.3   Intersecting circles.

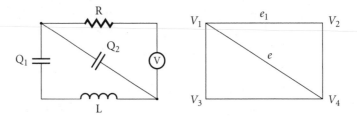

Figure 6.4   Electrical circuit and its underlying graph G.

Let G be a graph[13] with n vertices labeled $v_1, v_2, \ldots, v_n$. Suppose that G has no multiple edges or loops. Such a graph is called **simple**. A **circuit** of G at a vertex v_j is a set of distinct edges connecting vertices $v_{j_1}, v_{j_2}, \ldots, v_{j_k}$ such that $v_{j_1} = v_j$ and $v_{j_k} = v_j$ and the rest of the vertices are distinct. In other words, a circuit is a closed path with no self-crossings and edges that are traced only once. For example, the edges e_1, e_2, and e form a circuit at vertex V_1 in the graph of Fig. 6.4.

One of Kirchhoff's formulas asks for the number of spanning trees in a graph. A **spanning tree** of a graph G consists of *all* the vertices of G along with some edges such that

1. The spanning tree has no circuits;
2. Any two vertices can be connected by edges that belong to the spanning tree.

The **tree matrix** of the graph G is the $n \times n$ matrix T with entries t_{ij} defined by

$$t_{ij} := \begin{cases} \text{if } i = j, & t_{ii} = \text{number of edges incident to vertex } v_i \\ \text{if } i \neq j, & t_{ij} = \begin{cases} -1 & \text{if } v_i \text{ and } v_j \text{ are adjacent} \\ 0 & \text{if } v_i \text{ and } v_j \text{ are not adjacent} \end{cases} \end{cases}$$

NOTE   A tree matrix A is *symmetric*, i.e., $A = A^T$.

---

[13] Recall that some basics of graph theory were discussed in Chapter 3.

■ EXAMPLE 32   The tree matrix of the graph in Fig. 6.4 is

$$T = \begin{bmatrix} 3 & -1 & -1 & -1 \\ -1 & 2 & 0 & -1 \\ -1 & 0 & 2 & -1 \\ -1 & -1 & -1 & 3 \end{bmatrix}$$

The following theorem, due to Kirchhoff, counts the number of spanning trees of a graph. It is known as the *matrix tree theorem*.

THEOREM 14   (Matrix Tree Theorem)

All the cofactors of the tree matrix T of G are equal and their common value is the number of spanning trees of G.

■ EXAMPLE 33   Consider the graph of Fig. 6.4. The number of spanning trees of the graph of Fig. 6.4 is obtained by computing only one of the equal cofactors.

$$C_{11} = (-1)^{1+1}M_{11} = \begin{vmatrix} 2 & 0 & -1 \\ 0 & 2 & -1 \\ -1 & -1 & 3 \end{vmatrix} = 8$$

Therefore, the number of spanning trees is 8. In this case we can easily write down all the spanning trees and count them (Fig. 6.5).

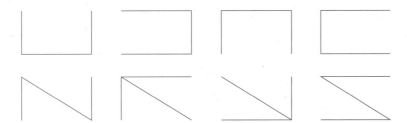

Figure 6.5   Spanning Trees of G.

You will appreciate the strength of Theorem 14 if you have to count the spanning trees of a graph like the one in Fig. 6.6. The tree matrix is

$$T = \begin{vmatrix} 2 & -1 & 0 & 0 & 0 & -1 \\ -1 & 4 & -1 & -1 & -1 & 0 \\ 0 & -1 & 4 & -1 & -1 & -1 \\ 0 & -1 & -1 & 3 & -1 & 0 \\ 0 & -1 & -1 & -1 & 4 & -1 \\ -1 & 0 & -1 & 0 & -1 & 3 \end{vmatrix}$$

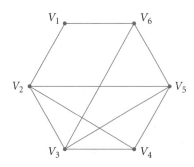

Figure 6.6

Computing the cofactor C_{11} yields $C_{11} = (-1)^{1+1}M_{11} = 115$. So, there are 115 spanning trees. Drawing and counting in this case is quite tedious.

Newton's Cows and Fields Problem

Suppose

1. a_1 cows graze b_1 fields bare in c_1 days;
2. a_2 cows graze b_2 fields bare in c_2 days;
3. a_3 cows graze b_3 fields bare in c_3 days.

Assuming that all the fields provide the same amount of grass, that the daily growth of the fields remains constant, and that the cows eat the same amount each day, what relation exists between the nine magnitudes $a_1, \ldots, c_3$?

Let x denote the initial amount of grass in each field, let y denote the daily growth, and let z denote the daily cow consumption. According to the assumptions, x, y, and z remain constant.

Consider Statement 1. In c_1 days, a_1 cows eat $a_1 c_1 z$ amount of grass. This amount equals the initial amount of grass, $b_1 x$, plus the amount that has grown in c_1 days, $b_1 c_1 y$. Hence $b_1 x + b_1 c_1 y = a_1 c_1 z$. Likewise, Statements 2 and 3 lead to $b_2 x + b_2 c_2 y = a_2 c_2 z$ and $b_3 x + b_3 c_3 y = a_3 c_3 z$, respectively. So we get the homogeneous system

$$b_1 x + b_1 c_1 y - a_1 c_1 z = 0$$

$$b_2 x + b_2 c_2 y - a_2 c_2 z = 0$$

$$b_3 x + b_3 c_3 y - a_3 c_3 z = 0$$

in unknowns x, y, z. By Theorem 4 the system has nontrivial solutions if and only if

$$\begin{vmatrix} a_1 c_1 & b_1 c_1 & b_1 \\ a_2 c_2 & b_2 c_2 & b_2 \\ a_3 c_3 & b_3 c_3 & b_3 \end{vmatrix} = 0$$

This is the required necessary condition that relates the seemingly unrelated nine numbers $a_1, \ldots, c_3$.

Resultants with CAS

Maple

```
> resultant(a*x^2+b*x+c,2*a*x+b,x);
```
$$4a^2c - b^2a$$

Mathematica

```
In[1]:=
    Resultant[a x^2+b x+c, 2a x+b, x]
Out[1]=
           2        2
       -(a b ) + 4 a  c
```

Exercises 6.5

In Exercises 1–4 find the equation of the line passing through P and Q.

1. $P(-1, 2)$ and $Q(1, 1)$

2. $P(2, 1)$ and $Q(1, -1)$

3. $P(0, 1)$ and $Q(6, -8)$

4. $P(-3, -5)$ and $Q(4, 7)$

In Exercises 5–6 determine whether the points P, Q, and R are on the same line.

5. $P(-2, 0)$, $Q(0, 1)$, and $R(2, 1)$

6. $P(-1, 2)$, $Q(0, 0)$, and $R(2, -1)$

In Exercises 7–10 find the equation, the center, and the radius of the circle passing through P, Q, and R.

7. $P(0, 0)$, $Q(-1, -1)$, and $R(0, -2)$

8. $P(2, 2)$, $Q(4, 0)$, and $R(6, 2)$

9. $P(5, 5)$, $Q(1, -1)$, and $R(0, 0)$

10. $P(7, 7)$, $Q(1, 1)$, and $R(-3, 2)$

In Exercises 11–13 find the equation of the parabola of the form $y = ax^2 + bx + c$ passing through P, Q, and R.

11. $P(0, 4)$, $Q(1, 3)$, and $R(-1, 9)$

12. $P(2, 0)$, $Q(0, 1)$, and $R(1, 0)$

13. $P(2, 2)$, $Q(3, 1)$, and $R(1, 7)$

In Exercises 14–16 find the equation of the plane passing through P, Q, and R.

14. $P(1, 1, 1)$, $Q(0, -1, 1)$, and $R(4, 3, -1)$

15. $P(-1, 1, 1)$, $Q(-1, 4, 3)$, and $R(4, 0, 2)$

16. $P(5, 4, 3)$, $Q(-1, 2, 2)$, and $R(4, 4, 4)$

17. Use the Sylvester resultant to solve the system

$$x^2 + y^2 - 1 = 0$$
$$x^2 + 2x + y^2 - 2y + 1 = 0$$

18. Use the Sylvester resultant to solve the system

$$x^2 + y^2 - 1 = 0$$
$$x^2 - 2x + y^2 - 1 = 0$$

19. Find the equation of the sphere passing through $(1, 2, 7)$, $(5, 2, 3)$, $(1, 6, 3)$, $(1, 2, -1)$. (*Hint:* Recall that the equation of a sphere is of the form $(x - a)^2 + (y - b)^2 + (z - c)^2 = r^2$ or $A(x^2 + y^2 + z^2) + Bx + Cy + Dz + E = 0$ with $A = 1$. Form a homogeneous system with unknowns A, B, C, D, E).

In Exercises 20–22 draw the underlying graphs of the given electrical circuits.

20. For Fig. 6.7

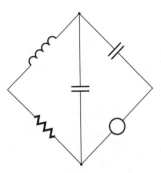

Figure 6.7

21. For Fig. 6.8

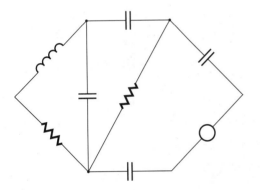

Figure 6.8

22. For Fig. 6.9

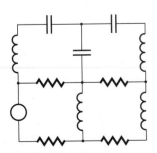

Figure 6.9

23. a. Draw all the spanning trees of the graph G_1 of Fig. 6.10 and count them.

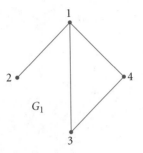

Figure 6.10

b. Write the tree matrix of the graph and use Theorem 14 to compute the number of spanning trees.

24. Write the tree matrix of the graph G_2 in Fig. 6.11 and use Theorem 14 to compute the number of the spanning trees.

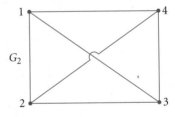

Figure 6.11

6.6 Miniprojects

1 ■ Determinants of Block Matrices

Let

$$A = \begin{bmatrix} 1 & 2 \\ 3 & 4 \end{bmatrix}, \qquad B = \begin{bmatrix} 5 & 6 \\ 7 & 8 \end{bmatrix}, \qquad C = \begin{bmatrix} 1 & 1 & 4 \\ 2 & 0 & -2 \\ 3 & 1 & 1 \end{bmatrix}$$

and let

$$D = \begin{bmatrix} 1 & 2 & 0 & 0 \\ 3 & 4 & 0 & 0 \\ 0 & 0 & 5 & 6 \\ 0 & 0 & 7 & 8 \end{bmatrix}, \qquad E = \begin{bmatrix} 1 & 2 & 0 & 0 & 0 \\ 3 & 4 & 0 & 0 & 0 \\ 0 & 0 & 1 & 1 & 4 \\ 0 & 0 & 2 & 0 & -2 \\ 0 & 0 & 3 & 1 & 1 \end{bmatrix}$$

Note that D is the block matrix formed by placing A and B along the block diagonal and padding the rest of the entries with zeros. E was formed from A and C in the same manner.

Problem A

Find a relation between

1. $\det(A)$, $\det(B)$ and $\det(D)$;
2. $\det(A)$, $\det(C)$ and $\det(E)$.

In general, let $A_1, \ldots, A_m$ denote m square matrices, not necessarily of the same size. Let A be the block diagonal matrix with $A_1, A_2, \ldots, A_m$ on the main diagonal.

$$A = \begin{bmatrix} A_1 & \mathbf{0} & \cdots & \mathbf{0} \\ \mathbf{0} & A_2 & & \mathbf{0} \\ \vdots & & \ddots & \vdots \\ \mathbf{0} & \cdots & \mathbf{0} & A_m \end{bmatrix}$$

Problem B

Show that the determinant of the block matrix A is the product of the determinants of the matrices on the diagonal.

$$\det(A) = \begin{vmatrix} A_1 & \mathbf{0} & \cdots & \mathbf{0} \\ \mathbf{0} & A_2 & & \mathbf{0} \\ \vdots & & \ddots & \vdots \\ \mathbf{0} & \cdots & \mathbf{0} & A_m \end{vmatrix} = \det(A_1)\det(A_2)\cdots\det(A_m)$$

(*Hint:*   First observe that, by careful row reduction, the determinant of any square matrix equals the determinant of an upper triangular matrix. Then, the determinant of an upper triangular matrix is the product of the main diagonal entries.)

Let A, B, C, and D be $n \times n$ matrices and let us form the $2n \times 2n$ block matrix

$$\mathbf{M} = \begin{bmatrix} A & B \\ C & D \end{bmatrix}$$

We can compute the determinant of $\mathbf{M}$ according to the following proposition, which we state without proof.

Proposition 15

If A, B, C, D commute with each other, i.e., if

$$AB = BA \qquad AC = CA \qquad AD = DA$$
$$BC = CB \qquad BD = DB \qquad CD = DC$$

then

$$\det \begin{bmatrix} A & B \\ C & D \end{bmatrix} = \det(AD - BC)$$

Problem C

1. Verify both the assumptions and the conclusion of Proposition 15 for

$$A = \begin{bmatrix} -1 & 0 \\ 0 & -1 \end{bmatrix}, \qquad B = \begin{bmatrix} 1 & 1 \\ 0 & 0 \end{bmatrix}, \qquad C = \begin{bmatrix} 4 & 1 \\ 0 & 3 \end{bmatrix}, \qquad D = \begin{bmatrix} 2 & 0 \\ 0 & 2 \end{bmatrix}$$

2. Find 2×2 matrices E, F, G, H where Proposition 15 does not apply.

2 ■ Vandermonde Determinants

Let

$$A = \begin{bmatrix} 1 & 1 & 1 \\ 2 & 3 & 5 \\ 4 & 9 & 25 \end{bmatrix}$$

Notice that the entries in each column are powers: $2^0 = 1$, $2^1 = 2$, $2^2 = 4$ for the first column, $3^0 = 1$, $3^1 = 3$, $3^2 = 9$ for the second column, and $5^0 = 1$, $5^1 = 5$, $5^2 = 25$ for the third column. A matrix with this property is called a *Vandermonde matrix*.

DEFINITION

An $n \times n$ matrix A_n is a **Vandermonde matrix** if there are numbers $x_1, x_2, \ldots, x_n$, such that

$$A_n = \begin{bmatrix} 1 & 1 & 1 & \cdots & 1 \\ x_1 & x_2 & x_3 & \cdots & x_n \\ x_1^2 & x_2^2 & x_3^2 & \cdots & x_n^2 \\ \vdots & \vdots & \vdots & \ddots & \vdots \\ x_1^{n-1} & x_2^{n-1} & x_3^{n-1} & \cdots & x_n^{n-1} \end{bmatrix}$$

There is a simple formula for the determinant of a Vandermonde matrix.

Proposition 16 (Vandermonde's Determinant)

The determinant V_n of the Vandermonde matrix A_n is given by

$$V_n = \det(A_n) = \prod_{1 \le i < j \le n} (x_j - x_i)$$

In other words,

$$V_n = (x_n - x_{n-1})(x_n - x_{n-2}) \cdots (x_n - x_2)(x_n - x_1) \times$$
$$(x_{n-1} - x_{n-2})(x_{n-1} - x_{n-3}) \cdots (x_{n-1} - x_2)(x_{n-1} - x_1) \times$$
$$\cdots\cdots\cdots\cdots\cdots\cdots\cdots\cdots\cdots\cdots\cdots\cdots\cdots\cdots\cdots$$
$$(x_4 - x_3)(x_4 - x_2)(x_4 - x_1) \times$$
$$(x_3 - x_2)(x_3 - x_1) \times$$
$$(x_2 - x_1)$$

Problem A

1. Verify Proposition 16 for the following Vandermonde matrices.

$$A = \begin{bmatrix} 1 & 1 & 1 \\ 2 & 3 & 5 \\ 4 & 9 & 25 \end{bmatrix}, \qquad B = \begin{bmatrix} 1 & 1 & 1 \\ 1 & -1 & 2 \\ 1 & 1 & 4 \end{bmatrix}$$

2. Use Proposition 16 to compute the determinant of the following Vandermonde matrices.

$$A = \begin{bmatrix} 1 & 1 & 1 \\ 10 & 11 & 12 \\ 100 & 121 & 144 \end{bmatrix}, \qquad B = \begin{bmatrix} 1 & 1 & 1 & 1 \\ 1 & 2 & 3 & 4 \\ 1 & 4 & 9 & 16 \\ 1 & 8 & 27 & 64 \end{bmatrix}$$

3. Use Proposition 16 to compute the determinant of the following matrices.

$$A = \begin{bmatrix} 1 & 5 & 25 \\ 1 & 9 & 81 \\ 1 & 12 & 144 \end{bmatrix}, \qquad B = \begin{bmatrix} 1 & 1 & 1 & 1 \\ 1 & 3 & 9 & 27 \\ 1 & 5 & 25 & 125 \\ 1 & 7 & 49 & 343 \end{bmatrix}$$

Problem B

Find a necessary and sufficient condition that a Vandermonde matrix has determinant zero.

Problem C

Prove Proposition 16.
    (*Hints*:

1. Perform the following operations

$$R_n - x_n R_{n-1} \rightarrow R_n$$

$$R_{n-1} - x_n R_{n-2} \rightarrow R_{n-1}$$

$$R_{n-2} - x_n R_{n-3} \rightarrow R_{n-2}$$

$$\vdots$$

$$R_2 - x_n R_1 \rightarrow R_2$$

to get a matrix whose ith column is

$$
\begin{bmatrix}
1 \\
x_i - x_n \\
x_i(x_i - x_n) \\
x_i^2(x_i - x_n) \\
\vdots \\
x_i^{n-2}(x_i - x_n)
\end{bmatrix}
\quad \text{if} \quad i < n \quad \text{and} \quad
\begin{bmatrix}
1 \\
0 \\
0 \\
0 \\
\vdots \\
0
\end{bmatrix}
\quad \text{if } i = n
$$

2. Expand the resulting determinant about the nth column to obtain the $(1, n)$ cofactor $C_{1,n} = (-1)^n M_{1,n}$.
3. Use Theorem 1 to factor the product

$$(x_1 - x_n)(x_2 - x_n) \cdots (x_{n-1} - x_n)$$

out of the minor $M_{1,n}$. The leftover determinant is just V_{n-1}. Continue this process.)

3 ■ Bezout Resultant

This project is a quick introduction to the increasingly important resultant of Bezout[14] (1774) of a system of two polynomials in one variable. We discuss *Cayley's statement* of the Bezout method.

Let $f(x)$ and $g(x)$ be two polynomials. We want to find a necessary condition that the system $f(x) = 0, g(x) = 0$ has a common solution. Let a be a second variable independent of x. By $f(a)$ and $g(a)$, we denote the polynomials $f(x)$ and $g(x)$, with x replaced by a. Consider the 2×2 determinant

$$\Delta(x, a) = \begin{vmatrix} f(x) & g(x) \\ f(a) & g(a) \end{vmatrix} = f(x)g(a) - g(x)f(a)$$

Note that Δ is zero for any common solution, x, of the system $f(x) = 0, g(x) = 0$. Moreover, $\Delta = 0$ if $x = a$. Therefore, $x - a$ divides $\Delta(x, a)$ exactly. Hence the quotient

$$\delta(x, a) = \frac{\Delta(x, a)}{x - a} = \frac{f(x)g(a) - g(x)f(a)}{x - a}$$

is zero for any solution of the original system. $\delta(x, a)$ is a polynomial in a and x. For any common zero of the system, say $x = x_0$, $\delta(x_0, a)$ is zero for all a; therefore, the coefficients of the powers of a in $\delta(x, a)$ are identically zero. Setting the coefficients of the powers of a in $\delta(x, a)$ equal to zero results in a homogeneous system in x. This system has nontrivial solutions if the determinant of the coefficient matrix is zero. This last

---

[14]Especially the multivariate version due to Dixon (1908).

determinant is called the **Bezout resultant** of the system. If the original system has a common zero, then the Bezout resultant is zero.

For example, let

$$f(x) = x^2 - 5x + 6$$

$$g(x) = x^2 + 2x - 8$$

Then

$$\delta(x, a) = \frac{1}{x - a} \begin{vmatrix} x^2 - 5x + 6 & x^2 + 2x - 8 \\ a^2 - 5a + 6 & a^2 + 2a - 8 \end{vmatrix}$$

$$= \frac{1}{x - a} \left[(x^2 - 5x + 6)(a^2 + 2a - 8) - (x^2 + 2x - 8)(a^2 - 5a + 6) \right]$$

which simplifies to

$$\delta(x, a) = 7ax - 14x - 14a + 28$$

For any common zero of the system, $\delta(x, a) = 0$ for all a. Hence, the coefficients of all powers of a must be zero. Because the coefficient of a^0 is $28 - 14x$ and the coefficient of a^1 is $-14 + 7x$, we have

$$28 - 14x = 0$$

$$-14 + 7x = 0$$

The determinant of the coefficient matrix of this system is the Bezout resultant. The system has a common solution, because the Bezout resultant is zero.

$$\begin{vmatrix} 28 & -14 \\ -14 & 7 \end{vmatrix} = 0$$

N O T E   If the two polynomials f and g are of the same degree, the Bezout and the Sylvester resultants are identical. The Bezout resultant is often preferred, because the size of the determinant Δ ($= \max(\deg(f), \deg(g))$) is much smaller than that of the Sylvester determinant ($= \deg(f) + \deg(g)$).

Problem A

The Bezout method can be used to eliminate one variable out of a system of two polynomial equations in two variables. Consider

$$x^2 + y^2 - 1 = 0$$

$$x^2 - 2x + y^2 - 2y + 1 = 0$$

as a system in y with coefficients polynomials in x:

$$y^2 + (x^2 - 1) = 0$$

$$y^2 - 2y + (x^2 - 2x + 1) = 0$$

1. Use the Bezout method to eliminate y.
2. Set the Bezout resultant equal to zero and solve for x.
3. Back substitute into the original system and find all the common zeros.

Problem B

Repeat the process of Problem A for the following system and conclude that there are no common solutions.

$$x^2 + y^2 - 1 = 0$$

$$x^2 - 6x + y^2 - 2y + 6 = 0$$

4 ■ Cross Products in $\mathbf{R}^n$

In this project we discuss an interesting generalization of the cross product[15] in $\mathbf{R}^n$.

Let $\mathbf{e}_1, \mathbf{e}_2, \ldots, \mathbf{e}_n$ denote the usual basis vectors of $\mathbf{R}^n$ and assume $n \geq 3$. Fix any $n - 3$ vectors $\mathbf{a}_1, \ldots, \mathbf{a}_{n-3}$ in $\mathbf{R}^n$ and define a product $\times$ on $\mathbf{R}^n$ by the determinant

$$\mathbf{u} \times \mathbf{v} = \begin{vmatrix} \mathbf{e}_1 & \cdots & \mathbf{e}_n \\ & \mathbf{a}_1 & \\ & \vdots & \\ & \mathbf{a}_{n-3} & \\ & \mathbf{u} & \\ & \mathbf{v} & \end{vmatrix}, \quad \mathbf{u}, \mathbf{v} \in \mathbf{R}^n$$

■ EXAMPLE 34   Let $n = 4$, $\mathbf{a}_1 = (a_1, a_2, a_3, a_4)$, $\mathbf{u} = (u_1, u_2, u_3, u_4)$, and $\mathbf{v} = (v_1, v_2, v_3, v_4)$. Then in $\mathbf{R}^4$ we have

$$\mathbf{u} \times \mathbf{v} = \begin{vmatrix} \mathbf{e}_1 & \mathbf{e}_2 & \mathbf{e}_3 & \mathbf{e}_4 \\ a_1 & a_2 & a_3 & a_4 \\ u_1 & u_2 & u_3 & u_4 \\ v_1 & v_2 & v_3 & v_4 \end{vmatrix}$$

$$= (a_2 u_3 v_4 - a_2 u_4 v_3 - u_2 a_3 v_4 + u_2 a_4 v_3 + v_2 a_3 u_4 - v_2 a_4 u_3)\mathbf{e}_1$$

$$+ (-v_4 u_3 a_1 + v_3 u_4 a_1 + v_4 a_3 u_1 - v_3 a_4 u_1 - u_4 a_3 v_1 + u_3 a_4 v_1)\mathbf{e}_2$$

$$+ (v_4 a_1 u_2 - u_4 a_1 v_2 - v_4 u_1 a_2 + a_4 u_1 v_2 + u_4 v_1 a_2 - a_4 v_1 u_2)\mathbf{e}_3$$

$$+ (-v_3 a_1 u_2 + u_3 a_1 v_2 + v_3 u_1 a_2 - a_3 u_1 v_2 - u_3 v_1 a_2 + a_3 v_1 u_2)\mathbf{e}_4$$

■ EXAMPLE 35   If $\mathbf{a}_1 = (1, 1, 1, 1)$, then

$$(2, 5, 10, 17) \times (3, 1, -1, -3) = (4, -12, 12, -4)$$

Problem A

1. If $\mathbf{a}_1 = (1, -1, 1, 1)$, find $(2, 5, 2, -3) \times (2, 3, -1, -3)$ in $\mathbf{R}^4$.
2. If $\mathbf{a}_1 = (1, 1, 1, 1, 1)$ and $\mathbf{a}_2 = (1, -1, 1, 0, 1)$, find $(2, 0, 2, -3, 1) \times (1, 0, 3, -1, -3)$ in $\mathbf{R}^5$.

---

[15]For more details on this subject, see Dittmer's article in the *American Mathematical Monthly* (November 1994) and the references cited there.

Problem B

Fix vectors $\mathbf{a}_1, \ldots, \mathbf{a}_{n-3}$ in $\mathbf{R}^n$. Use properties of determinants[16] to prove that for any vectors $\mathbf{u}$, $\mathbf{v}$, and $\mathbf{w}$ in $\mathbf{R}^n$, the following properties hold:

1. $\mathbf{u} \times \mathbf{v} = -\mathbf{v} \times \mathbf{u}$
2. $\mathbf{u} \times (\mathbf{v} + \mathbf{w}) = \mathbf{u} \times \mathbf{v} + \mathbf{u} \times \mathbf{w}$
3. $(\mathbf{u} + \mathbf{v}) \times \mathbf{w} = \mathbf{u} \times \mathbf{w} + \mathbf{v} \times \mathbf{w}$
4. $k(\mathbf{u} \times \mathbf{v}) = (k\mathbf{u}) \times \mathbf{v} = \mathbf{u} \times (k\mathbf{v})$
5. $\mathbf{0} \times \mathbf{u} = \mathbf{u} \times \mathbf{0} = \mathbf{0}$
6. $\mathbf{u} \times \mathbf{u} = \mathbf{0}$

Problem C

The generalized cross product is **nonassociative**; i.e., for any choice of nonzero $\mathbf{a}_i$s there are n-vectors $\mathbf{u}$, $\mathbf{v}$, and $\mathbf{w}$ such that

$$\mathbf{u} \times (\mathbf{v} \times \mathbf{w}) \neq (\mathbf{u} \times \mathbf{v}) \times \mathbf{w}$$

Prove this claim for $n = 4$.

6.7 Computer Exercises

This computer section helps you learn how to compute determinants with the software you use to do the computational parts of the exercises. It also helps review some of the basic material of this chapter. Note that an exercise designated as [S] requires symbolic manipulation.[17]

Let

$$A = \begin{bmatrix} 6 & 7 & 1 \\ 6 & -7 & 2 \\ 6 & 7 & 3 \end{bmatrix}, \qquad B = \begin{bmatrix} \frac{1}{3} & \frac{1}{4} & \frac{1}{5} \\ \frac{1}{4} & \frac{1}{4} & \frac{1}{5} \\ \frac{1}{5} & \frac{1}{5} & \frac{1}{5} \end{bmatrix}$$

$$C = \begin{bmatrix} 1 & 3 & 5 \\ 7 & 9 & 11 \\ 13 & 15 & 17 \end{bmatrix}, \qquad D = \begin{bmatrix} a & b & c \\ d & e & f \\ g & h & i \end{bmatrix}$$

1. Compute the determinants of A, B, and C. Which of these matrices are invertible?

2. Find $\det(B)$ by cofactor expansion about the second column.

3. Compute (a) $\det(A^5) - (\det(A))^5$, (b) $\det(A) - \det(A^T)$, (c) $\det(AB) - \det(A)\det(B)$. Explain in each case why the answer should be zero.

4. Compute (a) $\det(5B) - 125\det(B)$, (b) $\det(B^{-1}) - 1/\det(B)$, (c) $\det(B^{-2}) - 1/\det(B)^2$. Explain in each case why the answer should be zero.

5. [S] Compute and compare $\det(D)$ and $\det(D^T)$. Repeat with $\det(D^T D)$ and $\det(D)^2$. Explain the comparisons.

---

[16] Even though the first row is a row of vectors in the cross product.
[17] Skip these exercises if symbolic evaluation is not available.

Let M_n be the following matrix sequence:

$$M_2 = \begin{bmatrix} 1 & 1 \\ 1 & 2 \end{bmatrix}, \qquad M_3 = \begin{bmatrix} 1 & 1 & 1 \\ 1 & 2 & 1 \\ 1 & 1 & 3 \end{bmatrix},$$

$$M_4 = \begin{bmatrix} 1 & 1 & 1 & 1 \\ 1 & 2 & 1 & 1 \\ 1 & 1 & 3 & 1 \\ 1 & 1 & 1 & 4 \end{bmatrix}, \ldots$$

6. Define M_n as a function in n. Use this function to compute $\det(M_2)$, $\det(M_3)$, $\det(M_4)$,.... Do you see a pattern for $\det(M_n)$?

7. Compute $\det(S^T S)$ for several 3×2 matrices S. Find a connection between $S^T S$ being invertible and the linear dependence or independence of the columns of S.

8. Solve the system (a) by matrix inversion and (b) by Cramer's rule. (c) Compute the adjoint of the coefficient matrix.

$$\begin{aligned} -85x - 55y - 37z &= -306 \\ -35x + 97y + 50z &= 309 \\ 79x + 56y + 49z &= 338 \end{aligned}$$

9. Let $A\mathbf{x} = \mathbf{b}$ be a square system. Write the code for two functions, `CramerDisplay` and `CramerSolve`, each having three arguments, A, $\mathbf{b}$, and i. `CramerDisplay` is to return the matrix A_i obtained from A by replacing the ith column with $\mathbf{b}$. `CramerSolve` is to solve for x_i by Cramer's rule. Test the code by displaying A_1, A_2, and A_3 and solving the preceding system.

10. If the appropriate command is available, find *all* permutations of $\{1, 2, 3, 4\}$. Check that the number of permutations found is the correct one.

11. If the appropriate command is available find the signs of the permutations: $\{1, 4, 2, 3\}$, $\{4, 2, 3, 1\}$.

12. If the appropriate command is available, find a random permutation of $\{1, 2, 3, 4\}$.

13. [S] Using determinants, define a function $f(x_1, y_1, x_2, y_2, x_3, y_3)$ that computes the equation in x and y of a circle passing through three points (x_1, y_1), (x_2, y_2), (x_3, y_3).

14. [S] Use f to find the equation of the circle C_1 through $(-1, 1)$, $(1, 1)$, $(2, 4)$.

15. [S] Find the point(s) with x-coordinate -2 on the circle C_1 above. Also show that C_1 has no points with x-coordinate -3.

16. Using determinants, define a function $g(x_1, y_1, x_2, y_2, x_3, y_3)$ that tests whether or not the four points with coordinates (x, y), (x_1, y_1), (x_2, y_2), (x_3, y_3) lie on the same circle.

17. Use g to check whether the points $A(-2, 2)$, $B(-2, 4)$, and $C(-1, 2)$ lie on the circle C_1.

18. Let

$$p_1 = x^4 - 3x^3 - 63x^2 - 85x + 150$$

$$p_2 = x^4 + 22x^3 - 103x^2 - 2740x + 5700$$

$$p_3 = x^5 - 4x^3 + 8x^2 - 12x - 1$$

Compute the resultants of the polynomial pairs (p_1, p_2), (p_1, p_3) and (p_2, p_3). Which pairs have a common solution?

19. Compute the number of spanning trees of the graph formed by a regular hexagon and all its diagonals.

Determinants and Permutations in Maple-Mathematica-MATLAB

Let A be any square matrix and let n be any positive integer. We have:

function	Maple	Mathematica	MATLAB
determinant of A	det(A);	Det[A]	det(A)
permutation of $\{1,\dots,n\}$	permute(n);	Permutations[n]	
sign of permutation		Signature[p]	
random permutation	randperm(n);		randperm(n)
resultant	resultant	Resultant	
Sylvester matrix	sylvester		

In Maple permute(n) returns all permutations of $\{1,2,\dots,n\}$. First you need to load the package combinat by typing with(combinat);. randperm(n) returns a random permutation of $\{1,2,\dots,n\}$. The commands resultant and sylvester are in the linalg package.

In Mathematica, Permutations[n] returns all permutations of $\{1,2,\dots,n\}$. Signature[$\{2,5,4,1,3\}$] returns the sign of permutation $\{2,5,4,1,3\}$.

In MATLAB randperm(n) returns a random permutation of $\{1,2,\dots,n\}$.

Selected Solutions with Maple

```
with(linalg):
A := matrix([[6,7,1],[6,-7,2],[6,7,3]]);                    # DATA
B := matrix([[1/3,1/4,1/5],[1/4,1/4,1/5],[1/5,1/5,1/5]]);
C := matrix([[1,3,5],[7,9,11],[13,15,17]]);
D1 := matrix([[a,b,c],[d,e,f],[g,h,i]]);      # D is used for differentiation.
# Exercises 1-4.
det(A); det(B); det(C);    # C is the only noninvertible, since det(C)=0.
-(1/4)*det(minor(B,1,2))+(1/4)*det(minor(B,2,2))-(1/5)*det(minor(B,3,2));
det(A^5)-det(A)^5; det(A)-det(transpose(A)); det(A &* B)-det(A) * det(B);
det(5*B)-125*det(B); det(inverse(B))-1/det(B); det(B^(-2))-1/det(B)^2;
# Exercise 5.
det(D1); det(transpose(D1));
det(transpose(D1) &* D1); det(D1)^2;
""-expand(");                       # Expand the difference to get 0.
# Exercise 6.
m:=proc(n) matrix(n,n, (i,j)->if i=j then i else 1 fi) end:    # M_n.
det(m(2)); det(m(3)); det(m(4));                    # Etc..
#  Pattern:   det (M_n) = (n-1)!
# Exercise 7 - Comment.
# (S^T)S is invertible only if the columns of S are linearly independent.
```

```
# Exercise 8
A:=matrix([[-85,-55,-37],[-35,97,50],[79,56,49]]);
b:=vector([-306,309,338]);
sol:=evalm(inverse(A)&*b);          # (a)
A1:=delcols(A,1..1);                # Deletes column 1 from A
A1:=augment(b,A1);                  # then adds b as a first column
A2:=delcols(A,1..2);                # the last column of A
A2:=augment(b,A2);                  # prepended by b
A22:=delcols(A,2..3);               # the first column of A
A2:=augment(A22,A2);                # joined by the columns of A22
A3:=delcols(A,3..3);                # deletes column 3 from A
A3:=augment(A3,b);                  # then adds b as a last column
x:=det(A1)/det(A);                  # (b)
y:=det(A2)/det(A);                  # (b)
z:=det(A3)/det(A);                  # (b)
adj:=adjoint(A);                    # (c)
# Exercise 9.
CramerDisplay := proc (A,b,i) local AA, j; AA:= copy(A);
                  for j from 1 to rowdim(A) do
                    AA[j,i]:=b[j] od:                  # Replacing the ith
                    evalm(AA)                          # column with b.
                  end:
CramerSolve := proc (A,b,i) det(CramerDisplay(A,b,i))/det(A) end:
CramerDisplay(A,b,1); CramerDisplay(A,b,2); CramerDisplay(A,b,3);
CramerSolve(A,b,1); CramerSolve(A,b,2); CramerSolve(A,b,3);
# Exercises 10,12.
with(combinat);               # Loading the combinat package.
permute(4);                   # The permutations of {1,2,3,4}.
nops(");                      # The number of computed permutations.
4!;                           # The expected answer.
randperm(4);                  # A random permutation of {1,2,3,4}.
# Exercise 13.
circleeqn := proc(x1,y1,x2,y2,x3,y3)
                det(matrix(4,4,[x^2+y^2,x,y,1,
                               x1^2+y1^2,x1,y1,1,
                               x2^2+y2^2,x2,y2,1,
                               x3^2+y3^2,x3,y3,1]))
             end:
# Exercise 14.
ce:=circleeqn(-1,1,1,1,2,4);     # The equation of the circle.
# Exercise 17.
subs(x=-2,y=2,ce);               # Substitution of the points into the
subs(x=-2,y=4,ce);               # equation of the circle.
subs(x=-1,y=4,ce);               # Last point not on the circle.
# Exercise 18 - Partial
p1:=x^4-3*x^3-63*x^2-85*x+150;
```

```
p2:=x^4+22*x^3-103*x^2-2740*x+5700;
resultant(p1,p2,x);           # We also need to declare the variable x.
```

Note   Maple includes a `circle` command that computes, among other things, the equation of a circle. This command is part of the geometry package, which should be loaded first.

Selected Solutions with Mathematica

```
A = {{6,7,1},{6,-7,2},{6,7,3}}                              (* DATA *)
B = {{1/3,1/4,1/5},{1/4,1/4,1/5},{1/5,1/5,1/5}}
C1 = {{1,3,5},{7,9,11},{13,15,17}}       (* C is used in differential eqn. contants. *)
D1 = {{a,b,c},{d,e,f},{g,h,i}}           (* D is used for differentiation.         *)
(* Exercises 1-4. *)
Det[A]
Det[B]
Det[C1]                            (* C is the only noninvertible, since Det[C]=0.*)
(* Watch for the reverse numbering in the Minors command.*)
-(1/4) Minors[B,2][[3,2]]+(1/4) Minors[B,2][[2,2]]-(1/5) Minors[B,2][[1,2]]
Det[MatrixPower[A, 5]]-Det[A]^5
Det[A]-Det[Transpose[A]]
Det[A.B]-Det[A] Det[B]
Det[5B]-125Det[B]
Det[Inverse[B]]-1/Det[B]
Det[MatrixPower[B,-2]]-1/Det[B]^2
(* Exercise 5. *)
Det[D1]
Det[Transpose[D1]]
Det[Transpose[D1].D1]
Expand[% - Det[D1]^2]                      (* Expand the difference to get 0. *)
(* Exercise 6. *)
m[n_]:= Table[If[i==j, i, 1], {i,1,n},{j,1,n}]                (* M_n. *)
{Det[m[2]],Det[m[3]],Det[m[4]],Det[m[5]]}                    (* Etc.. *)
(*  Pattern:   det (M_n) = (n-1)!   *)
(* Exercise 7 - Comment. *)
(* (S^T)S is invertible only if the columns of S are linearly independent.*)
(* Exercise 8. *)
<<LinearAlgebra'MatrixManipulation' (* Loads the matrix manipulation package*)
A={{-85,-55,-37},{-35,97,50},{79,56,49}}
b={{-306},{309},{338}}
sol=Inverse[A] . b                    (* (a) *)
A1=TakeColumns[A,{2,3}]               (* Deletes column 1 from A         *)
A1=AppendRows[b,A1]                   (* then adds b as a first column   *)
A2=TakeColumns[A,{3}]                 (* takes column 3 from A           *)
A2=AppendRows[b,A2]                   (* adds b as a first column.       *)
A22=TakeColumns[A,{1}]                (* takes column 1 from A           *)
```

```
A2=AppendRows[A22,A2]              (* joins the columns of A22 and A2 *)
A3=TakeColumns[A,{1,2}]            (* deletes column 3 from A          *)
A3=AppendRows[A3,b]                (* then adds b as a last column     *)
x=Det[A1]/Det[A]                   (* (b) *)
y=Det[A2]/Det[A]                   (* (b) *)
z=Det[A3]/Det[A]                   (* (b) *)
adj=Det[A] Inverse[A]              (* (c) *)
(* Exercise 9. *)
CramerDisplay[A_,b_,i_] := Module[{AA=A, j},
                    For[j=1,j<=Length[A],j++,     (* Replacing the ith *)
                        AA[[j,i]]=b[[j,1]]]; AA]   (* column with b.    *)
CramerSolve[A_,b_,i_] := Det[CramerDisplay[A,b,i]]/Det[A]
CramerDisplay[A,b,1] (* Also CramerDisplay[A,b,2] and CramerDisplay[A,b,3]*)
CramerSolve[A,b,1] (* Also CramerSolve[A,b,2] and CramerSolve[A,b,3]*)
(* Exercises 10,11. *)
Permutations[{1,2,3,4}]            (* The permutations of {1,2,3,4}.*)
Length[%]                          (* The number of computed permutations. *)
4!                                 (* The expected answer. *)
Signature[{1,4,2,3}]               (* Sign. *)
Signature[{4,2,3,1}]
(* Exercise 13. *)
Clear[x,y,z]                       (* Clear the values from Exer. 8.    *)
circleeqn[x1_,y1_,x2_,y2_,x3_,y3_]:=Det[{{x^2+y^2,x,y,1},
                                {x1^2+y1^2,x1,y1,1},
                                {x2^2+y2^2,x2,y2,1},
                                {x3^2+y3^2,x3,y3,1}}]
(* Exercise 14. *)
ce=circleeqn[-1,1,1,1,2,4]         (* The equation of the circle. *)
(* Exercise 17. *)
ce /. {x->-2,y->2}                 (* Substitution of the points into the *)
ce /. {x->-2,y->4}                 (* equation of the circle. *)
ce /. {x->-1,y->2}                 (* Last point not on the circle. *)
(* Exercise 18. - Partial *)
p1=x^4-3x^3-63x^2-85x+150
p2=x^4+22x^3-103x^2-2740x+5700
Resultant[p1,p2,x]       (* We also need to declare the variable x.*)
```

Selected Solutions with MATLAB

Note   Any command line designated as (ST) requires the symbolic toolbox.

```
A = [6 7 1; 6 -7 2; 6 7 3]                     % DATA
B = [1/3 1/4 1/5; 1/4 1/4 1/5; 1/5 1/5 1/5]
C = [1 3 5; 7 9 11; 13 15 17]
D = sym('[a b c; d e f; g h i]')               % (ST)
% Exercises 1-4.
```

```
det(A), det(B), det(C)    % C is the only noninvertible, since det(C)=0.
-(1/4)*det(B([2 3],[1 3]))+(1/4)*det(B([1 3],[1 3]))-(1/5)*det(B([1 2],[1 3]))
det(A^5)-det(A)^5
det(A)-det(A.')
det(A*B)-det(A)*det(B)
det(5*B)-125*det(B)
det(inv(B))-1/det(B)
det(B^(-2))-1/det(B)^2
% Exercise 5.
determ(D)                           % (ST)
determ(transpose(D))                % (ST)
d1=determ(symmul(transpose(D),D))   % (ST)
d2=expand(sympow(determ(D),2))      % (ST)  First square then expand symbolically.
symsub(d1,d2)                        % (ST)  The difference is zero.
% Exercise 6.
% Create a script file named m.m having the following lines
function a = m(n)
for i=1:n,
  for j=1:n,
    if i==j
      a(i,j)=i;
      else a(i,j)=1;
    end
  end
end
% Then type
det(m(2)), det(m(3)), det(m(4)), det(m(5))          % Etc..
% Pattern: det (M_n) = (n-1)!
% Exercise 7 - Comment.
% (S^T)S is invertible only if the columns of S are linearly independent.
% Exercise 8.
A=[-85 -55 -37; -35 97 50; 79 56 49]
b=[-306;309;338]
sol=A\b                       % (a)
A1=[b A(:,2:3)]               % column b and columns 2 and 3 of A
A2=[A(:,1) b A(:,3)]          % column 1 of A column b column 3 of A
A3=[A(:,1:2) b]               % columns 2 and 3 of A and column b
x=det(A1)/det(A)              % (b)
y=det(A2)/det(A)              % (b)
z=det(A3)/det(A)              % (b)
adj=det(A)*inv(A)             % (c)
% Exercise 9.
% In a file called CramerD.m type and save the code:
function [B] = CramerD (A,b,i)
              B = [A(:,1:i-1) b A(:,i+1:length(A))];
        end
```

```
% In a file called CramerS.m type and save the code:
function [B] = CramerS (A,b,i)
                   B = det([A(:,1:i-1) b A(:,i+1:length(A))])/det(A);
                   end
% Then in MATLAB session type:
CramerD(A,b,1),CramerD(A,b,2),CramerD(A,b,3)
CramerS(A,b,1),CramerS(A,b,2),CramerS(A,b,3)
% Exercise 12.
randperm(4)                        % A random permutation of {1,2,3,4}.
% Exercises 16,17.
% Create a script file called "circ.m" with the following contents
function [a] = circ(x,y,x1,y1,x2,y2,x3,y3)    % x,y are used as
             a = det([x^2+y^2 x y 1;          % arguments of the function
                  x1^2+y1^2 x1 y1 1;          % because MATLAB will not
                  x2^2+y2^2 x2 y2 1;          % accept them as symbolic
                  x3^2+y3^2 x3 y3 1]);        % variables
% then return to your MATLAB session and type
circ(-2,2,-1,1,1,1,2,4)            % evaluation of the function
circ(-2,4,-1,1,1,1,2,4)            % at all four points
circ(-1,2,-1,1,1,1,2,4)            % (-1,2) is not on the circle since
                                   % the answer is nonzero.
% Exercise 18.  Find the det. of the manually entered the Sylvester matrix.
det([1 -3 -63 -85 150 0 0 0; 0 1 -3 -63 -85 150 0 0; 0 0 1 -3 -63 -85 150 0;...
0 0 0 1 -3 -63 -85 150; 1 22 -103 -2740 5700 0 0 0;0 1 22 -103 -2740 5700 0 0;...
0 0 1 22 -103 -2740 5700 0; 0 0 0 1 22 -103 -2740 5700])        %%%% Etc..
```

7

Eigenvalues and Eigenvectors

*No human investigation can be called real
science if it cannot be demonstrated mathematically.*

—Leonardo Da Vinci (1452–1519)

Introduction

Eigenvalues and eigenvectors are among the most useful topics of linear algebra. They are used in several areas of mathematics, physics, mechanics, electrical and nuclear engineering, hydrodynamics and aerodynamics, etc. In fact, it is rather odd to find an applied area of science where eigenvalues are never used.

Eigenvalues of matrices, as strange as it may seem, appeared in the literature before matrices did. This is because of the odd fact that, to paraphrase Cayley, the theory of matrices was well developed (via the theory of determinants) before matrices were even defined. According to Morris Kline,[1] eigenvalues arose originally in the context of quadratic forms and in celestial mechanics (the movement of the planets), where they were known as *characteristic roots of the secular equation*. As early as the 1740s Euler used eigenvalues implicitly to describe quadratic forms in three variables geometrically. These functions (studied in Chapter 8) are of the form

$$q(x, y, z) = ax^2 + by^2 + cz^2 + dxy + exz + fyz$$

In the 1760s Lagrange studied a system of six differential equations for the motion of the (only six known) planets and derived from that a sixth-degree polynomial equation, the roots of which were the eigenvalues of a 6×6 matrix (the definitions are discussed in Section 7.1). Cauchy in the 1820s realized the importance of eigenvalues for determining the "principal axes" of an n-variable quadratic form. He also applied his discoveries

---

[1] In *Mathematical Thought from Ancient to Modern Times* (Fair Lawn, N.J.: Oxford University Press, 1972).

to the theory of planetary motion. It was Cauchy who in 1840 first used the terms **characteristic values** and **characteristic equation** for the eigenvalues and for the basic polynomial equation they satisfy.

> **CONVENTION**
> Unless stated otherwise, all matrices in this chapter are square.

7.1   Eigenvalues and Eigenvectors

Reader's Goal for This Section

To know the definitions and how to compute eigenvalues and eigenvectors

If A is an $n \times n$ matrix, then $A\mathbf{v}$ is usually unrelated to the n-vector $\mathbf{v}$. A very interesting case arises when $A\mathbf{v}$ happens to be proportional (parallel) to $\mathbf{v}$. So, geometrically, $\mathbf{v}$ and $A\mathbf{v}$ are on the same line through the origin. In such a case we call $\mathbf{v}$ an *eigenvector* of A and the proportionality constant an *eigenvalue* of A.

DEFINITION   Let A be an $n \times n$ matrix. A nonzero vector $\mathbf{v}$ is an **eigenvector** of A if for some scalar λ

$$A\mathbf{v} = \lambda\mathbf{v} \tag{7.1}$$

The scalar λ (which may be zero) is called an **eigenvalue** of A corresponding to (or, *associated* with) the eigenvector $\mathbf{v}$. Eigenvalues are also known as **characteristic**, or **proper, values** (*Eigen* means *proper* in German) or even as **latent roots**.

■ **EXAMPLE 1**   Let

$$A = \begin{bmatrix} 2 & 2 \\ 2 & -1 \end{bmatrix}$$

Show that $\begin{bmatrix} 2 \\ 1 \end{bmatrix}$ and $\begin{bmatrix} 1 \\ -2 \end{bmatrix}$ are eigenvectors of A. What are the corresponding eigenvalues?

SOLUTION   Because

$$\begin{bmatrix} 2 & 2 \\ 2 & -1 \end{bmatrix} \begin{bmatrix} 2 \\ 1 \end{bmatrix} = \begin{bmatrix} 6 \\ 3 \end{bmatrix} = 3 \begin{bmatrix} 2 \\ 1 \end{bmatrix}$$

$(2, 1)$ is an eigenvector with corresponding eigenvalue $\lambda = 3$. Also,

$$\begin{bmatrix} 2 & 2 \\ 2 & -1 \end{bmatrix} \begin{bmatrix} 1 \\ -2 \end{bmatrix} = \begin{bmatrix} -2 \\ 4 \end{bmatrix} = -2 \begin{bmatrix} 1 \\ -2 \end{bmatrix}$$

So, $(1, -2)$ is an eigenvector with corresponding eigenvalue $\lambda = -2$.

Note that any nonzero scalar multiple of an eigenvector **v** is also an eigenvector, because if **w** = *r***v**, then

$$A\mathbf{w} = A(r\mathbf{v}) = rA\mathbf{v}$$

$$= r(\lambda\mathbf{v}) = \lambda(r\mathbf{v})$$

$$= \lambda\mathbf{w}$$

Furthermore, **v** and **w** have the same eigenvalue.

So, in Example 1 we see that the nonzero vectors on the lines l_1 and l_2 determined by $(2, 1)$ and $(1, -2)$ are eigenvectors of A. The linear transformation $A\mathbf{x}$ stretches the vectors of l_1 by a factor $\lambda = 3$. The vectors along l_2 are reflected about the origin and then stretched by a factor of 2. (Fig. 7.1).

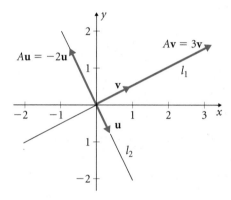

Figure 7.1

■ **EXAMPLE 2**   Find the eigenvalues and eigenvectors of A geometrically in each case.

(a) $A = \begin{bmatrix} 0 & 1 \\ 1 & 0 \end{bmatrix}$ (*Hint*: Recall that $A\mathbf{x}$ is the reflection of **x** about the line $y = x$.)

(b) A is the standard matrix of the rotation by $30°$ in $\mathbf{R}^3$ about the z-axis in the positive direction.

SOLUTION

(a) The only vectors that remain on the same line after rotation are all vectors along the lines $y = x$ and $y = -x$. These without the origin are the only eigenvectors. For **v** along $y = x$ we have $A\mathbf{v} = 1\mathbf{v}$, so **v** is an eigenvector with corresponding eigenvalue 1. For **v** along $y = -x$, $A\mathbf{v} = -1\mathbf{v}$, so **v** is an eigenvector with corresponding eigenvalue -1 (Fig. 7.2(a)).

(b) The only vectors that remain on the same line after rotation are all vectors along the z-axis (Fig. 7.2(b)). These without the origin are the only eigenvectors. The corresponding eigenvalue is 1.

REMARKS

1. Example 2 shows us that eigenvalues and eigenvectors are very *closely related to linear transformations*. In Part (b) we did not even have to display the matrix in order to compute its eigenvalues and eigenvectors. A geometric understanding of the corresponding matrix transformation was sufficient.

2. There are severe limitations in trying to find eigenvalues and eigenvectors geometrically. If A has size greater than 3, our geometric intuition abandons us. Besides, there are many 3-dimensional transformations that are too complicated to explain geometrically.

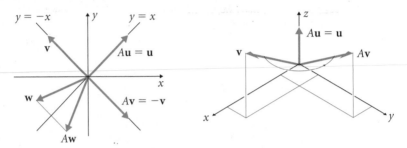

(a) Reflection about the line $y = x$.   **(b)** Rotation about the z-axis.

Figure 7.2   Visualizing eigenvectors through the linear transformation.

Computation of Eigenvalues and Eigenvectors

Let us now discuss how to compute eigenvalues and eigenvectors in general. Because

$$A\mathbf{v} = \lambda\mathbf{v} \Rightarrow A\mathbf{v} = \lambda I\mathbf{v}$$

$$\Rightarrow A\mathbf{v} - \lambda I\mathbf{v} = \mathbf{0}$$

$$\Rightarrow (A - \lambda I)\mathbf{v} = \mathbf{0}$$

we see that $\mathbf{v}$ is an eigenvector if and only if it is a nontrivial solution of the homogeneous system $(A - \lambda I)\mathbf{v} = \mathbf{0}$. In this case $\mathbf{v}$ is a nonzero vector of the *null space* of $A - \lambda I$. The system has a nontrivial solution if and only if the determinant of the coefficient matrix is zero. Thus, λ is an eigenvalue of A if and only if $\det(A - \lambda I) = 0$. We have proved the following theorem.

THEOREM 1

Let A be a square matrix.

1. A scalar λ is an eigenvalue of A if and only if

$$\det(A - \lambda I) = 0 \qquad\qquad \text{characteristic equation} \tag{7.2}$$

2. A vector $\mathbf{v}$ is an eigenvector of A corresponding to an eigenvalue λ if and only if $\mathbf{v}$ is a nontrivial solution of the system

$$(A - \lambda I)\mathbf{v} = \mathbf{0} \tag{7.3}$$

Equation (7.2) is called the **characteristic equation** of A and *is one of the most important equations in all of mathematics.* Also, $\det(A - \lambda I)$ is a polynomial of degree n in λ and is called the **characteristic polynomial** of A. The matrix $A - \lambda I$ is called the **characteristic matrix** of A. If an eigenvalue λ is a root of multiplicity k of the characteristic polynomial, we say that λ has **algebraic multiplicity** k. The null space of $A - \lambda I$, denoted by E_λ, is called the **eigenspace** of A corresponding to the eigenvalue λ.

$$E_\lambda = \text{Null}(A - \lambda I)$$

So, E_λ is a *subspace* of $\mathbf{R}^n$ and it consists of eigenvectors of A and the zero vector. The dimension of E_λ is called the **geometric multiplicity** of λ. It is a basic fact that *the geometric multiplicity never exceeds the algebraic* (see Exercise 38).

According to Theorem 1 and the preceding remarks we have the following.

Algorithm

(Computation of Eigenvalues, Eigenvectors, and Bases of Eigenspaces)

INPUT: $n \times n$ matrix A.

1. Compute the characteristic polynomial $\det(A - \lambda I)$.
2. Find the eigenvalues of A by solving $\det(A - \lambda I) = 0$ for λ.
3. For each eigenvalue λ_i solve the homogeneous system $(A - \lambda_i I)\mathbf{v} = \mathbf{0}$ by completely reducing the augmented matrix

$$[A - \lambda_i I : \mathbf{0}]$$

   The nontrivial solutions are the eigenvectors of A corresponding to λ_i.
4. Write the general solution of $(A - \lambda_i I)\mathbf{v} = \mathbf{0}$ in Step 3 as a linear combination of vectors with coefficients the free variables. These vectors form a basis for E_{λ_i}.

OUTPUT: Eigenvalues $\lambda_1, \ldots, \lambda_n$ of A; An eigenvector basis for each E_{λ_i}.

In Examples 3–7 we compute the eigenvalues and the eigenvectors and find bases for each eigenspace of the given matrix A.

■ EXAMPLE 3

$$A = \begin{bmatrix} 5 & 1 \\ 1 & 5 \end{bmatrix}$$

SOLUTION   The characteristic equation is

$$\det(A - \lambda I) = \begin{vmatrix} 5 - \lambda & 1 \\ 1 & 5 - \lambda \end{vmatrix} = \lambda^2 - 10\lambda + 24 = 0$$

Hence, the eigenvalues are

$$\lambda_1 = 4, \qquad \lambda_2 = 6$$

To find the corresponding eigenvectors, we have the following. For $\lambda_1 = 4$, we solve the system with augmented matrix $[A - \lambda_1 I : \mathbf{0}]$.

$$[A - 4I : \mathbf{0}] = \begin{bmatrix} 1 & 1 & : & 0 \\ 1 & 1 & : & 0 \end{bmatrix} \sim \begin{bmatrix} 1 & 1 & : & 0 \\ 0 & 0 & : & 0 \end{bmatrix}$$

to get $(-r, r)$ for any $r \in \mathbf{R}$. So, all eigenvectors corresponding to $\lambda_1 = 4$ are of the form $(-r, r)$, $r \neq 0$ and the eigenspace E_4 is

$$E_4 = \left\{ \begin{bmatrix} -r \\ r \end{bmatrix}, r \in \mathbf{R} \right\} = \text{Span} \left\{ \begin{bmatrix} -1 \\ 1 \end{bmatrix} \right\}$$

The eigenvector $\mathbf{v}_1 = \begin{bmatrix} -1 \\ 1 \end{bmatrix}$ defines the basis $\{\mathbf{v}_1\}$ for E_4.

For $\lambda_2 = 6$, we solve the system with augmented matrix $[A - \lambda_2 I : \mathbf{0}]$.

$$[A - 6I : \mathbf{0}] = \begin{bmatrix} -1 & 1 & : & 0 \\ 1 & -1 & : & 0 \end{bmatrix} \sim \begin{bmatrix} -1 & 1 & : & 0 \\ 0 & 0 & : & 0 \end{bmatrix}$$

to get (r, r) for any $r \in \mathbf{R}$. Hence, the eigenvectors corresponding to $\lambda_2 = 6$ are of the form (r, r), $r \neq 0$ and the eigenspace E_6 is

$$E_6 = \left\{ \begin{bmatrix} r \\ r \end{bmatrix}, r \in \mathbf{R} \right\} = \text{Span} \left\{ \begin{bmatrix} 1 \\ 1 \end{bmatrix} \right\}$$

The eigenvector $\mathbf{v}_2 = \begin{bmatrix} 1 \\ 1 \end{bmatrix}$ defines the basis $\{\mathbf{v}_2\}$ for E_6.

We see that the eigenspaces are the lines $y = -x$ and $y = x$. The effect of the linear transformation $A\mathbf{x}$ along these two lines is **stretching** by a factor of 4 along $y = -x$ and by a factor of 6 along $y = x$.

▪ EXAMPLE 4

$$A = \begin{bmatrix} 0 & 0 & 1 \\ 0 & 1 & 0 \\ 0 & 0 & 1 \end{bmatrix}$$

SOLUTION   The characteristic equation is

$$\det(A - \lambda I) = \begin{vmatrix} -\lambda & 0 & 1 \\ 0 & 1 - \lambda & 0 \\ 0 & 0 & 1 - \lambda \end{vmatrix} = -\lambda(1 - \lambda)^2 = 0$$

Hence, the eigenvalues are

$$\lambda_1 = 0, \qquad \lambda_2 = \lambda_3 = 1$$

For $\lambda_1 = 0$ we have

$$[A - 0I : \mathbf{0}] = \begin{bmatrix} 0 & 0 & 1 & : & 0 \\ 0 & 1 & 0 & : & 0 \\ 0 & 0 & 1 & : & 0 \end{bmatrix} \sim \begin{bmatrix} 0 & 1 & 0 & : & 0 \\ 0 & 0 & 1 & : & 0 \\ 0 & 0 & 0 & : & 0 \end{bmatrix}$$

The general solution is $(r, 0, 0)$ for $r \in \mathbf{R}$. Hence,

$$E_0 = \left\{ \begin{bmatrix} r \\ 0 \\ 0 \end{bmatrix}, r \in \mathbf{R} \right\} = \text{Span} \left\{ \begin{bmatrix} 1 \\ 0 \\ 0 \end{bmatrix} \right\}$$

and eigenvector $\mathbf{v}_1 = \begin{bmatrix} 1 \\ 0 \\ 0 \end{bmatrix}$ defines the basis $\{\mathbf{v}_1\}$ of E_0.

For $\lambda_2 = \lambda_3 = 1$ (with algebraic multiplicity 2), we have

$$[A - 1I : \mathbf{0}] = \begin{bmatrix} -1 & 0 & 1 & : & 0 \\ 0 & 0 & 0 & : & 0 \\ 0 & 0 & 0 & : & 0 \end{bmatrix}$$

with general solution (r, s, r) for $r \in \mathbf{R}$. Because $(r, s, r) = r(1, 0, 1) + s(0, 1, 0)$,

$$E_1 = \left\{ \begin{bmatrix} r \\ s \\ r \end{bmatrix}, r \in \mathbf{R} \right\} = \text{Span} \left\{ \begin{bmatrix} 1 \\ 0 \\ 1 \end{bmatrix}, \begin{bmatrix} 0 \\ 1 \\ 0 \end{bmatrix} \right\}$$

The spanning eigenvectors $\mathbf{v}_2 = \begin{bmatrix} 1 \\ 0 \\ 1 \end{bmatrix}$, $\mathbf{v}_3 = \begin{bmatrix} 0 \\ 1 \\ 0 \end{bmatrix}$ are also linearly independent. Hence, $\{\mathbf{v}_2, \mathbf{v}_3\}$ is a basis for E_1. The geometric multiplicity of $\lambda = 1$ is 2.

■ EXAMPLE 5

$$A = \begin{bmatrix} 1 & -1 & 0 \\ 0 & -4 & 2 \\ 0 & 0 & -2 \end{bmatrix}$$

SOLUTION We have

$$\det(A - \lambda I) = \begin{vmatrix} 1 - \lambda & -1 & 0 \\ 0 & -4 - \lambda & 2 \\ 0 & 0 & -2 - \lambda \end{vmatrix} = -(\lambda - 1)(\lambda + 2)(\lambda + 4) = 0$$

Hence, the eigenvalues are

$$\lambda_1 = 1, \qquad \lambda_2 = -2, \qquad \lambda_3 = -4$$

By row-reducing $[A - 1I : \mathbf{0}]$, $[A - (-2)I : \mathbf{0}]$, and $[A - (-4)I : \mathbf{0}]$, we get

$$E_1 = \text{Span} \left\{ \begin{bmatrix} 1 \\ 0 \\ 0 \end{bmatrix} \right\}, \qquad E_{-2} = \text{Span} \left\{ \begin{bmatrix} \frac{1}{3} \\ 1 \\ 1 \end{bmatrix} \right\}, \qquad E_{-4} = \text{Span} \left\{ \begin{bmatrix} \frac{1}{5} \\ 1 \\ 0 \end{bmatrix} \right\}$$

The spanning eigenvectors define bases for the corresponding eigenspaces.

■ EXAMPLE 6

$$A = \begin{bmatrix} 1 & 0 & 3 \\ 1 & -1 & 2 \\ -1 & 1 & -2 \end{bmatrix}$$

SOLUTION   We have

$$\det(A - \lambda I) = \begin{vmatrix} 1 - \lambda & 0 & 3 \\ 1 & -1 - \lambda & 2 \\ -1 & 1 & -2 - \lambda \end{vmatrix} = -\lambda^3 - 2\lambda^2 = -\lambda^2(2 + \lambda) = 0$$

Hence, the eigenvalues are

$$\lambda_1 = \lambda_2 = 0, \qquad \lambda_3 = -2$$

Row reduction of $[A - 0I : \mathbf{0}]$ and $[A - (-2)I : \mathbf{0}]$ yields

$$E_0 = \text{Span} \left\{ \begin{bmatrix} -3 \\ -1 \\ 1 \end{bmatrix} \right\}, \qquad E_{-2} = \text{Span} \left\{ \begin{bmatrix} -1 \\ -1 \\ 1 \end{bmatrix} \right\}$$

and the spanning eigenvectors define bases for the corresponding eigenspaces. Note that although the algebraic multiplicity of $\lambda = 0$ is 2, the geometric multiplicity is only 1.

■ EXAMPLE 7   (Complex Eigenvalues)

$$A = \begin{bmatrix} 0 & -1 \\ 1 & 0 \end{bmatrix}$$

SOLUTION   There are no real eigenvalues, because

$$\begin{vmatrix} 0 - \lambda & -1 \\ 1 & 0 - \lambda \end{vmatrix} = \lambda^2 + 1 = 0 \Rightarrow \lambda = \pm i$$

If we accept complex solutions, we can compute the (now complex) eigenvectors

$$\begin{bmatrix} -i & -1 & : & 0 \\ 1 & -i & : & 0 \end{bmatrix} \sim \begin{bmatrix} 1 & -i & : & 0 \\ 0 & 0 & : & 0 \end{bmatrix}$$

$$\begin{bmatrix} i & -1 & : & 0 \\ 1 & i & : & 0 \end{bmatrix} \sim \begin{bmatrix} 1 & i & : & 0 \\ 0 & 0 & : & 0 \end{bmatrix}$$

to get basic eigenvectors $(i, 1)$ and $(-i, 1)$ for $\lambda = i$ and $\lambda = -i$, respectively. We may also write

$$E_i = \text{Span} \left\{ \begin{bmatrix} i \\ 1 \end{bmatrix} \right\}, \qquad E_{-i} = \text{Span} \left\{ \begin{bmatrix} -i \\ 1 \end{bmatrix} \right\}$$

where the scalars used in the spans are *complex* numbers.

Although the roots of the characteristic equation can be complex numbers, we are mostly interested in real roots and thus in *real* eigenvalues.

A (Rare) Trick

It is usually hard or impossible to solve the characteristic equation exactly. However, if the characteristic polynomial has integer coefficients we may sometimes use a trick outlined in the next example.

■ EXAMPLE 8   Find the eigenvalues of a matrix with characteristic polynomial

$$p(\lambda) = -\lambda^3 - 2\lambda^2 + 3\lambda + 6$$

SOLUTION $p(\lambda)$ has *integer* coefficients. If it happens to have an integer root, this root should be a divisor of the constant term (this is a theorem from algebra). The divisors of 6 are ± 1, ± 2, ± 3, ± 6. If we substitute $\lambda = -2$ into p, we get zero. So -2 is a root; hence, $\lambda + 2$ evenly divides p. By using long division we see that

$$p(\lambda) = -(\lambda + 2)(\lambda^2 - 3)$$

Thus,

$$\lambda_1 = -2, \lambda_2 = \sqrt{3}, \lambda_3 = -\sqrt{3}$$

are the eigenvalues.

On Eigenvalues of Invertible Matrices

Because a square matrix A is invertible if and only if

$$\det(A) \neq 0 \Leftrightarrow \det(A - 0I) \neq 0$$

we have the following.

THEOREM 2   A square matrix A is invertible if and only if 0 is **not** an eigenvalue of A.

Eigenvalues of Triangular Matrices

If $A = [a_{ij}]$ is a triangular matrix, so is $A - \lambda I$. Hence, in this case

$$\det(A - \lambda I) = (a_{11} - \lambda)(a_{22} - \lambda) \cdots (a_{nn} - \lambda)$$

So we have Theorem 3.

THEOREM 3   The eigenvalues of a triangular matrix are its diagonal entries.

Fast Calculation of $A^k\mathbf{x}$

Let us now discuss a very interesting application that carries over to various mathematics, physics, and engineering problems (mainly through differential equations and dynamical systems).

Suppose we want to compute the matrix product $A^k \mathbf{x}$ for an n-vector $\mathbf{x}$ and an $n \times n$ matrix A. This can be tedious, especially for large k or n. Now let us **assume** that $\mathbf{x}$ can be written as a linear combination of eigenvectors $\mathbf{v}_1, \ldots, \mathbf{v}_m$ of A, say,

$$\mathbf{x} = c_1 \mathbf{v}_1 + \cdots + c_m \mathbf{v}_m$$

If $\lambda_1, \ldots, \lambda_m$ are the corresponding eigenvalues, then

$$Ax = A(c_1 \mathbf{v}_1 + \cdots + c_m \mathbf{v}_m)$$
$$= c_1 A \mathbf{v}_1 + \cdots + c_m A \mathbf{v}_m$$
$$= c_1 \lambda_1 \mathbf{v}_1 + \cdots + c_m \lambda_m \mathbf{v}_m$$

We repeat this process by using Ax in place of $\mathbf{x}$ and iterate $k - 1$ times to get

$$A^k \mathbf{x} = c_1 \lambda_1^k \mathbf{v}_1 + \cdots + c_m \lambda_m^k \mathbf{v}_m \tag{7.4}$$

Note that once we know the c_is, λ_is, and the $\mathbf{v}_i$s, computing $A^k \mathbf{x}$ becomes very easy, because the right-hand side of (7.4) involves no matrix multiplications.

■ EXAMPLE 9   Compute

$$\begin{bmatrix} 2 & 2 \\ 2 & -1 \end{bmatrix}^8 \begin{bmatrix} -6 \\ -8 \end{bmatrix}$$

SOLUTION $(1, -2)$ and $(2, 1)$ are the eigenvectors of the matrix, with eigenvalues -2 and 3. We have

$$\begin{bmatrix} -6 \\ -8 \end{bmatrix} = 2 \begin{bmatrix} 1 \\ -2 \end{bmatrix} - 4 \begin{bmatrix} 2 \\ 1 \end{bmatrix}$$

Hence, by (7.4), the given product equals

$$2 \cdot (-2)^8 \begin{bmatrix} 1 \\ -2 \end{bmatrix} - 4 \cdot 3^8 \begin{bmatrix} 2 \\ 1 \end{bmatrix} = \begin{bmatrix} -51{,}976 \\ -27{,}268 \end{bmatrix}$$

WEAKNESS OF THE METHOD   This method applies only if $\mathbf{x}$ can be written as a linear combination of eigenvectors of A. This assumption may fail for some A and $\mathbf{x}$.

Eigenvalues of Linear Transformations

We may also define eigenvalues and eigenvectors for linear transformations. If V is a vector space and $T : V \to V$ is a linear transformation, then a nonzero vector $\mathbf{v}$ is an **eigenvector** of T if

$$T(\mathbf{v}) = \lambda \mathbf{v}$$

for some (possibly zero) scalar λ. Just as before, we call λ an **eigenvalue** of T, and we say that the eigenvector $\mathbf{v}$ belongs to (or corresponds to, or is associated with) λ.

■ EXAMPLE 10   Let V be any vector space and let r be a fixed scalar. Find the eigenvalues and eigenvectors of the homothety

$$T : V \longrightarrow V, \qquad T(\mathbf{v}) = r\mathbf{v}$$

SOLUTION   Because $T(\mathbf{v}) = r\mathbf{v}$, any nonzero vector is an eigenvector, with corresponding eigenvalue r.

■ EXAMPLE 11   (Requires Calculus)   Let $V = C^1[a, b]$ be vector space of all real-valued functions in x defined on $[a, b]$ and differentiable. The differentiation operator $\frac{d}{dx} : V \longrightarrow V$, is a linear transformation. If r is a fixed scalar, then the function e^{rx} is in V. Show that e^{rx} is an eigenvector of $\frac{d}{dx}$. Find the corresponding eigenvalue.

SOLUTION   Because

$$\frac{d}{dx}(e^{rx}) = re^{rx}$$

it follows that e^{rx} is an eigenvector of $\frac{d}{dx}$ and r is the corresponding eigenvalue.

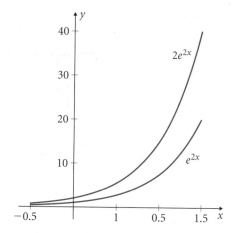

e^{2x} is an eigenvector of $\frac{d}{dx}$ with eigenvalue 2.

■ EXAMPLE 12   Find the eigenvalues and eigenvectors of

$$T : \mathbf{R}^2 \longrightarrow \mathbf{R}^2, T(x, y) = (x + y, x + y)$$

SOLUTION   We seek numbers λ and nonzero vectors (x, y) such that

$$\begin{bmatrix} x + y \\ x + y \end{bmatrix} = \lambda \begin{bmatrix} x \\ y \end{bmatrix}$$

or

$$\left(\begin{bmatrix} 1 & 1 \\ 1 & 1 \end{bmatrix} - \lambda I \right) \begin{bmatrix} x \\ y \end{bmatrix} = \begin{bmatrix} 0 \\ 0 \end{bmatrix}$$

Thus the problem reduces to finding the eigenvectors and eigenvalues of the matrix $\begin{bmatrix} 1 & 1 \\ 1 & 1 \end{bmatrix}$, which is the standard matrix of T. We see now that 0 and 2 are the only eigenvalues, and $(-1, 1)$ and $(1, 1)$ are the corresponding eigenvectors.

In general we have the following theorem.

THEOREM 4

Let V be a finite-dimensional vector space. Let $T : V \rightarrow V$ be a linear transformation with matrix A with respect to some basis $\mathcal{B} = \{v_1, \ldots, v_n\}$ of V. Then

$$T(\mathbf{v}) = \lambda \mathbf{v} \Leftrightarrow A[\mathbf{v}]_{\mathcal{B}} = \lambda [\mathbf{v}]_{\mathcal{B}} \qquad (7.5)$$

Hence,

1. λ is an eigenvalue of T if and only if it is an eigenvalue of A;
2. $\mathbf{v}$ is an eigenvector of T if and only if $[\mathbf{v}]_{\mathcal{B}}$ is an eigenvector of A.

PROOF   By Theorem 13, Section 5.4, we have

$$A[\mathbf{v}]_{\mathcal{B}} = [T(\mathbf{v})]_{\mathcal{B}}$$

Also $[\lambda \mathbf{v}]_{\mathcal{B}} = \lambda [\mathbf{v}]_{\mathcal{B}}$ by Theorem 18, Section 4.5. Hence,

$$T(\mathbf{v}) = \lambda \mathbf{v} \Leftrightarrow [T(\mathbf{v})]_{\mathcal{B}} = [\lambda \mathbf{v}]_{\mathcal{B}}$$
$$\Leftrightarrow A[\mathbf{v}]_{\mathcal{B}} = [\lambda \mathbf{v}]_{\mathcal{B}}$$
$$\Leftrightarrow A[\mathbf{v}]_{\mathcal{B}} = \lambda [\mathbf{v}]_{\mathcal{B}}$$

This proves (7.5), which implies Statements 1 and 2.

Eigenvalues and Eigenvectors with CAS

Maple

```
> with(linalg):
> eigenvects([[3,3,3],[2,2,2],[1,1,1]]);
```

$$[6, 1, \{[3, 2, 1]\}], [0, 2, \{[-1, 0, 1], [-1, 1, 0]\}]$$

Mathematica

```
In[1]:= Eigensystem[{{3,3,3},{2,2,2},{1,1,1}}]

Out[1]= {{0, 0, 6},
         {{-1, 0, 1}, {-1, 1, 0}, {3, 2, 1}}}
```

MATLAB

```
>> [V,D]=eig([3 3 3; 2 2 2; 1 1 1])
V =
        0.8018      0.8018     -0.7193
        0.5345     -0.5345      0.0250
        0.2673     -0.2673      0.6943
D =
        6.0000           0           0
             0      0.0000           0
             0           0           0
```

Exercises 7.1

Let

$$A = \begin{bmatrix} 3 & -2 \\ -3 & 2 \end{bmatrix}, \quad \mathbf{u} = \begin{bmatrix} -1 \\ 1 \end{bmatrix}, \quad \mathbf{v} = \begin{bmatrix} 2 \\ 3 \end{bmatrix}$$

1. Show that $\mathbf{u}$ is an eigenvector of A. What is the corresponding eigenvalue?

2. Show that any nonzero scalar multiple $b\mathbf{u}$ of $\mathbf{u}$ is also an eigenvector of A.

3. Show that $\mathbf{v}$ is an eigenvector of A. What is the corresponding eigenvalue?

4. Why is $c\mathbf{v}$ an eigenvector of A for any $c \neq 0$ scalar?

5. Is $\mathbf{u} + \mathbf{v}$ an eigenvector of A?

6. What is wrong with the following reasoning? Because $b\mathbf{u}$ and $c\mathbf{v}$ are eigenvectors of A, so is $b\mathbf{u} + c\mathbf{v}$. Hence, any 2-vector $\neq \mathbf{0}$ is an eigenvector of A.

7. If we row-reduce A to echelon form, are $\mathbf{u}$ and $\mathbf{v}$ still eigenvectors?

In Exercises 8–9 find which of the given vectors are eigenvectors of the matrix $A = \begin{bmatrix} 1 & 2 \\ 1 & 2 \end{bmatrix}$. If a vector is an eigenvector, find the corresponding eigenvalue.

8. $\begin{bmatrix} 1 \\ 1 \end{bmatrix}, \begin{bmatrix} 1 \\ -1 \end{bmatrix}, \begin{bmatrix} 0 \\ 0 \end{bmatrix}, \begin{bmatrix} 2 \\ 2 \end{bmatrix}$

9. $\begin{bmatrix} 1 \\ -2 \end{bmatrix}, \begin{bmatrix} -2 \\ -1 \end{bmatrix}, \begin{bmatrix} -2 \\ 1 \end{bmatrix}, \begin{bmatrix} 1 \\ -\frac{1}{2} \end{bmatrix}$

10. Which of $\mathbf{e}_1$ and $\mathbf{e}_2$ are eigenvectors of A?

$$A = \begin{bmatrix} 7 & 0 & 0 \\ 0 & -2 & 4 \\ 0 & 6 & 0 \end{bmatrix}$$

11. Show that $-6, 4, 7$ are eigenvalues of A of Exercise 10.

In Exercises 12–15 find the eigenvalues and eigenvectors of each matrix *geometrically.*

12. $\begin{bmatrix} 1 & 0 \\ 0 & -1 \end{bmatrix}$ (reflection with respect to the *x*-axis)

13. $\begin{bmatrix} 1 & 0 \\ 0 & 0 \end{bmatrix}$ (projection onto the *x*-axis)

14. $\begin{bmatrix} 3 & 0 \\ 0 & 3 \end{bmatrix}$ (stretching by a factor of 3 away from the origin)

15. $\begin{bmatrix} 0 & -1 \\ 1 & 0 \end{bmatrix}$ (rotation by 90° about the origin)

For the matrices of Exercises 16–23 find each.

(a) The characteristic polynomial.
(b) The eigenvalues.
(c) Bases of eigenvectors for all eigenspaces.
(d) The algebraic and geometric multiplicity of each eigenvalue.

16. a. $\begin{bmatrix} 3 & 2 \\ 3 & 2 \end{bmatrix}$ b. $\begin{bmatrix} 3 & 6 \\ 9 & 0 \end{bmatrix}$ c. $\begin{bmatrix} 1 & 11 \\ 11 & 1 \end{bmatrix}$

17. a. $\begin{bmatrix} -2 & 4 \\ 6 & 0 \end{bmatrix}$ **b.** $\begin{bmatrix} 0 & -9 \\ 1 & -6 \end{bmatrix}$ **c.** $\begin{bmatrix} 0 & 7 \\ 5 & 2 \end{bmatrix}$

18. a. $\begin{bmatrix} -2 & 17 \\ 17 & -2 \end{bmatrix}$ **b.** $\begin{bmatrix} a & b \\ b & a \end{bmatrix}$

19. a. $\begin{bmatrix} 0 & 0 & 1 \\ 0 & 1 & 0 \\ 1 & 0 & 0 \end{bmatrix}$ **b.** $\begin{bmatrix} 1 & 1 & 0 \\ 0 & 2 & 0 \\ 0 & 0 & 3 \end{bmatrix}$

20. a. $\begin{bmatrix} 0 & 0 & 1 \\ 0 & 2 & 0 \\ 4 & 0 & 0 \end{bmatrix}$ **b.** $\begin{bmatrix} 1 & 0 & 0 \\ 0 & 1 & 1 \\ 0 & 0 & 1 \end{bmatrix}$

21. a. $\begin{bmatrix} 0 & 2 & 0 \\ 2 & 0 & 0 \\ 0 & 0 & 3 \end{bmatrix}$ **b.** $\begin{bmatrix} 1 & 0 & 0 \\ 0 & 0 & 1 \\ 0 & 1 & 0 \end{bmatrix}$

22. a. $\begin{bmatrix} 1 & 2 & 3 \\ 1 & 2 & 3 \\ 1 & 2 & 3 \end{bmatrix}$ **b.** $\begin{bmatrix} 0 & -1 & 0 \\ -4 & 0 & 0 \\ 0 & 0 & 2 \end{bmatrix}$

23. a. $\begin{bmatrix} 1 & 0 & 0 \\ 0 & 2 & 1 \\ -3 & 0 & 3 \end{bmatrix}$ **b.** $\begin{bmatrix} 1 & 0 & 0 \\ 0 & 2 & 1 \\ -4 & 0 & 3 \end{bmatrix}$

24. Find a 3×3 matrix with three distinct eigenvalues.

25. Find a 3×3 matrix with only two distinct eigenvalues.

In Exercises 26–27 find the eigenvalues of the matrices without calculating.

26. a. $\begin{bmatrix} 1 & 0 \\ 0 & 2 \end{bmatrix}$ **b.** $\begin{bmatrix} -3 & 1 \\ 0 & 13 \end{bmatrix}$

27. a. $\begin{bmatrix} 1 & 0 \\ 7 & -2 \end{bmatrix}$ **b.** $\begin{bmatrix} 0 & 8 \\ 2 & 0 \end{bmatrix}$

In Exercises 28–29 without calculating find one eigenvector and the corresponding eigenvalue of the given matrix.

28. $\begin{bmatrix} 2 & 3 & 4 \\ 2 & 3 & 4 \\ 2 & 3 & 4 \end{bmatrix}$

29. $\begin{bmatrix} a & b & c \\ a & b & c \\ a & b & c \end{bmatrix}$

30. Why can't an $n \times n$ matrix have more than n distinct eigenvalues?

31. Prove that any square matrix A and its transpose A^T have the same characteristic polynomial. Conclude that they have the same eigenvalues.

32. Show that an $n \times n$ matrix A has 0 as an eigenvalue if and only if Null(A) $\neq \{0\}$. In this case, show that Null(A) $= E_0$.

33. Let $\mathbf{v}$ be an eigenvector of a matrix A with corresponding eigenvalue 2. Find one solution of the system $A\mathbf{x} = \mathbf{v}$.

Let $\mathbf{v}$ be an eigenvector of A with corresponding eigenvalue λ.

34. (Power) Show that $\mathbf{v}$ is also an eigenvector of A^k with corresponding eigenvalue λ^k.

35. (Inverse) If A is invertible, show that $\mathbf{v}$ is also an eigenvector of A^{-1} with corresponding eigenvalue λ^{-1}.

36. (Shift of Origin) Let $\mathbf{v}$ be an eigenvector of matrix A with eigenvalue λ and let c be any scalar. Show that $\mathbf{v}$ is an eigenvector of $A - cI$ with corresponding eigenvalue $\lambda - c$.

37. (Similar Matrices) Let A and B be $n \times n$ matrices similar to each other. So, there is an invertible matrix P such that $P^{-1}AP = B$. Prove the following.

   a. A and B have the same characteristic polynomial.

   b. A and B have the same eigenvalues.

   c. If $\mathbf{v}$ is an eigenvector of B with eigenvalue λ, then $P\mathbf{v}$ is an eigenvector of A with eigenvalue λ.

   d. If $\mathbf{u}$ is an eigenvector of A with eigenvalue λ, then $P^{-1}\mathbf{u}$ is an eigenvector of B with eigenvalue λ.

38. Let r be an eigenvalue of an $n \times n$ matrix A. Prove that the geometric multiplicity of r is less than or equal to the algebraic multiplicity of r. (*Hint:* Extend a basis of eigenvectors of r to a basis of $\mathbf{R}^n$. Let A' be the matrix of $T(\mathbf{x}) = A\mathbf{x}$ relative to this basis. Then $A = P^{-1}A'P$ for some invertible matrix P. Use Exercise 37.)

39. Let A be square and let $\mathbf{v} \in$ Null(A) such that $\mathbf{v} \neq \mathbf{0}$. Show that $\mathbf{v}$ is an eigenvector of A. What is its eigenvalue?

40. Find matrices A and B such that the eigenvalues of $A + B$ are not the sums of the eigenvalues of A and B.

41. Find matrices A and B such that the eigenvalues of AB are not the products of the eigenvalues of A and B.

42. Prove that the standard basis n-vectors $\mathbf{e}_1, \ldots, \mathbf{e}_n$ are eigenvectors of any diagonal $n \times n$ matrix A. What are the corresponding eigenvalues?

43. Suppose that an $n \times n$ matrix A has every nonzero n-vector as an eigenvector. Show that A is a scalar matrix.

44. Prove that if A is **nilpotent** (i.e., if $A^k = \mathbf{0}$ for some positive integer k), then 0 is its only eigenvalue. (*Hint:*

Consider nonzero $\mathbf{v}$ such that $A\mathbf{v} = \lambda\mathbf{v}$. If $\lambda \neq 0$, then $A\mathbf{v} \neq \mathbf{0}$. Hence, $A^2\mathbf{v} = \lambda A\mathbf{v} \neq \mathbf{0}$. Continue.)

45. Show that if A is nilpotent (see Exercise 44), then the geometric multiplicity of 0 equals the nullity of A.

The **trace**, $\mathrm{tr}(A)$, of a square matrix $A = [a_{ij}]$ is the sum of its diagonal elements:

$$\mathrm{tr}(A) = a_{11} + a_{22} + \cdots + a_{nn}$$

46. Let $\lambda_1, \ldots, \lambda_n$ be all the eigenvalues (repeated if multiple) of an $n \times n$ matrix A. Prove that

$$\mathrm{tr}(A) = \lambda_1 + \lambda_2 \cdots + \lambda_n$$

and

$$\det(A) = \lambda_1\lambda_2\cdots\lambda_n$$

(*Hint:* $\det(A - \lambda I) = (-1)^n(\lambda - \lambda_1)\cdots(\lambda - \lambda_n)$. Let $\lambda = 0$ to prove the second claim. For the first, note that the coefficient of λ^{n-1} can be determined from the product $(a_{11} - \lambda)\cdots(a_{nn} - \lambda)$.)

Companion Matrix of Polynomial

Let $p(x)$ be the polynomial

$$p(x) = x^n + a_{n-1}x^{n-1} + \cdots + a_0$$

The following $n \times n$ matrix is called the **companion matrix** of p:

$$C(p) = \begin{bmatrix} 0 & 1 & 0 & \cdots & 0 \\ 0 & 0 & 1 & \cdots & 0 \\ \vdots & \vdots & \vdots & \vdots & \vdots \\ 0 & 0 & 0 & \cdots & 1 \\ -a_0 & -a_1 & -a_2 & \cdots & -a_{n-1} \end{bmatrix}$$

47. Find the companion matrix $C(p)$ of $p(x) = x^2 + 2x - 15$ and then find the characteristic polynomial of $C(p)$.

48. Show that the companion matrix $C(p)$ of $p(x) = x^2 + ax + b$ has characteristic polynomial $\lambda^2 + a\lambda + b$.

49. Find a nondiagonal matrix with eigenvalues 4, -5. (*Hint:* Use Exercise 48.)

50. Show that the companion matrix $C(p)$ of $p(x) = x^3 + ax^2 + bx + c$ has characteristic polynomial $-(\lambda^3 + a\lambda^2 + b\lambda + c)$. Also show that for any eigenvalue λ of $C(p)$, the vector $(1, \lambda, \lambda^2)$ is an eigenvector of $C(p)$.

51. Find a nontriangular matrix with eigenvalues 4, -5, and -2. (*Hint:* Use Exercise 50.)

52. Find a matrix with eigenvectors $(1, 2, 4)$, $(1, 3, 9)$, and $(1, 4, 16)$. (*Hint:* Consider the companion matrix of $(x - 2)(x - 3)(x - 4)$.)

53. Show (by induction) that the companion matrix $C(p)$ of $p(x) = x^n + a_{n-1}x^{n-1} + \cdots + a_0$ has characteristic polynomial $(-1)^n p(\lambda)$.

Eigenvalues of Linear Transformations

54. Find the eigenvalues and eigenvectors of the projection p of $\mathbf{R}^3$ onto the xy-plane.

55. Find the eigenvalues and eigenvectors of $T : P_2 \to P_2$, $T(a + bx) = b + ax$.

56. (**Requires Calculus**) Find the eigenvalues and eigenvectors of differentiation $\frac{d}{dx} : P_2 \to P_2$.

7.2 Diagonalization

Reader's Goals for This Section

1. To know which matrices can be diagonalized and how to diagonalize them.
2. To know how to compute A^n efficiently if A can be diagonalized.
3. To know how to diagonalize a linear transformation.

The ideas and methods of this section are very useful in differential equations, in dynamical systems, in Markov processes, in the study of curves and surfaces, in graph theory, and in many other areas.

Anyone engaged in matrix arithmetic prefers diagonal matrices because they are easy to calculate with. This is most notable in matrix multiplication. For example, a diagonal matrix D does not "mix" the components of $\mathbf{x}$ in the product $D\mathbf{x}$,

$$\begin{bmatrix} 2 & 0 \\ 0 & 3 \end{bmatrix} \begin{bmatrix} a \\ b \end{bmatrix} = \begin{bmatrix} 2a \\ 3b \end{bmatrix}$$

and does not mix rows of A in a product DA (or columns in AD),

$$\begin{bmatrix} 2 & 0 \\ 0 & 3 \end{bmatrix} \begin{bmatrix} a & b & c \\ d & e & f \end{bmatrix} = \begin{bmatrix} 2a & 2b & 2c \\ 3d & 3e & 3f \end{bmatrix}$$

Moreover, it is very easy to compute the powers D^k:

$$\begin{bmatrix} 2 & 0 \\ 0 & 3 \end{bmatrix}^k = \begin{bmatrix} 2^k & 0 \\ 0 & 3^k \end{bmatrix}$$

We study matrices that can be transformed to diagonal matrices, and we try to take advantage of the easy arithmetic. We develop criteria to identify these matrices and explore their basic properties. All this is done by using eigenvalues and eigenvectors.

Diagonalization

If an $n \times n$ matrix A is similar to a *diagonal* matrix D, it is called **diagonalizable**. We also say that A **can be diagonalized**. This means that there exists an invertible $n \times n$ matrix P such that $P^{-1}AP$ is a diagonal matrix D.

$$P^{-1}AP = D$$

The process of finding matrices P and D as described is called **diagonalization**, and we say that P and D **diagonalize** A.

■ EXAMPLE 13   Show that

$$A = \begin{bmatrix} 0 & 1 \\ 0 & 1 \end{bmatrix}$$

is diagonalized by $P = \begin{bmatrix} 1 & 1 \\ 0 & 1 \end{bmatrix}$ and $D = \begin{bmatrix} 0 & 0 \\ 0 & 1 \end{bmatrix}$.

SOLUTION   This is true because

$$\begin{bmatrix} 1 & 1 \\ 0 & 1 \end{bmatrix}^{-1} \begin{bmatrix} 0 & 1 \\ 0 & 1 \end{bmatrix} \begin{bmatrix} 1 & 1 \\ 0 & 1 \end{bmatrix} = \begin{bmatrix} 0 & 0 \\ 0 & 1 \end{bmatrix}$$

It is worth noticing that $P' = \begin{bmatrix} 1 & 1 \\ 1 & 0 \end{bmatrix}$ and $D' = \begin{bmatrix} 1 & 0 \\ 0 & 0 \end{bmatrix}$ also diagonalize A. (Why?)

■ EXAMPLE 14   Show that

$$B = \begin{bmatrix} 0 & 1 \\ 0 & 0 \end{bmatrix}$$

is not diagonalizable.

SOLUTION   If B could be diagonalized by

$$P = \begin{bmatrix} a & b \\ c & d \end{bmatrix}, \quad ad - cb \neq 0 \quad \text{and} \quad D = \begin{bmatrix} e & 0 \\ 0 & f \end{bmatrix}$$

then $PD = BP$. So,

$$\begin{bmatrix} a & b \\ c & d \end{bmatrix} \begin{bmatrix} e & 0 \\ 0 & f \end{bmatrix} = \begin{bmatrix} 0 & 1 \\ 0 & 0 \end{bmatrix} \begin{bmatrix} a & b \\ c & d \end{bmatrix}$$

or

$$\begin{bmatrix} ae & bf \\ ce & df \end{bmatrix} = \begin{bmatrix} c & d \\ 0 & 0 \end{bmatrix}$$

Hence, $ce = 0$. If $c \neq 0$, then $e = 0$. So, $ae = a0 = 0 = c$. Thus, c must be zero. Similarly, $d = 0$. But then $ad - cb = 0$. So, P would be noninvertible. We conclude that A cannot be diagonalized.

In Examples 13 and 14 we saw that

1. *Not* all square matrices can be diagonalized;
2. The matrices P and D that diagonalize a matrix A are *not unique*.

Diagonalizing a Square Matrix

Let us see now when a square matrix A is diagonalizable and how to compute matrices P and D that diagonalize it.

First, it is worth noticing that if D is a diagonal matrix with diagonal entries $\lambda_1, \ldots, \lambda_n$, then for $i = 1, \ldots, n$,

$$D\mathbf{e}_i = \lambda_i \mathbf{e}_i$$

Hence, the standard basis vectors $\mathbf{e}_1, \ldots, \mathbf{e}_n$ are eigenvectors of D. In particular, the eigenvectors of D are *linearly independent*. More generally, we have Theorem 5.

THEOREM 5

(Criterion for Diagonalization)

Let A be an $n \times n$ matrix.

1. A is diagonalizable if and only if it has n linearly independent eigenvectors.
2. If A is diagonalizable with $P^{-1}AP = D$, then the columns of P are eigenvectors of A and the diagonal entries of D are the corresponding eigenvalues.
3. If $\{\mathbf{v}_1, \ldots, \mathbf{v}_n\}$ are linearly independent eigenvectors of A with corresponding eigenvalues $\lambda_1, \ldots, \lambda_n$, then A can be diagonalized by

$$P = [\mathbf{v}_1 \mathbf{v}_2 \cdots \mathbf{v}_n] \quad \text{and} \quad D = \begin{bmatrix} \lambda_1 & \cdots & 0 \\ \vdots & \ddots & \vdots \\ 0 & \cdots & \lambda_n \end{bmatrix}$$

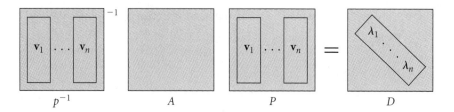

$$\underset{P^{-1}}{}\qquad \underset{A}{}\qquad \underset{P}{}\qquad \underset{D}{}$$

PROOF   Let P be any matrix with columns any n-vectors $\mathbf{v}_1, \ldots, \mathbf{v}_n$ and let D be any diagonal matrix with diagonal entries $\lambda_1, \ldots, \lambda_n$. Then

$$AP = A[\mathbf{v}_1 \mathbf{v}_2 \cdots \mathbf{v}_n] = [A\mathbf{v}_1 A\mathbf{v}_2 \cdots A\mathbf{v}_n] \tag{7.6}$$

and

$$[\lambda_1 \mathbf{v}_1 \lambda_2 \mathbf{v}_2 \cdots \lambda_n \mathbf{v}_n] = [\mathbf{v}_1 \mathbf{v}_2 \cdots \mathbf{v}_n] \begin{bmatrix} \lambda_1 & \cdots & 0 \\ \vdots & \ddots & \vdots \\ 0 & \cdots & \lambda_n \end{bmatrix} = PD \tag{7.7}$$

If A is diagonalizable, with $P^{-1}AP = D$, then $AP = PD$. Hence, $A\mathbf{v}_i = \lambda_i \mathbf{v}_i$, $i = 1, \ldots n$ by (7.6) and (7.7). So, the λ_is are eigenvalues and the $\mathbf{v}_i$s are corresponding eigenvectors. This proves Part 2 and the direct implication in Part 1.

Suppose that A has n linearly independent eigenvectors, say, $\mathbf{v}_1, \ldots, \mathbf{v}_n$ (the columns of P). If $\lambda_1, \ldots, \lambda_n$ are the corresponding eigenvalues, then $A\mathbf{v}_i = \lambda_i \mathbf{v}_i$, $i = 1, \ldots n$. If D is diagonal with diagonal entries $\lambda_1, \ldots, \lambda_n$, then $AP = PD$ by (7.6) and (7.7). Because P is square with linearly independent columns, it is invertible. Hence, $P^{-1}AP = D$ and A is diagonalizable. This proves Part 3 and the converse implication in Part 1.

THEOREM 6   An $n \times n$ matrix A is diagonalizable if and only if $\mathbf{R}^n$ has a basis of eigenvectors of A.

PROOF   This statement is true because n linearly independent n-vectors form a basis of $\mathbf{R}^n$.

■ **EXAMPLE 15**   Redo Example 14 using Theorem 5.

SOLUTION   We can easily verify that the only eigenvalue of B is 0 and $E_0 = \text{Span}\{(1, 0)\}$. So B does not have two linearly independent eigenvectors; hence, it is not diagonalizable, by Part 1 of Theorem 5. (Note how easy it was to reach the same conclusion this time.)

In Examples 16–18 we determine whether the given matrix is diagonalizable. If it is, we find P and D that diagonalize it.

■ EXAMPLE 16

$$A = \begin{bmatrix} 0 & 0 & 1 \\ 0 & 1 & 0 \\ 0 & 0 & 1 \end{bmatrix}$$

SOLUTION   In Example 4, Section 7.1, we found

$$\lambda_1 = 0, \qquad \lambda_2 = \lambda_3 = 1$$

and

$$E_0 = \text{Span} \left\{ \begin{bmatrix} 1 \\ 0 \\ 0 \end{bmatrix} \right\}, \qquad E_1 = \text{Span} \left\{ \begin{bmatrix} 1 \\ 0 \\ 1 \end{bmatrix}, \begin{bmatrix} 0 \\ 1 \\ 0 \end{bmatrix} \right\}$$

A has three linearly independent eigenvectors, so it is diagonalizable, and we may take

$$P = \begin{bmatrix} 1 & 1 & 0 \\ 0 & 0 & 1 \\ 0 & 1 & 0 \end{bmatrix}, \qquad D = \begin{bmatrix} 0 & 0 & 0 \\ 0 & 1 & 0 \\ 0 & 0 & 1 \end{bmatrix}$$

■ EXAMPLE 17

$$A = \begin{bmatrix} 1 & -1 & 0 \\ 0 & -4 & 2 \\ 0 & 0 & -2 \end{bmatrix}$$

SOLUTION   In Example 5, Section 7.1 we found

$$\lambda_1 = 1, \qquad \lambda_2 = -2, \qquad \lambda_3 = -4$$

and

$$E_1 = \text{Span} \left\{ \begin{bmatrix} 1 \\ 0 \\ 0 \end{bmatrix} \right\}, \qquad E_{-2} = \text{Span} \left\{ \begin{bmatrix} \frac{1}{3} \\ 1 \\ 1 \end{bmatrix} \right\}, \qquad E_{-4} = \text{Span} \left\{ \begin{bmatrix} \frac{1}{5} \\ 1 \\ 0 \end{bmatrix} \right\}$$

A has three linearly independent eigenvectors, so it is diagonalizable and

$$P = \begin{bmatrix} 1 & \frac{1}{3} & \frac{1}{5} \\ 0 & 1 & 1 \\ 0 & 1 & 0 \end{bmatrix}, \qquad D = \begin{bmatrix} 1 & 0 & 0 \\ 0 & -2 & 0 \\ 0 & 0 & -4 \end{bmatrix}$$

■ EXAMPLE 18

$$A = \begin{bmatrix} 1 & 0 & 3 \\ 1 & -1 & 2 \\ -1 & 1 & -2 \end{bmatrix}$$

SOLUTION   In Example 6, Section 7.1, we found

$$\lambda_1 = \lambda_2 = 0, \qquad \lambda_3 = -2$$

and

$$E_0 = \text{Span}\left\{ \begin{bmatrix} -3 \\ -1 \\ 1 \end{bmatrix} \right\}, \qquad E_{-2} = \text{Span}\left\{ \begin{bmatrix} -1 \\ -1 \\ 1 \end{bmatrix} \right\}$$

This time A has at most $2(<3)$ linearly independent eigenvectors, so it is *not* diagonalizable.

$n=3$

THEOREM 7

Let $\lambda_1, \ldots, \lambda_l$ be any distinct eigenvalues of an $n \times n$ matrix A.

1. Any corresponding eigenvectors $\mathbf{v}_1, \ldots, \mathbf{v}_l$ are linearly independent.
2. If $\mathcal{B}_1, \ldots, \mathcal{B}_l$ are bases for the corresponding eigenspaces, then $\mathcal{B} = \mathcal{B}_1 \cup \cdots \cup \mathcal{B}_l$ is linearly independent.
3. Let l be the number of *all* distinct eigenvalues of A. Then A is diagonalizable if and only if $\mathcal{B}$ (as in Part 2) has exactly n elements.

PROOF

1. If the vs are not linearly independent, let $\mathbf{v}_k$ be the first that can be written as a linear combination of the preceding ones. Hence,

$$\mathbf{v}_k = a_1\mathbf{v}_1 + \cdots + a_{k-1}\mathbf{v}_{k-1} \tag{7.8}$$

for scalars $a_1, \ldots, a_{k-1}$ not all zero (because $\mathbf{v} \neq \mathbf{0}$ as an eigenvector) and linearly independent $\mathbf{v}_1, \ldots, \mathbf{v}_{k-1}$. We multiply on the left by A to get

$$A\mathbf{v}_k = A(a_1\mathbf{v}_1 + \cdots + a_{k-1}\mathbf{v}_{k-1})$$
$$= a_1 A\mathbf{v}_1 + \cdots + a_{k-1}A\mathbf{v}_{k-1}$$

Hence,

$$\lambda_k\mathbf{v}_k = a_1\lambda_1\mathbf{v}_1 + \cdots + a_{k-1}\lambda_{k-1}\mathbf{v}_{k-1} \tag{7.9}$$

If we now multiply (7.8) by $-\lambda_k$ and add it to (7.9), we get

$$a_1(\lambda_1 - \lambda_k)\mathbf{v}_1 + \cdots + a_{k-1}(\lambda_{k-1} - \lambda_k)\mathbf{v}_{k-1} = \mathbf{0}$$

Therefore, $a_1(\lambda_1 - \lambda_k) = 0, \ldots, a_{k-1}(\lambda_{k-1} - \lambda_k) = 0$, because $\mathbf{v}_1, \ldots, \mathbf{v}_{k-1}$ are linearly independent. One of the as, say, a_i, is nonzero, so $\lambda_i - \lambda_k = 0$ or $\lambda_i = \lambda_k$, which contradicts the assumption that the λ's are distinct. We conclude that none of the eigenvectors can be written as a linear combination of the preceding, so they are linearly independent.
2. To save in writing we consider only two distinct eigenvalues, λ_1, λ_2, and two bases, $\mathcal{B}_1 = \{\mathbf{u}_1, \ldots, \mathbf{u}_p\}$ and $\mathcal{B}_2 = \{\mathbf{w}_1, \ldots, \mathbf{w}_q\}$, for E_{λ_1} and E_{λ_2}. The idea is the same for the general case. We shall show that $\mathcal{B} = \mathcal{B}_1 \cup \mathcal{B}_2$ is linearly independent. Let

$$c_1\mathbf{u}_1 + \cdots + c_p\mathbf{u}_p + d_1\mathbf{w}_1 + \cdots + d_q\mathbf{w}_q = 0$$

Then $\mathbf{u} = c_1\mathbf{u}_1 + \cdots + c_p\mathbf{u}_p$ is either an eigenvector of λ_1 or zero. Likewise, $\mathbf{w} = d_1\mathbf{w}_1 + \cdots + d_q\mathbf{w}_q$ is either an eigenvector of λ_2 or zero. In addition, $\mathbf{u} + \mathbf{v} = \mathbf{0}$. But if both $\mathbf{u}$ and $\mathbf{w}$ were eigenvectors, then they would be linearly independent, by Part 1. This contradicts that their sum is zero. We conclude that

$$\mathbf{u} = \mathbf{w} = \mathbf{0}$$

Hence, $c_1 = \cdots = c_p = 0$ and $d_1 = \cdots = d_q = 0$, because $\mathcal{B}_1$ and $\mathcal{B}_2$ are linearly independent. Therefore, $\mathcal{B}$ is linearly independent.

3. If $\mathcal{B}$ has n vectors, they are linearly independent, by Part 2. Hence, A is diagonalizable. Conversely, if A is diagonalizable, then it has n linearly independent eigenvectors. If exactly n_i of these eigenvectors correspond to eigenvalue λ_i, then $\mathcal{B}_i$ has at least n_i elements, because the eigenvectors are linearly independent. We conclude that $\mathcal{B}$ has at least n and, hence, exactly n elements.

We can now draw some interesting corollaries from Theorem 7.

THEOREM 8   Any $n \times n$ matrix A with n distinct eigenvalues is diagonalizable.

PROOF   By Theorem 5 it suffices to show that the corresponding eigenvectors are linearly independent. But this is guaranteed by Part 1 of Theorem 7.

CAUTION   *A diagonalizable matrix need not have distinct eigenvalues,* as Example 16 shows.

THEOREM 9 A is diagonalizable if and only if for each eigenvalue λ the geometric and algebraic multiplicities of λ are equal.

PROOF   Exercise.

Theorem 7 allows us the use of the following diagonalization process already illustrated in Examples 16–18. This time we do not need to prove that $\mathcal{B}$ is linearly independent, because it is guaranteed by the theorem.

Algorithm

(Diagonalization Process)

INPUT: $n \times n$ matrix A.

1. Compute bases $\mathcal{B}_1, \ldots, \mathcal{B}_l$ for all the eigenspaces of A. Form the union $\mathcal{B} = \mathcal{B}_1 \cup \cdots \cup \mathcal{B}_l$.
2. If $\mathcal{B}$ has less than n elements, stop: A is *not* diagonalizable.
3. If $\mathcal{B}$ has n elements, then A is diagonalizable.
4. A can be diagonalized by P with columns the elements of $\mathcal{B}$ and D with diagonal entries the corresponding eigenvalues.

OUTPUT: P and D that diagonalize A.

Powers of Diagonalizable Matrices

As we know, computing the powers A^k can be quite tedious. However, if A is diagonalizable and we have computed P and D, then $A = PDP^{-1}$. So,

$$A^2 = (PDP^{-1})(PDP^{-1}) = PD^2P^{-1}$$

We iterate to get $A^k = PD^kP^{-1}$. Because finding D^k amounts to raising the diagonal entries of D to only the kth power, we see that A^k is easy to compute *given P, P^{-1}, and D.*

If A happens to be invertible, then 0 is not an eigenvalue of A, by Theorem 2. Therefore, D^{-1} exists and

$$A^{-1} = (PDP^{-1})^{-1} = PD^{-1}P^{-1}$$

Again, we may iterate to get $A^{-k} = PD^{-k}P^{-1}$.

THEOREM 10

If A is diagonalized by P and D, then for $k = 0, 1, 2, \ldots$,

$$A^k = PD^kP^{-1} \tag{7.10}$$

If, in addition, A is invertible, then (7.10) is also valid for $k = -1, -2, -3, \ldots$.

■ **EXAMPLE 19**   Find a formula for A^k, $k = 0, 1, 2, \ldots$, where

$$A = \begin{bmatrix} 1 & 0 & 1 \\ 0 & 2 & 0 \\ 3 & 0 & 3 \end{bmatrix}$$

SOLUTION A has eigenvalues $0, 2, 4$, and the corresponding basic eigenvectors $(-1, 0, 1), (0, 1, 0), (1, 0, 3)$ are linearly independent. Hence,

$$A^k = \begin{bmatrix} -1 & 0 & 1 \\ 0 & 1 & 0 \\ 1 & 0 & 3 \end{bmatrix} \begin{bmatrix} 0 & 0 & 0 \\ 0 & 2 & 0 \\ 0 & 0 & 4 \end{bmatrix}^k \begin{bmatrix} -1 & 0 & 1 \\ 0 & 1 & 0 \\ 1 & 0 & 3 \end{bmatrix}^{-1}$$

$$= \begin{bmatrix} -1 & 0 & 1 \\ 0 & 1 & 0 \\ 1 & 0 & 3 \end{bmatrix} \begin{bmatrix} 0 & 0 & 0 \\ 0 & 2^k & 0 \\ 0 & 0 & 4^k \end{bmatrix} \begin{bmatrix} -\frac{3}{4} & 0 & \frac{1}{4} \\ 0 & 1 & 0 \\ \frac{1}{4} & 0 & \frac{1}{4} \end{bmatrix}$$

Therefore,

$$\begin{bmatrix} 1 & 0 & 1 \\ 0 & 2 & 0 \\ 3 & 0 & 3 \end{bmatrix}^k = \begin{bmatrix} 4^{k-1} & 0 & 4^{k-1} \\ 0 & 2^k & 0 \\ 3 \cdot 4^{k-1} & 0 & 3 \cdot 4^{k-1} \end{bmatrix}$$

An Important Change of Variables

Let us now discuss an idea that is the core of most applications of diagonalization. Let A be a diagonalizable matrix, diagonalized by P and D. Quite often a matrix equation

$$f(A, \mathbf{x}) = \mathbf{0}$$

can be substantially simplified if we replace **x** by the new vector **y** such that

$$\mathbf{x} = P\mathbf{y} \quad \text{or} \quad \mathbf{y} = P^{-1}\mathbf{x} \tag{7.11}$$

and replace A with PDP^{-1} to get an equation of the form

$$g(D, \mathbf{y}) = \mathbf{0}$$

that involves the diagonal matrix D and the new vector **y**.

The change in variables (7.11) is *very important*. It applies among other things to differential equations and to dynamical systems.

The following example shows us *how* to use (7.11) (although this time there is no computational advantage in the method).

■ EXAMPLE 20   Solve the system for (x, y) by diagonalizing the coefficient matrix.

$$\begin{bmatrix} 5 & 1 \\ 1 & 5 \end{bmatrix} \begin{bmatrix} x \\ y \end{bmatrix} = \begin{bmatrix} a \\ b \end{bmatrix}$$

SOLUTION   Let A be the coefficient matrix and let $\mathbf{a} = (a, b)$. Then $D = \begin{bmatrix} 4 & 0 \\ 0 & 6 \end{bmatrix}$, $P = \begin{bmatrix} -1 & 1 \\ 1 & 1 \end{bmatrix}$, and $P^{-1} = \dfrac{1}{2} \begin{bmatrix} -1 & 1 \\ 1 & 1 \end{bmatrix}$.

Now consider the new variable vector $\mathbf{y} = (x', y')$ defined by $\mathbf{y} = P\mathbf{x}$. We have

$$A\mathbf{x} = \mathbf{a} \Leftrightarrow PA\mathbf{x} = P\mathbf{a}$$

$$\Leftrightarrow PAP^{-1}\mathbf{y} = P\mathbf{a}$$

$$\Leftrightarrow D\mathbf{y} = P\mathbf{a}$$

The last equation is the "diagonal" system

$$\begin{bmatrix} 4x' \\ 6y' \end{bmatrix} = \begin{bmatrix} -a + b \\ a + b \end{bmatrix}$$

which is easy to solve for **y**. We get $\mathbf{y} = \left(-\frac{1}{4}a + \frac{1}{4}b, \frac{1}{6}a + \frac{1}{6}b\right)$. Therefore,

$$\mathbf{x} = P^{-1}\mathbf{y} = \frac{1}{24} \begin{bmatrix} 5a - b \\ -a + 5b \end{bmatrix}$$

Diagonalization of Linear Transformations

Diagonalization is closely related to linear transformations. Let A be an $n \times n$ diagonalizable matrix, diagonalized by P, with columns $\mathbf{v}_1, \ldots, \mathbf{v}_n$, and by D with diagonal entries $\lambda_1, \ldots, \lambda_n$. Consider the matrix transformation T defined by A,

$$T : \mathbf{R}^n \rightarrow \mathbf{R}^n, \qquad T(\mathbf{x}) = A\mathbf{x}$$

By Theorems 5 and 6 we know that the $\mathbf{v}_i$s are eigenvectors and the λ_is are eigenvalues of A and that the set $\mathcal{B} = \{\mathbf{v}_1, \ldots, \mathbf{v}_n\}$ forms a basis for $\mathbf{R}^n$. By Theorem 17, Section 5.4, the matrix of T with respect to this new basis $\mathcal{B}$ is

$$P^{-1}AP = D$$

Because P is the transition matrix from $\mathcal{B}$ to the standard basis by Theorem 20, Section 4.5. So, *the matrix of T with respect to a basis of eigenvectors of A is diagonal.*

■ EXAMPLE 21   Let

$$A = \begin{bmatrix} 1 & 0.5 \\ 0.5 & 1 \end{bmatrix}$$

Find a basis $\mathcal{B}$ of $\mathbf{R}^2$ such that the matrix D of $T(\mathbf{x}) = A\mathbf{x}$ with respect to it is diagonal. Also find the transition matrix P from $\mathcal{B}$ to the standard basis.

SOLUTION   The eigenvalues of the matrix are 0.5 and 1.5 with eigenvectors $(-1, 1)$ and $(1, 1)$. Hence,

$$D = \begin{bmatrix} 0.5 & 0 \\ 0 & 1.5 \end{bmatrix} \quad \text{and} \quad P = \begin{bmatrix} -1 & 1 \\ 1 & 1 \end{bmatrix}$$

Let $T : V \to V$ be a linear transformation of a *finite-dimensional* vector space V. It is desirable to have a basis $\mathcal{B}$ of V such that the matrix D of T with respect to $\mathcal{B}$ is diagonal. In that case T would be easy to evaluate. If such a basis $\mathcal{B}$ exists, we say that T is **diagonalizable** and that $\mathcal{B}$ **diagonalizes** T. It turns out that the vectors of $\mathcal{B}$ are eigenvectors of T. More precisely, we have the following theorem (whose proof is left as an exercise).

THEOREM 11
Let $T : V \to V$ be a linear transformation of a finite-dimensional vector space V. Then

1. T is diagonalizable if and only if V has a basis of eigenvectors of T;
2. T is diagonalizable if and only if the matrix of T with respect to any basis of V is diagonalizable;
3. If T is diagonalized by $\mathcal{B}$, then the vectors of $\mathcal{B}$ are eigenvectors of T.

■ EXAMPLE 22   Show that the linear transformation $T : P_2 \to P_2$, $T(a + bx) = b + ax$ is diagonalizable. Find a basis $\mathcal{B}$ of P_2 that diagonalizes T. Evaluate T using $\mathcal{B}$.

SOLUTION   Because

$$T(1) = 0 \cdot 1 + 1 \cdot x, \qquad T(x) = 1 \cdot 1 + 0 \cdot x$$

the matrix of T with respect to the standard basis $\{1, x\}$ is $A = \begin{bmatrix} 0 & 1 \\ 1 & 0 \end{bmatrix}$. A is diagonalizable (why?), so is T by Theorem 11. The eigenvectors of T can be obtained from those of A (by Theorem 4, Section 7.1) and are found to be $1 + x$ and $-1 + x$. These are linearly independent in P_2; hence, $\mathcal{B} = \{1 + x, -1 + x\}$ diagonalizes T. The corresponding eigenvalues are 1 and -1. We may use $\mathcal{B}$ to evaluate T as follows:

$$T(a + bx) = T\left(\frac{a+b}{2}(1+x) + \frac{b-a}{2}(-1+x)\right)$$

$$= \frac{a+b}{2}T(1+x) + \frac{b-a}{2}T(-1+x)$$

$$= \frac{a+b}{2} \cdot 1 \cdot (1+x) + \frac{b-a}{2} \cdot (-1) \cdot (-1+x)$$

$$= b + ax$$

▪ EXAMPLE 23   (Requires Calculus)   Show that differentiation $\frac{d}{dx} : P_2 \rightarrow P_2$ is not diagonalizable.

SOLUTION   Because

$$\frac{d}{dx}(1) = 0 = 0 \cdot 1 + 0 \cdot x, \qquad \frac{d}{dx}(x) = 1 = 1 \cdot 1 + 0 \cdot x$$

the matrix of $\frac{d}{dx}$ with respect to the standard basis $\{1, x\}$ is

$$\begin{bmatrix} 0 & 1 \\ 0 & 0 \end{bmatrix}$$

which is not diagonalizable. So, $\frac{d}{dx}$ is not diagonalizable by Theorem 11. Hence, P_2 has no basis of eigenvectors of $\frac{d}{dx}$.

Exercises 7.2

In Exercises 1–5 diagonalize the matrix A if it is diagonalizable; i.e., if possible, find P invertible and D diagonal such that $P^{-1}AP = D$.

1. $A = \begin{bmatrix} -2 & 5 \\ 5 & -2 \end{bmatrix}$

2. $A = \begin{bmatrix} -2 & 0 \\ 5 & -2 \end{bmatrix}$

3. $A = \begin{bmatrix} 1 & 0 & 2 \\ 0 & -2 & 5 \\ 0 & 5 & -2 \end{bmatrix}$

4. $A = \begin{bmatrix} 1 & 0 & 0 \\ 0 & -2 & 5 \\ 2 & 0 & -2 \end{bmatrix}$

5. $A = \begin{bmatrix} -4 & 0 & 0 \\ 0 & 2 & 1 \\ 2 & -1 & 0 \end{bmatrix}$

In Exercises 6–9 verify that S is a linearly independent set of eigenvectors of A. Diagonalize A using S.

6. $S = \{(-2, 2), (5, 5)\}$ and $A = \begin{bmatrix} -3 & 6 \\ 6 & -3 \end{bmatrix}$

7. $S = \{(10, 0), (6, 5)\}$ and $A = \begin{bmatrix} -1 & 6 \\ 0 & 4 \end{bmatrix}$

8. $S = \{(-2, 0, 0), (0, -3, 3), (0, 2, 3)\}$ and

$$A = \begin{bmatrix} 7 & 0 & 0 \\ 0 & -2 & 4 \\ 0 & 6 & 0 \end{bmatrix}$$

9. $S = \{(0, 2, 0), (1, 0, 2), (1, 0, -2)\}$ and

$$A = \begin{bmatrix} 0 & 0 & 1 \\ 0 & 2 & 0 \\ 4 & 0 & 0 \end{bmatrix}$$

In Exercises 10–14 S is a linearly independent set of eigenvectors of A and E is the set of the corresponding eigenvalues. Find A.

10. $S = \{(-1, 1), (1, 1)\}, E = \{-10, 12\}$

11. $S = \{(1, 0, 0), (1, 1, 0), (0, 0, 1)\}, E = \{1, 2, 3\}$

12. $S = \{(0, 1, 0), (1, 0, 1), (-1, 0, 1)\}, E = \{1, 1, -1\}$

13. $S = \{(1, 1, 1), (-3, 0, 1), (-2, 1, 0)\}, E = \{6, 0, 0\}$

14. $S = \{(-1, 1, 0), (1, 1, 0), (0, 0, 1)\}, E = \{-2, 2, 3\}$

15. Suppose that a 3×3 matrix has A eigenvalues 3, 0, -7. Is A diagonalizable? Why or why not.

16. Suppose that a 3×3 matrix A is upper triangular with diagonal entries 2, 1, -5. Show that A is diagonalizable. What is D?

17. Find a basis of $\mathbf{R}^3$ that consists of eigenvectors of
$$\begin{bmatrix} 1 & 2 & 2 \\ 1 & 2 & 2 \\ 1 & 2 & 2 \end{bmatrix}.$$

18. Find a basis of $\mathbf{R}^3$ that consists of eigenvectors of
$$\begin{bmatrix} 1 & 2 & 2 \\ 0 & 0 & 0 \\ 1 & 2 & 2 \end{bmatrix}.$$

19. Show that A is diagonalizable.
$$A = \begin{bmatrix} 2 & 3 & 4 \\ 2 & 3 & 4 \\ 2 & 3 & 4 \end{bmatrix}$$

20. Show that A is not diagonalizable.
$$A = \begin{bmatrix} 2 & 3 & -5 \\ 2 & 3 & -5 \\ 2 & 3 & -5 \end{bmatrix}$$

21. Show that A is not diagonalizable.
$$A = \begin{bmatrix} 5 & 1 & 0 \\ 0 & 5 & 1 \\ 0 & 0 & 5 \end{bmatrix}$$

22. Show that A is not diagonalizable.
$$A = \begin{bmatrix} 2 & 1 & 0 & 0 \\ 0 & 2 & 1 & 0 \\ 0 & 0 & 2 & 1 \\ 0 & 0 & 0 & 2 \end{bmatrix}$$

23. Show that A is diagonalizable if and only if $a + b + c \neq 0$.
$$A = \begin{bmatrix} a & b & c \\ a & b & c \\ a & b & c \end{bmatrix}$$

We assume that at least one of a, b, c is nonzero, so A is nonzero.

24. For which values of a is the matrix diagonalizable? Consider only diagonalization by matrices with *real* entries.
$$\begin{bmatrix} 1 & 0 & 0 \\ 0 & 2 & 1 \\ 0 & 0 & a \end{bmatrix}$$

25. For which values of a is the matrix diagonalizable?
$$\begin{bmatrix} 1 & 0 & 0 \\ 0 & 0 & 1 \\ 0 & a & 0 \end{bmatrix}$$

26. Show that the matrix is diagonalizable for all real values of a.
$$\begin{bmatrix} 1 & 0 & 0 \\ 0 & 2 & 1 \\ a & 0 & 0 \end{bmatrix}$$

27. Use diagonalization to compute A^6 and A^9, where
$$A = \begin{bmatrix} 0 & 8 \\ 2 & 0 \end{bmatrix}$$

28. Find a formula for A^n if A is as in Exercise 27.

29. Use diagonalization to compute
$$\begin{bmatrix} 2 & 2 & 2 \\ 1 & 1 & 1 \\ 2 & 2 & 2 \end{bmatrix}^7$$

30. Use diagonalization to compute
$$\begin{bmatrix} -2 & 0 & 3 \\ 0 & 1 & 0 \\ 3 & 0 & -2 \end{bmatrix}^{-n}$$

31. Prove that for $k = 0, 1, 2, \ldots,$
$$\begin{bmatrix} 1 & 1 \\ 1 & 1 \end{bmatrix}^{k+1} = \begin{bmatrix} 2^k & 2^k \\ 2^k & 2^k \end{bmatrix}$$

32. Prove the identity
$$\begin{bmatrix} 1 & 1 \\ 1 & 0 \end{bmatrix}^k \begin{bmatrix} 1 \\ 0 \end{bmatrix} = \frac{1}{\sqrt{5}} \begin{bmatrix} r_1^{k+1} - r_2^{k+1} \\ r_1^k - r_2^k \end{bmatrix} \quad (7.12)$$

where $r_1 = (1 + \sqrt{5})/2$, $r_2 = (1 - \sqrt{5})/2$, and k is a positive integer.

33. Prove Theorem 9.

Diagonalization of Linear Transformations

34. (Requires Calculus) Show that differentiation $\frac{d}{dx}$: $P_3 \to P_3$ is not a diagonalizable linear transformation.

In Exercises 35–41 determine whether the given transformation is diagonalizable. If it is, find a basis that diagonalizes it.

35. $T : P_3 \to P_3, \, T(p(x)) = p(0)$

36. $T : P_2 \to P_2, \, T(p(x)) = p(x + 1)$

37. Reflection of $\mathbf{R}^2$ about the x-axis

38. Reflection of $\mathbf{R}^2$ about the line $y = -x$

39. Projection of $\mathbf{R}^2$ onto the x-axis

40. Reflection of $\mathbf{R}^3$ about the xy-plane

41. Projection of $\mathbf{R}^3$ onto the xy-plane

42. Prove Theorem 11.

7.3   Approximations of Eigenvalues and Eigenvectors

Reader's Goal for This Section

To see how eigenvalues and eigenvectors are really computed in practice.

Up to now we have computed eigenvalues by solving the characteristic equation and then finding the corresponding eigenvectors. For large matrices, this is *not* a practical method even if we accept approximate solutions. The main problem is that the computation of the determinant of a large characteristic matrix is very *expensive*. Also, it may be hard to solve the resulting high-degree characteristic equation.

In practice, we use much more efficient methods that work backward: first they approximate an eigenvector and then the corresponding eigenvalue.

In this section we discuss two iterative such methods, the *power method* and the *inverse power method*. Another efficient method based on the QR decomposition of a matrix is discussed in Chapter 8.

The Power Method

The power method approximates an eigenvector of a matrix A by computing the power products

$$A^k \mathbf{x}, \qquad k = 1, 2, \ldots$$

starting with any n-vector $\mathbf{x}$. It is *often* true that as k grows, $A^k \mathbf{x}$ becomes parallel to an eigenvector of A. To illustrate this, consider Example 24.

■ EXAMPLE 24   Let

$$A = \begin{bmatrix} 8 & 7 \\ 1 & 2 \end{bmatrix} \text{ and } \mathbf{x} = \begin{bmatrix} 1 \\ 2 \end{bmatrix}$$

First find the eigenvalues and eigenvectors of A. Then determine the direction of $A^k \mathbf{x}$ for large k and relate it to the direction of the eigenvectors.

SOLUTION A has eigenvalues 1 and 9 with corresponding eigenvectors $\begin{bmatrix} -1 \\ 1 \end{bmatrix}$ and $\begin{bmatrix} 7 \\ 1 \end{bmatrix}$. Because

$$\begin{bmatrix} 1 \\ 2 \end{bmatrix} = \frac{13}{8} \begin{bmatrix} -1 \\ 1 \end{bmatrix} + \frac{3}{8} \begin{bmatrix} 7 \\ 1 \end{bmatrix}$$

we have by (7.4), Section 7.1,

$$A^k \mathbf{x} = \frac{13}{8} \cdot 1^k \begin{bmatrix} -1 \\ 1 \end{bmatrix} + \frac{3}{8} \cdot 9^k \begin{bmatrix} 7 \\ 1 \end{bmatrix} = \frac{1}{8} \begin{bmatrix} -13 + 21 \cdot 9^k \\ 13 + 3 \cdot 9^k \end{bmatrix}$$

To find the long-term direction of $A^k\mathbf{x}$, we compute the ratio of its components as $k \rightarrow \infty$:

$$\lim_{k \rightarrow \infty} \frac{-13 + 21 \cdot 9^k}{13 + 3 \cdot 9^k} = \lim_{k \rightarrow \infty} \frac{-13/9^k + 21}{13/9^k + 3} = \frac{21}{3} = 7$$

because $\lim_{k \rightarrow \infty}(1/9)^k = 0$. So, $A^k\mathbf{x}$ tends to become parallel to $\begin{bmatrix} 7 \\ 1 \end{bmatrix}$, the eigenvector with the eigenvalue of maximum absolute value. For example, for $k = 6$, we have

$$A^6\mathbf{x} = \begin{bmatrix} 1,395,031 \\ 199,292 \end{bmatrix} \qquad \text{with component ratio} \cong 6.9999$$

The solution in this example suggests that the direction of $A^k\mathbf{x}$ approaches the direction of the eigenvector with the largest absolute value.

Let A be an $n \times n$ matrix with eigenvalues $\lambda_1, \ldots, \lambda_n$. One eigenvalue, say, λ_1, is called the **dominant eigenvalue** if

$$|\lambda_1| > |\lambda_2| \geq \cdots \geq |\lambda_n|$$

For example, if A has eigenvalues 1, 4, -5, then -5 is the dominant eigenvalue. If A has eigenvalues 1, 4, -4, then there is *no* dominant eigenvalue, because there are two eigenvalues with maximum absolute value $|4| = |-4|$.

Note that a dominant eigenvalue cannot be zero. Also, it must be *real* if the matrix is real, because complex eigenvalues of a real matrix come in conjugate pairs and as we know, conjugate numbers have the same absolute values.

What makes matrices with a dominant eigenvalue special is the following theorem that is the backbone of the power method.

THEOREM 12   Let A be an $n \times n$ diagonalizable matrix with basic eigenvectors $\mathbf{v}_1, \ldots, \mathbf{v}_n$ and corresponding eigenvalues $\lambda_1, \ldots, \lambda_n$ and a dominant eigenvalue, say, λ_1. Let $\mathbf{x}$ be a vector that is a linear combination of the $\mathbf{v}$s,

$$\mathbf{x} = c_1\mathbf{v}_1 + \cdots + c_n\mathbf{v}_n$$

so that $c_1 \neq 0$. Then as k grows, a scalar multiple of $A^k\mathbf{x}$ approaches a scalar multiple of $\mathbf{v}_1$. In particular, the direction of $A^k\mathbf{x}$ approaches that of $\mathbf{v}_1$.

PROOF   By (7.4), Section 7.1,

$$A^k \mathbf{x} = c_1 \lambda_1^k \mathbf{v}_1 + \cdots + c_n \lambda_n^k \mathbf{v}_n$$

Because $\lambda_1 \neq 0$ the equation implies

$$\frac{1}{\lambda_1^k} A^k \mathbf{x} = c_1 \mathbf{v}_1 + c_2 \left(\frac{\lambda_2}{\lambda_1} \right)^k \mathbf{v}_2 + \cdots + c_n \left(\frac{\lambda_n}{\lambda_1} \right)^k \mathbf{v}_n \qquad (7.13)$$

Using the fact that $r^k \to 0$ as $k \to \infty$ if $|r| < 1$, we have

$$\frac{1}{\lambda_1^k} A^k \mathbf{x} \to c_1 \mathbf{v}_1 \qquad \text{as } k \to \infty$$

because $|\lambda_2/\lambda_1|, \ldots, |\lambda_n/\lambda_1| < 1$. Hence, $A^k \mathbf{x}$ scaled by $1/\lambda_1^k$ becomes parallel to $\mathbf{v}_1$ for large k, provided that $c_1 \neq 0$.

So, to find an approximation of an eigenvector of the dominant eigenvalue starting with an initial vector $\mathbf{x}$, we compute $A^k \mathbf{x}$ by the following iteration. Let $\mathbf{x}_0, \mathbf{x}_1, \mathbf{x}_2, \ldots$ be defined by

$$\mathbf{x} = \mathbf{x}_0, \qquad \mathbf{x}_1 = A\mathbf{x}_0, \qquad \mathbf{x}_2 = A^2 \mathbf{x}_0 = A\mathbf{x}_1, \ldots,$$

So, $\mathbf{x}_k = A^k \mathbf{x}$ and

$$\mathbf{x}_{k+1} = A\mathbf{x}_k, \qquad k = 0, 1, \ldots \qquad (7.14)$$

To approximate the corresponding eigenvalue λ_1, we observe that if $\mathbf{x}_k$ is an (approximate) eigenvector, then $\mathbf{x}_{k+1} = A\mathbf{x}_k \cong \lambda_1 \mathbf{x}_k$. Hence, λ_1 can be computed by taking the quotient of two corresponding components of $\mathbf{x}_{k+1}$ and $\mathbf{x}_k$.

Let us utilize Theorem 12 to find an eigenvector of a matrix by reworking Example 24.

■ EXAMPLE 25   Approximate the dominant eigenvalue of

$$A = \begin{bmatrix} 8 & 7 \\ 1 & 2 \end{bmatrix}$$

and a corresponding eigenvector by using the iteration (7.14) with $\mathbf{x}_0 = \begin{bmatrix} 1 \\ 2 \end{bmatrix}$.

SOLUTION   We need $\mathbf{x}_k$ for several values of k. The following table was compiled by using Maple. It displays the values of k, $\mathbf{x}_k$, the quotient d_k between the components of each $\mathbf{x}_k$, and the quotient l_k between the first components of $\mathbf{x}_k$ and $\mathbf{x}_{k-1}$.

k	0	1	2	3	4	5
$\mathbf{x}_k$	$\begin{bmatrix} 1 \\ 2 \end{bmatrix}$	$\begin{bmatrix} 22 \\ 5 \end{bmatrix}$	$\begin{bmatrix} 211 \\ 32 \end{bmatrix}$	$\begin{bmatrix} 1912 \\ 275 \end{bmatrix}$	$\begin{bmatrix} 17{,}221 \\ 2462 \end{bmatrix}$	$\begin{bmatrix} 155{,}002 \\ 22{,}145 \end{bmatrix}$
d_k	0.5	4.4	6.5938	6.9527	6.9947	6.9994
l_k		22.0	9.5909	9.0616	9.0068	9.0008

It is clear that the ratio d_k of the components of $\mathbf{x}_k$ tends to 7. So we get a scalar multiple of $(7, 1)$ as an approximate eigenvector. Because the values of l_k approach 9, so does the corresponding eigenvalue. Thus, we have found the largest absolute value eigenvalue and a corresponding eigenvector. In Fig. 7.3 we display only $\mathbf{x}_0, \mathbf{x}_1, \mathbf{x}_2$ and the line l through $\mathbf{0}$ and $(7, 1)$. The remaining $\mathbf{x}$'s are way out and the lines are too close to display.

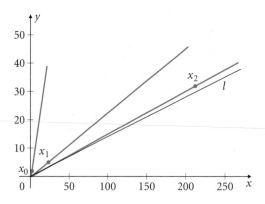

Figure 7.3 The power method.

There is a problem with the calculation of Example 25. The components of $\mathbf{x}_k$ grow very fast, almost out of control, whereas all we need is either $(7, 1)$ or some manageable scalar multiple of it. Because we are mainly interested in the direction of $(7, 1)$, we can scale each $\mathbf{x}_k$ to keep the numbers small. One way is to make $\mathbf{x}_k$ a unit vector by multiplying by $1/\|\mathbf{x}_k\|$. An easier way is to scale $\mathbf{x}_k$ so that its largest entry is 1. In Example 25, we start with $\mathbf{x}_0$ and scale it:

$$\mathbf{y}_0 = \frac{1}{2}\begin{bmatrix} 1 \\ 2 \end{bmatrix} = \begin{bmatrix} 0.5 \\ 1.0 \end{bmatrix}$$

Then we compute

$$\mathbf{x}_1 = \begin{bmatrix} 8 & 7 \\ 1 & 2 \end{bmatrix}\begin{bmatrix} 0.5 \\ 1.0 \end{bmatrix} = \begin{bmatrix} 11.0 \\ 2.5 \end{bmatrix}$$

and scale it:

$$\mathbf{y}_1 = \frac{1}{11.0}\begin{bmatrix} 11.0 \\ 2.5 \end{bmatrix} = \begin{bmatrix} 1.0 \\ 0.22727 \end{bmatrix}$$

and so on. In this setting we may approximate the eigenvalue λ_1 as follows. Suppose we have reached a scaled $\mathbf{y}_k$. Then we compute $\mathbf{x}_{k+1} = A\mathbf{y}_k$; before we scale it we save the component that corresponds to the component with 1 in $\mathbf{y}_k$. This component is an approximation for λ_1. (Why?) In the following table we display the $\mathbf{x}_k$s, $\mathbf{y}_k$s, and the components l_k that approximate λ_1 for A and $\mathbf{x}_0$ of Example 25.

k	0	1	2	3	4	5
$\mathbf{x}_k$	$\begin{bmatrix} 1 \\ 2 \end{bmatrix}$	$\begin{bmatrix} 11.0 \\ 2.5 \end{bmatrix}$	$\begin{bmatrix} 9.5909 \\ 1.4545 \end{bmatrix}$	$\begin{bmatrix} 9.0616 \\ 1.3033 \end{bmatrix}$	$\begin{bmatrix} 9.0068 \\ 1.2877 \end{bmatrix}$	$\begin{bmatrix} 9.0008 \\ 1.2859 \end{bmatrix}$
$\mathbf{y}_k$	$\begin{bmatrix} 0.5 \\ 1.0 \end{bmatrix}$	$\begin{bmatrix} 1.0 \\ 0.22727 \end{bmatrix}$	$\begin{bmatrix} 1.0 \\ 0.15165 \end{bmatrix}$	$\begin{bmatrix} 1.0 \\ 0.14383 \end{bmatrix}$	$\begin{bmatrix} 1.0 \\ 0.14297 \end{bmatrix}$	$\begin{bmatrix} 1.0 \\ 0.14287 \end{bmatrix}$
l_k	2	11.0	9.5909	9.0616	9.0068	9.0008

We see this time that the numbers are easier to handle. (In Fig. 7.4 we display only $\mathbf{y}_0$, $\mathbf{y}_1$, $\mathbf{y}_7$. The remaining ys are too close to $\mathbf{y}_7$, which in turn is too close to the line l through $\mathbf{0}$ and $(7, 1)$.) Let us review the steps of this modified power method.

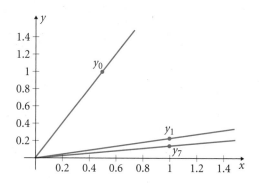

Figure 7.4   The (modified) power method.

Algorithm 1

(Power Method—Maximum Entry 1)

Let A be $n \times n$ diagonalizable matrix with a dominant eigenvalue. Let $\mathbf{x}_0$ be any vector. Let k be the desired number of iterations.

INPUT: $A, \mathbf{x}_0, k$.

1. Let l_0 be the component of $\mathbf{x}_0$ of largest absolute value.
2. Set $\mathbf{y}_0 = (1/l_0)\mathbf{x}_0$.
3. For $i = 1, \ldots, k$
    (a) Compute $A\mathbf{y}_{i-1}$. Let $\mathbf{x}_i = A\mathbf{y}_{i-1}$;
    (b) Let l_i be the component of $\mathbf{x}_i$ of largest absolute value;
    (c) Let $\mathbf{y}_i = (1/l_i)\mathbf{x}_i$.

OUTPUT: $\mathbf{y}_k$ and l_k.
For most $\mathbf{x}_0$, l_k approximates the dominant eigenvalue of A, and $\mathbf{y}_k$ the corresponding eigenvector.

REMARKS

1. We said that the power method applies for almost all $\mathbf{x}_0$, because we do not know in advance whether the coefficient c_1 of $\mathbf{v}_1$ in terms of the eigenvector is $\ne 0$, as Theorem 12 requires. However, in practice the method seems to work for any initial vector, because computer rounding errors will likely replace zeros with small floating-point numbers.
2. The power method is a **self-correcting method**. If at any point our $\mathbf{x}_i$ has been miscalculated, we can still go on, as if we were starting at this vector for the first time.
3. How slow or fast the iteration converges depends on the ratio $|\lambda_2/\lambda_1|$, where λ_2 is the eigenvalue with the second largest absolute value in Equation (7.13). If $|\lambda_2/\lambda_1|$ is close to 0, the convergence is very fast. This was the case in Example 25, where $|\lambda_2/\lambda_1| = \frac{1}{9} \cong 0.11111$. If $|\lambda_2/\lambda_1|$ is close to 1, the convergence is slow.
4. The power method works even when we have a **repeated dominant eigenvalue**.

$$\lambda_1 = \cdots = \lambda_r \quad \text{and} \quad |\lambda_1| > |\lambda_{r+1}| \ge \cdots \ge |\lambda_n|$$

For example, suppose A has eigenvalues 4, 1, -5, -5 (so $\lambda_1 = \lambda_2 = -5$, $\lambda_3 = 4$, $\lambda_4 = 1$). This is easily verified, because (7.13) of Theorem 12 would then be

$$(1/\lambda_1^k)A^k\mathbf{x} = \sum_{i=1}^{r} c_i\mathbf{v}_i + \sum_{i=r+1}^{n} c_i(\lambda_i/\lambda_1)^k\mathbf{v}_i$$

so $(1/\lambda_1^k)A^k\mathbf{x}$ approaches $\sum_{i=1}^{r} c_i\mathbf{v}_i$, which is an eigenvector of λ_1.
5. The power method works even when A is nondiagonalizable.[2] The convergence is usually slow in this case. Careful assumptions and proofs are fairly complicated.

Rayleigh Quotients (or Rayleigh-Ritz Method)

There is a common variation of the power method where we normalize by dividing by the *norm* of the vector; then we use the **Rayleigh quotient** $r(x)$,

$$r(x) = \frac{\mathbf{x}^T A\mathbf{x}}{\mathbf{x}^T\mathbf{x}}$$

to approximate the eigenvalues. This works because if $\mathbf{x}$ is an eigenvector, then $\mathbf{x}^T A\mathbf{x} = \mathbf{x}^T \lambda\mathbf{x} = \lambda(\mathbf{x}^T\mathbf{x})$. To illustrate using the data of Example 25, we let

$$\mathbf{y}_0 = \frac{\mathbf{x}_0}{\|\mathbf{x}_0\|} = \frac{1}{\sqrt{5}}\begin{bmatrix} 1 \\ 2 \end{bmatrix} = \begin{bmatrix} 0.44721 \\ 0.89443 \end{bmatrix}$$

Then we compute

$$\mathbf{x}_1 = A\mathbf{y}_0 = \begin{bmatrix} 9.8387 \\ 2.2361 \end{bmatrix}$$

Next we compute the Rayleigh quotient

$$r(\mathbf{y}_0) = \frac{\mathbf{y}_0^T A\mathbf{y}_0}{\mathbf{y}_0^T \mathbf{y}_0} = \mathbf{y}_0 \cdot \mathbf{x}_1$$

---

[2]Numerical analysts call a nondiagonalizable matrix *defective*.

because $\mathbf{y}_0^T \mathbf{y}_0 = \mathbf{y}_0 \cdot \mathbf{y}_0 = \|\mathbf{y}_0\|^2 = 1$. We continue the same way

$$\mathbf{y}_1 = \frac{\mathbf{x}_1}{\|\mathbf{x}_1\|}, \qquad \mathbf{x}_2 = A\mathbf{y}_1, \qquad r(\mathbf{y}_1) = \mathbf{y}_1 \cdot \mathbf{x}_2, \cdots$$

The Rayleigh quotients

$$r(\mathbf{y}_0), \qquad r(\mathbf{y}_1), \qquad r(\mathbf{y}_2), \ldots,$$

approach the dominant eigenvalue. In our case we have

k	0	1	2	3	4	5
$\mathbf{y}_k$	$\begin{bmatrix} 0.44721 \\ 0.89443 \end{bmatrix}$	$\begin{bmatrix} 0.97509 \\ 0.22162 \end{bmatrix}$	$\begin{bmatrix} 0.9887 \\ 0.14994 \end{bmatrix}$	$\begin{bmatrix} 0.98981 \\ 0.14236 \end{bmatrix}$	$\begin{bmatrix} 0.98994 \\ 0.14152 \end{bmatrix}$	$\begin{bmatrix} 0.98995 \\ 0.14143 \end{bmatrix}$
$\mathbf{x}_{k+1}$	$\begin{bmatrix} 9.8387 \\ 2.2361 \end{bmatrix}$	$\begin{bmatrix} 9.3521 \\ 1.4183 \end{bmatrix}$	$\begin{bmatrix} 8.9592 \\ 1.2886 \end{bmatrix}$	$\begin{bmatrix} 8.915 \\ 1.2745 \end{bmatrix}$	$\begin{bmatrix} 8.9102 \\ 1.273 \end{bmatrix}$	$\begin{bmatrix} 8.9096 \\ 1.2728 \end{bmatrix}$
$r(\mathbf{y}_k)$	6.4	9.4335	9.0512	9.0056	9.0007	9.0001

Algorithm 2

(Rayleigh Quotients, or Rayleigh-Ritz Method)

Let A be an $n \times n$ diagonalizable matrix with a dominant eigenvalue. Let $\mathbf{x}_0$ be any vector. Let k be the desired number of iterations.

INPUT: $A, \mathbf{x}_0, k$.
For $i = 0, \ldots, k - 1$,

1. Let $\mathbf{y}_i = (1/\|\mathbf{x}_i\|)\mathbf{x}_i$;
2. Let $\mathbf{x}_{i+1} = A\mathbf{y}_i$;
3. Let $r_i = \mathbf{y}_i \cdot \mathbf{x}_{i+1}$.

OUTPUT: $\mathbf{y}_{k-1}$ and r_{k-1}.
For most $\mathbf{x}_0$, r_k approximates the dominant eigenvalue of A, and $\mathbf{y}_k$ the corresponding eigenvector.

For **symmetric** matrices this method is very efficient and requires fewer iterations[3] to achieve the same accuracy. For example, consider the matrix $\begin{bmatrix} 2 & 2 \\ 2 & -1 \end{bmatrix}$, with eigenvalues 3 and -2. The power method is slow, because $|(-2)/3|$ is close to 1. To get $\lambda = 3.0$ to five decimal places it takes 33 iterations by the first power method and only 18 iterations by using Rayleigh quotients, if we start at $\mathbf{x}_0 = (1, 2)$. (For this calculation we used Mathematica.)

---

[3] It takes almost half the number of iterations.

Origin Shifts

Now that we know how to compute the dominant eigenvalue, we can use a simple trick to *find the eigenvalue farthest from the dominant one* (if there is one).

As we saw in Exercise 36, Section 7.1, if λ is an eigenvalue of A with corresponding eigenvector $\mathbf{v}$, then for any scalar c the matrix $A - cI$ has eigenvalue $\lambda - c$ and corresponding eigenvector $\mathbf{v}$, because

$$(A - cI)\mathbf{v} = A\mathbf{v} - c\mathbf{v} = (\lambda - c)\mathbf{v}$$

So, if $\lambda_1, \lambda_2, \ldots, \lambda_n$ are the eigenvalues of A, then $0, \lambda_2 - \lambda_1, \ldots, \lambda_n - \lambda_1$ are the eigenvalues of $B = A - \lambda_1 I$.

We can combine this observation with the power method to compute the eigenvalue λ_r farthest from the dominant eigenvalue, say, λ_1. First we compute λ_1. Now $\lambda_r - \lambda_1$ is the dominant eigenvalue of B, so we may use any power method to compute it. Finally, we add λ_1 to get λ_r.

To illustrate, let us compute the smallest of the eigenvalues of A in Example 25 given that we have computed the dominant eigenvalue to be $\lambda_1 = 9$. We form the matrix

$$B = A - 9I = \begin{bmatrix} 8 - 9 & 7 \\ 1 & 2 - 9 \end{bmatrix} = \begin{bmatrix} -1 & 7 \\ 1 & -7 \end{bmatrix}$$

Then we compute the dominant eigenvalue of B by, say, the power method with Rayleigh quotients to get a very fast convergence, starting at $\mathbf{x}_0 = (1, 2)$:

k	0	1	2
y_k	$\begin{bmatrix} 0.44721 \\ 0.89443 \end{bmatrix}$	$\begin{bmatrix} 0.7071 \\ -0.7071 \end{bmatrix}$	$\begin{bmatrix} -0.70711 \\ 0.70711 \end{bmatrix}$
$r(y_k)$	-2.6	-7.9998	-8.0001

Hence, the dominant eigenvalue of B is -8. Therefore, the eigenvalue of A farthest from $\lambda_1 = 9$ is $9 - 8 = 1$. The corresponding eigenvector is $\begin{bmatrix} -0.70711 \\ 0.70711 \end{bmatrix}$, or $\begin{bmatrix} -1 \\ 1 \end{bmatrix}$.

Inverse Power Method

In this paragraph we show how to use the power method to find *the eigenvalue closest to the origin* (if there is one).

The inverse power method is based on the observation that if λ is an eigenvalue of A with corresponding eigenvector $\mathbf{v}$, then λ^{-1} is an eigenvalue of A^{-1} with the same eigenvector (Exercise 35, Section 7.1). Thus, to find the eigenvalue of A closest to the origin we need to compute only the dominant eigenvalue of A^{-1}. The computation of A^{-1} is expensive, but we can avoid this by just solving the system

$$A\mathbf{x}_{k+1} = \mathbf{x}_k$$

for $\mathbf{x}_{k+1}$. Here is a version of this method that uses Rayleigh quotients.

Algorithm 3

(Inverse Power Method)

INPUT: $A, \mathbf{x}_0, k.$
For $i = 0, \ldots, k - 1,$

1. Let $\mathbf{y}_i = (1/\|\mathbf{x}_i\|)\mathbf{x}_i;$
2. Solve the system $A\mathbf{z} = \mathbf{y}_i$ for $\mathbf{z};$
3. Let $\mathbf{x}_{i+1} = \mathbf{z};$
4. Let $r_i = \mathbf{y}_i \cdot \mathbf{x}_{i+1}.$

OUTPUT: $\mathbf{y}_{k-1}$, and $r_{k-1}.$
For most $\mathbf{x}_0$, r_k^{-1} approximates the eigenvalue of A closest to the origin. $\mathbf{y}_k$ is the corresponding eigenvector.

To illustrate, suppose we want the eigenvalue of A in Example 25 closest to the origin. Then for $\mathbf{x}_0 = (1, 2)$, we have

$$\mathbf{y}_0 = \frac{\mathbf{x}_0}{\|\mathbf{x}_0\|} = \frac{1}{\sqrt{5}} \begin{bmatrix} 1 \\ 2 \end{bmatrix} = \begin{bmatrix} 0.44721 \\ 0.89443 \end{bmatrix}$$

Then we row-reduce the system

$$\begin{bmatrix} 8 & 7 & 0.44721 \\ 1 & 2 & 0.89443 \end{bmatrix} \sim \begin{bmatrix} 1 & 0 & -0.59629 \\ 0 & 1 & 0.74536 \end{bmatrix}$$

and set

$$\mathbf{x}_1 = \begin{bmatrix} -0.59629 \\ 0.74536 \end{bmatrix} \quad \text{and} \quad r_0 = \mathbf{x}_1 \cdot \mathbf{y}_0 = 0.40001$$

We continue the same way to get

$$\mathbf{y}_1 = \begin{bmatrix} -0.62469 \\ 0.78087 \end{bmatrix}, \qquad \mathbf{y}_2 = \begin{bmatrix} -0.69893 \\ 0.71519 \end{bmatrix}$$

and

$$r_1 = 1.0623, \qquad r_2 = 1.0075$$

If we stop now, the eigenvalue closest to the origin is approximately $1/r_2 = 1/1.0075 = 0.99256$, which is already close to 1, the true eigenvalue. The corresponding eigenvector is $\mathbf{y}_2$, which is parallel to $\begin{bmatrix} -1 \\ 1.0233 \end{bmatrix}$. This is close to the true eigenvector $\begin{bmatrix} -1 \\ 1 \end{bmatrix}$.

Algorithm 3 is more efficiently used with an LU factorization of A. In this case Step (2) is replaced by:

(2a) Solve $L\mathbf{y} = \mathbf{y}_i.$
(2b) Solve $L\mathbf{z} = \mathbf{y}.$

Shifted Inverse Power Method

Finally, we combine the inverse iteration with an origin shift to compute *the eigenvalue closest to a given number*. For example, if a number μ is closest to the eigenvalue λ than to any other eigenvalue, then $1/(\lambda - \mu)$ is a dominant eigenvalue of $(A - \mu I)^{-1}$, which we can compute by a power method. Just as with the inverse power method, it is more efficient to solve the system

$$(A - \mu I)\mathbf{x}_{k+1} = \mathbf{x}_k$$

rather than using the inverse matrix. We have the following.

Algorithm 4

(Shifted Inverse Power Method)

INPUT: $A, \mathbf{x}_0, k$. Initial guess μ for eigenvalue.
For $i = 0, \ldots, k - 1$,

1. Let $\mathbf{y}_i = (1/\|\mathbf{x}_i\|)\mathbf{x}_i$;
2. Solve the system $(A - \mu I)\mathbf{z} = \mathbf{y}_i$ for $\mathbf{z}$;
3. Let $\mathbf{x}_{i+1} = \mathbf{z}$;
4. Let $r_i = \mathbf{y}_i \cdot \mathbf{x}_{i+1}$.

OUTPUT: $\mathbf{y}_{k-1}$, and r_{k-1}.
For most $\mathbf{x}_0$, $\mu + r_k^{-1}$ approximates the eigenvalue of A closest to μ. $\mathbf{y}_k$ is the corresponding eigenvector.

This iteration works extremely well if the given number is an initial guess for an eigenvalue that we want to compute. The better the guess, the faster the convergence of the iteration.

In Example 25 suppose we have an initial guess $\mu = 2$ for one of the eigenvalues of A. Then for $\mathbf{x}_0 = (1, 2)$, we have

$$\mathbf{y}_0 = \frac{\mathbf{x}_0}{\|\mathbf{x}_0\|} = \begin{bmatrix} 0.44721 \\ 0.89443 \end{bmatrix}$$

Then we row-reduce the system $[A - 2I : \mathbf{y}_0]$,

$$\begin{bmatrix} 6 & 7 & 0.44721 \\ 1 & 0 & 0.89443 \end{bmatrix} \sim \begin{bmatrix} 1 & 0 & 0.89443 \\ 0 & 1 & -0.70277 \end{bmatrix}$$

and set

$$\mathbf{x}_1 = \begin{bmatrix} 0.89443 \\ -0.70277 \end{bmatrix} \quad \text{and} \quad r_0 = \mathbf{x}_1 \cdot \mathbf{y}_0 = -0.22858$$

We continue the same way to get

$$\mathbf{y}_1 = \begin{bmatrix} 0.78631 \\ -0.61782 \end{bmatrix}, \quad \mathbf{y}_2 = \begin{bmatrix} -0.69347 \\ 0.72049 \end{bmatrix}$$

and

$$r_1 = -0.88237, \qquad r_2 = -1.016$$

If we stop at this point, the eigenvalue closest to 2 is $2 + 1/r_2 = 2 + 1/(-1.016) = 1.0157$, which is already pretty close to 1, the true eigenvalue.

Again, Algorithm 4 is typically used with an LU factorization of A.

Application to Roots of Polynomials

Numerical approximations of eigenvalues can be so efficient that instead of computing the eigenvalues using the characteristic equation, we compute the roots of the characteristic equation or *any* polynomial equation by approximating the eigenvalues of the **companion matrix** . In Exercises 47–53, Section 7.1, we defined the companion matrix $C(p)$ of a monic polynomial $p(x)$ and saw that its eigenvalues are the roots of $p(x)$.

■ EXAMPLE 26   (Roots as Eigenvalues)   Approximate one root of $p(x) = x^3 - 15x^2 + 59x - 45$ using the initial guess $x = 10$.

SOLUTION   We apply the shifted inverse power method to find the eigenvalue of $C(p)$ closest to 10.

$$C(p) = \begin{bmatrix} 0 & 1 & 0 \\ 0 & 0 & 1 \\ 45 & -59 & 15 \end{bmatrix}$$

Starting with a unit initial vector, say, $\mathbf{y}_0 = \left(\frac{2}{3}, \frac{2}{3}, \frac{1}{3}\right)$, we reduce $[C(p) - 10I : \mathbf{y}_0]$ to get $\mathbf{x}_1$, then compute $\mathbf{y}_0 \cdot \mathbf{x}_1$ to get $r(y_0)$, and so on. After four iterations we have

$r(y_0)$	$r(y_1)$	$r(y_2)$	$r(y_3)$
0.17778	-1.2099	-1.0303	-1.0051

Hence, an approximate root of $p(x)$ is $10 + (-1.0051)^{-1} = 9.0051$, which is already close to the true root $x = 9$. In a similar manner we may obtain the other two roots, 1 and 5.

NOTE   Computing roots of polynomials as approximate eigenvalues of the companion matrix is often the method of choice for numerical software packages. For example, MATLAB's command `roots` is based on this method.

Approximate Eigenvalues with CAS

```
Maple

> with(linalg):
> evalf(eigenvals(matrix(
        [[1,1,1],[1,1,0],[1,0,0]])));

            [-.8019377352, .5549681326, 2.246979605]
```

Mathematica

```
In[1]:=
    N[Eigenvalues[{{1,1,1},{1,1,0},{1,0,0}}]]
Out[1]=
                              -16                            -17
    {2.24698+1.11022 |10    I,-0.801938-5.55112 10    I,
                              -16
        0.554958-1.11022 10    I
```

MATLAB

```
>> eig([1 1 1; 1 1 0; 1 0 0])
ans =
    0.5550
   -0.8019
    2.2478
```

Exercises 7.3   (Calculator Recommended)

In Exercises 1–2 use the information on the unspecified 2×2 matrix A and 2-vector $\mathbf{x}_0$ to do the following.

(a) Estimate an eigenvalue of A.
(b) Estimate an eigenvector with maximum entry 1.

1. $A^4\mathbf{x}_0 = \begin{bmatrix} 937 \\ 938 \end{bmatrix}$, $A^5\mathbf{x}_0 = \begin{bmatrix} 4687 \\ 4688 \end{bmatrix}$

2. $A^5\mathbf{x}_0 = \begin{bmatrix} 1561 \\ -1564 \end{bmatrix}$, $A^6\mathbf{x}_0 = \begin{bmatrix} -7811 \\ 7814 \end{bmatrix}$

Let $\mathbf{u} = \begin{bmatrix} 2 \\ 1 \end{bmatrix}$ and $\mathbf{v} = \begin{bmatrix} 8 \\ 4 \end{bmatrix}$. In Exercises 3–5 find an eigenvector and an eigenvalue of some (unspecified) 2×2 matrix A if for some (unspecified) 2-vector $\mathbf{x}$ we have

3. $A^6\mathbf{x} = \mathbf{u}$ and $A^7\mathbf{x} = \mathbf{v}$

4. $A^6\mathbf{x} = \mathbf{u}$ and $A^8\mathbf{x} = \mathbf{v}$

5. $A^{-6}\mathbf{x} = \mathbf{u}$ and $A^{-7}\mathbf{x} = \mathbf{v}$

In Exercises 6–19 apply the power method (Algorithm 1) with $k = 4$ to do the following.

(a) Approximate the dominant eigenvalue and a corresponding eigenvector for the given matrix.

(b) Compare the approximate eigenvalue with the true one.

6. $\begin{bmatrix} 3 & 2 \\ 2 & 3 \end{bmatrix}$

7. $\begin{bmatrix} -3 & 2 \\ 2 & -3 \end{bmatrix}$

8. $\begin{bmatrix} 4 & 3 \\ 3 & 4 \end{bmatrix}$

9. $\begin{bmatrix} -4 & 3 \\ 3 & -4 \end{bmatrix}$

10. $\begin{bmatrix} 5 & 4 \\ 4 & 5 \end{bmatrix}$

11. $\begin{bmatrix} -5 & 4 \\ 4 & -5 \end{bmatrix}$

12. $\begin{bmatrix} -8 & 7 \\ 1 & -2 \end{bmatrix}$

13. $\begin{bmatrix} -6 & 5 \\ 2 & -3 \end{bmatrix}$

14. $\begin{bmatrix} -7 & 6 \\ 1 & -2 \end{bmatrix}$

15. $\begin{bmatrix} 7 & 6 \\ 4 & 5 \end{bmatrix}$

16. $\begin{bmatrix} 10 & 9 \\ 2 & 3 \end{bmatrix}$

17. $\begin{bmatrix} -10 & 9 \\ 2 & -3 \end{bmatrix}$

18. $\begin{bmatrix} -7 & 6 \\ 4 & -5 \end{bmatrix}$

19. $\begin{bmatrix} -7 & 6 \\ 7 & -8 \end{bmatrix}$

In Exercises 20–24 use the Rayleigh-Ritz method (Algorithm 2) with $k = 4$ to approximate the dominant eigenvalue of the matrix.

20. The matrix of Exercise 6
21. The matrix of Exercise 7
22. The matrix of Exercise 8
23. The matrix of Exercise 9
24. The matrix of Exercise 10

In Exercises 25–34 use the inverse power method (Algorithm 3) with $k = 4$ to approximate the eigenvalue closest to the origin and compare your answer with the true eigenvalue for each matrix.

25. The matrix of Exercise 6
26. The matrix of Exercise 7
27. The matrix of Exercise 8
28. The matrix of Exercise 9
29. The matrix of Exercise 10
30. The matrix of Exercise 11
31. The matrix of Exercise 12
32. The matrix of Exercise 13
33. The matrix of Exercise 14
34. The matrix of Exercise 15

In Exercises 35–39 use the shifted inverse power method on the companion matrix to approximate the root closest to r of each polynomial $p(x)$.

35. $p(x) = x^2 - 5x + 4, r = 5$
36. $p(x) = x^2 - 8x + 12, r = 7$
37. $p(x) = x^2 - 7x + 6, r = 5$
38. $p(x) = x^2 - 3x - 4, r = 5$
39. $p(x) = x^2 - 8x - 9, r = 10$

7.4 Applications to Dynamical Systems

Reader's Goal for This Section

To study one of the important applications of eigenvalues.

Discrete Dynamical Systems

Eigenvalues and eigenvectors are used to solve many problems of physics, mathematics, and engineering. Here we use them to study discrete dynamical systems,[4] first introduced in Section 2.8.

A **dynamical system**, or **difference equation**, is an equation involving a time-dependent vector quantity $\mathbf{x}(t)$. In a *discrete* dynamical system the time variable is an integer, and we write $\mathbf{x}_k$ for $x(t)$. A **first-order discrete homogeneous dynamical system** is a vector equation of the form

$$\mathbf{x}_{k+1} = A\mathbf{x}_k \tag{7.15}$$

where A is a fixed square matrix of size matching that of the vector $\mathbf{x}_k$. We consider only A's with *real* entries that do *not* depend on k.

[4]Once again, we recommend James T. Sandefur's *Discrete Dynamical Systems, Theory and Applications* (Oxford: Clarendon Press, 1990).

Equation (7.15) gives the next value of $\mathbf{x}$ in terms of the current. We can compute $\mathbf{x}_k$ by repeated applications of (7.15),

$$\mathbf{x}_k = A\mathbf{x}_{k-1} = A^2\mathbf{x}_{k-2} = \cdots$$

Hence,

$$\mathbf{x}_k = A^k\mathbf{x}_0 \tag{7.16}$$

Equation (7.16) is called the **solution** of the dynamical system, and it gives $\mathbf{x}_k$ in terms of an initial vector $\mathbf{x}_0$.

The calculation of $\mathbf{x}_k$ by (7.16) has a main flow: computing A^k can be tedious. Besides, we are often interested in the *long-term behavior* of the system. That is, we want the limit vector

$$\lim_{k \to \infty} \mathbf{x}_k = \lim_{k \to \infty} A^k\mathbf{x}_0$$

if it exists. But computing $A^k\mathbf{x}_0$ for large k becomes a serious task.

Suppose we can write the vector $\mathbf{x}_0$ as a linear combination of eigenvectors $\mathbf{v}_1, \ldots, \mathbf{v}_n$ of A, say,

$$\mathbf{x}_0 = c_1\mathbf{v}_1 + \cdots + c_n\mathbf{v}_n$$

Let $\lambda_1, \ldots, \lambda_n$ be the corresponding eigenvalues. Then by (7.4), Section 7.1,

$$A^k\mathbf{x}_0 = c_1\lambda_1^k\mathbf{v}_1 + \cdots + c_k\lambda_n^k\mathbf{v}_n$$

Hence, the solution of the system can be simplified to

$$\mathbf{x}_k = c_1\lambda_1^k\mathbf{v}_1 + \cdots + c_k\lambda_n^k\mathbf{v}_n \tag{7.17}$$

Equation (7.17) involves *no matrix powers*, and its right-hand side is easy to compute, provided we know the c_is, the $\mathbf{v}_i$s, and the λ_is.

In the special case where the matrix A is *diagonalizable*, this method applies to *any* initial n-vector $\mathbf{x}_0$. For then $\mathbf{R}^n$ has a basis of eigenvectors of A; hence, any n-vector can be written as a linear combination in them. We have proved the following.

THEOREM 13   Let A be an $n \times n$ diagonalizable matrix with linearly independent eigenvectors $\mathbf{v}_1, \ldots, \mathbf{v}_n$ and corresponding eigenvalues $\lambda_1, \ldots, \lambda_n$. Then the solution of the dynamical system $\mathbf{x}_{k+1} = A\mathbf{x}_k$ is given by

$$\mathbf{x}_k = c_1\lambda_1^k\mathbf{v}_1 + \cdots + c_n\lambda_n^k\mathbf{v}_n$$

where the coefficients $c_1, \ldots c_n$ are such that

$$\mathbf{x}_0 = c_1\mathbf{v}_1 + \cdots + c_n\mathbf{v}_n$$

Long-Term Behavior of Dynamical Systems

What happens if k grows large in a dynamical system? Theorem 12, Section 7.3, tells us that if A is diagonalizable with a dominant eigenvalue λ_1 and corresponding eigenvector

v_1 and if

$$\mathbf{x}_0 = c_1\mathbf{v}_1 + \cdots + c_n\mathbf{v}_n, \qquad \text{such that } c_1 \neq 0$$

then as k grows, the direction of $\mathbf{x}_k$ approaches that of $\mathbf{v}_1$.

To illustrate, consider the case with $A = \begin{bmatrix} -2 & -2 \\ -2 & 1 \end{bmatrix}$ (eigenvalues $-3, 2$) and $\mathbf{x}_0 = \begin{bmatrix} 5 \\ 15 \end{bmatrix}$. We know that $A^k\mathbf{x}_0 = \mathbf{x}_{k+1}$ will have direction approaching the direction of $\begin{bmatrix} 2 \\ 1 \end{bmatrix}$, the eigenvector corresponding to $\lambda = -3$.

For example, for $k = 11$ we have

$$\mathbf{x}_{11} = \begin{bmatrix} -1{,}423{,}320 \\ -696{,}300 \end{bmatrix} \qquad \text{with component ratio} \cong 2.0441$$

In Fig. 7.5 we sketch the points $\mathbf{x}_0, \dots, \mathbf{x}_k$ for $k = 2, 3, 6$. Consecutive points are joined by a straight-line segment.

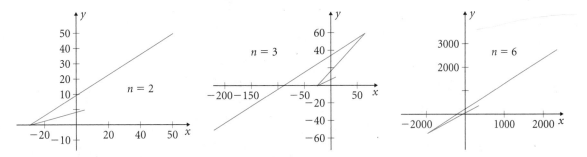

Figure 7.5

Attractors, Repellers, and Saddle Points

In the following examples we discuss the long-term behavior and plot some of the solutions of the dynamical system for the given matrices A. All plots include $(\pm 1, \pm 1)$, $(\pm 1, \pm 2)$, $(\pm 2, \pm 1)$ as starting points. Consecutive points $\mathbf{x}_k, \mathbf{x}_{k+1}$ are joined by straight-line segments. The resulting polygonal lines are the **trajectories** of the solutions.

■ EXAMPLE 27 Study the solutions of the dynamical systems defined by the following matrices.

(a) $\begin{bmatrix} 2 & 0 \\ 0 & 3 \end{bmatrix}$

(b) $\begin{bmatrix} .2 & 0 \\ 0 & .3 \end{bmatrix}$

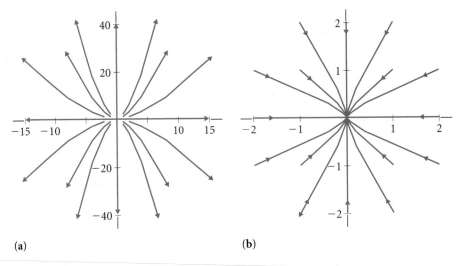

(a)   (b)

Figure 7.6

SOLUTION

(a) The eigenvalues are 3 and 2, with corresponding eigenvectors $(0, 1)$ and $(1, 0)$. Hence, if

$$\mathbf{x}_0 = c_1(0, 1) + c_2(1, 0) = (c_2, c_1)$$

then

$$\mathbf{x}_k = c_1 3^k(0, 1) + c_2 2^k(1, 0) = \left(2^k c_2, 3^k c_1\right) \tag{7.18}$$

So the components of $\mathbf{x}_k$ go to $\pm\infty$, depending on the signs of c_1 and c_2. By Theorem 12 the direction of $\mathbf{x}_k$ approaches that of $(0, 1)$ if $c_1 \neq 0$. In the long run, all trajectories become parallel to the y-axis and move *away* from the origin, except for the points of the x-axis (where $c_1 = 0$). These points remain on the x-axis and move toward $\pm\infty$, depending on whether $\mathbf{x}_0$ is > 0 or < 0. (Why?) In Fig. 7.6(a) we see trajectories up to $k = 2$.

(b) The second matrix has the same eigenvectors and eigenvalues 0.3 and 0.2. So,

$$\mathbf{x}_k = (0.2^k c_2, 0.3^k c_1) \quad \text{where} \quad \mathbf{x}_0 = (c_2, c_1)$$

as before. Hence, both components of $\mathbf{x}_k$ go to zero. By Theorem 12 the direction of $\mathbf{x}_k$ approaches that of $(0, 1)$ for large k. So, the trajectories become parallel to the y-axis and move toward the origin, except for the points of the x-axis (where $c_1 = 0$). These points remain on the x-axis and move toward the origin. (Why?) (Fig. 7.6(b), $k = 2$.)

■ EXAMPLE 28   Study the solutions of the dynamical systems defined by the following matrices.

(a) $\begin{bmatrix} 2.5 & 0.5 \\ 0.5 & \cdot 2.5 \end{bmatrix}$

(b) $\begin{bmatrix} 0.5 & 0.1 \\ 0.1 & 0.5 \end{bmatrix}$

SOLUTION

(a) The eigenvalues of A are 3 and 2, with corresponding eigenvectors $(1, 1)$ and $(-1, 1)$. Hence, for

$$\mathbf{x}_0 = c_1(1, 1) + c_2(-1, 1) = (c_1 - c_2, c_1 + c_2)$$

we have

$$\mathbf{x}_k = c_1 3^k(1, 1) + c_2 2^k(-1, 1) = \left(c_1 3^k - c_2 2^k, c_1 3^k + c_2 2^k\right) \qquad (7.19)$$

As k becomes large, $\mathbf{x}_k$ goes to (∞, ∞) if $c_1 > 0$ and to $(-\infty, -\infty)$ if $c_1 < 0$. So if $c_1 \neq 0$, eventually all the trajectories end up either in the *first* or the *third* quadrant. The long-term direction of $\mathbf{x}_k$ is parallel to $(1, 1)$, by Theorem 12, with the exception of the points $\mathbf{x}_0$ with $c_1 = 0$. These are the points of the line l through the origin and $(-1, 1)$. They remain on l, and their distance from the origin increases with k, because

$$\mathbf{x}_k = \left(-c_2 2^k, c_2 2^k\right) = 2^k c_2(-1, 1)$$

for $c_1 = 0$. (Fig. 7.7(a), $k = 2$.)

(b) For the second matrix we have the same eigenvectors and eigenvalues 0.6 and 0.4. The trajectories again tend to become parallel to $(1, 1)$ but move toward the origin, except for the points of the line l (where $c_1 = 0$). These points remain on l and also move toward the origin. (Fig. 7.7(b), $k = 4$.)

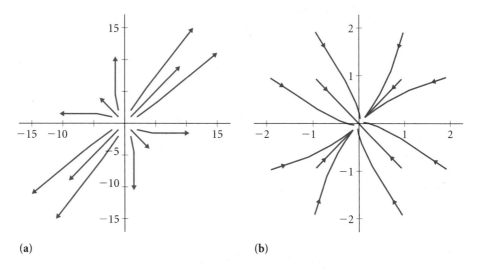

(a)   (b)

Figure 7.7

REMARKS

1. In Examples 27 and 28 we saw that if all eigenvalues have absolute value less than 1, all trajectories approach the origin. We say that the origin is an **attractor**. This is true in general, because each term of $\mathbf{x}_k = c_1\lambda_1^k\mathbf{v}_1 + \cdots + c_n\lambda_n^k\mathbf{v}_n$ would go to zero if all $|\lambda_i| < 1$. If, on the other hand, all eigenvalues have absolute value greater than 1, then all trajectories move away from the origin. We then say that the origin is a **repeller**. If, finally, some trajectories move toward the origin and some move away from it, we say that the origin is a **saddle point**.

2. The trajectories become parallel to the eigenvector with the largest-absolute-value eigenvalue, with the exception of the points along the line of the other eigenvector that remain on that line. The same is true (with zigzagging) if one or both eigenvalues are negative. (Try a few trajectories for $\begin{bmatrix} 2 & 2 \\ 2 & -1 \end{bmatrix}$ and $\begin{bmatrix} -1 & -2 \\ 1 & -4 \end{bmatrix}$.)

3. The graphs of Example 28 are similar to those of Example 27, where the matrices were diagonal. The roles of the axes are now played by the **eigenspaces**. The graphs for the nondiagonal matrices can be obtained from the those of the diagonal ones (or vice versa) by the **change of variables**, discussed in Section 7.1.

In fact, if A is any $n \times n$ diagonalizable matrix and P is the matrix with columns the elements of a basis $\mathcal{B} = \{\mathbf{v}_1, \ldots, \mathbf{v}_n\}$ of eigenvectors of A, we define new variables $\mathbf{y}_k$ such that

$$\mathbf{x}_k = P\mathbf{y}_k \tag{7.20}$$

So, $\mathbf{y}_k = [\mathbf{x}_k]_{\mathcal{B}}$. In this case the dynamical system $\mathbf{x}_{k+1} = A\mathbf{x}_k$ can be rewritten as

$$P\mathbf{y}_{k+1} = AP\mathbf{y}_k \Rightarrow \mathbf{y}_{k+1} = P^{-1}AP\mathbf{y}_k$$

Therefore,

$$\mathbf{y}_{k+1} = D\mathbf{y}_k \tag{7.21}$$

where D is, as usual, the diagonal matrix with diagonal entries the eigenvalues of A. The change of variables (7.20) transformed $\mathbf{x}_{k+1} = A\mathbf{x}_k$ into a **diagonal**, or **uncoupled**, dynamical system (7.21). The advantage in working with uncoupled systems is that the ith component of $\mathbf{y}_{k+1}$ depends only on the ith component of $\mathbf{y}_k$.

To illustrate, if $A = \begin{bmatrix} 2.5 & 0.5 \\ 0.5 & 2.5 \end{bmatrix}$, $P = \begin{bmatrix} -1 & 1 \\ 1 & 1 \end{bmatrix}$, $D = \begin{bmatrix} 2 & 0 \\ 0 & 3 \end{bmatrix}$, and $\mathbf{x}_k$ is as in (7.19), then

$$\mathbf{y}_k = P^{-1}\mathbf{x}_k = \begin{bmatrix} -1 & 1 \\ 1 & 1 \end{bmatrix}^{-1} \begin{bmatrix} c_1 3^k - c_2 2^k \\ c_1 3^k + c_2 2^k \end{bmatrix} = \begin{bmatrix} c_2 2^k \\ c_1 3^k \end{bmatrix}$$

which is Formula (7.18) of Example 27.

Let us now look at the case where one eigenvalue has absolute value less than 1 and one greater than 1.

▪ EXAMPLE 29   Study the solutions of (7.15) for

$$A = \begin{bmatrix} 0.5 & 0 \\ 0 & 1.5 \end{bmatrix} \quad \text{and} \quad A = \begin{bmatrix} 1 & 0.5 \\ 0.5 & 1 \end{bmatrix}$$

SOLUTION   The eigenvalues for both matrices are 0.5 and 1.5. The corresponding eigenvectors v_1 and v_2 are $(0, 1)$ and $(1, 0)$ for the first matrix and $(1, 1)$ and $(-1, 1)$ for the second. Hence,

$$\mathbf{x}_k = c_1(1.5)^k(0, 1) + c_2(0.5)^k(1, 0), \quad \text{where} \quad \mathbf{x}_0 = c_1(0, 1) + c_2(1, 0)$$

for the first matrix and

$$\mathbf{x}_k = c_1(1.5)^k(1, 1) + c_2(0.5)^k(-1, 1), \quad \text{where} \quad \mathbf{x}_0 = c_1(1, 1) + c_2(-1, 1)$$

for the second. In each case as $k \to \infty$, $(0.5)^k \to 0$, and $(1.5)^k \to \infty$, so if c_1 is not zero, all trajectories become parallel to v_1. If c_1 is zero, we have vectors along the direction of v_2, and their trajectories go to zero. Fig. 7.8(a) displays trajectories for the first system $(k = 4)$ and Fig. 7.8(b), for the second $(k = 3)$. This time the origin is a *saddle* point.

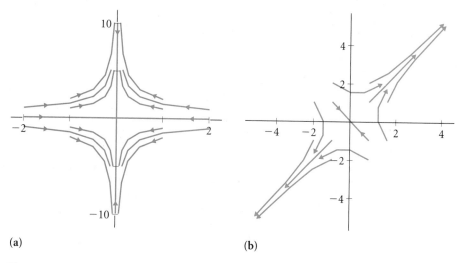

(a)   (b)

Figure 7.8

Repeated Eigenvalue

If our 2×2 matrix has only one eigenvalue λ with two linearly independent eigenvectors v_1 and v_2, then for $\mathbf{x}_0 = c_1 v_1 + c_2 v_2$,

$$\mathbf{x}_k = c_1 \lambda^k v_1 + c_2 \lambda^k v_2$$
$$= \lambda^k (c_1 v_1 + c_2 v_2)$$
$$= \lambda^k \mathbf{x}_0$$

Hence, x_k and x_0 are on the same line. Figure 7.9(a) and (b) illustrate this for the matrices $\begin{bmatrix} 2 & 0 \\ 0 & 2 \end{bmatrix}$ and $\begin{bmatrix} 0.2 & 0 \\ 0 & 0.2 \end{bmatrix}$. For the first matrix the origin is a repeller, and for the second, it is an attractor.

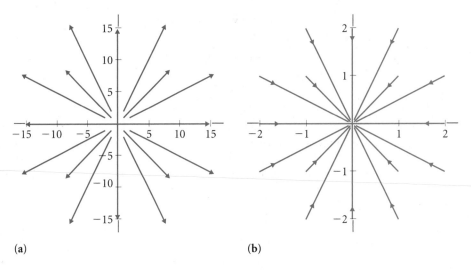

(a)   (b)

Figure 7.9

Complex Eigenvalues

If A has complex eigenvalues, the trajectories typically spiral around the origin toward or away from it (depending on whether the magnitude of the eigenvalues is greater than 1 or less than 1). Or, they circle around it.

■ EXAMPLE 30   Study the solutions of (7.15) for

$$A = \begin{bmatrix} 1 & -1 \\ 1 & 1 \end{bmatrix} \quad \text{and} \quad A = \begin{bmatrix} 0 & 1 \\ -1 & 1 \end{bmatrix}$$

SOLUTION

(a) The eigenvalues of the first matrix are $1 + i, 1 - i$, with corresponding eigenvectors $v_1 = (i, 1)$ and $v_2 = (-i, 1)$. Hence, if $x_0 = c_1 v_1 + c_2 v_2$, then x_k is of the form

$$x_k = c_1(1 + i)^k \begin{bmatrix} i \\ 1 \end{bmatrix} + c_2(1 - i)^k \begin{bmatrix} -i \\ 1 \end{bmatrix} \tag{7.22}$$

Note that since the components of x_k are real, the right-hand side of (7.22) must be a real number. Let us look at the trajectory that starts at $(1, 1)$. Because

$$(1, 1) = c_1 \begin{bmatrix} i \\ 1 \end{bmatrix} + c_2 \begin{bmatrix} -i \\ 1 \end{bmatrix}$$

implies $c_1 = \frac{1}{2} - \frac{1}{2}i$, and $c_2 = \frac{1}{2} + \frac{1}{2}i$, we have

$$\mathbf{x}_k = \left(\frac{1}{2} - \frac{1}{2}i\right)(1+i)^k \begin{bmatrix} i \\ 1 \end{bmatrix} + \left(\frac{1}{2} + \frac{1}{2}i\right)(1-i)^k \begin{bmatrix} -i \\ 1 \end{bmatrix} \tag{7.23}$$

Hence, for $k = 1, 2, 3, 4, \ldots$ we get

$$\begin{bmatrix} 0 \\ 2 \end{bmatrix}, \begin{bmatrix} -2 \\ 2 \end{bmatrix}, \begin{bmatrix} -4 \\ 0 \end{bmatrix}, \begin{bmatrix} -4 \\ -4 \end{bmatrix}, \ldots$$

These vectors are of increasing magnitude, hence the trajectories spiral away from the origin (Fig. 7.10(a)). This spiral behavior can be in fact predicted from (7.23), but we shall skip the details.

(b) For the second matrix a similar calculation yields

$$\mathbf{x}_k = c_1 \left(\frac{1}{2} + \frac{1}{2}i\sqrt{3}\right)^k \begin{bmatrix} \frac{1}{2} - \frac{1}{2}i\sqrt{3} \\ 1 \end{bmatrix}$$
$$+ c_2 \left(\frac{1}{2} - \frac{1}{2}i\sqrt{3}\right)^k \begin{bmatrix} \frac{1}{2} + \frac{1}{2}i\sqrt{3} \\ 1 \end{bmatrix} \tag{7.24}$$

and for $\mathbf{x}_0 = (1, 1)$, it is easy to show that $c_1 = \frac{1}{2} + \frac{1}{6}i\sqrt{3}$ and $c_2 = \frac{1}{2} - \frac{1}{6}i\sqrt{3}$. In this case for $k = 0, \ldots, 6$ we get

$$\begin{bmatrix} 1 \\ 1 \end{bmatrix}, \begin{bmatrix} 1 \\ 0 \end{bmatrix}, \begin{bmatrix} 0 \\ -1 \end{bmatrix}, \begin{bmatrix} -1 \\ -1 \end{bmatrix}, \begin{bmatrix} -1 \\ 0 \end{bmatrix}, \begin{bmatrix} 0 \\ 1 \end{bmatrix}, \begin{bmatrix} 1 \\ 1 \end{bmatrix}, \ldots$$

Notice that $\mathbf{x}_6$ is the same as $\mathbf{x}_0$. Hence, $\mathbf{x}_7$ is the same as $\mathbf{x}_1$, and so on. This time we have a 6-**cycle**, i.e., the vectors are repeated every 6 time units (Fig. 7.10(b)). So, for

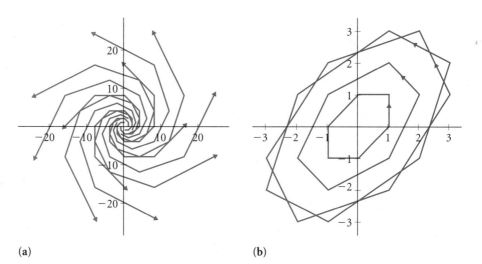

(a)   (b)

Figure 7.10

$$k = 0, 1, 2, \ldots$$

$$\mathbf{x}_{k+6} = \mathbf{x}_k$$

Also, we may write

$$\mathbf{x}_k = \mathbf{x}_r$$

where r is the remainder of the division of k by 6. For example,

$$\mathbf{x}_{44} = \mathbf{x}_2 = \begin{bmatrix} 0 \\ -1 \end{bmatrix}$$

This cyclical behavior is due to the fact the eigenvalues are sixth roots of 1, i.e., $\left(\frac{1}{2} \pm \frac{1}{2}i\sqrt{3}\right)^6 = 1$. So we see from (7.24) that the values of $\mathbf{x}_k$ are repeated any time k is incremented by 6. This is true for any choices of c_1 and c_2.

Note that there is nothing special about 6-cycles. We can also have cases with 2-, 3-, 4-, ... cycles.

A Population Growth Problem

Let us apply our methods to the insect population problem of Section 2.8.

A population of insects is divided into three age groups, A, B, and C. Group A consists of insects 0–1 wk old, group B consists of insects 1–2 wk old, and group C consists of insects 2–3 wk old. Suppose the groups have A_k, B_k, and C_k number of insects at the end of the kth week. We want to study how A, B, C change over time, given the following two conditions:

1. (Survival Rate) Only 10% of age group A survive a week. Hence,

$$B_{k+1} = \frac{1}{10}A_k \tag{7.25}$$

And only 40% of age group B survive a week. So,

$$C_{k+1} = \frac{2}{5}B_k \tag{7.26}$$

2. (Birth Rate) Each insect from group A has $\frac{2}{5}$ offspring, each insect from group B has 4 offspring, and each insect from group C has 5 offspring. In year $k + 1$ the insects of group A are offspring of insects in year k. Hence,

$$A_{k+1} = \frac{2}{5}A_k + 4B_k + 5C_k \tag{7.27}$$

■ PROBLEM   If there are initially 1000 insects in each age group, what is the population distribution in the long run?

SOLUTION   Equations (7.25), (7.26), and (7.27) can be rewritten as the dynamical system

$$\mathbf{x}_{k+1} = A\mathbf{x}_k$$

with

$$A = \begin{bmatrix} \frac{2}{5} & 4 & 5 \\ \frac{1}{10} & 0 & 0 \\ 0 & \frac{2}{5} & 0 \end{bmatrix} \quad \text{and} \quad \mathbf{x}_k = \begin{bmatrix} A_k \\ B_k \\ C_k \end{bmatrix}, \qquad k = 0, 1, 2, \dots,$$

Therefore,

$$\mathbf{x}_{k+1} = A^k \mathbf{x}_0$$

The eigenvalues of A are found to be $\lambda = 1, r/10, \bar{r}/10$, where $r = -3 - i\sqrt{11}$ and $\bar{r} = -3 + i\sqrt{11}$ (the conjugate of r) and the corresponding eigenvectors are $(50, 5, 2)$, $(r^2, r, 4)$, and $(\bar{r}^2, \bar{r}, 4)$. Hence, if

$$\mathbf{x}_0 = c_1 \begin{bmatrix} 50 \\ 5 \\ 2 \end{bmatrix} + c_2 \begin{bmatrix} r^2 \\ r \\ 4 \end{bmatrix} + c_3 \begin{bmatrix} \bar{r}^2 \\ \bar{r} \\ 4 \end{bmatrix} \tag{7.28}$$

then

$$\mathbf{x}_k = c_1 1^k \begin{bmatrix} 50 \\ 5 \\ 2 \end{bmatrix} + c_2 \left(\frac{r}{10}\right)^k \begin{bmatrix} r^2 \\ r \\ 4 \end{bmatrix} + c_3 \left(\frac{\bar{r}}{10}\right)^k \begin{bmatrix} \bar{r}^2 \\ \bar{r} \\ 4 \end{bmatrix} \tag{7.29}$$

Note that $|r/10| = |\bar{r}/10| = 1/\sqrt{5} < 1$. Hence, the positive numbers $|r/10|^k$ and $|\bar{r}/10|^k$ go to 0 as $k \to \infty$. Thus, the complex numbers

$$\left(\frac{r}{10}\right)^k \to 0, \qquad \left(\frac{\bar{r}}{10}\right)^k \to 0 \quad \text{as} \quad k \to \infty$$

Therefore, for large k (7.29) reduces to

$$\mathbf{x}_k \cong c_1 \begin{bmatrix} 50 \\ 5 \\ 2 \end{bmatrix} \tag{7.30}$$

So, for any given initial vector $\mathbf{x}_0 = (A_0, B_0, C_0)$, it suffices to compute c_1 from (7.28) and substitute into (7.30) to find $\mathbf{x}_k$ (for large k). We can solve for c_1 by, say, Cramer's rule,

$$c_1 = \begin{vmatrix} A_0 & r^2 & \bar{r}^2 \\ B_0 & r & \bar{r} \\ C_0 & 4 & 4 \end{vmatrix} \begin{vmatrix} 50 & r^2 & \bar{r}^2 \\ 5 & r & \bar{r} \\ 2 & 4 & 4 \end{vmatrix}^{-1}$$

to get

$$c_1 = \frac{1}{90}(A_0 + 6B_0 + 5C_0)$$

after (a rather lengthy) simplification. Hence, for $\mathbf{x}_0 = (1000, 1000, 1000)$ we have

$$\mathbf{x}_k \cong \frac{1}{90} \cdot 12 \cdot 1000 \cdot \begin{bmatrix} 50 \\ 5 \\ 2 \end{bmatrix} \cong \begin{bmatrix} 6666.66 \\ 666.66 \\ 266.66 \end{bmatrix}$$

Therefore, under the given survival and birth rates the numbers of insects in age groups A, B, and C approach 6666.66, 666.66, and 266.66, respectively. In Fig. 7.11 we see that the trajectory spirals to the point with coordinates these numbers.

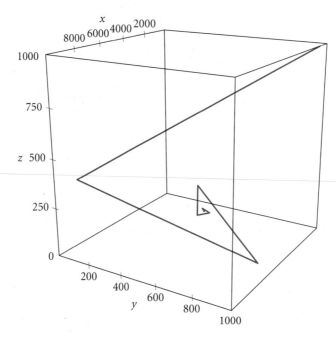

Figure 7.11

Exercises 7.4

In Exercises 1–9 suppose that a matrix A has eigenvectors $\mathbf{v}_1 = \begin{bmatrix} -1 \\ 1 \end{bmatrix}$ and $\mathbf{v}_2 = \begin{bmatrix} 1 \\ 1 \end{bmatrix}$ with corresponding given eigenvalues λ_1 and λ_2. Consider the dynamical system $\mathbf{x}_{k+1} = A\mathbf{x}_k$ with initial vector is $\mathbf{x}_0 = \begin{bmatrix} 1 \\ 4 \end{bmatrix}$.

(a) Find a formula for $\mathbf{x}_k$.
(b) Compute $A\mathbf{x}_0$ and $A^2\mathbf{x}_0$.
(c) Indicate whether the origin is an attractor, repeller, or neither.

1. $\lambda_1 = 1, \lambda_2 = 5$
2. $\lambda_1 = 2, \lambda_2 = 10$
3. $\lambda_1 = -7, \lambda_2 = -1$
4. $\lambda_1 = 1, \lambda_2 = 9$
5. $\lambda_1 = \frac{1}{2}, \lambda_2 = \frac{5}{2}$

6. $\lambda_1 = 2, \lambda_2 = 14$
7. $\lambda_1 = -13, \lambda_2 = -1$
8. $\lambda_1 = -\frac{1}{6}, \lambda_2 = \frac{5}{6}$
9. $\lambda_1 = -\frac{1}{10}, \lambda_2 = \frac{1}{2}$

In Exercises 10–15 consider the dynamical system $\mathbf{x}_{k+1} = A\mathbf{x}_k$ with the given matrix A and $\mathbf{x}_0 = \begin{bmatrix} 1 \\ 1 \end{bmatrix}$. Find $\mathbf{x}_5$ by using (a) $A^5\mathbf{x}_0$, (b) eigenvalues.

10. $A = \begin{bmatrix} 7 & 6 \\ 4 & 5 \end{bmatrix}$

11. $A = \begin{bmatrix} -3 & 2 \\ 2 & -3 \end{bmatrix}$

12. $A = \begin{bmatrix} -6 & 5 \\ 2 & -3 \end{bmatrix}$

13. $A = \begin{bmatrix} \frac{3}{10} & \frac{2}{10} \\ \frac{2}{10} & \frac{3}{10} \end{bmatrix}$

14. $A = \begin{bmatrix} \frac{3}{2} & 1 \\ 1 & \frac{3}{2} \end{bmatrix}$

15. $A = \begin{bmatrix} 5 & 4 \\ 4 & 5 \end{bmatrix}$

In Exercises 16–18 use eigenvalues and eigenvectors to compute $\mathbf{x}_1$, $\mathbf{x}_2$, $\mathbf{x}_3$ for the dynamical system $\mathbf{x}_{k+1} = A\mathbf{x}_k$ with the given matrix A and $\mathbf{x}_0 = \begin{bmatrix} 1 \\ 1 \end{bmatrix}$. Draw the trajectory through $\mathbf{x}_0, \ldots, \mathbf{x}_3$ and determine whether the origin is an attractor, repeller, or neither.

16. $A = \begin{bmatrix} 2 & -2 \\ 2 & 2 \end{bmatrix}$

17. $A = \begin{bmatrix} \frac{1}{2} & -\frac{1}{2} \\ \frac{1}{2} & \frac{1}{2} \end{bmatrix}$

18. $A = \begin{bmatrix} 0 & \frac{1}{2} \\ -2 & 1 \end{bmatrix}$

19. Show that all the solutions of the dynamical system
$$\mathbf{x}_{k+1} = \begin{bmatrix} 0 & 2 \\ -\frac{1}{2} & 1 \end{bmatrix} \mathbf{x}_k \text{ are 6-cycles.}$$

20. Show that all the solutions of the dynamical system
$$\mathbf{x}_{k+1} = \begin{bmatrix} \sqrt{3} & 1 \\ -1 & 0 \end{bmatrix} \mathbf{x}_k \text{ are 12-cycles.}$$

21. The kth generation of an animal population consists of A_k females and B_k males. Suppose that the next generation depends on the current one according to:
$$A_{k+1} = 0.8A_k + 0.7B_k$$
$$B_{k+1} = 0.2A_k + 0.3B_k$$

Write this dynamical system in matrix notation. If initially there were 100 females and 300 males, what is the approximate population (a) right after the third generation? (b) In the long run? Which gender will eventually dominate?

22. Repeat Exercise 21 for the following dependencies:
$$A_{k+1} = 0.7B_k$$
$$B_{k+1} = A_k + 0.3B_k$$

7.5 Applications to Markov Chains

Reader's Goal for This Section

To see how eigenvalues can be used to study the long term behavior of Markov chains.

One of the most interesting applications of eigenvalues is in computing advanced stages of Markov chains studied in Section 3.5. Recall that in a Markov chain, the next state of a system depends only on its current state. To illustrate, we revisit the study of smokers versus nonsmokers of that section.

Suppose that the probability a smoker will continue smoking a year later is 65%, whereas the probability a nonsmoker will continue not smoking is 85%. This information was tabulated by using the stochastic matrix of transition probabilities:

$$A = \begin{bmatrix} 0.65 & 0.15 \\ 0.35 & 0.85 \end{bmatrix}$$

For example, the entry 0.35 means that a smoker has a 35% chance of quitting a year later, and 0.15 means that a nonsmoker has a 15% chance of picking up smoking.

■ **EXAMPLE 31**   What are the percentages of smokers and nonsmokers in the long run if $100p$ percent are initially smokers and $100q$ percent are nonsmokers?

SOLUTION   First we note that $p + q = 1$. Recall from Section 3.5 that in k years the new percentages can be computed as

$$\begin{bmatrix} 0.65 & 0.15 \\ 0.35 & 0.85 \end{bmatrix}^{k} \begin{bmatrix} p \\ q \end{bmatrix}$$

So, we need the value of this vector as k goes to ∞. Diagonalization of A yields

$$A^{k} = \begin{bmatrix} 3 & -1 \\ 7 & 1 \end{bmatrix} \begin{bmatrix} 1 & 0 \\ 0 & \frac{1}{2} \end{bmatrix}^{k} \begin{bmatrix} 3 & -1 \\ 7 & 1 \end{bmatrix}^{-1}$$

Because $\left(\frac{1}{2}\right)^{k}$ approaches 0 as $k \to \infty$, we have

$$\lim_{k\to\infty} A^{k} = \begin{bmatrix} 3 & -1 \\ 7 & 1 \end{bmatrix} \begin{bmatrix} 1 & 0 \\ 0 & 0 \end{bmatrix} \begin{bmatrix} 3 & -1 \\ 7 & 1 \end{bmatrix}^{-1} = \begin{bmatrix} 0.3 & 0.3 \\ 0.7 & 0.7 \end{bmatrix}$$

Therefore,

$$\lim_{k\to\infty} A^{k} \begin{bmatrix} p \\ q \end{bmatrix} = \begin{bmatrix} 0.3 & 0.3 \\ 0.7 & 0.7 \end{bmatrix} \begin{bmatrix} p \\ q \end{bmatrix}$$

$$= \begin{bmatrix} 0.3p + 0.3q \\ 0.7p + 0.7q \end{bmatrix} = \begin{bmatrix} 0.3 \\ 0.7 \end{bmatrix}$$

because $p + q = 1$. Hence, in the long run the smokers will be 30% versus 70% of nonsmokers. This is true for any starting percentage vector (p, q) with $p + q = 1$.

A vector whose components are all nonnegative and add up to 1 is called a **probability** vector. For example,

$$\begin{bmatrix} 0.3 \\ 0.7 \end{bmatrix}, \qquad \begin{bmatrix} 0 \\ 1 \end{bmatrix}, \qquad \begin{bmatrix} 0.2 \\ 0.4 \\ 0.4 \end{bmatrix}$$

are probability vectors. However, $\begin{bmatrix} 2 \\ -1 \end{bmatrix}, \begin{bmatrix} 0.8 \\ 0.1 \end{bmatrix}$ are not, because $-1 < 0$ and since $0.1 + 0.8 \neq 1$. In Example 31 we showed that for any probability vector **v**, the limit of $A^{k}\mathbf{v}$ is $(0.3, 0.7)$ as $k \to \infty$.

As a second example, recall the Army-Navy games of Section 3.5. The probability that Army wins one year and Navy wins the next year is 70%; The probability that Navy wins one year and Army wins the next year is 30%. This can be expressed by the doubly stochastic matrix of transition probabilities:

$$B = \begin{bmatrix} 0.7 & 0.3 \\ 0.3 & 0.7 \end{bmatrix}$$

■ EXAMPLE 32   (Army-Navy Games)   Given that Navy won this year's game, what is the probability that it wins in the long run? What if Navy lost this year?

SOLUTION   Diagonalization of B yields

$$B^k = \begin{bmatrix} 1 & -1 \\ 1 & 1 \end{bmatrix} \begin{bmatrix} 1 & 0 \\ 0 & \frac{2}{5} \end{bmatrix}^k \begin{bmatrix} 1 & -1 \\ 1 & 1 \end{bmatrix}^{-1}$$

Because $\left(\frac{2}{5}\right)^k$ approaches 0 as $k \rightarrow \infty$, we have

$$\lim_{k \to \infty} B^k = \begin{bmatrix} 1 & -1 \\ 1 & 1 \end{bmatrix} \begin{bmatrix} 1 & 0 \\ 0 & 0 \end{bmatrix} \begin{bmatrix} 1 & -1 \\ 1 & 1 \end{bmatrix}^{-1} = \begin{bmatrix} 0.5 & 0.5 \\ 0.5 & 0.5 \end{bmatrix}$$

Therefore, for any p and q such that $p + q = 1$, we have

$$\lim_{k \to \infty} B^k \begin{bmatrix} p \\ q \end{bmatrix} = \begin{bmatrix} 0.5 & 0.5 \\ 0.5 & 0.5 \end{bmatrix} \begin{bmatrix} p \\ q \end{bmatrix}$$

$$= \begin{bmatrix} 0.5p + 0.5q \\ 0.5p + 0.5q \end{bmatrix} = \begin{bmatrix} 0.5 \\ 0.5 \end{bmatrix}$$

So, in the long run about 50% of the games will be won by Navy, whether it won this year ($p = 1, q = 0$) or lost ($p = 0, q = 1$).

Limits of Stochastic Matrices

We have just seen how to use diagonalization to find power limits of stochastic matrices, but is it clear that these limits always exist?

For example, consider the stochastic matrix $B = \begin{bmatrix} 0 & 1 \\ 1 & 0 \end{bmatrix}$. Then

$$B^2 = I, \qquad B^3 = B, \qquad B^4 = I, \qquad B^5 = B, \ldots$$

and clearly $\lim_{k \to \infty} B^k$ does not exist, even though B is diagonalizable.

QUESTION   When are we guaranteed that such a limit exists?

The key to answering this question lies in the following definition: a stochastic matrix A is called **regular** if some power A^k (k positive integer) consists of strictly positive entries.

$A = \begin{bmatrix} 0.5 & 1 \\ 0.5 & 0 \end{bmatrix}$ is regular, because

$$A^2 = \begin{bmatrix} 0.75 & 0.5 \\ 0.25 & 0.5 \end{bmatrix}$$

has only positive entries. On the other hand, B is not regular, because all its powers have *some* zero entries.

■ EXAMPLE 33   Show that A is regular.

$$A = \begin{bmatrix} 0.5 & 0.5 & 0 \\ 0.5 & 0 & 1 \\ 0 & 0.5 & 0 \end{bmatrix}$$

SOLUTION   It is easy to see that A^4 is the first power with only positive entries.

The following theorem answers the question. Its proof can be found in the book *Finite Markov Chains*, by J. G. Kemeny and J. L. Snell (New York: Springer-Verlag, 1976).

THEOREM 14

Let A be a regular $n \times n$ stochastic matrix. Then, as $k \to \infty$, A^k approaches an $n \times n$ matrix L of the form

$$L = \begin{bmatrix} \mathbf{v} & \mathbf{v} & \cdots & \mathbf{v} \end{bmatrix}$$

where $\mathbf{v}$ is a probability n-vector with all entries greater than 0.

So, for any regular stochastic matrix the limit of powers L exists. However, computing L using limits is quite inefficient. A much more efficient method is a consequence of our next result.

THEOREM 15

If A is a regular stochastic matrix and L and $\mathbf{v}$ are as in Theorem 14, then

1. For any initial probability vector $\mathbf{x}_0$, $A^k\mathbf{x}_0$ approaches $\mathbf{v}$ as $k \to \infty$, i.e.,

$$\lim_{k \to \infty}(A^k\mathbf{x}_0) = \mathbf{v}$$

2. $\mathbf{v}$ is the only probability vector that satisfies

$$A\mathbf{v} = \mathbf{v}$$

So, $\mathbf{v}$ is an eigenvector of A with eigenvalue $\lambda = 1$.

PROOF

1. Let $\mathbf{x}_0 = (x_1, \ldots, x_n)$. By Theorem 14,

$$\lim_{k \to \infty}(A^k\mathbf{x}_0) = \left(\lim_{k \to \infty} A^k\right)\mathbf{x}_0 = L\mathbf{x}_0$$

$$= x_1\mathbf{v} + \cdots + x_n\mathbf{v}$$

$$= (x_1 + \cdots + x_n)\mathbf{v} = \mathbf{v}$$

because $x_1 + \cdots + x_n = 1$.

2. We have

$$\mathbf{v} = \lim_{k \to \infty}(A^k\mathbf{x}_0) = \lim_{k \to \infty}(A^{k+1}\mathbf{x}_0) = A\lim_{k \to \infty}(A^k\mathbf{x}_0) = A\mathbf{v}$$

The proof of uniqueness of $\mathbf{v}$ is left as an exercise.

A nonzero vector $\mathbf{v}$ that satisfies $A\mathbf{v} = \mathbf{v}$ is called a **steady-state vector** (or **equilibrium**) of A. Let us see now how to compute $\mathbf{v}$ without limits. Because $\mathbf{v}$ is an eigenvector of A with eigenvalues 1, we just solve the system

$$(A - I)\mathbf{x} = \mathbf{0}$$

and pick out the solution whose entries add up to 1.

■ EXAMPLE 34   Find **v** and L for

$$A = \begin{bmatrix} 0.5 & 1 \\ 0.5 & 0 \end{bmatrix}$$

SOLUTION   We have

$$[A - I : \mathbf{0}] = \begin{bmatrix} -0.5 & 1 & 0 \\ 0.5 & -1 & 0 \end{bmatrix} \sim \begin{bmatrix} 1 & -2 & 0 \\ 0 & 0 & 0 \end{bmatrix}$$

So, the solution is $(2r, r)$, $r \in \mathbf{R}$. We want

$$2r + r = 1$$

Hence, $r = \frac{1}{3}$. So,

$$\mathbf{v} = \begin{bmatrix} \frac{2}{3} \\ \frac{1}{3} \end{bmatrix} \quad \text{and} \quad L = \begin{bmatrix} \frac{2}{3} & \frac{2}{3} \\ \frac{1}{3} & \frac{1}{3} \end{bmatrix}$$

(We may also use diagonalization to get L as the limit of A^k.)

NOTE   The proof of Part 1 of Theorem 15 shows that if A is regular, then for *any* initial vector $\mathbf{x}_0$ (not necessarily probability vector), as $k \rightarrow \infty$

$$A^k \mathbf{x}_0 \rightarrow r\mathbf{v}$$

where $r = x_1 + \cdots + x_n$. Thus, for any initial vector the dynamical system $\mathbf{x}_k = A^k \mathbf{x}_0$ has a limit, namely, $r\mathbf{v}$, that is a steady-state vector of A and that we can easily compute.

Exercises 7.5

1. Which of the following are probability vectors?

$$\begin{bmatrix} \frac{1}{3} \\ \frac{1}{2} \end{bmatrix}, \quad \begin{bmatrix} \frac{1}{2} \\ \frac{1}{2} \end{bmatrix}, \quad \begin{bmatrix} \frac{2}{3} \\ -\frac{1}{2} \end{bmatrix}, \quad \begin{bmatrix} \frac{1}{3} \\ \frac{2}{3} \end{bmatrix}$$

2. Show that the stochastic matrices are regular.

   a. $\begin{bmatrix} \frac{1}{2} & 1 \\ \frac{1}{2} & 0 \end{bmatrix}$ b. $\begin{bmatrix} 0 & \frac{1}{2} \\ 1 & \frac{1}{2} \end{bmatrix}$

3. Show that the matrices are not regular.

   a. $\begin{bmatrix} \frac{1}{2} & 0 \\ \frac{1}{2} & 1 \end{bmatrix}$ b. $\begin{bmatrix} 1 & \frac{1}{2} \\ 0 & \frac{1}{2} \end{bmatrix}$

4. Which of the matrices are regular?

   a. $\begin{bmatrix} 0.2 & 0.5 & 0 \\ 0.2 & 0.5 & 1 \\ 0.6 & 0 & 0 \end{bmatrix}$ b. $\begin{bmatrix} 0 & 1 & 0 \\ 0 & 0 & 1 \\ 1 & 0 & 0 \end{bmatrix}$

   c. $\begin{bmatrix} 0 & 0.5 & 1 \\ 0 & 0.5 & 0 \\ 1 & 0 & 0 \end{bmatrix}$ d. $\begin{bmatrix} 0 & 0.5 & 1 \\ 0.5 & 0.5 & 0 \\ 0.5 & 0 & 0 \end{bmatrix}$

5. Find the steady-state vectors of the matrices of Exercise 2.

6. Find the steady-state vectors of the matrices.

   a. $\begin{bmatrix} 0.6 & 0.5 \\ 0.4 & 0.5 \end{bmatrix}$ b. $\begin{bmatrix} 0.7 & 0.5 \\ 0.3 & 0.5 \end{bmatrix}$

7. Find the steady-state vectors of the matrices.

   a. $\begin{bmatrix} 0.2 & 0.5 & 0.5 \\ 0.2 & 0.5 & 0 \\ 0.6 & 0 & 0.5 \end{bmatrix}$ b. $\begin{bmatrix} 0.4 & 0.2 & 0.2 \\ 0.4 & 0.4 & 0.3 \\ 0.2 & 0.4 & 0.5 \end{bmatrix}$

8. Show that $\begin{bmatrix} 1 & \frac{1}{2} \\ 0 & \frac{1}{2} \end{bmatrix}$ has a steady-state vector even though it is nonregular.

9. Prove the uniqueness of the steady-state vector claimed in Theorem 15.

10. (**Demographics**) In a certain city a resident has a 40% chance to remain in the city 1 year later and 60% chance to move to the suburban area around it. A sub-

urban resident has a 20% chance to move to a city 1 year later. Write a stochastic matrix of transition of probabilities for this situation. What is the long-run distribution of a population living in this city and its surrounding suburban areas?

11. (**Economics**) Currently there are three investment plans available for the employees of a company: A, B, and C. An employee can only use one plan at a time and may switch from one plan to another only at the end of each year. The probability that someone in Plan A will continue with A is 20%, switch to Plan B is 50%, switch to Plan C is 50% and so on. The transition of probabilities matrix M for the employees that participate is given below.

This year

		A	B	C
Next	A	0.2	0.5	0.5
year	B	0.2	0.5	0
	C	0.6	0	0.5

Show that M is regular and find its equilibrium. Find the most and the least popular plans in the long run.

12. (**Psychology**) A psychologist places 40 rats in a box with three colored compartments: a blue (B), a green

(G), and a red (R). Each compartment has doors that lead to the other ones, as shown in the accompanying figure. The rats constantly move towards a door, so that the probability that they will stay in one compartment is 0. A rat in B has $\frac{3}{4}$ probability to go to G and $\frac{1}{4}$ probability to go to R, according to the distribution of doors. Likewise, a rat in R has $\frac{1}{2}$ probability to go to G and $\frac{1}{2}$ probability to go to B. So the transition of probabilities matrix A is of the form:

$$A = \begin{bmatrix} 0 & * & \frac{1}{2} \\ \frac{3}{4} & 0 & \frac{1}{2} \\ \frac{1}{4} & * & 0 \end{bmatrix}$$

Replace the asterisks in A with the correct probabilities. Show that A is regular and compute its steady-state vector. What is the long-run distribution of rats? What is the probability that a given rat will be in G in the long run?

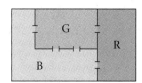

7.6   Miniprojects

1 ▪ The Cayley-Hamilton Theorem

If A is a square matrix of scalar entries and $p(x)$ is a polynomial, say,

$$p(x) = a_0 + a_1 x + \cdots + a_k x^k$$

then we denote by $p(A)$ the matrix

$$p(A) = a_0 I + a_1 A + \cdots + a_k A^k$$

For example, if $A = \begin{bmatrix} 2 & -2 \\ 1 & 4 \end{bmatrix}$ and $p(x) = 1 - 3x + x^2$, then

$$p(A) = \begin{bmatrix} 1 & 0 \\ 0 & 1 \end{bmatrix} - 3 \begin{bmatrix} 2 & -2 \\ 1 & 4 \end{bmatrix} + \begin{bmatrix} 2 & -2 \\ 1 & 4 \end{bmatrix}^2 = \begin{bmatrix} -3 & -6 \\ 3 & 3 \end{bmatrix}$$

In this project you are to verify and prove the following important theorem.

THEOREM 16   (Cayley-Hamilton)

Every square matrix satisfies its characteristic equation. So, if $p(x)$ is the characteristic polynomial of A, then

$$p(A) = \mathbf{0}$$

To illustrate, it is easy to see that $p(x) = 10 - 6x + x^2$ is the characteristic polynomial of the matrix A. Then

$$p(A) = 10 \begin{bmatrix} 1 & 0 \\ 0 & 1 \end{bmatrix} - 6 \begin{bmatrix} 2 & -2 \\ 1 & 4 \end{bmatrix} + \begin{bmatrix} 2 & -2 \\ 1 & 4 \end{bmatrix}^2 = \begin{bmatrix} 0 & 0 \\ 0 & 0 \end{bmatrix}$$

Problem A

Verify the Cayley-Hamilton theorem for the following matrices.

1. $\begin{bmatrix} 2 & 3 \\ -1 & 4 \end{bmatrix}$,   2. $\begin{bmatrix} -5 & 6 \\ 8 & 1 \end{bmatrix}$,   3. $\begin{bmatrix} -1 & -1 & 0 \\ 1 & -\frac{3}{2} & \frac{3}{2} \\ 0 & 1 & -1 \end{bmatrix}$.

Next you are going to prove the Cayley-Hamilton theorem for the special case where A is diagonalizable. Just follow the instructions.

Problem B

1. Let $\{\mathbf{v}_1, \dots, \mathbf{v}_k\}$ span $\mathbf{R}^n$ and let B be an $n \times n$ matrix such that

$$B\mathbf{v}_1 = \mathbf{0}, \dots, B\mathbf{v}_k = \mathbf{0}$$

Show that B is the zero matrix.
2. Let λ be an eigenvalue of a square matrix A with corresponding eigenvector $\mathbf{v}$. Show that for any positive integer k

$$A^k \mathbf{v} = \lambda^k \mathbf{v}$$

3. Let A be a diagonalizable matrix with characteristic polynomial $p(x)$. Prove the Cayley-Hamilton theorem for A as follows: Show that for any eigenvector $\mathbf{v}$ of A the vector $p(A)\mathbf{v}$ is zero (using Part 2). Then use Part 1 to conclude that $p(A) = \mathbf{0}$.

Next you will be led to proving the Cayley-Hamilton theorem for any square matrix.

If B is an $n \times n$ matrix with entries polynomials in x, then there are unique matrices $B_0, B_1, \dots, B_k$ with scalar entries such that

$$B = B_0 + B_1 x + \cdots + B_k x^k$$

For example,

$$\begin{bmatrix} 1 + x - 3x^2 & -1 + x \\ 2 + 5x & -6x + x^2 \end{bmatrix} = \begin{bmatrix} 1 & -1 \\ 2 & 0 \end{bmatrix} + \begin{bmatrix} 1 & 1 \\ 5 & -6 \end{bmatrix} x + \begin{bmatrix} -3 & 0 \\ 0 & 1 \end{bmatrix} x^2$$

Much of the matrix arithmetic for ordinary matrices extends to matrices with polynomial entries. In particular, the following formula generalizes Theorem 10, Section 6.3, to matrices B with polynomial entries.

$$\text{Adj}(B)B = \det(B)I_n$$

In the next few lines we show how to prove the Cayley-Hamilton theorem for any 2×2 matrix A. If the characteristic polynomial of A is

$$p(\lambda) = a + b\lambda + \lambda^2$$

then consider the matrix

$$B = A - \lambda I$$

Because the maximum degree in λ of the elements of $\text{Adj}(B)$ is 1, there are unique matrices B_0 and B_1 with scalar entries such that

$$\text{Adj}(B) = B_0 + B_1\lambda$$

Hence,

$$\text{Adj}(B)B = (B_0 + B_1\lambda)(A - \lambda I) = B_0A + (-B_0 + B_1A)\lambda - B_1\lambda^2$$

On the other hand,

$$\text{Adj}(B)B = \det(B)I = p(\lambda)I = aI + bI\lambda + I\lambda^2$$

Therefore, by uniqueness,

$$aI = B_0A, \qquad -B_0 + B_1A = bI, \qquad -B_1 = I$$

So,

$$p(A) = aI + bA + A^2 = B_0A + (-B_0 + B_1A)A + A^2 = \mathbf{0}$$

Problem C

Following the above steps prove the Cayley-Hamilton theorem for any square matrix.

2 ■ The Fibonacci Numbers (Part II)

In this project we return to the Fibonacci numbers $f_0, f_1, f_2, \ldots$ of Section 3.6 to study their long-term behavior using eigenvalues. These numbers arise when we count male-female pairs of rabbits, that reproduce monthly and create another male-female pair. Recall that

$$f_0 = 1, \qquad f_1 = 1, \qquad f_k = f_{k-1} + f_{k-2}, \qquad k \geq 2$$

To simplify notation we add an extra number $f_{-1} = 0$ to this list. So, the first few terms are

$$f_{-1} = 0, \qquad f_0 = 1, \qquad f_1 = 1, \qquad f_2 = 2, \qquad f_3 = 3, \qquad f_4 = 5, \qquad f_5 = 8, \ldots$$

$$\text{Let } \mathbf{x}_k = \begin{bmatrix} f_k \\ f_{k-1} \end{bmatrix}, k = 0, 1, \ldots, \text{ and let } A = \begin{bmatrix} 1 & 1 \\ 1 & 0 \end{bmatrix}.$$

Problem

1. Show that $\mathbf{x}_{k+1} = A\mathbf{x}_k$.
2. Prove (or use if proved already) the identity of Exercise 32, Section 7.2,

$$\begin{bmatrix} 1 & 1 \\ 1 & 0 \end{bmatrix}^k \begin{bmatrix} 1 \\ 0 \end{bmatrix} = \frac{1}{\sqrt{5}} \begin{bmatrix} r_1^{k+1} - r_2^{k+1} \\ r_1^k - r_2^k \end{bmatrix} \tag{7.31}$$

where r_1 and r_2 are the eigenvalues of A.
3. Conclude that for $k = -1, 0, 1, 2, \ldots$,

$$f_k = \frac{1}{\sqrt{5}} \left(\left(\frac{1+\sqrt{5}}{2} \right)^{k+1} - \left(\frac{1-\sqrt{5}}{2} \right)^{k+1} \right) \tag{7.32}$$

4. Without expansion find the integer

$$\frac{1}{\sqrt{5}} \left(\left(\frac{1+\sqrt{5}}{2} \right)^{8} - \left(\frac{1-\sqrt{5}}{2} \right)^{8} \right)$$

5. By noting that $r_2^k \to 0$ as $k \to \infty$ (since $|r_2| < 1$), conclude that

$$\lim_{k \to \infty} \frac{f_k}{f_{k-1}} = \frac{1+\sqrt{5}}{2} \cong 1.618$$

The number $r_1 = (1 + \sqrt{5})/2 \cong 1.618$ is called (**Plato's**) **golden mean**. It was recognized by the Greeks for its interesting properties. Artists know that the side ratio of the most eye-pleasing rectangles is $r_1 : 1$.

3 ■ Transition of Probabilities (Part II)

We now return to Project 3 of Section 3.6 to answer a few more questions.

Problem A

A group of people buys cars every 4 years from one of three automobile manufacturers, A, B and C. The transition of probabilities of switching from one manufacturer to another is given by the matrix

$$R = \begin{bmatrix} 0.5 & 0.4 & 0.6 \\ 0.3 & 0.4 & 0.3 \\ 0.2 & 0.2 & 0.1 \end{bmatrix}$$

1. Use eigenvalues to compute $\lim_{n \to \infty} R^n$.
2. Will one of the manufacturers eventually dominate the market no matter what the initial sales are?

Problem B

Consider the stochastic matrix of transition probabilities

$$T = \begin{bmatrix} \frac{1}{3} & \frac{1}{2} \\ \frac{2}{3} & \frac{1}{2} \end{bmatrix}$$

expressing the flow of customers from and to markets A and B after one purchase. Recall that a **market equilibrium** is a vector of shares (a, b) that remains the same from one purchase to the next.

1. Show that a market equilibrium is an eigenvector of the transition probabilities matrix. What is the corresponding eigenvalue?
2. Show that T has a market equilibrium.
3. Compute $\lim_{n \to \infty} T^n$.
4. Will one of the markets eventually dominate the other?

7.7 Computer Exercises

In this section you practice with the commands of your software that compute eigenvalues and eigenvectors. Using these you explore further topics. An exercise designated as [S] requires symbolic manipulation.[5]

Let

$$A = \begin{bmatrix} 3 & 3 & 4 & 4 \\ 3 & 3 & 4 & 4 \\ 5 & 5 & 6 & 6 \\ 5 & 5 & 6 & 6 \end{bmatrix} \qquad B = \begin{bmatrix} a & a & 1 \\ a & a & a \\ 1 & a & a \end{bmatrix}$$

$$C = \begin{bmatrix} 0.2 & 0.3 & 0.8 \\ 0.2 & 0.3 & 0.1 \\ 0.6 & 0.4 & 0.1 \end{bmatrix} \qquad R = \begin{bmatrix} 0.2 & 0 & 0.8 \\ 0 & 0 & 0.2 \\ 0.8 & 1 & 0 \end{bmatrix}$$

1. Without computing, find one eigenvalue of A. Then use your program to compute all the eigenvalues and basic eigenvectors numerically and, if possible, exactly. Confirm your answer by showing that the computed eigenvalues satisfy the characteristic equation and that the computed basic eigenvectors are indeed eigenvectors of A.
2. Diagonalize A by finding D and P. Then verify your answer by showing that $A = PDP^{-1}$.
3. [S] Compute all the values of a such that B has zero as an eigenvalue.
4. Find the roots of the polynomial $p(x) = x^5 - 15x^3 + 36x + 74$ directly and by computing the eigenvalues of the companion matrix.
5. Find an approximation to four decimal places for the limit of the stochastic matrix C, $\lim_{n \to \infty} C^n$ (a) directly, by computing C^n for large n, and (b) by using eigenvalues.
6. Show that R is a regular matrix. Is it true that if an $n \times n$ matrix S is regular, then S^n should have only nonzero entries? By examining regular matrices and their powers, form a conjecture on the smallest positive integer k such that S^k consists of only nonzero entries.
7. Verify the Hamilton-Cayley theorem (discussed in Project 1) for A.

---

[5] Skip these exercises if symbolic manipulation is not available.

8. Let A_n be the $n \times n$ matrix with entries 1. Find a formula for its eigenvalues and basic eigenvectors.

$$A_2 = \begin{bmatrix} 1 & 1 \\ 1 & 1 \end{bmatrix}, \qquad A_3 = \begin{bmatrix} 1 & 1 & 1 \\ 1 & 1 & 1 \\ 1 & 1 & 1 \end{bmatrix}, \dots$$

9. Let B_n be the $n \times n$ matrix with diagonal entries n and remaining entries 1. Find a formula for its eigenvalues and basic eigenvectors.

$$B_2 = \begin{bmatrix} 2 & 1 \\ 1 & 2 \end{bmatrix}, \qquad B_3 = \begin{bmatrix} 3 & 1 & 1 \\ 1 & 3 & 1 \\ 1 & 1 & 3 \end{bmatrix}, \dots$$

The following exercise is modeled after a known example in population dynamics attributed to H. Bernadelli, P. H. Leslie, and E. G. Leslie.

10. A species of beetles lives 3 years. Let A, B, and C be the 0–1-year-old, 1–2-year-old, and 2–3-year-old females, respectively. No female from group A produces offspring. Each female in group B produces 8 females and each female in group C produces 24 females. Suppose that only $\frac{1}{4}$ from group A survive to group B and only $\frac{1}{6}$ of group B survive to group C. If A_k, B_k, and C_k are the numbers of females in A, B, and C after k years, find a matrix M that $M(A_k, B_k, C_k)$ is $(A_{k+1}, B_{k+1}, C_{k+1})$. If $A_0 = 100$, $B_0 = 40$, and $C_0 = 20$, use eigenvalues and eigenvectors to determine whether the species will become extinct.

Selected Solutions with Maple

Commands charpoly, charmat, companion, eigenvals, Eigenvals, eigenvects.

```
# Exercise 1
# The matrix is noninvertible (repeated rows) so 0 is an eigenvalue.
with(linalg):
A := matrix([[3,3,4,4],[3,3,4,4],[5,5,6,6],[5,5,6,6]]);
evas:= eigenvals(A);           # Eigenvalues exactly,
evalf(");                      # and numerically.
eves := eigenvects(A);         # Eigenvectors exactly,
evalf(");                      # and numerically.
# May also use the inert version:
# Eigenvals(A); evalf(");  for the approximation.
p:=charpoly(A,x);              # The characteristic polynomial of A.
subs(x=evas[4],p);             # Substituting, for example, the 4th
simplify(");                   # eigenvalue and simplifying to get 0.
e := evas[3];                  # Pick an eigenvalue and the
v:=eves[1][3][1];              # corresponding eigenvector.
evalm(A &* v - e*v);           # Compute Av-ev and simplify to
map(simplify,");               # get the zero vector.
# Exercise 3
B := matrix([[a,a,1],[a,a,a],[1,a,a]]);
eigenvals(B);                  # Symbolic eigenvalues.
solve("[1]);solve(""[2]);solve("""[3]); # Set them = 0 and solve. So a=0,1.
```

```
# Exercise 4
p:=x^5-15*x^3+36*x+74;
solve(p);                     # The roots cannot be computed exactly.
allvalues(");                 # But allvalues will approximate.
eigenvals(companion(p,x));    # The eigenvalues of the TRANSPOSE of the
allvalues(");                 # companion matrix approximated. Same answer.
# Exercise 5
C := matrix([[.2,.3,.8],[.2,.3,.1],[.6,.4,.1]]);
evalm(C&^80);       # Yields identical columns. Limit to displayed accuracy.
v:=eigenvects(C);
DD:=diag(v[1][1],v[2][1],v[3][1]);     # DD since D is already used by Maple.
P:=concat(v[1][3][1],v[2][3][1],v[3][3][1]);     # P.
evalm(P&*DD&^80&*inverse(P));              # Same answer to 6 decimal places.
# Exercise 6 - Partial
R := matrix([[.2,0,.8],[0,0,.2],[.8,1,0]]);
evalm(R&^2);evalm(R&^3);evalm(R&^4);    # R^4 has all entries nonzero so R is regular.
# Exercise 7
subs(x=A,charpoly(A,x)); # Substitute the matrix in the characteristic polynomial.
evalm(");                 # Evaluate to get a zero matrix.
# Exercise 8 - Partial
An := proc(n) local i, j; matrix(n,n, (i,j) -> 1) end:    # Define A_n.
eigenvects(An(2));eigenvects(An(3));              # Etc..
# Exercise 9 - Partial
Bn:=proc(n) subs(nn=n,matrix(n,n,(i,j)->if i<>j then 1 else nn fi))end:
eigenvects(Bn(2));eigenvects(Bn(3));              # Etc..
# NOTE: proc(n) matrix(n,n,(i,j)->if i<>j then 1 else n fi)) end: fails
# to pass the correct n on the diagonal because of Maple's scoping rules.
# Exercise 10 - Partial
M := matrix([[0,8,24],[1/4,0,0],[0,1/6,0]]);    # The correct matrix.
```

Selected Solutions with Mathematica

Commands   Eigenvalues, Eigenvectors, Eigensystem.

```
(* Exercise 1 *)
(* The matrix is noninvertible (repeated rows) so 0 is an eigenvalue. *)
A = {{3,3,4,4},{3,3,4,4},{5,5,6,6},{5,5,6,6}}
evas = Eigenvalues[A]              (* Eigenvalues exactly,          *)
N[%]                               (* and numerically.             *)
Eigenvectors[A]                    (* Eigenvectors exactly.        *)
eves = Simplify[%]                 (* Answer needs simplification. *)
N[%]                               (* Eigenvectors numerically.    *)
Eigensystem[A] // Simplify         (* Or, eigenvals and eigenvects together.*)
p=Det[A-x*IdentityMatrix[4]]       (* Characteristic polynomial.    *)
p /. x->evas[[4]]                  (* Substituting, for example, the 4th*)
Simplify[%]                        (* eigenvalue and simplifying to get 0.*)
```

```
e = evas[[3]]                      (* Pick an eigenvalue and the    *)
v = eves[[3]]                      (* corresponding eigenvector.    *)
A . v - e v                        (* Compute Av-ev and simplify to *)
Simplify[%]                        (* get the zero vector.          *)
(* Exercise 3  *)
B = {{a,a,1},{a,a,a},{1,a,a}}
ee = Eigenvalues[B] // Simplify      (* Symbolic eigenvalues.       *)
Solve[ee[[1]]==0,a]                  (* Set them = 0 and solve to   *)
Solve[ee[[2]]==0,a]                  (* get a=0,1.                  *)
Solve[ee[[3]]==0,a]
(* Exercise 4  *)
p1 =x^5-15x^3+36x+74
Solve[p1==0,x]                    (* The roots cannot be computed    *)
N[%]                              (* exactly but approximately.      *)
Eigenvalues[N[{{0,1,0,0,0},{0,0,1,0,0},    (* The numerical eigenvalues *)
{0,0,0,1,0},{0,0,0,0,1},{-74,-36,0,15,0}}]]  (* of the  companion matrix. *)
(* Exercise 5 *)
CC = {{.2,.3,.8},{.2,.3,.1},{.6,.4,.1}}       (* C is used by Mathematica.  *)
MatrixPower[CC,80] (*Yields identical columns. Limit to displayed accuracy.*)
v=Eigensystem[CC]                          (* Next form D and P.          *)
DD=DiagonalMatrix[{v[[1,1]],v[[1,2]],v[[1,3]]}] (* D is used for differentiation.*)
P=Transpose[v[[2]]]                             (* P.                       *)
P . MatrixPower[DD,80] . Inverse[P]  (* Same answer to >15 decimal places. *)
(* Exercise 6 - Partial  *)
R = {{.2,0,.8},{0,0,.2},{.8,1,0}}
For[i=1,i<=4,i++,Print[MatrixPower[R,i]]] (* All entries of R^4 are not 0
                                 so R is regular.*)
(* Exercise 7  *)
p /. {Power->MatrixPower,x->A} (* Substitute A into the char. poly. and    *)
(* convert Power to  MatrixPower to a get the zero matrix, as expected.   *)
(* Exercise 8 - Partial  *)
An[n_]:= Table[1, {i,1,n},{j,1,n}]
Eigensystem[An[2]]                      (* Then Eigensystem[An[[3]]], etc.. *)
(* Exercise 9 - Partial     *)
Bn[n_]:= Table[If[i==j, n, 1], {i,1,n},{j,1,n}]
Eigensystem[Bn[2]]                      (* Then Eigensystem[Bn[[3]]], etc.. *)
(* Exercise 10 - Partial  *)
M = {{0,8,24},{1/4,0,0},{0,1/6,0}}    (* The correct matrix. Etc..        *)
```

Selected Solutions with MATLAB

Commands balance, eig, poly, polyeig, polyval, polyvalm, qz, and from the symbolic toolbox: charpoly, eigensys, sym.

```
% Exercise 1
% The matrix is noninvertible (repeated rows) so 0 is an eigenvalue.
```

```
A= [3 3 4 4; 3 3 4 4; 5 5 6 6; 5 5 6 6]; eig(A) % Numerical eigenvalues. Also may
[eves,evas]=eig(A)          % use the [,] format. eves is the matrix with columns
% the eigenvectors and evas is the diagonal matrix with diagonal the eigenvalues.
eigensys(A)                 % (ST) The eigenvalues symbolically.
[AA,BB]= eigensys(A)        % (ST) The eigenvectors and eigenvalues symbolically.
p = poly(A)                 % The characteristic polynomial of A.
roots(p)                    % All the eigenvalues are obtained as roots of p.
polyval(p,evas(1,1))        % Also evaluate p at an eigenvalue. Zero to 11 dec. places.
e = evas(1,1),v = eves(:,1) % Pick an eigenvalue and the corresponding eigenvector.
A * v - e * v                    % The answer is zero to 13 decimal places.
% Exercise 3
B = sym('[a a 1; a a a; 1 a a]'), eigensys(B)  % (ST) Symbolic matrix and eigenvalues.
solve('a-1')                          % (ST) copy and solve each of the
solve('a+1/2-1/2*(8*a^2+1)^(1/2)')    % (ST) symbolic eigenvalues (returned
solve('a+1/2+1/2*(8*a^2+1)^(1/2)')    % (ST)    as strings) to get a =0,1.
% Exercise 4
p1=[1 0 -15 0 36 74], roots(p1)      % roots finds the evals of the companion!
eig([0 1 0 0 0; 0 0 1 0 0; 0 0 0 1 0; 0 0 0 0 1; -74 -36 0 15 0]) % Same.
% Exercise 5
C = [.2 .3 .8; .2 .3 .1; .6 .4 .1]; C^80  % Yields identical columns.
[P,D] = eig(C)                       % P and D.
P * D^80 * P^(-1)                    % Same answer by diagonalization.
% Exercise 6 - Partial
R = [.2 0 .8; 0 0 .2; .8 1 0]
R^2,R^3,R^4            % R^4 has all entries nonzero so R is regular.
% Exercise 7
polyvalm(poly(A),A)     % Substitute A in the characterstic poly poly(A) and
% evaluate in a matrix sense to get a near zero matrix polyval does not work here.
% Exercise 8 - Partial
[P2,D2] = eig(ones(2)),[P3,D3] = eig(ones(3))  % An is just ones(n). However, the
[P2,D2] = eigensys(ones(2)) % (ST) exact calculation helps find a pattern easier.
% Exercise 9 - Partial
% To define Bn create a script file named Bn.m having the following lines:
function a = B(n)
for i=1:n,
  for j=1:n,
    if i==j
        a(i,j) = n;
      else a(i,j) =1;
    end
  end
end                     % Now type
[P2,D2] = eig(Bn(2)),[P3,D3] = eig(Bn(3))          % Etc.. However, the
[P2,D2] = eigensys(Bn(2)) % (ST) exact calculation helps find a patern easier.
% Exercise 10 - Partial
M = [0 8 24; 1/4 0 0; 0 1/6 0]                % The correct matrix. Etc..
```

Dot and Inner Products

$\mathbf{8}$

The knowledge of which geometry aims is the knowledge of the eternal.
—Plato (ca. 429–347 B.C.), *Republic*, VII, 52.

Introduction

In this chapter we study many useful properties of the dot product of $\mathbf{R}^n$. The material is of interest to both theoretically and applications inclined readers. We also study vector spaces that come equipped with an "inner product," a generalization of the dot product for abstract vectors. Inner products are widely used from theoretical analysis to applied signal processing.

Least Squares

One of the most interesting applications in this chapter is the method of *least squares*. Often in trying to understand experimental data, we wish to find a line or a curve that best "fits" (best describes) these data. For example, suppose a linear algebra instructor kept statistics (displayed next) of the percentage of B grades given over a period of six semesters.

Semester	1	2	3	4	5	6
Percentage of B's	0.20	0.25	0.20	0.30	0.45	0.40

If the instructor wishes to draw a straight line that comes close to the points of the table, there are many choices. However, there is a straight line that best fits these data in a certain sense. In Section 8.4 we see that this line is $y = 0.13333 + 0.05x$ (Fig. 8.1).

The method of least squares was discovered by Karl Friedrich Gauss and was used to solve an astronomical problem. In 1801 the asteroid *Ceres* had been observed for more than a month before it vanished when it came close to the much brighter sun. Based on the available observations, astronomers wanted to approximate the orbit of Ceres so that

505

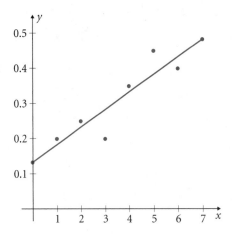

Figure 8.1   Line that best fits the data.

it could be observed again when it moved away from the sun. Gauss, using least squares, impressed the scientific community by predicting the correct time (about 10 months later) and location of the asteroid.

8.1   Orthogonal Sets and Matrices

Reader's Goals for This Section

1. To know the definitions and basic properties of orthogonal and orthonormal sets.
2. To know the definition and basic properties of orthogonal matrices.

In Chapter 2 we defined two vectors as orthogonal if their dot product was zero. In this section we study entire sets of vectors that are pairwise orthogonal. Such sets are called orthogonal and share many interesting properties that make them very useful in computations.

First, we prove an identity that we use several times in this chapter. For any $m \times n$ matrix A, n-vector $\mathbf{u}$, and m-vector $\mathbf{v}$, we have

$$(A\mathbf{u}) \cdot \mathbf{v} = \mathbf{u} \cdot (A^T\mathbf{v}) \qquad (8.1)$$

PROOF

$$(A\mathbf{u}) \cdot \mathbf{v} = (A\mathbf{u})^T\mathbf{v} = (\mathbf{u}^TA^T)\mathbf{v} = \mathbf{u}^T(A^T\mathbf{v}) = \mathbf{u} \cdot (A^T\mathbf{v})$$

Orthogonal Sets

We say that a set of n-vectors $\{\mathbf{v}_1, \ldots, \mathbf{v}_k\}$ is **orthogonal** if any two distinct vectors in it are orthogonal. This means that

$$\mathbf{v}_i \cdot \mathbf{v}_j = 0 \qquad \text{if } i \neq j$$

If $S = \{\mathbf{v}_1, \mathbf{v}_2, \mathbf{v}_3\} \subseteq \mathbf{R}^3$ is orthogonal, then all possible pairs of distinct vectors: $\{\mathbf{v}_1, \mathbf{v}_2\}, \{\mathbf{v}_1, \mathbf{v}_3\}, \{\mathbf{v}_2, \mathbf{v}_3\}$ must be orthogonal. Hence, S forms a right-angled coordinate frame (Fig. 8.2).

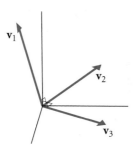

Figure 8.2   **Orthogonal vectors.**

■ **EXAMPLE 1**   Show that $S = \{\mathbf{v}_1, \mathbf{v}_2, \mathbf{v}_3\} \subseteq \mathbf{R}^4$ is orthogonal, where

$$\mathbf{v}_1 = \begin{bmatrix} 2 \\ 2 \\ 4 \\ 0 \end{bmatrix}, \qquad \mathbf{v}_2 = \begin{bmatrix} 0 \\ 2 \\ -1 \\ 1 \end{bmatrix}, \qquad \mathbf{v}_3 = \begin{bmatrix} -2 \\ 0 \\ 1 \\ 1 \end{bmatrix}$$

SOLUTION   This statement is true because

$$\mathbf{v}_1 \cdot \mathbf{v}_2 = 2 \cdot 0 + 2 \cdot 2 + 4 \cdot (-1) + 0 \cdot 1 \quad = 0$$
$$\mathbf{v}_1 \cdot \mathbf{v}_3 = 2 \cdot (-2) + 2 \cdot 0 + 4 \cdot 1 + 0 \cdot 1 \quad = 0$$
$$\mathbf{v}_2 \cdot \mathbf{v}_3 = 0 \cdot (-2) + 2 \cdot 0 + (-1) \cdot 1 + 1 \cdot 1 = 0$$

One of the most important properties of orthogonal sets is summarized in the following theorem. Recall that to compute the coefficients of a vector $\mathbf{u}$ as a linear combination in $\mathbf{v}_1, \ldots, \mathbf{v}_k$ may result in a tedious row reduction. If, however, $\{\mathbf{v}_1, \ldots, \mathbf{v}_k\}$ is orthogonal, then there is an *easy* formula for the coefficients.

THEOREM 1

Let $S = \{\mathbf{v}_1, \ldots, \mathbf{v}_k\}$ be an orthogonal set of nonzero vectors. If $\mathbf{u}$ is in Span(S) with

$$\mathbf{u} = c_1 \mathbf{v}_1 + \cdots + c_k \mathbf{v}_k \tag{8.2}$$

then

$$c_i = \frac{\mathbf{u} \cdot \mathbf{v}_i}{\mathbf{v}_i \cdot \mathbf{v}_i}, \qquad i = 1, \ldots, k \tag{8.3}$$

PROOF   For a fixed $i = 1, \ldots, k$ we take the dot product of each side of (8.2) with $\mathbf{v}_i$

$$\mathbf{u} \cdot \mathbf{v}_i = (c_1 \mathbf{v}_1 + \cdots + c_k \mathbf{v}_k) \cdot \mathbf{v}_i$$
$$= c_1 (\mathbf{v}_1 \cdot \mathbf{v}_i) + \cdots + c_k (\mathbf{v}_k \cdot \mathbf{v}_i)$$
$$= c_i (\mathbf{v}_i \cdot \mathbf{v}_i)$$

508   Chapter 8 ■ Dot and Inner Products

because $v_i \cdot v_j = 0$ for $i \neq j$, by orthogonality. Hence, $c_i = (u \cdot v_i)/(v_i \cdot v_i)$, as claimed. Note that $v_i \cdot v_i = \|v_i\|^2 \neq 0$, because $v_i \neq 0$.

The scalars $c_i = (u \cdot v_i)/(v_i \cdot v_i)$ can be defined for *any* n-vector u (not just one in Span(S) only), and they are often called the **Fourier coefficients** of u with respect to S.

As an *important* consequence of Theorem 1 we have the following.

THEOREM 2   Any orthogonal set $S = \{v_1, \ldots, v_k\}$ of nonzero n-vectors is linearly independent.

PROOF   Set a linear combination equal to 0:

$$c_1 v_1 + \cdots + c_k v_k = 0$$

By Theorem 1 with $u = 0$,

$$c_i = \frac{0 \cdot v_i}{v_i \cdot v_i} = 0, \qquad i = 1, \ldots, k$$

Hence, S is linearly independent.

We see that an orthogonal set of nonzero vectors is a **basis** for its span and that the coefficients c_i of (8.2) are **uniquely** determined by (8.3).

If a basis of a subspace V of $\mathbf{R}^n$ is an orthogonal set, we call it an **orthogonal basis**. Orthogonal bases are very useful, because the coordinates of vectors can be computed easily by using (8.3).

■ **EXAMPLE 2**   Show that the set $\mathcal{B} = \{v_1, v_2, v_3\}$ is an orthogonal basis of $\mathbf{R}^3$. Write u as a linear combination of v_1, v_2, v_3.

$$v_1 = \begin{bmatrix} 1 \\ -2 \\ 3 \end{bmatrix}, \quad v_2 = \begin{bmatrix} -2 \\ 2 \\ 2 \end{bmatrix}, \quad v_3 = \begin{bmatrix} \frac{5}{7} \\ \frac{4}{7} \\ \frac{1}{7} \end{bmatrix}, \quad u = \begin{bmatrix} 12 \\ -6 \\ 6 \end{bmatrix}$$

SOLUTION   It is easy to see that $\mathcal{B}$ is orthogonal. Hence, $\mathcal{B}$ is a linearly independent set of three 3-vectors, by Theorem 2. So, it is an orthogonal basis of $\mathbf{R}^3$. Let $u = c_1 v_1 + c_2 v_2 + c_3 v_3$. Then

$$u \cdot v_1 = 42 \qquad u \cdot v_2 = -24 \qquad u \cdot v_3 = 6$$
$$v_1 \cdot v_1 = 14 \qquad v_2 \cdot v_2 = 12 \qquad v_3 \cdot v_3 = 6/7$$

So, by (8.3)

$$u = \frac{u \cdot v_1}{v_1 \cdot v_1} v_1 + \frac{u \cdot v_2}{v_2 \cdot v_2} v_2 + \frac{u \cdot v_3}{v_3 \cdot v_3} v_3$$
$$= \frac{42}{14} v_1 + \frac{-24}{12} v_2 + \frac{6}{6/7} v_3$$
$$= 3v_1 - 2v_2 + 7v_3$$

This calculation is much easier than the row reduction of the matrix

$$[\mathbf{v}_1 \; \mathbf{v}_2 \; \mathbf{v}_3 \; \mathbf{u}]$$

THEOREM 3

Let $\mathcal{B}$ be an orthogonal basis of a subspace V of $\mathbf{R}^n$. If a vector $\mathbf{u}$ of V is orthogonal to each vector of $\mathcal{B}$, then $\mathbf{u} = \mathbf{0}$.

PROOF Exercise. (*Hint:* Use Theorem 1.)

An interesting property of orthogonal sets is related to matrices and their transposes.

THEOREM 4

If the columns of an $m \times n$ matrix A form an orthogonal set, then $A^T A$ is an $n \times n$ diagonal matrix. More precisely, if $A = [\mathbf{v}_1 \cdots \mathbf{v}_n]$, then

$$A^T A = \begin{bmatrix} \|\mathbf{v}_1\|^2 & 0 & \cdots & 0 \\ 0 & \|\mathbf{v}_2\|^2 & \cdots & 0 \\ \vdots & \vdots & \ddots & \vdots \\ 0 & 0 & \cdots & \|\mathbf{v}_n\|^2 \end{bmatrix} \tag{8.4}$$

Conversely, if (8.4) is valid then the columns of A form an orthogonal set.

PROOF Let c_{ij} be the (i, j) entry of $A^T A$, and let $\mathbf{r}_i$ be the ith row of A^T. Hence, $\mathbf{r}_i = \mathbf{v}_i$ as n-vectors. By definition of matrix multiplication,

$$c_{ij} = \mathbf{r}_i \cdot \mathbf{v}_j = \mathbf{v}_i \cdot \mathbf{v}_j = \begin{cases} \|\mathbf{v}_i\|^2 & \text{if } i = j \\ 0 & \text{if } i \neq j \end{cases}$$

because $\mathbf{v}_i \cdot \mathbf{v}_i = \|\mathbf{v}_i\|^2$ and $\mathbf{v}_i \cdot \mathbf{v}_j = 0$, by orthogonality. The proof of the converse is left as an exercise.

To illustrate, let A have columns the vectors of Example 1:

$$A^T A = \begin{bmatrix} 2 & 2 & 4 & 0 \\ 0 & 2 & -1 & 1 \\ -2 & 0 & 1 & 1 \end{bmatrix} \begin{bmatrix} 2 & 0 & -2 \\ 2 & 2 & 0 \\ 4 & -1 & 1 \\ 0 & 1 & 1 \end{bmatrix} = \begin{bmatrix} 24 & 0 & 0 \\ 0 & 6 & 0 \\ 0 & 0 & 6 \end{bmatrix}$$

The diagonal entries are clearly $\|\mathbf{v}_1\|^2$, $\|\mathbf{v}_2\|^2$, and $\|\mathbf{v}_3\|^2$, respectively.

Orthonormal Sets

We say that a set of vectors is **orthonormal** if it is orthogonal and consists of *unit* vectors (Fig. 8.3). Thus, $\{\mathbf{v}_1, \ldots, \mathbf{v}_k\}$ is orthonormal if

$$\mathbf{v}_i \cdot \mathbf{v}_j = 0, \qquad i \neq j \qquad \text{and} \qquad \|\mathbf{v}_i\| = 1, \qquad i = 1, \ldots, k$$

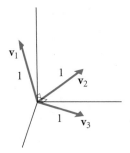

Figure 8.3   Orthonormal vectors.

Because $\|\mathbf{v}_i\| = 1 \Leftrightarrow \|\mathbf{v}_i\|^2 = \mathbf{v}_i \cdot \mathbf{v}_i = 1$, we have

$$\{\mathbf{v}_1, \ldots, \mathbf{v}_k\} \quad \text{orthonormal} \quad \Leftrightarrow \quad \mathbf{v}_i \cdot \mathbf{v}_j = \begin{cases} 1, & \text{if } i = j \\ 0, & \text{if } i \neq j \end{cases} \tag{8.5}$$

■ EXAMPLE 3   The standard basis of $\mathbf{R}^n$ is orthonormal.

■ EXAMPLE 4   The set $S = \{\mathbf{v}_1, \mathbf{v}_2\}$ is orthonormal, where

$$\mathbf{v}_1 = \begin{bmatrix} \dfrac{1}{\sqrt{2}} \\ \dfrac{1}{\sqrt{2}} \end{bmatrix}, \qquad \mathbf{v}_2 = \begin{bmatrix} -\dfrac{1}{\sqrt{2}} \\ \dfrac{1}{\sqrt{2}} \end{bmatrix}$$

Because $\mathbf{v}_1 \cdot \mathbf{v}_2 = 0$ and $\|\mathbf{v}_1\| = 1$, $\|\mathbf{v}_2\| = 1$.

We may normalize any orthogonal set of nonzero vectors to get an orthonormal set:

$$\{\mathbf{v}_1, \ldots, \mathbf{v}_k\} \quad \text{orthogonal} \quad \Rightarrow \quad \left\{ \frac{\mathbf{v}_1}{\|\mathbf{v}_1\|}, \ldots, \frac{\mathbf{v}_k}{\|\mathbf{v}_k\|} \right\} \quad \text{orthonormal}$$

This is because each vector $\mathbf{v}_i / \|\mathbf{v}_i\|$ is unit and, for $i \neq j$,

$$\frac{\mathbf{v}_i}{\|\mathbf{v}_i\|} \cdot \frac{\mathbf{v}_j}{\|\mathbf{v}_j\|} = \frac{\mathbf{v}_i \cdot \mathbf{v}_j}{\|\mathbf{v}_i\| \, \|\mathbf{v}_j\|} = 0$$

Normalizing $S = \{\mathbf{v}_1, \mathbf{v}_2, \mathbf{v}_3\}$ of Example 2 yields the orthonormal $S' = \{\mathbf{u}_1, \mathbf{u}_2, \mathbf{u}_3\}$,

$$\mathbf{u}_1 = \begin{bmatrix} \dfrac{1}{\sqrt{14}} \\ -\dfrac{2}{\sqrt{14}} \\ \dfrac{3}{\sqrt{14}} \end{bmatrix}, \qquad \mathbf{u}_2 = \begin{bmatrix} -\dfrac{1}{\sqrt{3}} \\ \dfrac{1}{\sqrt{3}} \\ \dfrac{1}{\sqrt{3}} \end{bmatrix}, \qquad \mathbf{u}_3 = \begin{bmatrix} \dfrac{5}{\sqrt{42}} \\ \dfrac{4}{\sqrt{42}} \\ \dfrac{1}{\sqrt{42}} \end{bmatrix} \tag{8.6}$$

An orthonormal set that is a basis of a subspace V of $\mathbf{R}^n$ is called an **orthonormal basis** of V. For example, S' is an orthonormal basis of $\mathbf{R}^3$. Written for orthonormal bases, Theorem 1 takes the following special form.

THEOREM 5

If $S = \{\mathbf{v}_1, \dots, \mathbf{v}_k\}$ is an orthonormal basis of the subspace V of $\mathbf{R}^n$, then each n-vector $\mathbf{u}$ of V can be uniquely written as

$$\mathbf{u} = (\mathbf{u} \cdot \mathbf{v}_1)\mathbf{v}_1 + \cdots + (\mathbf{u} \cdot \mathbf{v}_k)\mathbf{v}_k \qquad (8.7)$$

So, computing the components of a vector with respect to an orthonormal basis is easy.

Along the same lines, we also have the very useful Bessel's inequality, which appears in many applications. It asserts that the square of the length of a vector is at least the sum of the squares of its Fourier coefficients.

THEOREM 6

(Bessel's Inequality)

Let $S = \{\mathbf{v}_1, \dots, \mathbf{v}_k\}$ be an orthonormal subset of $\mathbf{R}^n$ (not necessarily a basis) and let $\mathbf{u}$ be any n-vector. Then

$$\|\mathbf{u}\|^2 \geq (\mathbf{u} \cdot \mathbf{v}_1)^2 + \cdots + (\mathbf{u} \cdot \mathbf{v}_k)^2 \qquad (8.8)$$

PROOF See Exercise 27.

Theorem 4 also has an important special case. The matrix $A = [\mathbf{v}_1 \cdots \mathbf{v}_n]$ has orthonormal columns if and only if each diagonal entry $\|\mathbf{v}_i\|^2$ of $A^T A$ is 1. This is equivalent to $A^T A = I$.

THEOREM 7

The columns of an $m \times n$ matrix A form an orthonormal set (hence, $m \geq n$) if and only if

$$A^T A = I_n$$

Orthogonal Matrices

A matrix A is called **orthogonal** if

1. It is square; and
2. Has *orthonormal* columns.

■ **EXAMPLE 5** I is orthogonal. So is the matrix with columns $\mathbf{u}_1, \mathbf{u}_2$, and $\mathbf{u}_3$ of (8.6).

Perhaps a better name for orthogonal matrix would be *orthonormal*, but this term is not used. Note that a *nonsquare* matrix with orthonormal columns is *not* called orthogonal.

Our first remark about orthogonal matrices is that they are **invertible**, because they are square with linearly independent columns. In fact, Theorem 7 for $m = n$ implies the following important result.

THEOREM 8

A square matrix A is orthogonal if and only if

$$A^T A = I \qquad \text{or} \qquad A^{-1} = A^T$$

So, the inverse of an orthogonal matrix is its transpose. No messy inversions here.

■ EXAMPLE 6   Show that A and B are orthogonal and compute their inverses.

$$A = \begin{bmatrix} 0 & 1 & 0 \\ 0 & 0 & 1 \\ 1 & 0 & 0 \end{bmatrix}, \qquad B = \begin{bmatrix} \cos\theta & -\sin\theta \\ \sin\theta & \cos\theta \end{bmatrix}$$

SOLUTION   The columns of A are clearly orthonormal, so A is orthogonal. Therefore,

$$A^{-1} = A^T = \begin{bmatrix} 0 & 0 & 1 \\ 1 & 0 & 0 \\ 0 & 1 & 0 \end{bmatrix}$$

The columns of B are orthonormal, because $\begin{bmatrix} \cos\theta \\ \sin\theta \end{bmatrix} \cdot \begin{bmatrix} -\sin\theta \\ \cos\theta \end{bmatrix} = 0$ and

$$\begin{bmatrix} \cos\theta \\ \sin\theta \end{bmatrix} \cdot \begin{bmatrix} \cos\theta \\ \sin\theta \end{bmatrix} = \cos^2\theta + \sin^2\theta = 1 = \begin{bmatrix} -\sin\theta \\ \cos\theta \end{bmatrix} \cdot \begin{bmatrix} -\sin\theta \\ \cos\theta \end{bmatrix}$$

Therefore,

$$B^{-1} = B^T = \begin{bmatrix} \cos\theta & \sin\theta \\ -\sin\theta & \cos\theta \end{bmatrix}$$

Note that A is a *permutation* matrix. In general, any permutation matrix is orthogonal.

■ EXAMPLE 7   Compute the inverse of $A = [\mathbf{u}_1 \ \mathbf{u}_2 \ \mathbf{u}_3]$ from (8.6).

SOLUTION   Since A is orthogonal,

$$A^{-1} = A^T = \begin{bmatrix} \dfrac{1}{\sqrt{14}} & -\dfrac{2}{\sqrt{14}} & \dfrac{3}{\sqrt{14}} \\[2mm] -\dfrac{1}{\sqrt{3}} & \dfrac{1}{\sqrt{3}} & \dfrac{1}{\sqrt{3}} \\[2mm] \dfrac{5}{\sqrt{42}} & \dfrac{4}{\sqrt{42}} & -\dfrac{1}{\sqrt{42}} \end{bmatrix}$$

The matrix transformation $A\mathbf{x}$ defined by an orthogonal matrix A is also called orthogonal. Orthogonal matrix transformations preserve dot products. Hence, they preserve lengths and angles. Conversely, if a matrix transformation preserves dot products, then its (standard) matrix is orthogonal.

THEOREM 9

Let A be an $n \times n$ matrix. The following statements are equivalent.

1. A is orthogonal.
2. $A\mathbf{u} \cdot A\mathbf{v} = \mathbf{u} \cdot \mathbf{v}$ for any n-vectors $\mathbf{u}$ and $\mathbf{v}$ (preservation of dot products).
3. $\|A\mathbf{v}\| = \|\mathbf{v}\|$ for any n-vector $\mathbf{v}$ (preservation of norms).

PROOF

$(1) \Rightarrow (2)$ If A is orthogonal, then $A^T A = I$. So by (8.1),

$$A\mathbf{u} \cdot A\mathbf{v} = \mathbf{u} \cdot (A^T A\mathbf{v}) = \mathbf{u} \cdot \mathbf{v}$$

$(2) \Rightarrow (1)$ Suppose $A\mathbf{u} \cdot A\mathbf{v} = \mathbf{u} \cdot \mathbf{v}$. In particular, $A\mathbf{e}_i \cdot A\mathbf{e}_j = \mathbf{e}_i \cdot \mathbf{e}_j$. But the standard basis is orthonormal. Hence,

$$A\mathbf{e}_i \cdot A\mathbf{e}_j = \mathbf{e}_i \cdot \mathbf{e}_j = \begin{cases} 1 & \text{if } i = j \\ 0 & \text{if } i \neq j \end{cases}$$

which shows that A is orthogonal, by (8.5), because $A\mathbf{e}_i$ is the ith column of A.

$(2) \Leftrightarrow (3)$ The proof of this equivalence is left as an exercise.

Theorem 9 has two interesting implications.

THEOREM 10

1. If A and B are $n \times n$ orthogonal, so is AB.
2. If A is orthogonal, so is A^{-1}.

PROOF

1. By Theorem 9 it suffices to show that AB preserves norms:

$$\|AB\mathbf{v}\| = \|A(B\mathbf{v})\| = \|B\mathbf{v}\| = \|\mathbf{v}\|$$

The proof of Part 2 is left as an exercise.

N O T E   Because the inverse—and hence the transpose—of an orthogonal matrix A is orthogonal, we conclude that *the rows of an orthogonal matrix are also orthonormal.*

The second implication of Theorem 9 is as follows.

THEOREM 11

If λ is an eigenvalue of an orthogonal matrix A, then $|\lambda| = 1$.

PROOF   If $\mathbf{v}$ an eigenvector of A, then by Part 3 of Theorem 9,

$$\|\mathbf{v}\| = \|A\mathbf{v}\| = \|\lambda\mathbf{v}\| = |\lambda|\|\mathbf{v}\|$$

Hence, $|\lambda| = 1$, because $\|\mathbf{v}\| \neq 0$.

Theorem 11 holds even for complex eigenvalues of A. For example, the eigenvalues

of $\begin{bmatrix} 0 & 1 & 0 \\ 0 & 0 & 1 \\ 1 & 0 & 0 \end{bmatrix}$ are 1, $-\frac{1}{2} + \frac{1}{2}i\sqrt{3}$, and $-\frac{1}{2} - \frac{1}{2}i\sqrt{3}$, and all three have absolute value

1. (Check.)

Exercises 8.1

In Exercises 1–4 show that the set of given n-vectors is orthogonal. Which of these sets form an orthogonal basis for $\mathbf{R}^n$?

1. $\begin{bmatrix} 1 \\ -2 \\ 1 \end{bmatrix}$, $\begin{bmatrix} 4 \\ 2 \\ 0 \end{bmatrix}$, $\begin{bmatrix} -1 \\ 2 \\ 5 \end{bmatrix}$

2. $\begin{bmatrix} 3 \\ -2 \\ 1 \end{bmatrix}$, $\begin{bmatrix} 1 \\ 2 \\ 1 \end{bmatrix}$, $\begin{bmatrix} -2 \\ -1 \\ 4 \end{bmatrix}$

3. $\begin{bmatrix} 1 \\ 1 \\ -1 \\ 1 \end{bmatrix}$, $\begin{bmatrix} 1 \\ 1 \\ 1 \\ -1 \end{bmatrix}$, $\begin{bmatrix} 0 \\ 0 \\ 1 \\ 1 \end{bmatrix}$

4. $\begin{bmatrix} 1 \\ 1 \\ -1 \\ 1 \end{bmatrix}$, $\begin{bmatrix} 1 \\ 1 \\ 1 \\ -1 \end{bmatrix}$, $\begin{bmatrix} 0 \\ 0 \\ 1 \\ 1 \end{bmatrix}$, $\begin{bmatrix} 1 \\ -1 \\ 0 \\ 0 \end{bmatrix}$

5. Give an example of a set of vectors $S = \{\mathbf{v}_1, \mathbf{v}_2, \mathbf{v}_3\}$ such that the pairs $\mathbf{v}_1, \mathbf{v}_2$ and $\mathbf{v}_2, \mathbf{v}_3$ are orthogonal, but S is not orthogonal.

In Exercises 6–9 show that each set of vectors forms an orthogonal basis for $\mathbf{R}^3$. Use Theorem 1 to express $\mathbf{u} = (1, 1, 1)$ as a linear combination in these vectors.

6. $\begin{bmatrix} 6 \\ 2 \\ 1 \end{bmatrix}$, $\begin{bmatrix} -1 \\ 3 \\ 0 \end{bmatrix}$, $\begin{bmatrix} -3 \\ -1 \\ 20 \end{bmatrix}$

7. $\begin{bmatrix} 6 \\ -1 \\ 1 \end{bmatrix}$, $\begin{bmatrix} 1 \\ 3 \\ -3 \end{bmatrix}$, $\begin{bmatrix} 0 \\ 1 \\ 1 \end{bmatrix}$

8. $\begin{bmatrix} 0 \\ 1 \\ 1 \end{bmatrix}$, $\begin{bmatrix} 4 \\ -1 \\ 1 \end{bmatrix}$, $\begin{bmatrix} 1 \\ 2 \\ -2 \end{bmatrix}$

9. $\begin{bmatrix} 1 \\ -2 \\ 1 \end{bmatrix}$, $\begin{bmatrix} 4 \\ 1 \\ -2 \end{bmatrix}$, $\begin{bmatrix} 3 \\ 6 \\ 9 \end{bmatrix}$

In Exercises 10–14 determine whether the given orthogonal set of vectors is orthonormal. If it is not, normalize the vectors to get an orthonormal set.

10. $\begin{bmatrix} \frac{1}{\sqrt{2}} \\ \frac{1}{\sqrt{2}} \end{bmatrix}$, $\begin{bmatrix} -\frac{1}{\sqrt{2}} \\ \frac{1}{\sqrt{2}} \end{bmatrix}$

11. $\begin{bmatrix} \frac{1}{\sqrt{2}} \\ \frac{2}{\sqrt{2}} \end{bmatrix}$, $\begin{bmatrix} -\frac{2}{\sqrt{2}} \\ \frac{1}{\sqrt{2}} \end{bmatrix}$

12. $\begin{bmatrix} 1 \\ 2 \\ 2 \end{bmatrix}$, $\begin{bmatrix} 2 \\ -2 \\ 1 \end{bmatrix}$, $\begin{bmatrix} 2 \\ 1 \\ -2 \end{bmatrix}$

13. $\begin{bmatrix} \frac{1}{3} \\ \frac{2}{3} \\ \frac{2}{3} \end{bmatrix}$, $\begin{bmatrix} \frac{2}{3} \\ -\frac{2}{3} \\ \frac{1}{3} \end{bmatrix}$, $\begin{bmatrix} \frac{2}{3} \\ \frac{1}{3} \\ -\frac{2}{3} \end{bmatrix}$

14. $\begin{bmatrix} 1 \\ 1 \\ -1 \\ 1 \end{bmatrix}$, $\begin{bmatrix} 1 \\ 1 \\ 1 \\ -1 \end{bmatrix}$, $\begin{bmatrix} 0 \\ 0 \\ 1 \\ 1 \end{bmatrix}$

In Exercises 15–16 show that $\mathcal{B}$ is an orthonormal basis for $\mathbf{R}^n$ (for the appropriate n). Use Theorem 5 to write $\mathbf{e}_1 \in \mathbf{R}^n$ as a linear combination in $\mathcal{B}$.

15. $\mathcal{B} = \left\{ \begin{bmatrix} \dfrac{2}{\sqrt{5}} \\ \dfrac{1}{\sqrt{5}} \end{bmatrix}, \begin{bmatrix} -\dfrac{1}{\sqrt{5}} \\ \dfrac{2}{\sqrt{5}} \end{bmatrix} \right\}$

16. $\mathcal{B} = \left\{ \begin{bmatrix} \frac{1}{3} \\ \frac{2}{3} \\ \frac{2}{3} \end{bmatrix}, \begin{bmatrix} \frac{2}{3} \\ -\frac{2}{3} \\ \frac{1}{3} \end{bmatrix}, \begin{bmatrix} \frac{2}{3} \\ \frac{1}{3} \\ -\frac{2}{3} \end{bmatrix} \right\}$

In Exercises 17–20 determine whether the given matrix is orthogonal. If the matrix is orthogonal, find its inverse.

17. $\begin{bmatrix} 0 & 1 \\ -1 & 0 \end{bmatrix}$

18. $\begin{bmatrix} 0 & 0 & 1 \\ 1 & 0 & 0 \\ 0 & 1 & 0 \\ 1 & 0 & 0 \end{bmatrix}$

19. $\begin{bmatrix} \dfrac{3}{\sqrt{14}} & \dfrac{1}{\sqrt{6}} & -\dfrac{2}{\sqrt{21}} \\ -\dfrac{2}{\sqrt{14}} & \dfrac{2}{\sqrt{6}} & -\dfrac{1}{\sqrt{21}} \\ \dfrac{1}{\sqrt{14}} & \dfrac{1}{\sqrt{6}} & \dfrac{4}{\sqrt{21}} \end{bmatrix}$

20. $\begin{bmatrix} \dfrac{1}{2} & \dfrac{1}{2} & 0 & \dfrac{1}{\sqrt{2}} \\ \dfrac{1}{2} & \dfrac{1}{2} & 0 & -\dfrac{1}{\sqrt{2}} \\ -\dfrac{1}{2} & \dfrac{1}{2} & \dfrac{1}{\sqrt{2}} & 0 \\ \dfrac{1}{2} & -\dfrac{1}{2} & \dfrac{1}{\sqrt{2}} & 0 \end{bmatrix}$

21. Suppose that the columns of an $m \times n$ matrix A form an orthonormal set. Why is $m \geq n$? If A only had orthogonal columns, would this imply that $m \geq n$?

22. Show that the rows of an $n \times n$ orthogonal matrix form a basis for $\mathbf{R}^n$.

23. Prove Theorem 3.

24. Complete the proof of Theorem 4.

25. Complete the proof of Theorem 9.

26. Complete the proof of Theorem 10.

27. Prove Bessel's inequality (Theorem 6). (*Hint:* Let $\mathbf{v} = \sum_{i=1}^{k} (\mathbf{u} \cdot \mathbf{v}_i)\mathbf{v}_i$ and let $\mathbf{r} = \mathbf{u} - \mathbf{v}$. Show that $\mathbf{r} \cdot \mathbf{v} = \mathbf{0}$. Then use the Pythagorean theorem to conclude that $\|\mathbf{u}\|^2 = \|\mathbf{v}\|^2 + \|\mathbf{r}\|^2$.)

8.2 Orthogonal Projections: Gram-Schmidt Process

Reader's Goals for This Section

1. To understand the concept and properties of an orthogonal projection.
2. To know how to find an orthonormal basis from a given basis of a subspace of $\mathbf{R}^n$.

In this section we study orthogonal *projections* and *complements*. These generalize basic notions of Chapter 2, such as projection of a vector along another vector, projection of a vector onto a plane, the normal vector to a plane and line, or plane perpendicular to a vector. We then use our new tools to construct an orthogonal basis out of an ordinary basis for any subspace of $\mathbf{R}^n$.

> **CONVENTION**
> All lines and planes in this section pass through the origin.

Orthogonal Complements

We know that a vector normal to a plane is orthogonal to any vector of that plane. In general, if an n-vector $\mathbf{u}$ is orthogonal to each vector of a subspace V of $\mathbf{R}^n$ we say that $\mathbf{u}$ is **orthogonal** to V. In practice, to check that $\mathbf{u}$ is orthogonal to V, we do not compute infinitely many dot products. Only the dot products of $\mathbf{u}$ with the elements of a *basis* of V (or of a finite spanning set).

THEOREM 12

The n-vector $\mathbf{u}$ is orthogonal to $V = \text{Span}\{\mathbf{v}_1, \ldots, \mathbf{v}_k\} \subseteq \mathbf{R}^n$ if and only if

$$\mathbf{u} \cdot \mathbf{v}_i = 0, \qquad i = 1, \ldots, k \qquad (8.9)$$

PROOF   If $\mathbf{u}$ is orthogonal to V, then (8.9) holds. Conversely, if we assume (8.9) and let $\mathbf{v}$ be any element of V, then there are scalars c_i such that

$$\mathbf{v} = c_1 \mathbf{v}_1 + \cdots + c_k \mathbf{v}_k$$

"Dotting" with $\mathbf{u}$ yields

$$\mathbf{u} \cdot \mathbf{v} = \mathbf{u} \cdot (c_1 \mathbf{v}_1 + \cdots + c_k \mathbf{v}_k)$$
$$= c_1(\mathbf{u} \cdot \mathbf{v}_1) + \cdots + c_k(\mathbf{u} \cdot \mathbf{v}_k)$$
$$= c_1 0 + \cdots + c_k 0 = 0$$

So, $\mathbf{u}$ and $\mathbf{v}$ are orthogonal. Hence, $\mathbf{u}$ is orthogonal to V, as asserted.

DEFINITION

The set of all n-vectors orthogonal to V is called the **orthogonal complement** of V and it is denoted by $V^\perp$ (read "V perp.").

▪ **EXAMPLE 8**   In $\mathbf{R}^3$, the orthogonal complement of a plane through $\mathbf{0}$ is the line through $\mathbf{0}$ perpendicular to the plane. Also, the orthogonal complement of a line through $\mathbf{0}$ is the plane through $\mathbf{0}$ perpendicular to the line (Fig. 8.4).

$$V^\perp = l$$
$$l^\perp = V$$
$$\mathbf{u} \cdot \mathbf{v} = 0$$

Figure 8.4   Orthogonal complement of a line and of a plane.

THEOREM 13

Let V be a subspace of $\mathbf{R}^n$. Then

1. $V^{\perp}$ is a subspace of $\mathbf{R}^n$;
2. $(V^{\perp})^{\perp} = V$.

PROOF

1. Let $\mathbf{u}_1$ and $\mathbf{u}_2$ be in $V^{\perp}$ and let $\mathbf{v}$ be any vector of V. Then $\mathbf{u}_1 \cdot \mathbf{v} = 0$ and $\mathbf{u}_2 \cdot \mathbf{v} = 0$. For any scalars c_1 and c_2,

$$(c_1\mathbf{u}_1 + c_2\mathbf{u}_2) \cdot \mathbf{v} = c_1\mathbf{u}_1 \cdot \mathbf{v} + c_2\mathbf{u}_2 \cdot \mathbf{v} = c_1 0 + c_2 0 = 0$$

Therefore, $c_1\mathbf{u}_1 + c_2\mathbf{u}_2$ is orthogonal to any vector of V; hence, it is in $V^{\perp}$. So, $V^{\perp}$ is a subspace of $\mathbf{R}^n$.

2. The proof is left as an exercise.

The notion of orthogonal complement allows us easily to express an important relation between the column space of a matrix and the null space of its transpose.

THEOREM 14

Let A be any $m \times n$ matrix.[1] Then

$$(\mathrm{Col}(A))^{\perp} = \mathrm{Null}(A^T)$$

By switching rows to columns, we also have

$$(\mathrm{Row}(A))^{\perp} = \mathrm{Null}(A)$$

PROOF   Let $A = [\mathbf{v}_1 \cdots \mathbf{v}_n]$. Then $\mathbf{u}$ is in $(\mathrm{Col}(A))^{\perp}$ if and only if $\mathbf{u}$ is orthogonal to any vector of $\mathrm{Col}(A)$. Equivalently, $\mathbf{u}$ is orthogonal to the columns of A (which span $\mathrm{Col}(A)$). So,

$$\mathbf{u} \in (\mathrm{Col}(A))^{\perp} \Leftrightarrow \mathbf{u} \cdot \mathbf{v}_1 = 0, \ldots, \mathbf{u} \cdot \mathbf{v}_n = 0$$

$$\Leftrightarrow \mathbf{v}_1^T \mathbf{u} = 0, \ldots, \mathbf{v}_n^T \mathbf{u} = 0$$

$$\Leftrightarrow A^T \mathbf{u} = 0$$

$$\Leftrightarrow \mathbf{u} \in \mathrm{Null}(A^T)$$

[1]Some authors call this theorem *the fundamental theorem of linear algebra*.

■ EXAMPLE 9   Verify Theorem 14 for

$$A = \begin{bmatrix} 1 & 2 \\ -2 & 0 \\ 1 & 4 \end{bmatrix}$$

SOLUTION   By Theorem 12 it suffices to show that each vector of some basis of Col(A) is orthogonal to each vector of some basis of Null(A^T). Because the columns of A are linearly independent, they form a basis for Col(A). By reducing $[A^T : \mathbf{0}]$ it is easy to show that $\{(4, 1, -2)\}$ is a basis for Null(A^T). We now check

$$\begin{bmatrix} 1 \\ -2 \\ 1 \end{bmatrix} \cdot \begin{bmatrix} 4 \\ 1 \\ -2 \end{bmatrix} = 0 = \begin{bmatrix} 2 \\ 0 \\ 4 \end{bmatrix} \cdot \begin{bmatrix} 4 \\ 1 \\ -2 \end{bmatrix}$$

as predicted by Theorem 14.

Orthogonal Projections

In Section 2.2 we saw that for given plane or space vectors $\mathbf{u}$ and $\mathbf{v}$ ($\neq \mathbf{0}$), it is possible to write $\mathbf{u}$ as a sum of two *orthogonal* vectors $\mathbf{u}_{pr}$ and $\mathbf{u}_c$,

$$\mathbf{u} = \mathbf{u}_{pr} + \mathbf{u}_c$$

such that $\mathbf{u}_c$ is along the line defined by $\mathbf{v}$. Here, $\mathbf{u}_{pr}$ is the **orthogonal projection** of $\mathbf{u}$ along $\mathbf{v}$ and $\mathbf{u}_c$ is the **component** of $\mathbf{u}$ **orthogonal** to $\mathbf{v}$. This applies to physics when we try to decompose a force vector into more manageable components.

In addition, we saw that $\mathbf{u}_{pr}$ and $\mathbf{u}_c$ can be nicely computed in terms of dot products as

$$\mathbf{u}_{pr} = \frac{\mathbf{u} \cdot \mathbf{v}}{\mathbf{v} \cdot \mathbf{v}} \mathbf{v} \tag{8.10}$$

and

$$\mathbf{u}_c = \mathbf{u} - \frac{\mathbf{u} \cdot \mathbf{v}}{\mathbf{v} \cdot \mathbf{v}} \mathbf{v} \tag{8.11}$$

(Fig. 8.5). The fact that $\mathbf{u}_{pr}$ and $\mathbf{u}_c$ are orthogonal follows from

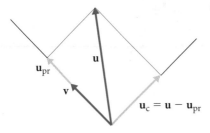

Figure 8.5   The orthogonal projection $\mathbf{u}_{pr}$ of $\mathbf{u}$ along $\mathbf{v}$.

$$\mathbf{u}_c \cdot \mathbf{v} = \mathbf{u} \cdot \mathbf{v} - \frac{\mathbf{u} \cdot \mathbf{v}}{\mathbf{v} \cdot \mathbf{v}} \mathbf{v} \cdot \mathbf{v}$$

$$= \mathbf{u} \cdot \mathbf{v} - \mathbf{u} \cdot \mathbf{v} = 0$$

Note that $\text{Span}\{\mathbf{v}, \mathbf{u}\} = \text{Span}\{\mathbf{v}, \mathbf{u}_c\}$ by (8.11).

It is also worth noticing that $\mathbf{u}_{pr}$ and $\mathbf{u}_c$ remain the same if we replace $\mathbf{v}$ with $c\mathbf{v}$, $c \neq 0$, because

$$\frac{\mathbf{u} \cdot c\mathbf{v}}{c\mathbf{v} \cdot c\mathbf{v}} c\mathbf{v} = \frac{c(\mathbf{u} \cdot \mathbf{v})}{c^2(\mathbf{v} \cdot \mathbf{v})} c\mathbf{v} = \frac{\mathbf{u} \cdot \mathbf{v}}{\mathbf{v} \cdot \mathbf{v}} \mathbf{v}$$

Thus, $\mathbf{u}_{pr}$ and $\mathbf{u}_c$ depend *only* on $\text{Span}\{\mathbf{v}\}$ and not on $\mathbf{v}$ itself.

The length of $\mathbf{u}_c$ is clearly the shortest distance from (the tip of) $\mathbf{u}$ to the line $l = \text{Span}\{\mathbf{v}\}$.

■ EXAMPLE 10   Find the shortest distance d from $\mathbf{u} = (1, -2, 3)$ to the line $\mathbf{p} = (1, 1, 1)t, t \in \mathbf{R}$.

SOLUTION   Let $\mathbf{v} = (1, 1, 1)$. Then $\mathbf{u} \cdot \mathbf{v} = 2$ and $\mathbf{v} \cdot \mathbf{v} = 3$. Hence,

$$\mathbf{u}_{pr} = \frac{\mathbf{u} \cdot \mathbf{v}}{\mathbf{v} \cdot \mathbf{v}} \mathbf{v} = \left(\frac{2}{3}, \frac{2}{3}, \frac{2}{3} \right)$$

So (Fig. 8.6),

$$d = \|\mathbf{u}_c\| = \|\mathbf{u} - \mathbf{u}_{pr}\| = \left\| \left(\frac{1}{3}, -\frac{8}{3}, \frac{7}{3} \right) \right\| = \frac{\sqrt{114}}{3}$$

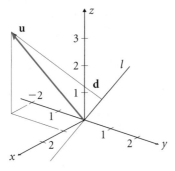

Figure 8.6   *The length of the component orthogonal to l is the shortest distance from l.*

Instead of projecting 2- or 3-vectors onto a line, let us see how we can, in general, project an n-vector onto a subspace V of $\mathbf{R}^n$ so that the main properties of the plane or space projections are preserved. In order to extend Equations (8.10) and (8.11) in an easy manner, we assume that V has an orthogonal basis. This is not a restriction, because—as we shall see shortly—an orthogonal basis always exists.

DEFINITION

Let $\mathbf{u}$ be an n-vector and let V be a subspace of $\mathbf{R}^n$ with an orthogonal basis $\mathcal{B} = \{\mathbf{v}_1, \ldots, \mathbf{v}_k\}$. Then the **orthogonal projection** of $\mathbf{u}$ onto V is the vector

$$\mathbf{u}_{\text{pr}} = \frac{\mathbf{u} \cdot \mathbf{v}_1}{\mathbf{v}_1 \cdot \mathbf{v}_1}\mathbf{v}_1 + \cdots + \frac{\mathbf{u} \cdot \mathbf{v}_k}{\mathbf{v}_k \cdot \mathbf{v}_k}\mathbf{v}_k \qquad (8.12)$$

The difference $\mathbf{u}_{\text{c}} = \mathbf{u} - \mathbf{u}_{\text{pr}}$ is called the **component of u orthogonal to** V.

$$\mathbf{u}_{\text{c}} = \mathbf{u} - \frac{\mathbf{u} \cdot \mathbf{v}_1}{\mathbf{v}_1 \cdot \mathbf{v}_1}\mathbf{v}_1 - \cdots - \frac{\mathbf{u} \cdot \mathbf{v}_k}{\mathbf{v}_k \cdot \mathbf{v}_k}\mathbf{v}_k \qquad (8.13)$$

$$\mathbf{u} = \mathbf{u}_{\text{pr}} + \mathbf{u}_{\text{c}} \qquad (8.14)$$

We see that $\mathbf{u}_{\text{c}}$ is orthogonal to all vectors of $\mathcal{B}$.

$$\mathbf{u}_{\text{c}} \cdot \mathbf{v}_1 = 0, \ldots, \mathbf{u}_{\text{c}} \cdot \mathbf{v}_k = 0 \qquad (8.15)$$

We have

$$\mathbf{u}_{\text{c}} \cdot \mathbf{v}_i = \mathbf{u} \cdot \mathbf{v}_i - \frac{\mathbf{u} \cdot \mathbf{v}_1}{\mathbf{v}_1 \cdot \mathbf{v}_1}\mathbf{v}_1 \cdot \mathbf{v}_i - \cdots - \frac{\mathbf{u} \cdot \mathbf{v}_k}{\mathbf{v}_k \cdot \mathbf{v}_k}\mathbf{v}_k \cdot \mathbf{v}_i$$

$$= \mathbf{u} \cdot \mathbf{v}_i - \frac{\mathbf{u} \cdot \mathbf{v}_i}{\mathbf{v}_i \cdot \mathbf{v}_i}\mathbf{v}_i \cdot \mathbf{v}_i$$

$$= \mathbf{u} \cdot \mathbf{v}_i - \mathbf{u} \cdot \mathbf{v}_i = 0$$

because $\mathbf{v}_j \cdot \mathbf{v}_i = 0$ for $j \neq i$. By Theorem 12 we see that $\mathbf{u}_{\text{c}}$ is orthogonal to V. Hence,

$$\mathbf{u}_{\text{pr}} \in V \qquad \text{and} \qquad \mathbf{u}_{\text{c}} \in V^{\perp} \qquad (8.16)$$

Equation (8.13) implies that

$$\text{Span}\{\mathbf{v}_1, \ldots, \mathbf{v}_k, \mathbf{u}\} = \text{Span}\{\mathbf{v}_1, \ldots, \mathbf{v}_k, \mathbf{u}_{\text{c}}\} \qquad (8.17)$$

Geometrically, if $n = 3$, we see that $\mathbf{u}_{\text{pr}}$ is the vector sum of the projections of $\mathbf{u}$ along $\mathbf{v}_1$ and $\mathbf{v}_2$ and lies in the plane $\text{Span}\{\mathbf{v}_1, \mathbf{v}_2\}$. Also, $\mathbf{u}_{\text{c}}$ is the normal to this plane such that $\mathbf{u} = \mathbf{u}_{\text{pr}} + \mathbf{u}_{\text{c}}$ (Fig. 8.7).

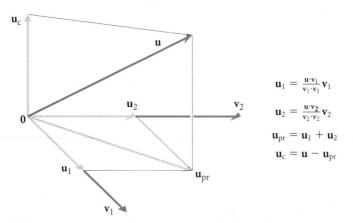

$$\mathbf{u}_1 = \frac{\mathbf{u} \cdot \mathbf{v}_1}{\mathbf{v}_1 \cdot \mathbf{v}_1}\mathbf{v}_1$$

$$\mathbf{u}_2 = \frac{\mathbf{u} \cdot \mathbf{v}_2}{\mathbf{v}_2 \cdot \mathbf{v}_2}\mathbf{v}_2$$

$$\mathbf{u}_{\text{pr}} = \mathbf{u}_1 + \mathbf{u}_2$$

$$\mathbf{u}_{\text{c}} = \mathbf{u} - \mathbf{u}_{\text{pr}}$$

Figure 8.7 The orthogonal projection $\mathbf{u}_{\text{pr}}$ of $\mathbf{u}$ onto $\text{Span}\{\mathbf{v}_1, \mathbf{v}_2\}$.

The vector $\mathbf{u}_{pr}$ satisfies a most interesting property. It is the closest point in V to $\mathbf{u}$, or the *best approximation* of $\mathbf{u}$ by vectors of V (Fig. 8.8).

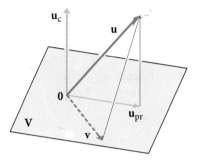

Figure 8.8   The orthogonal projection $\mathbf{u}_{pr}$ is the only vector of V closest to $\mathbf{u}$.

THEOREM 15

(Best Approximation)

With the preceding notation,

$$\|\mathbf{u}_c\| = \|\mathbf{u} - \mathbf{u}_{pr}\| < \|\mathbf{u} - \mathbf{v}\|$$

for any vector $\mathbf{v}$ of V other than $\mathbf{u}_{pr}$.

PROOF   The vectors $\mathbf{u}_{pr} - \mathbf{v}$ and $\mathbf{u} - \mathbf{u}_{pr}$ are orthogonal, because the first is in V and the second is in $V^{\perp}$. Hence, by the Pythagorean theorem for n-vectors (Section 2.2), we have

$$\|\mathbf{u} - \mathbf{u}_{pr}\|^2 + \|\mathbf{u}_{pr} - \mathbf{v}\|^2 = \|(\mathbf{u} - \mathbf{u}_{pr}) + (\mathbf{u}_{pr} - \mathbf{v})\|^2$$
$$= \|\mathbf{u} - \mathbf{v}\|^2$$

Therefore, $\|\mathbf{u} - \mathbf{u}_{pr}\| < \|\mathbf{u} - \mathbf{v}\|$, because $\mathbf{u}_{pr} - \mathbf{v} \neq \mathbf{0}$.

■ EXAMPLE 11   Find the vector in the plane spanned by the orthogonal vectors $\mathbf{v}_1 = (-1, 4, 1)$ and $\mathbf{v}_2 = (5, 1, 1)$ that best approximates $\mathbf{u} = (1, -1, 2)$.

SOLUTION   The vector we need is $\mathbf{u}_{pr}$. Because

$$\mathbf{u} \cdot \mathbf{v}_1 = -3 \qquad \mathbf{u} \cdot \mathbf{v}_2 = 6$$
$$\mathbf{v}_1 \cdot \mathbf{v}_1 = 18 \qquad \mathbf{v}_2 \cdot \mathbf{v}_2 = 27$$

we have

$$\mathbf{u}_{pr} = \frac{\mathbf{u} \cdot \mathbf{v}_1}{\mathbf{v}_1 \cdot \mathbf{v}_1} \mathbf{v}_1 + \frac{\mathbf{u} \cdot \mathbf{v}_2}{\mathbf{v}_2 \cdot \mathbf{v}_2} \mathbf{v}_2$$

$$= \frac{-3}{18}(-1, 4, 1) + \frac{6}{27}(5, 1, 1)$$

$$= \left(\frac{23}{18}, -\frac{4}{9}, \frac{1}{18} \right)$$

The best approximation theorem implies that the orthogonal projection $\mathbf{u}_{pr}$ is unique and *does not depend on the orthogonal basis* of V we used to compute it. Another orthogonal basis $\mathcal{B}'$ would produce the same $\mathbf{u}_{pr}$ and $\mathbf{u}_c$. In fact, we shall see shortly that $\mathbf{u}_{pr}$ and $\mathbf{u}_c$ depend only on $\mathbf{u}$ and V.

The Gram-Schmidt Process

In this paragraph we describe a *very important* method, called the **Gram-Schmidt process**,[2] which allows us to "orthogonalize" any basis $\mathcal{B}$ of any subspace V of $\mathbf{R}^n$; i.e., we transform $\mathcal{B}$ into a new basis of V that has orthogonal vectors.

Let V be any subspace of $\mathbf{R}^n$ and let $\mathcal{B} = \{\mathbf{v}_1, \ldots, \mathbf{v}_k\}$ be any basis of V. We want to gradually replace the vectors $\mathbf{v}_1, \ldots, \mathbf{v}_k$ with vectors $\mathbf{u}_1, \ldots, \mathbf{u}_k$ that are orthogonal and still form a basis of V. First, we replace the set $\{\mathbf{v}_1, \mathbf{v}_2\}$ with an orthogonal set $\{\mathbf{u}_1, \mathbf{u}_2\}$ such that $\text{Span}\{\mathbf{v}_1, \mathbf{v}_2\} = \text{Span}\{\mathbf{u}_1, \mathbf{u}_2\}$. We simply let $\mathbf{u}_1$ be $\mathbf{v}_1$ and let $\mathbf{u}_2$ be the component of $\mathbf{v}_2$ orthogonal to $\mathbf{v}_1$. By (8.15) $\{\mathbf{u}_1, \mathbf{u}_2\}$ is orthogonal. By (8.13) $\{\mathbf{v}_1, \mathbf{v}_2\}$ and $\{\mathbf{u}_1, \mathbf{u}_2\}$ have the same span. Also,

$$\mathbf{u}_1 = \mathbf{v}_1$$

$$\mathbf{u}_2 = \mathbf{v}_2 - \frac{\mathbf{v}_2 \cdot \mathbf{u}_1}{\mathbf{u}_1 \cdot \mathbf{u}_1} \mathbf{u}_1$$

We continue the same way orthogonalizing the set $\{\mathbf{u}_1, \mathbf{u}_2, \mathbf{v}_3\}$. We replace $\mathbf{v}_3$ with $\mathbf{u}_3$, the component of $\mathbf{v}_2$ orthogonal to $\text{Span}\{\mathbf{u}_1, \mathbf{u}_2\}$. Then $\{\mathbf{u}_1, \mathbf{u}_2, \mathbf{u}_3\}$ is orthogonal and spans $\text{Span}\{\mathbf{v}_1, \mathbf{v}_2, \mathbf{v}_3\}$. In addition,

$$\mathbf{u}_3 = \mathbf{v}_3 - \frac{\mathbf{v}_3 \cdot \mathbf{u}_1}{\mathbf{u}_1 \cdot \mathbf{u}_1} \mathbf{u}_1 - \frac{\mathbf{v}_3 \cdot \mathbf{u}_2}{\mathbf{u}_2 \cdot \mathbf{u}_2} \mathbf{u}_2$$

(Fig. 8.9). By induction, we continue until all of $\mathcal{B}$ is replaced with $\{\mathbf{u}_1, \ldots, \mathbf{u}_k\}$, which is

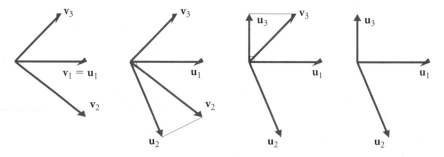

Figure 8.9   Stages of the Gram-Schmidt Process.

[2] Named after the Danish mathematician and actuary **Jörgen Pedersen Gram** (1850–1916) and the German mathematician **Erhard Schmidt** (1876–1959).

orthogonal, and spans the span of $\mathcal{B}$, which is all of V. This is the entire process. If we want to obtain an orthonormal basis, we just normalize $\{\mathbf{u}_1, \ldots, \mathbf{u}_k\}$. Let us summarize.

THEOREM 16

(Gram-Schmidt Process)

Any subspace V of $\mathbf{R}^n$ has at least one orthogonal basis and at least one orthonormal basis. If $\mathcal{B} = \{\mathbf{v}_1, \ldots, \mathbf{v}_k\}$ is any basis of V, then $\mathcal{B}' = \{\mathbf{u}_1, \ldots, \mathbf{u}_k\}$ is an orthogonal basis, where

$$\mathbf{u}_1 = \mathbf{v}_1$$

$$\mathbf{u}_2 = \mathbf{v}_2 - \frac{\mathbf{v}_2 \cdot \mathbf{u}_1}{\mathbf{u}_1 \cdot \mathbf{u}_1} \mathbf{u}_1$$

$$\mathbf{u}_3 = \mathbf{v}_3 - \frac{\mathbf{v}_3 \cdot \mathbf{u}_1}{\mathbf{u}_1 \cdot \mathbf{u}_1} \mathbf{u}_1 - \frac{\mathbf{v}_3 \cdot \mathbf{u}_2}{\mathbf{u}_2 \cdot \mathbf{u}_2} \mathbf{u}_2$$

$$\vdots$$

$$\mathbf{u}_k = \mathbf{v}_k - \frac{\mathbf{v}_k \cdot \mathbf{u}_1}{\mathbf{u}_1 \cdot \mathbf{u}_1} \mathbf{u}_1 - \frac{\mathbf{v}_k \cdot \mathbf{u}_2}{\mathbf{u}_2 \cdot \mathbf{u}_2} \mathbf{u}_2 \cdots - \frac{\mathbf{v}_k \cdot \mathbf{u}_{k-1}}{\mathbf{u}_{k-1} \cdot \mathbf{u}_{k-1}} \mathbf{u}_{k-1}$$

and

$$\text{Span}\{\mathbf{v}_1, \ldots, \mathbf{v}_i\} = \text{Span}\{\mathbf{u}_1, \ldots, \mathbf{u}_i\}, \qquad i = 1, \ldots, k$$

An orthonormal basis $\mathcal{B}''$ is obtained by normalizing $\mathcal{B}'$:

$$\mathcal{B}'' = \left\{ \frac{\mathbf{u}_1}{\|\mathbf{u}_1\|}, \ldots, \frac{\mathbf{u}_k}{\|\mathbf{u}_k\|} \right\}$$

■ **EXAMPLE 12**   Find an orthogonal and an orthonormal basis of $\mathbf{R}^3$ by applying the Gram-Schmidt process to the basis $\mathcal{B} = \{\mathbf{v}_1, \mathbf{v}_2, \mathbf{v}_3\}$, where

$$\mathbf{v}_1 = \begin{bmatrix} 1 \\ -1 \\ 1 \end{bmatrix}, \qquad \mathbf{v}_2 = \begin{bmatrix} -2 \\ 3 \\ -1 \end{bmatrix}, \qquad \mathbf{v}_3 = \begin{bmatrix} 1 \\ 2 \\ -4 \end{bmatrix}$$

SOLUTION   Let $\mathbf{u}_1 = \mathbf{v}_1$. Because $\mathbf{v}_2 \cdot \mathbf{u}_1 = -6$ and $\mathbf{u}_1 \cdot \mathbf{u}_1 = 3$, we have

$$\mathbf{u}_2 = \mathbf{v}_2 - \frac{\mathbf{v}_2 \cdot \mathbf{u}_1}{\mathbf{u}_1 \cdot \mathbf{u}_1} \mathbf{u}_1$$

$$= \begin{bmatrix} -2 \\ 3 \\ -1 \end{bmatrix} - \frac{-6}{3} \begin{bmatrix} 1 \\ -1 \\ 1 \end{bmatrix} = \begin{bmatrix} 0 \\ 1 \\ 1 \end{bmatrix}$$

Because $\mathbf{v}_3 \cdot \mathbf{u}_1 = -5$, $\mathbf{v}_3 \cdot \mathbf{u}_2 = -2$, and $\mathbf{u}_2 \cdot \mathbf{u}_2 = 2$, we have

$$\mathbf{u}_3 = \mathbf{v}_3 - \frac{\mathbf{v}_3 \cdot \mathbf{u}_1}{\mathbf{u}_1 \cdot \mathbf{u}_1} \mathbf{u}_1 - \frac{\mathbf{v}_3 \cdot \mathbf{u}_2}{\mathbf{u}_2 \cdot \mathbf{u}_2} \mathbf{u}_2$$

$$= \begin{bmatrix} 1 \\ 2 \\ -4 \end{bmatrix} - \frac{-5}{3} \begin{bmatrix} 1 \\ -1 \\ 1 \end{bmatrix} - \frac{-2}{2} \begin{bmatrix} 0 \\ 1 \\ 1 \end{bmatrix} = \begin{bmatrix} \frac{8}{3} \\ \frac{4}{3} \\ -\frac{4}{3} \end{bmatrix}$$

So the orthogonal basis is $\mathcal{B}' = \{\mathbf{u}_1, \mathbf{u}_2, \mathbf{u}_3\}$, where

$$\mathbf{u}_1 = \begin{bmatrix} 1 \\ -1 \\ 1 \end{bmatrix}, \qquad \mathbf{u}_2 = \begin{bmatrix} 0 \\ 1 \\ 1 \end{bmatrix}, \qquad \mathbf{u}_3 = \begin{bmatrix} \frac{8}{3} \\ \frac{4}{3} \\ -\frac{4}{3} \end{bmatrix}$$

We finally normalize to get an orthonormal basis $\mathcal{B}''$:

$$\mathcal{B}'' = \left\{ \begin{bmatrix} \frac{1}{\sqrt{3}} \\ -\frac{1}{\sqrt{3}} \\ \frac{1}{\sqrt{3}} \end{bmatrix}, \begin{bmatrix} 0 \\ \frac{1}{\sqrt{2}} \\ \frac{1}{\sqrt{2}} \end{bmatrix}, \begin{bmatrix} \frac{2}{\sqrt{6}} \\ \frac{1}{\sqrt{6}} \\ -\frac{1}{\sqrt{6}} \end{bmatrix} \right\}$$

The Gram-Schmidt process theorem asserts that orthogonal bases exist. This combined with Theorem 3 of Section 8.1 yields the important observation that if $\mathbf{v}$ is in *both* V and $V^\perp$, then $\mathbf{v} = \mathbf{0}$.

THEOREM 17

Let V be any subspace of $\mathbf{R}^n$. Then

$$V \cap V^\perp = \{\mathbf{0}\}$$

The definition of the vectors $\mathbf{u}_{\mathrm{pr}}$ and $\mathbf{u}_c$ initially assumed the existence of an orthogonal basis, but now that we know that orthogonal bases exist, we conclude that $\mathbf{u}_{\mathrm{pr}}$ and $\mathbf{u}_c$ are *always* defined, for any $\mathbf{u}$ and any V. The best approximation theorem implies that $\mathbf{u}_{\mathrm{pr}}$ and $\mathbf{u}_c$ do not depend on the particular orthogonal basis that constructed them. So, they only depend on $\mathbf{u}$ and V.

THEOREM 18

(Orthogonal Decomposition)

Let $\mathbf{u}$ be any n-vector and V be any subspace of $\mathbf{R}^n$. Then $\mathbf{u}$ always has an orthogonal projection $\mathbf{u}_{\mathrm{pr}}$ onto V and a component $\mathbf{u}_c$ orthogonal to V.

$$\mathbf{u} = \mathbf{u}_{\mathrm{pr}} + \mathbf{u}_c, \qquad \text{with} \qquad \mathbf{u}_{\mathrm{pr}} \in V, \mathbf{u}_c \in V^\perp \qquad (8.18)$$

$\mathbf{u}_{\mathrm{pr}}$ and $\mathbf{u}_c$ can be computed by (8.12) and (8.13) if an orthogonal basis is known. Furthermore, the decomposition (8.18) is unique. In other words, if

$$\mathbf{u} = \mathbf{v} + \mathbf{v}^\perp \quad \text{with} \quad \mathbf{v} \in V, \qquad \mathbf{v}^\perp \in V^\perp$$

then

$$\mathbf{v} = \mathbf{u}_{\text{pr}} \quad \text{and} \quad \mathbf{v}^{\perp} = \mathbf{u}_{\text{c}}$$

PROOF   The only part of these claims that still needs proof is uniqueness. We have

$$\mathbf{u} = \mathbf{u}_{\text{pr}} + \mathbf{u}_{\text{c}} = \mathbf{v} + \mathbf{v}^{\perp} \Rightarrow \mathbf{u}_{\text{pr}} - \mathbf{v} = \mathbf{v}^{\perp} - \mathbf{u}_{\text{c}}$$

This common vector is zero, by Theorem 17, because $\mathbf{u}_{\text{pr}} - \mathbf{v} \in V$ and $\mathbf{v}^{\perp} - \mathbf{u}_{\text{c}} \in V^{\perp}$. Therefore, $\mathbf{v} = \mathbf{u}_{\text{pr}}$ and $\mathbf{v}^{\perp} = \mathbf{u}_{\text{c}}$, as asserted.

The unique decomposition (8.18) of $\mathbf{u}$ into a summand in V and one in $V^{\perp}$ is called the **orthogonal decomposition** of $\mathbf{u}$ with respect to V.

Note that in the special case when $\mathbf{u}$ is already in V, then $\mathbf{u}_{\text{pr}} = \mathbf{u}$ and $\mathbf{u}_{\text{c}} = \mathbf{0}$. This follows from $\mathbf{u} = \mathbf{u} + \mathbf{0}$ and the uniqueness of the decomposition, because $\mathbf{u} \in V$ and $\mathbf{0} \in V^{\perp}$.

$$\text{If} \quad \mathbf{u} \in V \quad \Rightarrow \quad \mathbf{u}_{\text{pr}} = \mathbf{u} \quad \text{and} \quad \mathbf{u}_{\text{c}} = \mathbf{0}$$

■ EXAMPLE 13   Find the orthogonal decomposition of $\mathbf{u} = (1, 1, 1)$ with respect to $V = \text{Span}\{\mathbf{v}_1, \mathbf{v}_2\}$, where $\mathbf{v}_1$ and $\mathbf{v}_2$ are as in Example 12.

SOLUTION   In Example 12 we orthogonalized $\{\mathbf{v}_1, \mathbf{v}_2\}$ and obtained $\{\mathbf{u}_1, \mathbf{u}_2\}$, with $\mathbf{u}_1 = (1, -1, 1)$ and $\mathbf{u}_2 = (0, 1, 1)$. Therefore,

$$\mathbf{u}_{\text{pr}} = \frac{\mathbf{u} \cdot \mathbf{u}_1}{\mathbf{u}_1 \cdot \mathbf{u}_1}\mathbf{u}_1 + \frac{\mathbf{u} \cdot \mathbf{u}_2}{\mathbf{u}_2 \cdot \mathbf{u}_2}\mathbf{u}_2 = \begin{bmatrix} \frac{1}{3} \\ \frac{2}{3} \\ \frac{4}{3} \end{bmatrix}$$

$$\mathbf{u}_{\text{c}} = \mathbf{u} - \mathbf{u}_{\text{pr}} = \begin{bmatrix} \frac{2}{3} \\ \frac{1}{3} \\ -\frac{1}{3} \end{bmatrix}$$

So,

$$\begin{bmatrix} 1 \\ 1 \\ 1 \end{bmatrix} = \begin{bmatrix} \frac{1}{3} \\ \frac{2}{3} \\ \frac{4}{3} \end{bmatrix} + \begin{bmatrix} \frac{2}{3} \\ \frac{1}{3} \\ -\frac{1}{3} \end{bmatrix}$$

where $(\frac{1}{3}, \frac{2}{3}, \frac{4}{3}) \in V$ and $(\frac{2}{3}, \frac{1}{3}, -\frac{1}{3}) \in V^{\perp}$. (To check: $\mathbf{u}_{\text{pr}} = \frac{7}{3}\mathbf{v}_1 + \mathbf{v}_2$ and $\mathbf{u}_{\text{c}} \cdot \mathbf{v}_1 = 0 = \mathbf{u}_{\text{c}} \cdot \mathbf{v}_2$.)

NOTE   The Gram-Schmidt process is not well suited for large numerical calculations. There is almost always a loss of orthogonality in the computed $\mathbf{u}_i$s. In practice, professionals use variants of the QR method (discussed in Section 8.3) and a version of the Gram-Schmidt process called the **modified Gram-Schmidt**, which has much better numerical properties.

Gram-Schmidt with CAS

Maple

```
> with(linalg):
> v1:=vector([1,-1,1]):v2:=vector([-2,3,-1]):
  v3:=vector([1,2,-4]):
> GramSchmidt({v1,v2,v3});
```

$$\left\{ [1,-1,1],[0,1,1],\left[\frac{8}{3},\frac{4}{3},\frac{-4}{3}\right] \right\}$$

Mathematica

```
In[1]:=
   <<LinearAlgebra'Orthogonalization'
   GramSchmidt[{{1,-1,1},{-2,3,-1},{1,2,-4}}]
Out[2]=

            1                1               1
  {{---------, -(---------), ---------},
     Sqrt[3]       Sqrt[3]     Sqrt[3]


                 1             1
      {0, ---------, ---------},
          Sqrt[2]    Sqrt[2]


          2         1               1
  {Sqrt[-], ---------, -(---------)}}}
          3   Sqrt[6]      Sqrt[6]
```

MATLAB

```
>> orth([1 -2 1; -1 3 2; 1 -1 -4])
ans=
    0.1133   -0.7713    0.6262
   -0.6577    0.4142    0.6292
    0.7447    0.4832    0.4684
```

Exercises 8.2

Orthogonal Projections

1. Find the projection of $\mathbf{u} = \begin{bmatrix} -2 \\ 1 \end{bmatrix}$ onto the line l

through $\mathbf{p} = \begin{bmatrix} 1 \\ 2 \end{bmatrix}$ and the origin.

2. Referring to Exercise 1, find the shortest distance from $\mathbf{u}$ to l.

3. Find the projection of $\mathbf{u} = (3,-1,2)$ onto the line $l = \{(1,1,-3)t, t \in \mathbf{R}\}$.

4. Referring to Exercise 3, find the shortest distance from $\mathbf{u}$ to l.

In Exercises 5–6 find the orthogonal decomposition of $\mathbf{u}$ with respect to V.

5. $\mathbf{u} = \begin{bmatrix} -2 \\ 1 \end{bmatrix}$, $V = \mathrm{Span}\left\{ \begin{bmatrix} 2 \\ 3 \end{bmatrix} \right\}$

6. $\mathbf{u} = \begin{bmatrix} 2 \\ 1 \\ 1 \end{bmatrix}$, $V = \mathrm{Span}\left\{ \begin{bmatrix} 1 \\ 1 \\ -4 \end{bmatrix}, \begin{bmatrix} 2 \\ 2 \\ 1 \end{bmatrix} \right\}$

In Exercises 7–8 write $\mathbf{u}$ as a sum of two orthogonal vectors, one in V and one in $V^{\perp}$.

7. $\mathbf{u} = \begin{bmatrix} 1 \\ 3 \end{bmatrix}$, $V = \mathrm{Span}\left\{ \begin{bmatrix} -2 \\ 2 \end{bmatrix} \right\}$

8. $\mathbf{u} = \begin{bmatrix} 0 \\ 1 \\ 1 \end{bmatrix}$, $V = \mathrm{Span}\left\{ \begin{bmatrix} 1 \\ 1 \\ -2 \end{bmatrix}, \begin{bmatrix} 2 \\ 0 \\ 1 \end{bmatrix} \right\}$

In Exercises 9–10 find the vector in the plane V that best approximates $\mathbf{u}$.

9. $V = \mathrm{Span}\left\{ \begin{bmatrix} 1 \\ -1 \\ 2 \end{bmatrix}, \begin{bmatrix} -2 \\ 2 \\ 2 \end{bmatrix} \right\}$, $\mathbf{u} = \begin{bmatrix} 1 \\ 1 \\ 1 \end{bmatrix}$

10. $V = \mathrm{Span}\left\{ \begin{bmatrix} 1 \\ 1 \\ -2 \end{bmatrix}, \begin{bmatrix} 2 \\ 0 \\ 1 \end{bmatrix} \right\}$, $\mathbf{u} = \begin{bmatrix} 2 \\ 1 \\ 1 \end{bmatrix}$

11. Complete the proof of Theorem 13.

12. Verify Theorem 14 for $A = \begin{bmatrix} 4 & 2 \\ -2 & 2 \\ 1 & 2 \end{bmatrix}$.

Gram-Schmidt Process

In Exercises 13–14 find an orthogonal and an orthonormal basis of $\mathbf{R}^3$ by applying the Gram-Schmidt process to the basis $\mathcal{B}$.

13. $\mathcal{B} = \left\{ \begin{bmatrix} 2 \\ -1 \\ 1 \end{bmatrix}, \begin{bmatrix} 0 \\ 3 \\ -1 \end{bmatrix}, \begin{bmatrix} 1 \\ 2 \\ 0 \end{bmatrix} \right\}$

14. $\mathcal{B} = \left\{ \begin{bmatrix} 1 \\ -2 \\ 1 \end{bmatrix}, \begin{bmatrix} 4 \\ 3 \\ -5 \end{bmatrix}, \begin{bmatrix} 1 \\ 2 \\ 3 \end{bmatrix} \right\}$

In Exercises 15–16 apply the Gram-Schmidt process to find an orthogonal basis for V.

15. $V = \mathrm{Span}\left\{ \begin{bmatrix} 4 \\ 2 \\ -1 \end{bmatrix}, \begin{bmatrix} 1 \\ 2 \\ 3 \end{bmatrix} \right\}$

16. $V = \mathrm{Span}\left\{ \begin{bmatrix} 3 \\ 0 \\ 1 \\ -1 \end{bmatrix}, \begin{bmatrix} 0 \\ 2 \\ -1 \\ 0 \end{bmatrix}, \begin{bmatrix} 2 \\ 2 \\ -2 \\ 2 \end{bmatrix} \right\}$

In Exercises 17–18 find the orthogonal decomposition of $\mathbf{u}$ with respect to V. In each case you first need to apply Gram-Schmidt to find an orthogonal basis of V.

17. $\mathbf{u} = \begin{bmatrix} 2 \\ 0 \\ 1 \end{bmatrix}$, $V = \mathrm{Span}\left\{ \begin{bmatrix} 5 \\ 1 \\ -4 \end{bmatrix}, \begin{bmatrix} 0 \\ 2 \\ 1 \end{bmatrix} \right\}$

18. $\mathbf{u} = \begin{bmatrix} 2 \\ 0 \\ 1 \\ 2 \end{bmatrix}$, $V = \mathrm{Span}\left\{ \begin{bmatrix} 1 \\ 1 \\ -1 \\ 1 \end{bmatrix}, \begin{bmatrix} 1 \\ -1 \\ 1 \\ 0 \end{bmatrix} \right\}$

Distance from a Vector to a Subspace

Let $\mathbf{u}$ be an n-vector and V be a subspace of $\mathbf{R}^n$. If $\mathbf{u}_{\mathrm{pr}}$ is the orthogonal projection of $\mathbf{u}$ onto V then, by the best approximation theorem $\mathbf{u}_c = \mathbf{u} - \mathbf{u}_{\mathrm{pr}}$ has the shortest length among all the vectors $\mathbf{u} - \mathbf{v}$ with $\mathbf{v}$ in V. We call the length $\|\mathbf{u}_c\|$ the **distance** from $\mathbf{u}$ to V. Note that if the distance from $\mathbf{u}$ to V is zero, then $\mathbf{u}$ is in V.

19. Find the distance from $(4, -4, 4)$ to

$$\mathrm{Span}\{(1, 1, -2), (2, 0, 1)\}$$

20. Find the distance from $(2, 0, 4)$ to

$$\mathrm{Span}\{(1, 1, -2), (2, 3, 1)\}$$

(You need an orthogonal basis.)

21. Find the distance from $(1, 2, 1, 2)$ to

$$\mathrm{Span}\{(1, 1, -2, 2), (2, 0, 1, 0)\}$$

8.3   The QR Factorization

Reader's Goals for This Section

1. To know how to find the QR factorization of matrices that have one.
2. To know how to apply the QR method to approximate eigenvalues.

In this section we study a very interesting topic that is particularly useful in applications, the QR factorization of a matrix. This factorization yields an important method for numerical approximation of eigenvalues and eigenvectors.

The QR Factorization

The orthogonalization of the columns of a matrix A may lead to a certain factorization of A that is very useful in numerical calculations, especially in approximating eigenvalues and eigenvectors, as we shall see shortly.

THEOREM 19   **(QR Factorization)**

Let A be an $m \times n$ matrix with linearly independent columns (hence, $m \geq n$). Then A can be factored as

$$A = QR$$

where Q is a matrix with orthonormal columns and R is an invertible upper triangular matrix.

PROOF   Let $\mathbf{v}_1, \ldots, \mathbf{v}_n$ be the columns of A and let $\mathbf{u}_1, \ldots, \mathbf{u}_n$ be the vectors obtained by orthonormalizing them, in such a way that $\mathrm{Span}\{\mathbf{v}_1, \ldots, \mathbf{v}_i\} = \mathrm{Span}\{\mathbf{u}_1, \ldots, \mathbf{u}_i\}$, $i = 1, \ldots, n$. For example, the Gram-Schmidt process followed by normalization will guarantee these conditions (see Section 8.2). Let

$$Q = [\mathbf{u}_1\, \mathbf{u}_2 \cdots \mathbf{u}_n]$$

Each $\mathbf{v}_i$ is a linear combination of $\mathbf{u}_1, \ldots, \mathbf{u}_i$ and, hence, a linear combination of $\mathbf{u}_1, \ldots, \mathbf{u}_n$ of the form

$$\mathbf{v}_i = r_{1i}\mathbf{u}_1 + \cdots + r_{ni}\mathbf{u}_n = Q \begin{bmatrix} r_{1i} \\ \vdots \\ r_{ni} \end{bmatrix}, \qquad i = 1, \ldots, n \qquad (8.19)$$

with

$$r_{i+1,i} = \cdots = r_{ni} = 0, \qquad i = 1, \ldots, n \qquad (8.20)$$

Therefore,

$$A = [\mathbf{v}_1 \cdots \mathbf{v}_n] = \begin{bmatrix} Q \begin{bmatrix} r_{11} \\ \vdots \\ 0 \end{bmatrix} \cdots Q \begin{bmatrix} r_{1n} \\ \vdots \\ r_{nn} \end{bmatrix} \end{bmatrix} = QR$$

where R is the matrix with (i, j) entry r_{ij}, $i, j = 1, \ldots, n$. Q and R are the matrices we wanted. Q has orthonormal columns and R is upper triangular by (8.20). R is also *invertible*, because the homogeneous system $R\mathbf{x} = \mathbf{0}$ has only the trivial solution. Otherwise, the system $QR\mathbf{x} = \mathbf{0}$ or $A\mathbf{x} = \mathbf{0}$ would have a nontrivial solution, so A would have linearly *dependent* columns.

NOTES

1. It is easy—but not necessary—to give formulas for Q and R based on the equations of the Gram-Schmidt process. What we do in practice is orthonormalize the columns of A to get Q. Then we compute R by

$$R = Q^T A$$

   because

$$Q^T A = Q^T (QR) = (Q^T Q)R = IR = R$$

   and $Q^T Q = I$, by Theorem 7, Section 8.1.
2. The matrix R can be so arranged that its diagonal entries are always **strictly positive**. If $r_{ii} < 0$ in (8.19), then we replace $\mathbf{u}_i$ with $-\mathbf{u}_i$. If we do this, then Q is unique, because when we orthonormalize, the $\mathbf{u}_i$s are unique up to sign.
3. The columns of Q form an **orthonormal basis** for Col(A). Furthermore,

$$\text{Span}\{\mathbf{v}_1, \ldots, \mathbf{v}_i\} = \text{Span}\{\mathbf{u}_1, \ldots, \mathbf{u}_i\}$$

   for $i = 1, \ldots, n$.
4. In the special case when A is square, Q is an **orthogonal** matrix.

■ EXAMPLE 14   Find the QR factorization of A.

$$A = \begin{bmatrix} 1 & 1 & 0 \\ 1 & -1 & 0 \\ 1 & 1 & 1 \\ 1 & 1 & 1 \end{bmatrix}$$

SOLUTION   First we note that the columns $\mathbf{v}_1, \mathbf{v}_2, \mathbf{v}_3$ of A are linearly independent; hence a QR factorization exists. Next, we need to "orthonormalize" $\{\mathbf{v}_1, \mathbf{v}_2, \mathbf{v}_3\}$. By the Gram-Schmidt process we have

$$\mathbf{u}_1 = \mathbf{v}_1 = (1, 1, 1, 1), \qquad \mathbf{u}_2 = \mathbf{v}_2 - \frac{\mathbf{v}_2 \cdot \mathbf{u}_1}{\mathbf{u}_1 \cdot \mathbf{u}_1} \mathbf{u}_1 = \left(\frac{1}{2}, -\frac{3}{2}, \frac{1}{2}, \frac{1}{2} \right)$$

$$\mathbf{u}_3 = \mathbf{v}_3 - \frac{\mathbf{v}_3 \cdot \mathbf{u}_1}{\mathbf{u}_1 \cdot \mathbf{u}_1} \mathbf{u}_1 - \frac{\mathbf{v}_3 \cdot \mathbf{u}_2}{\mathbf{u}_2 \cdot \mathbf{u}_2} \mathbf{u}_2 = \left(-\frac{2}{3}, 0, \frac{1}{3}, \frac{1}{3} \right)$$

and we form the matrix

$$Q = \begin{bmatrix} \dfrac{\mathbf{u}_1}{\|\mathbf{u}_1\|} & \dfrac{\mathbf{u}_2}{\|\mathbf{u}_2\|} & \dfrac{\mathbf{u}_3}{\|\mathbf{u}_3\|} \end{bmatrix}$$

$$= \begin{bmatrix} \dfrac{1}{2} & \dfrac{\sqrt{3}}{6} & -\dfrac{\sqrt{6}}{3} \\[2ex] \dfrac{1}{2} & -\dfrac{\sqrt{3}}{2} & 0 \\[2ex] \dfrac{1}{2} & \dfrac{\sqrt{3}}{6} & \dfrac{\sqrt{6}}{6} \\[2ex] \dfrac{1}{2} & \dfrac{\sqrt{3}}{6} & \dfrac{\sqrt{6}}{6} \end{bmatrix}$$

Because, $R = Q^T A$, we have

$$R = \begin{bmatrix} \dfrac{1}{2} & \dfrac{1}{2} & \dfrac{1}{2} & \dfrac{1}{2} \\[2ex] \dfrac{\sqrt{3}}{6} & -\dfrac{\sqrt{3}}{2} & \dfrac{\sqrt{3}}{6} & \dfrac{\sqrt{3}}{6} \\[2ex] -\dfrac{\sqrt{6}}{3} & 0 & \dfrac{\sqrt{6}}{6} & \dfrac{\sqrt{6}}{6} \end{bmatrix} \begin{bmatrix} 1 & 1 & 0 \\ 1 & -1 & 0 \\ 1 & 1 & 1 \\ 1 & 1 & 1 \end{bmatrix}$$

$$= \begin{bmatrix} 2 & 1 & 1 \\[2ex] 0 & \sqrt{3} & \dfrac{\sqrt{3}}{3} \\[2ex] 0 & 0 & \dfrac{\sqrt{6}}{3} \end{bmatrix}$$

The QR Method for Eigenvalues

The QR decomposition can be used to approximate the eigenvalues of a square matrix.[3] The resulting algorithm, called the **QR method**, adds a very important tool to the numerical approximations of eigenvalues we studied in Chapter 7. In contrast with the other methods, the QR method will find *all* the eigenvalues of a matrix. It is also used to solve linear systems. Also, variants of it are used in orthonormalization to replace the numerically unstable Gram-Schmidt process.

Let us first outline the method. We start out with an invertible $n \times n$ matrix A. We compute the QR factorization of $A = QR$, and then we form the matrix $A_1 = RQ$. We observe that A and A_1 are similar, because

$$Q^{-1}AQ = Q^{-1}(QR)Q = RQ = A_1$$

Hence, they have the same eigenvalues. Now we continue by finding the QR factorization $R_1 Q_1$ of A_1 and forming the matrix $A_2 = Q_1 R_1$ that has the same eigenvalues as A. We

---

[3]The QR method as it stands today was introduced in 1961 by J. G. F. Francis and independently by V. N. Kublanovskaya. The original idea, however, is due to H. Rutishauser (1958), who used the LU factorization of a matrix to compute eigenvalues and called the iterations *LR transformations*.

iterate to get a sequence of matrices

$$A,\ A_1,\ A_2,\ A_3,\dots$$

It turns out that if A has n eigenvalues of different magnitudes, then this sequence approaches an upper triangular matrix $\widehat{R}$ similar to A. Hence, the diagonal entries of $\widehat{R}$ are *all* the eigenvalues of A.

The following algorithm-theorem, whose proof we omit, is the core of the QR method just described.

Algorithm

(The QR Method)

INPUT: $n \times n$ matrix invertible matrix A with eigenvalues $\lambda_1, \dots, \lambda_n$, such that

$$|\lambda_1| < |\lambda_2| < \cdots < |\lambda_n|$$

1. Set $A_0 = A$.
2. For $i = 1, 2, \dots, k-1$
   (a) Find the QR decomposition of A_i, say, $A_i = Q_i R_i$.
   (b) Let $A_{i+1} = R_i Q_i$.

OUTPUT: A_k, which approximates a triangular matrix $\widehat{R}$ with diagonal entries all the eigenvalues of A.

Numerical Concerns with Gram-Schmidt and QR

As we pointed out in Section 8.2, when we apply the Gram-Schmidt process in numerical calculations, we do not exactly get an orthonormal basis. This is due to the accumulated round-off error, which may be huge for this process. As a result, the triangular matrix R in the QR decomposition is *not* exactly triangular after a numerical calculation. Entries that were supposed to be zero are often very small numbers. In the following example, we computed all factorizations exactly; then each answer was approximated and truncated to four decimal places. This was done only to show that the matrices R_i are indeed upper triangular and the Q_i are (almost) orthogonal. In practice we go straight to the approximations.

■ EXAMPLE 15   Find the eigenvalues of

$$A = \begin{bmatrix} 9 & 8 \\ 1 & 2 \end{bmatrix}$$

by using the QR method.

SOLUTION   First we find the QR factorization of $A = A_0$.

$$Q_0 = \begin{bmatrix} \dfrac{9}{\sqrt{82}} & -\dfrac{1}{\sqrt{82}} \\[3mm] \dfrac{1}{\sqrt{82}} & \dfrac{9}{\sqrt{82}} \end{bmatrix} \cong \begin{bmatrix} 0.9938 & -0.1104 \\ 0.1104 & 0.9938 \end{bmatrix}$$

$$R_0 = \begin{bmatrix} \sqrt{82} & \dfrac{37}{41}\sqrt{82} \\[3mm] 0 & \dfrac{5}{41}\sqrt{82} \end{bmatrix} \cong \begin{bmatrix} 9.0553 & 8.1719 \\ 0 & 1.1043 \end{bmatrix}$$

Hence,

$$A_1 = R_0 Q_0 = \begin{bmatrix} 9.9024 & 7.1219 \\ 0.1219 & 1.0975 \end{bmatrix}$$

Now we repeat by factoring A_1,

$$Q_1 = \begin{bmatrix} 0.9999 & -0.0123 \\ 0.0123 & 0.9999 \end{bmatrix}, \qquad R_1 = \begin{bmatrix} 9.9031 & 7.1349 \\ 0 & 1.0097 \end{bmatrix}$$

$$A_2 = R_1 Q_1 = \begin{bmatrix} 9.9903 & 7.0124 \\ 0.0124 & 1.0096 \end{bmatrix}$$

and factoring A_2,

$$Q_2 = \begin{bmatrix} 0.9999 & -0.0012 \\ 0.0012 & 0.9999 \end{bmatrix} \qquad R_2 = \begin{bmatrix} 9.9903 & 7.0136 \\ 0 & 1.0009 \end{bmatrix}$$

$$A_3 = \begin{bmatrix} 9.9990 & 7.0012 \\ 0.0012 & 1.0009 \end{bmatrix}$$

We see that A_3 is almost diagonal with eigenvalues close to 10 and 1, which are the true eigenvalues of A.

QR with CAS

Maple

```
> with(linalg):
> R:=QRdecomp(matrix([[1,-2],[-1,3]]),Q='q');
```

$$R = \begin{bmatrix} \sqrt{2} & -\dfrac{5}{2}\sqrt{2} \\[2mm] 0 & \dfrac{1}{2}\sqrt{2} \end{bmatrix}$$

```
> evalm(q);
```

$$\begin{bmatrix} \dfrac{1}{2}\sqrt{2} & \dfrac{1}{2}\sqrt{2} \\[2mm] -\dfrac{1}{2}\sqrt{2} & \dfrac{1}{2}\sqrt{2} \end{bmatrix}$$

Mathematica

```
In[1]:=
    QRDecomposition[{{1.,-2},{-1,3}}]
Out[1]=
{{{-0.707107, 0.707107},
   {-0.707107, -0.707107}},
  {{-1.41421, 3.53553}, {0, -0.707107}}}
```

MATLAB

```
>> [Q,R]=qr([1 -2; -1 3])
  Q =
     -0.7071      0.7071
      0.7071      0.7071
  R =
     -1.4142      3.5355
          0      0.7071
```

Exercises 8.3

In Exercises 1–7 find a QR factorization of A.

1. $A = \begin{bmatrix} 0 & -2 \\ 1 & 3 \end{bmatrix}$

2. $A = \begin{bmatrix} 1 & -2 \\ 1 & 1 \end{bmatrix}$

3. $A = \begin{bmatrix} 0 & 0 & 4 \\ 0 & 1 & 2 \\ -1 & 0 & 2 \end{bmatrix}$

4. $A = \begin{bmatrix} 1 & 0 & 4 \\ 0 & 1 & 0 \\ -1 & 0 & 2 \end{bmatrix}$

5. $A = \begin{bmatrix} 1 & -2 & 1 \\ 0 & 1 & 2 \\ -1 & 0 & 1 \end{bmatrix}$

6. $A = \begin{bmatrix} 1 & -2 \\ 1 & 4 \\ 1 & 0 \\ -1 & 2 \end{bmatrix}$

7. $A = \begin{bmatrix} 1 & -2 \\ 1 & 0 \\ 1 & 2 \\ -1 & 4 \end{bmatrix}$

In Exercises 8–10 find the triangular matrix R such that $A = QR$, given that Q was obtained from A by orthonormalizing its columns.

8. $A = \begin{bmatrix} 1 & 2 \\ 1 & 1 \\ -1 & 0 \end{bmatrix}$, $Q = \begin{bmatrix} \frac{1}{\sqrt{3}} & \frac{1}{\sqrt{2}} \\ \frac{1}{\sqrt{3}} & 0 \\ -\frac{1}{\sqrt{3}} & \frac{1}{\sqrt{2}} \end{bmatrix}$

9. $A = \begin{bmatrix} 1 & -1 \\ 1 & 0 \\ 1 & -1 \\ -1 & -2 \end{bmatrix}$, $Q = \begin{bmatrix} \frac{1}{2} & -\frac{1}{\sqrt{6}} \\ \frac{1}{2} & 0 \\ \frac{1}{2} & -\frac{1}{\sqrt{6}} \\ -\frac{1}{2} & -\frac{\sqrt{6}}{3} \end{bmatrix}$

10. $A = \begin{bmatrix} 1 & -1 & 1 \\ 0 & 1 & -1 \\ -1 & 1 & 1 \end{bmatrix}$, $Q = \begin{bmatrix} \frac{1}{\sqrt{2}} & 0 & \frac{1}{\sqrt{2}} \\ 0 & 1 & 0 \\ -\frac{1}{\sqrt{2}} & 0 & \frac{1}{\sqrt{2}} \end{bmatrix}$

11. Find a QR factorization for an orthogonal matrix A.

12. Prove that A is invertible if and only if $A = QR$ for some orthogonal matrix Q and some upper triangular matrix R with nonzero main diagonal entries.

13. Let A have linearly independent columns and let $A = QR$ be a QR factorization. Show that

$$A\mathbf{x} = \mathbf{b} \text{ is consistent} \quad \Leftrightarrow$$

$$Q\mathbf{y} = \mathbf{b} \text{ is consistent}$$

Conclude that A and Q have the same column space.

14. Let A be a matrix with nonzero columns that form an orthogonal set. If $A = QR$ is a QR factorization, show that R is diagonal.

In Exercises 15–17 find $A_1, A_2,$ and A_3 of the QR method. Use A_3 to estimate the eigenvalues of A. What is the error in each case?

15. $A = \begin{bmatrix} 8 & 7 \\ 1 & 2 \end{bmatrix}$

16. $A = \begin{bmatrix} 10 & 8 \\ 1 & 3 \end{bmatrix}$

17. $A = \begin{bmatrix} 12 & 11 \\ 2 & 3 \end{bmatrix}$

8.4   Least Squares

Reader's Goal for This Section

To know how to solve basic least squares problems.

In this section we study a topic of considerable interest in applications, the method of least squares, discussed in the introduction.

A Least Squares Problem

In trying to solve a problem, we sometimes get data points and seek a function with a graph passing through these points. Usually, the nature of the problem dictates the kind of function we need. For example, if a car moves at a constant speed and we measure the distance $s(t)$ covered every minute, we expect the graph of $s(t)$ to be a straight line. A high-degree polynomial or an exponential function would be unsuitable for this situation.

Suppose our problem suggests a straight line and we have the points $(1, 2), (2, 4),$ $(3, 3)$. Let $y = b + mx$ be the equation of this line. We want to find the slope m and the y-intercept b. Because the line should pass through the three points, we have

$$2 = b + m \cdot 1, \qquad 4 = b + m \cdot 2, \qquad 3 = b + m \cdot 3$$

Unfortunately, the resulting linear system in unknowns m and b,

$$\begin{bmatrix} 1 & 1 \\ 1 & 2 \\ 1 & 3 \end{bmatrix} \begin{bmatrix} b \\ m \end{bmatrix} = \begin{bmatrix} 2 \\ 4 \\ 3 \end{bmatrix}$$

is easily seen to be *inconsistent*. So, our problem cannot be solved exactly.[4] The next best thing then is to try to find the straight line that best "fits" these points.

Best fitting may have different meanings, depending on what aspects of the solution we need emphasized. In this case, suppose we want our best line to be such that if δ_1, δ_2 and δ_3 are the errors in the y-direction,

$$\delta_1 = 2 - b - m \cdot 1, \qquad \delta_2 = 4 - b - m \cdot 2, \qquad \delta_3 = 3 - b - m \cdot 3$$

then the number

$$\delta_1^2 + \delta_2^2 + \delta_3^2$$

is minimum (Fig. 8.10). A solution for m and b that minimizes this sum of the squares of the errors is called a **least squares solution**.

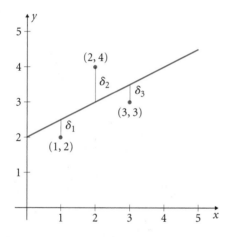

Figure 8.10   Least squares: minimize $\delta_1^2 + \delta_2^2 + \delta_3^2$.

We may express all this in vector notation. If Δ is the error vector

$$\Delta = (\delta_1, \delta_2, \delta_3)$$

we want to minimize $\delta_1^2 + \delta_2^2 + \delta_3^2 = \|\Delta\|^2$ or, equivalently, minimize $\|\Delta\|$.

■ EXAMPLE 16   Find which of the lines yields the smallest least squares error for the points $(1, 2), (2, 4), (3, 3)$.

(a) $y = 2x$
(b) $y = 3$
(c) $y = 0.5x + 2$

---

[4]Note that the quadratic $-\frac{3}{2}x^2 + \frac{13}{2}x - 3$ passes through these points, but that is not what we need.

SOLUTION   We have the following data.

	$y = 2x$ $(m = 2,\ b = 0)$	$y = 3$ $(m = 0,\ b = 3)$	$y = 0.5x + 2$ $(m = 0.5,\ b = 2)$
δ_1	$2 - 0 - 1 \cdot 2 = 0$	$2 - 3 - 1 \cdot 0 = -1$	$2 - 2 - 1 \cdot 0.5 = -0.5$
δ_2	$4 - 0 - 2 \cdot 2 = 0$	$4 - 3 - 2 \cdot 0 = 1$	$4 - 2 - 2 \cdot 0.5 = 1$
δ_2	$3 - 0 - 3 \cdot 2 = -3$	$3 - 3 - 3 \cdot 0 = 0$	$3 - 2 - 3 \cdot 0.5 = -0.5$
$\lVert \Delta \rVert^2$	$0^2 + 0^2 + (-3)^2 = 9$	$(-1)^2 + 1^2 + 0^2 = 2$	$(-0.5)^2 + (-0.5)^2 + 1 = 1.5$

We see that the line $y = 0.5x + 2$ yields the smallest least squares error of the three lines. In fact, as we shall see, this line yields the smallest error compared with *any* line.

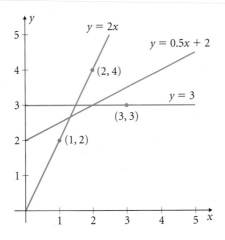

Figure 8.11

Solution of the Least Squares Problem

Let us now see how to find the least squares solution for the points shown in Fig. 8.11 and in general. Suppose we have an inconsistent linear system

$$A\mathbf{x} = \mathbf{b} \tag{8.21}$$

where A is an $m \times n$ matrix. Because for any n-vector $\mathbf{x}$, the product $A\mathbf{x}$ is never $\mathbf{b}$, the resulting error Δ,

$$\Delta = \mathbf{b} - A\mathbf{x}$$

is a nonzero m-vector for all n-vectors $\mathbf{x}$. Solving the least squares problem for (8.21) amounts to finding an n-vector $\tilde{\mathbf{x}}$ such that the length of $\Delta = \mathbf{b} - A\tilde{\mathbf{x}}$ is minimum. Then $\tilde{\mathbf{x}}$ would be our *least squares solution*.

Least squares problem: Find $\widetilde{\mathbf{x}}$ such that $\|\mathbf{b} - A\widetilde{\mathbf{x}}\|$ is minimum.

As $\mathbf{x}$ varies, $A\mathbf{x}$ generates $\mathrm{Col}(A)$. So, $\|\mathbf{b} - A\widetilde{\mathbf{x}}\|$ is minimum only if $A\widetilde{\mathbf{x}}$ is the orthogonal projection $\mathbf{b}_{\mathrm{pr}}$ of $\mathbf{b}$ onto $\mathrm{Col}(A)$, by the best approximation theorem (Section 8.2). Hence,

$$\|\Delta\| = \min \Leftrightarrow A\widetilde{\mathbf{x}} = \mathbf{b}_{\mathrm{pr}} \Leftrightarrow \mathbf{b} - A\widetilde{\mathbf{x}} = \mathbf{b}_c$$

We conclude that a least squares solution $\widetilde{\mathbf{x}}$ for $A\mathbf{x} = \mathbf{b}$ **always exists** (because $\mathbf{b}_{\mathrm{pr}}$ does).

THEOREM 20

For any $m \times n$ matrix A and any m-vector $\mathbf{b}$, there is a least squares solution $\widetilde{\mathbf{x}}$ of $A\mathbf{x} = \mathbf{b}$. In addition, if $\mathbf{b}_{\mathrm{pr}}$ is the orthogonal projection of $\mathbf{b}$ onto $\mathrm{Col}(A)$, then

$$A\widetilde{\mathbf{x}} = \mathbf{b}_{\mathrm{pr}} \tag{8.22}$$

Because $\mathbf{b}_c = \mathbf{b} - A\widetilde{\mathbf{x}}$ is orthogonal to $\mathrm{Col}(A)$, $A\mathbf{x}$ and $\mathbf{b} - A\widetilde{\mathbf{x}}$ must be orthogonal for any n-vector $\mathbf{x}$. Hence,

$$
\begin{aligned}
(\mathbf{b} - A\widetilde{\mathbf{x}}) \cdot A\mathbf{x} &= 0 \\
\Leftrightarrow A^T(\mathbf{b} - A\widetilde{\mathbf{x}}) \cdot \mathbf{x} &= 0 \\
\Leftrightarrow A^T(\mathbf{b} - A\widetilde{\mathbf{x}}) &= \mathbf{0} \qquad \text{by Theorem 3, Section 8.1} \\
\Leftrightarrow A^T\mathbf{b} - A^T A\widetilde{\mathbf{x}} &= \mathbf{0} \\
\Leftrightarrow A^T A\widetilde{\mathbf{x}} &= A^T\mathbf{b}
\end{aligned}
$$

So, $\widetilde{\mathbf{x}}$ is a least squares solution if and only if it satisfies the system $A^T A\widetilde{\mathbf{x}} = A^T\mathbf{b}$. This system is known as the **normal equations for $\widetilde{\mathbf{x}}$**. We know now how to find $\widetilde{\mathbf{x}}$. In addition, we may use $\|\Delta\| = \|\mathbf{b} - A\widetilde{\mathbf{x}}\|$ to compute the **least squares error** involved.

THEOREM 21

(Least Squares Solutions)

Let A be an $m \times n$ matrix. Then there are always least squares solutions $\widetilde{\mathbf{x}}$ of $A\mathbf{x} = \mathbf{b}$. Furthermore,

1. $\widetilde{\mathbf{x}}$ is a least squares solution of $A\mathbf{x} = \mathbf{b}$ if and only if $\widetilde{\mathbf{x}}$ is a solution of the normal equations

$$A^T A\widetilde{\mathbf{x}} = A^T\mathbf{b} \tag{8.23}$$

The least squares error $\|\Delta\|$ is then given by

$$\|\Delta\| = \|\mathbf{b} - A\widetilde{\mathbf{x}}\|$$

2. A has linearly independent columns if and only if $A^T A$ is invertible. In this case the least squares solution is unique, and it can be computed by

$$\widetilde{\mathbf{x}} = \left(A^T A\right)^{-1} A^T\mathbf{b}$$

PROOF   We need to prove only the last part of the theorem. First, we show that A and $A^T A$ have the same null space. Indeed, if $\mathbf{v} \in \mathrm{Null}(A)$, then $A\mathbf{v} = \mathbf{0}$, so $A^T A\mathbf{v} = \mathbf{0}$, which implies $\mathbf{v} \in \text{Null}(A^T A)$. Therefore, $\text{Null}(A) \subseteq \text{Null}(A^T A)$. On the other hand,

$$\mathbf{v} \in \text{Null}(A^T A)$$
$$\Rightarrow A^T A \mathbf{v} \quad = \mathbf{0}$$
$$\Rightarrow A^T A \mathbf{v} \cdot \mathbf{v} = \mathbf{0} \cdot \mathbf{v} = 0$$
$$\Rightarrow A \mathbf{v} \cdot A \mathbf{v} = 0$$
$$\Rightarrow \|A \mathbf{v}\|^2 \quad = 0$$
$$\Rightarrow A \mathbf{v} \quad = \mathbf{0}$$
$$\Rightarrow \mathbf{v} \in \text{Null}(A)$$

Hence, $\text{Null}(A^T A) \subseteq \text{Null}(A)$, and the two null spaces are equal. So, $\text{Nullity}(A) = \text{Nullity}(A^T A)$. But by the dimension theorem,

$$\text{Rank}(A) + \text{Nullity}(A) = n = \text{Rank}(A^T A) + \text{Nullity}(A^T A)$$

Hence, $\text{Rank}(A) = \text{Rank}(A^T A)$. If A has linearly independent columns, then $\text{Rank}(A) = n$. So, $\text{Rank}(A^T A) = n$. Therefore, the $n \times n$ matrix $A^T A$ has linearly independent columns. Hence, $A^T A$ is invertible. Conversely, if $A^T A$ is invertible, then $\text{Rank}(A^T A) = n$; hence, $\text{Rank}(A) = n$. So, A has linearly independent columns. If $A^T A$ is invertible, then $A^T A \widetilde{\mathbf{x}} = A^T \mathbf{b}$ implies the unique solution $\widetilde{\mathbf{x}} = (A^T A)^{-1} A^T \mathbf{b}$. ■

■ EXAMPLE 17   Solve the least squares problem and compute the least squares error for the system $A\mathbf{x} = \mathbf{b}$,

$$\begin{bmatrix} 1 & 1 \\ 1 & 2 \\ 1 & 3 \end{bmatrix} \begin{bmatrix} x_1 \\ x_2 \end{bmatrix} = \begin{bmatrix} 2 \\ 4 \\ 3 \end{bmatrix}$$

Use the solution to find the straight line that yields the smallest least squares error for the points $(1, 2), (2, 4), (3, 3)$ discussed in the introduction of this section.

SOLUTION   By Theorem 21 it suffices to solve the normal equations $A^T A \widetilde{\mathbf{x}} = A^T \mathbf{b}$:

$$\begin{bmatrix} 1 & 1 & 1 \\ 1 & 2 & 3 \end{bmatrix} \begin{bmatrix} 1 & 1 \\ 1 & 2 \\ 1 & 3 \end{bmatrix} \widetilde{\mathbf{x}} = \begin{bmatrix} 1 & 1 & 1 \\ 1 & 2 & 3 \end{bmatrix} \begin{bmatrix} 2 \\ 4 \\ 3 \end{bmatrix}$$

or

$$\begin{bmatrix} 3 & 6 \\ 6 & 14 \end{bmatrix} \widetilde{\mathbf{x}} = \begin{bmatrix} 9 \\ 19 \end{bmatrix}$$

The solution of this system yields the least squares solution

$$\widetilde{\mathbf{x}} = \begin{bmatrix} 2 \\ \frac{1}{2} \end{bmatrix}$$

with least squares error (Fig. 8.12)

$$\|\Delta\| = \|\mathbf{b} - A\tilde{\mathbf{x}}\|$$

$$= \left\| \begin{bmatrix} 2 \\ 4 \\ 3 \end{bmatrix} - \begin{bmatrix} 1 & 1 \\ 1 & 2 \\ 1 & 3 \end{bmatrix} \begin{bmatrix} 2 \\ \frac{1}{2} \end{bmatrix} \right\|$$

$$= \left\| \begin{bmatrix} -\frac{1}{2} \\ 1 \\ -\frac{1}{2} \end{bmatrix} \right\| = \frac{\sqrt{6}}{2}$$

Because $\tilde{\mathbf{x}} = (2, 0.5)$, the slope of the least squares line is 0.5 and its y-intercept is 2, so its equation is $y = 0.5x + 2$. (This is exactly the line sketched in Fig. 8.10.)

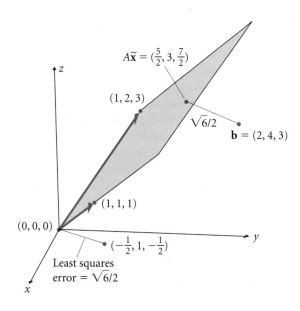

Figure 8.12 Least squares error.

■ EXAMPLE 18 Find the least squares line for the grades data our linear algebra instructor had collected in the introduction of this chapter. Moreover, find the expected percentage of Bs after the tenth semester.

Semester	1	2	3	4	5	6
Percentage of Bs	0.20	0.25	0.20	0.35	0.45	0.40

SOLUTION   Let

$$A = \begin{bmatrix} 1 & 1 \\ 1 & 2 \\ 1 & 3 \\ 1 & 4 \\ 1 & 5 \\ 1 & 6 \end{bmatrix}, \qquad \mathbf{b} = \begin{bmatrix} 0.20 \\ 0.25 \\ 0.20 \\ 0.35 \\ 0.45 \\ 0.40 \end{bmatrix}$$

Hence,

$$A^T A = \begin{bmatrix} 1 & 1 & 1 & 1 & 1 & 1 \\ 1 & 2 & 3 & 4 & 5 & 6 \end{bmatrix} \begin{bmatrix} 1 & 1 \\ 1 & 2 \\ 1 & 3 \\ 1 & 4 \\ 1 & 5 \\ 1 & 6 \end{bmatrix} = \begin{bmatrix} 6 & 21 \\ 21 & 91 \end{bmatrix}$$

and

$$A^T \mathbf{b} = \begin{bmatrix} 1 & 1 & 1 & 1 & 1 & 1 \\ 1 & 2 & 3 & 4 & 5 & 6 \end{bmatrix} \begin{bmatrix} 0.20 \\ 0.25 \\ 0.20 \\ 0.35 \\ 0.45 \\ 0.40 \end{bmatrix} = \begin{bmatrix} 1.85 \\ 7.35 \end{bmatrix}$$

Therefore, the normal equations are

$$\begin{bmatrix} 6 & 21 \\ 21 & 91 \end{bmatrix} \begin{bmatrix} \widetilde{b} \\ \widetilde{m} \end{bmatrix} = \begin{bmatrix} 1.85 \\ 7.35 \end{bmatrix}$$

But

$$\begin{bmatrix} 6 & 21 & 1.85 \\ 21 & 91 & 7.35 \end{bmatrix} \sim \begin{bmatrix} 1 & 0 & 0.13333 \\ 0 & 1 & 0.05 \end{bmatrix}$$

Hence, $\widetilde{m} = 0.05$ and $\widetilde{b} = 0.13333$. Therefore, the line is

$$y = 0.13333 + 0.05x$$

For $x = 10$ we get $y = 0.13333 + 0.05 \cdot 10 = 0.63333$. So, roughly 63.3% of B grades are expected after the tenth semester if this trend of grading performance continues.

If A does not have linearly independent columns, then there are several least squares solutions.

■ EXAMPLE 19   Find all least squares solutions of the system:

$$x - y = 1$$
$$x - y = 5$$

SOLUTION $A = \begin{bmatrix} 1 & -1 \\ 1 & -1 \end{bmatrix}$ and $\mathbf{b} = \begin{bmatrix} 1 \\ 5 \end{bmatrix}$. Hence,

$$A^T A \widetilde{\mathbf{x}} = \begin{bmatrix} 2 & -2 \\ -2 & 2 \end{bmatrix} \begin{bmatrix} \widetilde{x} \\ \widetilde{y} \end{bmatrix} = A^T \mathbf{b} = \begin{bmatrix} 6 \\ -6 \end{bmatrix}$$

Reduction of the augmented matrix,

$$\begin{bmatrix} 2 & -2 & 6 \\ -2 & 2 & -6 \end{bmatrix} \sim \begin{bmatrix} 1 & -1 & 3 \\ 0 & 0 & 0 \end{bmatrix}$$

yields $\widetilde{x} = 3 + r, \widetilde{y} = r, r \in \mathbf{R}.$

Least Squares with QR Factorization

The least squares solutions discussed previously suffer from a frequent problem. The matrix $A^T A$ of the normal equations is usually **ill-conditioned**, which means that a small numerical error in a row reduction causes a large error in the solution.

Usually, Gauss elimination for $A^T A$ of size $n \geq 5$ does not yield any good approximate solutions. An answer to this serious problem is to use the QR factorization of A. The idea behind this approach is that because orthogonal matrices preserve lengths, they should preserve the length of the error vector as well.

Let A have linearly independent columns and let $A = QR$ be a QR factorization, as discussed in Section 8.3. Then for $\widetilde{\mathbf{x}}$ a least squares solution of $A\mathbf{x} = \mathbf{b}$, we have

$$
\begin{aligned}
A^T A \widetilde{\mathbf{x}} &= A^T \mathbf{b} \\
\Leftrightarrow (QR)^T (QR) \widetilde{\mathbf{x}} &= (QR)^T \mathbf{b} \\
\Leftrightarrow R^T Q^T Q R \widetilde{\mathbf{x}} &= R^T Q^T \mathbf{b} \\
\Leftrightarrow R^T R \widetilde{\mathbf{x}} &= R^T Q^T \mathbf{b} \qquad \text{because } Q^T Q = I \\
\Leftrightarrow R \widetilde{\mathbf{x}} &= Q^T \mathbf{b} \qquad \text{because } R^T \text{ is invertible}
\end{aligned}
$$

Note that $R\widetilde{\mathbf{x}} = Q^T \mathbf{b}$ is equivalent to $\widetilde{\mathbf{x}} = R^{-1} Q^T \mathbf{b}$. However, instead of inverting R, it is easier to use back-substitution on the system $R\widetilde{\mathbf{x}} = Q^T \mathbf{b}$. We have proved the following theorem.

THEOREM 22   If A is an $m \times n$ matrix with linearly independent columns and if $A = QR$ is a QR factorization, then the unique least squares solution $\widetilde{\mathbf{x}}$ of $A\mathbf{x} = \mathbf{b}$ is, theoretically, given by

$$\widetilde{\mathbf{x}} = R^{-1} Q^T \mathbf{b}$$

and it is usually computed by solving the system

$$R\widetilde{\mathbf{x}} = Q^T \mathbf{b}$$

■ EXAMPLE 20   Find the least squares solution for $A\mathbf{x} = \mathbf{b}$ by using QR factorization.

$$A = \begin{bmatrix} 2 & 2 & 6 \\ 1 & 4 & -3 \\ 2 & -4 & 9 \end{bmatrix}, \qquad \mathbf{b} = \begin{bmatrix} 1 \\ -1 \\ 4 \end{bmatrix}$$

SOLUTION By applying the method of Section 8.3, we have

$$A = QR = \begin{bmatrix} \frac{2}{3} & \frac{1}{3} & -\frac{2}{3} \\ \frac{1}{3} & \frac{2}{3} & \frac{2}{3} \\ \frac{2}{3} & -\frac{2}{3} & \frac{1}{3} \end{bmatrix} \begin{bmatrix} 3 & 0 & 9 \\ 0 & 6 & -6 \\ 0 & 0 & -3 \end{bmatrix}$$

But

$$Q^T\mathbf{b} = \begin{bmatrix} \frac{2}{3} & \frac{1}{3} & \frac{2}{3} \\ \frac{1}{3} & \frac{2}{3} & -\frac{2}{3} \\ -\frac{2}{3} & \frac{2}{3} & \frac{1}{3} \end{bmatrix} \begin{bmatrix} 1 \\ -1 \\ 4 \end{bmatrix} = \begin{bmatrix} 3 \\ -3 \\ 0 \end{bmatrix}$$

Hence, by Theorem 22, $\tilde{\mathbf{x}}$ can be computed by solving $R\tilde{\mathbf{x}} = Q^T\mathbf{b}$:

$$\begin{bmatrix} 3 & 0 & 9 \\ 0 & 6 & -6 \\ 0 & 0 & -3 \end{bmatrix} \tilde{\mathbf{x}} = \begin{bmatrix} 3 \\ -3 \\ 0 \end{bmatrix}$$

Using back-substitution we easily get $\tilde{\mathbf{x}} = \begin{bmatrix} 1 \\ -\frac{1}{2} \\ 0 \end{bmatrix}$.

Exercises 8.4

In Exercises 1–3 use the normal equations to find the least squares solution for $A\mathbf{x} = \mathbf{b}$.

1. $A = \begin{bmatrix} 1 & -2 \\ -1 & 1 \\ 1 & 2 \end{bmatrix}, \mathbf{b} = \begin{bmatrix} -1 \\ 2 \\ 1 \end{bmatrix}$

2. $A = \begin{bmatrix} -1 & 1 \\ 2 & 2 \\ -1 & 0 \end{bmatrix}, \mathbf{b} = \begin{bmatrix} 5 \\ -4 \\ 1 \end{bmatrix}$

3. $A = \begin{bmatrix} 2 & 1 \\ -1 & 1 \\ 2 & 2 \\ -1 & 0 \end{bmatrix}, \mathbf{b} = \begin{bmatrix} 1 \\ 2 \\ 0 \\ 1 \end{bmatrix}$

4. Solve the system $\begin{bmatrix} 1 & -2 \\ 1 & 0 \end{bmatrix} \mathbf{x} = \begin{bmatrix} 3 \\ 5 \end{bmatrix}$. Also find its least squares solution. Why did you get the same answer?

5. Write your own 3×3 system whose solution coincides with its least squares solution.

6. Write your own 3×2 inconsistent system and compute its least squares solution.

In Exercises 7–9 find the least squares line for the given points. In each case draw the points and the line.

7. $(-1, -2), (0, 1), (1, 5)$

8. $(1, 2), (2, 3), (4, 4)$

9. $(-1, -1), (1, 2), (3, 2), (4, 4)$

10. Find all the least squares solutions for the system

$$x - y = 0$$
$$x - y = 1$$
$$x - y = 2$$

11. Find all the least squares solutions for $A\mathbf{x} = \mathbf{b}$, where

$$A = \begin{bmatrix} 1 & 1 & 1 \\ -2 & 0 & 2 \end{bmatrix}, \qquad \mathbf{b} = \begin{bmatrix} 2 \\ 2 \end{bmatrix}$$

In Exercises 12–14 use the given QR factorization to find the least squares solution for $A\mathbf{x} = \mathbf{b}$.

12. $A = \begin{bmatrix} -2 & 0 \\ 1 & -1 \\ 2 & -1 \end{bmatrix}, Q = \begin{bmatrix} -\frac{2}{3} & -\frac{2}{3} \\ \frac{1}{3} & -\frac{2}{3} \\ \frac{2}{3} & -\frac{1}{3} \end{bmatrix},$

$R = \begin{bmatrix} 3 & -1 \\ 0 & 1 \end{bmatrix}$, and $\mathbf{b} = \begin{bmatrix} 1 \\ 2 \\ 4 \end{bmatrix}$

13. $A = \begin{bmatrix} 0 & 3 \\ 0 & 4 \\ 5 & 10 \end{bmatrix}, Q = \begin{bmatrix} 0 & \frac{3}{5} \\ 0 & \frac{4}{5} \\ 1 & 0 \end{bmatrix},$

$R = \begin{bmatrix} 5 & 10 \\ 0 & 5 \end{bmatrix}$, and $\mathbf{b} = \begin{bmatrix} 3 \\ 0 \\ -4 \end{bmatrix}$

14. $A = \begin{bmatrix} 1 & 1 \\ -1 & 1 \\ 1 & 1 \\ 1 & -1 \end{bmatrix}, Q = \begin{bmatrix} \frac{1}{2} & \frac{1}{2} \\ -\frac{1}{2} & \frac{1}{2} \\ \frac{1}{2} & \frac{1}{2} \\ \frac{1}{2} & -\frac{1}{2} \end{bmatrix},$

$R = \begin{bmatrix} 2 & 0 \\ 0 & 2 \end{bmatrix}$, and $\mathbf{b} = \begin{bmatrix} 1 \\ 2 \\ 4 \\ -1 \end{bmatrix}$

Least Squares when A Has Orthogonal Columns

If a matrix A has nonzero orthogonal columns, then it is easy to find the least squares solution for $A\mathbf{x} = \mathbf{b}$.

15. Let A be an $m \times n$ matrix with **nonzero** orthogonal columns $\mathbf{a}_i$ and let $\tilde{\mathbf{x}}$ be the least squares solution of the system $A\mathbf{x} = \mathbf{b}$. Show that

$$\tilde{\mathbf{x}} = \begin{bmatrix} \dfrac{\mathbf{b} \cdot \mathbf{a}_1}{\mathbf{a}_1 \cdot \mathbf{a}_1} \\ \dfrac{\mathbf{b} \cdot \mathbf{a}_2}{\mathbf{a}_2 \cdot \mathbf{a}_2} \\ \vdots \\ \dfrac{\mathbf{b} \cdot \mathbf{a}_n}{\mathbf{a}_n \cdot \mathbf{a}_n} \end{bmatrix} \qquad (8.24)$$

(*Hint:* A^TA has a very special form under the assumptions on A.)

16. How does (8.24) simplify when A has orthonormal columns?

17. Use Exercise 15 to find $\tilde{\mathbf{x}}$ if

$$A = \begin{bmatrix} -2 & -2 \\ 1 & -2 \\ 2 & -1 \end{bmatrix}, \qquad \mathbf{b} = \begin{bmatrix} 1 \\ -3 \\ 5 \end{bmatrix}$$

18. Use Exercise 15 to find $\tilde{\mathbf{x}}$ if

$$A = \begin{bmatrix} 1 & 1 \\ -1 & 1 \\ 1 & 1 \\ 1 & -1 \end{bmatrix}, \qquad \mathbf{b} = \begin{bmatrix} 1 \\ 2 \\ 4 \\ -1 \end{bmatrix}$$

19. Use Exercise 15 to find $\tilde{\mathbf{x}}$ if

$$A = \begin{bmatrix} -\frac{2}{3} & -\frac{2}{3} \\ \frac{1}{3} & -\frac{2}{3} \\ \frac{2}{3} & -\frac{1}{3} \end{bmatrix}, \qquad \mathbf{b} = \begin{bmatrix} 1 \\ -3 \\ 5 \end{bmatrix}$$

8.5 Orthogonalization of Symmetric Matrices

Reader's Goal for This Section

To know how to orthogonally diagonalize a symmetric matrix.

In this section we study some very interesting properties of symmetric matrices. Recall that A is **symmetric** if any entry off the main diagonal has a mirror image about the main diagonal. In other words,

$$A^T = A$$

In particular, A is square.

For example, $\begin{bmatrix} 1 & 3 \\ 3 & -2 \end{bmatrix}$ and $\begin{bmatrix} -1 & -1 & 1 \\ -1 & 2 & 4 \\ 1 & 4 & 2 \end{bmatrix}$ are symmetric, but $\begin{bmatrix} 1 & -3 \\ 3 & -2 \end{bmatrix}$ is not.

In Section 7.2 we saw that if an $n \times n$ matrix A is diagonalizable (i.e., similar to a diagonal matrix), then we can define an invertible matrix P whose columns are n linearly independent eigenvectors of A and a diagonal matrix D with diagonal entries the corresponding eigenvalues, such that

$$P^{-1}AP = D$$

Here we are interested in those diagonalizable matrices for which P can be *orthogonal*.

DEFINITION

We say that A is **orthogonally diagonalizable** if it can be diagonalized by an invertible matrix Q and a diagonal matrix D, so that Q is orthogonal. So, $Q^{-1}AQ = D$, or, equivalently,

$$Q^T AQ = D$$

because $Q^{-1} = Q^T$.

In general, we say that two $n \times n$ matrices A and B are **orthogonally similar** if there is an orthogonal matrix Q such that $Q^{-1}AQ = B$, or

$$Q^T AQ = B$$

So, an orthogonally diagonalizable matrix is one that is orthogonally similar to a diagonal matrix.

THEOREM 23   An orthogonally diagonalizable matrix is symmetric.

PROOF   If A is orthogonally diagonalizable, then $Q^{-1}AQ = D$ for some orthogonal matrix Q. So

$$A = QDQ^{-1} = QDQ^T$$

which implies

$$A^T = (QDQ^T)^T = Q^{TT}D^TQ^T = QDQ^T = A$$

because $D^T = D$ (because D is diagonal). Hence, A is symmetric.   ■

What is totally surprising (and amazing) is that the converse of this theorem is true. That is, *any symmetric matrix is orthogonally diagonalizable*. This statement is not easy to prove.

An interesting property of symmetric matrices is the following.

THEOREM 24   A real symmetric matrix has only real eigenvalues.

PROOF (OPTIONAL)   The proof requires a few easy facts on conjugates of complex numbers (discussed in Appendix A).

Preparation: The (**complex**) **conjugate** of $a + bi$ is the complex number $a - bi$ and is denoted by $\overline{a + bi}$. For example,

$$\overline{1 + i} = 1 - i, \qquad \overline{-i} = i, \qquad \overline{1} = 1$$

Because conjugation changes the sign of only the imaginary part, we see that a number is real if and only if it is equal to its conjugate.

$$z \text{ is real} \quad \Leftrightarrow \quad \overline{z} = z \tag{8.25}$$

Complex conjugation satisfies the following basic properties:

$$\overline{z_1 + z_2} = \overline{z_1} + \overline{z_2}, \qquad \overline{z_1 - z_2} = \overline{z_1} - \overline{z_2}, \qquad \overline{z_1 z_2} = \overline{z_1}\,\overline{z_2} \tag{8.26}$$

The complex conjugate $\overline{A}$ of matrix A is the matrix with entries the complex conjugates of the entries of A. It is easy to use the above properties to show that for compatible matrices A and B,

$$\overline{A + B} = \overline{A} + \overline{B}, \qquad \overline{A - B} = \overline{A} - \overline{B}, \qquad \overline{AB} = \overline{A}\,\overline{B} \tag{8.27}$$

Main Proof: Let A be a symmetric matrix and let $\mathbf{v}$ be an eigenvector with corresponding eigenvalue λ. Then $A\mathbf{v} = \lambda\mathbf{v}$. Also, $\overline{A} = A$, because A has real entries, by (8.26). Hence, by (8.27),

$$\overline{A\mathbf{v}} = \overline{\lambda\mathbf{v}} \Rightarrow \overline{A}\,\overline{\mathbf{v}} = \overline{\lambda}\,\overline{\mathbf{v}}$$

$$\Rightarrow A\overline{\mathbf{v}} = \overline{\lambda}\,\overline{\mathbf{v}}$$

$$\Rightarrow (A\overline{\mathbf{v}})^T = (\overline{\lambda}\,\overline{\mathbf{v}})^T$$

$$\Rightarrow \overline{\mathbf{v}}^T A^T = \overline{\lambda}\,\overline{\mathbf{v}}^T$$

$$\Rightarrow \overline{\mathbf{v}}^T A = \overline{\lambda}\,\overline{\mathbf{v}}^T$$

$$\Rightarrow (\overline{\mathbf{v}}^T A)\mathbf{v} = (\overline{\lambda}\,\overline{\mathbf{v}}^T)\mathbf{v}$$

$$\Rightarrow \overline{\mathbf{v}}^T (A\mathbf{v}) = \overline{\lambda}(\overline{\mathbf{v}}^T\mathbf{v})$$

$$\Rightarrow \overline{\mathbf{v}}^T (\lambda\mathbf{v}) = \overline{\lambda}(\overline{\mathbf{v}}^T\mathbf{v})$$

$$\Rightarrow (\lambda - \overline{\lambda})(\overline{\mathbf{v}}^T\mathbf{v}) = 0$$

But, $\overline{\mathbf{v}}^T\mathbf{v} \neq 0$, because if $\mathbf{v} = (a_1 + ib_1, \ldots, a_n + ib_n)$, then $\overline{\mathbf{v}} = (a_1 - ib_1, \ldots, a_n - ib_n)$, and so

$$\overline{\mathbf{v}}^T\mathbf{v} = (a_1^2 + b_1^2) + \cdots + (a_n^2 + b_n^2) \neq 0$$

because $\mathbf{v} \neq \mathbf{0}$. Therefore, $\lambda - \overline{\lambda} = 0$ or $\lambda = \overline{\lambda}$. Hence, the eigenvalue λ is real by (8.26).

THEOREM 25   Any two eigenvectors of a symmetric matrix A that correspond to different eigenvalues are orthogonal.

PROOF   Let $\mathbf{v}_1$ and $\mathbf{v}_2$ be two eigenvectors of A with corresponding eigenvalues λ_1 and λ_2 such that $\lambda_1 \neq \lambda_2$. Because $A\mathbf{v}_1 = \lambda_1\mathbf{v}_1$, $A\mathbf{v}_2 = \lambda_2\mathbf{v}_2$, and $A^T = A$, we have

$$\begin{aligned}
\lambda_1\mathbf{v}_1 \cdot \mathbf{v}_2 &= (A\mathbf{v}_1) \cdot \mathbf{v}_2 \\
&= \mathbf{v}_1 \cdot (A^T\mathbf{v}_2) \\
&= \mathbf{v}_1 \cdot A\mathbf{v}_2 \\
&= \mathbf{v}_1 \cdot \lambda_2\mathbf{v}_2 \\
&= \lambda_2\mathbf{v}_1 \cdot \mathbf{v}_2
\end{aligned}$$

Hence,

$$(\lambda_1 - \lambda_2)\mathbf{v}_1 \cdot \mathbf{v}_2 = 0$$

But $\lambda_1 - \lambda_2 \neq 0$; so $\mathbf{v}_1 \cdot \mathbf{v}_2 = 0$, as asserted.

▪ EXAMPLE 21   Verify Theorems 24 and 25 for A and B.

$$A = \begin{bmatrix} -1 & -1 & 1 \\ -1 & 2 & 4 \\ 1 & 4 & 2 \end{bmatrix}, \qquad B = \begin{bmatrix} 0 & 3 & 3 \\ 3 & 0 & 3 \\ 3 & 3 & 0 \end{bmatrix}$$

SOLUTION

1. It is easy to see that the eigenvalues of A are $-3, 0, 6$, and the corresponding eigenvectors are

$$\mathbf{v}_1 = \begin{bmatrix} -1 \\ -1 \\ 1 \end{bmatrix}, \qquad \mathbf{v}_2 = \begin{bmatrix} 2 \\ -1 \\ 1 \end{bmatrix}, \qquad \mathbf{v}_3 = \begin{bmatrix} 0 \\ 1 \\ 1 \end{bmatrix}$$

Note that $\mathbf{v}_1 \cdot \mathbf{v}_2 = \mathbf{v}_1 \cdot \mathbf{v}_3 = \mathbf{v}_2 \cdot \mathbf{v}_3 = 0$. So, the eigenvalues are all real and the eigenvectors orthogonal, as claimed by Theorems 24 and 25.

2. B has characteristic polynomial $(x - 6)(x + 3)^2$ and, hence, real eigenvalues, -3 and 6. Also,

$$E_{-3} = \text{Span}\left\{ \begin{bmatrix} -1 \\ 1 \\ 0 \end{bmatrix}, \begin{bmatrix} -1 \\ 0 \\ 1 \end{bmatrix} \right\}, \qquad E_6 = \text{Span}\left\{ \begin{bmatrix} 1 \\ 1 \\ 1 \end{bmatrix} \right\}$$

For the eigenvectors, $(-1, 1, 0) \cdot (1, 1, 1) = 0$ and $(-1, 0, 1) \cdot (1, 1, 1) = 0$. Note, however, that $(-1, 1, 0) \cdot (-1, 0, 1) = 1 \neq 0$.

REMARK   Although eigenvectors that correspond to *different* eigenvalues are orthogonal, eigenvectors that correspond to the *same* eigenvalue do *not have to be orthogonal*, as we saw in the last example.

■ EXAMPLE 22   Orthogonally diagonalize matrices A and B of Example 21.

SOLUTION

1. In Example 21 we found basic eigenvectors of A that were already orthogonal. If we normalize them, they will remain orthogonal. Hence,

$$Q = \begin{bmatrix} -\dfrac{1}{\sqrt{3}} & \dfrac{2}{\sqrt{6}} & 0 \\ -\dfrac{1}{\sqrt{3}} & -\dfrac{1}{\sqrt{6}} & \dfrac{1}{\sqrt{2}} \\ \dfrac{1}{\sqrt{3}} & \dfrac{1}{\sqrt{6}} & \dfrac{1}{\sqrt{2}} \end{bmatrix}$$

is orthogonal with columns eigenvectors of A. So, Q orthogonally diagonalizes A. It is not hard to check that

$$Q^T A Q = \begin{bmatrix} -3 & 0 & 0 \\ 0 & 0 & 0 \\ 0 & 0 & 6 \end{bmatrix}$$

2. For B, because the eigenvectors $(-1, 1, 0)$ and $(-1, 0, 1)$ were not orthogonal, we may apply Gram-Schmidt to orthogonalize them. We easily get $(-1, 1, 0)$ and $(-\frac{1}{2}, -\frac{1}{2}, 1)$. It is very important to note that since the Gram-Schmidt process did not alter the span of the original vectors, the new vector $(-\frac{1}{2}, -\frac{1}{2}, 1)$ is still in E_{-3}. (Check.) Thus, it is an eigenvector of A corresponding to -3. Hence, it must be orthogonal to $(1, 1, 1)$. Because $(-1, 1, 0)$, $(-\frac{1}{2}, -\frac{1}{2}, 1)$ and $(1, 1, 1)$ are now mutually orthogonal, all we need is to normalize them to get an orthogonal matrix Q that orthogonally diagonalizes B:

$$Q = \begin{bmatrix} -\dfrac{1}{\sqrt{2}} & -\dfrac{1}{\sqrt{6}} & \dfrac{1}{\sqrt{3}} \\ \dfrac{1}{\sqrt{2}} & -\dfrac{1}{\sqrt{6}} & \dfrac{1}{\sqrt{3}} \\ 0 & \dfrac{2}{\sqrt{6}} & \dfrac{1}{\sqrt{3}} \end{bmatrix}$$

Again, it is not hard to check that

$$Q^T B Q = \begin{bmatrix} -3 & 0 & 0 \\ 0 & -3 & 0 \\ 0 & 0 & 6 \end{bmatrix}$$

N O T E   In Example 22 we saw how to orthogonally diagonalize a symmetric matrix in practice: If necessary, we apply the Gram-Schmidt process to orthogonalize eigenvectors that correspond to the same eigenvalue. Then we normalize the orthogonal eigenvectors and use them as columns for Q. What we have not done yet is to show that any symmetric matrix is diagonalizable. To prove this basic fact, we use a classical theorem due to Schur (1909). Its proof is rather hard. It is included in the end of the section for the interested reader.

THEOREM 26   **(Schur's Decomposition)**

Any real square matrix A with only real eigenvalues is orthogonally similar to an upper triangular matrix T. So, there exists an orthogonal matrix Q and an upper triangular matrix T such that $Q^T A Q = T$. Equivalently,

$$A = QTQ^T$$

We have finally arrived at a major result on symmetric matrices, the spectral theorem.

THEOREM 27   **(The Spectral Theorem)**

A square matrix is real symmetric if and only if it is orthogonally diagonalizable.

PROOF   We already know that if A is orthogonally diagonalizable, it has to be symmetric, by Theorem 23. Conversely, let A be symmetric. Then by Theorem 24, A has real eigenvalues. Hence, by Schur's decomposition theorem, there is an orthogonal matrix Q and an upper triangular matrix D such that

$$Q^T A Q = D$$

Because $A^T = A$, we have

$$D^T = (Q^T A Q)^T = Q^T A^T Q^{TT} = Q^T A Q = D$$

We see that D is symmetric and upper triangular, so it is diagonal. Therefore, A is orthogonally diagonalizable.

Now that we know that symmetric matrices are orthogonally diagonalizable, we can describe the procedure applied in Example 22 to find Q and D.

Algorithm   **(Diagonalization of a Symmetric Matrix)**

INPUT: $n \times n$ symmetric matrix A.

1. Compute all the eigenvalues of A. Let $\lambda_1, \ldots, \lambda_k$ be all the *distinct* ones. (They are all real and some are possibly multiple.)
2. Find a basis $\mathcal{B}_i$ of eigenvectors for each eigenspace E_{λ_i}, $i = 1, \ldots, k$. (The union $\mathcal{B}_1 \cup \cdots \cup \mathcal{B}_k$ is a basis of eigenvectors of A, because A is symmetric and, hence, diagonalizable.)

3. Apply, if necessary, the Gram-Schmidt process to each $\mathcal{B}_i$ to get orthogonal sets $\mathcal{B}_i'$. (So, each $\mathcal{B}_i'$ is automatically linearly independent. Because $\text{Span}(\mathcal{B}_i) = \text{Span}(\mathcal{B}_i')$, each $\mathcal{B}_i'$ forms an orthogonal basis for E_{λ_i}.)
4. Let $\mathbf{u}_1, \ldots, \mathbf{u}_n$ be the vectors of $\mathcal{B}_1', \ldots, \mathcal{B}_k'$. They form an orthogonal basis of eigenvectors of A. (Because the vectors from the same $\mathcal{B}_i'$ are orthogonal and vectors from different $\mathcal{B}_i'$'s are eigenvectors corresponding to distinct eigenvalues, they, too, are orthogonal.)
5. Let $\mathbf{v}_1, \ldots, \mathbf{v}_n$ be the normalizations of the $\mathbf{u}_i$s. These form an orthonormal basis of eigenvectors of A.
6. Let $Q = [\mathbf{v}_1 \cdots \mathbf{v}_n]$. Q is orthogonal.
7. Let D be the diagonal matrix with diagonal entries the corresponding eigenvalues, in the same order, and with repeated entries for multiple eigenvalues.
8. Q and D orthogonally diagonalize A.

OUTPUT: Q orthogonal and D diagonal such that $Q^T A Q = D$.

Proof of Schur's Decomposition Theorem and Example

Let $\mathbf{v}_1$ be a unit eigenvector of A and let λ_1 be the corresponding (real) eigenvalue. By the Gram-Schmidt process, there is an orthogonal matrix $Q_1 = [\mathbf{v}_1 \cdots \mathbf{v}_n]$ with first column $\mathbf{v}_1$. Then

$$Q_1^T A Q_1 = \begin{bmatrix} \mathbf{v}_1^T \\ \vdots \\ \mathbf{v}_n^T \end{bmatrix} [A\mathbf{v}_1 \cdots A\mathbf{v}_n] = \begin{bmatrix} \mathbf{v}_1^T \\ \vdots \\ \mathbf{v}_n^T \end{bmatrix} [\lambda_1 \mathbf{v}_1 \cdots A\mathbf{v}_n] = \begin{bmatrix} \lambda_1 & \vdots & * \\ \cdots & \vdots & \cdots \\ \mathbf{0} & \vdots & A_1 \end{bmatrix}$$

because $\lambda_1 \mathbf{v}_1^T \mathbf{v}_1 = \lambda_1 \mathbf{v}_1 \cdot \mathbf{v}_1 = \lambda_1$ (because $\mathbf{v}_1$ is unit) and $\lambda_1 \mathbf{v}_i^T \mathbf{v}_1 = \lambda_1 \mathbf{v}_1 \cdot \mathbf{v}_i = \lambda_1 0 = 0$, for $i \neq 1$, by orthogonality. Now, if λ_2 is an eigenvalue of A_1, it is also an eigenvalue of A; hence, λ_2 is real. We apply the same procedure to the $(n-1) \times (n-1)$ matrix A_1 to find orthogonal matrix Q_2 such that

$$Q_2^T A_1 Q_2 = \begin{bmatrix} \lambda_2 & \vdots & * \\ \cdots & \vdots & \cdots \\ \mathbf{0} & \vdots & A_2 \end{bmatrix}$$

The matrix $Q_2' = \begin{bmatrix} 1 & \vdots & \mathbf{0} \\ \cdots & \vdots & \cdots \\ \mathbf{0} & \vdots & Q_2 \end{bmatrix}$ has the property

$$
Q_2'^T(Q_1^T A Q_1)Q_2' =
\begin{bmatrix}
1 & \vdots & \mathbf{0} \\
\cdots & \vdots & \cdots \\
\mathbf{0} & \vdots & Q_2^T
\end{bmatrix}
\begin{bmatrix}
\lambda_1 & \vdots & * \\
\cdots & \vdots & \cdots \\
\mathbf{0} & \vdots & A_1
\end{bmatrix}
\begin{bmatrix}
1 & \vdots & \mathbf{0} \\
\cdots & \vdots & \cdots \\
\mathbf{0} & \vdots & Q_2
\end{bmatrix}
$$

$$
=
\begin{bmatrix}
\lambda_1 & \vdots & * \\
\cdots & \vdots & \cdots \\
\mathbf{0} & \vdots & Q_2^T A_1 Q_2
\end{bmatrix}
=
\begin{bmatrix}
\lambda_1 & * & \vdots & * \\
0 & \lambda_2 & \vdots & * \\
\cdots & \cdots & \vdots & \cdots \\
0 & 0 & \vdots & A_2
\end{bmatrix}
$$

Continuing the same way, after $n - 1$ steps we obtain an orthogonal matrix Q,

$$
Q = Q_1
\begin{bmatrix}
1 & \vdots & \mathbf{0} \\
\cdots & \vdots & \cdots \\
\mathbf{0} & \vdots & Q_2
\end{bmatrix}
\begin{bmatrix}
I_2 & \vdots & \mathbf{0} \\
\cdots & \vdots & \cdots \\
\mathbf{0} & \vdots & Q_3
\end{bmatrix}
\cdots
\begin{bmatrix}
I_{n-2} & \vdots & \mathbf{0} \\
\cdots & \vdots & \cdots \\
\mathbf{0} & \vdots & Q_{n-1}
\end{bmatrix}
$$

such that $Q^T A Q$ is upper triangular:

$$
Q^T A Q =
\begin{bmatrix}
\lambda_1 & * & * & * \\
0 & \lambda_2 & * & * \\
0 & 0 & \ddots & * \\
\vdots & & & \\
0 & 0 & \cdots & \lambda_n
\end{bmatrix}
$$

This completes the proof of Schur's decomposition theorem.

■ EXAMPLE 23   Find the Schur decomposition of

$$
A = \begin{bmatrix} 9 & 8 \\ 1 & 2 \end{bmatrix}
$$

SOLUTION   Both eigenvalues $1, 10$ of A are real, and 1 has unit eigenvector $\mathbf{v}_1 = \begin{bmatrix} -1/\sqrt{2} \\ 1/\sqrt{2} \end{bmatrix}$. By the proof of the theorem we need an orthogonal matrix $Q_1 = [\mathbf{v}_1 \ \mathbf{v}_2]$.

We may take $Q_1 = \begin{bmatrix} -1/\sqrt{2} & 1/\sqrt{2} \\ 1/\sqrt{2} & 1/\sqrt{2} \end{bmatrix}$. (Check.) Then $T = Q_1^T A Q_1 = \begin{bmatrix} 1 & -7 \\ 0 & 10 \end{bmatrix}$.

So,

$$
\begin{bmatrix} 9 & 8 \\ 1 & 2 \end{bmatrix}
=
\begin{bmatrix} -\dfrac{1}{\sqrt{2}} & \dfrac{1}{\sqrt{2}} \\ \dfrac{1}{\sqrt{2}} & \dfrac{1}{\sqrt{2}} \end{bmatrix}
\begin{bmatrix} 1 & -7 \\ 0 & 10 \end{bmatrix}
\begin{bmatrix} -\dfrac{1}{\sqrt{2}} & \dfrac{1}{\sqrt{2}} \\ \dfrac{1}{\sqrt{2}} & \dfrac{1}{\sqrt{2}} \end{bmatrix}
$$

Exercises 8.5

In Exercises 1–4 determine which matrices are symmetric.

1. $A = \begin{bmatrix} 0 & 2 \\ -2 & 3 \end{bmatrix}$

2. $A = \begin{bmatrix} 0 & -2 \\ -2 & 3 \end{bmatrix}$

3. $A = \begin{bmatrix} 1 & -3 & 2 \\ -3 & 0 & 1 \\ 2 & -1 & 5 \end{bmatrix}$

4. $A = \begin{bmatrix} 1 & 3 & 0 \\ 3 & 0 & 1 \\ 0 & 1 & 5 \end{bmatrix}$

5. Find a 2×2 matrix A that is both symmetric and orthogonal. Compute A^2.

6. Show that if a matrix A is both symmetric and orthogonal, then $A^2 = I$. Conclude that $A = A^{-1}$.

7. Without computing, determine the inverse of
$$\begin{bmatrix} \frac{3}{5} & \frac{4}{5} \\ \frac{4}{5} & -\frac{3}{5} \end{bmatrix}.$$

In Exercises 8–15 orthogonally diagonalize the matrices.

8. $A = \begin{bmatrix} 3 & -1 \\ -1 & 3 \end{bmatrix}$

9. $A = \begin{bmatrix} -1 & 2 \\ 2 & -1 \end{bmatrix}$

10. $A = \begin{bmatrix} -4 & 2 \\ 2 & -4 \end{bmatrix}$

11. $A = \begin{bmatrix} 6 & -4 \\ -4 & 6 \end{bmatrix}$

12. $A = \begin{bmatrix} 2 & 0 & -1 \\ 0 & 0 & 0 \\ -1 & 0 & 2 \end{bmatrix}$

13. $A = \begin{bmatrix} 1 & 0 & 0 \\ 0 & 2 & -2 \\ 0 & -2 & 2 \end{bmatrix}$

14. $A = \begin{bmatrix} 5 & -4 & 0 \\ -4 & 3 & 4 \\ 0 & 4 & 1 \end{bmatrix}$

15. $A = \begin{bmatrix} 1 & -1 & 0 & 0 \\ -1 & 1 & 0 & 0 \\ 0 & 0 & 2 & -2 \\ 0 & 0 & -2 & 2 \end{bmatrix}$

16. Suppose that A and B are $n \times n$ and orthogonally diagonalizable with real entries and let c be any real scalar. Use the spectral theorem to show that the following are also orthogonally diagonalizable.

a. $A + B$　　　　　**b.** $A - B$

c. cA　　　　　　**d.** A^2

17. Show that if A is real symmetric, then A^2 has nonnegative eigenvalues. (*Hint:* Use the spectral theorem and write $Q^T A^2 Q$ as $(Q^T A Q)(Q^T A Q)$.)

18. Show that the geometric and algebraic multiplicities of each eigenvalue of a symmetric matrix are equal.

In Exercises 19–22 find the Schur decomposition of A.

19. $A = \begin{bmatrix} 9 & 8 \\ 2 & 3 \end{bmatrix}$

20. $A = \begin{bmatrix} 6 & 5 \\ 1 & 2 \end{bmatrix}$

21. $A = \begin{bmatrix} 10 & 9 \\ 2 & 3 \end{bmatrix}$

22. $A = \begin{bmatrix} 1 & 0 & 0 \\ 0 & 4 & 3 \\ 0 & 1 & 2 \end{bmatrix}$

Spectral Decomposition

23. Suppose $A = QDQ^T$ is an orthogonal diagonalization of A. If $Q = [v_1 \cdots v_n]$ and D has diagonal entries $\lambda_1, \ldots, \lambda_n$, prove that
$$A = \lambda_1 v_1 v_1^T + \cdots + \lambda_n v_n v_n^T$$
This is called the **spectral decomposition** of A.

24. Find the spectral decomposition of $A = \begin{bmatrix} 12 & 3 \\ 3 & 4 \end{bmatrix}$.

25. Find the spectral decomposition of $A = \begin{bmatrix} 13 & 4 \\ 4 & 7 \end{bmatrix}$.

8.6　Quadratic Forms and Conic Sections

Reader's Goals for This Section

1. To know the definition and how to diagonalize a quadratic form.
2. To know how to identify a conic section not in standard position.

In this section we use the diagonalization of symmetric matrices (discussed in Section 8.5) to study *quadratic expressions*, such as

$$ax^2 + by^2 + cxy + dx + ey + f$$

or

$$ax^2 + by^2 + cz^2 + dxy + exz + fyz + gx + hy + kz + l$$

Actually, we are interested in only the quadratic (or *principal*) terms of these sums:

$$ax^2 + by^2 + cxy \qquad (8.28)$$

and

$$ax^2 + by^2 + cz^2 + dxy + exz + fyz \qquad (8.29)$$

Expressions of the form (8.28) and (8.29) are called **quadratic forms**, and they are used in a wide variety of problems in mathematics, physics, mechanics, economics, statistics, robotics, and image processing and in many industrial applications.

Quadratic forms can be written as matrix products $\mathbf{x}^T A \mathbf{x}$. For example,

$$3x^2 + 7y^2 - 2xy = [\, x \quad y \,] \begin{bmatrix} 3 & -1 \\ -1 & 7 \end{bmatrix} \begin{bmatrix} x \\ y \end{bmatrix}$$

or even

$$3x^2 + 7y^2 - 2xy = [\, x \quad y \,] \begin{bmatrix} 3 & -2 \\ 0 & 7 \end{bmatrix} \begin{bmatrix} x \\ y \end{bmatrix}$$

We prefer the first decomposition over the second, because the first square matrix is symmetric. In fact, when we write a quadratic form as $\mathbf{x}^T A \mathbf{x}$, we may always assume that A is symmetric. A is called the **associated** matrix of the form. So, we have

$$ax^2 + by^2 + cxy = [\, x \quad y \,] \begin{bmatrix} a & \frac{1}{2}c \\ \frac{1}{2}c & b \end{bmatrix} \begin{bmatrix} x \\ y \end{bmatrix}$$

and

$$ax^2 + by^2 + cz^2 + dxy + exz + fyz = [\, x \quad y \quad z \,] \begin{bmatrix} a & \frac{1}{2}d & \frac{1}{2}e \\ \frac{1}{2}d & b & \frac{1}{2}f \\ \frac{1}{2}e & \frac{1}{2}f & c \end{bmatrix} \begin{bmatrix} x \\ y \\ z \end{bmatrix}$$

The main reason we want A to be symmetric is that symmetric matrices can be *orthogonally diagonalized*. In general, we may define quadratic forms in n variables using only symmetric matrices.

DEFINITION

A **quadratic form** (in n variables) is a function $q : \mathbf{R}^n \to \mathbf{R}$ of the form

$$q(\mathbf{x}) = \mathbf{x}^T A \mathbf{x}$$

for some *symmetric* $n \times n$ matrix A and any n-vector $\mathbf{x}$. We say that A is the matrix **associated** with q.

■ EXAMPLE 24   Show that the dot product in $\mathbf{R}^n$ defines a quadratic form q by

$$q(\mathbf{x}) = \mathbf{x} \cdot \mathbf{x} \qquad \text{or} \qquad q(\mathbf{x}) = \|\mathbf{x}\|^2$$

What is the associated matrix?

SOLUTION   Because

$$q(\mathbf{x}) = \mathbf{x} \cdot \mathbf{x} = \mathbf{x}^T\mathbf{x} = \mathbf{x}^T I \mathbf{x}$$

we see that q is a quadratic form with associated matrix I.

Quadratic forms in two variables, $q(x, y) = ax^2 + by^2 + cxy$, are closely related to **conic** sections, studied in algebra and analytic geometry. In fact, if $c = 0$, then the **cross** term cxy is zero, and the equation $q(x, y) = ax^2 + by^2 = 1$ represents, in general, an *ellipse* or a *hyperbola*.[5]

■ EXAMPLE 25   Describe the conic sections defined by each equation.

(a) $x^2 + 2y^2 = 1$
(b) $2x^2 + y^2 = 1$
(c) $x^2 - 2y^2 = 1$
(d) $-x^2 + 2y^2 = 1$

SOLUTION   Equations (a) and (b) represent ellipses and (c) and (d) are hyperbolas (Fig. 8.13).

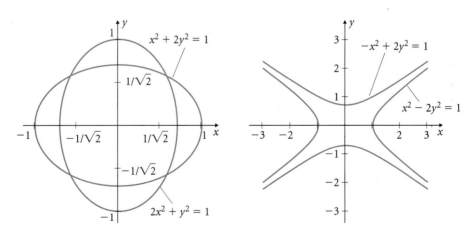

Figure 8.13

---

[5]Parabolas (such as $y = 2x^2$ or $y^2 = 3x$) are also conic sections but we since they contain non-quadratic terms we do not include them in our study.

In general (but not always), a two-variable quadratic form with no cross term represents an ellipse or a hyperbola in **standard position** (Figs. 8.13 and 8.14). This means that the principal axes of these conics are the x- and y-axes. In this case the matrix of the quadratic form is **diagonal**.

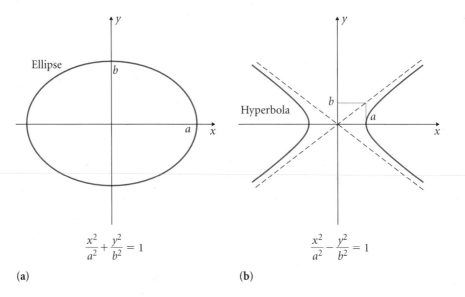

$$\frac{x^2}{a^2} + \frac{y^2}{b^2} = 1$$

(a)

$$\frac{x^2}{a^2} - \frac{y^2}{b^2} = 1$$

(b)

Figure 8.14

What happens if our quadratic form $q(x, y) = ax^2 + by^2 + cxy$ *does* have a cross term (so $c \neq 0$)? We use a change of variables, so that with respect to the new variables x', y' we have no cross term: $q(x', y') = a'x'^2 + b'y'^2$. Then q in the new coordinate system can be identified as a conic section. This is our next topic.

Diagonalization of Quadratic Forms

Let $q(\mathbf{x}) = \mathbf{x}^T A \mathbf{x}$ be any quadratic form in n variables. Because A is symmetric we can orthogonally diagonalize it, say, by Q orthogonal and D diagonal. Recall from Section 8.5 that Q has as columns n orthonormal eigenvectors of A, and D is diagonal with diagonal entries the corresponding eigenvalues. Using the familiar (from Chapter 7) change of variables,

$$\mathbf{x} = Q\mathbf{y}, \quad \text{or} \quad \mathbf{y} = Q^{-1}\mathbf{x} = Q^T\mathbf{x} \tag{8.30}$$

we have

$$q(\mathbf{x}) = \mathbf{x}^T A \mathbf{x}$$
$$= (Q\mathbf{y})^T A Q\mathbf{y}$$
$$= \mathbf{y}^T Q^T A Q\mathbf{y}$$
$$= \mathbf{y}^T D \mathbf{y}$$

So, q has the same values as a quadratic form in the new variables $\mathbf{y}$ with matrix D. In the new variables there are no cross terms, because D is diagonal. This process is called **diagonalization** of q. Because the diagonal entries of D are the eigenvalues of A, we have proved Theorem 28.

THEOREM 28

(Principal Axes)

Let A be an $n \times n$ symmetric matrix orthogonally diagonalized by Q and D. Then the change of variables $\mathbf{x} = Q\mathbf{y}$ transforms the quadratic form $q(\mathbf{x}) = \mathbf{x}^T A\mathbf{x}$ into the form $\mathbf{y}^T D\mathbf{y}$, which has no cross terms. In fact, if $\lambda_1, \ldots, \lambda_n$ are the eigenvalues of A and if $\mathbf{y} = (y_1, \ldots, y_n)$, then

$$q(\mathbf{x}) = q(\mathbf{y}) = \mathbf{y}^T D\mathbf{y} = \lambda_1 y_1^2 + \cdots + \lambda_n y_n^2$$

■ EXAMPLE 26 Write $q(x, y) = x^2 - 2xy + y^2$ without cross terms. Compute $q(1, -1)$ by using the old and the new variables.

SOLUTION   The matrix of q is $A = \begin{bmatrix} 1 & -1 \\ -1 & 1 \end{bmatrix}$ with basic eigenvectors $\begin{bmatrix} 1 \\ 1 \end{bmatrix}$ and $\begin{bmatrix} -1 \\ 1 \end{bmatrix}$ and corresponding eigenvalues 0 and 2. Because the eigenvectors are already orthogonal, we need only to normalize them to get Q. We have

$$Q = \begin{bmatrix} \dfrac{1}{\sqrt{2}} & -\dfrac{1}{\sqrt{2}} \\ \dfrac{1}{\sqrt{2}} & \dfrac{1}{\sqrt{2}} \end{bmatrix} \quad \text{and} \quad D = \begin{bmatrix} 0 & 0 \\ 0 & 2 \end{bmatrix}$$

Consider next new variables $\mathbf{y} = (x', y')$ such that $\mathbf{x} = (x, y) = Q\mathbf{y}$.

$$\mathbf{y} = \begin{bmatrix} x' \\ y' \end{bmatrix} = Q^T \begin{bmatrix} x \\ y \end{bmatrix}$$

$$= \begin{bmatrix} \dfrac{1}{\sqrt{2}} & -\dfrac{1}{\sqrt{2}} \\ \dfrac{1}{\sqrt{2}} & \dfrac{1}{\sqrt{2}} \end{bmatrix}^T \begin{bmatrix} x \\ y \end{bmatrix}$$

$$= \begin{bmatrix} \dfrac{x}{\sqrt{2}} + \dfrac{y}{\sqrt{2}} \\ -\dfrac{x}{\sqrt{2}} + \dfrac{y}{\sqrt{2}} \end{bmatrix}$$

and

$$q(\mathbf{y}) = \begin{bmatrix} x' & y' \end{bmatrix} \begin{bmatrix} 0 & 0 \\ 0 & 2 \end{bmatrix} \begin{bmatrix} x' \\ y' \end{bmatrix} = 2y'^2$$

Hence, $q(\mathbf{y}) = 2y'^2$ without cross terms. To evaluate q at $(1, -1)$ we have, in the old variables,

$$q(1, -1) = 1^2 - 2 \cdot 1 \cdot (-1) + 1^2 = 4$$

In the new variables, first we find

$$\mathbf{y} = \begin{bmatrix} x' \\ y' \end{bmatrix} = Q^T \begin{bmatrix} 1 \\ -1 \end{bmatrix} = \begin{bmatrix} 0 \\ -\dfrac{2}{\sqrt{2}} \end{bmatrix}$$

and then

$$\begin{bmatrix} 0 & -\dfrac{2}{\sqrt{2}} \end{bmatrix} \begin{bmatrix} 0 & 0 \\ 0 & 2 \end{bmatrix} \begin{bmatrix} 0 \\ -\dfrac{2}{\sqrt{2}} \end{bmatrix} = 4$$

We are interested in the *orthogonal* diagonalization of quadratic forms because the change of variables is then done by an **orthogonal transformation** $Q\mathbf{x}$ (meaning Q is orthogonal). So, the lengths (norms) and angles of any vectors mapped are preserved. Therefore, the shapes of the curves, surfaces, solids, etc., are also preserved in the new coordinates.

Applications of Quadratic Forms to Geometry

The principal axes theorem can be used for $n = 2, 3$ to identify conic sections or even quadric surfaces with cross terms.

Conic Sections: Ellipses and Hyperbolas

▪ EXAMPLE 27   Use diagonalization to identify the conic sections $q_1(x, y) = 1$ and $q_2(x, y) = 1$, where

$$q_1(x, y) = 2x^2 + 2y^2 - 2xy$$
$$q_2(x, y) = -x^2 - y^2 + 6xy$$

SOLUTION   Let $A_1 = \begin{bmatrix} 2 & -1 \\ -1 & 2 \end{bmatrix}$ and $A_2 = \begin{bmatrix} -1 & 3 \\ 3 & -1 \end{bmatrix}$ be the corresponding matrices. A_1 and A_2 are easily diagonalized by Q_1, D_1 and Q_2, D_2, where

$$Q_1 = \begin{bmatrix} \dfrac{1}{\sqrt{2}} & -\dfrac{1}{\sqrt{2}} \\ \dfrac{1}{\sqrt{2}} & \dfrac{1}{\sqrt{2}} \end{bmatrix} = Q_2$$

and

$$D_1 = \begin{bmatrix} 1 & 0 \\ 0 & 3 \end{bmatrix}, \qquad D_2 = \begin{bmatrix} 2 & 0 \\ 0 & -4 \end{bmatrix}$$

Therefore,

$$q_1(x', y') = [\, x' \;\; y' \,] \begin{bmatrix} 1 & 0 \\ 0 & 3 \end{bmatrix} \begin{bmatrix} x' \\ y' \end{bmatrix} = x'^2 + 3y'^2$$

and

$$q_2(x', y') = [\, x' \;\; y' \,] \begin{bmatrix} 2 & 0 \\ 0 & -4 \end{bmatrix} \begin{bmatrix} x' \\ y' \end{bmatrix} = 2x'^2 - 4y'^2$$

So, in the new variables we have an ellipse, $x'^2 + 3y'^2 = 1$, and a hyperbola, $2x'^2 - 4y'^2 = 1$. To sketch these curves we need to know which vectors map to $(1, 0)$ and $(0, 1)$ in x' and y'. Because

$$Q_1 \begin{bmatrix} 1 \\ 0 \end{bmatrix} = Q_2 \begin{bmatrix} 1 \\ 0 \end{bmatrix} = \begin{bmatrix} \dfrac{1}{\sqrt{2}} \\ \dfrac{1}{\sqrt{2}} \end{bmatrix}$$

$$Q_1 \begin{bmatrix} 0 \\ 1 \end{bmatrix} = Q_2 \begin{bmatrix} 0 \\ 1 \end{bmatrix} = \begin{bmatrix} \dfrac{1}{\sqrt{2}} \\ -\dfrac{1}{\sqrt{2}} \end{bmatrix}$$

we see that $(1, 0)$ and $(0, 1)$ in the new systems are $(1/\sqrt{2}, 1/\sqrt{2})$ and $(1/\sqrt{2}, -1/\sqrt{2})$ in the old. Hence, the ellipse and parabola are rotated $45°$ from the standard position (Fig. 8.15).

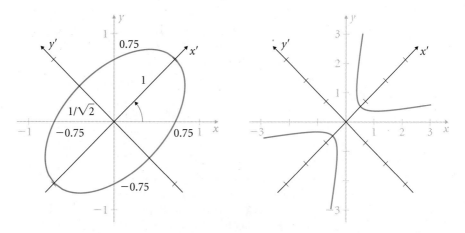

Figure 8.15 Conics in non-standard position.

▪ **EXAMPLE 28**   Use diagonalization to identify the conic sections $q_1(x, y) = 1$ and $q_2(x, y) = 1$, where

$$q_1(x, y) = 2x^2 + 2y^2 + 2xy$$

$$q_2(x, y) = 2x^2 + 2y^2 - 4xy$$

SOLUTION   It is easy to see that q_1 and q_2 are diagonalized, respectively, by

$$Q_1 = \begin{bmatrix} -\dfrac{1}{\sqrt{2}} & \dfrac{1}{\sqrt{2}} \\ \dfrac{1}{\sqrt{2}} & \dfrac{1}{\sqrt{2}} \end{bmatrix}, \qquad D_1 = \begin{bmatrix} 1 & 0 \\ 0 & 3 \end{bmatrix}$$

and

$$Q_2 = \begin{bmatrix} \dfrac{1}{\sqrt{2}} & \dfrac{1}{\sqrt{2}} \\ \dfrac{1}{\sqrt{2}} & -\dfrac{1}{\sqrt{2}} \end{bmatrix}, \qquad D_2 = \begin{bmatrix} 0 & 0 \\ 0 & 4 \end{bmatrix}$$

Hence,

$$q_1(x', y') = \begin{bmatrix} x' & y' \end{bmatrix} \begin{bmatrix} 1 & 0 \\ 0 & 3 \end{bmatrix} \begin{bmatrix} x' \\ y' \end{bmatrix} = x'^2 + 3y'^2$$

and

$$q_2(x', y') = \begin{bmatrix} x' & y' \end{bmatrix} \begin{bmatrix} 0 & 0 \\ 0 & 4 \end{bmatrix} \begin{bmatrix} x' \\ y' \end{bmatrix} = 4y'^2$$

So, $q_1(x', y') = x'^2 + 3y'^2 = 1$ is an ellipse in the $x'y'$–system, just as in Example 27. This time, however, $Q_1(1, 0) = (-1/\sqrt{2}, 1/\sqrt{2})$ and $Q_1(0, 1) = (1/\sqrt{2}, 1/\sqrt{2})$, so the positive x'-axis is the half-line at an angle of $135°$ and the positive y'-axis is the half-line at an angle of $45°$. In fact, since

$$Q_1 = \begin{bmatrix} 0 & 1 \\ 1 & 0 \end{bmatrix} \begin{bmatrix} \dfrac{1}{\sqrt{2}} & \dfrac{1}{\sqrt{2}} \\ -\dfrac{1}{\sqrt{2}} & \dfrac{1}{\sqrt{2}} \end{bmatrix}$$

the transformation defined by Q_1 is a rotation by $-45°$ (the second matrix) followed by a reflection about $y = x$ (the first matrix) (Fig. 8.16).

Because $q_2(x', y') = 4y'^2 = 1$, we have $y' = \pm\frac{1}{2}$, so this time we do not get an ellipse or a hyperbola but two straight lines in the $x'y'$–system (Fig. 8.16). This is an example of a **degenerate** quadratic form.

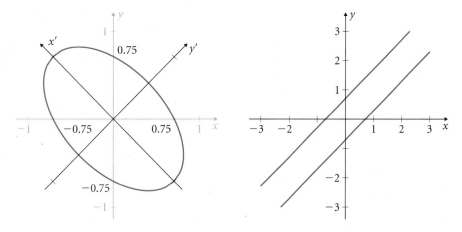

Figure 8.16   (a) Rotation followed by reflection, (b) degenerate form: two parallel lines.

Quadric Surfaces: Ellipsoids

We may apply these methods to identify quadric surfaces. For example, an equation of the form

$$\frac{x^2}{a^2} + \frac{y^2}{b^2} + \frac{z^2}{c^2} = 1$$

for $a, b, c > 0$ is an ellipsoid in standard position. The cross sections of such a surface with coordinates planes are ellipses. Figure 8.17(a) shows the ellipsoid $3x^2 + 6y^2 + 9z^2 = 1$.

■ EXAMPLE 29   Identify the quadric surface $5x^2 + 6y^2 + 7z^2 + 4xy + 4yz = 1$.

SOLUTION   The matrix of $q(x, y, z) = 5x^2 + 6y^2 + 7z^2 + 4xy + 4yz$ is

$$A = \begin{bmatrix} 5 & 2 & 0 \\ 2 & 6 & 2 \\ 0 & 2 & 7 \end{bmatrix}$$

with eigenvalues $2, 6, 9$ and corresponding eigenvectors, $(2, -2, 1)$, $(2, 1, -2)$, and $(1, 2, 2)$. These are already orthogonal, so we normalize them to get

$$Q = \frac{1}{3} \begin{bmatrix} 2 & 2 & 1 \\ -2 & 1 & 2 \\ 1 & -2 & 2 \end{bmatrix}, \qquad D = \begin{bmatrix} 3 & 0 & 0 \\ 0 & 6 & 0 \\ 0 & 0 & 9 \end{bmatrix}$$

Using the change of variables $\mathbf{y} = Q^T \mathbf{x}$, we get

$$q(x, y, z) = \begin{bmatrix} x' & y' & z' \end{bmatrix} \begin{bmatrix} 3 & 0 & 0 \\ 0 & 6 & 0 \\ 0 & 0 & 9 \end{bmatrix} \begin{bmatrix} x' \\ y' \\ z' \end{bmatrix} = 3x'^2 + 6y'^2 + 9z'^2$$

Therefore, $q(x, y, z) = 1$ takes the form $3x'^2 + 6y'^2 + 9z'^2 = 1$ in the new system. Clearly, the graph is an ellipsoid in $x'y'z'$-coordinates. Figure 8.17(b) shows this ellipsoid as somewhat turned and titled, compared with the one with the same equation in standard position. We leave it to the reader to locate and draw the new axes.

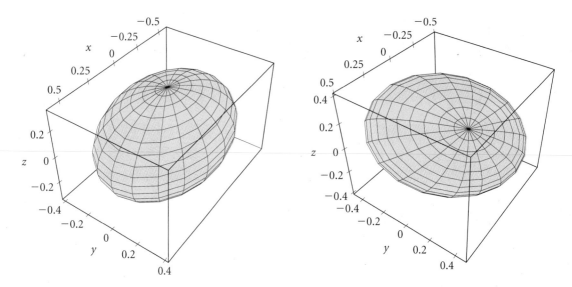

Figure 8.17   Ellipsoids in: (a) standard and (b) non-standard positions.

Positive and Negative Definite Quadratic Forms

In this section we classify quadratic forms according to their possible values. Because the values of any quadratic form $q(\mathbf{x}) = \mathbf{x}^T A\mathbf{x}$ are real numbers, they can be greater than 0, less than 0, or equal to 0. So, we may classify q according to whether its values are *always* positive on nonzero vectors or *always* negative. (We have to exclude $\mathbf{0}$, because $q(\mathbf{0}) = 0$.)

DEFINITION

Let $q(\mathbf{x}) = \mathbf{x}^T A\mathbf{x}$ be quadratic form with A symmetric.

1. If $q(\mathbf{x}) > 0$ for all $\mathbf{x} \neq \mathbf{0}$, then q is called **positive definite.**
2. If $q(\mathbf{x}) < 0$ for all $\mathbf{x} \neq \mathbf{0}$, then q is called **negative definite**.
3. If $q(\mathbf{x})$ takes on both positive and negative values, then q is called **indefinite**.

We also use the same terminology for the associated symmetric matrix A. So, for example, A is **positive definite** if $q(\mathbf{x}) = \mathbf{x}^T A\mathbf{x}$ is a positive definite quadratic form.

In addition to these basic types of forms, we have **positive** and **negative semidefinite** quadratic forms and symmetric matrices, according to whether $q(\mathbf{x}) \geq 0$ or $q(\mathbf{x}) \leq 0$ for all $\mathbf{x} \neq \mathbf{0}$.

The principal axes theorem can be easily used to identify the type of a quadratic form by looking at the signs of the eigenvalues of its matrix.

THEOREM 29   A quadratic form $q(\mathbf{x}) = \mathbf{x}^T A \mathbf{x}$ with A symmetric is

1. Positive definite if and only if all the eigenvalues of A are > 0;
2. Negative definite if and only if all the eigenvalues of A are < 0;
3. Indefinite if and only if A has positive and negative eigenvalues.

PROOF   Exercise.

■ EXAMPLE 30   (Relativity)   Show that the quadratic form q, used in the theory of relativity to define distance in space-time, is indefinite.

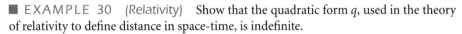

$$q(\mathbf{x}) = [\,x \quad y \quad z \quad t\,] \begin{bmatrix} 1 & 0 & 0 & 0 \\ 0 & 1 & 0 & 0 \\ 0 & 0 & 1 & 0 \\ 0 & 0 & 0 & -1 \end{bmatrix} \begin{bmatrix} x \\ y \\ z \\ t \end{bmatrix} = x^2 + y^2 + z^2 - t^2$$

SOLUTION   By Theorem 29, q is indefinite, because its matrix has both positive and negative eigenvalues.

NOTE   The appearance of the signs in the formula of a quadratic form may be deceiving sometimes. For example, we may be inclined to say that the form $q(\mathbf{x}) = x^2 + y^2 + 10xy$ is positive definite, but it is actually indefinite, because the eigenvalues of $\begin{bmatrix} 1 & 5 \\ 5 & 1 \end{bmatrix}$ are -4 and 6. In fact, $q(1, -1) = -8 < 0$.

Exercises 8.6

In this section all unspecified matrices have real entries.
In Exercises 1–4 evaluate the quadratic form $q(\mathbf{x}) = \mathbf{x}^T A \mathbf{x}$ for the given A and $\mathbf{x}$.

1. $A = \begin{bmatrix} -2 & 2 \\ 2 & 3 \end{bmatrix}$, $\mathbf{x} = \begin{bmatrix} x \\ y \end{bmatrix}$

2. $A = \begin{bmatrix} 4 & 7 \\ 7 & 3 \end{bmatrix}$, $\mathbf{x} = \begin{bmatrix} 1 \\ -1 \end{bmatrix}$

3. $A = \begin{bmatrix} 1 & -3 & 2 \\ -3 & 0 & 1 \\ 2 & 1 & 5 \end{bmatrix}$, $\mathbf{x} = \begin{bmatrix} x \\ y \\ z \end{bmatrix}$

4. $A = \begin{bmatrix} 1 & -3 & 2 \\ -3 & 0 & 1 \\ 2 & 1 & 5 \end{bmatrix}$, $\mathbf{x} = \begin{bmatrix} 1 \\ 2 \\ 3 \end{bmatrix}$

In Exercises 5–12 find the symmetric matrix A of the quadratic form.

5. $q(x, y) = 3x^2 - 6xy + 3y^2$

6. $q(x, y) = -x^2 + 10xy - y^2$

7. $q(x, y) = -4x^2 + 2xy - 4y^2$

8. $q(x, y) = 6x^2 - 2xy + 6y^2$

9. $q(x, y, z) = 2x^2 + 2xz + 2z^2$

10. $q(x, y, z) = x^2 + 2y^2 + 8yz + 2z^2$

11. $q(x, y, z) = 5x^2 - 8xy + 3y^2 + 12yz + z^2$

12. $q(x, y, z, w) = x^2 + 2xy + y^2 + 2z^2 + 4zw + 2w^2$

In Exercises 13–19 orthogonally diagonalize the quadratic form. Use a change of variables to rewrite the form without cross terms.

13. $q(x, y) = 3x^2 - 2xy + 3y^2$

14. $q(x, y) = -x^2 + 4xy - y^2$

15. $q(x, y) = -4x^2 + 4xy - 4y^2$

16. $q(x, y) = 6x^2 - 8xy + 6y^2$

17. $q(x, y, z) = 2x^2 - 2xz + 2z^2$

18. $q(x, y, z) = x^2 + 2y^2 - 4yz + 2z^2$

19. $q(x, y, z) = 5x^2 - 8xy + 3y^2 + 8yz + z^2$

20. Identify the conic section $3x^2 - 2xy + 3y^2 = 1$.

21. Identify the conic section $5x^2 - 8xy + 5y^2 = 1$.

22. Identify the quadric surface $6x^2 + 8xy + 4xz + 10y^2 + 12yz + 11z^2 = 1$.

23. Prove Theorem 29.

24. Show that the quadratic form $q(x, y) = ax^2 + bxy + cy^2$ is positive definite if and only if $a > 0$ and $b^2 - 4ac < 0$.

25. Show that if A is symmetric, then the quadratic form $q(\mathbf{x}) = \mathbf{x}^T A^2 \mathbf{x}$ is positive semidefinite. (*Hint:* Use the spectral theorem, Section 8.5, and write $Q^T A^2 Q$ as $(Q^T A Q)(Q^T A Q)$.)

26. Let $A = P^T P$, where P is an invertible matrix. Show that A is positive definite. (*Hint:* A is symmetric and semidefinite, because $\mathbf{x}^T A \mathbf{x} = \mathbf{x}^T P^T P \mathbf{x} = (P\mathbf{x})^T P\mathbf{x} = \|P\mathbf{x}\|^2 \geq 0$. Then show that if $\mathbf{x} \neq \mathbf{0}$, then $\mathbf{x}^T A \mathbf{x} > 0$.)

27. Let A be a positive definite matrix (hence, A is symmetric). Show that there exists an invertible matrix P such that $A = P^T P$. (*Hint:* Use the spectral theorem, Section 8.5, and write A as QDQ^T with Q orthogonal and D diagonal. Then find a matrix B such $D = B^T B$. What should P be?)

28. Prove that A is positive definite if and only if there exists an invertible matrix P such that $A = P^T P$.

Completion of the Square

The familiar completion of square

$$A^2 + bA = \left(A^2 + 2\left(\frac{b}{2}\right)A + \frac{b^2}{4} \right) - \frac{b^2}{4}$$

$$= \left(A + \frac{b}{2} \right)^2 - \frac{b^2}{4}$$

can be used to convert a 2-variable quadratic form into one without cross terms.

29. Let $q(x, y) = ax^2 + bxy + cy^2$. If $a \neq 0$, complete the square to write q in the form $aX^2 + By^2$ for some constant B and a new variable X that depends on x and y.

30. Apply the formula from Exercise 29 to write $q(x, y) = 3x^2 - 2xy + 3y^2$ without cross terms.

31. Referring to Exercise 29, if $a = 0$ and $c \neq 0$, can you still complete the square and write q without cross terms?

32. Referring to Exercise 29, if $a = 0$ and $c = 0$, can you write q without cross terms?

8.7   The Singular Value Decomposition (SVD)

Reader's Goals for This Section

1. To know how to compute the singular value decomposition of a matrix.
2. To realize the theoretical and practical importance of this method.

We have seen that factorizations of matrices into factors with special properties can be very useful. Examples include the LU, QR, diagonalization, orthogonal diagonalization, Schur decomposition, etc. A factorization is of particular interest if some of the factors are orthogonal matrices. The reason is that orthogonal transformations preserve norms and angles. In particular, they preserve the lengths of the error vectors that are inevitable in numerical calculations.

In this section we study one of the most important factorizations that applies to any $m \times n$ matrix A, the **singular value decomposition**, or **SVD**. This method is of both theoretical and applied interest and is rather old.[6] In fact, it is so useful that it deserves much more attention and credit. Among the many applications is the most reliable estimation of the rank of a matrix.

Our goal is to factor any $m \times n$ matrix A in the form

$$A = U\Sigma V^T$$

where U is $m \times m$, V is $n \times n$, and they are both orthogonal. Also, Σ is an $m \times n$ matrix with a diagonal upper left block of positive entries of decreasing magnitude and the remaining entries 0. So,

$$\Sigma = \begin{bmatrix} D & \vdots & \mathbf{0} \\ \cdots & \cdots & \cdots \\ \mathbf{0} & \vdots & \mathbf{0} \end{bmatrix}, \qquad \text{where } D = \begin{bmatrix} \sigma_1 & \cdots & 0 \\ \vdots & \ddots & \vdots \\ 0 & \cdots & \sigma_r \end{bmatrix} \qquad (8.31)$$

and

$$\sigma_1 \geq \sigma_2 \geq \cdots \geq \sigma_r > 0, \qquad r \leq m, n$$

Here are some examples for Σ with $r = 2$:

$$\begin{bmatrix} 6 & 0 \\ 0 & 3 \\ 0 & 0 \end{bmatrix}, \qquad \begin{bmatrix} 9 & 0 & 0 \\ 0 & 3 & 0 \end{bmatrix}, \qquad \begin{bmatrix} 9 & 0 & 0 \\ 0 & 9 & 0 \\ 0 & 0 & 0 \end{bmatrix}, \qquad \begin{bmatrix} 9 & 0 & 0 & 0 \\ 0 & 9 & 0 & 0 \\ 0 & 0 & 0 & 0 \end{bmatrix}$$

The corresponding Ds are

$$\begin{bmatrix} 6 & 0 \\ 0 & 3 \end{bmatrix}, \qquad \begin{bmatrix} 9 & 0 \\ 0 & 3 \end{bmatrix}, \qquad \begin{bmatrix} 9 & 0 \\ 0 & 9 \end{bmatrix}, \qquad \begin{bmatrix} 9 & 0 \\ 0 & 9 \end{bmatrix}$$

Singular Values; Finding V, Σ, and U

First we define V then find the σ_is along the diagonal of D to form Σ. Consider the $n \times n$ symmetric matrix $A^T A$. By the spectral theorem, $A^T A$ is orthogonally diagonalizable and has real eigenvalues, say, $\lambda_1, \ldots, \lambda_n$. Let $\mathbf{v}_1, \ldots, \mathbf{v}_n$ be the corresponding eigenvectors so that they form an orthonormal basis of $\mathbf{R}^n$. V is simply

$$V = [\mathbf{v}_1 \; \mathbf{v}_2 \cdots \mathbf{v}_n]$$

Next, we observe that all the eigenvalues are nonnegative (so $A^T A$ is positive semidefinite). Because $(A^T A)\mathbf{v}_i = \lambda_i \mathbf{v}_i$ and $\|\mathbf{v}_i\| = 1$, we have

---

[6]It is certainly mentioned in E. Beltrami's "Sulle Funzioni Bilineari," *Giornale di Matematische 11* (1873), pp. 98–106.

$$0 \le \|A\mathbf{v}_i\|^2$$
$$= (A\mathbf{v}_i)^T A\mathbf{v}_i$$
$$= \mathbf{v}_i^T A^T A\mathbf{v}_i$$
$$= \mathbf{v}_i^T \lambda_i \mathbf{v}_i$$
$$= \lambda_i \|\mathbf{v}_i\|^2$$
$$= \lambda_i$$

Hence, $\lambda_i \ge 0$ for $i = 1,\ldots,n$. By renumbering, if necessary, we order the λs from largest to smallest and take their square roots, which we call σ:

$$\sigma_1 = \sqrt{\lambda_1} \ge \cdots \ge \sigma_n = \sqrt{\lambda_n} \ge 0$$

So,

$$\sigma_i = \|A\mathbf{v}_i\|, \qquad i = 1,\ldots,n \qquad (8.32)$$

The numbers $\sigma_1,\ldots,\sigma_n$ are called the **singular values** of A, and they carry important information about A. Let r be the positive integer such that

$$\sigma_1 \ge \sigma_2 \ge \cdots \ge \sigma_r > 0 \qquad \text{and} \qquad \sigma_{r+1} = \cdots = \sigma_n = 0$$

So, $\sigma_1,\ldots,\sigma_r$ are the nonzero singular values of A ordered by magnitude. These are the diagonal entries of D in Σ.

■ EXAMPLE 31   Compute V and Σ in each case.

(a) $A = \begin{bmatrix} 2 & 4 \\ 1 & -4 \\ -2 & 2 \end{bmatrix}$

(b) $A = \begin{bmatrix} -2 & 1 & 2 \\ 6 & 6 & 3 \end{bmatrix}$

(c) $A = \begin{bmatrix} 0 & 6 & 6 \\ -6 & -3 & 0 \\ 6 & 0 & -3 \end{bmatrix}$

SOLUTION   We need $A^T A$, its eigenvalues, and the corresponding basic eigenvectors.

	$A^T A$	Eigenvalues	Eigenvectors
(a)	$\begin{bmatrix} 9 & 0 \\ 0 & 36 \end{bmatrix}$	$36, 9$	$\begin{bmatrix} 0 \\ 1 \end{bmatrix}, \begin{bmatrix} 1 \\ 0 \end{bmatrix}$
(b)	$\begin{bmatrix} 40 & 34 & 14 \\ 34 & 37 & 20 \\ 14 & 20 & 13 \end{bmatrix}$	$81, 9, 0$	$\begin{bmatrix} 2 \\ 2 \\ 1 \end{bmatrix}, \begin{bmatrix} -2 \\ 1 \\ 2 \end{bmatrix}, \begin{bmatrix} 1 \\ -2 \\ 2 \end{bmatrix}$

$A^T A$	Eigenvalues	Eigenvectors

(c) $\begin{bmatrix} 72 & 18 & -18 \\ 18 & 45 & 36 \\ -18 & 36 & 45 \end{bmatrix}$ $81, 81, 0$ $\begin{bmatrix} -2 \\ 0 \\ 1 \end{bmatrix}, \begin{bmatrix} 2 \\ 1 \\ 0 \end{bmatrix}, \begin{bmatrix} 1 \\ -2 \\ 2 \end{bmatrix}$

(a) The singular values of A are $\sigma_1 = \sqrt{36} = 6$ and $\sigma_2 = \sqrt{9} = 3$. The eigenvectors are already orthonormal, so

$$V = \begin{bmatrix} 0 & 1 \\ 1 & 0 \end{bmatrix} \quad \text{and} \quad \Sigma = \begin{bmatrix} 6 & 0 \\ 0 & 3 \\ 0 & 0 \end{bmatrix}$$

(b) The singular values of A are $\sigma_1 = 9$, $\sigma_2 = 3$, and $\sigma_3 = 0$. The eigenvectors are orthogonal and need normalization. So,

$$V = \begin{bmatrix} \frac{2}{3} & -\frac{2}{3} & \frac{1}{3} \\ \frac{2}{3} & \frac{1}{3} & -\frac{2}{3} \\ \frac{1}{3} & \frac{2}{3} & \frac{2}{3} \end{bmatrix} \quad \text{and} \quad \Sigma = \begin{bmatrix} 9 & 0 & 0 \\ 0 & 3 & 0 \end{bmatrix}$$

(c) The singular values of A are $\sigma_1 = 9$, $\sigma_2 = 9$, and $\sigma_3 = 0$. We now need to orthonormalize the eigenvectors. If we orthogonalize the first two that belong to E_{81}, we get $(-2, 0, 1)$, $(\frac{2}{5}, 1, \frac{4}{5})$ by a one-step Gram-Schmidt process. So, normalizing the orthogonal set $\{(-2, 0, 1), (\frac{2}{5}, 1, \frac{4}{5}), (1, -2, 2)\}$ yields

$$V = \begin{bmatrix} -\dfrac{2}{\sqrt{5}} & \dfrac{2}{3\sqrt{5}} & \dfrac{1}{3} \\ 0 & \dfrac{5}{3\sqrt{5}} & -\dfrac{2}{3} \\ \dfrac{1}{\sqrt{5}} & \dfrac{4}{3\sqrt{5}} & \dfrac{2}{3} \end{bmatrix} \quad \text{and} \quad \Sigma = \begin{bmatrix} 9 & 0 & 0 \\ 0 & 9 & 0 \\ 0 & 0 & 0 \end{bmatrix}$$

NOTE Σ is unique, because it is determined by the ordered singular values. The computation of V, however, involves choices, so V is not unique. In (c) of Example 31, instead of the eigenvectors $\mathbf{v}_1 = (-2, 0, 1)$, $\mathbf{v}_2 = (2, 1, 0)$, we could have used the linear combinations $\mathbf{v}_1 + 2\mathbf{v}_2 = (2, 2, 1)$ and $2\mathbf{v}_1 + \mathbf{v}_2 = (-2, 1, 2)$. These are already orthogonal, so after normalization we get a different V:

$$V = \begin{bmatrix} \frac{2}{3} & -\frac{2}{3} & \frac{1}{3} \\ \frac{2}{3} & \frac{1}{3} & -\frac{2}{3} \\ \frac{1}{3} & \frac{2}{3} & \frac{2}{3} \end{bmatrix}$$

Finally, we come to the definition of U. This is done in two steps:

1. First, we form

$$\mathbf{u}_i = \frac{1}{\sigma_i} A\mathbf{v}_i \quad \text{for } i = 1, \ldots, r \tag{8.33}$$

These vectors are *orthonormal*:

$$\mathbf{u}_i \cdot \mathbf{u}_j = \frac{1}{\sigma_i \sigma_j}(A\mathbf{v}_i \cdot A\mathbf{v}_j)$$

$$= \frac{1}{\sigma_i \sigma_j}(A^T A\mathbf{v}_i) \cdot \mathbf{v}_j$$

$$= \frac{\lambda_i}{\sigma_i \sigma_j}\mathbf{v}_i \cdot \mathbf{v}_j$$

$$= \begin{cases} 0 & i \neq j \\ 1 & i = j \end{cases} \tag{8.34}$$

For $i \neq j$, the $\mathbf{v}_i$s are orthogonal, so $\mathbf{v}_i \cdot \mathbf{v}_j = 0$. Also, for $i = j$, we have $\lambda_i/\sigma_i^2 = 1$, by the definition of the singular values, and $\mathbf{v}_i \cdot \mathbf{v}_i = 1$, because $\mathbf{v}_i$ is unit.

2. Next we extend the set $\{\mathbf{u}_1, \ldots, \mathbf{u}_r\}$ to an orthonormal basis $\{\mathbf{u}_1, \ldots, \mathbf{u}_m\}$ of $\mathbf{R}^m$. This is necessary only if $r < m$. We define

$$U = [\mathbf{u}_1 \ \mathbf{u}_2 \cdots \mathbf{u}_m]$$

■ EXAMPLE 32   Find an SVD for A of Example 31(b).

SOLUTION   Because $\sigma_1 = 9$, $\sigma_2 = 3$, $\mathbf{v}_1 = (\frac{2}{3}, \frac{2}{3}, \frac{1}{3})$, and $\mathbf{v}_2 = (-\frac{2}{3}, \frac{1}{3}, \frac{2}{3})$, we have

$$\mathbf{u}_1 = \frac{1}{9}\begin{bmatrix} -2 & 1 & 2 \\ 6 & 6 & 3 \end{bmatrix}\begin{bmatrix} \frac{2}{3} \\ \frac{2}{3} \\ \frac{1}{3} \end{bmatrix} = \begin{bmatrix} 0 \\ 1 \end{bmatrix}$$

$$\mathbf{u}_2 = \frac{1}{3}\begin{bmatrix} -2 & 1 & 2 \\ 6 & 6 & 3 \end{bmatrix}\begin{bmatrix} -\frac{2}{3} \\ \frac{1}{3} \\ \frac{2}{3} \end{bmatrix} = \begin{bmatrix} 1 \\ 0 \end{bmatrix}$$

Because $m = r = 2$, $\{\mathbf{u}_1, \mathbf{u}_2\}$ needs no extension. So,

$$U = \begin{bmatrix} 0 & 1 \\ 1 & 0 \end{bmatrix}$$

An SVD of $A = U\Sigma V^T$ is then

$$\begin{bmatrix} -2 & 1 & 2 \\ 6 & 6 & 3 \end{bmatrix} = \begin{bmatrix} 0 & 1 \\ 1 & 0 \end{bmatrix}\begin{bmatrix} 9 & 0 & 0 \\ 0 & 3 & 0 \end{bmatrix}\begin{bmatrix} \frac{2}{3} & -\frac{2}{3} & \frac{1}{3} \\ \frac{2}{3} & \frac{1}{3} & -\frac{2}{3} \\ \frac{1}{3} & \frac{2}{3} & \frac{2}{3} \end{bmatrix}^T$$

One way to extend an orthonormal set $S = \{\mathbf{u}_1, \ldots, \mathbf{u}_r\}$ to an orthonormal basis $\mathcal{B} = \{\mathbf{u}_1, \ldots, \mathbf{u}_m\}$ is outlined by the following steps:

1. Form $S' = \{\mathbf{u}_1, \ldots, \mathbf{u}_r, \mathbf{e}_1, \ldots, \mathbf{e}_m\}$ and find the pivot columns of the matrix with columns these vectors.
2. Form the subset S'' of S' that consists of the pivot columns. S'' is a basis of $\mathbf{R}^m$.
3. Apply the Gram-Schmidt to S'' and normalize the resulting vectors to get $\mathcal{B}$.

■ EXAMPLE 33   (Extension to Orthonormal Basis)   Find an SVD for A of Example 31(a).

SOLUTION   To find U we have

$$\mathbf{u}_1 = \frac{1}{6} \begin{bmatrix} 2 & 4 \\ 1 & -4 \\ -2 & 2 \end{bmatrix} \begin{bmatrix} 0 \\ 1 \end{bmatrix} = \begin{bmatrix} \frac{2}{3} \\ -\frac{2}{3} \\ \frac{1}{3} \end{bmatrix}$$

$$\mathbf{u}_2 = \frac{1}{3} \begin{bmatrix} 2 & 4 \\ 1 & -4 \\ -2 & 2 \end{bmatrix} \begin{bmatrix} 1 \\ 0 \end{bmatrix} = \begin{bmatrix} \frac{2}{3} \\ \frac{1}{3} \\ -\frac{2}{3} \end{bmatrix}$$

Now we need to extend $\{\mathbf{u}_1, \mathbf{u}_2\}$ to an orthonormal basis $\{\mathbf{u}_1, \mathbf{u}_2, \mathbf{u}_3\}$ of $\mathbf{R}^3$. Because

$$\begin{bmatrix} \frac{2}{3} & \frac{2}{3} & 1 & 0 & 0 \\ -\frac{2}{3} & \frac{1}{3} & 0 & 1 & 0 \\ \frac{1}{3} & -\frac{2}{3} & 0 & 0 & 1 \end{bmatrix} \sim \begin{bmatrix} 1 & 0 & 0 & -2 & -1 \\ 0 & 1 & 0 & -1 & -2 \\ 0 & 0 & 1 & 2 & 2 \end{bmatrix}$$

the first three columns are pivot columns, so $\{\mathbf{u}_1, \mathbf{u}_2, (1, 0, 0)\}$ forms a basis of $\mathbf{R}^3$. Gram-Schmidt and normalization now yield $\mathbf{u}_3 = (\frac{1}{3}, \frac{2}{3}, \frac{2}{3})$. So,

$$U = \begin{bmatrix} \frac{2}{3} & \frac{2}{3} & \frac{1}{3} \\ -\frac{2}{3} & \frac{1}{3} & \frac{2}{3} \\ \frac{1}{3} & -\frac{2}{3} & \frac{2}{3} \end{bmatrix}$$

An SVD of $A = U\Sigma V^T$ is then

$$\begin{bmatrix} 2 & 4 \\ 1 & -4 \\ -2 & 2 \end{bmatrix} = \begin{bmatrix} \frac{2}{3} & \frac{2}{3} & \frac{1}{3} \\ -\frac{2}{3} & \frac{1}{3} & \frac{2}{3} \\ \frac{1}{3} & -\frac{2}{3} & \frac{2}{3} \end{bmatrix} \begin{bmatrix} 6 & 0 \\ 0 & 3 \\ 0 & 0 \end{bmatrix} \begin{bmatrix} 0 & 1 \\ 1 & 0 \end{bmatrix}^T$$

We leave it to the reader to verify the following SVD for A of Example 31(c) and to find another one based on the first V.

$$\begin{bmatrix} 0 & 6 & 6 \\ -6 & -3 & 0 \\ 6 & 0 & -3 \end{bmatrix} = \begin{bmatrix} \frac{2}{3} & \frac{2}{3} & \frac{1}{3} \\ -\frac{2}{3} & \frac{1}{3} & \frac{2}{3} \\ \frac{1}{3} & -\frac{2}{3} & \frac{2}{3} \end{bmatrix} \begin{bmatrix} 9 & 0 & 0 \\ 0 & 9 & 0 \\ 0 & 0 & 0 \end{bmatrix} \begin{bmatrix} \frac{2}{3} & -\frac{2}{3} & \frac{1}{3} \\ \frac{2}{3} & \frac{1}{3} & -\frac{2}{3} \\ \frac{1}{3} & \frac{2}{3} & \frac{2}{3} \end{bmatrix}^T$$

We have almost proved the basic theorem of this section.

THEOREM 30

Let A be any $m \times n$ matrix and let $\sigma_1, \ldots, \sigma_r$ be all its nonzero singular values. Then there are orthogonal matrices U ($m \times m$) and V ($n \times n$) and·an $m \times n$ matrix Σ of the form (8.31) such that

$$A = U\Sigma V^T$$

PROOF U, V, and Σ (of the indicated sizes) have been already explicitly defined. Moreover, U and V are orthogonal by construction. It remains to show only that $A = U\Sigma V^T$. It suffices to show that $AV = U\Sigma$, because $V^T = V^{-1}$. By (8.33),

$$\sigma_i \mathbf{u}_i = A\mathbf{v}_i \qquad \text{for } i = 1, \ldots, r$$

and by (8.32), $\|A\mathbf{v}_i\| = \sigma_i = 0$ for $i = r + 1, \ldots, n$. So,

$$A\mathbf{v}_i = \mathbf{0} \qquad \text{for } i = r + 1, \ldots, n \tag{8.35}$$

Therefore,

$$AV = [A\mathbf{v}_1 \cdots A\mathbf{v}_n]$$

$$= [A\mathbf{v}_1 \cdots A\mathbf{v}_r \; \mathbf{0} \cdots \mathbf{0}]$$

$$= [\sigma_1 \mathbf{u}_1 \cdots \sigma_r \mathbf{u}_r \; \mathbf{0} \cdots \mathbf{0}]$$

$$= [\mathbf{u}_1 \cdots \mathbf{u}_m] \begin{bmatrix} \sigma_1 & & 0 & \vdots & \\ & \ddots & & \vdots & 0 \\ 0 & & \sigma_r & \vdots & \\ \cdots & \cdots & \cdots & \cdots & \cdots \\ & & 0 & \vdots & 0 \end{bmatrix}$$

$$= U\Sigma$$

This concludes the proof of the theorem.

U, Σ, V, and r (the number of nonzero singular values) provide important information about A.

THEOREM 31

Let V, Σ, U be singular value decomposition matrices for an $m \times n$ matrix A. Let $\sigma_1, \ldots, \sigma_r$ be all the nonzero singular values of A.

1. The rank of A is r.
2. $\{\mathbf{u}_1, \ldots, \mathbf{u}_r\}$ is an orthonormal basis for $\text{Col}(A)$.
3. $\{\mathbf{u}_{r+1}, \ldots, \mathbf{u}_m\}$ is an orthonormal basis for $\text{Null}(A^T)$.
4. $\{\mathbf{v}_1, \ldots, \mathbf{v}_r\}$ is an orthonormal basis for $\text{Row}(A)$.
5. $\{\mathbf{v}_{r+1}, \ldots, \mathbf{v}_n\}$ is an orthonormal basis for $\text{Null}(A)$.

PROOF

1 and 2. Let $\mathcal{B} = \{\mathbf{u}_1, \ldots, \mathbf{u}_r\}$. Then $\mathcal{B}$ is orthonormal (thus, linearly independent) by (8.34) and a subset of $\text{Col}(A)$ by (8.33). Because $\{\mathbf{v}_1, \ldots, \mathbf{v}_n\}$ is a basis of $\mathbf{R}^n$, the set $\{A\mathbf{v}_1, \ldots, A\mathbf{v}_n\}$ spans $\text{Col}(A)$. Therefore, $\{A\mathbf{v}_1, \ldots, A\mathbf{v}_r\}$ spans $\text{Col}(A)$, by (8.35). So, the dimension of $\text{Col}(A)$ is $\leq r$; thus, it is exactly r, because $\mathcal{B}$ is a linearly independent subset with r elements. Hence, $\mathcal{B}$ is an orthonormal basis of $\text{Col}(A)$ and $\text{Rank}(A) = r$.

3. By 2, $\{\mathbf{u}_{r+1}, \ldots, \mathbf{u}_m\}$ is an orthonormal basis for the orthogonal complement of $\text{Col}(A)$. The claim now follows from $(\text{Col}(A))^\perp = \text{Null}(A^T)$ of Theorem 14, Section 8.2.

5. $\{v_{r+1}, \ldots, v_n\}$ is an orthonormal subset of Null(A), by (8.35). But by the rank theorem, the nullity of A is $n - \text{Rank}(A) = n - r$. Hence, the dimension of Null(A) is $n - r$, so $\{v_{r+1}, \ldots, v_n\}$ is an orthonormal basis.

4. By 5, $\{v_1, \ldots, v_r\}$ is an orthonormal basis for the orthogonal complement of Null(A). But, $(\text{Null}(A))^{\perp} = (\text{Row}(A)^{\perp})^{\perp} = \text{Row}(A)$, by Theorem 14, Section 8.2. And the claim follows.

On the Numerical Computation of Rank

One of the most important applications of the SVD is in the computation of the rank of a matrix, using Theorem 31. Numerical reduction of large matrices often yields the wrong rank due to the accumulation of round-off errors. Entries that should have been zero could be replaced by small numbers. This is propagated, repeated, and magnified during the reduction. So, Gauss elimination can be unreliable in the computation of the rank. On the other hand, when we factor a matrix by using SVD, it can be shown that most of the round-off errors occur in the computation of Σ. So, we usually discard very small values for σ_i as 0s and count the remaining σ_is to get the rank.

Pseudoinverse

Let A be any $m \times n$ matrix. We may use the SVD of A to define an $n \times m$ matrix A^+ such that in the special case when A is invertible (so $m = n$) $A^+ = A^{-1}$. The matrix A^+ has several interesting properties, and it gives an optimal solution to the least squares problem studied in Section 8.4.

DEFINITION

Let $A = U\Sigma V^T$ be an SVD for an $m \times n$ matrix A. The **pseudoinverse**, or **Moore-Penrose inverse**, of A is the $n \times m$ matrix A^+ given by

$$A^+ = V\Sigma^+ U^T \tag{8.36}$$

where Σ^+ is the $n \times m$ matrix

$$\Sigma^+ = \begin{bmatrix} D^{-1} & \vdots & 0 \\ \cdots & \cdots & \cdots \\ 0 & \vdots & 0 \end{bmatrix}$$

D is, as before, the $r \times r$ diagonal with diagonal entries the positive singular values $\sigma_1 \geq \cdots \geq \sigma_r > 0$ of A.

■ EXAMPLE 34   Compute the pseudoinverse of

$$A = \begin{bmatrix} 2 & 0 & 0 \\ 0 & 0 & -6 \end{bmatrix}$$

SOLUTION   It is easy to find an SVD for A. For instance,

$$\begin{bmatrix} 2 & 0 & 0 \\ 0 & 0 & -6 \end{bmatrix} = \begin{bmatrix} 0 & 1 \\ -1 & 0 \end{bmatrix} \begin{bmatrix} 6 & 0 & 0 \\ 0 & 2 & 0 \end{bmatrix} \begin{bmatrix} 0 & 1 & 0 \\ 0 & 0 & 1 \\ 1 & 0 & 0 \end{bmatrix}^T$$

Hence, $\Sigma^+ = \begin{bmatrix} \frac{1}{6} & 0 \\ 0 & \frac{1}{2} \\ 0 & 0 \end{bmatrix}$, so

$$A^+ = \begin{bmatrix} 0 & 1 & 0 \\ 0 & 0 & 1 \\ 1 & 0 & 0 \end{bmatrix} \begin{bmatrix} \frac{1}{6} & 0 \\ 0 & \frac{1}{2} \\ 0 & 0 \end{bmatrix} \begin{bmatrix} 0 & 1 \\ -1 & 0 \end{bmatrix}^T$$

$$= \begin{bmatrix} \frac{1}{2} & 0 \\ 0 & 0 \\ 0 & -\frac{1}{6} \end{bmatrix}$$

Note that if A is $n \times n$ invertible, then $n = r$ and $\Sigma = D$. So, Σ is $n \times n$ and invertible. Moreover, $\Sigma\Sigma^+ = I_n$. Therefore,

$$AA^+ = AV\Sigma^+U^T = U\Sigma V^T V\Sigma^+U^T = U\Sigma\Sigma^+U^T = UU^T = I$$

This holds only if A is invertible. Hence, $A^+ = A^{-1}$, in this case.

Penrose showed that A^+ is the *unique* matrix B that satisfies the **Moore-Penrose conditions**:

1. $ABA = A$
2. $BAB = B$
3. $(AB)^T = AB$
4. $(BA)^T = BA$

It is instructive to verify these conditions for the pair (A, A^+) of Example 34. The verification for any pair (A, A^+) is discussed in the exercises. Although we do not prove it, we *use* the uniqueness part of Penrose's statement. So, if we can show that the pair (A, B) satisfies the conditions, then B is the unique pseudoinverse of A. So, $B = A^+$.

A very important use of the pseudoinverse A^+ is in the solution of the least squares problem, our next topic.

SVD and Least Squares

Recall from Section 8.4 that a least squares solution for the possibly inconsistent system $Ax = \mathbf{b}$ is a vector $\widetilde{\mathbf{x}}$ that minimizes the length of the error vector $\Delta = \mathbf{b} - A\widetilde{\mathbf{x}}$,

$$\|\Delta\| = \|\mathbf{b} - A\widetilde{\mathbf{x}}\| = \min$$

The vector $\widetilde{\mathbf{x}}$ in not necessarily unique. If A is $m \times n$ with rank $r < n$, then its nullity is greater than or equal to 1. In this case any vector of the form $\widetilde{\mathbf{x}} + \mathbf{z}$ with $\mathbf{z} \neq \mathbf{0}$ in Null(A)

will also be a least squares solution. Because

$$\mathbf{b} - A(\widetilde{\mathbf{x}} + \mathbf{z}) = \mathbf{b} - A\widetilde{\mathbf{x}} - A\mathbf{z} = \mathbf{b} - A\widetilde{\mathbf{x}}$$

If, however, we demand that $\widetilde{\mathbf{x}}$ has also minimum length, then such solution is unique and can be computed by using the Moore-Penrose inverse of A.

THEOREM 32

The least squares problem $A\mathbf{x} = \mathbf{b}$ has a unique least squares solution $\widetilde{\mathbf{x}}$ of minimal length given by

$$\widetilde{\mathbf{x}} = A^+\mathbf{b}$$

PROOF   Let $\mathbf{x}$ be an n-vector and let $\mathbf{y} = (y_1, \ldots, y_n)$ be $V^T\mathbf{x}$. The matrix U^T is orthogonal, because U is. Hence, $\|U^T\mathbf{z}\| = \|\mathbf{z}\|$ for any m-vector $\mathbf{z}$. We have

$$\|\mathbf{b} - A\mathbf{x}\| = \|\mathbf{b} - U\Sigma V^T\mathbf{x}\| = \|U^T\mathbf{b} - \Sigma V^T\mathbf{x}\| = S_1 + S_2$$

where

$$S_1 = (\mathbf{u}_1^T\mathbf{b} - \sigma_1 y_1)^2 + \cdots + (\mathbf{u}_r^T\mathbf{b} - \sigma_r y_r)^2$$
$$S_2 = (\mathbf{u}_{r+1}^T\mathbf{b})^2 + \cdots + (\mathbf{u}_m^T\mathbf{b})^2$$

because Σ has only r nonzero entries located at the upper left $r \times r$ block.

Because the sum S_2 is fixed, $\|\mathbf{b} - A\mathbf{x}\|$ is minimized if the sum S_1 is minimum. In fact, if we could choose $\mathbf{x} = (x_1, \ldots, x_n)$ such that

$$\mathbf{u}_i^T\mathbf{b} = \sigma_i y_i, \qquad i = 1, \ldots, r$$

then S_1 would be 0. So, all we need is an $\mathbf{x}$ of the form

$$\mathbf{x} = \left(V\frac{\mathbf{u}_1^T\mathbf{b}}{\sigma_1}, \ldots, V\frac{\mathbf{u}_r^T\mathbf{b}}{\sigma_r}, *, \ldots, * \right)$$

Any such $\mathbf{x}$ would be a least squares solution, because it minimizes $\|\mathbf{b} - A\mathbf{x}\|$. To get such an $\mathbf{x}$ of least magnitude, we have to set the last $n - r$ coordinates equal to 0. Hence,

$$\widetilde{\mathbf{x}} = \left(V\frac{\mathbf{u}_1^T\mathbf{b}}{\sigma_1}, \ldots, V\frac{\mathbf{u}_r^T\mathbf{b}}{\sigma_r}, 0, \ldots, 0 \right)$$

is the only least squares solution of minimal length. Moreover, we may rewrite $\widetilde{\mathbf{x}}$ as

$$\widetilde{\mathbf{x}} = V\Sigma^+U^T\mathbf{b} \Rightarrow \widetilde{\mathbf{x}} = A^+\mathbf{b}$$

■ **EXAMPLE 35**   Find the minimum length least squares solution of

$$\begin{bmatrix} 2 & 0 & 0 \\ 0 & 0 & -6 \end{bmatrix} \mathbf{x} = \begin{bmatrix} 1 \\ 2 \end{bmatrix}$$

SOLUTION   If A is the coefficient matrix, then, by Theorem 32 and Example 34,

$$\widetilde{x} = A^+ \begin{bmatrix} 1 \\ 2 \end{bmatrix} = \begin{bmatrix} \frac{1}{2} & 0 \\ 0 & 0 \\ 0 & -\frac{1}{6} \end{bmatrix} \begin{bmatrix} 1 \\ 2 \end{bmatrix} = \begin{bmatrix} \frac{1}{2} \\ 0 \\ -\frac{1}{3} \end{bmatrix}$$

The Polar Decomposition of a Square Matrix

An interesting and useful consequence of the SVD for a **square** matrix A is the **polar decomposition** of A.

THEOREM 33   **(Polar Decomposition)**

Any square matrix A can be factored as

$$A = PQ \tag{8.37}$$

where P is positive semidefinite and Q is orthogonal.

PROOF   If A is $n \times n$, then so are U, Σ, and V in an SVD of A. Therefore,

$$A = U\Sigma V^T = U\Sigma(U^T U)V^T = (U\Sigma U^T)UV^T$$

and we let $P = U\Sigma U^T$ and $Q = UV^T$. The verification that P is positive semidefinite and Q is orthogonal is left as an exercise.

The polar decomposition is analogous to writing a complex number z in polar form $z = re^{i\theta}$, where $r \geq 0$ is the magnitude of z and θ is its argument, with $|e^{i\theta}| = 1$. P plays the role of r and Q, the role of $e^{i\theta}$.

■ EXAMPLE 36   Find the polar decomposition of

$$A = \begin{bmatrix} -2 & 0 \\ 0 & -5 \end{bmatrix}$$

SOLUTION   From the SVD of A,

$$\begin{bmatrix} -2 & 0 \\ 0 & -5 \end{bmatrix} = \begin{bmatrix} 0 & -1 \\ 1 & 0 \end{bmatrix} \begin{bmatrix} 5 & 0 \\ 0 & 2 \end{bmatrix} \begin{bmatrix} 0 & 1 \\ -1 & 0 \end{bmatrix}^T$$

we set

$$P = \begin{bmatrix} 0 & -1 \\ 1 & 0 \end{bmatrix} \begin{bmatrix} 5 & 0 \\ 0 & 2 \end{bmatrix} \begin{bmatrix} 0 & -1 \\ 1 & 0 \end{bmatrix}^T = \begin{bmatrix} 2 & 0 \\ 0 & 5 \end{bmatrix}$$

and

$$Q = \begin{bmatrix} 0 & -1 \\ 1 & 0 \end{bmatrix} \begin{bmatrix} 0 & 1 \\ -1 & 0 \end{bmatrix}^T = \begin{bmatrix} -1 & 0 \\ 0 & -1 \end{bmatrix}$$

SVD with CAS

Maple

```
> evalf(Svd(matrix([[1,2],[3,4]]),U,V));
```

$$[5.464985704, .3659661906]$$

```
> evalf(evalm(U));evalf(evalm((V));
```

$$\begin{bmatrix} -.4045535848 & -.9145142957 \\ -.9145142957 & .4045535848 \end{bmatrix}$$

$$\begin{bmatrix} -.5760484368 & .8174155605 \\ -.8174155605 & -.5760484368 \end{bmatrix}$$

Mathematica

```
In[1]:=
    {U,S,V}=SingularValues[{{1.,2},{3,4}}]
Out[1]=
    {{{-0.404554, -0.914514},
      {-0.914514, 0.404554}},
     {5.46499, 0.365966},
     {{-0.576048, -0.817416},
      {0.817416, -0.576048}}}
```

MATLAB

```
>> [U,S,V]=svd([1 2; 3 4])
U =
     0.4046      0.9145
     0.9145     -0.4046
S =
     5.4650           0
          0      0.3660
V =
     0.5760     -0.8174
     0.8174      0.5760
```

Exercises 8.7

In Exercises 1–3 find the singular values of the matrix.

1. $\begin{bmatrix} 0 & 0 \\ 0 & -2 \\ 3 & 0 \end{bmatrix}$

2. $\begin{bmatrix} -2 & 0 & 0 \\ 0 & 0 & 5 \end{bmatrix}$

3. $\begin{bmatrix} 1 & 0 & 1 \\ 0 & 1 & 0 \\ 1 & 0 & 1 \end{bmatrix}$

In Exercises 4–11 find an SVD for the matrix.

4. $\begin{bmatrix} -2 & 0 \\ 0 & 0 \\ 0 & 5 \end{bmatrix}$

5. $\begin{bmatrix} -2 & 0 & 0 \\ 0 & 0 & 5 \end{bmatrix}$

6. $\begin{bmatrix} 2 & 0 & 0 \\ 0 & 4 & 0 \\ 0 & 0 & 6 \end{bmatrix}$

7. $\begin{bmatrix} 0 & 0 & 1 \\ 0 & 2 & 0 \\ 3 & 0 & 0 \end{bmatrix}$

8. $\begin{bmatrix} 1 & 0 & 1 \\ 0 & 1 & 0 \\ 1 & 0 & 1 \end{bmatrix}$

9. $\begin{bmatrix} 2 & 0 & 4 \\ 0 & 4 & 0 \\ 4 & 0 & 8 \end{bmatrix}$

10. $\begin{bmatrix} 1 & 6 & -4 \\ -2 & 6 & 2 \\ 2 & 3 & 4 \end{bmatrix}$

11. $\begin{bmatrix} 2 & 6 & -4 \\ -4 & 6 & 2 \\ 4 & 3 & 4 \end{bmatrix}$

In Exercises 12–13 find an SVD by working with the transpose of the matrix.

12. $\begin{bmatrix} 2 & -4 & 4 \\ 6 & 6 & 3 \\ -4 & 2 & 4 \end{bmatrix}$

13. $\begin{bmatrix} 2 & 1 & -2 \\ 0 & 0 & 0 \\ 6 & -6 & 3 \end{bmatrix}$

In Exercises 14–15 compute the pseudoinverses and verify the Moore-Penrose properties.

14. a. $\begin{bmatrix} -2 & 0 \\ 0 & 0 \\ 0 & 5 \end{bmatrix}^+$ **b.** $\begin{bmatrix} -2 & 0 & 0 \\ 0 & 0 & 5 \end{bmatrix}^+$

15. a. $\begin{bmatrix} 2 & 0 & 0 \\ 0 & 4 & 0 \\ 0 & 0 & 6 \end{bmatrix}^+$ **b.** $\begin{bmatrix} 0 & 0 & 1 \\ 0 & 2 & 0 \\ 3 & 0 & 0 \end{bmatrix}^+$

In Exercises 16–17 compute and compare A^+ and A^{-1}.

16. $\begin{bmatrix} 2 & 0 & 0 \\ 0 & 4 & 0 \\ 0 & 0 & 6 \end{bmatrix}$

17. $\begin{bmatrix} 0 & 0 & 1 \\ 0 & 2 & 0 \\ 3 & 0 & 0 \end{bmatrix}$

18. Let A be any matrix. Prove that the pair (A, A^+) satisfies the Moore-Penrose conditions. (*Hint:* Verify the conditions for (Σ, Σ^+) first.)

In Exercises 19–20 prove the identities by verifying the Moore-Penrose conditions.

19. $\begin{bmatrix} -3 & 0 \\ 0 & 0 \\ 0 & 4 \end{bmatrix}^+ = \begin{bmatrix} -\frac{1}{3} & 0 & 0 \\ 0 & 0 & \frac{1}{4} \end{bmatrix}$

20. $\begin{bmatrix} -2 & 6 \\ 1 & 6 \\ 2 & 3 \end{bmatrix}^+ = \begin{bmatrix} -\frac{2}{9} & \frac{1}{9} & \frac{2}{9} \\ \frac{2}{27} & \frac{2}{27} & \frac{1}{27} \end{bmatrix}$

21. Show that $A^{++} = A$. (*Hint:* Verify the Moore-Penrose conditions for (A^+, A).)

22. Show that $(A^T)^+ = (A^+)^T$. Conclude that if A is symmetric so is A^+. (*Hint:* Verify the Moore-Penrose conditions for $(A^T, (A^+)^T)$.)

In Exercises 23–25 solve the least squares problem for $A\mathbf{x} = \mathbf{b}$ by using A^+.

23. $A = \begin{bmatrix} -2 & 0 \\ 0 & 0 \\ 0 & 5 \end{bmatrix}$, $\mathbf{b} = \begin{bmatrix} 1 \\ 2 \\ 3 \end{bmatrix}$

24. $A = \begin{bmatrix} -2 & 0 & 0 \\ 0 & 0 & 5 \end{bmatrix}$, $\mathbf{b} = \begin{bmatrix} 1 \\ 2 \end{bmatrix}$

25. $A = \begin{bmatrix} -2 & 6 \\ 1 & 6 \\ 2 & 3 \end{bmatrix}$, $\mathbf{b} = \begin{bmatrix} 1 \\ 2 \\ 3 \end{bmatrix}$

In Exercises 26–27 compute the polar decomposition for each of the matrices.

26. a. $\begin{bmatrix} -2 & 0 \\ 0 & 3 \end{bmatrix}$ **b.** $\begin{bmatrix} -2 & 0 \\ 0 & -3 \end{bmatrix}$

27. a. $\begin{bmatrix} 1 & -1 \\ 1 & 1 \end{bmatrix}$ **b.** $\begin{bmatrix} 1 & 6 & -4 \\ -2 & 6 & 2 \\ 2 & 3 & 4 \end{bmatrix}$

8.8 Inner Products

Reader's Goals for This Section

1. To know the definition and basic properties of inner products.
2. To find more properties of inner products analogous to those for dot products.

In this section we discuss a very useful generalization of the dot product for vector spaces. Although the dot product in $\mathbf{R}^n$ and its properties are at the core of many theoretical and applied results, working directly with n-vectors can sometimes be restrictive. This is quite apparent when we deal with vector spaces of polynomials or functions, because these sets have their own natural notations.

We introduce a "dot product," called an *inner product*, that works with general vectors. To define it we use the basic properties of the dot product outlined by Theorem 4, Section 2.2. Because the remaining properties of the dot product were proved using this theorem, the properties of the inner product have proofs *identical* to the analogous ones for the dot product. So, the material of this section will seem quite familiar.

DEFINITION

An **inner product** on a (real) vector space V is a function that to each pair of vectors **u** and **v** of V associates a real number $\langle \mathbf{u}, \mathbf{v} \rangle$ satisfying the following properties, or **axioms**.

For any vectors **u**, **v**, **w** and any scalar c,

1. $\langle \mathbf{u}, \mathbf{v} \rangle = \langle \mathbf{v}, \mathbf{u} \rangle$ (**Symmetry axiom**)
2. $\langle \mathbf{u} + \mathbf{w}, \mathbf{v} \rangle = \langle \mathbf{u}, \mathbf{v} \rangle + \langle \mathbf{w}, \mathbf{v} \rangle$ (**Additivity axiom**)
3. $\langle c\mathbf{u}, \mathbf{v} \rangle = c\langle \mathbf{u}, \mathbf{v} \rangle$ (**Homogeneity axiom**)
4. $\langle \mathbf{u}, \mathbf{u} \rangle \geq 0$, and $\langle \mathbf{u}, \mathbf{u} \rangle = 0$ if and only if $\mathbf{u} = \mathbf{0}$ (**Positivity axiom**)

Any real vector space with an inner product is called an **inner product space**.

To prove that a vector space is an inner product space, first we must have a function that associates a number to each pair of vectors. Then we must verify the four axioms for this function. Usually, the part $\langle \mathbf{u}, \mathbf{u} \rangle = 0 \Rightarrow \mathbf{u} = \mathbf{0}$ of the positivity axiom is the hardest to verify.

The axioms of an inner product imply the following additional basic properties.

THEOREM 34

Let **u**, **v**, and **w** be any vectors in an inner product space and let c be any scalar.

1. $\langle \mathbf{u}, \mathbf{v} + \mathbf{w} \rangle = \langle \mathbf{u}, \mathbf{v} \rangle + \langle \mathbf{u}, \mathbf{w} \rangle$
2. $\langle \mathbf{u}, c\mathbf{v} \rangle = c\langle \mathbf{u}, \mathbf{v} \rangle$
3. $\langle \mathbf{u} - \mathbf{w}, \mathbf{v} \rangle = \langle \mathbf{u}, \mathbf{v} \rangle - \langle \mathbf{w}, \mathbf{v} \rangle$
4. $\langle \mathbf{u}, \mathbf{v} - \mathbf{w} \rangle = \langle \mathbf{u}, \mathbf{v} \rangle - \langle \mathbf{u}, \mathbf{w} \rangle$
5. $\langle \mathbf{0}, \mathbf{v} \rangle = \langle \mathbf{v}, \mathbf{0} \rangle = 0$

PROOF   We prove the first property and leave the remaining as exercises.

$$\begin{aligned}
\langle \mathbf{u}, \mathbf{v} + \mathbf{w} \rangle &= \langle \mathbf{v} + \mathbf{w}, \mathbf{u} \rangle && \text{by symmetry} \\
&= \langle \mathbf{v}, \mathbf{u} \rangle + \langle \mathbf{w}, \mathbf{u} \rangle && \text{by additivity} \\
&= \langle \mathbf{u}, \mathbf{v} \rangle + \langle \mathbf{u}, \mathbf{w} \rangle && \text{by symmetry}
\end{aligned}$$

■ EXAMPLE 37   Let $\mathbf{u} = (u_1, \ldots, u_n)$ and $\mathbf{v} = (v_1, \ldots, v_n)$ be any n-vectors. Show that the dot product of $\mathbf{R}^n$,

$$\langle \mathbf{u}, \mathbf{v} \rangle = \mathbf{u} \cdot \mathbf{v} = \mathbf{u}^T \mathbf{v} = u_1 v_1 + \cdots + u_n v_n$$

is an inner product.

SOLUTION   All axioms hold by Theorem 4, Section 2.2.

■ EXAMPLE 38   Let $\mathbf{u} = (u_1, u_2)$ and $\mathbf{v} = (v_1, v_2)$ be any 2-vectors. Show that

$$\langle \mathbf{u}, \mathbf{v} \rangle = 3u_1 v_1 + 4u_2 v_2$$

defines an inner product in $\mathbf{R}^2$.

SOLUTION

Symmetry: Symmetry holds because

$$\langle \mathbf{u}, \mathbf{v} \rangle = 3u_1 v_1 + 4u_2 v_2 = 3v_1 u_1 + 4v_2 u_2 = \langle \mathbf{v}, \mathbf{u} \rangle$$

Additivity: If $\mathbf{w} = (w_1, w_2)$, then

$$\begin{aligned}
\langle \mathbf{u} + \mathbf{w}, \mathbf{v} \rangle &= 3(u_1 + w_1)v_1 + 4(u_2 + w_2)v_2 \\
&= (3u_1 v_1 + 4u_2 v_2) + (3w_1 v_1 + 4w_2 v_2) \\
&= \langle \mathbf{u}, \mathbf{v} \rangle + \langle \mathbf{w}, \mathbf{v} \rangle
\end{aligned}$$

Homogeneity: For any scalar c,

$$\begin{aligned}
\langle c\mathbf{u}, \mathbf{v} \rangle &= 3(cu_1)v_1 + 4(cu_2)v_2 \\
&= c(3u_1 v_1 + 4u_2 v_2) \\
&= c\langle \mathbf{u}, \mathbf{v} \rangle
\end{aligned}$$

Positivity:

$$\langle \mathbf{u}, \mathbf{u} \rangle = 3u_1 u_1 + 4u_2 u_2 = 3u_1^2 + 4u_2^2 \geq 0$$

which also implies

$$\langle \mathbf{u}, \mathbf{u} \rangle = 0 \Leftrightarrow (u_1 = 0 \quad \text{and} \quad u_2 = 0) \Leftrightarrow \mathbf{u} = \mathbf{0}$$

Hence, all axioms hold and $\langle , \rangle$ defines an inner product.

We have just found an inner product of $\mathbf{R}^2$ other than the dot product. So, a vector space may have *several different inner products*.

Example 38 is a special case of Example 39.

■ EXAMPLE 39   (Weighted Dot Product)   Let $w_1, \ldots, w_n$ be any *positive* numbers and let $\mathbf{u} = (u_1, \ldots, u_n)$ and $\mathbf{v} = (v_1, \ldots, v_n)$ be any n-vectors. Show that

$$\langle \mathbf{u}, \mathbf{v} \rangle = w_1 u_1 v_1 + \cdots + w_n u_n v_n \tag{8.38}$$

defines an inner product in $\mathbf{R}^n$.

SOLUTION   All axioms are verified just as in Example 38.

The inner product of Example 39 is called the **weighted dot product** of $\mathbf{R}^n$ **with weights** $w_1, \ldots, w_n$. It is important that all weights $w_1, \ldots, w_n$ are positive. Otherwise, the positivity axiom may fail. Formula (8.38) may also be written in matrix notation as

$$\langle \mathbf{u}, \mathbf{v} \rangle = \mathbf{u}^T W \mathbf{v}, \qquad \text{where } W = \begin{bmatrix} w_1 & \cdots & 0 \\ \vdots & \ddots & \vdots \\ 0 & \cdots & w_n \end{bmatrix}$$

■ EXAMPLE 40   Let

$$A = \begin{bmatrix} a_1 & a_2 \\ a_3 & a_4 \end{bmatrix} \quad \text{and} \quad B = \begin{bmatrix} b_1 & b_2 \\ b_3 & b_4 \end{bmatrix}$$

be 2×2 matrices with real entries (i.e., elements of M_{22}). It is easy to show that

$$\langle A, B \rangle = a_1 b_1 + a_2 b_2 + a_3 b_3 + a_4 b_4$$

defines an inner product in M_{22}.

For example, if $A = \begin{bmatrix} 1 & 2 \\ -3 & 4 \end{bmatrix}$ and $B = \begin{bmatrix} 0 & -2 \\ -2 & 1 \end{bmatrix}$, then

$$\langle A, B \rangle = 1 \cdot 0 + 2 \cdot (-2) + (-3) \cdot (-2) + 4 \cdot 1 = 6$$

■ EXAMPLE 41   Let $p(x) = a_0 + a_1 x + \cdots + a_n x^n$ and $q(x) = b_0 + b_1 x + \cdots + b_n x^n$ be polynomials in P_n. It is easy to show that

$$\langle p, q \rangle = a_0 b_0 + a_1 b_1 + \cdots + a_n b_n$$

defines an inner product in P_n.

As an example, if $p(x) = 1 - x^2$ and $q(x) = -2x + x^2$, then

$$\langle p, q \rangle = 1 \cdot 0 + 0 \cdot (-2) + (-1) \cdot 1 = -1$$

■ EXAMPLE 42   Let $r_0, r_1, \ldots, r_n$ be $n + 1$ distinct real numbers and let $p(x)$ and $q(x)$ be any polynomials in P_n. Show that

$$\langle p, q \rangle = p(r_0) q(r_0) + \cdots + p(r_n) q(r_n)$$

defines an inner product in P_n.

SOLUTION   Axioms 1–3 are easily verified. For the positivity axiom we have

$$\langle p, p \rangle = p(r_0)^2 + \cdots + p(r_n)^2 \geq 0$$

and

$$\langle p, p \rangle = 0 \Leftrightarrow p(r_0) = 0, \ldots, p(r_n) = 0 \Leftrightarrow p = \mathbf{0}$$

because the polynomial p has degree at most n, so if it has more than n roots, it has to be the zero polynomial.

For example, let $r_0 = -2, r_1 = 0, r_2 = 1, p(x) = 1 - x^2$, and $q(x) = -2x + x^2$. Because

$$p(-2) = -3, \quad q(-2) = 8, \quad p(0) = 1, \quad q(0) = 0, \quad p(1) = 0, \quad q(1) = -1$$

we have

$$\langle p, q \rangle = p(-2)q(-2) + p(0)q(0) + p(1)q(1) = -24$$

Our next example generalizes the weighted dot product, and it is an important source of inner products.

■ EXAMPLE 43   Let A be any **positive definite** (hence, symmetric) $n \times n$ matrix. Show that for any n-vectors $\mathbf{u}$ and $\mathbf{v}$,

$$\langle \mathbf{u}, \mathbf{v} \rangle = \mathbf{u}^T A \mathbf{v}$$

defines an inner product of $\mathbf{R}^n$.

SOLUTION   We need to verify the four axioms.

Symmetry: Because $A = A^T$, we have

$$\langle \mathbf{u}, \mathbf{v} \rangle = \mathbf{u}^T A \mathbf{v} = \mathbf{u} \cdot A \mathbf{v} = A^T \mathbf{u} \cdot \mathbf{v}$$
$$= A \mathbf{u} \cdot \mathbf{v} = \mathbf{v} \cdot A \mathbf{u} = \mathbf{v}^T A \mathbf{u} = \langle \mathbf{v}, \mathbf{u} \rangle$$

Additivity:

$$\langle \mathbf{u} + \mathbf{w}, \mathbf{v} \rangle = (\mathbf{u} + \mathbf{w})^T A \mathbf{v}$$
$$= \mathbf{u}^T A \mathbf{v} + \mathbf{w}^T A \mathbf{v} = \langle \mathbf{u}, \mathbf{v} \rangle + \langle \mathbf{w}, \mathbf{v} \rangle$$

Homogeneity:

$$\langle c\mathbf{u}, \mathbf{v} \rangle = (c\mathbf{u})^T A \mathbf{v} = c\mathbf{u}^T A \mathbf{v} = c\langle \mathbf{u}, \mathbf{v} \rangle$$

Positivity: Because A is positive definite, so is the quadratic form $q(\mathbf{u}) = \mathbf{u}^T A \mathbf{u}$. Hence,

$$\langle \mathbf{u}, \mathbf{u} \rangle = \mathbf{u}^T A \mathbf{u} > 0 \qquad \text{for all } \mathbf{u} \neq \mathbf{0}$$

This verifies the last axiom.

■ EXAMPLE 44   Show that

$$\langle \mathbf{u}, \mathbf{v} \rangle = 6u_1v_1 - 2u_2v_1 - 2u_1v_2 + 3u_2v_2$$

defines an inner product in $\mathbf{R}^2$.

SOLUTION $\langle \mathbf{u}, \mathbf{v} \rangle$ can be written in the form $\mathbf{u}^T A \mathbf{v}$:

$$\langle \mathbf{u}, \mathbf{v} \rangle = \begin{bmatrix} u_1 & u_2 \end{bmatrix} \begin{bmatrix} 6 & -2 \\ -2 & 3 \end{bmatrix} \begin{bmatrix} v_1 \\ v_2 \end{bmatrix}$$

Because $\begin{bmatrix} 6 & -2 \\ -2 & 3 \end{bmatrix}$ has positive eigenvalues (2 and 7), it is positive definite by Theorem 29, Section 8.6. Hence, $\langle \mathbf{u}, \mathbf{v} \rangle$ defines as inner product of $\mathbf{R}^2$, by Example 43.

Note that if A is not positive definite, $\mathbf{u}^T A \mathbf{v}$ may not define an inner product. For example,

$$\langle \mathbf{u}, \mathbf{v} \rangle = \begin{bmatrix} u_1 & u_2 \end{bmatrix} \begin{bmatrix} 2 & -2 \\ -2 & 2 \end{bmatrix} \begin{bmatrix} v_1 \\ v_2 \end{bmatrix} = 2u_1v_1 - 2u_2v_1 - 2u_1v_2 + 2u_2v_2$$

is *not* an inner product, because

$$\langle (1, 1), (1, 1) \rangle = 0$$

$\begin{bmatrix} 2 & -2 \\ -2 & 2 \end{bmatrix}$ has eigenvalues 0 and 4, so it is not positive definite (it is positive semidefinite).

■ EXAMPLE 45   (Requires Calculus)   Let $f(x)$ and $g(x)$ be in $C[a, b]$, the vector space of the continuous real-valued functions defined on $[a, b]$. Then

$$\langle f, g \rangle = \int_a^b f(x)g(x)\, dx$$

defines an inner product on $C[a, b]$.

SOLUTION

Symmetry: We have

$$\langle f, g \rangle = \int_a^b f(x)g(x)\, dx = \int_a^b g(x)f(x)\, dx = \langle g, f \rangle$$

Additivity:

$$\langle f + h, g \rangle = \int_a^b (f(x) + h(x))g(x)\, dx$$

$$= \int_a^b (f(x)g(x) + h(x)g(x))\, dx$$

$$= \int_a^b f(x)g(x) + \int_a^b h(x)g(x)\, dx$$

$$= \langle f, g \rangle + \langle h, g \rangle$$

Homogeneity:

$$\langle cf, g \rangle = \int_a^b cf(x)\, dx = c \int_a^b f(x)\, dx = c\langle f, g \rangle$$

Positivity: For any function $f(x)$ of $C[a, b]$, $f(x)^2 \geq 0$. Hence,

$$\langle f, f \rangle = \int_a^b f(x)^2\, dx \geq 0$$

Let $g(x) = f(x)^2$. Then g is nonnegative and continuous, so by a theorem of calculus,

$$\int_a^b g(x)\, dx = 0 \Leftrightarrow g = \mathbf{0}$$

($\mathbf{0}$ is the zero function, i.e., $\mathbf{0}(x) = 0$ for all $x \in [a, b]$.) Thus,

$$\langle f, f \rangle = \int_a^b f(x)^2\, dx = 0 \Leftrightarrow f = \mathbf{0}$$

This concludes the proof of the positivity axiom.

Length and Orthogonality

In an inner product space we can define lengths, distances, and orthogonal vectors using formulas identical to those for the dot product.

DEFINITION

Let V be an inner product space. Two vectors $\mathbf{u}$ and $\mathbf{v}$ are called **orthogonal** if their inner product is zero.

$$\mathbf{u} \text{ and } \mathbf{v} \text{ are orthogonal if } \langle \mathbf{u}, \mathbf{v} \rangle = 0$$

The **norm** (or **length**, or **magnitude**) of $\mathbf{v}$ is the nonnegative number $\|\mathbf{v}\|$ defined by:

$$\|\mathbf{v}\| = \sqrt{\langle \mathbf{v}, \mathbf{v} \rangle} \tag{8.39}$$

The positive square root is defined, because $\langle \mathbf{v}, \mathbf{v} \rangle \geq 0$, by the positivity axiom. Equivalently,

$$\|\mathbf{v}\|^2 = \langle \mathbf{v}, \mathbf{v} \rangle \tag{8.40}$$

We also define the **distance** between two vectors $\mathbf{u}$ and $\mathbf{v}$ by

$$d(\mathbf{u}, \mathbf{v}) = \|\mathbf{u} - \mathbf{v}\| \tag{8.41}$$

Note that

$$d(\mathbf{0}, \mathbf{v}) = d(\mathbf{v}, \mathbf{0}) = \|\mathbf{v}\|$$

A vector with norm 1 is called a **unit** vector. The set $\mathbf{S}$ of all unit vectors of V is called the **unit circle** or the **unit sphere**.

$$\mathbf{S} = \{\mathbf{v}, \mathbf{v} \in V \text{ and } \|\mathbf{v}\| = 1\} \tag{8.42}$$

The unit sphere consists of all vectors of V of distance 1 from the origin. This is how the unit circle and sphere are defined in $\mathbf{R}^2$ and $\mathbf{R}^3$ with respect to the ordinary (dot product) norm, thus justifying the names. Note, however, that a unit circle in $\mathbf{R}^2$ may **not** have the graph of a circle in the Cartesian coordinate system.

■ **EXAMPLE 46** For the inner product $\langle \mathbf{u}, \mathbf{v} \rangle = 3u_1 v_1 + 4u_2 v_2$ of Example 38, do the following.

(a) Compute $\|(-2, 1)\|$.
(b) Compute $d(\mathbf{e}_1, \mathbf{e}_2)$.
(c) Show that $(4, 3)$ and $(1, -1)$ are orthogonal.
(d) Describe and sketch a graph of the unit circle.

SOLUTION

(a) We have

$$\|(-2, 1)\|^2 = \langle (-2, 1), (-2, 1) \rangle = 3(-2)(-2) + 4 \cdot 1 \cdot 1 = 16$$

Hence,

$$\|(-2, 1)\| = \sqrt{16} = 4$$

(b) Because

$$\|(1, 0) - (0, 1)\|^2 = \|(1, -1)\|^2 = 3 \cdot 1 \cdot 1 + 4 \cdot (-1) \cdot (-1) = 7$$

we have

$$d(\mathbf{e}_1, \mathbf{e}_2) = \sqrt{7}$$

(c) $(4, 3)$ and $(1, -1)$ are orthogonal with respect to this inner product (but not with respect to the dot product) because

$$\langle (4, 3), (1, -1) \rangle = 3 \cdot 4 \cdot 1 + 4 \cdot 3 \cdot (-1) = 0$$

(d) Because $\|(v_1, v_2)\| = 1$ is equivalent to $\|(v_1, v_2)\|^2 = 1$, we have

$$\mathbf{S} = \{(v_1, v_2), \quad \text{such that} \quad 3v_1^2 + 4v_2^2 = 1\} \subseteq \mathbf{R}^2$$

Thus the unit sphere (circle) with respect to this inner product looks like an **ellipse** in the coordinate system equipped with the ordinary dot product, angles, and distances (Fig. 8.18). ■

Basic Identities and Inequalities

The axioms can be combined with the identity of the following theorem to generalize familiar identities from Section 2.2.

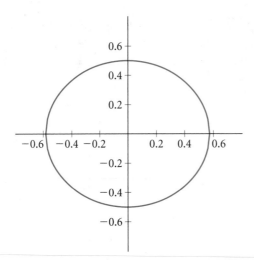

Figure 8.18   The unit circle for the inner product $3u_1v_1 + 4u_2v_2$.

THEOREM 35

Let V be an inner product space. For any vectors $\mathbf{u}$ and $\mathbf{v}$ of V we have

$$\|\mathbf{u} + \mathbf{v}\|^2 = \|\mathbf{u}\|^2 + \|\mathbf{v}\|^2 + 2\langle \mathbf{u}, \mathbf{v} \rangle \qquad (8.43)$$

PROOF   We have (explain which axioms are used)

$$\|\mathbf{u} + \mathbf{v}\|^2 = \langle \mathbf{u} + \mathbf{v}, \mathbf{u} + \mathbf{v} \rangle$$
$$= \langle \mathbf{u}, \mathbf{u} + \mathbf{v} \rangle + \langle \mathbf{v}, \mathbf{u} + \mathbf{v} \rangle$$
$$= \langle \mathbf{u}, \mathbf{u} \rangle + \langle \mathbf{u}, \mathbf{v} \rangle + \langle \mathbf{v}, \mathbf{u} \rangle + \langle \mathbf{v}, \mathbf{v} \rangle$$
$$= \langle \mathbf{u}, \mathbf{u} \rangle + 2\langle \mathbf{u}, \mathbf{v} \rangle + \langle \mathbf{v}, \mathbf{v} \rangle$$
$$= \|\mathbf{u}\|^2 + \|\mathbf{v}\|^2 + 2\langle \mathbf{u}, \mathbf{v} \rangle$$

Replacing $\mathbf{v}$ with $-\mathbf{v}$ in (8.43) yields

$$\|\mathbf{u} - \mathbf{v}\|^2 = \|\mathbf{u}\|^2 + \|\mathbf{v}\|^2 - 2\langle \mathbf{u}, \mathbf{v} \rangle \qquad (8.44)$$

Next, we have a generalization of the parallelogram law (See Exercise 22, Section 2.2).

THEOREM 36

(Parallelogram Law)

Let V be an inner product space. For any vectors $\mathbf{u}$ and $\mathbf{v}$ of V we have

$$\|\mathbf{u} + \mathbf{v}\|^2 + \|\mathbf{u} - \mathbf{v}\|^2 = 2\|\mathbf{u}\|^2 + 2\|\mathbf{v}\|^2$$

PROOF   Add Equations (8.43) and (8.44).

The identity of the following theorem gives the inner product in terms of the norm. (See Exercise 23, Section 2.2.)

THEOREM 37

(Polarization Identity)

Let V be an inner product space. For any vectors $\mathbf{u}$ and $\mathbf{v}$ of V we have

$$\langle \mathbf{u}, \mathbf{v} \rangle = \frac{1}{4}\|\mathbf{u} + \mathbf{v}\|^2 - \frac{1}{4}\|\mathbf{u} - \mathbf{v}\|^2$$

PROOF   Subtract Equation (8.44) from Equation (8.43) and solve for $\langle \mathbf{u}, \mathbf{v} \rangle$.

We also have a generalization of the Pythagorean theorem of Section 2.2.

THEOREM 38

(Pythagorean Theorem)

Let V be an inner product space. The vectors $\mathbf{u}$ and $\mathbf{v}$ of V are orthogonal if and only if

$$\|\mathbf{u} + \mathbf{v}\|^2 = \|\mathbf{u}\|^2 + \|\mathbf{v}\|^2$$

PROOF   Exercise.

The Cauchy-Schwarz-Bunyakovsky Inequality (CSBI)

One of the most useful consequences of the axioms is a generalization of the Cauchy-Schwarz inequality, which we call the Cauchy-Schwarz-Bunyakovsky inequality (CSBI).

THEOREM 39

(Cauchy-Schwarz-Bunyakovsky Inequality)

$$|\langle \mathbf{u}, \mathbf{v} \rangle| \leq \|\mathbf{u}\|\,\|\mathbf{v}\|$$

Furthermore, equality holds if and only if $\mathbf{u}$ and $\mathbf{v}$ are scalar multiples of each other.

PROOF   By Theorem 34 $\langle u, u \rangle \geq 0$

$$0 \leq \langle x\mathbf{u} + \mathbf{v}, x\mathbf{u} + \mathbf{v} \rangle = x^2 \langle \mathbf{u}, \mathbf{u} \rangle + 2x\langle \mathbf{u}, \mathbf{v} \rangle + \langle \mathbf{v}, \mathbf{v} \rangle \qquad (8.45)$$

for all scalars x. This is a quadratic polynomial $p(x) = ax^2 + bx + c$ with $a = \langle \mathbf{u}, \mathbf{u} \rangle$, $b = 2\langle \mathbf{u}, \mathbf{v} \rangle$, and $c = \langle \mathbf{v}, \mathbf{v} \rangle$. Because $a \geq 0$ and $p(x) \geq 0$ for all x, the graph of $p(x)$ is a parabola in the upper half-plane that opens upward. Hence, the parabola either is above the x-axis, in which case $p(x)$ has two complex roots, or is tangent to the x-axis, in which case $p(x)$ has a repeated real root. Therefore, $b^2 - 4ac \leq 0$. So,

$$\left(2\langle \mathbf{u}, \mathbf{v} \rangle\right)^2 - 4\langle \mathbf{u}, \mathbf{u} \rangle\langle \mathbf{v}, \mathbf{v} \rangle \leq 0 \qquad \text{or} \qquad 4\langle \mathbf{u}, \mathbf{v} \rangle^2 - 4\|\mathbf{u}\|^2\|\mathbf{v}\|^2 \leq 0$$

which implies the CSBI.

Equality holds if and only if $b^2 - 2ac = 0$ or if and only if $p(x)$ has a double real root, say, r. Hence, by Equation (8.45) with $x = r$, we have

$$\langle r\mathbf{u} + \mathbf{v}, r\mathbf{u} + \mathbf{v} \rangle = 0$$
$$\Leftrightarrow \qquad \|r\mathbf{u} + \mathbf{v}\| = 0$$
$$\Leftrightarrow \qquad r\mathbf{u} + \mathbf{v} = \mathbf{0}$$
$$\Leftrightarrow \qquad \mathbf{v} = -r\mathbf{u}$$

So, $\mathbf{v}$ is a scalar product of $\mathbf{u}$. This proves the last claim of the theorem.

■ **EXAMPLE 47**   Verify the CSBI for Example 42 with $r_0 = -2, r_1 = 0, r_2 = 1$, $p(x) = 1 - x^2$, and $q(x) = -2x + x^2$.

SOLUTION   We have

$$\langle p, q \rangle = p(-2)q(-2) + p(0)q(0) + p(1)q(1) = -24$$
$$\langle p, p \rangle = p(-2)^2 + p(0)^2 + p(1)^2 = 10$$
$$\langle q, q \rangle = q(-2)^2 + q(0)^2 + q(1)^2 = 65$$

Hence,

$$\|p\| = \sqrt{10} \qquad \|q\| = \sqrt{65}$$

and

$$|\langle p, q \rangle| = |-24| = 24 \le \sqrt{10} \cdot \sqrt{65} \cong 25.495$$

verifies the CSBI.

On the History of the CSBI

Cauchy[7] is credited with the inequality for vectors. Schwarz[8] is credited with the inequality for the integral inner products, as in Example 45. However, Bunyakovsky[9] proved and published the Schwarz inequality in a monograph 25 years before Schwarz.

---

[7]**Augustin Louis Cauchy** (1789–1857) was born in Paris, France, and died near Paris. He did important work in differential equations, infinite series, determinants, probability, permutation groups, and mathematical physics. In 1814 he published a memoir that became the foundation of the theory of complex functions. His work is known for its rigor. He published 789 papers and held positions at Faculté des Sciences, the Collège de France, and the École Polytechnique in Paris. Many terms and theorems bear his name. He was a faithful royalist and spent time in Switzerland, Turin, and Prague after refusing to take an oath of allegiance. He returned to Paris in 1838 and regained his position at the Academy of Sciences. In 1848 he regained his chair at the Sorbonne, which he held until his death.

[8]**Karl Herman Amandus Schwarz** (1843–1921) was born in Hermsdorf, Poland (now Germany), and died in Berlin, Germany. He studied chemistry at Berlin, but he changed to mathematics, where he obtained a doctorate. He held academic positions at Halle, Zurich, and Göttingen. He succeeded Weierstrass at Berlin and taught there until 1917. He worked on calculus of variation and on minimal surfaces. His memoir for Weierstrass's 70th birthday contains, among other important topics, the inequality for integrals now known as the Schwarz inequality.

[9]**Viktor Yakovlevich Bunyakovsky** (1804–1889) was born in Bar, Ukraine, and died in St. Petersburg, Russia. He was a professor at St. Petersburg between 1846 and 1880. He published more than

As an application of Theorem 35 and the CSBI, we have the useful *triangle inequality*.

THEOREM 40   (The Triangle Inequality)

$$\|u + v\| \le \|u\| + \|v\|$$

$$|u+v| \le |u| + |v|$$

PROOF

$$\|u + v\|^2 = \|u\|^2 + \|v\|^2 + 2\langle u, v \rangle \qquad \text{By Theorem 35}$$
$$\le \|u\|^2 + \|v\|^2 + 2\|u\|\,\|v\| \qquad \text{By CSBI}$$
$$= (\|u\| + \|v\|)^2$$

Therefore, $\|u + v\| \le \|u\| + \|v\|$.

The Gram-Schmidt Process

The Gram-Schmidt Process can be easily extended to general inner products, thus establishing the existence of orthogonal bases for finite-dimensional inner product spaces. The formulas are the same as before, except that we replace the dot product with a general inner product.

■ EXAMPLE 48   (Generalized Gram-Schmidt)   Find an orthogonal basis of P_2 starting with $1, x, x^2$ and using the inner product of Example 42 with $r_0 = 0, r_1 = 1$, $r_2 = 2$.

SOLUTION   Let $p_1 = 1$. Because

$$\langle 1, 1 \rangle = 1^2 + 1^2 + 1^2 = 3, \qquad \langle x, 1 \rangle = 0 \cdot 1 + 1 \cdot 1 + 2 \cdot 1 = 3$$

we let

$$p_2 = x - \frac{\langle x, 1 \rangle}{\langle 1, 1 \rangle} 1 = x - 1$$

Likewise,

$$\langle x - 1, x - 1 \rangle = 2, \qquad \langle x^2, 1 \rangle = 5, \qquad \langle x^2, x - 1 \rangle = 4$$

Set

$$p_3 = x^2 - \frac{\langle x^2, 1 \rangle}{\langle 1, 1 \rangle} 1 - \frac{\langle x^2, x - 1 \rangle}{\langle x - 1, x - 1 \rangle} (x - 1)$$

$$= x^2 - 2x + \frac{1}{3}$$

---

150 papers in mathematics and in mechanics. He did important work in number theory and he proved and published the Schwarz inequality in 1859, 25 years before Schwarz. He also worked in geometry and hydrostatistics.

Therefore,

$$\{p_1, p_2, p_3\} = \left\{1, x - 1, x^2 - 2x + \frac{1}{3}\right\}$$

is an orthogonal basis of P_2 with respect to the given inner product.

The best approximation theorem (Section 8.2) also generalizes easily to inner product spaces. We leave it to the reader to write out the details. This generalization is particularly useful when we try to approximate a function by using other functions. The kind of approximation depends on the inner product we use.

■ **EXAMPLE 49** (Generalized Best Approximation)   Referring to Example 48 and its solution, find a polynomial $\widetilde{p}$ in $P_1 = \text{Span}\{1, x\} \subseteq P_2$ that best approximates $p(x) = 2x^2 - 1$.

SOLUTION   By the solution of Example 48, $\{p_0, p_1\} = \{1, x - 1\}$ is an orthogonal basis of P_1. Because

$$\langle 2x^2 - 1, 1\rangle = (-1) \cdot 1 + 1 \cdot 1 + 7 \cdot 1 = 7$$

$$\langle 2x^2 - 1, x - 1\rangle = (-1) \cdot (-1) + 1 \cdot 0 + 7 \cdot 1 = 8$$

we have

$$\widetilde{p} = p_{\text{pr}} = \frac{\langle p, p_0\rangle}{\langle p_0, p_0\rangle} p_0 + \frac{\langle p, p_1\rangle}{\langle p_1, p_1\rangle} p_1$$

$$= \frac{7}{3} + \frac{8}{2}(x - 1) = 4x - \frac{5}{3}$$

Hence, $4x - \frac{5}{3}$ of P_1 best approximates $2x^2 - 1$ with respect to the given inner product (Fig. 8.19).

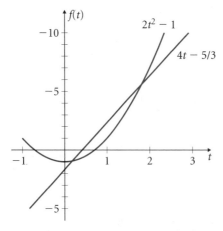

Figure 8.19 $4t - \frac{5}{3}$ is the best linear polynomial approximating $2t^2 - 1$ when distance is measured at 0, 1, 2.

Exercises 8.8

Let $\mathbf{u} = \begin{bmatrix} 1 \\ -2 \end{bmatrix}$ and $\mathbf{v} = \begin{bmatrix} 4 \\ 3 \end{bmatrix}$. In Exercises 1–2:

(a) Compute $\langle \mathbf{u}, \mathbf{v} \rangle$, $\|\mathbf{u}\|$, $\|\mathbf{v}\|$, $\|\mathbf{u} + \mathbf{v}\|$;
(b) Compute the distance $d(\mathbf{u}, \mathbf{v})$;
(c) Verify the CSBI for $\mathbf{u}$ and $\mathbf{v}$;
(d) Verify the triangle inequality for $\mathbf{u}$ and $\mathbf{v}$;
(e) Verify the polarization identity for $\mathbf{u}$ and $\mathbf{v}$.

1. The inner product is that of Example 38.
2. The inner product is that of Example 39 with $W = \begin{bmatrix} 2 & 0 \\ 0 & 5 \end{bmatrix}$.
3. Referring to Exercise 1, find a vector orthogonal to $\mathbf{u}$.
4. Referring to Exercise 2, find a vector orthogonal to $\mathbf{u}$.

Consider the inner product of Example 40 and let
$$A = \begin{bmatrix} 1 & 0 \\ 0 & 1 \end{bmatrix}, B = \begin{bmatrix} -1 & 0 \\ 0 & 1 \end{bmatrix}, C = \begin{bmatrix} -\frac{1}{2} & \frac{1}{2} \\ \frac{1}{2} & \frac{1}{2} \end{bmatrix}.$$

5. Find the orthogonal pairs between A, B, and C.
6. For the orthogonal pairs found in Exercise 5, verify the Pythagorean theorem.
7. Construct orthonormal pairs out of the orthogonal ones found in Exercise 5.
8. Which of A, B, and C are unit "vectors"?

In Exercises 9–15 determine whether, for the given matrix A, the function

$$f(\mathbf{u}, \mathbf{v}) = \mathbf{u}^T A \mathbf{v}, \qquad \mathbf{u}, \mathbf{v} \in \mathbf{R}^2$$

defines an inner product of $\mathbf{R}^2$ as follows: Refer to Example 43 and check to see if A is positive definite (hence, symmetric). If A is not positive definite, find an axiom of the definition of inner product that fails.

9. $A = \begin{bmatrix} 1 & 0 \\ 0 & 2 \end{bmatrix}$
10. $A = \begin{bmatrix} 1 & 0 \\ 0 & -2 \end{bmatrix}$
11. $A = \begin{bmatrix} 1 & -1 \\ -1 & 2 \end{bmatrix}$
12. $A = \begin{bmatrix} 1 & 2 \\ 2 & 1 \end{bmatrix}$
13. $A = \begin{bmatrix} 2 & 2 \\ 2 & 2 \end{bmatrix}$
14. $A = \begin{bmatrix} -1 & 2 \\ 2 & 2 \end{bmatrix}$
15. $A = \begin{bmatrix} 8 & 7 \\ 1 & 2 \end{bmatrix}$

16. Use a 3×3 positive definite matrix of your choice to define your own inner product in $\mathbf{R}^3$.

17. If A is a positive definite (thus symmetric) $n \times n$ matrix, show that
$$f(\mathbf{u}, \mathbf{v}) = \mathbf{v}^T A \mathbf{u}, \qquad \mathbf{u}, \mathbf{v} \in \mathbf{R}^n$$
defines an inner product in $\mathbf{R}^n$.

In Exercises 18–20 let $p(x) = 1 - 2x^2$, $q(x) = -2x + x^2$ in P_2.

(a) Compute $\langle p, q \rangle$, $\|p\|$, $\|q\|$, $\|p + q\|$.
(b) Compute the distance $d(q, p)$.
(c) Verify the CSBI for p and q.
(d) Verify the triangle inequality for p and q.

18. The inner product in P_2 is that of Example 41.
19. The inner product in P_2 is that of Example 42 with $r_0 = -3, r_1 = 0, r_2 = 2$.
20. (**Requires Calculus**) The inner product is that of Example 45 with $a = -1$ and $b = 1$.
21. Referring to Exercise 18, find a vector orthogonal to p.
22. Referring to Exercise 19, find a vector orthogonal to p.
23. Referring to Exercise 20, find a vector orthogonal to p.
24. Consider $\mathbf{R}^2$ equipped with the inner product of Example 44. Is the standard basis $\{\mathbf{e}_1, \mathbf{e}_2\}$ an orthogonal basis?
25. Complete the proof of Theorem 34.
26. Complete the proof of Theorem 38.
27. Suppose that $T : \mathbf{R}^n \rightarrow \mathbf{R}^n$ is an invertible linear transformation. Show that the assignment
$$\langle \mathbf{u}, \mathbf{v} \rangle = T(\mathbf{u}) \cdot T(\mathbf{v}), \qquad \mathbf{u}, \mathbf{v} \in \mathbf{R}^n$$
defines an inner product in $\mathbf{R}^n$.

Projections

NOTE

1. We compute orthogonal projections with respect to inner products using the same formula as before, where the dot product is replaced with the given inner product.
2. The distance from a vector to $\mathbf{u}$, a subspace W in an inner product space, is the norm $\|\mathbf{u}_c\|$ of the orthogonal component $\mathbf{u}_c$ of $\mathbf{u}$ with respect to W.

28. Consider $\mathbf{R}^2$ equipped with the inner product of Example 44. Find the orthogonal projection of $(1, 1)$ onto the line l through $\mathbf{0}$ and $\mathbf{e}_1$.

29. Referring to Exercise 28, find the distance from $(1, 1)$ to l.

30. Referring to Example 48, find the orthogonal projection (thus, the best approximation) of $x^2 + x + 1$ with respect to the subspace $W = \text{Span}\{p, q\}$ of P_2, where

$$p = x - 1, \qquad q = x^2 - 2x + \frac{1}{3}$$

   (Note that $\{p, q\}$ is an orthogonal basis of W.)

31. Referring to Exercise 30, find the distance from $x^2 + x + 1$ to W.

Gram-Schmidt

32. Consider $\mathbf{R}^2$ equipped with the inner product of Example 44. Apply the Gram-Schmidt process to the standard basis $\{\mathbf{e}_1, \mathbf{e}_2\}$ to find an orthogonal basis for this inner product.

33. Apply the Gram-Schmidt process to find an orthogonal basis of P_2, starting with $1, x, x^2$ and using the inner product of Example 42 with $r_0 = -2, r_1 = 0, r_2 = 2$.

Exercises 34–35 require calculus.

34. Show that the first four **Legendre** polynomials form an orthogonal basis for P_3 for the inner product of Example 45 with $a = -1$ and $b = 1$.

$$\mathcal{L} = \left\{1, x, \frac{3}{2}x^2 - \frac{1}{2}, \frac{5}{2}x^3 - \frac{3}{2}x\right\}$$

35. Use the Legendre polynomials and the inner product of Exercise 34 to find an orthonormal basis for P_3.

8.9   Applications and Additional Topics

Reader's Goal for This Section

To see some of the applications of dot products.

In this section we discuss a few of the many applications of the dot products. Not only are the applications of orthogonality, least squares, etc., very interesting, but they can also be fun. Let us begin with how the NFL rates quarterbacks.

The NFL Rating of Quarterbacks

We present the essence of an interesting article by Roger W. Johnson, published in *The College Mathematics Journal*, 5 (November 1993). In this article Johnson finds the formula for the rating of a quarterback.[10] Although we follow Johnson's exposition almost to the letter, we use slightly more recent data, partly to see if the old formula is still valid.

Tables 8.1 and 8.2 were taken from *The Sports Illustrated 1995 Sports Almanac* article by Peter King. Table 8.1 shows the 1993 NFL individual leading passers for the American Football Conference.

---

[10]Given a table of all-time leading passers through the 1989–1990 season.

TABLE 8.1	Player	Att.	Comp.	Yards	TD	Int.	Rating
1993 American	Elway	551	348	4030	25	10	92.8
Football	Montana	298	181	2144	13	7	87.4
Conference	Testaverde	230	130	1797	14	9	85.7
	Esiason	473	288	3421	16	11	84.5
	Mitchell	233	133	1773	12	8	84.2
	Hostetler	419	236	3242	14	10	82.5
	Kelly	470	288	3382	18	18	79.9
	O'Donnell	486	270	3208	14	7	79.5
	George	407	234	2526	8	6	76.3
	DeBerg	227	136	1707	7	10	75.3

Table 8.2 displays the same information for the National Football Conference.

TABLE 8.2	Player	Att.	Comp.	Yards	TD	Int.	Rating
1993 National	Young	462	314	4023	29	16	101.5
Football	Aikman	392	271	3100	15	6	99.0
Conference	Simms	400	247	3038	15	9	88.3
	Brister	309	181	1905	14	5	84.9
	Hebert	430	263	2978	24	17	84.0
	Buerlein	418	258	3164	18	17	82.5
	McMahon	331	200	1967	9	8	76.2
	Favre	522	318	3303	19	24	72.2
	Harbaugh	325	200	2002	7	11	72.1
	Wilson	388	221	2457	12	15	70.1

Given the attempts (Att.), completions (Comp.), yards, touchdowns (TD), interceptions (Int.), and ratings, we want to find a formula for the rating. As Johnson points out, it is known that this rating depends on the percentages of the completions, touchdowns, and interceptions and also on the average gain per pass attempt (computed in Tables 8.3 and 8.4). However, there seems to be no published formula of the ratings. Let us assume that the rating depends *linearly* on these four quantities and a constant, say,

$$\text{Rating} = x_1 + x_2 \,(\%\ \text{Comp.}) + x_3 \,(\%\ \text{TD}) + x_4 \,(\%\ \text{Int.}) + x_5 \,(\text{Yards/Att.}) \quad (8.46)$$

We want to compute the unknown coefficients $\mathbf{x} = (x_1, \ldots, x_5)$ using Tables 8.3 and 8.4. This yields a system of 20 equations (10 per table) and 5 unknowns. Let A be the coefficient matrix of this system. Then

$$A\mathbf{x} = \mathbf{b} \quad (8.47)$$

where $\mathbf{b}$ is the vector of all the ratings. Note that A has a first column of 1s; the remaining columns are the percentages of the tables.

It seems reasonable to use the data from only five players to get a square system in $\mathbf{x}$. However, if we compare the solutions using the first five players of the American conference and the last five players of the National conference, the two answers differ

TABLE 8.3
American Football
Conference

Player	% Comp.	% TDs	% Int.	Yd/Att.
Elway	63.1579	4.5372	1.8149	7.3140
Montana	60.7383	4.3624	2.3490	7.1946
Testaverde	56.5217	6.0870	3.9130	7.8130
Esiason	60.8879	3.3827	2.3256	7.2326
Mitchell	57.0815	5.1502	3.4335	7.6094
Hostetler	56.3246	3.3413	2.3866	7.7375
Kelly	61.2766	3.8298	3.8298	7.1957
O'Donnell	55.5556	2.8807	1.4403	6.6008
George	57.4939	1.9656	1.4742	6.2064
DeBerg	59.9119	3.0837	4.4053	7.5198

TABLE 8.4
National Football
Conference

Player	% Comp.	% TDs	% Int.	Yd/Att.
Young	67.9654	6.2771	3.4632	8.7078
Aikman	69.1327	3.8265	1.5306	7.9082
Simms	61.7500	3.7500	2.2500	7.5950
Brister	58.5761	4.5307	1.6181	6.1650
Hebert	61.1628	5.5814	3.9535	6.9256
Buerlein	61.7225	4.3062	4.0670	7.5694
McMahon	60.4230	2.7190	2.4169	5.9426
Favre	60.9195	3.6398	4.5977	6.3276
Harbaugh	61.5385	2.1538	3.3846	6.1600
Wilson	56.9588	3.0928	3.8660	6.3325

by $(0.82, -0.04, 0, -0.2, 0.28)$ (rounded to two decimal places). Clearly System (8.47) is inconsistent.

So, we have to find an optimal solution by using least squares. The normal equations are

$$A^T A\mathbf{x} = A^T \mathbf{b}$$

Using MATLAB (bank notation format), we get the system

$$
\begin{bmatrix}
20.00 & 1{,}209.10 & 78.50 & 58.52 & 142.06 \\
1{,}209.10 & 73{,}333.88 & 4{,}766.20 & 3{,}536.22 & 8{,}609.83 \\
78.50 & 4{,}766.20 & 335.31 & 236.62 & 568.22 \\
58.52 & 3{,}536.22 & 236.62 & 192.91 & 417.78 \\
142.06 & 8{,}609.83 & 568.22 & 417.78 & 1{,}019.51
\end{bmatrix}
\mathbf{x} =
\begin{bmatrix}
1{,}658.90 \\
100{,}652.76 \\
6{,}634.23 \\
4{,}794.16 \\
11{,}871.57
\end{bmatrix}
$$

and its solution

$$\mathbf{x} = (2.0589, 0.8321, 3.3178, -4.1666, 4.1884)$$

Therefore, the formula for the ratings is

$$\text{Rating} = 2.0589 + 0.8321\,(\% \text{ Comp.}) + 3.3178\,(\% \text{ TD})$$

$$-4.1666\,(\% \text{ Int.}) + 4.1884\,(\text{Yd/Att.})$$

In fact, if we compute the product $A\mathbf{x}$, we get all the ratings to the displayed accuracy.

These coefficients with the four decimal places are not very handy. There is a rational approximation found in Johnson's article,

$$\text{Rating} = \frac{1}{24}(50 + 20\,(\%\ \text{Comp.}) + 80\,(\%\ \text{TD}) - 100\,(\%\ \text{Int.}) + 100\,(\text{Yd/Att.}))$$

that yields the same accuracy. It seems reasonable to assume that this is the correct formula for the ratings.

REMARK   We used data from both conferences to get a more accurate least squares approximation.

Trend Analysis and Least Squares Polynomials

In Section 8.4 we saw how to find a least squares line fitting planar data points. However, not all sets of data can be satisfactorily approximated by straight lines. Often we have to use quadratic or cubic polynomials or even more complicated functions. The question then arises: Which function is suitable in a given situation and how do we compute it? This is the subject of a **trend analysis**. Let us explore the polynomial fitting of data.

We seek a polynomial $q(x)$ of degree at most $n - 1$ that best approximates a set of points $(a_1, b_1), \ldots, (a_m, b_m)$. Let

$$q(x) = \alpha_0 + \alpha_1 x + \cdots + \alpha_{n-1} x^{n-1}$$

Evaluation of q at the x-coordinates of the points may not quite yield the corresponding y-coordinates. Suppose that the errors are $\delta_1, \ldots, \delta_m$. We have

$$b_1 = \alpha_0 + \alpha_1 a_1 + \cdots + \alpha_{n-1} a_1^{n-1} + \delta_1$$
$$b_2 = \alpha_0 + \alpha_1 a_2 + \cdots + \alpha_{n-1} a_2^{n-1} + \delta_2$$
$$\vdots$$
$$b_m = \alpha_0 + \alpha_1 a_m + \cdots + \alpha_{n-1} a_m^{n-1} + \delta_m$$

In matrix notation,

$$\mathbf{b} = A\alpha + \Delta$$

where

$$\mathbf{b} = \begin{bmatrix} b_1 \\ \vdots \\ b_m \end{bmatrix}, \qquad \alpha = \begin{bmatrix} \alpha_0 \\ \vdots \\ \alpha_{n-1} \end{bmatrix}, \qquad \Delta = \begin{bmatrix} \delta_1 \\ \vdots \\ \delta_m \end{bmatrix}.$$

and A is the $m \times n$ coefficient matrix

$$A = \begin{bmatrix} 1 & a_1 & a_1^2 & \cdots & a_1^{n-1} \\ \vdots & \vdots & \vdots & \ddots & \vdots \\ 1 & a_m & a_m^2 & \cdots & a_m^{n-1} \end{bmatrix}.$$

The goal is to find a vector α that minimizes the length of the error vector $\|\Delta\| = \|\mathbf{b} - A\alpha\|$. As we saw in Section 8.4, this amounts to solving for α the normal equations

$$A^T A \alpha = A^T \mathbf{b}$$

We know that least squares solutions always exist, but what about uniqueness? After all we want *one* polynomial that best fits the data. The answer is in the next theorem.

THEOREM 41

Let the data points $(a_1, b_1), \ldots, (a_m, b_m)$ have all distinct x-coordinates (i.e., all a_is are different). For any positive integer $n \le m$, there is a *unique* polynomial

$$q(x) = \alpha_0 + \alpha_1 x + \cdots + \alpha_{n-1} x^{n-1}$$

that minimizes $\|\Delta\|$.

PROOF   Exercise.

The unique polynomial of Theorem 41 is called the **least squares polynomial** of degree $n - 1$ for these points. A very interesting special case is when A is square, so $m = n$.

THEOREM 42

Let the n data points $(a_1, b_1), \ldots, (a_n, b_n)$ have all distinct x-coordinates. Then $\Delta = \mathbf{0}$ and the unique least squares polynomial

$$q(x) = \alpha_0 + \alpha_1 x + \cdots + \alpha_{n-1} x^{n-1}$$

actually *passes* through all the points. So,

$$q(a_i) = b_i, \qquad i = 1, \ldots, n$$

PROOF   Exercise.

The unique polynomial of Theorem 42 is called the **interpolating polynomial** for the points. Note that if A is square, its transpose

$$A^T = \begin{bmatrix} 1 & 1 & \cdots & 1 \\ a_1 & a_2 & \cdots & a_n \\ \vdots & \vdots & \ddots & \vdots \\ a_1^{n-1} & a_2^{n-1} & \cdots & a_n^{n-1} \end{bmatrix}$$

is the Vandermonde matrix studied in Section 6.6.

■ EXAMPLE 50   Find the least squares (a) line and (b) quadratic for the C grades data of our linear algebra instructor.

Semester	1	2	3	4	5	6
Percentage of Cs	0.20	0.25	0.25	0.35	0.35	0.30

SOLUTION

(a) To compute the least squares line, let

$$A = \begin{bmatrix} 1 & 1 \\ 1 & 2 \\ 1 & 3 \\ 1 & 4 \\ 1 & 5 \\ 1 & 6 \end{bmatrix}, \quad b = \begin{bmatrix} 0.20 \\ 0.25 \\ 0.25 \\ 0.35 \\ 0.35 \\ 0.30 \end{bmatrix}$$

Then,

$$A^T A \alpha = \begin{bmatrix} 6 & 21 \\ 21 & 91 \end{bmatrix} \begin{bmatrix} \alpha_0 \\ \alpha_1 \end{bmatrix} = A^T b = \begin{bmatrix} 1.7 \\ 6.4 \end{bmatrix}$$

with solution $\alpha = (0.19333, 0.0257)$. So, the least squares line is

$$y = 0.19333 + 0.0257x$$

(b) To compute the least squares quadratic, let

$$A = \begin{bmatrix} 1 & 1 & 1 \\ 1 & 2 & 4 \\ 1 & 3 & 9 \\ 1 & 4 & 16 \\ 1 & 5 & 25 \\ 1 & 6 & 36 \end{bmatrix}$$

The normal equations are

$$A^T A \alpha = \begin{bmatrix} 6 & 21 & 91 \\ 21 & 91 & 441 \\ 91 & 441 & 2275 \end{bmatrix} \begin{bmatrix} \alpha_0 \\ \alpha_1 \\ \alpha_2 \end{bmatrix} = A^T b = \begin{bmatrix} 1.7 \\ 6.4 \\ 28.6 \end{bmatrix}$$

with unique solution $\alpha = (0.11, 0.0882, -0.0089)$. Hence, the least squares quadratic is

$$y = 0.11 + 0.0882x - 0.0089x^2$$

Figure 8.20 shows the two least squares curves and the points. If the quadratic is the better approximation (it seems to be), then the number of Cs is expected to decrease overall.

Continuous Least Squares (Requires Calculus)

We are by now experienced in computing least squares fittings for a finite set of points (the discrete case). What do we do if we have an infinite continuous set of points and want to find a curve fitting this set?

Suppose we want the straight line $y = b + mx$ that best approximates the function $f(x) = x^2$ on the interval $[0, 1]$.

Unless we pick finitely many points, we cannot use the ordinary dot product anymore. We can, however, use the inner product for continuous functions of Example 45, Section 8.8.

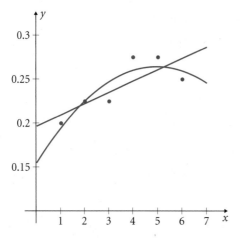

Figure 8.20   Least squares line and quadratic for the same data.

We need to find b and m that minimize the error vector

$$\Delta = x^2 - (b + mx) \tag{8.48}$$

using the integral inner product. Equivalently, we have to minimize

$$\|\Delta\|^2 = \int_0^1 \left(x^2 - (b + mx)\right)^2 dx$$

Equation (8.48) can be written in vector notation as

$$\Delta = x^2 - A\mathbf{x} \tag{8.49}$$

where

$$A = \begin{bmatrix} 1 & x \end{bmatrix} \qquad \text{and} \qquad \mathbf{x} = \begin{bmatrix} b \\ m \end{bmatrix}$$

We know from Section 8.4 that if $\widetilde{\mathbf{x}} = \begin{bmatrix} \widetilde{b} \\ \widetilde{m} \end{bmatrix}$ is a least squares solution for the system $A\mathbf{x} = x^2$, then $A\widetilde{\mathbf{x}}$ must be the unique projection of x^2 onto $\mathrm{Col}(A)$. And we have a formula for the projection, provided we use an orthogonal basis of $\mathrm{Col}(A)$. It is easy to see that $\{1, x\}$ yields the orthogonal basis $\{1, x - \frac{1}{2}\}$, using Gram-Schmidt process. So,

$$A\widetilde{\mathbf{x}} = x^2_{\mathrm{pr}} = \frac{\langle x^2, 1 \rangle}{\langle 1, 1 \rangle} 1 + \frac{\langle x^2, x - \frac{1}{2} \rangle}{\langle x - \frac{1}{2}, x - \frac{1}{2} \rangle} \left(x - \frac{1}{2} \right)$$

$$= -\frac{1}{6} + x$$

since

$$\int_0^1 x^2 \cdot 1 \, dx = \frac{1}{3} \qquad \int_0^1 x^2 \cdot \left(x - \frac{1}{2} \right) dx = \frac{1}{12}$$

$$\int_0^1 1 \cdot 1 \, dx = 1 \qquad \int_0^1 \left(x - \frac{1}{2} \right) \cdot \left(x - \frac{1}{2} \right) dx = \frac{1}{12}$$

Therefore,

$$A\widetilde{\mathbf{x}} = \begin{bmatrix} 1 & x \end{bmatrix} \begin{bmatrix} \widetilde{b} \\ \widetilde{m} \end{bmatrix} = -\frac{1}{6} + x$$

Hence, $\widetilde{b} = -\frac{1}{6}$ and $\widetilde{m} = 1$. So, the least squares line that best approximates x^2 on $[0, 1]$ is

$$y = -\frac{1}{6} + x$$

(see Fig. 8.21). Note that the answer strongly depends on the interval we choose.

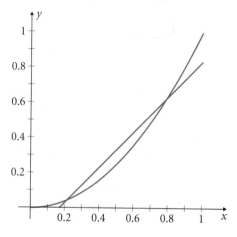

Figure 8.21   The least squares line for x^2 on $[0, 1]$.

Actually, we do not need to orthogonalize. We can use the corresponding "normal equations"

$$A^T A \widetilde{\mathbf{x}} = A^T x^2$$

The matrix multiplication this time is a funny one. It uses the current inner product, not the dot product. So, by $A^T A$ we mean

$$A^T A = \begin{bmatrix} 1 \\ x \end{bmatrix} \begin{bmatrix} 1 & x \end{bmatrix} = \begin{bmatrix} \langle 1, 1 \rangle & \langle 1, x \rangle \\ \langle x, 1 \rangle & \langle x, x \rangle \end{bmatrix}$$

and $A^T x^2$ is

$$A^T x^2 = \begin{bmatrix} 1 \\ x \end{bmatrix} \begin{bmatrix} x^2 \end{bmatrix} = \begin{bmatrix} \langle 1, x^2 \rangle \\ \langle x, x^2 \rangle \end{bmatrix}$$

Computing the corresponding integrals of the entries, we get the system

$$\begin{bmatrix} 1 & \frac{1}{2} \\ \frac{1}{2} & \frac{1}{3} \end{bmatrix} \begin{bmatrix} \widetilde{b} \\ \widetilde{m} \end{bmatrix} = \begin{bmatrix} \frac{1}{3} \\ \frac{1}{4} \end{bmatrix}$$

with the same solution $\widetilde{b} = -\frac{1}{6}$, $\widetilde{m} = 1$, as before.

We leave the justification of these actions to the reader.

Fourier Series and Polynomials (Requires Calculus)

In many applications we need to analyze a function (such as one representing a sound wave) in terms of its periodicity. Most functions, however, are not periodic, so we try to approximate them using periodic functions such as sine and cosine. This idea goes back to Euler. However, it flourished with the work of Fourier.[11]

Let $\mathcal{B}$ be the set of the following trigonometric functions defined on $[-\pi, \pi]$.

$$\mathcal{B} = \{1, \cos x, \cos 2x, \dots, \cos nx, \ \sin x, \sin 2x, \dots, \sin nx\}$$

A **trigonometric polynomial** is a linear combination of elements of $\mathcal{B}$.

$$p(x) = a_0 + a_1 \cos x + \cdots + a_n \cos nx + b_1 \sin x + \cdots + b_n \sin nx$$

If a_n and b_n are not both zero, we say that $p(x)$ has **order** n.

It is a basic fact that any function $f(x)$ in $C[-\pi, \pi]$ can be approximated by a trigonometric polynomial. By approximated we mean that $f(x)$ and some $p(x)$ are close with respect to the norm defined by integral inner product of Example 45, Section 8.8.

Let $T_n[-\pi, \pi]$ be the subspace of $C[-\pi, \pi]$ that consists of all trigonometric polynomials of order at most n. Then $T_n[-\pi, \pi] = \mathrm{Span}(\mathcal{B})$. Our first basic fact is the following.

THEOREM 43 $\mathcal{B}$ is an orthogonal basis of $T_n[-\pi, \pi]$.

PROOF   It is clear that $\mathcal{B}$ spans $T_n[-\pi, \pi]$. We leave as an exercise the fact that $\mathcal{B}$ is linearly independent. To show that $\mathcal{B}$ is orthogonal, we need to show any two distinct functions are orthogonal, i.e.,

1. $\langle 1, \cos nx \rangle = 0, n = 1, 2, \dots$
2. $\langle 1, \sin nx \rangle = 0, n = 1, 2, \dots$
3. $\langle \cos mx, \cos nx \rangle = 0, m \neq n$
4. $\langle \cos mx, \sin nx \rangle = 0, m, n = 1, 2, \dots$
5. $\langle \sin mx, \sin nx \rangle = 0, m \neq n$

---

[11] **Jean-Baptiste Joseph Fourier** (1768–1830) French mathematician and physicist who became famous for his solution of the heat equation. He introduced the Fourier series, which became a fundamental tool in mathematical physics. He followed Napoleon to Egypt and later was made a baron by him.

To prove the third identity, we have

$$\langle \cos mx, \cos nx \rangle = \int_{-\pi}^{\pi} \cos mx \cos nx \, dx$$

$$= \frac{1}{2} \int_{-\pi}^{\pi} (\cos(m+n)x + \cos(m-n)x) \, dx$$

$$= \frac{1}{2} \left[\frac{\sin(m+n)x}{m+n} + \frac{\sin(m-n)x}{m-n} \right]_{-\pi}^{\pi} = 0$$

In the second step we used a trigonometric identity and in the last, the fact that $\sin k\pi = 0$ for any integer k. The rest of the identities are proved similarly.

It is easy to compute the norms of the functions of $\mathcal{B}$. For example, using the half-angle formula,

$$\| \cos kx \|^2 = \langle \cos kx, \cos kx \rangle$$

$$= \int_{-\pi}^{\pi} \cos^2 kx \, dx$$

$$= \frac{1}{2} \int_{-\pi}^{\pi} (1 + \cos 2kx) \, dx$$

$$= \frac{1}{2} \left[x + \frac{\sin 2kx}{2k} \right]_{-\pi}^{\pi} = \pi$$

Similarly, we compute $\|1\|^2$ and $\|\sin kx\|^2$ to get

$$\|1\| = \sqrt{2\pi}, \qquad \| \cos kx \| = \sqrt{\pi}, \qquad \| \sin kx \| = \sqrt{\pi}$$

Now, to approximate f we need just the orthogonal projection f_{pr} of f onto $T_n[-\pi, \pi]$ using the orthogonal basis $\mathcal{B}$. Suppose

$$f_{\text{pr}}(x) = a_0 + a_1 \cos x + \cdots + a_n \cos nx + b_1 \sin x + \cdots + b_n \sin nx \qquad (8.50)$$

then the Fourier coefficients are given (just as in the case of the dot product) by

$$a_0 = \frac{\langle f, 1 \rangle}{\langle 1, 1 \rangle}, \qquad a_k = \frac{\langle f, \cos kx \rangle}{\langle \cos kx, \cos kx \rangle}, \qquad b_k = \frac{\langle f, \sin kx \rangle}{\langle \sin kx, \sin kx \rangle}$$

Therefore, for $k \geq 1$,

$$a_0 = \frac{1}{2\pi} \int_{-\pi}^{\pi} f(x) \, dx$$

$$a_k = \frac{1}{\pi} \int_{-\pi}^{\pi} f(x) \cos kx \, dx \qquad (8.51)$$

$$b_k = \frac{1}{\pi} \int_{-\pi}^{\pi} f(x) \sin kx \, dx$$

These are the famous **Euler's formulas** that Fourier used to solve the heat equation. The trigonometric polynomial that approximates f given by (8.50) and (8.51) is called the nth-order **Fourier polynomial** (or **Fourier approximation**) of f on the interval $[-\pi, \pi]$.

▪ EXAMPLE 51   Find the nth-order Fourier polynomial of $f(x) = x$ on $[-\pi, \pi]$.

SOLUTION   We have

$$a_0 = \frac{1}{2\pi} \int_{-\pi}^{\pi} x \, dx = \frac{1}{2\pi} \left. \frac{x^2}{2} \right|_{-\pi}^{\pi} dx = 0$$

For $k \geq 1$, by integration by parts,

$$a_k = \frac{1}{\pi} \int_{-\pi}^{\pi} x \cos kx \, dx = \frac{1}{\pi} \left[\frac{\cos kx}{k^2} + \frac{x \sin kx}{k} \right]_{-\pi}^{\pi} = 0$$

$$b_k = \frac{1}{\pi} \int_{-\pi}^{\pi} x \sin kx \, dx = \frac{1}{\pi} \left[\frac{\sin kx}{k^2} - \frac{x \cos kx}{k} \right]_{-\pi}^{\pi} = \frac{2(-1)^{k+1}}{k}$$

because $\cos k\pi = (-1)^k$ for any integer k. Therefore, the Fourier approximation p_n of f is

$$p_n(x) = 2 \sin x - \sin 2x + \frac{2}{3} \sin 3x - \cdots + \frac{2(-1)^{n+1}}{n} \sin nx$$

Figure 8.22 shows $f(x) = x$ sketched with $p_2(x)$ and $p_3(x)$ on $[-\pi, \pi]$.

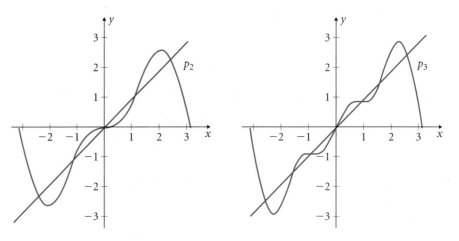

Figure 8.22   The Fourier approximations of orders 2 and 3 for x on $[-\pi, \pi]$.

As n grows, the polynomials p_n get closer to f. Taking the limit as $n \to \infty$ yields an infinite series. We write

$$f(x) = a_0 + \sum_{n=1}^{\infty} (a_n \cos nx + b_n \sin nx)$$

The right side is called the **Fourier series** of f on $[-\pi, \pi]$.

Wavelets (Requires Calculus)

Our last, but not least, application of inner products is in the theory of wavelets,[12] which is becoming quite significant these days. This theory targets many of the problems that Fourier polynomials were designed to solve. These are usually problems involving waves, frequencies, amplitudes etc. In many cases the results from using wavelets are much more favorable compared with those using Fourier analysis. We illustrate some of the highlights of the theory. Additional information is supplied in Project 1, Section 8.10.

First we define the **mother wavelet** $\psi(x)$ by

$$\psi(x) = \psi_{0,0}(x) = \begin{cases} 1, & \text{if } 0 \leq x \leq \frac{1}{2} \\ -1, & \text{if } \frac{1}{2} < x \leq 1 \\ 0, & \text{otherwise} \end{cases}$$

See Fig. 8.23.

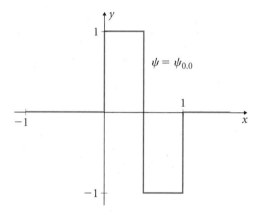

Figure 8.23   The mother wavelet.

Next, for any pair of integers m and n we define the **Haar** (or **basic**) **wavelets** $\psi_{m,n}(x)$ in terms of the mother wavelet by

$$\psi_{m,n}(x) = 2^{-m/2}\psi(2^{-m}x - n)$$

As we see in Project 1 this is equivalent to the full definition

$$\psi_{m,n}(x) = \begin{cases} 2^{-m/2}, & \text{if } 2^m n \leq x \leq 2^m(n + \frac{1}{2}) \\ -2^{-m/2}, & \text{if } 2^m(n + \frac{1}{2}) < x \leq 2^m(n + 1) \\ 0, & \text{otherwise} \end{cases}$$

Figure 8.24 shows the basic wavelets $\psi_{-2,-3}$, $\psi_{0,1}$, $\psi_{1,2}$, and $\psi_{2,2}$.

The interval $I_{m,n} = [2^m n, 2^m(n + 1)]$, which is the only set over which $\psi_{m,n}$ is nonzero, is called the **support** of the wavelet. In general, the support of a function f is

---

[12]The first author is in debt to professor P. R. Turner for allowing him to read and use his notes in this topic.

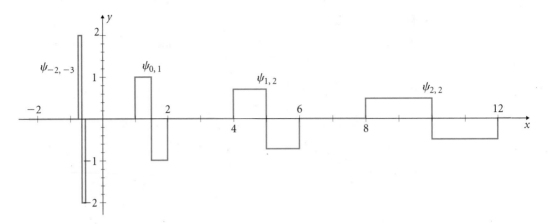

Figure 8.24   Some basic wavelets.

the set of points x such that $f(x) \neq 0$. For example, the support of $\psi_{-2,-3}$ is $[-\frac{3}{4}, -\frac{1}{2}]$, whereas that of $\psi_{2,2}$ is $[8, 12]$.

For functions f and g we consider the usual inner product, except that we integrate over the *entire real line*,

$$\langle f, g \rangle = \int_{-\infty}^{\infty} f(x) g(x)\, dx \tag{8.52}$$

The indefinite integral is not always defined. However, it can be proved that it is defined for functions with finite norm,

$$\| f \| = \left(\int_{-\infty}^{\infty} f(x)^2\, dx \right)^{1/2} < \infty \tag{8.53}$$

The set of functions that satisfies this condition, denoted by L_2, is a vector space under the usual addition and scalar multiplication of functions. It is also an inner product space with (8.52) as the defining inner product. The functions of L_2 are called L_2-**functions**. The basic wavelets are L_2-functions.

The first interesting fact is that all the basic wavelets are units, *i.e.*,

$$\| \psi_{m,n} \| = 1$$

for all integers m and n. Because

$$\| \psi_{m,n} \|^2 = \int_{-\infty}^{\infty} \psi_{m,n}(x)^2\, dx$$

$$= \int_{-\infty}^{2^m n} \psi_{m,n}(x)^2\, dx + \int_{2^m n}^{2^m (n+1/2)} \psi_{m,n}(x)^2\, dx$$

$$+ \int_{2^m (n+1/2)}^{2^m (n+1)} \psi_{m,n}(x)^2\, dx + \int_{2^m (n+1)}^{\infty} \psi_{m,n}(x)^2\, dx$$

$$= 0 + \int_{2^m n}^{2^m(n+1/2)} 2^{-m} \, dx + \int_{2^m(n+1/2)}^{2^m(n+1)} 2^{-m} \, dx + 0$$

$$= 2^{-m} x \big|_{2^m n}^{2^m(n+1/2)} + 2^{-m} x \big|_{2^m(n+1/2)}^{2^m(n+1)}$$

$$= \frac{1}{2} + \frac{1}{2} = 1$$

Also any two basic wavelets are *orthogonal*. So, for $(m_1, n_1) \neq (m_2, n_2)$

$$\langle \psi_{m_1,n_1}, \psi_{m_2,n_2} \rangle = \int_{-\infty}^{\infty} \psi_{m_1,n_1}(x) \, \psi_{m_2,n_2}(x) \, dx = 0 \tag{8.54}$$

The proof of this basic fact is discussed in Project 1. We have Theorem 44.

THEOREM 44 All the basic wavelets $\psi_{m,n}$ form an orthonormal set.

Just as functions can be approximated by trigonometric polynomials, they can also be approximated by linear combinations of basic wavelets. It turns out that this is the case for all functions of L_2. If f is any L_2-function and V is the span of finitely many Haar wavelets, then the projection f_{pr} of f onto V is a linear combination

$$f_{\mathrm{pr}}(x) = \sum_{m,n} c_{m,n} \psi_{m,n}(x)$$

where m and n take on values from two finite sets. The coefficients $c_{m,n}$ are computed as usual by

$$c_{m,n} = \frac{\langle f, \psi_{m,n} \rangle}{\langle \psi_{m,n}, \psi_{m,n} \rangle} = \int_{-\infty}^{\infty} f(x) \psi_{m,n}(x) \, dx$$

because $\langle \psi_{m,n}, \psi_{m,n} \rangle = \|\psi_{m,n}\|^2 = 1$. The integral this time is not indefinite, since the support of $\psi_{m,n}$ is a finite interval. We have

$$c_{m,n} = \int_{2^m n}^{2^m(n+1)} f(x) \psi_{m,n}(x) \, dx$$

$$= \int_{2^m n}^{2^m(n+1/2)} f(x) 2^{-m/2} \, dx + \int_{2^m(n+1/2)}^{2^m(n+1)} f(x)(-2^{-m/2}) \, dx$$

So, we may write

$$c_{m,n} = A_{m,n} - B_{m,n}$$

$$A_{m,n} = 2^{-m/2} \int_{2^m n}^{2^m(n+1/2)} f(x) \, dx \tag{8.55}$$

$$B_{m,n} = 2^{-m/2} \int_{2^m(n+1/2)}^{2^m(n+1)} f(x) \, dx$$

Formulas (8.55) yield the coefficients c_{mn} of f_{pr} as a linear combination of $\psi_{m,n}$. These are analogous to Formulas (8.51) that computed the coefficients in the trigonometric polynomial approximation of f. To properly approximate a function f we need to take the coefficients of *all* Haar wavelets $\psi_{m,n}$ into account, infinitely many of which may be nonzero. So, just as with the Fourier series, we write f as an infinite series in terms of $\psi_{m,n}$.

$$f(x) = \sum_{m=-\infty}^{\infty} \sum_{n=-\infty}^{\infty} c_{m,n}\psi_{m,n}(x)$$

■ EXAMPLE 52   Let

$$f(x) = \begin{cases} 1, & 0 \le x \le 1 \\ 0, & \text{otherwise} \end{cases}$$

(Fig. 8.25), and let V_k be the span of Haar wavelets

$$V_k = \text{Span}\{\psi_{1,0}, \psi_{2,0}, \ldots, \psi_{k,0}\}$$

Approximate f by computing f_{pr} with respect to V_k.

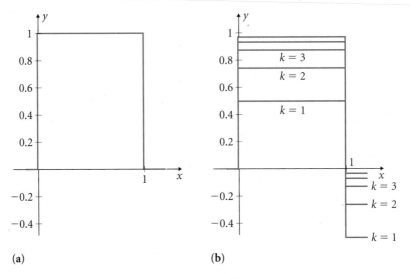

Figure 8.25   Approximation by Haar wavelets.

SOLUTION   Let $c_{1,0}, \ldots, c_{k,0}$ be scalars such that for all x

$$f_{pr}(x) = c_{1,0}\psi_{1,0}(x) + \cdots + c_{k,0}\psi_{k,0}(x)$$

In Project 1 we see that for $m = 1, \ldots, k$,

$$c_{m,0} = 2^{-m/2}$$

Therefore,

$$f_{pr}(x) = 2^{-1/2}\psi_{1,0}(x) + 2^{-2/2}\psi_{2,0}(x) + \cdots + 2^{-k/2}\psi_{k,0}(x)$$

Figure 8.25(b) shows the graphs of f_{pr} for $k = 1, \ldots, 5$. It is clear that as k grows, f_{pr} approaches f very fast.

Exercises 8.9

1. **(Needs Computer)** Compute the NFL ratings using Table 8.1 only. Is the answer satisfactory?

2. **(Needs Computer)** Repeat Exercise 1 with Table 8.2.

In Exercises 3–5 find the least squares quadratic $q(x)$ passing through the given points. Then evaluate $q(6)$.

3. $(1, -3), (2, 0), (3, 4), (4, 13), (5, 20)$

4. $(1, -1), (2, 0), (3, 1), (4, 4), (5, 8)$

5. $(-2, 1), (-1, 2), (0, 4), (1, 6), (2, 9)$

In Exercises 6–7 find the least squares cubic $q(x)$ passing through the points. Then evaluate $q(3)$.

6. $(-2, -8), (-1, -2), (0, 0), (1, 8), (2, 12)$

7. $(-2, -2), (-1, 0), (0, 1), (1, -3), (2, 7)$

8. Prove Theorem 41.

9. Prove Theorem 42.

10. Find the least squares line that approximates $f(x) = x^2$ on $[0, 2]$.

11. Find the least squares line that approximates $f(x) = x^2$ on $[1, 2]$.

12. Find the least squares line that approximates $f(x) = x^3$ on $[0, 1]$.

13. Find the least squares line that approximates $f(x) = x^3$ on $[0, 2]$.

In Exercises 14–15 use the normal equations approach with $A = \begin{bmatrix} 1 & x & x^2 \end{bmatrix}$.

14. Find the least squares quadratic that approximates $f(x) = x^3$ on $[0, 1]$.

15. Find the least squares quadratic that approximates $f(x) = x^3$ on $[0, 2]$.

From the proof of Theorem 43 prove the following.

16. Relation 1

17. Relation 2

18. Relation 4

19. Relation 5

In Exercises 20–22 find the Fourier coefficients a_0, a_n, and b_n of $f(x)$.

20. $f(x) = \begin{cases} -1, & \text{if } -\pi < x < 0 \\ 1, & \text{if } 0 < x < \pi \end{cases}$

21. $f(x) = \begin{cases} 0, & \text{if } -\pi < x < 0 \\ 1, & \text{if } 0 < x < \pi \end{cases}$

22. $f(x) = \begin{cases} 0, & \text{if } -\pi < x < 0 \\ 1, & \text{if } 0 < x < \dfrac{\pi}{2} \\ 0, & \text{if } \dfrac{\pi}{2} < x < \pi \end{cases}$

In Exercises 23–27 show that the given set is orthogonal using the integral inner product on the given interval.

23. $\{\sin(x), \sin(2x), \ldots, \sin(nx)\}$, $[0, \pi]$

24. $\{1, \cos(x), \cos(2x), \ldots, \cos(nx)\}$, $[0, 2\pi]$

25. $\{\sin(\pi x), \sin(2\pi x), \ldots, \sin(n\pi x)\}$, $[-1, 1]$

26. $\{1, \cos(\pi x), \cos(2\pi x), \ldots, \cos(n\pi x)\}$, $[-1, 1]$

27. $\{1, \cos(\pi x), \cos(2\pi x), \ldots, \cos(n\pi x)\}$, $[0, 2]$

28. Let

$$f(x) = \begin{cases} 1, & \text{if } -1 \leq x \leq 0 \\ 0, & \text{otherwise} \end{cases}$$

Write f_{pr} as

$$f_{\mathrm{pr}}(x) = \sum_{m=1}^{k} c_{m,-1}\psi_{m,-1}(x)$$

and show that

$$c_{m,-1} = -2^{-m/2}$$

Sketch the graphs of f and f_{pr} for (a) $k = 2$, (b) $k = 3$.

29. Let

$$f(x) = \begin{cases} -1, & \text{if } 0 \leq x \leq 1 \\ 0, & \text{otherwise} \end{cases}$$

Write f_{pr} as

$$f_{\mathrm{pr}}(x) = \sum_{m=1}^{k} c_{m,0}\psi_{m,0}(x)$$

and show that

$$c_{m,0} = -2^{-m/2}$$

Sketch the graphs of f and f_{pr} for (a) $k = 2$, (b) $k = 3$.

8.10   Miniprojects

1 ■ Wavelets

In this project you are guided to prove some claims made in wavelet theory of Section 8.9.

Problem A

Prove that the definition of the basic wavelets in terms of the mother wavelet

$$\psi_{m,n}(x) = 2^{-m/2}\psi(2^{-m}x - n)$$

is equivalent to the full definition

$$\psi_{m,n}(x) = \begin{cases} 2^{-m/2}, & \text{if } 2^m n \leq x \leq 2^m\left(n + \dfrac{1}{2}\right) \\ -2^{-m/2}, & \text{if } 2^m\left(n + \dfrac{1}{2}\right) < x \leq 2^m(n + 1) \\ 0, & \text{otherwise} \end{cases}$$

Problem B

Use the following steps to prove that the basic wavelets are *orthogonal*, i.e., for $(m_1, n_1) \neq (m_2, n_2)$

$$\langle \psi_{m_1,n_1}, \psi_{m_2,n_2} \rangle\rangle = \int_{-\infty}^{\infty} \psi_{m_1,n_1}(x)\,\psi_{m_2,n_2}(x)\,dx = 0 \qquad (8.56)$$

Recall that $I_{m,n} = [2^m n, 2^m(n + 1)]$ is the support of $\psi_{m,n}$.

1. If $m_1 = m_2$ and $n_1 \neq n_2$, show that the intersection $I_{m_1,n_1} \cap I_{m_2,n_2}$ contains at most one point.
2. If $m_1 > m_2$, then show that either $I_{m_1,n_1} \cap I_{m_2,n_2}$ contains at most one point or I_{m_2,n_2} is contained in I_{m_1,n_1}.
3. If $(m_1, n_1) \neq (m_2, n_2)$ and if I_{m_2,n_2} is contained in I_{m_1,n_1} show that I_{m_2,n_2} is contained either in $[2^{m_1} n_1, 2^{m_1}(n_1 + 1/2)]$ or in $[2^{m_1}(n_1 + 1/2), 2^{m_1}(n_1 + 1)]$.
4. Use Parts 1, 2, and 3 to prove (8.56) for $(m_1, n_1) \neq (m_2, n_2)$.

Problem C

Let

$$f(x) = \begin{cases} 1, & 0 \leq x \leq 1 \\ 0, & \text{otherwise} \end{cases}$$

Let $V_k = \text{Span}\{\psi_{1,0}, \psi_{2,0}, \dots, \psi_{k,0}\}$ and let V any span of Haar wavelets containing V_k. Using the following steps, show that the projection f_{pr} with respect to V is given by

$$f_{\text{pr}}(x) = 2^{-1/2}\psi_{1,0}(x) + 2^{-2/2}\psi_{2,0}(x) + \cdots + 2^{-k/2}\psi_{k,0}(x)$$

First, let

$$c_{m,n} = \int_{-\infty}^{\infty} f(x)\psi_{m,n}(x)\,dx = \int_0^1 \psi_{m,n}(x)\,dx$$

1. If $m \geq 0$, $n \neq 0$, show that $c_{m,n} = 0$.
2. If $m = 0$, $n = 0$, show that $c_{0,0} = 0$.
3. If $m \leq 0$, $n \neq 0$, show that the intersection $I_{m,n} \cap [0,1]$ is either at most one point or is $I_{m,n}$. Conclude that $c_{m,n} = 0$.
4. If $m > 0$, $n = 0$ show that

$$c_{m,0} = 2^{-m/2}$$

The reason we are interested in f is that if we can show that f can be approximated by wavelets, then so can all piecewise constant functions. These functions are *dense* in L_2, i.e., they can approximate any L_2-function. So, the basic wavelets would approximate any L_2-function. The relation between f and the $\psi_{k,0}$s is a strong one. It can be proved that for all x,

$$f(x) = \sum_{m=1}^{\infty} c_{m,0}\psi_{m,0}(x)$$

2 ■ Complex Inner Products

NOTE   This paragraph requires material from Appendix A. Some familiarity with complex arithmetic is necessary.

In this project you are to study inner products over *complex* vector spaces. Complex vector spaces are vector spaces with scalars complex numbers. They were discussed in Section 4.8, Project 2.

DEFINITION

> An (complex) **inner product** on a complex vector space V is a function that, to each pair of vectors $\mathbf{u}$ and $\mathbf{v}$ of V, associates a complex number $\langle \mathbf{u}, \mathbf{v} \rangle$ satisfying the following properties or **axioms**.
>
> For any vectors $\mathbf{u}, \mathbf{v}, \mathbf{w}$ and any complex scalar c,
>
> 1. $\langle \mathbf{u}, \mathbf{v} \rangle = \overline{\langle \mathbf{v}, \mathbf{u} \rangle}$;
> 2. $\langle \mathbf{u} + \mathbf{w}, \mathbf{v} \rangle = \langle \mathbf{u}, \mathbf{v} \rangle + \langle \mathbf{w}, \mathbf{v} \rangle$;
> 3. $\langle c\mathbf{u}, \mathbf{v} \rangle = c\langle \mathbf{u}, \mathbf{v} \rangle$;
> 4. $\langle \mathbf{u}, \mathbf{u} \rangle > 0$ if $\mathbf{u} \neq \mathbf{0}$.

Problem A

Let $\mathbf{u} = (u_1, \ldots, u_n)$ and $\mathbf{v} = (v_1, \ldots, v_n)$ be in $\mathbf{C}^n$. Prove that the (complex) **dot** product

$$\mathbf{u} \cdot \mathbf{v} = u_1\overline{v_1} + \cdots + u_n\overline{v_n}$$

defines a complex inner product on $\mathbf{C}^n$.

Problem B

1. Using a complex inner product, define the notions of

    (a) Orthogonal pair of vectors;
    (b) Orthogonal set;
    (c) Orthonormal set.

2. Provide examples for these concepts.
3. Prove that the columns of a unitary matrix (see Appendix A for a definition) form an orthonormal set with respect to the complex dot product of Problem A.

Problem C

Let $\mathbf{u}$, $\mathbf{v}$, and $\mathbf{w}$ be any vectors in a complex inner product space and let c be any complex scalar. Prove the following properties:

1. $\langle \mathbf{u}, \mathbf{v} + \mathbf{w} \rangle = \langle \mathbf{u}, \mathbf{v} \rangle + \langle \mathbf{u}, \mathbf{w} \rangle$
2. $\langle \mathbf{u}, c\mathbf{v} \rangle = \bar{c} \langle \mathbf{u}, \mathbf{v} \rangle$

3 ■ The Pauli Spin and Dirac Matrices

NOTE   This paragraph requires material from Appendix A. The definitions of Hermitian and unitary matrices are in Appendix A. Some familiarity with complex arithmetic is necessary.

In this project we explore the basic properties of certain matrices with complex entries that play an important role in nuclear physics and quantum mechanics.

W. Pauli[13] introduced the following three matrices, known as **Pauli spin matrices** to compute the electron spin.

$$\sigma_x = \begin{bmatrix} 0 & 1 \\ 1 & 0 \end{bmatrix} \qquad \sigma_y = \begin{bmatrix} 0 & -i \\ i & 0 \end{bmatrix} \qquad \sigma_z = \begin{bmatrix} 1 & 0 \\ 0 & -1 \end{bmatrix}$$

In 1927 P. A. M. Dirac,[14] working in quantum mechanics, generalized Pauli's spin matrices to the following, known as **Dirac matrices**.

---

[13]**Wolfgang Joseph Pauli** (1900–1958) was born in Vienna, Austria. He studied physics in Münich and did postgraduate work with Bohr in Copenhagen. He taught in Hamburg and Zurich. He is known for his work in elementary particles, such as the prediction of the neutrino particle and the discovery of the exclusion principle that bears his name. He also used three 2×2 complex matrices, called the Pauli spin matrices, to describe his theory of spin of elementary particles in 1927.
[14]**Paul Adrien Maurice Dirac** was born in Bristol, England in 1902. In 1926 he obtained a doctorate from Cambridge University. He then studied under Bohr in Copenhagen and under Born in Göttingen. In 1932 he became the Lucasian Professor of mathematics at Cambridge (the chair once held by Newton). He won the Nobel Prize in Physics in 1933. He is known for his work on quantum mechanics, elementary particles, and the theory of antimatter.

$$\alpha_x = \begin{bmatrix} 0 & 0 & 0 & 1 \\ 0 & 0 & 1 & 0 \\ 0 & 1 & 0 & 0 \\ 1 & 0 & 0 & 0 \end{bmatrix}, \qquad \alpha_y = \begin{bmatrix} 0 & 0 & 0 & -i \\ 0 & 0 & i & 0 \\ 0 & -i & 0 & 0 \\ i & 0 & 0 & 0 \end{bmatrix}$$

$$\alpha_z = \begin{bmatrix} 0 & 0 & 1 & 0 \\ 0 & 0 & 0 & -1 \\ 1 & 0 & 0 & 0 \\ 0 & -1 & 0 & 0 \end{bmatrix}, \qquad \beta = \begin{bmatrix} 1 & 0 & 0 & 0 \\ 0 & 1 & 0 & 0 \\ 0 & 0 & -1 & 0 \\ 0 & 0 & 0 & -1 \end{bmatrix}$$

The first three Dirac matrices are block matrices in the Pauli spin matrices.

$$\alpha_x = \begin{bmatrix} \mathbf{0} & \sigma_x \\ \sigma_x & \mathbf{0} \end{bmatrix}, \qquad \alpha_y = \begin{bmatrix} \mathbf{0} & \sigma_y \\ \sigma_y & \mathbf{0} \end{bmatrix}, \qquad \alpha_z = \begin{bmatrix} \mathbf{0} & \sigma_z \\ \sigma_z & \mathbf{0} \end{bmatrix}$$

A square matrix A is **involutory** if $A^{-1} = A$, or equivalently, if $A^2 = I$. For example, $-I$ is involutory.

Problem A

Show that the Pauli spin matrices and the Dirac matrices are

1. Hermitian;
2. Unitary;
3. Involutory.

Problem B

Show that the Pauli spin matrices satisfy the relations

$$\sigma_x \sigma_y = i\sigma_z \qquad \sigma_y \sigma_z = i\sigma_x \qquad \sigma_z \sigma_x = i\sigma_y$$

and that they *anticommute*, i.e.,

$$\sigma_x \sigma_y = -\sigma_y \sigma_x \qquad \sigma_y \sigma_z = -\sigma_z \sigma_y \qquad \sigma_x \sigma_z = -\sigma_z \sigma_x$$

Problem C

Show that the Dirac matrices satisfy the relations

$$\alpha_x \beta = -\beta \alpha_x \qquad \alpha_y \beta = -\beta \alpha_y \qquad \alpha_z \beta = -\beta \alpha_z$$

and that they *anticommute*, i.e.,

$$\alpha_x \alpha_y = -\alpha_y \alpha_x \qquad \alpha_y \alpha_z = -\alpha_z \alpha_y \qquad \alpha_x \alpha_z = -\alpha_z \alpha_x$$

8.11   Computer Exercises

This computer section is designed to familiarize you with your program's basic commands on the material of Chapter 8. It will also help you review this material.

Let

$$A = \begin{bmatrix} 1 & 4 & 69 \\ 2 & -3 & 28 \\ -3 & 2 & -37 \\ 4 & 2 & -59 \end{bmatrix}, \quad B = \begin{bmatrix} 1 & 2 & 3 \\ 4 & 5 & 6 \\ 7 & 8 & 9 \end{bmatrix},$$

$$C = \begin{bmatrix} 1 & 2 & 3 & 4 \\ 2 & 2 & 3 & 4 \\ 3 & 3 & 3 & 4 \\ 4 & 4 & 4 & 4 \end{bmatrix}, \quad S = \begin{bmatrix} -1 & -1 & 1 \\ -1 & 2 & 4 \\ 1 & 4 & 2 \end{bmatrix}.$$

1. Show that A has orthogonal columns by: (a) using the dot product, (b) verifying Relation (8.4) of Theorem 4, Section 8.1.

2. Let A_1 be the matrix obtained by adding $e_4 \in \mathbf{R}^4$ as a last column to A. Apply Gram-Schmidt process to the columns of A_1 to orthogonalize them. Form a matrix A_2 with these columns and verify relation (8.4).

3. Orthonormalize the columns of matrix A_2 of Exercise 2 to obtain an orthogonal matrix A_3. Verify that A_3 is orthogonal.

4. Verify Bessel's inequality with the set of orthonormal vectors consisting of the first three columns of A_3 of Exercise 3 and with $u = e_4 \in \mathbf{R}^4$.

5. Find a matrix whose columns are orthogonal and span the column space of B.

6. Write a short program that computes the orthogonal projection of a vector u onto the span of a set S of orthogonal vectors.

7. Test your program of Exercise 6 with $u = (1, -1, 2)$ and $S = \{(-1, 4, 1), (5, 1, 1)\}$ (see Example 11, Section 8.2).

8. Modify your program of Exercise 6 so that it computes orthogonal projection of a vector onto the span of any finite set of linearly independent but not necessarily orthogonal vectors.

9. Test your program of Exercise 8 with $u = (1, 1, 1)$ and $S = \{(1, -1, 2), (-1, 1, 4)\}$.

10. Compute the QR factorization of C and verify your answer.

11. Find the least squares line through the 10 points of x^2 with x-coordinates $0, 1, \ldots, 9$. On the same graph, plot this line and x^2 over $[0, 9]$.

12. Find and plot the least squares quadratic through the points $(-2, 2), (-1, 0), (2, 4), (3, 7), (4, 9)$.

13. Orthogonally diagonalize the symmetric matrix S.

14. Find a Schur decomposition for B.

15. Find the singular values and an SVD for A. Verify that $A = U\Sigma V^T$. Repeat with C.

16. Compute A^+ and C^+. Verify the Moore-Penrose properties for (A, A^+) and (C, C^+).

17. Consider the inner product of Example 45, Section 8.8, on $[0, 1]$. Use your program to compute the values $\langle x^2, \sin(x) \rangle$, $\|x^2\|$, $\|\sin(x)\|$. Verify the CSB inequality.

18. Verify the CSB inequality using the inner product of Example 42, Section 8.8, with $r_0 = -10$, $r_1 = 3, r_2 = 15$ for $u = x^2 + x - 1$ and $v = -x^2 + 2x$.

19. Write and test a function with arguments u, v, and w that computes the weighted dot product $\langle u, v \rangle$ with weight vector w (see Example 39, Sec. 8.8).

20. Use your program to verify the quarterback rating formula of Sec. 8.9.

Selected Solutions with Maple

Commands definite, dotprod, GramSchmidt, innerprod, leastsqrs, norm, normalize, orthog, QRdecomp, singularvals.

```
with(linalg):
A := matrix([[1,4,69],[2,-3,28],[-3,2,-37],[4,2,-59]]);  # All data.
B := matrix([[1,2,3],[4,5,6],[7,8,9]]);
C := matrix([[1,2,3,4],[2,2,3,4],[3,3,3,4],[4,4,4,4]]);
DD := matrix([[-1,-1,1],[-1,2,4],[1,4,2]]);  # D is already used by Maple.
# Exercise 1
dotprod(col(A,1),col(A,2));      # Etc.. Repeat with the other two pairs.
evalm(transpose(A)&*A);      #(A^T)A is diagonal with diag. entries norms^2 of
norm(col(A,1),2)^2;norm(col(A,2),2)^2;norm(col(A,3),2)^2;   # the columns of A.
# Exercise 2
e4:=vector([0,0,0,1]);                          # e_4.
A1 := concat(A,e4);                             # A1. Then Gram-Schmidt on
GramSchmidt([col(A1,1),col(A1,2),col(A1,3),col(A1,4)]);   # the columns of A1.
A2 := concat("[1]","[2]","[3]","[4]");          # A2.
evalm(transpose(A2) &* A2);                     # Etc.. Same verification.
# Exercise 3
seq(col(A2,i)/norm(col(A2,i),2), i=1..4);   # All columns of A2 orthonormalized
A3:=concat(");                              # and put in a matrix.
orthog(A3);                                 # Test for orthogonality.
evalm(transpose(A3) &* A3);                 # Another way. (A3^T)A3 yields I.
# Exercise 4                                # e_4 is defined above.
norm(e4)^2-(dotprod(e4,col(A3,1))^2 \       # LHS-RHS of Bessel >0. OK.
+ dotprod(e4,col(A3,2))^2 + dotprod(e4,col(A3,3))^2);
# Exercise 5
GramSchmidt([col(B,1),col(B,2),col(B,3)]); # GS on the columns of B to form
B1:=concat("[1]","[2]");                    # a new matrix B1 whose columns
concat(B1,B);                               # span Col(B) since, the first
rref(");                             # 2 columns of [B1,B] are pivot columns.
# Exercise 6
proj := proc(u,lis) local i, s;      # proj for projection.
s := [seq(0,i=1..nops(u))];  # 0 list converted to vector in the loop.
for i from 1 to nops(lis) do
s := evalm(s + (dotprod(u,lis[i])/dotprod(lis[i],lis[i])*lis[i])) od:
evalm(s) end:
# Exercise 7
proj([1,-1,2],[[-1,4,1],[5,1,1]]);    # Testing with Example 11.
# Exercise 10
R := QRdecomp(C, Q='q');   # QR factorization.
evalm(C - q &* R);          # The difference is the zero matrix,
evalm(transpose(q)&*q);     # and Q is orthogonal, since (Q^T)Q=I.
# Exercise 11
A := matrix(10,2,[1,0,1,1,1,2,1,3,1,4,1,5,1,6,1,7,1,8,1,9]);
```

```
b := vector([0,1,4,9,16,25,36,49,64,81]);  # No need for normal
leastsqrs(A,b);                 # equations, leastsqrs does it in one step.
plot({x^2,9*x-12},x=0..9); # The least squares line and x^2 plotted.
# Exercise 13
eigsys:=eigenvects(S);                       # Eigenvalues and eigenvectors.
D1 := diag(eigsys[1][1],eigsys[2][1],eigsys[3][1]);    # Diagonal of eigenvalues.
eves := eigsys[1][3][1],eigsys[2][3][1],eigsys[3][3][1];    # The eigvecs are
Q:=concat(eves[1]/norm(eves[1],2),eves[2]/norm(eves[2],2),    # automatically
eves[3]/norm(eves[3],2));              # orthogonal since they corresp. to distinct
orthog(");                             # eigvals. We divide by the norms to get Q.
evalm(transpose(Q) &* S &* Q);         # which is orthogonal and (Q^T)SQ=D1. OK.
# Exercise 14
# As of now there is no one-step Schur decomposition available, but you may
# try to compute one using the steps of the proof of Schur's theorem.
# Exercise 15 - Partial
singularvals(A);evalf(");              # The singular values exactly and approximately.
sv := evalf(Svd(A,U,V));               # sv are also the numerical sing. vals. and U and V.
evalm(U); evalm(V);                    # Looking at U and V and checking their
evalm(transpose(U) &* U); evalm(transpose(V) &* V);        # orthogonality.
evalm(transpose(U) &* A &* V);         # Checking that U'AV is Sigma - the answer form sv.
# Exercise 16 - Partial              # There is no 1-step pseudoinverse but we compute
diag(sv[1]^(-1),sv[2]^(-1),sv[3]^(-1));    # easily. In the notation of Exercise 15,
sigplus:=concat(",[0,0,0]);            # we form sigma plus and mulitply to get
psA := evalm(V &* sigplus &* transpose(U));    # the pseudoinverse.
evalm(A &* psA &* A - A);              # Checking all
evalm(psA &* A &* psA - psA);          # Moore-Penrose
evalm( transpose(A &* psA) - A &* psA);    # conditions. All matrices
evalm( transpose(psA &* A) - psA &* A);    # are approx. zero.
# Exercise 17
int(x^2*sin(x),x=0..1);                # <x^2, sin(x)>.
sqrt(int(x^2,x=0..1));                 # Norm(x^2).
sqrt(int(sin(x),x=0..1));              # Norm(sin(x)).
evalf(""*"-""");           # Norm(x^2)*Norm(sin(x))-<x^2, sin(x)> is >0. OK.
```

Selected Solutions with Mathematica

Commands   Dot, Outer, QRDecomposition, SchurDecomposition, SingularValues. And from the LinearAlgebra'Orthogonalization' package: GramSchmidt, InnerProduct, Normalize, Normalized, Projection.

```
A = {{1,4,69},{2,-3,28},{-3,2,-37},{4,2,-59}} (* All data. *)
B = {{1,2,3},{4,5,6},{7,8,9}}
CC = {{1,2,3,4},{2,2,3,4},{3,3,3,4},{4,4,4,4}}(* C is already used by Mathematica. *)
DD = {{-1,-1,1},{-1,2,4},{1,4,2}}             (* D is already used by Mathematica. *)
(* Exercise 1 *)
AT=Transpose[A];               (* We can access the columns easier by transposition.
```

```
Dot[AT[[1]],AT[[2]]]              (* Etc.. Repeat with the other two pairs.*)
AT . A    (* (A^T)A is diagonal with diag. entries norms^2 of the columns of A.*)
sqnorm[lis_]:=Plus @@ (lis^2) (* This little function computes the squares of
                               norms of vectors. *)
sqnorm[AT[[1]]]                   (* Norm^2 of column 1 of A, etc.. *)
(* Another way: <<LinearAlgrebra`Orthogonalization` then
(* AT[[1]][[1]]/Normalize[AT[[1]]][[1]] ,    Etc.. *)
(* Exercises 2 and 3 *)
e4 = {{0},{0},{0},{1}}                    (* e_4  *)
<<LinearAlgebra`MatrixManipulation`       (* The matrix manipulation package. *)
A1 = AppendRows[A,e4]; TA1=Transpose[A1]  (* A1 and its transpose.         *)
<<LinearAlgebra`Orthogonalization`        (* Orthogonalization package.   *)
A2 = Transpose[GramSchmidt[TA1]]          (* Gram-Schmidt on the columns of A.*)
Transpose[A2] . A2   (* The matrix is already orthogonal! GS normalizes too.*)
(* Note: GramSchmidt[Vec_list, Normalized->False] is GS without normalization. *)
(* Exercise 4 *)
ee4 = {0,0,0,1}                (* LHS-RHS of Bessel >0. OK. Look up *)
sqnorm[ee4] - Sum[Dot[ee4,Flatten[TakeColumns[A2,{i}]]]^2,{i,1,3}] (* Flatten. *)
(* Exercise 5 *)   (* By row reducing B^T we see that B has dependent columns *)
GramSchmidt[{Transpose[B][[1]],Transpose[B][[2]]}] (* Since Mathematica computes
B1 = Transpose[%]   (* the GS of independent vectors we use only the first 2 cols. *)
AppendRows[B1,B]    (* The cols. of B1 span Col(B) since, the first 2 cols. of *)
RowReduce[%]        (* [B1,B] are pivot columns.                        *)
(* Exercise 6 *)
proj[u_,lis_] :=
Sum[Dot[u,lis[[i]]]/Dot[lis[[i]],lis[[i]]]*lis[[i]],{i,1,Length[lis]}]
(* Exercise 7 *)
proj[{1,-1,2},{{-1,4,1},{5,1,1}}]    (* Testing with Example 11.     *)
(* Exercise 10 *)       (* QR factorization.                        *)
QRDecomposition[N[CC]]  (* Need to numerically evaluate CC first.   *)
Q = %[[1]]; R= %[[2]];  (* Warning: Q is such that  (Q^T)R is CC.   *)
Transpose[Q] . R - CC   (* The difference is approx. the zero matrix,*)
Transpose[Q] . Q        (* and Q is orthogonal, since (Q^T)Q=I.     *)
(* Exercise 11 *)
A = {{1,0},{1,1},{1,2},{1,3},{1,4},{1,5},{1,6},{1,7},{1,8},{1,9}}
AT = Tranpose[A]
b = {{0},{1},{4},{9},{16},{25},{36},{49},{64},{81}}
LinearSolve[AT.A,AT.b]     (* Solve the normal equations to get 9x-12.*)
Plot[{x^2,9x-12},{x,0,9}]  (* The least squares line and x^2 plotted. *)
(* Exercise 13 *)
D1=DiagonalMatrix[Eigenvalues[S]]    (* Diag. matrix with diag. entries the *)
eves=Eigenvectors[S]   (* eigenvalues which are distinct so the eigenvectors *)
<<LinearAlgebra`Orthogonalization`   (* are already orthogonal, so they    *)
Q=Transpose[Map[Normalize, eves]]    (* only need normalization. Q and D1 *)
Transpose[Q] . S . Q                 (* orthogonally diagonalize S.      *)
Transpose[Q] . Q                     (* Testing Q for orthogonality.     *)
```

```
(* Exercise 14 *)
Eigenvalues[B]              (* First we check the eigenvalues. All real. OK. *)
sd = SchurDecomposition[N[B]];        (* Schur Decomposition.              *)
Q=sd[[1]]                   (* Q and *)
T=sd[[2]]                   (* T. *)
Transpose[Q] . Q            (* Q is orthogonal. *)
Q . T . Transpose[Q]        (* Checking and got B.*)
(* Exercise 15  - Partial *)
{Ut,sig,V}=SingularValues[N[A]]     (* The SVD has a slightly different notation here.*)
                            (* U is not 4X4 so not orthogonal but 3X4 and      *)
                            (* sigma is square with diagonal the sing. values.*)
Transpose[Ut] . Ut          (* Ut is not orthogonal but ((Ut)^T)Ut = I_3       *)
Transpose[V] . V            (* V is orthogonal.                                *)
Transpose[Ut] . DiagonalMatrix[sig] . V   (* The product yields A.            *)
(* Exercise 16  - Partial *)
psA = PseudoInverse[A]      (* One step computation of the pseudoinverse.      *)
N[A]                        (* Approximate.                                    *)
A . psA . A - A                 (* Checking all                                *)
psA . A . psA - psA             (* Moore-Penrose                               *)
Transpose[A . psA] - A . psA    (* conditions. All matrices                    *)
Transpose[psA . A] - psA . A    (* are zero.                                   *)
(* Exercise 17 *)
Integrate[x^2 Sin[x],{x,0,1}]       (* <x^2, sin(x)>.                          *)
Sqrt[Integrate[x^2,{x,0,1}]]        (* Norm(x^2).                              *)
Sqrt[Integrate[Sin[x],{x,0,1}]]     (* Norm(sin(x)).                           *)
N[%% %- %%%]                (* Norm(x^2)*Norm(sin(x))-<x^2, sin(x)> is >0. OK. *)
```

Selected Solutions with MATLAB

Commands lscov, nnls, orth, norm, normest, pinv, qr, qrdelete, qrinsert, rcond, schur, svd. And from the symbolic toolbox: singvals.

```
A = [1 4 69; 2 -3 28; -3 2 -37; 4 2 -59]       % All data.
B = [1 2 3; 4 5 6; 7 8 9]
C = [1 2 3 4; 2 2 3 4; 3 3 3 4; 4 4 4 4]
D = [-1 -1 1; -1 2 4; 1 4 2]
% Exercise 1
dot(A(:,1), A(:,2))                % Etc.. Repeat with the other two pairs.
A.'*A              % (A^T)A is diagonal with diag. entries norms^2 of
norm(A(:,1))^2,norm(A(:,2))^2,norm(A(:,3))^2       % the columns of A.
% Exercises 2 and 3
e4 = [0;0;0;1]                % e4.
A1 = [A e4]                   % A1.
A2=orth(A1)                   % Gram-Schmidt on the columns of A.
A2.'*A2            % The matrix is already orthogonal! GS normalizes too.
% Exercise 4                         % LHS-RHS of Bessel >0. OK.
```

```
norm(e4)^2-(dot(e4,A2(:,1))^2+dot(e4,A2(:,2))^2+dot(e4,A2(:,3))^2)
% Exercise 5
B1 = orth(B)        % GS on the columns of B to form a new matrix B1
[B1 B]              % whose columns span Col(B) since, the first 2
rref(ans)           % columns of [B1,B] are pivot columns.
% Exercise 6
function [A] = proj(u,lis)          % In an m-file called proj.m type
       [m,n]=size(lis);             % the code on the left for
       s = zeros(1,n);              % the orthogonal projection.
       for i=1:m
          s =  s + dot(u,lis(i,:))/dot(lis(i,:),lis(i,:))*lis(i,:);
       end
       A = s;
end
% Exercise 7
proj([1 -1 2],[-1 4 1; 5 1 1])
% Exercise 10
[Q,R] = qr(C)           % QR factorization.
C - Q*R                 % The difference is approx. the zero matrix,
Q.' * Q                 % and Q is orthogonal, since (Q^T)Q=I.
% Exercise 11
A = [1 0; 1 1; 1 2; 1 3; 1 4; 1 5; 1 6; 1 7; 1 8; 1 9]
b = [0;1;4;9;16;25;36;49;64;81] % No need for normal equations lscov does it
lscov(A,b,diag(ones(10,1)))     % in one step. Look it up. See also nnls.
x = 0:.1:9;                % Plotting the least squares line
y1 = 9*x-12;               % 9x-12 and x^2 in one graph
y2 = x.^2;                 % on [0,9].
plot(x,y1,x,y2);
% Exercise 13
[D1,Q]=eig(D)   % The orthogonal normalization is done in 1 step!
                % D1 is diagonal with the eigvals on the diagonal.
Q' * Q          % and Q is orthgonal.
Q' * D * Q      % The product is D as it should.
% Exercise 14
eig[B]                  % First we check the eigenvalues. All real. OK.
[Q, T] = schur(B)       % Schur Decomposition, Q and T.
Q' * Q                  % Q is orthogonal.
Q * T * Q'              % Checking and got B.
% Exercise 15  - Partial
[U,S,V] = svd(A)            % One step SVD.         Checking...
U' * U, V' * V             % U and V are orthogonal.
U*S*V'                     % The product is A.
% Exercise 16  - Partial
psA = pinv(A)              % One step computation of the pseudoinverse.
A * psA * A - A            % Checking all
psA * A * psA - psA        % Moore-Penrose
```

```
(A * psA)' - A * psA              % conditions. All matrices
(psA * A)' - psA * A              % are approximately zero.
% Exercise 17
ff = eval(int('x^2*sin(x)',0,1))              % (ST)   <x^2, sin(x)>.
f1 = sqrt(eval(int('x^2',0,1)))               % (ST)   Norm(x^2).
f2 = sqrt(eval(int('sin(x)',0,1)))            % (ST)   Norm(sin(x)).
f1*f2-ff                        % Norm(x^2)*Norm(sin(x))-<x^2, sin(x)> is >0. OK.
```

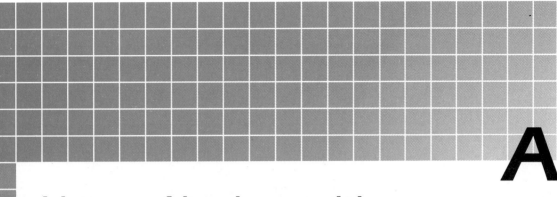

A

Linear Algebra with Complex Numbers

Today most areas of mathematics, physics, and engineering use complex numbers. Complex numbers were discovered by Cardano[1] and first mentioned in his book *Ars magna*[2] (A.D. 1545). It was Gauss, however—who according to G. H. Hardy—"was the first mathematician to use complex numbers in a really confident and scientific way."

In this appendix we *outline* the mechanics of doing linear algebra with complex numbers. All processes are the same as before, except that we use complex arithmetic.

Arithmetic with Complex Numbers

The **imaginary unit**—i, or $\sqrt{-1}$—is defined by the property

$$i^2 = -1$$

Hence,

$$i^3 = -i, \qquad i^4 = 1, \qquad i^5 = i$$

■ EXAMPLE 1   Compute i^{1246}.

SOLUTION $i^{1246} = i^{311 \cdot 4 + 2} = (i^4)^{311} \, i^2 = 1^{311}(-1) = -1$ ■

---

[1]**Girolamo Cardano** (1501–1576) was born in Pavia, Italy, the illegitimate son of a lawyer. After a childhood in extreme poverty and poor health, he studied Medicine and eventually was appointed professor of mathematics in Milan. He became famous as a mathematician, physician, and astrologer. In his book, *Ars magna*, the complete solutions of the cubic and quadric equations appeared for the first time.
[2]There Cardano divides 10 into parts whose product is 40. The answer is $5 + 15i$ and $5 - 15i$, as we can see by solving the resulting quadratic equation. Cardano writes this answer as 5 p: Rm: 15 and 5 m: Rm: 15.

A **complex number** z is an expression of the form $z = a + bi$, where both a and b are real numbers. The set of all complex numbers is denoted by **C**. The **real part**, $\text{Re}(z)$, of z is a. The **imaginary part**, $\text{Im}(z)$, of z is b. If $b = 0$ then z is a real number. If $a = 0$ then z is pure imaginary. The **complex conjugate** of z is $\bar{z} = a - ib$.

■ EXAMPLE 2

$$\text{Re}(1 - 2i) = 1, \qquad \text{Im}(5 - 2i) = -2, \qquad \overline{1 - i} = 1 + i, \qquad \overline{-3} = -3$$

Two complex numbers are **equal** if their respective real and imaginary parts are equal. For example, $5 + xi = y - 4i$ if and only if $y = 5$ and $x = -4$.

The **absolute value** $|z| = |a + bi|$ of a complex number z is the nonnegative real number $\sqrt{a^2 + b^2}$.

■ EXAMPLE 3

$$|-2 + 3i| = \sqrt{(-2)^2 + 3^2} = \sqrt{13}$$

Note that

$$z\bar{z} = |z|^2$$

The **sum**, **difference**, and **product** of complex numbers is, as in real numbers, with the following provisions: All powers of i are calculated. Terms are collected so that the final result in the form $a + ib$ for real a and b.

■ EXAMPLE 4

$$(1 - 2i) - (2 + 3i)(-1 + i) = (1 - 2i) - (-5 - i) = 6 - i$$

The **quotient**, z/w, of two complex numbers $z = a + bi$ and $w = c + di$ with $c + di \neq 0$, is the number

$$\frac{z}{w} = \frac{z\bar{w}}{w\bar{w}} = \frac{ac + bd}{c^2 + d^2} + \frac{bc - ad}{c^2 + d^2}i$$

It is easy to verify that $w(\frac{z}{w}) = z$.

■ EXAMPLE 5

$$\frac{2 + 3i}{1 + 2i} = \frac{(2 + 3i)(1 - 2i)}{(1 + 2i)(1 - 2i)} = \frac{8 - i}{5} = \frac{8}{5} - \frac{1}{5}i$$

The following properties are left as exercises.

$$z + \bar{z} = 2\text{Re}(z) \qquad z - \bar{z} = 2\text{Im}(z)i$$

$$\overline{z + w} = \bar{z} + \bar{w}, \qquad \overline{z - w} = \bar{z} - \bar{w}, \qquad \overline{zw} = \bar{z}\,\bar{w}, \qquad \overline{z/w} = \bar{z}/\bar{w}$$

Geometric Interpretation of Complex Numbers

Every complex number $z = a + ib$ can be represented by the vector (or point) (a, b) in the plane. The x- and the y-axes in this context are called the **real** and the **imaginary** axes, respectively. The opposite $-z$ of z is the reflection of z with respect to the origin, and the conjugate $\bar{z}$ is the reflection with respect to the real axis. Addition of two complex numbers corresponds to vector addition in $\mathbf{R}^2$, and scalar multiplication by a **real** number corresponds to scalar multiplication in $\mathbf{R}^2$ (Fig. A.1).

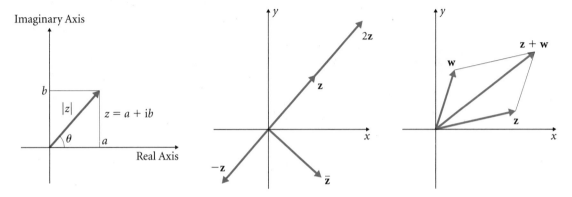

Figure A.1   Complex numbers as 2-vectors.

Geometrically, the absolute value $|z|$ is the length of vector z. The angle θ between the positive real axis and the vector (a, b) representing $z = a + ib$ is called the **argument** of z. Because

$$a = |z| \cos \theta, \qquad b = |z| \sin \theta$$

we have

$$z = |z| (\cos \theta + i \sin \theta) \qquad (A.1)$$

Equation (A.1) is called a **polar representation** of z.

■ EXAMPLE 6   Find the polar representation of $-1 + i$.

SOLUTION   First, $|-1 + i| = \sqrt{2}$. The argument of $-1 + i$ can be computed from

$$\sqrt{2} \cos \theta = -1, \qquad \sqrt{2} \sin \theta = 1$$

which imply $\theta = 3\pi/4$. Hence,

$$-1 + i = \sqrt{2} \left(\cos \frac{3\pi}{4} + i \sin \frac{3\pi}{4} \right)$$

Polar representations of complex numbers are very useful, especially in questions related to complex multiplication and division. For example, if $w = |w| (\cos \phi + i \sin \phi)$, then the product zw and the quotient z/w have polar representations:

$$zw = |z|\,|w|\,(\cos(\theta + \phi) + i\sin(\theta + \phi))$$

$$\frac{z}{w} = \frac{|z|}{|w|}\,(\cos(\theta - \phi) + i\sin(\theta - \phi))$$

These identities can be proved by using the standard trigonometric identities expressing the sine and cosine of the sum or difference of two angles. Also we may easily compute the polar representations of powers:

$$z^n = |z|^n\,(\cos n\theta + i\sin n\theta)$$

■ EXAMPLE 7   Write $(-1 + i)^{10}$ in the form $a + ib$.

SOLUTION   We have

$$(-1 + i)^{10} = \left(\sqrt{2}\right)^{10}\left(\cos\left(10 \cdot \frac{3\pi}{4}\right) + i\sin\left(10 \cdot \frac{3\pi}{4}\right)\right)$$

$$= 32\left(\cos\frac{15\pi}{2} + i\sin\frac{15\pi}{2}\right)$$

$$= -32i$$

Systems of Complex Numbers

Solving systems with complex numbers is the same as before.

■ EXAMPLE 8   Solve the system for z and w.

$$3iz + 4w = 5 + 15i$$

$$(5 - i)z + (3 - 4i)w = 24 + 5i$$

SOLUTION   By Gauss elimination,

$$\begin{bmatrix} 3i & 4 & 5 + 15i \\ 5 - i & 3 - 4i & 24 + 5i \end{bmatrix} \sim \begin{bmatrix} 3i & 4 & 5 + 15i \\ 0 & \frac{13}{3} + \frac{8}{3}i & \frac{2}{3} + \frac{55}{3}i \end{bmatrix}$$

$$\begin{bmatrix} 3i & 4 & 5 + 15i \\ 0 & 1 & 2 + 3i \end{bmatrix} \sim \begin{bmatrix} 3i & 0 & -3 + 3i \\ 0 & 1 & 2 + 3i \end{bmatrix} \sim \begin{bmatrix} 1 & 0 & 1 + i \\ 0 & 1 & 2 + 3i \end{bmatrix}$$

Thus the solution to the system is $z = 1 + i$ and $w = 2 + 3i$.

Matrices of Complex Numbers

All matrix arithmetic and theorems we have discussed extend to matrices whose entries are complex numbers.

■ EXAMPLE 9 Compute $A^TA - (1 + i)B$, where

$$A = \begin{bmatrix} i & 2 \\ 1 - i & 2i \\ 2 & -i \end{bmatrix}, \qquad B = \begin{bmatrix} 1 & 0 \\ -i & 1 - i \end{bmatrix}$$

SOLUTION

$$\begin{bmatrix} 3 - 2i & 2 + 2i \\ 2 + 2i & -1 \end{bmatrix} - \begin{bmatrix} 1 + i & 0 \\ 1 - i & 2 \end{bmatrix} = \begin{bmatrix} 2 - 3i & 2 + 2i \\ 1 + 3i & -3 \end{bmatrix}$$

■ EXAMPLE 10 Compute A^{-1} by row reduction.

$$A = \begin{bmatrix} i & 1 \\ 1 & 1 - i \end{bmatrix}$$

SOLUTION $[A : I]$ reduces as

$$\begin{bmatrix} i & 1 & 1 & 0 \\ 0 & 1 & i & 1 \end{bmatrix} \sim \begin{bmatrix} i & 0 & 1 - i & -1 \\ 0 & 1 & i & 1 \end{bmatrix} \sim \begin{bmatrix} 1 & 0 & -1 - i & i \\ 0 & 1 & i & 1 \end{bmatrix}$$

So, $A^{-1} = \begin{bmatrix} -1 - i & i \\ i & 1 \end{bmatrix}$.

The **complex conjugate** $\overline{A}$ of a matrix A is the matrix whose entries are the complex conjugates of the corresponding entries of A.

■ EXAMPLE 11

$$\overline{\begin{bmatrix} i & 0 \\ -1 & 1 + i \end{bmatrix}} = \begin{bmatrix} -i & 0 \\ -1 & 1 - i \end{bmatrix}$$

It is easy to verify the following properties:

$$\overline{\overline{A}} = A, \qquad \overline{A \pm B} = \overline{A} \pm \overline{B}, \qquad \overline{AB} = \overline{A}\,\overline{B}, \qquad \overline{A^{-1}} = \left(\overline{A}\right)^{-1}, \qquad \overline{A^T} = \left(\overline{A}\right)^T$$

The **real part**, Re(A), and the **imaginary part**, Im(A), of a matrix A are the matrices whose entries are the real and imaginary parts of the corresponding entries of A.

■ EXAMPLE 12 Let

$$A = \begin{bmatrix} i & 2 \\ 1 - i & -i \end{bmatrix}$$

Then

$$\text{Re}(A) = \begin{bmatrix} 0 & 2 \\ 1 & 0 \end{bmatrix}, \qquad \text{Im}(A) = \begin{bmatrix} 1 & 0 \\ -1 & -1 \end{bmatrix}$$

Computing determinants is the same as with real entries:

■ EXAMPLE 13

$$\begin{vmatrix} i & 1+i & 1 \\ 2i & 0 & i \\ -2i & 2 & -i \end{vmatrix} = -(1+i)\begin{vmatrix} 2i & i \\ -2i & -i \end{vmatrix} - 2\begin{vmatrix} i & 1 \\ 2i & i \end{vmatrix} = 2 + 4i$$

Some Special Square Matrices

Certain types of matrices, now called Hermitian, were introduced by the French mathematician Hermite.[3] These matrices are useful in engineering, mathematics, and physics, especially in atomic physics.

DEFINITION

Let A be a square matrix. A is **Hermitian** if $\overline{A}^T = A$. A is **skew-Hermitian** if $\overline{A}^T = -A$.

■ EXAMPLE 14 A is Hermitian and B is skew-Hermitian.

$$A = \begin{bmatrix} -1 & 2i & 3 \\ -2i & -2 & -i \\ 3 & i & -3 \end{bmatrix}, \qquad B = \begin{bmatrix} 0 & 2-i \\ -2-i & i \end{bmatrix}$$

The following theorem summarizes the basic properties of Hermitian and skew-Hermitian matrices. Its proof is left as an exercise.

THEOREM 1

1. The main diagonal of a Hermitian matrix consists of real numbers.
2. The main diagonal of a skew-Hermitian matrix consists of 0s or pure imaginary numbers.
3. A matrix that is both Hermitian and skew-Hermitian is a zero matrix.
4. If A and B are Hermitian, so are $A + B$, $A - B$, and cA for any real scalar c.
5. If A and B are skew-Hermitian so are $A + B$, $A - B$, and cA for any real scalar c.

We also have the following.

THEOREM 2

Every square matrix A can be written uniquely as a sum of a Hermitian matrix S and a skew-Hermitian matrix R. More precisely,

$$A = H + R \quad \text{with} \quad H = \frac{1}{2}\left(A + \overline{A}^T\right), \qquad R = \frac{1}{2}\left(A - \overline{A}^T\right)$$

[3]**Charles Hermite** was born in Dieuze, Lorraine, France, in 1822. He studied mathematics on his own. He is known for many beautiful, though relatively technical, results. He proved that the number e is transcendental and that the roots of arbitrary fifth-degree polynomials may be expressed in terms of elliptic modular functions.

The proof is also left as an exercise.

▪ EXAMPLE 15

$$A = \begin{bmatrix} 4 & 1-i \\ -1-3i & i \end{bmatrix} = \begin{bmatrix} 4 & i \\ -i & 0 \end{bmatrix} + \begin{bmatrix} 0 & 1-2i \\ -1-2i & i \end{bmatrix} = H + R$$

DEFINITION   A square matrix A is **unitary** if $\overline{A}^T = A^{-1}$, or equivalently, if $\overline{A}^T A = I$.

▪ EXAMPLE 16   Show that the matrices are unitary.

$$A = \begin{bmatrix} \frac{1}{2} & -\frac{\sqrt{3}}{2}i \\ -\frac{\sqrt{3}}{2}i & \frac{1}{2} \end{bmatrix}, \qquad B = \begin{bmatrix} 0 & i & 0 \\ 0 & 0 & i \\ 1 & 0 & 0 \end{bmatrix}$$

SOLUTION

$$\overline{A}^T A = \begin{bmatrix} \frac{1}{2} & \frac{\sqrt{3}}{2}i \\ \frac{\sqrt{3}}{2}i & \frac{1}{2} \end{bmatrix} \begin{bmatrix} \frac{1}{2} & -\frac{\sqrt{3}}{2}i \\ -\frac{\sqrt{3}}{2}i & \frac{1}{2} \end{bmatrix} = I_2$$

Likewise, $\overline{B}^T B = I_3$.

▪ EXAMPLE 17   Compute the inverse of the unitary matrix.

$$A = \begin{bmatrix} \frac{1}{\sqrt{2}} & \frac{i}{\sqrt{2}} & 0 \\ \frac{i}{\sqrt{2}} & \frac{1}{\sqrt{2}} & 0 \\ 0 & 0 & 1 \end{bmatrix}$$

SOLUTION

$$A^{-1} = \overline{A}^T = \begin{bmatrix} \frac{1}{\sqrt{2}} & -\frac{i}{\sqrt{2}} & 0 \\ -\frac{i}{\sqrt{2}} & \frac{1}{\sqrt{2}} & 0 \\ 0 & 0 & 1 \end{bmatrix}$$

REMARK   If a square matrix is real, then $\overline{A} = A$. In this case the relations $\overline{A}^T = A$ and $\overline{A}^T = A^{-1}$ reduce to $A^T = A$ and $A^T = A^{-1}$. Thus,

- A real Hermitian matrix is symmetric;
- A real unitary matrix is orthogonal.

Note for a *real* skew-Hermitian matrix A, we have $\overline{A} = -A$. Such a matrix is called **skew-symmetric**.

B

Linear Algebra Commands

In this appendix we have collected **most** of the linear algebra commands in Maple, Mathematica, and MATLAB.

Maple

Nearly all of Maple's linear algebra commands are in the package linalg, which can be loaded by:

$$\text{with(linalg);}$$

The linalg Package Commands

GramSchmidt	JordanBlock	LUdecomp	QRdecomp	addcol
addrow	adjoint	angle	augment	backsub
band	basis	bezout	blockmatrix	charmat
charpoly	cholesky	col	coldim	colspace
colspan	companion	cond	copyinto	crossprod
curl	definite	delcols	delrows	det
diag	diverge	dotprod	eigenvals	eigenvects
entermatrix	equal	exponential	extend	ffgausselim
fibonacci	forwardsub	frobenius	gausselim	gaussjord
geneqns	genmatrix	grad	hadamard	hermite
hessian	hilbert	htranspose	ihermite	indexfunc
innerprod	intbasis	inverse	ismith	issimilar
iszero	jacobian	jordan	kernel	laplacian
leastsqrs	linsolve	matadd	matrix	minor
minpoly	mulcol	multiply	norm	normalize

orthog	permanent	pivot	potential	randmatrix
randvector	rank	references	row	rowdim
rowspace	rowspan	scalarmul	singularvals	smith
stack	submatrix	subvector	sumbasis	swapcol
swaprow	sylvester	toeplitz	trace	transpose
vandermonde	vecpotent	vectdim	vector	wronskian

See also   array, evalm, identity, list, with, range, table, &*, ^, and &^.

Mathematica

Matrix Operations Commands

Det	Dot	Eigensystem	Eigenvalues
Eigenvectors	Inverse	LatticeReduce	LinearProgramming
LinearSolve	MatrixExp	MatrixPower	Minors
NullSpace	Outer	PseudoInverse	QRDecomposition
RowReduce	SchurDecomposition	SingularValues	Transpose

Linear Algebra Packages

Package Name	Commands
LinearAlgebra`Cholesky`	CholeskyDecomposition
LinearAlgebra`CrossProduct`	Cross
LinearAlgebra`GaussianElimination`	LU, LUFactor, LUSolve
LinearAlgebra`MatrixManipulation`	AppendColumns, AppendRows, BlockMatrix, HankelMatrix, HilbertMatrix, LowerDiagonalMatrix, SquareMatrixQ, SubMatrix, TakeColumns, TakeMatrix, TakeRows, TridiagonalMatrix, UpperDiagonalMatrix, ZeroMatrix
LinearAlgebra`Orthogonalization`	GramSchmidt, InnerProduct, Normalize, Normalized, Projection
LinearAlgebra`Tridiagonal`	TridiagonalSolve

See also   Array, ColumnForm, DiagonalMatrix, Dimensions, IdentityMatrix, Length, List, MatrixForm, MatrixPower, Range, Table, and . (Dot).
    A package, say, LinearAlgebra`Cholesky`, can be loaded by

<<LinearAlgebra`Cholesky`

MATLAB

Numerical Linear Algebra

\	/	balance	cdf2rdf	chol
cond	det	diag	eig	eye
fliplr	flipud	hess	inv	linspace
logspace	lscov	lu	meshgrid	nnls
norm	null	ones	orth	pinv
poly	polyeig	qr	qrdelete	qrinsert
qz	rand	randn	rank	rcond
reshape	rot90	rref	rsf2csf	schur
svd	trace	tril	triu	zeros

Symbolic Math Toolbox (Linear Algebra and Operations)

charpoly	colspace	determ	eigensys
inverse	jordan	linsolve	nullspace
numeric	singvals	solve	subs
sym	sym2poly	symadd	symdiv
symmul	symop	sympow	symsize
symsub	symvar	transpose	

See also   poly, polyval, polyvalm, and roots.

Answers to Selected Exercises

Chapter 1

Section 1.1

1. (a)–(e) Linear (f) Nonlinear
Only (c) is homogeneous.

3. (a1) x, y (a2) y (a3) x (b1) x, y (b2) None (b3) x, y
(c1) x, y, z (c2) x (c3) y, z (d1) x, y, z (d2) x (d3) y, z
(e1) x, y, z, w, t (e2) x (e3) y, z, w, t

5. Only S is on the plane.

7. (a) If $a = -2$, infinitely many solutions; if $a = 2$, no solutions; if $a \neq \pm 2$, the unique solution $x = 1/(a - 2)$

(b) If $a = \pm 2$, no solutions; if $a \neq \pm 2$, unique solution $x = 3/(a^2 - 4)$

(c) If $a = \pm 2$, infinitely many solutions; if $a \neq \pm 2$, the unique solution $x = 0$

(d) If $a = 0$, infinitely many solutions given by $x = r, y = t$; if $a \neq 0$, infinitely many solutions given by $x = 3 + ar, y = r$

9. The standard form is

$$2x + 4z = -1$$
$$-x + 2z + 2w = 2$$
$$-2x - z + 3w = -3$$
$$y + z + t - w = 4$$

(a)–(c) The coefficient matrix and the vector of constants are:

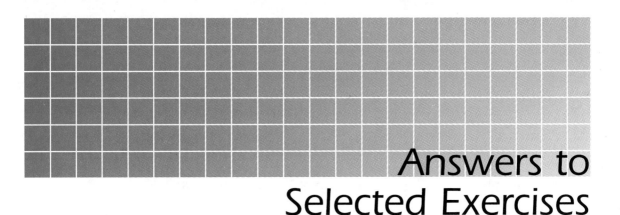

$$\begin{bmatrix} 2 & 0 & 4 & 0 & 0 \\ -1 & 0 & 2 & 0 & 2 \\ -2 & 0 & -1 & 0 & 3 \\ 0 & 1 & 1 & 1 & -1 \end{bmatrix}, \quad \begin{bmatrix} -1 \\ 2 \\ -3 \\ 4 \end{bmatrix}$$

The augmented matrix is

$$\begin{bmatrix} 2 & 0 & 4 & 0 & 0 & : & -1 \\ -1 & 0 & 2 & 0 & 2 & : & 2 \\ -2 & 0 & -1 & 0 & 3 & : & -3 \\ 0 & 1 & 1 & 1 & -1 & : & 4 \end{bmatrix}$$

(d) The associated homogeneous system is

$$2x + 4z = 0$$
$$-x + 2z + 2w = 0$$
$$-2x - z + 3w = 0$$
$$y + z + t - w = 0$$

11. General solution: $(-2t - 2r - s, t, 2r, s/2, s, r)$ for $r, s, t \in \mathbf{R}$

13. $x_6 = r, x_5 = r, x_3 = r, x_4 = s, x_2 = t, x_1 = 1 + t - 6r + 5s$ for all r, s, t

15. $P(\frac{1}{6}, \frac{5}{6})$

17. In the second system the first equation is a multiple of the first equation of the first system. The last equation is the sum of all three equations of the first system.

19. $x = -1, y = 2, z = 2$

21. No solutions

23. $(r - 1, 2r + 3, r), r \in \mathbf{R}$

25. $(-\frac{r}{3} - 9, \frac{r}{3} + 3, -6, r), r \in \mathbf{R}$

27. $(1 - r, r, 1 - r, r), r \in \mathbf{R}$

29. $(\frac{37}{4}, \frac{17}{4}, \frac{11}{4})$

31. $(1, 0, 3)$

33. $(3, -2, -4)$

35. $\theta = (2k - 1)\pi$, where k is any integer

41. 32

43. 16 days

45. The angles are $100°, 80°, 100°, 80°$

47. $\frac{14}{3}$ and $\frac{20}{3}$ or 6 and 4.

49. 14 rows

51. The boats traveled 24 miles

Section 1.2

1. (a) Not echelon form (b) Reduced row echelon form (c) Not echelon form (d) Row echelon form but not reduced row echelon form

3. (a) Row echelon form but not reduced row echelon form (b) Row echelon form but not reduced row echelon form (c) Reduced row echelon form (d) Not echelon form

5. (a) Not row echelon form (b) Row echelon form but not reduced row echelon form (c) Reduced row echelon form (d) Reduced row echelon form

9. Both A and B have I_3 as a reduced row echelon form. Hence, $A \sim I$ and $B \sim I$. Therefore, $A \sim I$ and $I \sim B$, by Exercise 8. Hence, $A \sim B$ by Exercise 7.

11. Yes. The second matrix is the reduced row echelon form of the first.

13. True, because the respective reduced row echelon forms, $\begin{bmatrix} 1 & 0 & 1 \\ 0 & 1 & 1 \end{bmatrix}$ and $\begin{bmatrix} 1 & 0 & -2 \\ 0 & 1 & 1 \end{bmatrix}$, are not equal.

15. Because $\cos\theta\cos\theta - (-\sin\theta)(\sin\theta) = \cos^2\theta + \sin^2\theta = 1 \neq 0$, the matrix reduces to I by Exercise 14.

17. (a) $\begin{bmatrix} 1 & 1 & 1 \\ 0 & 1 & -1 \\ 0 & 0 & 2 \end{bmatrix}$ and $\begin{bmatrix} 1 & 0 & 0 \\ 0 & 1 & 0 \\ 0 & 0 & 1 \end{bmatrix}$

(b) $\begin{bmatrix} 1 & 0 & 1 & 1 \\ 0 & -1 & 0 & -1 \\ 0 & 0 & 0 & -1 \end{bmatrix}$ and $\begin{bmatrix} 1 & 0 & 1 & 0 \\ 0 & 1 & 0 & 0 \\ 0 & 0 & 0 & 1 \end{bmatrix}$

19. $\begin{bmatrix} 1 & 0 & 0 & -1 & 0 \\ 0 & 1 & 0 & 0 & 0 \\ 0 & 0 & 1 & 0 & 1 \\ 0 & 0 & 0 & 0 & 0 \end{bmatrix}$

21. $\begin{bmatrix} 1 & 4 & 0 & 5 & 0 & 6 \\ 0 & 0 & 1 & 4 & 0 & 4 \\ 0 & 0 & 0 & 0 & 1 & 2 \end{bmatrix}$

23. $x = 1, y = 2, z = -2, w = -4$

25. $x_1 = -4t - 5s, x_2 = 6 - 2t - 6s, x_3 = t, x_4 = s, x_5 = -3$

27. No solution

29. $x = 1, y = 2, z = -2, w = -4$

31. $x = r_1 - 2, y = -r_2 + 1, z = r_1 - r_2 - 4, w = r_2, t = r_1$

33. $x = 0, y = 0, z = 0, w = 0$

35. $x = \frac{1}{2}a + \frac{1}{2}b, y = \frac{1}{2}a - \frac{1}{2}b, z = -b$

37. $x_1 = r - 1, x_2 = r, x_3 = 1, x_4 = -2$

39. $x_1 = 6 + r, x_2 = r - 1, x_3 = 4, x_4 = r, x_5 = -1$

41. $x_1 = 0, x_2 = 0, x_3 = 0, x_4 = 0, x_5 = 0$

43. The systems are equivalent because both augmented matrices have the same reduced echelon form:
$$\begin{bmatrix} 1 & 4 & 0 & 5 & 0 & 6 \\ 0 & 0 & 1 & 4 & 0 & 4 \\ 0 & 0 & 0 & 0 & 1 & 2 \end{bmatrix}$$

45. (a) The last column is nonpivot, so there are solutions. Because the third column is nonpivot there are infinitely many solutions. (b) The last column is pivot, so there are no solutions.

47. (a) If the last column is pivot, there are no solutions. Otherwise, there are infinitely many solutions. (b) No solutions (c) If the last column is pivot, there are no solutions. Otherwise, there is exactly one solution. (d) No solutions

53. (a) If $a = 4$, there are infinitely many solutions. If $a \neq 4$, there is exactly one solution.

(b) If $a = 3$, there are infinitely many solutions. If $a = -3$, there are no solutions. If $a \neq -3, 3$, there is only one solution.

Section 1.3

1. Yes

3. No

5. $5x + y = 14, x - 2y = -6$

7. $x = 1.99, y = 3.97$; the exact solution is $x = 2, y = 4$.

9. $x = 1.0011, y = 4.9990, z = -2.0022$

11. $x = 2.0176, y = -1.9952, z = 2.9936$

13. $x = 3.9984, y = -1.0003, z = 2.0005$

15. (a) The first five Gauss-Seidel iterates are $(2, 0)$, $(2, -2), (0, -2), (0, 0), (2, 0)$. Because the first and last iterates are identical, these values are repeated as k grows larger. Thus the iteration diverges.

(b) The fourth and fifth Gauss-Seidel iterates are $(-0.5977, 0.5977)$ and $(-0.6006, 0.6006)$. So the iteration converges to $(-0.6, 0.6)$ to at least two decimal places.

17. $x = 3.1, y = 1.1$

19. $x = 2, y = -1, z = 1$

Section 1.4

1. The relation is $B = 3A - 1000$, which is linear.

3. 80,000 yen, 900 francs and 1,200 marks

5. Boston 30°, New York 36°, Montreal 24°

7. The volumes of the solutions containing A, B, and C are 2.0 cm³, 3.5 cm³, 1.8 cm³.

9. $i_1 = \frac{3}{2} = 1.5, i_2 = 1, i_3 = \frac{1}{2} = 0.5$ A.

11. $x_1 = \frac{10}{7} = x_3$ and $x_2 = \frac{12}{7}$

13. The system is balanced if $w_1 = 8r, w_2 = 2r, w_3 = 5r,$ $w_4 = r$, where r is any positive real number.

15. $y = x^3 + 2x^2 - 3x + 1$

17. $a = \frac{1}{2}, b = -\frac{3}{2},$ and $c = 1$

19. $A = -\frac{1}{3}$ and $B = \frac{1}{3}$

Chapter 2

Section 2.1

1. (a) $\begin{bmatrix} 1 \\ -1 \end{bmatrix}$ (b) $\begin{bmatrix} -5 \\ -9 \end{bmatrix}$

3. (a) $\begin{bmatrix} 4 \\ -9 \\ 4 \end{bmatrix}$ (b) $\begin{bmatrix} 9 \\ -10 \\ 7 \end{bmatrix}$

5. (a) $\begin{bmatrix} -39 \\ 32 \end{bmatrix}$ (b) $\begin{bmatrix} 40 \\ 5 \end{bmatrix}$

7.

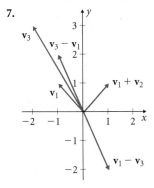

9.

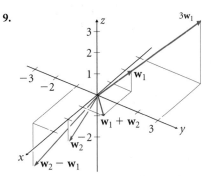

11. (a) $a = 0, b = 0$ (b) No values

13. $x = \frac{3}{2}a + 2b$

15. $x = -a - 3b, y = -2a - 4b$

17. (a) $PQ = (0, -2)$ (b) $PQ = (-3, 2, 2)$

(a)

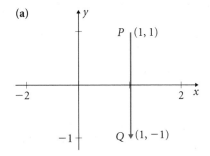

(b)

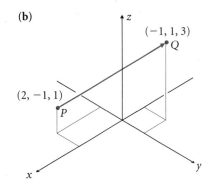

19. Yes

21. No

23. No

25. Yes

27. It is.

29. No, it is not.

31. It is.

33. True, by Theorem 2, because $\begin{bmatrix} 1 & 1 & 0 \\ 1 & 0 & -1 \\ 1 & -1 & 1 \end{bmatrix}$ has three pivots.

35. $\begin{bmatrix} 0 \\ 0 \\ 1 \end{bmatrix}$

37. $\mathbf{R}^2$

39. $\left\{ \begin{bmatrix} r \\ s \\ 0 \end{bmatrix}, \ r, s \in \mathbf{R} \right\}$

41. $\mathbf{R}^3$

43. Any nonzero k

45. (a) $x \begin{bmatrix} 1 \\ -1 \\ 0 \end{bmatrix} + y \begin{bmatrix} 2 \\ 1 \\ 1 \end{bmatrix} + z \begin{bmatrix} -1 \\ 0 \\ -1 \end{bmatrix} = \begin{bmatrix} 1 \\ 0 \\ 2 \end{bmatrix}$

(b) $x \begin{bmatrix} 1 \\ -1 \\ 0 \end{bmatrix} + y \begin{bmatrix} -2 \\ 0 \\ 1 \end{bmatrix} + z \begin{bmatrix} -1 \\ 2 \\ 0 \end{bmatrix} = \begin{bmatrix} 1 \\ -1 \\ 2 \end{bmatrix}$

47. (a) No   (b) Yes

51. (a) True (b) True (c) False (d) True (e) False (f) False (g) False (h) True

53. $(10, 20)$ mi/h (if the eastbound and northbound directions are along the positive x- and y-axes).

55. (a) $(1100, 1800, 2400)$; a total of $1100, $1800, $2400 is spent for food per trip for first, second, and third class, respectively.

(b) $(150, 100, 200)$; the airline spends $150, $100, $200 more for plane 3 than for plane 2 for food per trip for first, second, and third class, respectively.

(c) $(4500, 7000, 9000)$; the airline spends $4500, $7000, and $9000 for 10 trips of plane 3 for food for first, second, and third class, respectively.

(d) $(8900, 14600, 19300)$; the airline spends $8900, $14,600, and $9300 in food for first, second, and third class, respectively for seven trips of the first plane, eight trips of the second and nine trips of the third.

Section 2.2

1. $\|\mathbf{u}\| = 3, \|\mathbf{v}\| = 5\sqrt{2}, \|\mathbf{w}\| = 2\sqrt{5},$
$\|\mathbf{u} + \mathbf{v}\| = \sqrt{19}, \|\mathbf{u} - \mathbf{v}\| = 3\sqrt{11},$
$\|\mathbf{u} - \mathbf{v} + \mathbf{w}\| = \sqrt{139}, \|\mathbf{d}\| = 3,$
$\|10\mathbf{d}\| = 30, \|\|\mathbf{d}\| \, \mathbf{d}\| = 9$

3. $\mathbf{u} \cdot \mathbf{v} = -20, \ \mathbf{w} \cdot \mathbf{u} = 0, \ \mathbf{u} \cdot (\mathbf{v} + \mathbf{w}) = -20,$
$\mathbf{v} \cdot \mathbf{u} + \mathbf{w} \cdot \mathbf{u} = -20,$
$\mathbf{d} \cdot \mathbf{d} = 9, \ (\mathbf{d} \cdot \mathbf{d})\mathbf{d} = \left(-9, -18, 9, 9\sqrt{3} \right)$

5. (a) $\left(-\frac{1}{3}, \frac{2}{3}, -\frac{2}{3} \right)$ (b) $\left(\frac{2\sqrt{2}}{5}, -\frac{3\sqrt{2}}{10}, \frac{1\sqrt{2}}{2} \right)$

(c) $\left(-\frac{2\sqrt{5}}{5}, -\frac{\sqrt{5}}{5}, 0 \right)$ (d) $\left(-\frac{1}{3}, -\frac{2}{3}, \frac{1}{3}, \frac{1}{3}\sqrt{3} \right)$

7. $\left(\frac{2}{3}, -\frac{4}{3}, \frac{4}{3} \right)$

9. $\left(-3, -6, 3, 3\sqrt{3} \right)$

11. (a), (b) and (d)

13. $\left(\frac{\sqrt{5}}{5}, 0, \frac{2\sqrt{5}}{5} \right)$

15. (a) 0.24871. (b) 0.3681 rad. (c) $\frac{2}{3}\pi$

17. (a) $\left(\frac{8}{5}, \frac{16}{5} \right)$ (b) $\left(\frac{33}{35}, \frac{64}{35}, \frac{9}{7} \right)$ (c) $\left(-2, -1, \frac{3}{10}, \frac{1}{10} \right)$

21. No

25. The xy-plane circle of radius 1 centered at the origin

29. $\|\mathbf{v}\| = \|c\mathbf{u}\| = |c| \, \|\mathbf{u}\| = c1 = c$, because $c > 0$ and $\mathbf{u}$ is unit. Hence,
$$\mathbf{u} = \frac{1}{c}\mathbf{v} = \frac{1}{\|\mathbf{v}\|}\mathbf{v} = (\cos\alpha, \cos\beta, \cos\gamma) \text{ by } (2.10).$$

31. (a) $(1, 1, 1) = \sqrt{3}\left(\frac{\sqrt{3}}{3}, \frac{\sqrt{3}}{3}, \frac{\sqrt{3}}{3} \right)$. Each direction cosine is $\frac{\sqrt{3}}{3}$ and each direction angle is 0.95532 rad.

(b) $(-1, 1, -1) = \sqrt{3}\left(-\frac{\sqrt{3}}{3}, \frac{\sqrt{3}}{3}, -\frac{\sqrt{3}}{3} \right)$ with direction cosines $-\frac{\sqrt{3}}{3}, \frac{\sqrt{3}}{3}, -\frac{\sqrt{3}}{3}$ and direction angles $2.1863, 0.95532,$ and 2.1863 rad.

(c) $(\sqrt{3}, 1, 0) = 2\left(\frac{\sqrt{3}}{2}, \frac{1}{2}, 0 \right)$ with direction cosines $\frac{\sqrt{3}}{2}, \frac{1}{2}, 0$ and direction angles $\frac{\pi}{6}, \frac{\pi}{3}, \frac{\pi}{2}$.

(d) $(2, -2, 0) = 2\sqrt{2}\left(\frac{\sqrt{2}}{2}, -\frac{\sqrt{2}}{2}, 0 \right)$ with direction cosines: $\frac{\sqrt{2}}{2}, -\frac{\sqrt{2}}{2}, 0$ and direction angles: $\frac{\pi}{4}, \frac{3\pi}{4}, \frac{\pi}{2}$.

Section 2.3

1. (a) False (b) True (c) True (d) True

3. (a) Yes (b) Yes (c) Yes (d) Yes (e) Yes (f) No (g) Yes

5. (a) Yes (b) No (c) No

7. (a) Yes (b) No

9. (a) Yes (b) No

11. (a) No (b) No

13. Yes

15. (a) False (b) True (c) False (d) True (e) True (f) False
(g) True

21. (a) No (b) Yes

23. $\left\{ \begin{bmatrix} 3 \\ 0 \\ -1 \end{bmatrix}, \begin{bmatrix} -1 \\ 4 \\ 0 \end{bmatrix} \right\}$

25. That the system $[A : \mathbf{b}]$ is consistent

27. All $x \neq -1$

29. The corresponding spans are the shaded planes in the following figures.

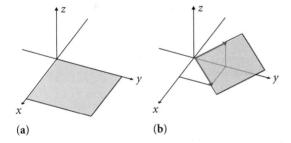

(a)   (b)

Section 2.4

1. Linearly independent

3. Linearly independent

5. Linearly independent

7. Linearly dependent

9. Linearly independent

11. Linearly dependent

13. Linearly dependent

15. Linearly dependent

17. Linearly independent

19. $a = -1, 2$

21. Yes

23. $c_1 = -1, c_2 = -1, c_3 = -2$

31. The intersection is a straight line through the origin if the planes are distinct or the common plane if the planes coincide.

Section 2.5

1. $A\mathbf{u} = \begin{bmatrix} -10 \\ -4 \\ 23 \end{bmatrix}$, $A\mathbf{v}$ and $A\mathbf{w}$ are undefined.

3. If $\mathbf{x} = (x_1, x_2)$ then the system is $-3x_1 - 2x_2 = 100$, $-x_1 = 200, 5x_1 - 3x_2 = 300$.

5. $\begin{bmatrix} 1 & -7 \\ -2 & 4 \end{bmatrix} \begin{bmatrix} x \\ y \end{bmatrix} = \begin{bmatrix} -5 \\ 0 \end{bmatrix}$

7. $n = 2$; the set is all of $\mathbf{R}^2$.

9. $n = 2$; the set is all of $\mathbf{R}^2$.

11. $(33, 44, 40)$; there were 33 students in 1999, 44 in spring 2000, and 40 in fall 2000.

13. 6 and 18

15. $\mathbf{u}$

17. The solution set of the system is
$S = \left\{ \begin{bmatrix} 10 + 3r \\ r \end{bmatrix}, r \in \mathbf{R} \right\}$. $\mathbf{p} = \begin{bmatrix} 10 \\ 0 \end{bmatrix}$ is a particular solution. The null space of the coefficient matrix is
$N = \left\{ \begin{bmatrix} 3 \\ 1 \end{bmatrix}, r \in \mathbf{R} \right\}$. Clearly $S = \mathbf{p} + N$, as claimed by Theorem 19.

Section 2.6

1. $(4, -3, -5)$, $(-10, 20, -20)$,
$(-80, -40, 0)$, $(6, -12, -10)$,
$(6, -12, -10)$, $(9, -18, -22)$

3. $-5 \qquad 5$
$\;\;\;5 \qquad -5$

5. $\|\mathbf{u} \times \mathbf{v}\|^2 = \|(-29, -18, -15)\|^2 = 1390$, and
$\|\mathbf{u}\|^2 \|\mathbf{v}\|^2 - (\mathbf{u} \cdot \mathbf{v})^2 = 26(61) - 196 = 1390$

7. $\dfrac{1}{\sqrt{2249}} (16, 12, 43)$

9. $\sqrt{6}$

11. 40

13. (a) They are. (b) They are not.

15. No

Section 2.7

1. P and Q

3. $(5, -4, 2), (2, -2, 1), (8, -6, 3)$

5. l_1 and l_2

7. The point of intersection is $(11, -8, 4)$

9. $l_1 \colon \dfrac{x-5}{-3} = \dfrac{y+4}{2} = \dfrac{z-2}{-1}$;

$l_2 \colon \dfrac{x-1}{6} = \dfrac{y-6}{-4} = \dfrac{z-8}{2}$;

$l_3 \colon \dfrac{x-5}{-1} = \dfrac{y+7}{-1} = \dfrac{z-11}{1}$;

$l_4 \colon \dfrac{x-14}{1} = \dfrac{y+2}{2} = \dfrac{z-13}{3}$

11. $x = 5 + 4t, y = -4 - 3t, z = 2 + t$

13. $x = 5 - 3t, y = -4 + 2t, z = 2 - t$

15. $\dfrac{x-3}{-4} = \dfrac{y+1}{3} = \dfrac{z+2}{7}$

17. P and Q

19. $-6(x - 5) + 4(y + 4) + 5(z - 2) = 0$

21. $-6x + 4y + 5z + 36 = 0$

23. $x + 2y + z + 1 = 0$

25. $x + y - z + 1 = 0$

27. $x - 6y + 2z - 21 = 0$

29. $\dfrac{6}{\sqrt{114}}$

31. $-x_1 + 3x_2 - 2x_3 + 8x_4 + 4x_5 + 3 = 0$

Section 2.8

1. Averaging once yields $(1, \frac{5}{2}, 5, \frac{9}{2}, \frac{5}{2}, 6, 5, \frac{11}{2})$, and averaging twice gives $(\frac{1}{2}, \frac{7}{4}, \frac{15}{4}, \frac{19}{4}, \frac{7}{2}, \frac{17}{4}, \frac{11}{2}, \frac{21}{4})$. In the following figure, plot I is the original, plot II is averaging once, and plot III is averaging twice.

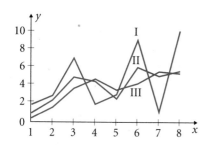

3. $\begin{bmatrix} Y_{k+1} \\ A_{k+1} \end{bmatrix} = \begin{bmatrix} 2 & 10 \\ \frac{4}{5} & 0 \end{bmatrix} \begin{bmatrix} Y_k \\ A_k \end{bmatrix}$;

initial condition: $\begin{bmatrix} Y_0 \\ A_0 \end{bmatrix} = \begin{bmatrix} 100 \\ 100 \end{bmatrix}$;

after three time units: $\begin{bmatrix} Y_3 \\ A_3 \end{bmatrix} = \begin{bmatrix} 16000 \\ 2560 \end{bmatrix}$

5. $\begin{bmatrix} Y_{k+1} \\ A_{k+1} \end{bmatrix} = \begin{bmatrix} 4 & 10 \\ \frac{1}{2} & 0 \end{bmatrix} \begin{bmatrix} Y_k \\ A_k \end{bmatrix}$;

initial condition: $\begin{bmatrix} Y_0 \\ A_0 \end{bmatrix} = \begin{bmatrix} 100 \\ 100 \end{bmatrix}$;

after three time units: $\begin{bmatrix} Y_3 \\ A_3 \end{bmatrix} = \begin{bmatrix} 31400 \\ 3050 \end{bmatrix}$

7. $\begin{bmatrix} Y_{k+1} \\ A_{k+1} \end{bmatrix} = \begin{bmatrix} 3 & 12 \\ \frac{1}{3} & 0 \end{bmatrix} \begin{bmatrix} Y_k \\ A_k \end{bmatrix}$;

initial condition: $\begin{bmatrix} Y_0 \\ A_0 \end{bmatrix} = \begin{bmatrix} 300 \\ 100 \end{bmatrix}$;

after three time units: $\begin{bmatrix} Y_3 \\ A_3 \end{bmatrix} = \begin{bmatrix} 30900 \\ 2500 \end{bmatrix}$

9. $\begin{bmatrix} A_{k+1} \\ B_{k+1} \\ C_{k+1} \end{bmatrix} = \begin{bmatrix} \frac{1}{4} & \frac{5}{2} & \frac{3}{2} \\ \frac{1}{4} & 0 & 0 \\ 0 & \frac{1}{3} & 0 \end{bmatrix} \begin{bmatrix} A_k \\ B_k \\ C_k \end{bmatrix}$;

initial condition: $\begin{bmatrix} A_0 \\ B_0 \\ C_0 \end{bmatrix} = \begin{bmatrix} 4800 \\ 4800 \\ 4800 \end{bmatrix}$;

after 6 weeks: $\begin{bmatrix} A_3 \\ B_3 \\ C_3 \end{bmatrix} = \begin{bmatrix} 15975 \\ 2625 \\ 1700 \end{bmatrix}$

11. 0; the point is on the plane.

13. $\dfrac{2\sqrt{14}}{7}$

15. $\dfrac{4\sqrt{146}}{73}$

19. $\left(0, \frac{3}{4}, 0\right)$

21. $\left(-\frac{2}{3}, \frac{29}{12}, -\frac{25}{12}\right)$

23. $\mathbf{R} \cdot \mathbf{d} = (1, 1, 1) \cdot (1, 9, -7) = 3$

25. $25 \cos 45° = \dfrac{25\sqrt{2}}{2}$ lb

Chapter 3

Section 3.1

1. A: rows $\begin{bmatrix} -1 & 0 \end{bmatrix}, \begin{bmatrix} 2 & 3 \end{bmatrix}, \begin{bmatrix} -2 & 1 \end{bmatrix}$;

columns $\begin{bmatrix} -1 \\ 2 \\ -2 \end{bmatrix}, \begin{bmatrix} 0 \\ 3 \\ 1 \end{bmatrix}$; size 3×2; $(2, 2)$ entry: 3; $(3, 1)$ entry: -2;

B: rows $\begin{bmatrix} -1 & 0 & -2 \end{bmatrix}, \begin{bmatrix} 2 & 2 & 1 \end{bmatrix}$;

columns $\begin{bmatrix} -1 \\ 2 \end{bmatrix}, \begin{bmatrix} 0 \\ 2 \end{bmatrix}, \begin{bmatrix} -2 \\ 1 \end{bmatrix}$; size 2×3; $(2, 2)$ entry: 2; $(2, 3)$ entry: 1.

3. **(a)** The system $x = 1$, $y - 2 = 0$, $x - y = 0$ is inconsistent. **(b)** The system $x + y = 1$, $-y + z = 1$, $x + z = 3$ is inconsistent.

5. **(c)** and **(e)** are impossible due to incompatible sizes.

(a) $\begin{bmatrix} 1 & 0 \\ 1 & 3 \end{bmatrix}$

(b) $\begin{bmatrix} -3 & -3 & 3 \\ -3 & -3 & 3 \end{bmatrix}$

(d) $\begin{bmatrix} -9 & 9 \\ 9 & -9 \\ -9 & 9 \end{bmatrix}$

(f) $\begin{bmatrix} -12 & 26 \\ -40 & -18 \end{bmatrix}$

7. **(a)** $\begin{bmatrix} -4 & \frac{1}{2} \\ \frac{1}{2} & -1 \end{bmatrix}$

(b) $\begin{bmatrix} \frac{17}{2} & -4 \\ -\frac{25}{2} & 8 \end{bmatrix}$

(c) $\begin{bmatrix} 4 & -16 \\ -52 & 32 \end{bmatrix}$

11. $\begin{bmatrix} 13 & 12 & 9 & 8 \end{bmatrix}$

13. 1

15. $A^8 = \begin{bmatrix} 1 & 8 \\ 0 & 1 \end{bmatrix}$; $A^n = \begin{bmatrix} 1 & n \\ 0 & 1 \end{bmatrix}$

19. $\begin{bmatrix} 1 & 1 \\ 0 & 0 \end{bmatrix}$

21. $A = \begin{bmatrix} 1 & 1 \\ 0 & 0 \end{bmatrix}$, $B = \begin{bmatrix} 1 & 0 \\ 1 & 0 \end{bmatrix}$

23. Because $AB = BA$, we have

$$(AB)^2 = (AB)(AB) = A(BA)B$$
$$= A(AB)B = (AA)(BB) = A^2B^2$$

27. $\begin{bmatrix} 1 & 2 \\ 4 & 5 \end{bmatrix}$, $\begin{bmatrix} 2 & 3 \\ 5 & 6 \end{bmatrix}$, $\begin{bmatrix} 1 & 3 \\ 4 & 6 \end{bmatrix}$

31. $\begin{bmatrix} 64 & 32 & 24 \\ 80 & 40 & 32 \\ 48 & 24 & 16 \end{bmatrix}$

Section 3.2

1. $\begin{bmatrix} \frac{1}{2} & -\frac{1}{2} \\ -\frac{1}{4} & \frac{3}{4} \end{bmatrix}$, $\begin{bmatrix} \frac{1}{2} & -\frac{5}{8} \\ -\frac{1}{2} & \frac{7}{8} \end{bmatrix}$

3. $\begin{bmatrix} 4 & -\frac{3}{2} \\ -1 & \frac{1}{2} \end{bmatrix}$

5. $A^{-1} = \dfrac{1}{a^2 + b^2} \begin{bmatrix} a & -b \\ b & a \end{bmatrix} = \begin{bmatrix} a & -b \\ b & a \end{bmatrix}$

7. By Theorem 7

$$(2A)^3 = 2^3 A^3$$
$$= 8 \begin{bmatrix} 1 & 1 \\ -5 & -2 \end{bmatrix} = \begin{bmatrix} 8 & 8 \\ -40 & -16 \end{bmatrix}$$

Hence,

$$(2A)^{-3} = \begin{bmatrix} 8 & 8 \\ -40 & -16 \end{bmatrix}^{-1}$$
$$= \begin{bmatrix} -\frac{1}{12} & -\frac{1}{24} \\ \frac{5}{24} & \frac{1}{24} \end{bmatrix}$$

9. $\begin{bmatrix} -1 & 0 & 0 \\ 0 & 1 & 0 \\ 0 & 0 & -1 \end{bmatrix}$, $\begin{bmatrix} \frac{1}{5} & 0 & 0 \\ 0 & \frac{1}{5} & 0 \\ 0 & 0 & \frac{1}{5} \end{bmatrix}$

13. **(a)** $\begin{bmatrix} -1 & -1 & 0 \\ 0 & -1 & 0 \\ 0 & 0 & 1 \end{bmatrix}$

(b) $\begin{bmatrix} -\frac{1}{2} & 0 & \frac{1}{4} \\ 0 & -1 & 1 \\ 0 & 0 & -\frac{1}{2} \end{bmatrix}$

(c) Noninvertible

(d) $\begin{bmatrix} -1 & -1 & 1 \\ 0 & -1 & 1 \\ -1 & -2 & 1 \end{bmatrix}$

15. $\begin{bmatrix} -1 & 0 & 1 & 2 \\ 0 & 0 & 1 & 1 \\ -1 & 1 & 1 & 2 \\ -1 & 1 & 1 & 1 \end{bmatrix}$

17. **(a)** $\begin{bmatrix} 1 & 1 & -1 \\ 1 & 0 & -1 \\ -1 & 1 & 0 \end{bmatrix} \begin{bmatrix} 1 \\ 2 \\ 3 \end{bmatrix} = \begin{bmatrix} -4 \\ -1 \\ -6 \end{bmatrix}$

(b) $\begin{bmatrix} 1 & 1 & -1 \\ 1 & 0 & -1 \\ -1 & 1 & 0 \end{bmatrix} \begin{bmatrix} 2 \\ 4 \\ -1 \end{bmatrix} = \begin{bmatrix} 7 \\ 3 \\ 2 \end{bmatrix}$

19. $A = \begin{bmatrix} \frac{1}{2} & \frac{1}{2} & -\frac{1}{2} \\ \frac{1}{2} & -\frac{1}{2} & \frac{1}{2} \\ 0 & 0 & 1 \end{bmatrix}$

21. $A^{-1} = \begin{bmatrix} 0 & 0 & \frac{1}{c_3} \\ 0 & \frac{1}{c_2} & 0 \\ \frac{1}{c_1} & 0 & 0 \end{bmatrix}$

$$B^{-1} = \begin{bmatrix} 0 & 0 & 0 & \frac{1}{c_4} \\ 0 & 0 & \frac{1}{c_3} & 0 \\ 0 & \frac{1}{c_2} & 0 & 0 \\ \frac{1}{c_1} & 0 & 0 & 0 \end{bmatrix}$$

23. $A^{-1} = \begin{bmatrix} 1 & 0 & 0 \\ 0 & -1 & 1 \\ 0 & 0 & 1 \end{bmatrix}$, $A^{-2} = I$. Therefore, $A^{-2k} = I$

and $A^{-(2k-1)} = A^{-1}$, for any positive integer k. In particular, $A^{-24} = I$ and $A^{-25} = A^{-1}$.

25. $A = \begin{bmatrix} 1 & 0 \\ 0 & 1 \end{bmatrix}$, $B = \begin{bmatrix} 0 & 2 \\ 1 & 0 \end{bmatrix}$

Section 3.3

1. A, C, D are elementary. The corresponding operations are $R_1 + R_2 \rightarrow R_1$, $R_1 - 2R_3 \rightarrow R_1$, and $R_2 \leftrightarrow R_4$.

3. $R_1 \leftrightarrow R_2, 2R_2 \rightarrow R_2, R_1 - 5R_3 \rightarrow R_1$, and $-R_1 + R_3 \rightarrow R_3$

5. (a) $R_1 \leftrightarrow R_3$ (b) $R_1 - 5R_3 \rightarrow R_1$
     (c) $2R_4 \rightarrow R_4$ (d) $-10R_1 + R_3 \rightarrow R_3$

7. $A = \begin{bmatrix} 1 & 0 \\ -\frac{1}{2} & 1 \end{bmatrix} \begin{bmatrix} 1 & 0 \\ 0 & \frac{1}{2} \end{bmatrix} \begin{bmatrix} 1 & 1 \\ 0 & 1 \end{bmatrix} \begin{bmatrix} 2 & 0 \\ 0 & 1 \end{bmatrix}$

$A^{-1} = \begin{bmatrix} \frac{1}{2} & 0 \\ 0 & 1 \end{bmatrix} \begin{bmatrix} 1 & -1 \\ 0 & 1 \end{bmatrix} \begin{bmatrix} 1 & 0 \\ 0 & 2 \end{bmatrix} \begin{bmatrix} 1 & 0 \\ \frac{1}{2} & 1 \end{bmatrix}$

9. $A = \begin{bmatrix} 1 & -2 \\ 0 & 1 \end{bmatrix} \begin{bmatrix} 3 & 0 \\ 0 & 1 \end{bmatrix} \begin{bmatrix} 1 & 0 \\ 0 & 3 \end{bmatrix}$

$B = \begin{bmatrix} 2 & 0 \\ 0 & 1 \end{bmatrix} \begin{bmatrix} 1 & 0 \\ 1 & 1 \end{bmatrix}$

$C = \begin{bmatrix} 1 & 0 & 0 \\ 0 & 1 & 0 \\ 0 & 1 & 1 \end{bmatrix} \begin{bmatrix} 1 & 0 & 0 \\ 0 & 1 & 0 \\ 1 & 0 & 1 \end{bmatrix}$

11. The matrix is singular, because its reduced row echelon form

$$\begin{bmatrix} 1 & 0 & 1 \\ 0 & 1 & 0 \\ 0 & 0 & 0 \end{bmatrix}$$

has a row of zeros.

13. This is true by Theorem 15, because the coefficient matrix is invertible. More precisely,

$$\begin{bmatrix} 1 & -1 \\ 1 & 2 \end{bmatrix}^{-1} = \begin{bmatrix} \frac{2}{3} & \frac{1}{3} \\ -\frac{1}{3} & \frac{1}{3} \end{bmatrix}$$

15. $A = \begin{bmatrix} 0 & 0 & 1 \\ 0 & 1 & 0 \\ 1 & 0 & 0 \end{bmatrix} \begin{bmatrix} c_3 & 0 & 0 \\ 0 & 1 & 0 \\ 0 & 0 & 1 \end{bmatrix}$

$\times \begin{bmatrix} 1 & 0 & 0 \\ 0 & c_2 & 0 \\ 0 & 0 & 1 \end{bmatrix} \begin{bmatrix} 1 & 0 & 0 \\ 0 & 1 & 0 \\ 0 & 0 & c_1 \end{bmatrix}$

$A^{-1} = \begin{bmatrix} 0 & 0 & c_3^{-1} \\ 0 & c_2^{-1} & 0 \\ c_1^{-1} & 0 & 0 \end{bmatrix}$

$= \begin{bmatrix} 1 & 0 & 0 \\ 0 & 1 & 0 \\ 0 & 0 & c_1^{-1} \end{bmatrix} \begin{bmatrix} 1 & 0 & 0 \\ 0 & c_2^{-1} & 0 \\ 0 & 0 & 1 \end{bmatrix}$

$\times \begin{bmatrix} c_3^{-1} & 0 & 0 \\ 0 & 1 & 0 \\ 0 & 0 & 1 \end{bmatrix} \begin{bmatrix} 0 & 0 & 1 \\ 0 & 1 & 0 \\ 1 & 0 & 0 \end{bmatrix}$

19. **(a)** False **(b)** True **(c)** False **(d)** False **(e)** True **(f)** True

21. $n \times m$

Section 3.4

1. $\mathbf{x} = \begin{bmatrix} -3 \\ 1 \end{bmatrix}$

3. $\mathbf{x} = \begin{bmatrix} 1 \\ -1 \\ -2 \end{bmatrix}$

5. $\mathbf{x} = \begin{bmatrix} 1 \\ 1 \\ 1 \end{bmatrix}$

7. $\begin{bmatrix} 1 & 0 \\ -5 & 1 \end{bmatrix} \begin{bmatrix} 2 & 1 \\ 0 & -7 \end{bmatrix}$

9. $\begin{bmatrix} 1 & 0 & 0 \\ -4 & 1 & 0 \\ 7 & -3 & 1 \end{bmatrix} \begin{bmatrix} -1 & 2 & 1 \\ 0 & 3 & -1 \\ 0 & 0 & -5 \end{bmatrix}$

11. $\begin{bmatrix} 1 & 0 & 0 & 0 \\ -3 & 1 & 0 & 0 \\ 0 & -2 & 1 & 0 \\ 5 & -1 & 0 & 1 \end{bmatrix} \begin{bmatrix} 4 & 1 & 1 & 2 \\ 0 & 2 & -1 & 2 \\ 0 & 0 & 3 & 2 \\ 0 & 0 & 0 & -1 \end{bmatrix}$

13. $\begin{bmatrix} 1 & 0 & 0 & 0 \\ -3 & 1 & 0 & 0 \\ 0 & -2 & 1 & 0 \\ 5 & -1 & 0 & 1 \end{bmatrix} \begin{bmatrix} 4 & 1 & 1 \\ 0 & 2 & -1 \\ 0 & 0 & 3 \\ 0 & 0 & 0 \end{bmatrix}$

15. $LU = \begin{bmatrix} 1 & 0 \\ 7 & 1 \end{bmatrix} \begin{bmatrix} 2 & 1 \\ 0 & -5 \end{bmatrix}$, $\mathbf{x} = \begin{bmatrix} -2 \\ 10 \end{bmatrix}$

17. $LU = \begin{bmatrix} 1 & 0 & 0 & 0 \\ 0 & 1 & 0 & 0 \\ -2 & 2 & 1 & 0 \\ 0 & 5 & 0 & 1 \end{bmatrix} \begin{bmatrix} 2 & 1 & 1 & 2 \\ 0 & 1 & -1 & 2 \\ 0 & 0 & 3 & 2 \\ 0 & 0 & 0 & 2 \end{bmatrix}$,

$\mathbf{x} = \begin{bmatrix} 1 \\ 1 \\ 0 \\ 0 \end{bmatrix}$

19. $\begin{bmatrix} 0 & 1 & 0 \\ 1 & 0 & 0 \\ 0 & 0 & 1 \end{bmatrix} A$

$= \begin{bmatrix} 1 & 0 & 0 \\ 0 & 1 & 0 \\ -2 & -1 & 1 \end{bmatrix} \begin{bmatrix} -1 & 2 & -4 \\ 0 & 1 & 1 \\ 0 & 0 & -6 \end{bmatrix}$

21. A $PA = LU$ factorization is given by

$\begin{bmatrix} 0 & 1 & 0 \\ 1 & 0 & 0 \\ 0 & 0 & 1 \end{bmatrix} A = \begin{bmatrix} 1 & 0 & 0 \\ 0 & 1 & 0 \\ 1 & -2 & 1 \end{bmatrix} \begin{bmatrix} 2 & 0 & 1 \\ 0 & 3 & -1 \\ 0 & 0 & -2 \end{bmatrix}$

Hence, the LU method on the system,

$$PA\mathbf{x} = P\mathbf{b} = \begin{bmatrix} -1 \\ -3 \\ -1 \end{bmatrix}$$

yields $\mathbf{x} = (-2, 0, 3)$.

23. A $PA = LU$ factorization is given by

$\begin{bmatrix} 0 & 0 & 1 \\ 0 & 1 & 0 \\ 1 & 0 & 0 \end{bmatrix} A = \begin{bmatrix} 1 & 0 & 0 \\ 0 & 1 & 0 \\ 0 & 1/2 & 1 \end{bmatrix} \begin{bmatrix} 2 & -5 & 1 \\ 0 & 2 & -4 \\ 0 & 0 & 3 \end{bmatrix}$

Hence, the LU method on the system

$$PA\mathbf{x} = P\mathbf{b} = \begin{bmatrix} -8 \\ 4 \\ 2 \end{bmatrix}$$

yields $\mathbf{x} = (1, 2, 0)$.

Section 3.5

1. $M_1 = \begin{bmatrix} 0 & 1 \\ 1 & 0 \end{bmatrix}$, $M_2 = \begin{bmatrix} 0 & 1 & 1 \\ 1 & 0 & 1 \\ 1 & 1 & 0 \end{bmatrix}$,

$M_3 = \begin{bmatrix} 0 & 0 & 0 \\ 0 & 0 & 0 \\ 0 & 0 & 0 \end{bmatrix}$

3. $A(G_1) = \begin{bmatrix} 0 & 1 \\ 1 & 0 \end{bmatrix}$, $A(G_2) = \begin{bmatrix} 0 & 1 & 1 \\ 1 & 0 & 1 \\ 1 & 1 & 0 \end{bmatrix}$,

$A(G_3) = \begin{bmatrix} 0 & 0 & 0 \\ 0 & 0 & 0 \\ 0 & 0 & 0 \end{bmatrix}$

5. $A(G_7) = \begin{bmatrix} 0 & 1 & 0 \\ 1 & 0 & 1 \\ 0 & 1 & 0 \end{bmatrix}$, $A(G_7)^2 = \begin{bmatrix} 1 & 0 & 1 \\ 0 & 2 & 0 \\ 1 & 0 & 1 \end{bmatrix}$

The number of walks of length 2 from **(a)** 1 to 1 is 1, **(b)** 1 to 2 is 0, **(c)** 1 to 3 is 1. These numbers agree with the $(1, 1), (1, 2), (1, 3)$ entries of $A(G_7)^2$.

7. $I(G_1) = \begin{bmatrix} 1 \\ 1 \end{bmatrix}$, $I(G_2) = \begin{bmatrix} 1 & 0 & 1 \\ 1 & 1 & 0 \\ 0 & 1 & 1 \end{bmatrix}$,

$I(G_3)$ does not exist.

9.

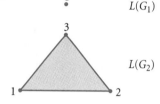

$L(G_1)$

$L(G_2)$

No graph $L(G_3)$

11. $A(L(G_7)) = \begin{bmatrix} 0 & 1 \\ 1 & 0 \end{bmatrix}$, $I(G_7) = \begin{bmatrix} 1 & 0 \\ 1 & 1 \\ 0 & 1 \end{bmatrix}$

Hence,

$I(G_7)^T I(G_7) - 2I_2$

$= \begin{bmatrix} 1 & 1 & 0 \\ 0 & 1 & 1 \end{bmatrix} \begin{bmatrix} 1 & 0 \\ 1 & 1 \\ 0 & 1 \end{bmatrix} - 2 \begin{bmatrix} 1 & 0 \\ 0 & 1 \end{bmatrix}$

$= \begin{bmatrix} 2 & 1 \\ 1 & 2 \end{bmatrix} - \begin{bmatrix} 2 & 0 \\ 0 & 2 \end{bmatrix}$

$= \begin{bmatrix} 0 & 1 \\ 1 & 0 \end{bmatrix} = A(L(G_7))$

13. Only C is stochastic.

15. G and H are doubly stochastic.

17. All powers of stochastic are doubly stochastic. Therefore, the given matrix is doubly stochastic, because

$\begin{bmatrix} \frac{1}{2} & \frac{1}{2} \\ \frac{1}{2} & \frac{1}{2} \end{bmatrix}$ is.

19. For A: $x = 0.8, y = 0.8$; for B: $x = \frac{3}{4}, y = \frac{1}{4}$.

21. (a) Because

$$(I - C)^{-1} = \left(\begin{bmatrix} 1 & 0 \\ 0 & 1 \end{bmatrix} - \begin{bmatrix} 0.5 & 0.4 \\ 0.1 & 0.6 \end{bmatrix} \right)^{-1}$$

$$= \begin{bmatrix} 0.5 & -0.4 \\ -0.1 & 0.4 \end{bmatrix}^{-1}$$

$$= \begin{bmatrix} 2.5 & 2.5 \\ 0.625 & 3.125 \end{bmatrix}$$

exists, C is productive.

(b) The production vector is computed by

$$X = (I - C)^{-1}D$$

$$= \begin{bmatrix} 2.5 & 2.5 \\ 0.625 & 3.125 \end{bmatrix} \begin{bmatrix} 10 \\ 20 \end{bmatrix} = \begin{bmatrix} 75 \\ 68.75 \end{bmatrix}$$

Chapter 4

Section 4.1

1. True, since both $\begin{bmatrix} a \\ 0 \end{bmatrix} + \begin{bmatrix} b \\ 0 \end{bmatrix} = \begin{bmatrix} a+b \\ 0 \end{bmatrix}$ and

$c\begin{bmatrix} a \\ 0 \end{bmatrix} = \begin{bmatrix} ca \\ 0 \end{bmatrix}$ are in V. Hence, V is a vector subspace

of $\mathbf{R}^2$.

3. We have

$$\begin{bmatrix} a \\ 0 \\ -2a \end{bmatrix} + \begin{bmatrix} b \\ 0 \\ -2b \end{bmatrix} = \begin{bmatrix} a+b \\ 0 \\ -2a-2b \end{bmatrix}$$

$$= \begin{bmatrix} a+b \\ 0 \\ -2(a+b) \end{bmatrix} \in V$$

and

$$c\begin{bmatrix} a \\ 0 \\ -2a \end{bmatrix} = \begin{bmatrix} ca \\ 0 \\ -2(ca) \end{bmatrix} \in V$$

Hence, V is a vector subspace of $\mathbf{R}^3$.

5. We have

$$\begin{bmatrix} a-b \\ b-c \\ c-d \\ d-a \end{bmatrix} + \begin{bmatrix} a'-b' \\ b'-c' \\ c'-d' \\ d'-a' \end{bmatrix} = \begin{bmatrix} a-b+a'-b' \\ b-c+b'-c' \\ c-d+c'-d' \\ d-a+d'-a' \end{bmatrix}$$

$$= \begin{bmatrix} (a+a')-(b+b') \\ (b+b')-(c+c') \\ (c+c')-(d+d') \\ (d+d')-(a+a') \end{bmatrix}$$

$$\in V$$

and

$$k\begin{bmatrix} a-b \\ b-c \\ c-d \\ d-a \end{bmatrix} = \begin{bmatrix} (ka)-(kb) \\ (kb)-(kc) \\ (kc)-(kd) \\ (kd)-(ka) \end{bmatrix} \in V$$

Hence, V is a vector subspace of $\mathbf{R}^4$.

7. Let V be the set of all 4-vectors with first three components zero and let $\mathbf{v}_1$ and $\mathbf{v}_2$ be in V. Then the sum $\mathbf{v}_1 + \mathbf{v}_2$ has the first three components zero, so it is in V. Also $c\mathbf{v}_1$ has the first three components zero, so it is in V. Hence, V is a vector subspace of $\mathbf{R}^4$.

9. Let $\mathbf{v}_1 = (1, -1, 0)$ and $\mathbf{v}_2 = (2, 0, 2)$ and let V be the set of all linear combinations of $\mathbf{v}_1$ and $\mathbf{v}_2$. Then if $\mathbf{u}_1 = c_1\mathbf{v}_1 + c_2\mathbf{v}_2$ and $\mathbf{u}_2 = k_1\mathbf{v}_1 + k_2\mathbf{v}_2$,

$$\mathbf{u}_1 + \mathbf{u}_2 = (c_1\mathbf{v}_1 + c_2\mathbf{v}_2) + (k_1\mathbf{v}_1 + k_2\mathbf{v}_2)$$
$$= (c_1 + k_1)\mathbf{v}_1 + (c_2 + k_2)\mathbf{v}_2 \in V$$

and

$$c\mathbf{u}_1 = c(c_1\mathbf{v}_1 + c_2\mathbf{v}_2) = (cc_1)\mathbf{v}_1 + (cc_2)\mathbf{v}_2 \in V$$
Hence, V is a subspace of $\mathbf{R}^3$.

11. Yes

13. No

15. No

17. Yes

19. No

21. Yes

23. No

25. No

27. Yes

29. No

31. $V_1 \cap V_2$ is either a line or a plane (if $V_1 = V_2$) through the origin; hence, it is a subspace of $\mathbf{R}^3$.

33. Yes

35. No

37. Yes

39. Yes

41. $\left\{ \begin{bmatrix} 1 \\ 2 \end{bmatrix}, \begin{bmatrix} -1 \\ 1 \end{bmatrix} \right\}$

43. $\left\{ \begin{bmatrix} 1 \\ 0 \\ 0 \end{bmatrix}, \begin{bmatrix} 0 \\ 1 \\ 0 \end{bmatrix}, \begin{bmatrix} -1 \\ 1 \\ 5 \end{bmatrix} \right\}$, or even the standard basis of $\mathbf{R}^3$

45. Yes

47. No

49. No

51. (a) No (b) No

53. (a) No (b) No

55. $\left\{ \begin{bmatrix} 1 \\ -1 \end{bmatrix}, \begin{bmatrix} -2 \\ 0 \end{bmatrix} \right\}$

57. $\left\{ \begin{bmatrix} -5 \\ -1 \\ 1 \end{bmatrix}, \begin{bmatrix} 4 \\ 1 \\ 7 \end{bmatrix} \right\}$

61. $x = \begin{bmatrix} -7 \\ 6 \\ 2 \end{bmatrix}$

63. $[x]_{\mathcal{B}} = \begin{bmatrix} 7 \\ -3 \end{bmatrix}$

65. $[x]_{\mathcal{B}} = \begin{bmatrix} -2 \\ 4 \\ -1 \end{bmatrix}$.

67. $[x]_{\mathcal{B}} = \begin{bmatrix} a - b \\ a + b \end{bmatrix}$

Section 4.2

1. If f and g are two polynomials of degree $\leq n$, so is the sum $f + g$ and so is the scalar product cf. So, Axioms (A1) and (M1) hold. Axioms (A2), (A3), (M2)–(M5) hold, because the corresponding properties hold for the coefficients of the polynomials. Now P_n contains the zero polynomial by definition, and certainly $f + 0 = 0 + f = f$. Also, $-f$ has degree $\leq n$, and $f + (-f) = 0$. Therefore, P_n is a vector space.

3. Let V be the given set and let $f = ap + bq$ and $f' = a'p + b'q$ be two elements of V. Then

$$f + f' = ap + bq + a'p + b'q$$
$$= (a + a')p + (b + b')q \in V$$

and

$$cf = c(ap + bq) = (ca)p + (cb)q \in V$$

Hence, (A1) and (M1) hold. The rest of the axioms hold, because they hold in the larger set P.

5. No, Axiom (M5) fails: $1(1, 1) = (0, 0) \neq (1, 1)$.

7. True, because the sum of two invertible matrices may not be invertible. $C = 0$ then $CM = 0$ $M = 0$

9. No, look at the leading coefficient of $2p$. not invertible.

11. No. (Show that for f and g in the set the sum $f + g$ is not in the set.)

13. Yes. This is the same as Exercise 8.

15. No. The sum $(1 + x^2) + (x + x^2) = 2x^2 + x + 1$ is not in the set.

17. Yes, because if V is the set, then

$$\begin{bmatrix} a & -b \\ b & a \end{bmatrix} + \begin{bmatrix} a' & -b' \\ b' & a' \end{bmatrix}$$
$$= \begin{bmatrix} a + a' & -(b + b') \\ b + b' & a + a' \end{bmatrix} \in V$$

and

$$c\begin{bmatrix} a & -b \\ b & a \end{bmatrix} = \begin{bmatrix} ca & -(cb) \\ cb & ca \end{bmatrix} \in V$$

19. No, the zero matrix is not in the set.

21. Yes

23. Yes

25. Yes. If f and g are even, so is $f + g$ and so is cf.

29. If $r = 0$, fine. If $r \neq 0$, then r^{-1} exists and

$$ru = rv \qquad \Rightarrow$$
$$r^{-1}(ru) = r^{-1}(rv) \Rightarrow$$
$$(r^{-1}r)u = (r^{-1}r)v \Rightarrow$$
$$1u = 1v \qquad \Rightarrow$$
$$u = v$$

Section 4.3

1. (a) Yes (b) No (c) No.

3. No

5. Yes

7. Yes

9. No

11. Yes

13. Yes

15. (a) False (b) True (c) False (d) True (e) True (f) False (g) True

19. No

21. Yes

23. (a) False (b) False (c) True (d) False

27. No. For example, take $p = x + 1, q = 1, r = -x - 1$.

29. No

31. In Exercise 18, Section 4.2, we saw that V is a vector subspace of M_{22}. If we take $a = 1, b = c = 0$, we see that $E_{11} - E_{22}$ is in V. Also from $a = c = 0$ and $b = 1$, we see that E_{12} is in V. Finally, taking $a = b = 0$, $c = 1$ shows that E_{21} is in V. $\mathcal{B}$ spans V, because

$$\begin{bmatrix} a & b \\ c & -a \end{bmatrix} = a \begin{bmatrix} 1 & 0 \\ 0 & -1 \end{bmatrix} + b \begin{bmatrix} 0 & 1 \\ 0 & 0 \end{bmatrix}$$
$$+ c \begin{bmatrix} 0 & 0 \\ 1 & 0 \end{bmatrix}$$

$\mathcal{B}$ is linearly independent, because the right-hand side linear combination, when set equal to zero implies that the left-hand side matrix is zero. Hence, $a = b = c = 0$. So $\mathcal{B}$ is a basis of V.

33. True, because

$$\begin{bmatrix} 1 & 0 & 0 & 3 \\ 0 & 1 & 0 & -3 \\ 0 & 0 & 2 & 0 \\ 0 & 0 & 0 & 1 \end{bmatrix}$$

has exactly four pivots.

35. (a) $\mathcal{B} = \{2 + x + 2x^2, x^2, 1 - x - x^2\}$
(b) $V = P_2$
37. (a) $\mathcal{B} = \{x^2, 1 + x, -1 + x^2\}$
(b) $V = P_2$
39. (a) $\mathcal{B} = \{-x + x^2, -5 + x, -x^2\}$
(b) $V = P_2$
41. $\{-x + x^2, x + x^2, 1\}$
43. $\{1 + x, -1 + x^2, x^2\}$
45. $\{-x + x^2, -5 + x, x^2\}$

Section 4.4

1. $\dim(V) = 1$
3. $\dim(V) = 1$
5. $\dim(V) = 2$
7. $\dim(V) = 1$
9. The dimension is 2.
11. The dimension is 3.
13. $\dim(V) = 3$
15. $\dim(V) = 2$
17. $\dim(V) = 2$
19. (a) False **(b)** True **(c)** False **(d)** True **(e)** False **(f)** True **(g)** False
21. The dimension is 2.
23. The dimension is 2.
25. $\{0\}$, $\mathbf{R}^4$, all lines through the origin, all planes through the origin, and all hyperplanes through the origin

Section 4.5

1. $p = -3 + 24x$
3. $p = (2a - 3b) + (2a + 3b)x$
5. $[p]_{\mathcal{B}} = \begin{bmatrix} -3 \\ -2 \end{bmatrix}$
7. $[p]_{\mathcal{B}} = \begin{bmatrix} a - b \\ a + b \end{bmatrix}$
9. $(1, -2, 1, -1)$
11. $\begin{bmatrix} 0 & 1 \\ 1 & 0 \end{bmatrix}$
13. $\begin{bmatrix} 0 & 0 & 1 \\ 1 & 0 & 0 \\ 0 & 1 & 0 \end{bmatrix}$
15. $\begin{bmatrix} 1 & 0 & \frac{1}{4} & 0 \\ 0 & \frac{1}{2} & 0 & \frac{1}{4} \\ 0 & 0 & \frac{1}{4} & 0 \\ 0 & 0 & 0 & \frac{1}{8} \end{bmatrix}$
17. $\begin{bmatrix} 1 & 0 & \frac{1}{2} & 0 \\ 0 & \frac{1}{2} & 0 & \frac{3}{4} \\ 0 & 0 & \frac{1}{4} & 0 \\ 0 & 0 & 0 & \frac{1}{8} \end{bmatrix}$
19. $\begin{bmatrix} 1 & \frac{1}{2} & \frac{1}{2} & \frac{1}{2} \\ 0 & 1 & 1 & \frac{3}{2} \\ 0 & 0 & 1 & \frac{3}{2} \\ 0 & 0 & 0 & 1 \end{bmatrix}$
21. The transition matrix is $\begin{bmatrix} 0 & -1 \\ -1 & 0 \end{bmatrix}$, and the new coordinates of $(1, 1)$ are given by $\begin{bmatrix} -1 \\ -1 \end{bmatrix}$.

Section 4.6

1. (a) Basis: $\left\{ \begin{bmatrix} 2 \\ 1 \end{bmatrix} \right\}$; the nullity is 1.

(b) Basis: $\left\{ \begin{bmatrix} 1 \\ 1 \\ 0 \end{bmatrix} \right\}$; the nullity is 1.

3. (a) Basis: $\left\{ \begin{bmatrix} -2 \\ 1 \\ 0 \\ 0 \\ 0 \\ 0 \end{bmatrix}, \begin{bmatrix} 1 \\ 0 \\ 1 \\ 0 \\ 0 \\ 0 \end{bmatrix}, \begin{bmatrix} 3 \\ 0 \\ 0 \\ 1 \\ 0 \\ 0 \end{bmatrix} \right\}$;
the nullity is 3.

(b) Basis: Empty. The nullity is 0.

5. (a) Basis: $\left\{ \begin{bmatrix} -1 \\ 1 \\ 1 \\ 0 \end{bmatrix}, \begin{bmatrix} 2 \\ 1 \\ 0 \\ 1 \end{bmatrix} \right\}$;

the nullity is 2.

(b) Basis: $\left\{ \begin{bmatrix} 1 \\ 1 \end{bmatrix} \right\}$; the nullity is 1.

7. Basis: $\left\{ \begin{bmatrix} -1 \\ -1 \\ 0 \\ 0 \\ 1 \end{bmatrix}, \begin{bmatrix} -4 \\ -2 \\ 1 \\ 0 \\ 0 \end{bmatrix} \right\}$; the nullity is 2.

Let N be the nullity, P be the number of pivot columns, and C be the number of columns of a matrix. Then:

9. (a) $N = 2, P = 2, 2 + 2 = 4 = C$

(b) $N = 1, P = 2, 1 + 2 = 3 = C$

11. $\mathbf{b} = \mathbf{0}$

13. $\mathbf{a}, \mathbf{b}, \mathbf{c}$ are in the column space.

15. $\mathbf{u}$ and $\mathbf{w}$ are in the column space.

17. None of the vectors is in the column space.

19. $\left\{ \begin{bmatrix} 2 \\ 0 \\ 0 \end{bmatrix}, \begin{bmatrix} 0 \\ 1 \\ 0 \end{bmatrix}, \begin{bmatrix} 0 \\ 1 \\ 1 \end{bmatrix} \right\}$

21. $\left\{ \begin{bmatrix} 1 \\ 0 \\ 0 \\ 0 \end{bmatrix}, \begin{bmatrix} -2 \\ -2 \\ 2 \\ 0 \end{bmatrix}, \begin{bmatrix} 1 \\ 2 \\ -4 \\ 0 \end{bmatrix}, \begin{bmatrix} 0 \\ 4 \\ 2 \\ 2 \end{bmatrix} \right\}$

23. $\left\{ \begin{bmatrix} 1 \\ 0 \\ 0 \\ 0 \end{bmatrix}, \begin{bmatrix} -2 \\ -2 \\ 0 \\ 0 \end{bmatrix}, \begin{bmatrix} 0 \\ 4 \\ 2 \\ 2 \end{bmatrix} \right\}$

25. The column space is $\mathbf{R}^2$.

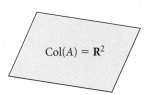

Col$(A) = \mathbf{R}^2$

27. The null space is spanned by $\{(0, 2, 1)\}$ and the column space is spanned by the set $\{(3, 0, 0), (0, 1, -2)\}$.

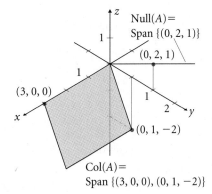

Null$(A) =$ Span $\{(0, 2, 1)\}$

Col$(A) =$ Span $\{(3, 0, 0), (0, 1, -2)\}$

29. $A = \begin{bmatrix} 1 & 1 \\ 1 & 1 \end{bmatrix}$ and $B = \begin{bmatrix} 1 & 1 \\ 0 & 0 \end{bmatrix}$; $A \sim B$, but Col(A) consists of all multiples of $\begin{bmatrix} 1 \\ 1 \end{bmatrix}$, whereas Col$(B)$ consists of all multiples of $\begin{bmatrix} 1 \\ 0 \end{bmatrix}$.

31. $\left\{ \begin{bmatrix} -1 \\ 2 \\ 3 \end{bmatrix}, \begin{bmatrix} 0 \\ 1 \\ 1 \end{bmatrix} \right\}$

33. $\left\{ \begin{bmatrix} 1 \\ -2 \\ -3 \\ 1 \end{bmatrix}, \begin{bmatrix} 0 \\ 1 \\ 2 \\ 0 \end{bmatrix} \right\}$

35. $\left\{ \begin{bmatrix} -1 \\ 0 \\ 1 \end{bmatrix}, \begin{bmatrix} 1 \\ 0 \\ 0 \end{bmatrix}, \begin{bmatrix} 0 \\ 1 \\ 0 \end{bmatrix} \right\}$

37. $\left\{ \begin{bmatrix} -1 \\ 0 \\ 1 \\ 0 \end{bmatrix}, \begin{bmatrix} 1 \\ -1 \\ 0 \\ 0 \end{bmatrix}, \begin{bmatrix} 1 \\ 0 \\ 0 \\ 0 \end{bmatrix}, \begin{bmatrix} 0 \\ 0 \\ 0 \\ 1 \end{bmatrix} \right\}$

39. Basis: $\left\{ \begin{bmatrix} 1 \\ 2 \\ 2 \\ -1 \end{bmatrix}, \begin{bmatrix} 0 \\ -1 \\ 2 \\ 3 \end{bmatrix} \right\}$; the rank is 2.

41. Basis: $\left\{ \begin{bmatrix} 1 \\ 2 \\ 2 \end{bmatrix}, \begin{bmatrix} 0 \\ -1 \\ 2 \end{bmatrix}, \begin{bmatrix} 0 \\ 0 \\ 1 \end{bmatrix} \right\}$; the rank is 3.

43. $\left\{ \begin{bmatrix} 1 \\ 0 \end{bmatrix}, \begin{bmatrix} 0 \\ 1 \end{bmatrix} \right\}$

45. $\left\{ \begin{bmatrix} 1 \\ 0 \\ -2 \end{bmatrix}, \begin{bmatrix} 0 \\ 1 \\ -4 \end{bmatrix} \right\}$

47. $B \sim \begin{bmatrix} 1 & 1 & 2 & 2 \\ 0 & 0 & -1 & 2 \\ 0 & 0 & 0 & 0 \end{bmatrix}$, $\text{rank}(A) = 2$

$B^T \sim \begin{bmatrix} 1 & 0 & 0 \\ 0 & -1 & 1 \\ 0 & 0 & 0 \\ 0 & 0 & 0 \end{bmatrix}$, $\text{rank}(B^T) = 2$

49. The nullity of B is 2 (a null space basis: $\{(-6, 0, 2, 1), (-1, 1, 0, 0)\}$). The rank is 2. $4 = 2 + 2$ equals the number of columns.

51. $B \sim \begin{bmatrix} 1 & 1 & 2 & 2 \\ 0 & 0 & -1 & 2 \\ 0 & 0 & 0 & 0 \end{bmatrix}$, so the rank of B is 2. And

$[B : b] \sim \begin{bmatrix} 1 & 1 & 2 & 2 & 1 \\ 0 & 0 & -1 & 2 & 0 \\ 0 & 0 & 0 & 0 & 0 \end{bmatrix}$, so the rank of

$[B : b]$ is again 2.

53. Yes. A has 450 columns and nullity 50, so its rank is 400, by the rank theorem. Hence, the column space has dimension 400. Therefore, the column space is all of $\mathbf{R}^{400}$. So, every 400-vector $\mathbf{b}$ is spanned by the columns of A. Therefore, $A\mathbf{x} = \mathbf{b}$ is consistent for all 400-vectors $\mathbf{b}$.

Section 4.7

1. (a) $\begin{bmatrix} 1 \\ 1 \\ 0 \end{bmatrix}$ (b) $\begin{bmatrix} 0 \\ 1 \\ 1 \end{bmatrix}$

(c) $\begin{bmatrix} 0 \\ 1 \\ 1 \end{bmatrix}$ (d) $\begin{bmatrix} 0 \\ 0 \\ 0 \end{bmatrix}$

3. $A^2 = \begin{bmatrix} 0 & 1 & 1 \\ 1 & 0 & 1 \\ 1 & 1 & 0 \end{bmatrix}$; $A^3 = \begin{bmatrix} 1 & 0 & 1 \\ 0 & 1 & 1 \\ 1 & 1 & 0 \end{bmatrix}$

5. No, because $\mathbf{u} + \mathbf{v} + \mathbf{w} = \mathbf{0}$.

7. $\{(1, 1, 1)\}$ is a basis of the null space of A over Z_2. The only elements of the null space over Z_2 are $(1, 1, 1)$ and $(0, 0, 0)$. Over $\mathbf{R}$ the null space of A is $\{\mathbf{0}\}$, so the only basis is the empty set.

9. Because the matrices commute, $AB = BA$. Hence,

$$(A + B)^2 = (A + B)(A + B)$$
$$= A^2 + AB + BA + B^2$$
$$= A^2 + 2AB + B^2$$
$$= A^2 + B^2$$

because $2AB$ is the zero matrix over Z_2.

11. 0101010

13. (a) $(1, 1, 1, 1)$ (b) $(0, 1, 1, 1)$

15. (a) $(0, 1, 1, 0)$ (b) $(0, 1, 1, 0)$

17. (a) $(1, 1, 1, 0)$ (b) $(1, 1, 1, 0)$

Chapter 5

Section 5.1

1. It is a transformation. Its domain is $\mathbf{R}^2$, and its codomain is $\mathbf{R}^2$.

3. It is a transformation. Its domain is $\mathbf{R} \times \mathbf{R}^+$ ($\mathbf{R}^+$ is the set of positive real numbers), and its codomain is $\mathbf{R}^2$.

5. Not a transformation (The matrix product is undefined.)

7. Not equal; they have different codomains.

9. (a) $n = 2$, $m = 2$ (b) $\mathbf{R}^2$ and $\mathbf{R}^2$ (c) $\{\mathbf{0}\}$ (d) $\mathbf{R}^2$

11. (a) $n = 3$, $m = 2$ (b) $\mathbf{R}^3$ and $\mathbf{R}^2$ (c) $\text{Span}\{(1, -1, 1)\}$ (d) $\mathbf{R}^2$

13. (a) $n = 3$, $m = 3$ (b) $\mathbf{R}^3$ and $\mathbf{R}^3$ (c) $\text{Span}\{(0, 1, 0), (1, 0, 1)\}$ (d) $\text{Span}\{(1, -1, 1)\}$

15. (a) $n = 4$, $m = 2$ (b) $\mathbf{R}^4$ and $\mathbf{R}^2$ (c) $\text{Span}\{(-1, -1, 1, 0), (2, 0, 0, 1)\}$ (d) $\mathbf{R}^2$

17. $\left\{ \begin{bmatrix} 1 \\ 0 \end{bmatrix}, \begin{bmatrix} 2 \\ 1 \end{bmatrix} \right\}$

19. $\left\{ \begin{bmatrix} 1 \\ 0 \end{bmatrix}, \begin{bmatrix} 2 \\ 1 \end{bmatrix} \right\}$

21. $\left\{ \begin{bmatrix} 1 \\ -2 \\ 0 \end{bmatrix}, \begin{bmatrix} 0 \\ 0 \\ -1 \end{bmatrix}, \begin{bmatrix} -2 \\ 1 \\ 4 \end{bmatrix} \right\}$

23. (a) $\begin{bmatrix} -1 \\ 0 \end{bmatrix}, \begin{bmatrix} 0 \\ -1 \end{bmatrix}, \begin{bmatrix} 1 \\ -2 \end{bmatrix}$; rotation by $180°$.

(b) $\begin{bmatrix} -1 \\ 0 \end{bmatrix}, \begin{bmatrix} 0 \\ 3 \end{bmatrix}, \begin{bmatrix} 1 \\ 6 \end{bmatrix}$; none

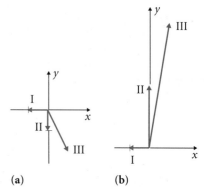

(a)   (b)

25. (a) $\begin{bmatrix} 1 \\ 0 \end{bmatrix}, \begin{bmatrix} -3 \\ 1 \end{bmatrix}, \begin{bmatrix} -7 \\ 2 \end{bmatrix}$; shear by a factor of 3 along the opposite x-direction.

(b) $\begin{bmatrix} 1 \\ -3 \end{bmatrix}, \begin{bmatrix} -3 \\ 1 \end{bmatrix}, \begin{bmatrix} -7 \\ 5 \end{bmatrix}$; none

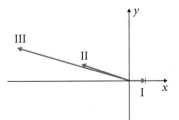

(a)

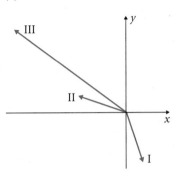

(b)

27. (a) $\begin{bmatrix} 1 \\ 3 \end{bmatrix}, \begin{bmatrix} 2 \\ 4 \end{bmatrix}, \begin{bmatrix} 3 \\ 5 \end{bmatrix}$; none

(b) $\begin{bmatrix} 3 \\ 3 \end{bmatrix}, \begin{bmatrix} 3 \\ 3 \end{bmatrix}, \begin{bmatrix} 3 \\ 3 \end{bmatrix}$; none

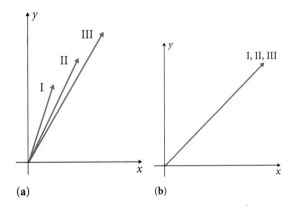

(a)   **(b)**

29. (a) $\begin{bmatrix} 1 \\ 0 \end{bmatrix}, \begin{bmatrix} 0 \\ 0 \end{bmatrix}, \begin{bmatrix} -1 \\ 0 \end{bmatrix}$; projection onto the x-axis

(b) $\begin{bmatrix} 3 \\ 0 \end{bmatrix}, \begin{bmatrix} 0 \\ 0 \end{bmatrix}, \begin{bmatrix} -3 \\ 0 \end{bmatrix}$; none

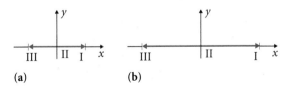

(a)   **(b)**

31. $T(x, y) = (-y, -x). A = \begin{bmatrix} 0 & -1 \\ -1 & 0 \end{bmatrix}$

33. $R(x, y) = \begin{bmatrix} (\cos \theta) x + (\sin \theta) y \\ -(\sin \theta) x + (\cos \theta) y \end{bmatrix}$

$A = \begin{bmatrix} \cos \theta & \sin \theta \\ -\sin \theta & \cos \theta \end{bmatrix}$

37. R_θ^x and R_θ^y with respective matrices:

$$\begin{bmatrix} 1 & 0 & 0 \\ 0 & \cos \theta & -\sin \theta \\ 0 & \sin \theta & \cos \theta \end{bmatrix}$$

and

$$\begin{bmatrix} \cos \theta & 0 & -\sin \theta \\ 0 & 1 & 0 \\ \sin \theta & 0 & \cos \theta \end{bmatrix}$$

Section 5.2

1. $\mathbf{R}^2$ and $\mathbf{R}^2$; linear

3. $\mathbf{R}^2$ and $\mathbf{R}^2$; nonlinear

5. $\mathbf{R}^2$ and $\mathbf{R}^2$; linear

7. $\mathbf{R}^2$ and $\mathbf{R}^2$; linear

9. $\mathbf{R}^2$ and $\mathbf{R}^3$; linear

11. $\mathbf{R}^3$ and $\mathbf{R}^2$; linear

13. $T\left(c_1 \begin{bmatrix} x_1 \\ y_1 \\ z_1 \end{bmatrix} + c_2 \begin{bmatrix} x_2 \\ y_2 \\ z_2 \end{bmatrix} \right)$

$= T \begin{bmatrix} c_1 x_1 + c_2 x_2 \\ c_1 y_1 + c_2 y_2 \\ c_1 z_1 + c_2 z_2 \end{bmatrix}$

$= \begin{bmatrix} c_1 x_1 + c_2 x_2 + c_1 y_1 + c_2 y_2 + c_1 z_1 + c_2 z_2 \\ c_1 x_1 + c_2 x_2 - c_1 y_1 - c_2 y_2 - c_1 z_1 - c_2 z_2 \end{bmatrix}$

$= c_1 \begin{bmatrix} x_1 + y_1 + z_1 \\ x_1 - y_1 - z_1 \end{bmatrix} + c_2 \begin{bmatrix} x_2 + y_2 + z_2 \\ x_2 - y_2 - z_2 \end{bmatrix}$

$$= c_1 T \left(\begin{bmatrix} x_1 \\ y_1 \\ z_1 \end{bmatrix} \right) + c_2 T \left(\begin{bmatrix} x_2 \\ y_2 \\ z_2 \end{bmatrix} \right)$$

15. Linear

17. Nonlinear

19. Nonlinear

21. Yes

23. No

25. $\begin{bmatrix} \frac{1}{2}x + \frac{1}{2}y + 2z \\ x - y - z \end{bmatrix}$ and $\begin{bmatrix} -\frac{95}{2} \\ 0 \end{bmatrix}$.

27. $-a - b + c + \left(2a + \frac{1}{2}b + \frac{1}{2}c \right) x$ and $\left(-\frac{95}{2} \right) x$

29. Because

$$T(c_1 X + c_2 Y) = C^{-1}(c_1 X + c_2 Y)C$$
$$= C^{-1}(c_1 XC + c_2 YC)$$
$$= C^{-1}(c_1 XC) + C^{-1}(c_2 YC)$$
$$= c_1(C^{-1}XC) + c_2(C^{-1}YC)$$
$$= c_1 T(X) + c_2 T(Y)$$

T is linear by Theorem 3.

31. $(-2, 2)$ is in the span of $\{(1, -1)\}$, so T cannot be specified for any vector not in that span. Examples of different linear transformations such that $T(1, -1) = (3, -1)$, so $T(-2, 2) = (-6, 2)$, are

$$T(x, y) = (-6x - 9y, 2x + 3y)$$
$$T(x, y) = (-3x - 6y, x + 2y)$$
$$T(x, y) = \left(-\frac{3}{2}x - \frac{9}{2}y, \frac{1}{2}x + \frac{3}{2}y \right)$$

33. $T(x, y) = (xy, x)$

35. $T(x, y) = (x, x)$, $\mathbf{v}_1 = (2, 0)$, $\mathbf{v}_2 = (1, 1)$

39. A has size $m \times n$ and $\mathbf{b}$ is an n-vector. T is nonlinear, because

$$T(\mathbf{0}) = A\mathbf{0} + \mathbf{b} = \mathbf{b} \neq \mathbf{0}$$

41. $A = \begin{bmatrix} 1 & -1 \\ 0 & 1 \\ 1 & 0 \end{bmatrix}$

43. $A = \begin{bmatrix} 0 & 0 \\ 0 & 0 \\ 0 & 0 \end{bmatrix}$

45. $A = \begin{bmatrix} -1 & 0 \\ 0 & 1 \end{bmatrix}$

47. $A = \begin{bmatrix} \cos\theta & -\sin\theta \\ \sin\theta & \cos\theta \end{bmatrix}$

49. They are of the form $T(x) = cx$ for some fixed scalar c.

Section 5.3

1. Kernel basis: $\{(1, 1, 1)\}$;
range basis: $\{(1, 0, -1), (-1, 1, 0)\}$;
nullity is 1 and rank is 2.
Dimension theorem: $1 + 2 = \dim(\mathbf{R}^3)$.

3. Kernel basis: $\{(2, 1)\}$;
range basis: $\{(1, 1, 2, 0)\}$;
nullity is 1 and rank is 1.
Dimension theorem: $1 + 1 = \dim(\mathbf{R}^2)$.

5. Kernel basis: the empty set;
range basis: $\{(1, 0, 0), (0, 1, 0), (0, 0, 1)\}$;
nullity is 0 and rank is 3.
Dimension theorem: $0 + 3 = \dim(\mathbf{R}^3)$.

7. Kernel basis: the empty set;
range basis: $\{(1, 0), (0, 1)\}$;
nullity is 0 and rank is 2.
Dimension theorem: $0 + 2 = \dim(\mathbf{R}^2)$.

9. Kernel basis: $\{1 + x - x^2 + x^3\}$;
range basis: $\{1, -1 + x, x^2\}$;
nullity is 1 and rank is 3.
Dimension theorem: $1 + 3 = \dim(P_3)$.

11. Kernel basis: the empty set;
range basis: $\{\mathbf{e}_1, \mathbf{e}_2, \mathbf{e}_3\}$

13. Kernel basis: $\{1, x\}$;
range basis: $\{1\}$.

15. $\dim(\text{Kernel}(T)) = 3$

17. The nullity is 2. The rank is 2.

19. The nullity is 2. The rank is 2.

21. The system $2x + y = 0$, $-3x + 4y = 0$ has only the trivial solution, so T is one-to-one by Theorem 9.

23. The system $2x - y = 0$, $x - y = 0$, $-x + y = 0$, $x - 2y = 0$ has only the trivial solution, so T is one-to-one by Theorem 9.

25. The system $2a - b = 0$, $-b + c = 0$, $-3a + c = 0$ has only the trivial solution, so T is one-to-one by Theorem 9.

27. Because the nullity is zero, the rank is 2, so the range is $\mathbf{R}^2$; hence, T is onto.

29. T is onto, because the system $x - y + z = a$, $-x + y + z = b$ can be solved for any a, b. For example, $x = \frac{1}{2}a - \frac{1}{2}b, y = 0, z = \frac{1}{2}a + \frac{1}{2}b$ is a solution, so for these choices (x, y, z) maps to (a, b).

31. T is onto, because the system $a + b = A, a + c = B$ can be solved for any A, B. For example, $a = B, b = A - B$, $c = 0$ is a solution, so for these choices $a + bx + cx^2$ maps to $A + Bx$.

33. The kernel is zero, because $a - 2b = 0$, $-2a + b = 0$ has only the trivial solution. So T is one-to-one. Hence, it is an isomorphism by Theorem 11.

35. The kernel is zero, because $x - y + z = 0$, $-x + y + z = 0$, $-y + z = 0$ has only the trivial solution. So T is one-to-one. Hence, it is an isomorphism by Theorem 11.

37. The kernel is zero, because $c = 0$, $b = 0$, $a - b = 0$ has only the trivial solution. So T is one-to-one. Hence, it is an isomorphism by Theorem 11.

39. Not isomorphism

41. Isomorphism

42. Not isomorphism

Section 5.4

1. $\begin{bmatrix} 1 & -1 \\ 0 & -2 \\ 2 & 2 \end{bmatrix}$

3. $\begin{bmatrix} 3 & -1 & 0 \\ 1 & 0 & -1 \\ 0 & 0 & 2 \end{bmatrix}$

5. **(a)** $A = \begin{bmatrix} 0 & 0 & -2 \\ 0 & 1 & 0 \\ 0 & 0 & 0 \end{bmatrix}$ **(b)** $A' = \begin{bmatrix} 0 & 0 & 0 \\ -2 & 0 & 0 \\ 1 & 1 & 1 \end{bmatrix}$

(c) Directly, $T(6x - 2x^2) = 4 + 6x$. Using A:

$$\begin{bmatrix} 0 & 0 & -2 \\ 0 & 1 & 0 \\ 0 & 0 & 0 \end{bmatrix} \begin{bmatrix} 0 \\ 6 \\ -2 \end{bmatrix} = \begin{bmatrix} 4 \\ 6 \\ 0 \end{bmatrix}$$

So, we get $4 + 6x$ in terms of the standard basis. Using A': First $6x - 2x^2$ in terms of $\mathcal{B}'$, we have

$$6x - 2x^2 = -2(-x + x^2) + 0(1 + x) + 4(x)$$

Then form the matrix product

$$\begin{bmatrix} 0 & 0 & 0 \\ -2 & 0 & 0 \\ 1 & 1 & 1 \end{bmatrix} \begin{bmatrix} -2 \\ 0 \\ 4 \end{bmatrix} = \begin{bmatrix} 0 \\ 4 \\ 2 \end{bmatrix}$$

to get the coordinates of the image with respect to $\mathcal{B}'$. So, $T(6x - 2x^2) = 0(-x + x^2) + 4(1 + x) + 2(x) = 4 + 6x$.

7. **(a)** $A = \begin{bmatrix} -1 & -1 \\ 1 & -1 \\ -3 & 1 \end{bmatrix}$

(b) Directly: $T(5 - 2x) = 5 - 5x + 2x^2$; using A:

$$5 - 2x = \frac{3}{2}(1 + x) - \frac{7}{2}(-1 + x)$$

So,

$$\begin{bmatrix} -1 & -1 \\ 1 & -1 \\ -3 & 1 \end{bmatrix} \begin{bmatrix} \frac{3}{2} \\ -\frac{7}{2} \end{bmatrix} = \begin{bmatrix} 2 \\ 5 \\ -8 \end{bmatrix}$$

Hence,

$$T(5 - 2x) = 2(-x + x^2) + 5(1 + x) - 8x$$
$$= 5 - 5x + 2x^2$$

9. The coordinate vector $\begin{bmatrix} 1 & -2 \\ -4 & 3 \end{bmatrix}$ with respect to $\mathcal{R}$ is $(2, 0, -4, -1)$. Hence,

$$\begin{bmatrix} 1 & 3 & 0 & 2 \\ 0 & 1 & 0 & 0 \\ 0 & -2 & 1 & -1 \\ 1 & 0 & 0 & 0 \end{bmatrix} \begin{bmatrix} 2 \\ 0 \\ -4 \\ -1 \end{bmatrix} = \begin{bmatrix} 0 \\ 0 \\ -3 \\ 2 \end{bmatrix}$$

and

$$T \begin{bmatrix} 1 & -2 \\ -4 & 3 \end{bmatrix} = -3 \begin{bmatrix} 0 & 0 \\ 1 & -1 \end{bmatrix} + 2 \begin{bmatrix} 1 & 0 \\ 0 & 1 \end{bmatrix}$$
$$= \begin{bmatrix} 2 & 0 \\ -3 & 5 \end{bmatrix}$$

13. $m = 2, n = 4$

$$T(a + bx + cx^2 + dx^3) = (-9a - b + 8c + 5d)$$
$$+ (11a + 3b - 10c + d)x$$

15. $A = \begin{bmatrix} 1 & 0 & 0 & 0 \\ 1 & 0 & -1 & 1 \\ 1 & -1 & 0 & 1 \\ 0 & -1 & 0 & 0 \end{bmatrix}$

A is invertible, so T is an isomorphism by Theorem 16.

Section 5.5

1. $(f + g)(x, y, z) = \begin{bmatrix} z \\ 3x - z \end{bmatrix}$

$(f + g)(-1, 2, 0) = \begin{bmatrix} 0 \\ -3 \end{bmatrix}$

$(-4f)(x, y, z) = \begin{bmatrix} -4x + 4y - 4z \\ -4x - 4y \end{bmatrix}$

$(-4f)(-1, 2, 0) = \begin{bmatrix} 12 \\ -4 \end{bmatrix}$

3. $(f + g)(x, y, z) = \begin{bmatrix} -x - y + z \\ 2x - 3y - z \end{bmatrix}$

$(f + g)(-1, 2, 0) = \begin{bmatrix} -1 \\ -8 \end{bmatrix}$

$$(-4f)(x, y, z) = \begin{bmatrix} 4x + 8y \\ -4x + 4z \end{bmatrix}$$

$$(-4f)(-1, 2, 0) = \begin{bmatrix} 12 \\ 4 \end{bmatrix}$$

5. Matrix(f) + Matrix(g)

$$= \begin{bmatrix} 1 & -1 & 1 \\ 1 & 1 & 0 \end{bmatrix} + \begin{bmatrix} -1 & 1 & 0 \\ 2 & -1 & -1 \end{bmatrix}$$

$$= \begin{bmatrix} 0 & 0 & 1 \\ 3 & 0 & -1 \end{bmatrix}$$

$$= \text{Matrix}(f + g) - 2\,\text{Matrix}(f)$$

$$= \begin{bmatrix} -2 & 2 & -2 \\ -2 & -2 & 0 \end{bmatrix} = \text{Matrix}(-2f)$$

7. Matrix(f) + Matrix(g)

$$= \begin{bmatrix} -1 & -2 & 0 \\ 1 & 0 & -1 \end{bmatrix} + \begin{bmatrix} 0 & 1 & 1 \\ 1 & -3 & 0 \end{bmatrix}$$

$$= \begin{bmatrix} -1 & -1 & 1 \\ 2 & -3 & -1 \end{bmatrix}$$

$$= \text{Matrix}(f + g) - 2\,\text{Matrix}(f)$$

$$= \begin{bmatrix} 2 & 4 & 0 \\ -2 & 0 & 2 \end{bmatrix} = \text{Matrix}(-2f)$$

9. $f \circ g(x, y) = (x + 4y, 3x - y)$, $f \circ g(-1, -3) = (-13, 0)$

11. Matrix(f) $\cdot$ Matrix(g) $= \begin{bmatrix} 1 & -1 \\ 1 & 1 \end{bmatrix} \begin{bmatrix} -5 & 1 \\ 1 & 3 \end{bmatrix}$

$$= \begin{bmatrix} -6 & -2 \\ -4 & 4 \end{bmatrix}$$

$$= \text{Matrix}(f \circ g)$$

13. Matrix(f) $\cdot$ Matrix(g) $= \begin{bmatrix} -1 & -2 & 1 & 0 \\ 1 & 0 & 0 & -1 \end{bmatrix}$

$$\cdot \begin{bmatrix} 0 & 1 \\ 1 & -3 \\ 1 & -1 \\ 1 & 0 \end{bmatrix} = \begin{bmatrix} -1 & 4 \\ -1 & 1 \end{bmatrix} = \text{Matrix}(f \circ g)$$

15. (a) The codomain of g is $\mathbf{R}^3$, whereas the domain of f is $\mathbf{R}^2$. $g \circ f$ is undefined.

    (b) The codomain of g is $\mathbf{R}^2$, whereas the domain of f is $\mathbf{R}^3$. $g \circ f$ is defined.

    (c) The codomain of g is $\mathbf{R}^3$, whereas the domain of f is $\mathbf{R}^2$. $g \circ f$ is defined.

17. $f^3(x, y) = (-2x - 2y, 2x - 2y)$. $f^3(1, -1) = (0, 4)$

19. f is invertible, because its standard matrix

$$\begin{bmatrix} -1 & 1 & 1 \\ 1 & 0 & -2 \\ 2 & -1 & 0 \end{bmatrix}, \text{ is invertible.}$$

21. f is invertible, because its standard matrix is invertible

with inverse $\begin{bmatrix} -\frac{1}{2} & 0 & \frac{1}{4} \\ 0 & -1 & 1 \\ 0 & 0 & -\frac{1}{2} \end{bmatrix}$. Hence,

$$f^{-1}(x, y, z) = \left(-\tfrac{1}{2}x + \tfrac{1}{4}z, -y + z, -\tfrac{1}{2}z\right)$$

23. f is invertible, because its standard matrix is invertible

with inverse $\begin{bmatrix} \frac{5}{3} & -\frac{7}{6} & \frac{1}{6} \\ -1 & 1 & 0 \\ -\frac{1}{3} & -\frac{1}{6} & \frac{1}{6} \end{bmatrix}$. Hence,

$$f^{-1}(x, y, z) = \left(\tfrac{5}{3}x - \tfrac{7}{6}y + \tfrac{1}{6}z, -x + y,\right.$$
$$\left. -\tfrac{1}{3}x - \tfrac{1}{6}y + \tfrac{1}{6}z\right)$$

25. The matrix is invertible, so f is invertible. f^{-1} is multiplication on the left by the inverse of the given matrix. So, $f^{-1} : \mathbf{R}^4 \to \mathbf{R}^4$ is given by

$$f^{-1}(\mathbf{x}) = \begin{bmatrix} -1 & 0 & 1 & 2 \\ 0 & 0 & 1 & 1 \\ -1 & 1 & 1 & 2 \\ -1 & 1 & 1 & 1 \end{bmatrix} \mathbf{x}$$

35. Because $f^{-1}\begin{bmatrix} x \\ y \end{bmatrix} = \begin{bmatrix} \frac{1}{5}x + \frac{2}{5}y \\ -\frac{2}{5}x + \frac{1}{5}y \end{bmatrix}$ and

$$f^{-2}\begin{bmatrix} x \\ y \end{bmatrix} = \begin{bmatrix} -\frac{3}{25}x + \frac{4}{25}y \\ -\frac{4}{25}x - \frac{3}{25}y \end{bmatrix}, f^{-2} \circ f^{-1} \text{ is given by}$$

$$f^{-2} \circ f^{-1}\begin{bmatrix} x \\ y \end{bmatrix} = \begin{bmatrix} -\frac{11}{125}x - \frac{2}{125}y \\ \frac{2}{125}x - \frac{11}{125}y \end{bmatrix}$$

which is also the formula for $f^{-3}(x, y)$.

Section 5.6

1. $T(\mathbf{0}) = \begin{bmatrix} 1 \\ -1 \end{bmatrix}$, $T(\mathbf{e}_1) = \begin{bmatrix} 2 \\ -1 \end{bmatrix}$, $T(\mathbf{e}_2) = \begin{bmatrix} 1 \\ 4 \end{bmatrix}$

3. $T(\mathbf{0}) = \begin{bmatrix} 1 \\ -1 \\ 0 \end{bmatrix}$, $T(\mathbf{e}_1) = \begin{bmatrix} 0 \\ 0 \\ 0 \end{bmatrix}$,

$$T(\mathbf{e}_2) = \begin{bmatrix} 3 \\ 1 \\ -4 \end{bmatrix}, \quad T(\mathbf{e}_3) = \begin{bmatrix} 1 \\ -2 \\ 1 \end{bmatrix}$$

5. $\begin{bmatrix} 1 & -1 \\ -1 & 1 \end{bmatrix} \mathbf{x} + \begin{bmatrix} 0 \\ -1 \end{bmatrix}$

7. $\begin{bmatrix} -1 & 3 & 0 \\ 1 & 0 & -1 \\ 1 & -5 & 1 \end{bmatrix} \mathbf{x} + \begin{bmatrix} 1 \\ 0 \\ -1 \end{bmatrix}$

9. $A = \begin{bmatrix} 0 & 5 \\ 2 & -8 \end{bmatrix}$, $\mathbf{b} = \begin{bmatrix} -1 \\ 1 \end{bmatrix}$

11. $A = \begin{bmatrix} 1 & -2 \\ 3 & 4 \end{bmatrix}$, $\mathbf{b} = \begin{bmatrix} -3 \\ 2 \end{bmatrix}$

13. $A = \begin{bmatrix} 1 & -2 & 3 \\ -4 & 5 & -6 \end{bmatrix}$, $\mathbf{b} = \begin{bmatrix} 1 \\ -1 \end{bmatrix}$

15. Because $T(\mathbf{0}) = \mathbf{b}$, $\mathbf{b}$ is determined by $T(\mathbf{0})$. Also, because $L(\mathbf{x}) = T(\mathbf{x}) - \mathbf{b} = A\mathbf{x}$ is a matrix transformation, it can be uniquely determined by $T(\mathbf{e}_1), \ldots, T(\mathbf{e}_n)$, so that T is uniquely determined by the values

$$T(\mathbf{e}_1), \ldots, T(\mathbf{e}_n), T(\mathbf{0})$$

17. The two sets are equal.

Chapter 6

Section 6.1

1. (a) -3, (b) -3

3. (a) -10, (b) -10

5. (a) 158, (b) -15

7. (a) 1, (b) 0

9. -80

11. 0

13. 76

15. (a) $\det(A) = -2 = \det(A^T)$

   (b) $\det(AB) = 4 = -2(-2) = \det(A)\det(B)$

   (c) $\det(A^{-1}) = -\frac{1}{2} = 1/\det(A)$

17. (a) $M_{11} = -30,$ $C_{11} = -30,$
    $M_{12} = 10,$ $C_{12} = -10,$
    $M_{13} = -35,$ $C_{13} = -35,$
    $M_{21} = 12,$ $C_{21} = -12,$
    $M_{22} = -20,$ $C_{22} = -20,$
    $M_{23} = 14,$ $C_{23} = -14,$
    $M_{31} = -2,$ $C_{31} = -2,$
    $M_{32} = -10,$ $C_{32} = 10,$
    $M_{33} = 11,$ $C_{33} = 11$

   (b) (b_1) $1(-30) - 2(-10) + 2(-35) = -80$
       (b_2) $3(-12) + 5(-20) - 4(-14) = -80$
       (b_3) $7(-2) + 0(10) - 6(11) = -80$
       (b_4) $1(-30) + 3(-12) + 7(-2) = -80$
       (b_5) $-2(-10) + 5(-20) + 0(10) = -80$
       (b_6) $2(-35) - 4(-14) - 6(11) = -80$

19. $\det(A^{-1}) = \frac{1}{2} = 1/\det(A)$

21. $\det(A^{-1}) = \begin{vmatrix} \dfrac{d}{ad-bc} & -\dfrac{b}{ad-bc} \\ -\dfrac{c}{ad-bc} & \dfrac{a}{ad-bc} \end{vmatrix}$

   $= \dfrac{ad-bc}{(ad-bc)^2} = \dfrac{1}{ad-bc} = \dfrac{1}{\det(A)}$

23. $\det(E_3) = r = \det(E_4)$

25. The determinant of an elementary matrix E is $1, r,$ or -1, according to whether E comes from elimination, scaling (of factor r), or interchange.

27. (a) $\det(E_3 A) = r(aei - afh - bdi + cdh + bfg - ceg) = \det(E_3)\det(A) = \det(A_3)$

   (b) $\det(E_4 A) = r(aei - afh - bdi + cdh + bfg - ceg) = \det(E_4)\det(A) = \det(A_4)$

29. (a) $\lambda = 2 \pm \sqrt{3}$ (b) $\lambda = 0, 11$

31. $\lambda = 0, 3$

33. $x = 1, x = 2$

35. $a = 2, b = 4$ or $a = -2, b = -4$

37. *Hint:* If T is invertible, then the images of $\mathbf{e}_1, \mathbf{e}_2,$ and $\mathbf{e}_3$ are linearly independent; hence, they define a parallelepiped. Its volume is $|\det(A)|$, by Section 2.6. What happens if T is not invertible?

Section 6.2

1. (a) -50 (b) 0

3. (a) -1 (b) -1

5. (a) 1 (b) 1

7. 24

9. Property 3 of Theorem 1 was used.

11. Property 4 of Theorem 4 was used.

13. Property 4 of Theorem 1 was used to modify the second row (did not change the determinant), and then Property 2 was used to scale the last row.

15. Theorem 7 was used.

17. $x = 0$ makes the second row zero and $x = 2$ makes the first and last rows proportional. So in each case the determinant should be 0.

19. (a) $-\begin{vmatrix} 2 & 2 & 4 \\ 0 & 2 & 0 \\ 0 & 0 & 1 \end{vmatrix} = -4$

   (b) $-1\begin{vmatrix} 1 & 1 & 0 \\ 0 & 1 & 1 \\ 0 & 0 & 1 \end{vmatrix} = -1$

21. $\begin{vmatrix} 2 & -4 & 2 & 8 \\ 0 & -1 & 2 & 1 \\ 0 & 0 & 7 & 0 \\ 0 & 0 & 0 & -2 \end{vmatrix} = 28$

23. $\begin{vmatrix} 1 & 0 & -1 & 2 & 1 \\ 0 & 2 & 6 & -2 & 6 \\ 0 & 0 & 3 & 0 & 3 \\ 0 & 0 & 0 & 5 & -1 \\ 0 & 0 & 0 & 0 & \frac{2}{5} \end{vmatrix} = 12$

25. (a) Expanding each side yields

$$ka_1b_2c_3 - ka_1b_3c_2 - ka_2b_1c_3$$
$$+ ka_3b_1c_2 + ka_2b_3c_1 - ka_3b_2c_1$$

(b) Expanding each side yields

$$a_1b_2c_3 - a_1b_3c_2 - a_2b_1c_3$$
$$+ a_3b_1c_2 + a_2b_3c_1 - a_3b_2c_1$$

29. $\det(C) = -1$, so C is invertible. $\det(D) = -1$, so D is invertible.

31. $\det(G) = 3$, so G is invertible. $\det(H) = 2$, so H is invertible.

33. $k = -2, 0, 2$

35. $\det(B^{-1}AB) = \det(B^{-1})\det(A)\det(B)$
$$= \det(B)^{-1}\det(A)\det(B)$$
$$= \det(A)$$

37. (a) -14 **(b)** -56
   (c) 7 **(d)** -2
   (e) -14 **(f)** -1134

39. $\begin{vmatrix} 1 & 1 & 1 \\ a & b & c \\ a^2 & b^2 & c^2 \end{vmatrix} = \begin{vmatrix} 1 & 1 & 1 \\ 0 & b-a & c-a \\ 0 & b^2-a^2 & c^2-a^2 \end{vmatrix}$

$$= \begin{vmatrix} 1 & 1 & 1 \\ 0 & b-a & c-a \\ 0 & 0 & (c-a)(c-b) \end{vmatrix}$$

$$= (b-a)(c-b)(c-a)$$

41. $\begin{vmatrix} 1 & 1 & 1 \\ a & b & c \\ a^3 & b^3 & c^3 \end{vmatrix}$

$$= \begin{vmatrix} 1 & 1 & 1 \\ 0 & b-a & c-a \\ 0 & b^3-a^3 & c^3-a^3 \end{vmatrix}$$

$$= \begin{vmatrix} 1 & 1 & 1 \\ 0 & b-a & c-a \\ 0 & 0 & (c-b)(c-a)(a+c+b) \end{vmatrix}$$

$$= (b-a)(c-b)(c-a)(a+b+c)$$

Section 6.3

1. (a) $\dfrac{1}{-2}\begin{bmatrix} 5 & -4 \\ -3 & 2 \end{bmatrix}$ **(b)** $\dfrac{1}{2}\begin{bmatrix} 4 & -2 \\ 3 & -1 \end{bmatrix}$

3. $\dfrac{1}{4}\begin{bmatrix} 0 & 2 & 2 \\ 2 & -2 & 0 \\ 2 & 0 & -2 \end{bmatrix}$

5. $\dfrac{1}{1}\begin{bmatrix} 1 & 0 & 0 \\ -2 & 1 & 0 \\ 5 & -4 & 1 \end{bmatrix}$

7. We get

$$A^{-1} = \begin{bmatrix} 0 & 1 & 0 & -1 \\ -1 & 1 & -1 & 1 \\ 3 & -3 & 2 & -1 \\ -1 & 0 & 0 & 1 \end{bmatrix}$$

by either method. Row reduction is much more efficient.

9. (a) $x = \frac{-2}{-2} = 1$, $y = \frac{0}{-2} = 0$
   (b) $x = \frac{2}{-2} = -1$, $z = \frac{-2}{-2} = 1$

11. $x = \frac{4}{1} = 4; y = \frac{-3}{1} = -3; z = \frac{1}{1} = 1$

13. $z = \frac{-21}{-101} = \frac{21}{101}$

15. *Hint:* Apply determinants to both sides of $A\,\mathrm{Adj}(A) = \det(A)I_n$.

Section 6.4

1. $(2,1,3,4)$ is odd with sign -1. $(1,4,2,3)$ is even with sign 1. $(1,5,2,4,3)$ is even with sign 1. $(1,4,3,5,2)$ is even with sign 1.

3. $(3,1,4,2)$ is odd with sign -1. $(4,2,1,3)$ is even with sign 1. $(3,4,2,1,5)$ is odd with sign -1. $(4,2,5,1,3)$ is even with sign 1.

5. (a) $-2(3)(4) = -24$ **(b)** $2(3)(4) = 24$

7. (a) $2(3)(4) - 5(3)(6) = -66$ **(b)** $-2(3)(-4) - 5(3)(-6) = 114$

9. $1(2)(3)(4)(5) = 120$

11. $\begin{bmatrix} 1 & 0 & 0 \\ 0 & 1 & 0 \\ 0 & 0 & 1 \end{bmatrix}$ $(1,2,3)$

$\begin{bmatrix} 1 & 0 & 0 \\ 0 & 0 & 1 \\ 0 & 1 & 0 \end{bmatrix}$ $(1,3,2)$

$$\begin{bmatrix} 0 & 1 & 0 \\ 0 & 0 & 1 \\ 1 & 0 & 0 \end{bmatrix} \quad (3,1,2)$$

$$\begin{bmatrix} 0 & 0 & 1 \\ 0 & 1 & 0 \\ 1 & 0 & 0 \end{bmatrix} \quad (3,2,1)$$

$$\begin{bmatrix} 0 & 1 & 0 \\ 1 & 0 & 0 \\ 0 & 0 & 1 \end{bmatrix} \quad (2,1,3)$$

$$\begin{bmatrix} 0 & 0 & 1 \\ 1 & 0 & 0 \\ 0 & 1 & 0 \end{bmatrix} \quad (2,3,1)$$

13. Let $p = (j_1, \ldots, j_n)$ be a permutation and let A be the corresponding permutation matrix. So, the ith row of A has an 1 on its j_ith column and 0s elsewhere. Hence, the determinant of A consists of only one term of the form $\operatorname{sign}(p)1 \cdots 1 = \operatorname{sign}(p)$.

Section 6.5

1. $x + 2y - 3 = 0$

3. $3x + 2y - 2 = 0$

5. No, they are not.

7. $x^2 + y^2 + 2y = 0$ or $x^2 + (y+1)^2 = 1$; the center is at $(0, -1)$ and the radius is 1.

9. $x^2 + y^2 - 6x - 4y = 0$ or $(x-3)^2 + (y-2)^2 = 13$. The center is at $(3,2)$ and the radius is $\sqrt{13}$.

11. $-2y + 4x^2 - 6x + 8 = 0$ or $y = 2x^2 - 3x + 4$.

13. $-2y + 4x^2 - 22x + 32 = 0$ or $y = 2x^2 - 11x + 16$

15. $x + 2y - 3z + 2 = 0$

17. First, we rewrite the system in unknown y and parametric coefficients in x.

$$y^2 + (x^2 - 1) = 0$$
$$y^2 + (x^2 - 2x - 1) = 0$$

The Sylvester resultant is

$$\begin{vmatrix} 1 & 0 & x^2 - 1 & 0 \\ 0 & 1 & 0 & x^2 - 1 \\ 1 & -2 & x^2 + 2x + 1 & 0 \\ 0 & 1 & -2 & x^2 + 2x + 1 \end{vmatrix} = 0$$

Hence, $8x^2 + 8x = 0$. So, $x = 0$ or $x = -1$. If $x = 0$, then $y = 1$ by the second equation, and the first equation is also satisfied. If $x = -1$, then $y = 0$ by the first equation and the second equation is also satisfied. So we have two solutions: $x = 0$, $y = 1$, and $x = -1, y = 0$.

19. From

$$\begin{vmatrix} x^2 + y^2 + z^2 & x & y & z & 1 \\ 54 & 1 & 2 & 7 & 1 \\ 38 & 5 & 2 & 3 & 1 \\ 46 & 1 & 6 & 3 & 1 \\ 6 & 1 & 2 & -1 & 1 \end{vmatrix} = 0$$

we get, after simplification,

$$x^2 + y^2 + z^2 - 2x - 4y - 6z - 2 = 0$$

or

$$(x-1)^2 + (y-2)^2 + (z-3)^2 = 16$$

Hence, the sphere has center $(1, 2, 3)$ and radius 4.

21.

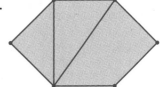

23. (a) We have the following three spanning trees:

(b) The tree matrix is

$$\begin{bmatrix} 3 & -1 & -1 & -1 \\ -1 & 1 & 0 & 0 \\ 0 & -1 & 2 & -1 \\ -1 & 0 & -1 & 2 \end{bmatrix}$$

and the cofactor

$$C_{11} = (-1)^2 \begin{vmatrix} 1 & 0 & 0 \\ -1 & 2 & -1 \\ 0 & -1 & 2 \end{vmatrix} = 3$$

Hence, there are three spanning trees.

Chapter 7

Section 7.1

1. $A\mathbf{u} = (-5, 5) = 5\mathbf{u}$. The eigenvalue is 5.

3. $A\mathbf{v} = (0, 0) = 0\mathbf{v}$. The eigenvalue is 0.

5. No, because $A(\mathbf{u} + \mathbf{v}) = A(1, 4) = (-5, 5)$, which is not a multiple of $(1, 4)$.

7. $(-1, 1)$ is not an eigenvector of the reduced row echelon form of A. On the other hand, $(2, 3)$ is.

9. Only $(-2, 1)$ and $(1, -\frac{1}{2})$ are eigenvectors. The corresponding eigenvalue for both is 0.

11. Because the characteristic polynomial is $-(\lambda - 7) \times (\lambda^2 + 2\lambda - 24)$, the eigenvalues are $7, 4, -6$.

13. Along the x-axis, the vectors remain the same, so they are eigenvectors with eigenvalue 1. Along the y-axis, the vectors go to the origin, so they are eigenvectors with eigenvalue 0.

15. There are no eigenvectors (with real entries). No nonzero vector remains in its own line after rotation.

17. **(a)** The characteristic polynomial is $\lambda^2 + 2\lambda - 24$. The eigenvalues are 4 and -6. The corresponding bases of eigenvectors are $\{(2, 3)\}$ and $\{(1, -1)\}$. All multiplicities are 1.

 (b) The characteristic polynomial is $\lambda^2 + 6\lambda + 9$. The only eigenvalue is -3. The corresponding basis of eigenvectors is $\{(3, 1)\}$. The algebraic multiplicity is 2. The geometric multiplicity is 1.

 (c) The characteristic polynomial is $\lambda^2 - 2\lambda - 35$. The eigenvalues are 7 and -5. The corresponding bases of eigenvectors are $\{(1, 1)\}$ and $\{(-7, 5)\}$. All multiplicities are 1.

19. **(a)** The characteristic polynomial is $-\lambda^3 + \lambda^2 + \lambda - 1$. The eigenvalues are 1, with basis of eigenvectors $\{(1, 0, 1), (0, 1, 0)\}$, and -1, with basis of eigenvectors $\{(-1, 0, 1)\}$. The algebraic and geometric multiplicity of 1 is 2. The algebraic and geometric multiplicity of -1 is 1.

 (b) The characteristic polynomial is $-(\lambda - 1) \times (\lambda - 2)(\lambda - 3)$. The eigenvalues are 1, 2, 3, with corresponding bases of eigenvectors $\{(1, 0, 0)\}$, $\{(1, 1, 0)\}, \{(0, 0, 1)\}$. All multiplicities are 1.

21. **(a)** The characteristic polynomial is $-\lambda^3 + 3\lambda^2 + 4\lambda - 12$. The eigenvalues are $2, 3$, and -2, with corresponding bases of eigenvectors $\{(1, 1, 0)\}$, $\{(0, 0, 1)\}$ and $\{(-1, 1, 0)\}$. All multiplicities are 1.

 (b) The characteristic polynomial is $-(\lambda - 1) \times (\lambda^2 - 1)$. The eigenvalues are 1, with basis of eigenvectors $\{(1, 0, 0), (0, 1, 1)\}$, and -1, with basis of eigenvectors $\{(0, -1, 1)\}$. The algebraic and geometric multiplicity of 1 is 2. The algebraic and geometric multiplicity of -1 is 1.

23. **(a)** The characteristic polynomial is $-(\lambda - 1) \times (\lambda - 2)(\lambda - 3)$. The eigenvalues are 1, 2, 3, with corresponding bases of eigenvectors $\{(2, -3, 3)\}$, $\{(0, 1, 0)\}, \{(0, 1, 1)\}$. All multiplicities are 1.

 (b) The characteristic polynomial is $-(\lambda - 1) \times (\lambda - 2)(\lambda - 3)$. The eigenvalues are 1, 2, 3, with corresponding bases of eigenvectors $\{(1, -2, 2)\}$, $\{(0, 1, 0)\}, \{(0, 1, 1)\}$. All multiplicities are 1.

25. $\begin{bmatrix} 1 & 1 & 1 \\ 0 & 2 & 1 \\ 0 & 0 & 1 \end{bmatrix}$ with eigenvalues are 1 and 2

27. **(a)** $1, -2$ **(b)** $-4, 4$

29. $(1, 1, 1)$ with eigenvalue $a + b + c$

31. *Hint:* First show that $\det((A - \lambda I)^T) = \det(A^T - \lambda I)$.

33. $\frac{1}{2}\mathbf{v}$

35. *Hint:* First show that if A is invertible, then $\lambda \neq 0$. Then left-multiply $A\mathbf{v} = \lambda\mathbf{v}$ by $\lambda^{-1}A^{-1}$.

37. *Hint:* First show that $\det(P^{-1}AP - \lambda I) = \det(A - \lambda I)$.

39. $A\mathbf{v} = \mathbf{0} = 0\mathbf{v}$, because $\mathbf{v} \in \text{Null}(A)$. Hence, $\mathbf{v} \neq \mathbf{0}$ is an eigenvector with eigenvalue 0.

41. Take $A = \begin{bmatrix} 1 & 1 \\ 0 & 1 \end{bmatrix}$ and $B = \begin{bmatrix} 0 & 1 \\ 1 & 0 \end{bmatrix}$.

43. *Hint:* First note that $\mathbf{e}_1, \ldots, \mathbf{e}_n$ are eigenvectors. If $\lambda_1, \ldots, \lambda_n$ are the corresponding eigenvalues, then $A\mathbf{e}_i = \lambda_i\mathbf{e}_i$. This shows that A is diagonal. Next show that all the λ_i's are equal.

45. *Hint:* 0 is the only eigenvalue of A. The geometric multiplicity of 0 is the dimension of E_0. Now use Exercise 32.

47. $C(p) = \begin{bmatrix} 0 & 1 \\ 15 & -2 \end{bmatrix}$. The characteristic polynomial is $\lambda^2 + 2\lambda - 15$.

49. By the last exercise it suffices to construct a monic polynomial with roots 4 and -5 and then take its companion matrix. So,

$$p(x) = (x - 4)(x + 5) = x^2 + x - 20$$

Hence, $C(p) = \begin{bmatrix} 0 & 1 \\ 20 & -1 \end{bmatrix}$ must have eigenvalues: $4, -5$. It does.

51. The polynomial $(x - 4)(x + 5)(x + 2) = x^3 + 3x^2 - 18x - 40$ has companion matrix

$$C(p) = \begin{bmatrix} 0 & 1 & 0 \\ 0 & 0 & 1 \\ 40 & 18 & -3 \end{bmatrix}$$

Its eigenvalues are the roots of its characteristic polynomial $-\lambda^3 - 3\lambda^2 + 18\lambda + 40$, which are 4, -5, and -2.

53. *Hint:* First verify the claim for $n = 2$. Then assume it is true for $n - 1$ and prove it for n.

55. First we find the eigenvalues and eigenvectors of the matrix $A = \begin{bmatrix} 0 & 1 \\ 1 & 0 \end{bmatrix}$ of T with respect to the standard basis $\{1, x\}$. The eigenvalues are 1 and -1, with corresponding basic eigenvectors $(1, 1)$ and $(-1, 1)$. Hence, T has eigenvalues 1 and -1 with corresponding basic eigenvectors $x + 1$ and $-x + 1$.

Section 7.2

1. $P = \begin{bmatrix} 1 & -1 \\ 1 & 1 \end{bmatrix}$, $D = \begin{bmatrix} 3 & 0 \\ 0 & -7 \end{bmatrix}$

3. $P = \begin{bmatrix} 1 & 1 & 1 \\ 1 & 0 & 4 \\ 1 & 0 & -4 \end{bmatrix}$, $D = \begin{bmatrix} 3 & 0 & 0 \\ 0 & 1 & 0 \\ 0 & 0 & -7 \end{bmatrix}$

5. Not diagonalizable

7. $A(10, 0) = (-10, 0)$, so $(10, 0)$ is an eigenvector with eigenvalue -1. $A(6, 5) = (24, 20)$, so $(6, 5)$ is an eigenvector with eigenvalue 4. Because $(10, 0)$ and $(6, 5)$ belong to different eigenvalues, they are linearly independent. A is diagonalizable, with $P = \begin{bmatrix} 10 & 6 \\ 0 & 5 \end{bmatrix}$ and $D = \begin{bmatrix} -1 & 0 \\ 0 & 4 \end{bmatrix}$.

9. Because
$$A \begin{bmatrix} 0 & 1 & 1 \\ 2 & 0 & 0 \\ 0 & 2 & -2 \end{bmatrix} = \begin{bmatrix} 0 & 2 & -2 \\ 4 & 0 & 0 \\ 0 & 4 & 4 \end{bmatrix}$$
we see that $(0, 2, 0), (1, 0, 2), (1, 0, -2)$ are eigenvectors with corresponding eigenvalues $2, 2, -2$. The first two eigenvectors are clearly linearly independent. So, S is linearly independent, because $(1, 0, -2)$ belongs to a different eigenvalue. A is diagonalizable, with
$$P = \begin{bmatrix} 0 & 1 & 1 \\ 2 & 0 & 0 \\ 0 & 2 & -2 \end{bmatrix} \text{ and } D = \begin{bmatrix} 2 & 0 & 0 \\ 0 & 2 & 0 \\ 0 & 0 & -2 \end{bmatrix}.$$

11. Let $P = \begin{bmatrix} 1 & 1 & 0 \\ 0 & 1 & 0 \\ 0 & 0 & 1 \end{bmatrix}$, $D = \begin{bmatrix} 1 & 0 & 0 \\ 0 & 2 & 0 \\ 0 & 0 & 3 \end{bmatrix}$. Then
$$A = PDP^{-1} = \begin{bmatrix} 1 & 1 & 0 \\ 0 & 2 & 0 \\ 0 & 0 & 3 \end{bmatrix}$$

13. Let $P = \begin{bmatrix} 1 & -3 & -2 \\ 1 & 0 & 1 \\ 1 & 1 & 0 \end{bmatrix}$, $D = \begin{bmatrix} 6 & 0 & 0 \\ 0 & 0 & 0 \\ 0 & 0 & 0 \end{bmatrix}$. Then
$$A = PDP^{-1} = \begin{bmatrix} 1 & 2 & 3 \\ 1 & 2 & 3 \\ 1 & 2 & 3 \end{bmatrix}$$

15. A has three distinct eigenvalues, so it is diagonalizable by Theorem 8.

17. $\{(1, 1, 1), (-2, 1, 0), (-2, 0, 1)\}$

19. A is diagonalizable, because it has three linearly independent eigenvectors, namely, $(1, 1, 1), (-3, 2, 0), (-2, 0, 1)$.

21. A is not diagonalizable, because its only eigenvalue 5 has only one (< 3) basic eigenvector, $(1, 0, 0)$.

25. The matrix is diagonalizable over the reals if and only if $a > 0$.

27. $A^6 = \begin{bmatrix} 2 & -2 \\ 1 & 1 \end{bmatrix} \begin{bmatrix} 4 & 0 \\ 0 & -4 \end{bmatrix}^6 \begin{bmatrix} 2 & -2 \\ 1 & 1 \end{bmatrix}^{-1}$
$= \begin{bmatrix} 4096 & 0 \\ 0 & 4096 \end{bmatrix} = \begin{bmatrix} 2^{12} & 0 \\ 0 & 2^{12} \end{bmatrix}$
$A^9 = \begin{bmatrix} 2 & -2 \\ 1 & 1 \end{bmatrix} \begin{bmatrix} 4 & 0 \\ 0 & -4 \end{bmatrix}^9 \begin{bmatrix} 2 & -2 \\ 1 & 1 \end{bmatrix}^{-1}$
$= \begin{bmatrix} 0 & 524,288 \\ 131,072 & 0 \end{bmatrix} = \begin{bmatrix} 0 & 2^{19} \\ 2^{17} & 0 \end{bmatrix}$

29. $P = \begin{bmatrix} 0 & 1 & 2 \\ 1 & 0 & 1 \\ -1 & -1 & 2 \end{bmatrix}$, $D = \begin{bmatrix} 0 & 0 & 0 \\ 0 & 0 & 0 \\ 0 & 0 & 5 \end{bmatrix}$ diagonalize the matrix. Hence,
$$PD^7P^{-1} = \begin{bmatrix} 2 \cdot 5^6 & 2 \cdot 5^6 & 2 \cdot 5^6 \\ 5^6 & 5^6 & 5^6 \\ 2 \cdot 5^6 & 2 \cdot 5^6 & 2 \cdot 5^6 \end{bmatrix}$$
$$= \begin{bmatrix} 31,250 & 31,250 & 31,250 \\ 15,625 & 15,625 & 15,625 \\ 31,250 & 31,250 & 31,250 \end{bmatrix}$$

31. The identity is true, because by diagonalizing the base matrix on the left we get
$$\begin{bmatrix} 1 & 1 \\ 1 & 1 \end{bmatrix}^{k+1} = \begin{bmatrix} 1 & -1 \\ 1 & 1 \end{bmatrix} \begin{bmatrix} 2 & 0 \\ 0 & 0 \end{bmatrix}^{k+1} \begin{bmatrix} 1 & -1 \\ 1 & 1 \end{bmatrix}^{-1}$$
$$= \begin{bmatrix} 2^k & 2^k \\ 2^k & 2^k \end{bmatrix}$$
for $k = 0, 1, 2, \ldots$.

35. Let $\mathcal{B} = \{1, x, x^2, x^3\}$ be the standard basis of P_3. Then the matrix of $T : P_3 \rightarrow P_3$,

$$T(a + bx + cx^2 + dx^3) = a$$

with respect to $\mathcal{B}$ is

$$A = \begin{bmatrix} 1 & 0 & 0 & 0 \\ 0 & 0 & 0 & 0 \\ 0 & 0 & 0 & 0 \\ 0 & 0 & 0 & 0 \end{bmatrix}$$

A is diagonalizable. Its eigenvalues are 0, 1, with corresponding bases of eigenvectors $\{e_2, e_3, e_4\}$ and $\{e_1\}$. Therefore, T is diagonalizable. $\mathcal{B}$ is a basis of P_3 that consists of eigenvectors of T.

37. Let T be the given reflection. Then the standard matrix of T is

$$A = \begin{bmatrix} 1 & 0 \\ 0 & -1 \end{bmatrix}$$

which has eigenvalues $1, -1$ and corresponding bases of eigenvectors $\{(1, 0)\}, \{(0, 1)\}$. Therefore, A is diagonalizable. Hence, T is diagonalizable. The standard basis $\{e_1, e_2\}$ is a basis of $\mathbf{R}^2$ that consists of eigenvectors of T.

39. Let T be the given projection. The standard matrix of T is

$$A = \begin{bmatrix} 1 & 0 \\ 0 & 0 \end{bmatrix}$$

which has eigenvalues $0, 1$ and corresponding bases of eigenvectors $\{(0, 1)\}$ and $\{(1, 0)\}$. Therefore, A is diagonalizable. Hence, T is diagonalizable. The standard basis $\{e_1, e_2\}$ is a basis of $\mathbf{R}^2$ that consists of eigenvectors of T.

41. Let T be the given projection. The standard matrix of T is

$$A = \begin{bmatrix} 1 & 0 & 0 \\ 0 & 1 & 0 \\ 0 & 0 & 0 \end{bmatrix}$$

which has eigenvalues $0, 1$ and corresponding bases of eigenvectors $\{(0, 0, 1)\}$ and $\{(1, 0, 0), (0, 1, 0)\}$. Therefore, A is diagonalizable. Hence, T is diagonalizable. The standard basis $\{e_1, e_2, e_3\}$ is a basis of $\mathbf{R}^3$ that consists of eigenvectors of T.

Section 7.3

1. 5 is an eigenvalue with eigenvector $(1, 1)$.

3. $(2, 1)$ is an eigenvector with corresponding eigenvalue 4.

5. $(2, 1)$ is an eigenvector with corresponding eigenvalue $\frac{1}{4}$.

7. For $\mathbf{x} = (1, 2)$ we have

$$A^3\mathbf{x} = \begin{bmatrix} 61 \\ -64 \end{bmatrix} \qquad A^4\mathbf{x} = \begin{bmatrix} -311 \\ 314 \end{bmatrix}$$

$$\lambda_{appr} = -5.0984 \qquad \lambda = -5$$

$$\mathbf{v}_{appr} = \begin{bmatrix} -0.99 \\ 1.0 \end{bmatrix} \qquad \mathbf{v} = \begin{bmatrix} -1 \\ 1 \end{bmatrix}$$

9. For $\mathbf{x} = (1, 2)$, we have

$$A^3\mathbf{x} = \begin{bmatrix} 170 \\ -173 \end{bmatrix} \qquad A^4\mathbf{x} = \begin{bmatrix} -1199 \\ 1202 \end{bmatrix}$$

$$\lambda_{appr} = -7.0529 \qquad \lambda = -7$$

$$\mathbf{v}_{appr} = \begin{bmatrix} -0.997 \\ 1.0 \end{bmatrix} \qquad \mathbf{v} = \begin{bmatrix} -1 \\ 1 \end{bmatrix}$$

11. For $\mathbf{x} = (1, 2)$, we have

$$A^3\mathbf{x} = \begin{bmatrix} 363 \\ -366 \end{bmatrix} \qquad A^4\mathbf{x} = \begin{bmatrix} -3279 \\ 3282 \end{bmatrix}$$

$$\lambda_{appr} = -9.0331 \qquad \lambda = -9$$

$$\mathbf{v}_{appr} = \begin{bmatrix} -0.999 \\ 1.0 \end{bmatrix} \qquad \mathbf{v} = \begin{bmatrix} -1 \\ 1 \end{bmatrix}$$

13. For $\mathbf{x} = (1, 2)$, we have

$$A^3\mathbf{x} = \begin{bmatrix} 364 \\ -148 \end{bmatrix} \qquad A^4\mathbf{x} = \begin{bmatrix} -2924 \\ 1172 \end{bmatrix}$$

$$\lambda_{appr} = -8.033 \qquad \lambda = -8$$

$$\mathbf{v}_{appr} = \begin{bmatrix} 1.0 \\ -0.4008 \end{bmatrix} \qquad \mathbf{v} = \begin{bmatrix} 1 \\ -\frac{2}{5} \end{bmatrix}$$

15. For $\mathbf{x} = (1, 2)$, we have

$$A^3\mathbf{x} = \begin{bmatrix} 2395 \\ 1598 \end{bmatrix} \qquad A^4\mathbf{x} = \begin{bmatrix} 26,353 \\ 17,570 \end{bmatrix}$$

$$\lambda_{appr} = 11.003 \qquad \lambda = 11$$

$$\mathbf{v}_{appr} = \begin{bmatrix} 1.0 \\ 0.66672 \end{bmatrix} \qquad \mathbf{v} = \begin{bmatrix} 1 \\ \frac{2}{3} \end{bmatrix}$$

17. For $\mathbf{x} = (1, 2)$, we have

$$A^3\mathbf{x} = \begin{bmatrix} 1412 \\ -316 \end{bmatrix} \qquad A^4\mathbf{x} = \begin{bmatrix} -16,964 \\ 3772 \end{bmatrix}$$

$$\lambda_{appr} = -12.014 \qquad \lambda = -12$$

$$\mathbf{v}_{appr} = \begin{bmatrix} 1.0 \\ -0.22235 \end{bmatrix} \qquad \mathbf{v} = \begin{bmatrix} 1 \\ -\frac{2}{9} \end{bmatrix}$$

19. For $\mathbf{x} = (1, 2)$, we have

$$A^3\mathbf{x} = \begin{bmatrix} 1265 \\ -1479 \end{bmatrix} \qquad A^4\mathbf{x} = \begin{bmatrix} -17,729 \\ 20,687 \end{bmatrix}$$

$$\lambda_{appr} = -14.015 \qquad \lambda = -14$$

$$\mathbf{v}_{appr} = \begin{bmatrix} -0.85701 \\ 1.0 \end{bmatrix} \qquad \mathbf{v} = \begin{bmatrix} -\frac{6}{7} \\ 1 \end{bmatrix}$$

21. Starting at $(1, 2)$, the first four iterations yield -1.4000, -3.9412, -4.9432, -4.9977. So, the dominant eigenvalue is -5.

23. Starting at $(1, 2)$, the first four iterations yield -1.6000, -6.0690, -6.9776, -6.9995. So, the dominant eigenvalue is -7.

25. Starting at $(1, 2)$, the first four iterations yield 0.2800, 0.7882, 0.9886, 0.9995. The true eigenvalue closest to the origin is 1 and is approximated by 0.9995.

27. Starting at $(1, 2)$, the first four iterations yield 0.2286, 0.8670, 0.9968, 0.9999. The true eigenvalue closest to the origin is 1 and is approximated by 0.9999.

29. Starting at $(1, 2)$, the first four iterations yield 0.2000, 0.9111, 0.9988, 1.0000. The true eigenvalue closest to the origin is 1 and is approximated to 4 decimal places by 1.0000.

31. Starting at $(1, 2)$, the first four iterations yield -1.1111, -1.0194, -1.0022, -1.0022. The true eigenvalue closest to the origin is -1 and is approximated by -1.0022.

33. Starting at $(1, 2)$, the first four iterations yield -1.1000, -1.0205, -1.0026, -1.0003. The true eigenvalue closest to the origin is -1 and is approximated by -1.0003.

35. Applying the shifted inverse power method to $\begin{bmatrix} 0 & 1 \\ -4 & 5 \end{bmatrix}$ starting at $(1, 0)$ yields -1.0139 after four iterations. So, the root nearest 5 is $5 + (-1.0139)^{-1} = 4.0138$.

37. Applying the shifted inverse power method to $\begin{bmatrix} 0 & 1 \\ -6 & 7 \end{bmatrix}$ starting at $(1, 0)$ yields 0.9780 after four iterations. So, the root nearest 5 is $5 + (0.9780)^{-1} = 6.0225$.

39. Applying the shifted inverse power method to $\begin{bmatrix} 0 & 1 \\ 9 & 8 \end{bmatrix}$ starting at $(1, 0)$ yields -1.0006 after four iterations. So, the root nearest 10 is $10 + (-1.0006)^{-1} = 9.0006$.

Section 7.4

1. (a) $\mathbf{x}_k = \frac{3}{2} \cdot 1^k \cdot \begin{bmatrix} -1 \\ 1 \end{bmatrix} + \frac{5}{2} \cdot 5^k \cdot \begin{bmatrix} 1 \\ 1 \end{bmatrix}$

   (b) $A\mathbf{x}_0 = \mathbf{x}_1 = \begin{bmatrix} 11 \\ 14 \end{bmatrix}$; $A^2\mathbf{x}_0 = \mathbf{x}_2 = \begin{bmatrix} 61 \\ 64 \end{bmatrix}$

   (c) Neither

3. (a) $\mathbf{x}_k = \frac{3}{2} \cdot (-7)^k \cdot \begin{bmatrix} -1 \\ 1 \end{bmatrix} + \frac{5}{2} \cdot (-1)^k \cdot \begin{bmatrix} 1 \\ 1 \end{bmatrix}$

   (b) $A\mathbf{x}_0 = \mathbf{x}_1 = \begin{bmatrix} 8 \\ -13 \end{bmatrix}$; $A^2\mathbf{x}_0 = \mathbf{x}_2 = \begin{bmatrix} -71 \\ 76 \end{bmatrix}$

   (c) Neither

5. (a) $\mathbf{x}_k = \frac{3}{2} \cdot \left(\frac{1}{2}\right)^k \begin{bmatrix} -1 \\ 1 \end{bmatrix} + \frac{5}{2} \cdot \left(\frac{5}{2}\right)^k \begin{bmatrix} 1 \\ 1 \end{bmatrix}$

   (b) $A\mathbf{x}_0 = \mathbf{x}_1 = \begin{bmatrix} \frac{11}{2} \\ 7 \end{bmatrix}$; $A^2\mathbf{x}_0 = \mathbf{x}_2 = \begin{bmatrix} \frac{61}{4} \\ 16 \end{bmatrix}$

   (c) Neither

7. (a) $\mathbf{x}_k = \frac{3}{2} \cdot (-13)^k \cdot \begin{bmatrix} -1 \\ 1 \end{bmatrix} + \frac{5}{2} \cdot (-1)^k \cdot \begin{bmatrix} 1 \\ 1 \end{bmatrix}$

   (b) $A\mathbf{x}_0 = \mathbf{x}_1 = \begin{bmatrix} 17 \\ -22 \end{bmatrix}$; $A^2\mathbf{x}_0 = \mathbf{x}_2 = \begin{bmatrix} -251 \\ 256 \end{bmatrix}$

   (c) Neither

9. (a) $\mathbf{x}_k = \frac{3}{2} \cdot \left(-\frac{1}{10}\right)^k \begin{bmatrix} -1 \\ 1 \end{bmatrix} + \frac{5}{2} \cdot \left(\frac{1}{2}\right)^k \begin{bmatrix} 1 \\ 1 \end{bmatrix}$

   (b) $A\mathbf{x}_0 = \mathbf{x}_1 = \begin{bmatrix} \frac{7}{5} \\ \frac{11}{10} \end{bmatrix}$; $A^2\mathbf{x}_0 = \mathbf{x}_2 = \begin{bmatrix} \frac{61}{100} \\ \frac{16}{25} \end{bmatrix}$

   (c) Attractor

11. (a) $\begin{bmatrix} -3 & 2 \\ 2 & -3 \end{bmatrix}^5 \begin{bmatrix} 1 \\ 1 \end{bmatrix} = \begin{bmatrix} -1 \\ -1 \end{bmatrix}$

   (b) $\mathbf{x}_5 = 0 \cdot (-5)^5 \begin{bmatrix} -1 \\ 1 \end{bmatrix} + 1 \cdot (-1)^5 \begin{bmatrix} 1 \\ 1 \end{bmatrix}$

$$= \begin{bmatrix} -1 \\ -1 \end{bmatrix}$$

13. (a) $\begin{bmatrix} \frac{3}{10} & \frac{2}{10} \\ \frac{2}{10} & \frac{3}{10} \end{bmatrix}^5 \begin{bmatrix} 1 \\ 1 \end{bmatrix} = \begin{bmatrix} \frac{1}{32} \\ \frac{1}{32} \end{bmatrix}$

   (b) $\mathbf{x}_5 = 0 \cdot \left(\frac{1}{10}\right)^5 \begin{bmatrix} -1 \\ 1 \end{bmatrix} + 1 \cdot \left(\frac{1}{2}\right)^5 \begin{bmatrix} 1 \\ 1 \end{bmatrix} = \begin{bmatrix} \frac{1}{32} \\ \frac{1}{32} \end{bmatrix}$

15. (a) $\begin{bmatrix} 5 & 4 \\ 4 & 5 \end{bmatrix}^5 \begin{bmatrix} 1 \\ 1 \end{bmatrix} = \begin{bmatrix} 59,049 \\ 59,049 \end{bmatrix}$

   (b) $\mathbf{x}_5 = 0 \cdot 1^5 \begin{bmatrix} -1 \\ 1 \end{bmatrix} + 1 \cdot 9^5 \begin{bmatrix} 1 \\ 1 \end{bmatrix} = \begin{bmatrix} 59,049 \\ 59,049 \end{bmatrix}$

17. $\mathbf{x}_1 = (0, 1)$, $\mathbf{x}_2 = (-\frac{1}{2}, \frac{1}{2})$, $\mathbf{x}_3 = (-\frac{1}{2}, 0)$. The origin is an attractor.

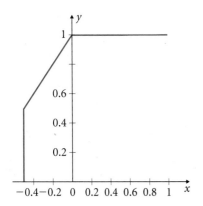

19. This is true, because both eigenvalues, $\frac{1}{2} \pm \frac{\sqrt{3}}{2}i$, are 6th roots of 1. I.e., $\left(\frac{1}{2} \pm \frac{\sqrt{3}}{2}i\right)^6 = 1$.

21. In matrix notation:

$$\begin{bmatrix} A_{k+1} \\ B_{k+1} \end{bmatrix} = \begin{bmatrix} 0.8 & 0.7 \\ 0.2 & 0.3 \end{bmatrix}^k \begin{bmatrix} A_k \\ B_k \end{bmatrix}$$

$$\mathbf{x}_0 = \begin{bmatrix} A_0 \\ B_0 \end{bmatrix} = \begin{bmatrix} 300 \\ 100 \end{bmatrix}$$

(a) After the third generation, there are about 311 females and 89 males.

(b) In the long run, there are 311.11 females and 88.889 males. So, the females will eventually dominate the population.

Section 7.5

1. Second and fourth

3. (a) The matrix is not regular, because it is lower triangular, so all its powers are also lower triangular; hence there are always entries that are 0.

(b) The matrix is not regular, because it is upper triangular, so all its powers are also upper triangular; hence there are always entries that are 0.

5. Let $A = \begin{bmatrix} \frac{1}{2} & 1 \\ \frac{1}{2} & 0 \end{bmatrix}$. Solving $[A - I : \mathbf{0}]$ yields $(2r, r)$.

But we want $2r + r = 1$. So, $r = \frac{1}{3}$. Hence, $\mathbf{v} = \begin{bmatrix} \frac{2}{3} \\ \frac{1}{3} \end{bmatrix}$ is the steady-state vector of A. Let $B = \begin{bmatrix} 0 & \frac{1}{2} \\ 1 & \frac{1}{2} \end{bmatrix}$. Solving $[B - I : \mathbf{0}]$ yields $(r/2, r)$. But we want $r/2 + r = 1$.

So, $r = \frac{2}{3}$. Hence, $\mathbf{v} = \begin{bmatrix} \frac{1}{3} \\ \frac{2}{3} \end{bmatrix}$ is the steady-state vector of B.

7. (a) $\left(\frac{5}{13}, \frac{2}{13}, \frac{6}{13}\right)$ **(b)** $\left(\frac{1}{4}, \frac{13}{36}, \frac{7}{18}\right)$

9. Assume that $\mathbf{u}$ is another probability vector such that $A\mathbf{u} = \mathbf{u}$. Then $\lim_{k \to \infty} A^k \mathbf{u} = \mathbf{v}$. And because $A\mathbf{u} = \mathbf{u}$, we have $A^k \mathbf{u} = \mathbf{u}$ for all k. Hence, $\mathbf{u} = \mathbf{v}$.

11. M is regular, because M^2 has only positive entries. The equilibrium of M is $\left(\frac{5}{13}, \frac{2}{13}, \frac{6}{13}\right)$. So, in the long run plan C is the most popular and plan B is the least popular.

Chapter 8

Section 8.1

1. $(1, -2, 1) \cdot (4, 2, 0) = 0$
$(1, -2, 1) \cdot (-1, 2, 5) = 0$
$(4, 2, 0) \cdot (-1, 2, 5) = 0$
This set forms an orthogonal basis of $\mathbf{R}^3$.

3. $(1, 1, -1, 1) \cdot (1, 1, 1, -1) = 0$
$(1, 1, -1, 1) \cdot (0, 0, 1, 1) = 0$
$(1, 1, 1, -1) \cdot (0, 0, 1, 1) = 0$
This set does not form an orthogonal basis of $\mathbf{R}^4$.

5. $\mathbf{v}_1 = (1, 1, 2)$, $\mathbf{v}_2 = (-1, 1, 0)$, $\mathbf{v}_3 = (0, 0, 1)$

7. Every pair of vectors has dot product zero; hence, the vectors are linearly independent. Thus, they form an orthogonal basis of $\mathbf{R}^3$.

$$(1, 1, 1) = \frac{3}{19}\mathbf{v}_1 + \frac{1}{19}\mathbf{v}_2 + 1\mathbf{v}_3$$

9. Every pair of vectors has dot product zero; hence, the vectors are linearly independent. Thus, they form an orthogonal basis of $\mathbf{R}^3$.

$$(1, 1, 1) = 0\mathbf{v}_1 + \frac{1}{7}\mathbf{v}_2 + \frac{1}{7}\mathbf{v}_3$$

11. Not orthonormal. Orthonormalization yields
$$\begin{bmatrix} 1/\sqrt{5} \\ 2/\sqrt{5} \end{bmatrix}, \begin{bmatrix} -2/\sqrt{5} \\ 1/\sqrt{5} \end{bmatrix}$$

13. Orthonormal

15. $\mathcal{B}$ is an orthonormal basis for $\mathbf{R}^2$, because both vectors are unit and their dot product is zero.

$$\mathbf{e}_1 = \frac{2}{\sqrt{5}}\mathbf{v}_1 - \frac{1}{\sqrt{5}}\mathbf{v}_2$$

17. Orthogonal; its inverse is $\begin{bmatrix} 0 & -1 \\ 1 & 0 \end{bmatrix}$.

19. Orthogonal; its inverse is

$$\begin{bmatrix} \dfrac{3}{\sqrt{14}} & -\dfrac{2}{\sqrt{14}} & \dfrac{1}{\sqrt{14}} \\[2mm] \dfrac{1}{\sqrt{6}} & \dfrac{2}{\sqrt{6}} & \dfrac{1}{\sqrt{6}} \\[2mm] -\dfrac{2}{\sqrt{21}} & -\dfrac{1}{\sqrt{21}} & \dfrac{4}{\sqrt{21}} \end{bmatrix}$$

21. No. We can possibly have zero columns.

Section 8.2

1. $\mathbf{u}_{pr} = \dfrac{(-2,1)\cdot(1,2)}{(1,2)\cdot(1,2)}(1,2) = (0,0)$

3. $\mathbf{u}_{pr} = \dfrac{(3,-1,2)\cdot(1,1,-3)}{(1,1,-3)\cdot(1,1,-3)}(1,1,-3)$

$= \left(-\frac{4}{11}, -\frac{4}{11}, \frac{12}{11}\right)$

5. $\mathbf{u} = \left(-\frac{2}{13}, -\frac{3}{13}\right) + \left(-\frac{24}{13}, \frac{16}{13}\right)$

7. $\mathbf{u} = (-1,1) + (2,2)$

9. $\mathbf{u}_{pr} = (0,0,1)$

13. Orthogonal:

$$\left\{ \begin{bmatrix} 2 \\ -1 \\ 1 \end{bmatrix}, \begin{bmatrix} \frac{4}{3} \\ \frac{7}{3} \\ -\frac{1}{3} \end{bmatrix}, \begin{bmatrix} -\frac{1}{11} \\ \frac{1}{11} \\ \frac{3}{11} \end{bmatrix} \right\}$$

Orthonormal:

$$\left\{ \begin{bmatrix} \frac{2}{\sqrt{6}} \\ \frac{-1}{\sqrt{6}} \\ \frac{1}{\sqrt{6}} \end{bmatrix}, \begin{bmatrix} \frac{4}{\sqrt{66}} \\ \frac{7}{\sqrt{66}} \\ -\frac{1}{\sqrt{66}} \end{bmatrix}, \begin{bmatrix} -\frac{1}{\sqrt{11}} \\ \frac{1}{\sqrt{11}} \\ \frac{3}{\sqrt{11}} \end{bmatrix} \right\}$$

15. $\left\{ \begin{bmatrix} 4 \\ 2 \\ -1 \end{bmatrix}, \begin{bmatrix} \frac{1}{21} \\ \frac{32}{21} \\ \frac{68}{21} \end{bmatrix} \right\}$

17. $\mathbf{u} = \begin{bmatrix} \frac{80}{103} \\ \frac{70}{103} \\ -\frac{37}{103} \end{bmatrix} + \begin{bmatrix} \frac{126}{103} \\ -\frac{70}{103} \\ \frac{140}{103} \end{bmatrix}$

19. $\|\mathbf{u}_c\| = \left\| \begin{bmatrix} \frac{8}{15} \\ -\frac{8}{3} \\ -\frac{16}{15} \end{bmatrix} \right\| = \frac{8}{15}\sqrt{30}$

21. $\|\mathbf{u}_c\| = \left\| \begin{bmatrix} -\frac{7}{10} \\ \frac{3}{2} \\ \frac{7}{5} \\ 1 \end{bmatrix} \right\| = \frac{1}{10}\sqrt{570}$

Section 8.3

1. $\begin{bmatrix} 0 & 1 \\ -1 & 0 \end{bmatrix} \begin{bmatrix} -1 & -3 \\ 0 & -2 \end{bmatrix}$

3. $\begin{bmatrix} 0 & 0 & -1 \\ 0 & -1 & 0 \\ 1 & 0 & 0 \end{bmatrix} \begin{bmatrix} -1 & 0 & 2 \\ 0 & -1 & -2 \\ 0 & 0 & -4 \end{bmatrix}$

5. $\begin{bmatrix} \frac{\sqrt{2}}{2} & -\frac{\sqrt{3}}{3} & \frac{\sqrt{6}}{6} \\[1mm] 0 & \frac{\sqrt{3}}{3} & \frac{\sqrt{6}}{3} \\[1mm] -\frac{\sqrt{2}}{2} & -\frac{\sqrt{3}}{3} & \frac{\sqrt{6}}{6} \end{bmatrix} \begin{bmatrix} \sqrt{2} & -\sqrt{2} & 0 \\ 0 & \sqrt{3} & 0 \\ 0 & 0 & \sqrt{6} \end{bmatrix}$

7. $\begin{bmatrix} \frac{1}{2} & -\frac{\sqrt{5}}{10} \\[1mm] \frac{1}{2} & \frac{\sqrt{5}}{10} \\[1mm] \frac{1}{2} & \frac{3\sqrt{5}}{10} \\[1mm] -\frac{1}{2} & \frac{3\sqrt{5}}{10} \end{bmatrix} \begin{bmatrix} 2 & -2 \\ 0 & 2\sqrt{5} \end{bmatrix}$

9. $\begin{bmatrix} 2 & 0 \\ 0 & \sqrt{6} \end{bmatrix}$

11. Let $Q = A$. Such a Q is acceptable, because A has orthogonal columns. Then $R = Q^T A = A^T A = I$, because A is orthonormal. So, $A = AI$ is a QR factorization of A.

15. Because,

$$A_1 = \begin{bmatrix} 8.8916 & 6.1382 \\ 0.13817 & 1.1075 \end{bmatrix}$$

$$A_2 = \begin{bmatrix} 8.9869 & 6.0159 \\ 0.015947 & 1.0121 \end{bmatrix}$$

$$A_3 = \begin{bmatrix} 8.9970 & 6.0013 \\ 0.0015697 & 1.0013 \end{bmatrix}$$

the estimated eigenvalues are 8.9970, 1.0013. The true eigenvalues are 9, 1. The errors are 0.003, −0.0013.

17. Because,

$$A_1 = \begin{bmatrix} 13.864 & 9.1909 \\ 0.19135 & 1.1367 \end{bmatrix}$$

$$A_2 = \begin{bmatrix} 13.992 & 9.0157 \\ 0.015813 & 1.0109 \end{bmatrix}$$

$$A_3 = \begin{bmatrix} 14.002 & 8.9943 \\ -0.0054328 & 0.99650 \end{bmatrix}$$

the estimated eigenvalues are $14.002, 0.99650$. The true eigenvalues are $14, 1$. The errors are $-0.002, 0.0035$.

Section 8.4

1. The normal equations

$$\begin{bmatrix} 3 & -1 \\ -1 & 9 \end{bmatrix} \widetilde{\mathbf{x}} = \begin{bmatrix} -2 \\ 6 \end{bmatrix}$$

yield $\widetilde{\mathbf{x}} = \begin{bmatrix} -\frac{6}{13} \\ \frac{8}{13} \end{bmatrix}$.

3. The normal equations

$$\begin{bmatrix} 10 & 5 \\ 5 & 6 \end{bmatrix} \widetilde{\mathbf{x}} = \begin{bmatrix} -1 \\ 3 \end{bmatrix}$$

yield $\widetilde{\mathbf{x}} = \begin{bmatrix} -\frac{3}{5} \\ 1 \end{bmatrix}$.

5. All we need is a system with invertible coefficient matrix. For example:

$$\begin{bmatrix} 2 & 1 & 0 \\ 0 & 1 & 1 \\ 0 & 0 & -3 \end{bmatrix} \mathbf{x} = \begin{bmatrix} 9 \\ 6 \\ 3 \end{bmatrix}$$

with solution $(1, 7, -1)$ by direct elimination or via the normal equations.

7. $y = \frac{7}{2}x + \frac{4}{3}$

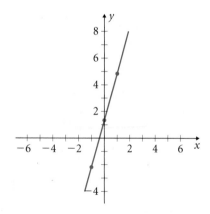

9. $y = \frac{51}{59}x + \frac{14}{59}$

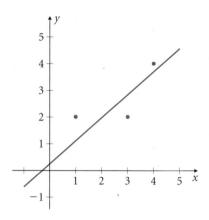

11. Solving the normal equations

$$\begin{bmatrix} 5 & 1 & -3 \\ 1 & 1 & 1 \\ -3 & 1 & 5 \end{bmatrix} \widetilde{\mathbf{x}} = \begin{bmatrix} -2 \\ 2 \\ 6 \end{bmatrix}$$

we get $\widetilde{\mathbf{x}} = (r - 1, -2r + 3, r), r \in \mathbf{R}$.

13. The system $R\widetilde{\mathbf{x}} = Q^T\mathbf{b}$,

$$\begin{bmatrix} 5 & 10 \\ 0 & 5 \end{bmatrix} \widetilde{\mathbf{x}} = \begin{bmatrix} -4 \\ \frac{9}{5} \end{bmatrix}$$

yields $\widetilde{\mathbf{x}} = (-\frac{38}{25}, \frac{9}{25})$.

17. By Exercise 15,

$$\widetilde{\mathbf{x}} = \begin{bmatrix} \frac{5}{9} \\ -\frac{1}{9} \end{bmatrix}$$

19. By Exercise 15,

$$\widetilde{\mathbf{x}} = \begin{bmatrix} \frac{5/3}{1} \\ \frac{-1/3}{1} \end{bmatrix} = \begin{bmatrix} \frac{5}{3} \\ -\frac{1}{3} \end{bmatrix}$$

Section 8.5

1. Not symmetric

3. Not symmetric

5. $A^2 = \begin{bmatrix} \frac{1}{\sqrt{2}} & \frac{1}{\sqrt{2}} \\ \frac{1}{\sqrt{2}} & -\frac{1}{\sqrt{2}} \end{bmatrix}^2 = \begin{bmatrix} 1 & 0 \\ 0 & 1 \end{bmatrix}$

7. $A^{-1} = \begin{bmatrix} \frac{3}{5} & \frac{4}{5} \\ \frac{4}{5} & -\frac{3}{5} \end{bmatrix}$ by Exercise 6

9. $Q = \begin{bmatrix} \frac{1}{\sqrt{2}} & -\frac{1}{\sqrt{2}} \\ \frac{1}{\sqrt{2}} & \frac{1}{\sqrt{2}} \end{bmatrix}$, $D = \begin{bmatrix} 1 & 0 \\ 0 & -3 \end{bmatrix}$

11. $Q = \begin{bmatrix} \frac{1}{\sqrt{2}} & -\frac{1}{\sqrt{2}} \\ \frac{1}{\sqrt{2}} & \frac{1}{\sqrt{2}} \end{bmatrix}$, $D = \begin{bmatrix} 2 & 0 \\ 0 & 10 \end{bmatrix}$

13. $Q = \begin{bmatrix} 0 & 1 & 0 \\ \frac{1}{\sqrt{2}} & 0 & -\frac{1}{\sqrt{2}} \\ \frac{1}{\sqrt{2}} & 0 & \frac{1}{\sqrt{2}} \end{bmatrix}$, $D = \begin{bmatrix} 0 & 0 & 0 \\ 0 & 1 & 0 \\ 0 & 0 & 4 \end{bmatrix}$

15. $Q = \begin{bmatrix} -\frac{1}{\sqrt{2}} & 0 & 0 & \frac{1}{\sqrt{2}} \\ \frac{1}{\sqrt{2}} & 0 & 0 & \frac{1}{\sqrt{2}} \\ 0 & -\frac{1}{\sqrt{2}} & \frac{1}{\sqrt{2}} & 0 \\ 0 & \frac{1}{\sqrt{2}} & \frac{1}{\sqrt{2}} & 0 \end{bmatrix}$

$D = \begin{bmatrix} 2 & 0 & 0 & 0 \\ 0 & 4 & 0 & 0 \\ 0 & 0 & 0 & 0 \\ 0 & 0 & 0 & 0 \end{bmatrix}$

17. *Hint:* Show that Q and D^2 orthogonally diagonalize A^2.

19. $Q = \begin{bmatrix} -\frac{1}{\sqrt{2}} & \frac{1}{\sqrt{2}} \\ \frac{1}{\sqrt{2}} & \frac{1}{\sqrt{2}} \end{bmatrix}$, $T = \begin{bmatrix} 1 & -6 \\ 0 & 11 \end{bmatrix}$

21. $Q = \begin{bmatrix} -\frac{1}{\sqrt{2}} & \frac{1}{\sqrt{2}} \\ \frac{1}{\sqrt{2}} & \frac{1}{\sqrt{2}} \end{bmatrix}$, $T = \begin{bmatrix} 1 & -7 \\ 0 & 12 \end{bmatrix}$

23. We have

$$A = QDQ^T$$

$$= [\mathbf{v}_1 \cdots \mathbf{v}_n] \begin{bmatrix} \lambda_1 & \cdots & 0 \\ \vdots & \ddots & \vdots \\ 0 & \cdots & \lambda_n \end{bmatrix} \begin{bmatrix} \mathbf{v}_1 \\ \vdots \\ \mathbf{v}_n \end{bmatrix}^T$$

$$= [\lambda_1\mathbf{v}_1 \cdots \lambda_n\mathbf{v}_n] \begin{bmatrix} \mathbf{v}_1 \\ \vdots \\ \mathbf{v}_n \end{bmatrix}^T$$

$$= \lambda_1\mathbf{v}_1\mathbf{v}_1^T + \cdots + \lambda_n\mathbf{v}_n\mathbf{v}_n^T$$

25. $5 \begin{bmatrix} \frac{1}{\sqrt{5}} \\ -\frac{2}{\sqrt{5}} \end{bmatrix} \begin{bmatrix} \frac{1}{\sqrt{5}} \\ -\frac{2}{\sqrt{5}} \end{bmatrix}^T + 15 \begin{bmatrix} \frac{2}{\sqrt{5}} \\ \frac{1}{\sqrt{5}} \end{bmatrix} \begin{bmatrix} \frac{2}{\sqrt{5}} \\ \frac{1}{\sqrt{5}} \end{bmatrix}^T$

Section 8.6

1. $q(\mathbf{x}) = -2x^2 + 3y^2 + 4xy$

3. $q(\mathbf{x}) = x^2 + 5z^2 - 6xy + 4xz + 2yz$

5. $\begin{bmatrix} 3 & -3 \\ -3 & 3 \end{bmatrix}$

7. $\begin{bmatrix} -4 & 1 \\ 1 & -4 \end{bmatrix}$

9. $\begin{bmatrix} 2 & 0 & 1 \\ 0 & 0 & 0 \\ 1 & 0 & 2 \end{bmatrix}$

11. $\begin{bmatrix} 5 & -4 & 0 \\ -4 & 3 & 6 \\ 0 & 6 & 1 \end{bmatrix}$

13. The matrix of the form is $\begin{bmatrix} 3 & -1 \\ -1 & 3 \end{bmatrix}$, which is orthogonally diagonalized by

$$Q = \begin{bmatrix} \frac{1}{\sqrt{2}} & -\frac{1}{\sqrt{2}} \\ \frac{1}{\sqrt{2}} & \frac{1}{\sqrt{2}} \end{bmatrix}, \quad D = \begin{bmatrix} 2 & 0 \\ 0 & 4 \end{bmatrix}$$

Let $\mathbf{y} = (x', y')$. Then

$$\mathbf{y} = Q^T\mathbf{x} = \begin{bmatrix} \frac{x}{\sqrt{2}} + \frac{y}{\sqrt{2}} \\ -\frac{x}{\sqrt{2}} + \frac{y}{\sqrt{2}} \end{bmatrix}$$

and

$$q(\mathbf{x}) = \mathbf{y}^T \begin{bmatrix} 2 & 0 \\ 0 & 4 \end{bmatrix} \mathbf{y}$$

$$= 2x'^2 + 4y'^2$$

15. The matrix of the form is $\begin{bmatrix} -4 & 2 \\ 2 & -4 \end{bmatrix}$, which is orthogonally diagonalized by

$$Q = \begin{bmatrix} \dfrac{1}{\sqrt{2}} & -\dfrac{1}{\sqrt{2}} \\ \dfrac{1}{\sqrt{2}} & \dfrac{1}{\sqrt{2}} \end{bmatrix}, \quad D = \begin{bmatrix} -2 & 0 \\ 0 & -6 \end{bmatrix}$$

Let $\mathbf{y} = (x',y')$. Then

$$\mathbf{y} = Q^T \mathbf{x} = \begin{bmatrix} \dfrac{x}{\sqrt{2}} + \dfrac{y}{\sqrt{2}} \\ -\dfrac{x}{\sqrt{2}} + \dfrac{y}{\sqrt{2}} \end{bmatrix}$$

and

$$q(\mathbf{x}) = \mathbf{y}^T \begin{bmatrix} -2 & 0 \\ 0 & -6 \end{bmatrix} \mathbf{y}$$
$$= -2x'^2 - 6y'^2$$

17. The matrix of the form is $\begin{bmatrix} 2 & 0 & -1 \\ 0 & 0 & 0 \\ -1 & 0 & 2 \end{bmatrix}$, which is orthogonally diagonalized by

$$Q = \begin{bmatrix} 0 & \dfrac{1}{\sqrt{2}} & -\dfrac{1}{\sqrt{2}} \\ 1 & 0 & 0 \\ 0 & \dfrac{1}{\sqrt{2}} & \dfrac{1}{\sqrt{2}} \end{bmatrix}$$

$$D = \begin{bmatrix} 0 & 0 & 0 \\ 0 & 1 & 0 \\ 0 & 0 & 3 \end{bmatrix}$$

Let $\mathbf{y} = (x',y',z')$. Then

$$\mathbf{y} = Q^T \mathbf{x} = \begin{bmatrix} y \\ \dfrac{x}{\sqrt{2}} + \dfrac{z}{\sqrt{2}} \\ -\dfrac{x}{\sqrt{2}} + \dfrac{z}{\sqrt{2}} \end{bmatrix}$$

and

$$q(\mathbf{x}) = \mathbf{y}^T \begin{bmatrix} 0 & 0 & 0 \\ 0 & 1 & 0 \\ 0 & 0 & 3 \end{bmatrix} \mathbf{y}$$
$$= y'^2 + 3z'^2$$

19. The matrix of the form is $\begin{bmatrix} 5 & -4 & 0 \\ -4 & 3 & 4 \\ 0 & 4 & 1 \end{bmatrix}$, which is orthogonally diagonalized by

$$Q = \begin{bmatrix} \dfrac{1}{3} & -\dfrac{2}{3} & \dfrac{2}{3} \\ \dfrac{2}{3} & \dfrac{2}{3} & \dfrac{1}{3} \\ -\dfrac{2}{3} & \dfrac{1}{3} & \dfrac{2}{3} \end{bmatrix}$$

$$D = \begin{bmatrix} -3 & 0 & 0 \\ 0 & 9 & 0 \\ 0 & 0 & 3 \end{bmatrix}$$

Let $\mathbf{y} = (x',y',z')$. Then

$$\mathbf{y} = Q^T \mathbf{x} = \begin{bmatrix} \frac{1}{3}x + \frac{2}{3}y - \frac{2}{3}z \\ -\frac{2}{3}x + \frac{2}{3}y + \frac{1}{3}z \\ \frac{2}{3}x + \frac{1}{3}y + \frac{2}{3}z \end{bmatrix}$$

and

$$q(\mathbf{x}) = \mathbf{y}^T \begin{bmatrix} -3 & 0 & 0 \\ 0 & 9 & 0 \\ 0 & 0 & 3 \end{bmatrix} \mathbf{y}$$
$$= -3x'^2 + 9y'^2 + 3z'^2$$

21. The matrix of the form is $\begin{bmatrix} 5 & -4 \\ -4 & 5 \end{bmatrix}$, which is orthogonally diagonalized by

$$Q = \begin{bmatrix} \dfrac{1}{\sqrt{2}} & -\dfrac{1}{\sqrt{2}} \\ \dfrac{1}{\sqrt{2}} & \dfrac{1}{\sqrt{2}} \end{bmatrix}, \quad D = \begin{bmatrix} 1 & 0 \\ 0 & 9 \end{bmatrix}$$

Let $\mathbf{y} = (x',y')$. Then

$$\mathbf{y} = Q^T \mathbf{x} = \begin{bmatrix} \dfrac{x}{\sqrt{2}} + \dfrac{y}{\sqrt{2}} \\ -\dfrac{x}{\sqrt{2}} + \dfrac{y}{\sqrt{2}} \end{bmatrix}$$

and

$$q(\mathbf{x}) = \mathbf{y}^T \begin{bmatrix} 1 & 0 \\ 0 & 9 \end{bmatrix} \mathbf{y}$$
$$= x'^2 + 9y'^2$$

Hence, the conic section is an ellipse.

25. *Hint:* First show that Q and D^2 orthogonally diagonalize A^2.

27. *Hint:* Consider $P = BQ^T$, where B is the diagonal matrix with diagonal entries the square roots of the diagonal entries of D.

29. We have

$$q(x,y) = ax^2 + bxy + cy^2$$
$$= a\left(x^2 + 2x\frac{b}{2a}y + \frac{c}{a}y^2 + \frac{b^2}{4a^2}y^2 - \frac{b^2}{4a^2}y^2\right)$$

$$= a\left(x + \frac{b}{2a}y\right)^2 + \left(c - \frac{b^2}{4a}\right)y^2$$

$$= aX^2 + By^2$$

where $X = x + (b/2a)y$ and $B = c - b^2/4a$.

31. Yes, as follows:

$$q(x, y) = bxy + cy^2$$

$$= c\left(y^2 + 2\frac{b}{2c}xy + \frac{b^2}{4c^2}x^2 - \frac{b^2}{4c^2}x^2\right)$$

$$= c\left(y + \frac{b}{2c}x\right)^2 + \frac{-b^2}{4c}x^2$$

$$= cY^2 + Ax^2$$

where $Y = y + (b/2c)x$ and $A = -b^2/4c^2$.

Section 8.7

1. $3, 2$

3. $2, 1, 0$

5. $\begin{bmatrix} 0 & -1 \\ 1 & 0 \end{bmatrix} \begin{bmatrix} 5 & 0 & 0 \\ 0 & 2 & 0 \end{bmatrix} \begin{bmatrix} 0 & 1 & 0 \\ 0 & 0 & 1 \\ 1 & 0 & 0 \end{bmatrix}^T$

7. $\begin{bmatrix} 0 & 0 & 1 \\ 0 & 1 & 0 \\ 1 & 0 & 0 \end{bmatrix} \begin{bmatrix} 3 & 0 & 0 \\ 0 & 2 & 0 \\ 0 & 0 & 1 \end{bmatrix} \begin{bmatrix} 1 & 0 & 0 \\ 0 & 1 & 0 \\ 0 & 0 & 1 \end{bmatrix}^T$

9.
$$\begin{bmatrix} \frac{1}{\sqrt{5}} & 0 & -\frac{2}{\sqrt{5}} \\ 0 & 1 & 0 \\ \frac{2}{\sqrt{5}} & 0 & \frac{1}{\sqrt{5}} \end{bmatrix} \begin{bmatrix} 10 & 0 & 0 \\ 0 & 4 & 0 \\ 0 & 0 & 0 \end{bmatrix}$$

$$\times \begin{bmatrix} \frac{1}{\sqrt{5}} & 0 & -\frac{2}{\sqrt{5}} \\ 0 & 1 & 0 \\ \frac{2}{\sqrt{5}} & 0 & \frac{1}{\sqrt{5}} \end{bmatrix}^T$$

11. $\begin{bmatrix} \frac{2}{3} & \frac{1}{3} & -\frac{2}{3} \\ \frac{2}{3} & -\frac{2}{3} & \frac{1}{3} \\ \frac{1}{3} & \frac{2}{3} & \frac{2}{3} \end{bmatrix} \begin{bmatrix} 9 & 0 & 0 \\ 0 & 6 & 0 \\ 0 & 0 & 6 \end{bmatrix} \begin{bmatrix} 0 & 1 & 0 \\ 1 & 0 & 0 \\ 0 & 0 & 1 \end{bmatrix}^T$

13. A^TA is messy, but AA^T is diagonal. An SVD of A^T is

$$\begin{bmatrix} \frac{2}{3} & \frac{2}{3} & \frac{1}{3} \\ -\frac{2}{3} & \frac{1}{3} & \frac{2}{3} \\ \frac{1}{3} & -\frac{2}{3} & \frac{2}{3} \end{bmatrix} \begin{bmatrix} 9 & 0 & 0 \\ 0 & 3 & 0 \\ 0 & 0 & 0 \end{bmatrix} \begin{bmatrix} 0 & 1 & 0 \\ 0 & 0 & 1 \\ 1 & 0 & 0 \end{bmatrix}^T$$

Hence, an SVD of A is obtained by transposition:

$$\begin{bmatrix} 0 & 1 & 0 \\ 0 & 0 & 1 \\ 1 & 0 & 0 \end{bmatrix} \begin{bmatrix} 9 & 0 & 0 \\ 0 & 3 & 0 \\ 0 & 0 & 0 \end{bmatrix} \begin{bmatrix} \frac{2}{3} & \frac{2}{3} & \frac{1}{3} \\ -\frac{2}{3} & \frac{1}{3} & \frac{2}{3} \\ \frac{1}{3} & -\frac{2}{3} & \frac{2}{3} \end{bmatrix}^T$$

15. (a) The verification of the Moore-Penrose properties is straightforward given:

$$A^+ = \begin{bmatrix} 0 & 0 & 1 \\ 0 & 1 & 0 \\ 1 & 0 & 0 \end{bmatrix} \begin{bmatrix} \frac{1}{6} & 0 & 0 \\ 0 & \frac{1}{4} & 0 \\ 0 & 0 & \frac{1}{2} \end{bmatrix} \begin{bmatrix} 0 & 0 & 1 \\ 0 & 1 & 0 \\ 1 & 0 & 0 \end{bmatrix}^T$$

$$= \begin{bmatrix} \frac{1}{2} & 0 & 0 \\ 0 & \frac{1}{4} & 0 \\ 0 & 0 & \frac{1}{6} \end{bmatrix}$$

(b) The verification of the Moore-Penrose properties is straightforward:

$$A^+ = \begin{bmatrix} 1 & 0 & 0 \\ 0 & 1 & 0 \\ 0 & 0 & 1 \end{bmatrix} \begin{bmatrix} \frac{1}{3} & 0 & 0 \\ 0 & \frac{1}{2} & 0 \\ 0 & 0 & 1 \end{bmatrix} \begin{bmatrix} 0 & 0 & 1 \\ 0 & 1 & 0 \\ 1 & 0 & 0 \end{bmatrix}^T$$

$$= \begin{bmatrix} 0 & 0 & \frac{1}{3} \\ 0 & \frac{1}{2} & 0 \\ 1 & 0 & 0 \end{bmatrix}$$

17. A^+ was computed in Exercise 15(**b**) and was found to be $\begin{bmatrix} 0 & 0 & \frac{1}{3} \\ 0 & \frac{1}{2} & 0 \\ 1 & 0 & 0 \end{bmatrix}$, which is also A^{-1}.

19. $AA^+A = \begin{bmatrix} -3 & 0 \\ 0 & 0 \\ 0 & 4 \end{bmatrix}$

$A^+AA^+ = \begin{bmatrix} -\frac{1}{3} & 0 & 0 \\ 0 & 0 & \frac{1}{4} \end{bmatrix}$

$\left(AA^+\right)^T = \begin{bmatrix} 1 & 0 & 0 \\ 0 & 0 & 0 \\ 0 & 0 & 1 \end{bmatrix}$

$AA^+ = \begin{bmatrix} 1 & 0 & 0 \\ 0 & 0 & 0 \\ 0 & 0 & 1 \end{bmatrix}$

$\left(A^+A\right)^T = I$

$A^+A = I$

21. It suffices to verify the Moore-Penrose conditions for (A^+, A), because then A would be the unique inverse of A^+. So, $A^{++} = A$. By Exercise 18 we have

$$A^+AA^+ = A^+, \quad AA^+A = A,$$
$$(A^+A)^T = A^TA, \quad (AA^+)^T = AA^+$$

Hence, the conditions hold for the pair (A^+, A).

23. $A^+\mathbf{b} = \begin{bmatrix} -\frac{1}{2} & 0 & 0 \\ 0 & 0 & \frac{1}{5} \end{bmatrix} \begin{bmatrix} 1 \\ 2 \\ 3 \end{bmatrix} = \begin{bmatrix} -\frac{1}{2} \\ \frac{3}{5} \end{bmatrix}$

25. $A^+\mathbf{b} = \begin{bmatrix} -\frac{2}{9} & \frac{1}{9} & \frac{2}{9} \\ \frac{2}{27} & \frac{2}{27} & \frac{1}{27} \end{bmatrix} \begin{bmatrix} 1 \\ 2 \\ 3 \end{bmatrix} = \begin{bmatrix} \frac{2}{3} \\ \frac{1}{3} \end{bmatrix}$

27. (a) $A = PQ = \begin{bmatrix} \sqrt{2} & 0 \\ 0 & \sqrt{2} \end{bmatrix} \begin{bmatrix} \frac{1}{\sqrt{2}} & -\frac{1}{\sqrt{2}} \\ \frac{1}{\sqrt{2}} & \frac{1}{\sqrt{2}} \end{bmatrix}$

(b) $A = PQ = \begin{bmatrix} 7 & 2 & 0 \\ 2 & 6 & 2 \\ 0 & 2 & 5 \end{bmatrix} \begin{bmatrix} \frac{1}{3} & \frac{2}{3} & -\frac{2}{3} \\ -\frac{2}{3} & \frac{2}{3} & \frac{1}{3} \\ \frac{2}{3} & \frac{1}{3} & \frac{2}{3} \end{bmatrix}$

Section 8.8

1. (a) $-12, \sqrt{19}, 2\sqrt{21}, \sqrt{79}$

(b) $\|(-3, -5)\| = \sqrt{127}$

(c) $\sqrt{19} \cdot 2\sqrt{21} \cong 39.95 \geq |-12| = 12$

(d) $\sqrt{79} \cong 8.8882 \leq \sqrt{19} + 2\sqrt{21} \cong 13.524$

(e) $\frac{1}{4}(79 - 127) = -12$

3. $\left(1, \frac{3}{8}\right)$

5. A and B are orthogonal. Also A, and C are orthogonal.

7. The orthonormal pairs are (A', B') and (A', C'), where

$$A' = \frac{1}{\|A\|}A = \begin{bmatrix} \frac{1}{\sqrt{2}} & 0 \\ 0 & \frac{1}{\sqrt{2}} \end{bmatrix}$$

$$B' = \frac{1}{\|B\|}B = \begin{bmatrix} -\frac{1}{\sqrt{2}} & 0 \\ 0 & \frac{1}{\sqrt{2}} \end{bmatrix}$$

$$C' = \frac{1}{\|C\|}C = \begin{bmatrix} -\frac{1}{2} & \frac{1}{2} \\ \frac{1}{2} & \frac{1}{2} \end{bmatrix}$$

9. It is, because A is positive definite.

11. It is, because A is positive definite.

13. It is not. f is not positive definite, because $f(\mathbf{u}, \mathbf{u}) = 0$ for $\mathbf{u} = (-1, 1)$.

15. It is not. f is not symmetric, because for $\mathbf{u} = (1, 2)$ and $\mathbf{v} = (2, 1)$, we have $f(\mathbf{u}, \mathbf{v}) = 31$, whereas $f(\mathbf{v}, \mathbf{u}) = 49$.

17. *Hint:* Check the four axioms.

19. (a) $-255, \sqrt{339}, 15, 3\sqrt{6}$

(b) $\|1 - 3x^2 + 2x\| = \sqrt{1074}$

(c) $|-255| = 255 \leq \sqrt{339} \cdot 15 \cong 276.18$

(d) $3\sqrt{6} \cong 7.3485 \leq \sqrt{339} + 15 \cong 33.412$

21. $2 + x^2$

23. $\frac{1}{5} + x^2$

29. Using the calculation $(1, 1)_{pr} = \left(\frac{2}{3}, 0\right)$ from the previous exercise, the distance from $(1, 1)$ to l is

$$\|(1, 1)_c\| = \|(1, 1) - (1, 1)_{pr}\|$$
$$= \|(1, 1) - \left(\frac{2}{3}, 0\right)\| = \|\left(\frac{1}{3}, 1\right)\|$$
$$= \sqrt{\frac{7}{3}}$$

31. Let $f(x) = x^2 + x + 1$. From the previous exercise we have $f_{pr} = x^2 + x - \frac{8}{3}$. Hence, the distance from $x^2 + x + 1$ to W is

$$\|f_c\| = \|f - f_{pr}\| = \left\|\frac{11}{3}\right\| = \frac{11}{\sqrt{3}}$$

33. Let $\mathbf{u}_1 = 1$. Then

$$\mathbf{u}_2 = x - \frac{0}{3} \cdot 1 = x$$

and

$$\mathbf{u}_3 = x^2 - \frac{8}{3}1 - \frac{0}{4}x$$
$$= x^2 - \frac{8}{3}$$

So, $\left\{1, x, x^2 - \frac{8}{3}\right\}$ is an orthogonal basis of P_2 for this inner product.

35. Because

$$\int_{-1}^{1} 1^2 \, dx = 2$$
$$\int_{-1}^{1} x^2 \, dx = \frac{2}{3}$$
$$\int_{-1}^{1} \left(\frac{3}{2}x^2 - \frac{1}{2}\right)^2 \, dx = \frac{2}{5}$$
$$\int_{-1}^{1} \left(\frac{5}{2}x^3 - \frac{3}{2}x\right)^2 \, dx = \frac{2}{7}$$

is an orthonormal basis for P_3 is

$$\mathcal{L}' = \left\{\frac{1}{2}, \frac{3}{2}x, \frac{15}{4}x^2 - \frac{5}{4}, \frac{35}{4}x^3 - \frac{21}{4}x\right\}$$

Section 8.9

1. This time we get the solution

$$\mathbf{x} = (1.9847, 0.8318, 3.3135, -4.1813, 4.2094)$$

which differs from the one discussed in the section by

$$(0.0745, 0.0003, 0.0044, 0.0147, -0.0211)$$

The answer is satisfactory, but we do not quite get the correct rational coefficients.

3. $q(x) = -\frac{22}{5} + \frac{23}{70}x + \frac{13}{14}x^2$; $q(6) = 31$

5. $q(x) = \frac{134}{35} + 2x + \frac{2}{7}x^2$; $q(6) = \frac{914}{35} \cong 26.114$

7. $q(x) = -\frac{34}{35} - \frac{11}{4}x + \frac{11}{14}x^2 + \frac{5}{4}x^3$; $q(3) = \frac{158}{5} \cong 31.6$

11. Because

$$\int_1^2 1\,dx = 1, \qquad \int_1^2 x\,dx = \frac{3}{2}$$
$$\int_1^2 x^2\,dx = \frac{7}{3}, \qquad \int_1^2 x^3\,dx = \frac{15}{4}$$

we have

$$\begin{bmatrix} 1 & \frac{3}{2} & \frac{7}{3} \\ \frac{3}{2} & \frac{7}{3} & \frac{15}{4} \end{bmatrix} \sim \begin{bmatrix} 1 & 0 & -\frac{13}{6} \\ 0 & 1 & 3 \end{bmatrix}$$

So, $y = -\frac{13}{6} + 3x$.

13. Because

$$\int_0^2 1\,dx = 2, \qquad \int_0^2 x\,dx = 2$$
$$\int_0^2 x^2\,dx = \frac{8}{3}, \qquad \int_0^2 x^3\,dx = 4$$
$$\int_0^2 x^4\,dx = \frac{32}{5}$$

we have

$$\begin{bmatrix} 2 & 2 & 4 \\ 2 & \frac{8}{3} & \frac{32}{5} \end{bmatrix} \sim \begin{bmatrix} 1 & 0 & -\frac{8}{5} \\ 0 & 1 & \frac{18}{5} \end{bmatrix}$$

So, $y = -\frac{8}{5} + \frac{18}{5}x$.

15. Because

$$\int_0^2 1\,dx = 2, \qquad \int_0^2 x\,dx = 2$$
$$\int_0^2 x^2\,dx = \frac{8}{3}, \qquad \int_0^2 x^3\,dx = 4$$
$$\int_0^2 x^4\,dx = \frac{32}{5}, \qquad \int_0^2 x^5\,dx = \frac{32}{3}$$

we have

$$\begin{bmatrix} 2 & 2 & \frac{8}{3} & 4 \\ 2 & \frac{8}{3} & 4 & \frac{32}{5} \\ \frac{8}{3} & 4 & \frac{32}{5} & \frac{32}{3} \end{bmatrix} \sim \begin{bmatrix} 1 & 0 & 0 & \frac{2}{5} \\ 0 & 1 & 0 & -\frac{12}{5} \\ 0 & 0 & 1 & 3 \end{bmatrix}$$

So, $y = \frac{2}{5} - \frac{12}{5}x + 3x^2$.

17. $\langle 1, \sin nx \rangle = \int_{-\pi}^{\pi} 1 \cdot \sin nx\,dx$

$$= -\frac{\cos nx}{n} \Big|_{-\pi}^{\pi} = 0$$

19. $\langle \sin mx, \sin nx \rangle = \int_{-\pi}^{\pi} \sin mx \sin nx\,dx$

$$= \frac{1}{2} \frac{\sin(m-n)x}{m-n}$$
$$- \frac{1}{2} \frac{\sin(m+n)x}{m+n} \Big|_{-\pi}^{\pi} = 0$$

21. $a_0 = \frac{1}{2}$, $a_n = 0$, $b_n = \dfrac{1-(-1)^n}{n\pi}$.

23. We have

$$\langle \sin nx, \sin mx \rangle = \int_0^{\pi} \sin nx \sin mx\,dx$$
$$= \frac{1}{2} \frac{\sin(n-m)x}{n-m}$$
$$- \frac{1}{2} \frac{\sin(n+m)x}{n+m} \Big|_0^{\pi} = 0$$

25. We have

$$\langle \sin n\pi x, \sin m\pi x \rangle = \int_{-1}^{1} \sin n\pi x \sin m\pi x\,dx$$
$$= \frac{1}{2} \frac{\sin(n\pi - m\pi)x}{n\pi - m\pi}$$
$$- \frac{1}{2} \frac{\sin(n\pi + m\pi)x}{n\pi + m\pi} \Big|_{-1}^{1} = 0$$

27. We have

$$\langle 1, \cos m\pi x \rangle = \int_0^2 \cos m\pi x\,dx$$
$$= \frac{\sin m\pi x}{m\pi} \Big|_0^2 = 0$$

and

$$\langle \cos n\pi x, \cos m\pi x \rangle = \int_0^2 \cos n\pi x \cos m\pi x\,dx$$
$$= \frac{1}{2} \frac{\sin(n\pi - m\pi)x}{n\pi - m\pi}$$
$$+ \frac{1}{2} \frac{\sin(n\pi + m\pi)x}{n\pi + m\pi} \Big|_0^2$$
$$= 0$$

29. Because for $m = 1, 2, \ldots$,

$$\psi_{m,0} = \begin{cases} 2^{-m/2} & 0 \le x \le 2^{m-1} \\ -2^{-m/2} & 2^{m-1} < x \le 2^m \end{cases}$$

we have

$$c_{m,0} = \int_{-\infty}^{\infty} f(x)\psi_{m,0}(x)\,dx$$
$$\int_0^1 (-1)\psi_{m,0}(x)\,dx = \int_0^1 (-2^{-m/2})\,dx = -2^{-m/2}$$

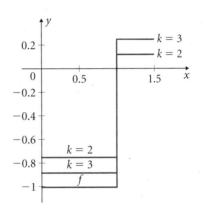

Index